SPORTS AND FITNESS NUTRITION

Robert E. C. Wildman, Ph.D., R.D.

Barry S. Miller, Ph.D.
University of Delaware

THOMSON

WADSWORTH

Australia • Canada • Mexico • Singapore • Spain
United Kingdom • United States

THOMSON

WADSWORTH

Publisher: Peter Marshall
Development Editor: Elizabeth Howe
Assistant Editor: Madinah Chang
Editorial Assistant: Elesha Feldman
Technology Project Manager: Travis Metz
Marketing Manager: Jennifer Somerville
Marketing Assistant: Melanie Banfield
Advertising Project Manager: Shemika Britt
Project Manager, Editorial Production: Sandra Craig
Print/Media Buyer: Doreen Suruki
Permissions Editor: Joohee Lee

Production: Robin Lockwood Productions
Text Design: Rokusek Design
Photo Research: Judy Mason
Copy Editor: Anita Wagner
Illustrations: Ralph Lao, Lotus Art
Cover Design: Bill Stanton
Cover Images: woman doing yoga: John Kelly/Getty Images; body-builder: Jim Cummins/Getty Images; cyclist: The Mooks/Getty Images; salad: Aaron Rezny/Corbis
Composition: UG / GGS Information Services, Inc.
Printer: Phoenix Color Corp

For more information about our products, contact us at:
Thomson Learning Academic Resource Center
1-800-423-0563
For permission to use material from this text, contact us by:
Phone: 1-800-730-2214
Fax: 1-800-730-2215
Web: http://www.thomsonrights.com

Library of Congress Control Number: 2003105866

ISBN 0-534-57564-1

Wadsworth/Thomson Learning
10 Davis Drive
Belmont, CA 94002-3098
USA

Asia
Thomson Learning
5 Shenton Way #01–01
UIC Building
Singapore 068808

Australia/New Zealand
Thomson Learning
102 Dodds Street
Southbank, Victoria 3006
Australia

Canada
Nelson
1120 Birchmount Road
Toronto, Ontario M1K 5G4
Canada

Europe/Middle East/Africa
Thomson Learning
High Holborn House
50/51 Bedford Row
London WC1R 4LR
United Kingdom

Latin America
Thomson Learning
Seneca, 53
Colonia Polanco
11560 Mexico D.F.
Mexico

Spain/Portugal
Paraninfo
Calle/Magallanes, 25
28015 Madrid, Spain

ABOUT THE AUTHORS

Robert E. C. Wildman Ph.D., R.D., received his B.S. in Dietetics and Nutrition from the University of Pittsburgh in 1988, M.S. in Food and Nutrition from The Florida State University in 1991, and Ph.D. from The Ohio State University in Human Nutrition in 1994. Dr. Wildman is Director of Nutrition for the Bally Total Fitness Corporation, Chicago, IL. He is also the author of *Advanced Human Nutrition* and *The Nutritionist: Food, Nutrition and Optimal Health* and editor of *The Handbook of Nutraceuticals and Functional Food.*

Barry S. Miller Ph.D., received his B.A. in Physical Education from California State University at Chico in 1987, M.S. in Physical Education/Biomechanics from CSU, Chico, in 1990, and a Ph.D. in Kinesiology with emphasis in Biomechanics from the University of Minnesota, Twin Cities in 1997. Dr. Miller is currently an assistant director of recreation and assistant professor of Health and Exercise Sciences at the University of Delaware. He plays an active role in training the Fitness Program's student staff in the subject matter contained in this text. He also serves as a reference to the many recreation participants and athletes on campus.

To my academic mentors—Bruce Rengers and Denis Medeiros.

To my coaches—Pro, Scully, Brown, McDonald, Long, LaRusso, Hamilton, Derr, Hutch, Warrick, Keithley, Miller, Killian, DeFrank, Stevenson, Demily and so many others!

Robert

To my academic mentors—Cheryl Maglischo and June Stoner.

To my many coaches—Fuller, Gilley, Frankie, Candaele, Moore, Marshall, Nunes, Pettigrew, Pickett, Stewart, and Metcalf.

Barry

CONTENTS

Chapter 7

ENERGY BALANCE, BODY WEIGHT AND COMPOSITION FOR SPORT AND FITNESS 195

Chapter 12

STRENGTH AND RESISTANCE EXERCISE AND TRAINING 374

Chapter 15

NUTRITION, EXERCISE, AND SPECIAL POPULATIONS   470

Personal Snapshot   470

Chapter Objectives   470

The following appendixes are available at Wadsworth Nutrition Resource Center: http://nutrition.wadsworth.com

Appendix B: Table of Food Composition
Appendix C: Web Sites
Appendix D: Answers to Study Questions

PREFACE

Sports and Fitness Nutrition is a complete resource for science majors and professionals from a variety of backgrounds. The chapters are designed to provide both foundation and applied information in the areas of nutrition, exercise, health, and fitness. This gives readers access to fundamental concepts in nutrition and exercise science if these areas have not been covered previously. The text material is sectioned into small, concise units of information to help readers find information new to them and reduce the time spent on material studied in other classes.

Readers of *Sports and Fitness Nutrition* will develop a solid basic and applied understanding of the relationship between nutrition and exercise and their application to health, fitness, and sport performance. The necessary skills for students using this book are completion of basic chemistry and biology courses and at least an introduction to human anatomy and physiology and to biochemistry. This text is well suited for college juniors and seniors, and as a main text for a graduate level course with appropriate supplemental materials. The book can also serve as a resource for health professionals such as personal trainers, pharmacists, physical therapists, chiropractors, and physicians.

■ OVERVIEW

The book begins with a survey of basic concepts in nutrition and exercise and their involvement in general health. Chapter 1 also looks at types of nutrition recommendations and at how nutrition assessment techniques are applied. Chapter 2 discusses the basic structure and function of skeletal muscle, including an overview of the energy systems that fuel physical activity. Chapter 3 describes basic exercise theory and concepts of fitness

Chapters 4–6 provide a sound overview of carbohydrates, protein, and fat and how exercise influences their metabolism, as well as adaptive processes that relate to these nutrients. Chapter 7 examines concepts related to body weight and composition and discusses diet strategies for changing weight and composition. Chapter 8 explores the fundamental importance of water in the body and the importance of sweating during exercise as a primary mechanism for heat removal. Chapters 9 and 10 close out the chapters related to essential nutrients with the roles of vitamins and minerals in exercise and health. Chapter 11 provides a foundation for understanding the purpose and composition of sport foods and popular sport supplements and other ergogenic efforts, such as doping agents.

Chapters 12 and 13 apply the preceding concepts to more specific discussions of strength training and cardiorespiratory (aerobic, endurance) training. Chapter 14 details the characteristics of specific sports and the types of training and nutrition that can maximize athletes' performance in these sports. Chapter 15 discusses exercise and nutrition for active populations not discussed in previous chapters, including children, teens, women during pregnancy and lactation, older populations, vegetarians, and people with diabetes mellitus.

■ FEATURES

Each chapter of *Sports and Fitness Nutrition* provides Key Points, an In Review list for each section, and a Conclusions section for easy review of text material. In addition, chapters include In Focus and In Practice features that offer practical information about applied areas associated with the material in that chapter. For instance, the Chapter 9 In Focus is on antioxidant supplementation for athletes, and the one in Chapter 15 explores eating disorders and the female athlete triad. The Chapter 5 In Focus is dedicated to the health risks of high-protein diets. The In Practice components give instructions for such topics as making a sport shake or gauging sweat rate to estimate water needs during exercise.

■ OUR APPROACH

The book is intended to provide a thorough understanding of the relationship between nutrition and exercise. Two important concepts are clearly developed in the book. First, nutrition and exercise cannot be separated. Nutrition dictates the magnitude and efficiency of physical performance and exercise, and chronic training affects the functions of metabolic systems.

Second, the field of sports and fitness nutrition has a long history. Ancient Greek and Roman athletes chose certain foods such as breads, figs, and meat in an attempt to improve physical performance. Elite athletes of the twentieth century sought similar foods to enhance performance. For instance, Finnish Olympic cyclist Kalevi Eskelinen and his teammates (Munich 1972) focused on foods such as potatoes, milk, bread, butter, meat, fish, and fruits, and they consumed liver from time to time to preserve blood hemoglobin levels. Bodybuilding legend Red Lerille (Mr. America 1960) emphasized protein in his diet and was able to achieve amazing mass and symmetry.

Today sports nutrition continues to be founded on natural carbohydrate- and protein-rich foods such as fish, poultry, lean meats, legumes, whole-grain products, and fresh fruits and vegetables. In addition, the field of sports nutrition has developed into big business; supplements and manufactured sport foods such as sport bars, drinks, and shakes have become popular among athletes and others. The field of sports nutrition has also expanded its boundaries to include fitness and health. Today the greater percentage of people who exercise do so to improve their fitness and health.

Sports nutrition is also sport- or activity-specific. For instance, the nutrition considerations for a middleweight wrestler are different from those for a football lineman or a sand volleyball player. Therefore, the characteristics of a sport (game duration, season length, travel) and its physical demands (speed, mass, power, endurance) need to be clearly defined to see what its nutrition requirements are.

■ INSTRUCTOR AND STUDENT RESOURCES

Instructors can find a test bank and lecture outlines on the Web at the Wadsworth Nutrition Resource Center: http://nutrition.wadsworth.com. Students will also find helpful resources at the Wadsworth Nutrition Resource Center including the answers to the end of the chapter study questions.

■ ACKNOWLEDGMENTS

We would like to acknowledge the efforts of the following individuals who played a role in the development of this book: research assistants Louis Ciliberti, M.S.; Mark Doyle, B.S.; Brenda Larrimer, M.S., R.P.T.; Jukka Eskelinen and Jeff Driban; and Dr. Kevin Tipton, who provided material advisement. We'd like to thank Sharon M. Nickols-Richardson Ph.D., R.D. for creating the Instructor materials for the Web site. We also thank the reviewers for their helpful comments and suggestions: Eldon Wayne Askew, University of Utah; Janine Baer, University of Dayton; Dan Benardot, Georgia State University; Becky Black, Long Beach City College; Christina Gayer Campbell, Montana State University; Alana D. Cline, University of Northern Colorado; Rachel A. Geik, Purdue University; Art Gilbert, University of California, Santa Barbara; Robert E. Keith, Auburn University; Alan Levine, Marywood University; Dorothy Chen-Maynard, Cal State San Bernardino; Jennifer McLean, Corning Community College; Chris Melby, Colorado State University; Steven Nizielski, Grand Valley State University; Shelly Nickols-Richardson, Virginia Polytechnic University; Nancy Rodriguez, University of Connecticut; Lorraine Sirota, Brooklyn College of the City University of New York; and Joan Thompson, Weber State University.

NUTRITION, EXERCISE, AND HEALTH

Bill Aron/PhotoEdit

Personal Snapshot

Susan recently visited her physician for a routine examination. She is 24 years old and has gained 10 lb during the past 5 years, which she attributes to lack of exercise and poor dietary choices while in college. Her physical examination and blood tests also reveal that she has high total cholesterol and elevated blood pressure. Because heart disease runs in her family she wants to do something about her health. However, the diet and exercise information from the Internet, bookstores, and TV infomercials has proved to be confusing and often conflicting. With so many organizations, products, and services out there, she is not sure which are credible and most appropriate. She is referred to a registered dietitian (R.D.) who helps her structure her diet to allow for changes in body weight and composition and provides a list of reliable Web sites and books. She also commits to a community health walking program and will join a local health club and begin working with a personal trainer.

Chapter Objectives

- Introduce current nutrition recommendations and diet planning tools.

- Discuss the most common nutrition assessment techniques and how they are most appropriately applied.

- Discuss the current health status of Americans and health promotion programs in the United States and other countries.

- Provide current estimates of nutrition status and exercise practices of the American population.

- Describe the characteristics and risk factors for heart disease and osteoporosis that are particularly influenced by diet and exercise.

How important are diet and exercise in the promotion of general health and prevention of diseases such as heart disease and osteoporosis? In a word—remarkably! In fact, the influence of diet and exercise on an individual's general health is recognized worldwide. Eating a healthy diet and exercising on a regular basis can each reduce risk factors for the major disease processes. When a healthy diet is combined with regular exercise, the reduction of risk factors is likely to be greater than the sum of the two.

This book is dedicated to describing the relationship between nutrition, physical activity, and athletic performance. In accordance, this chapter provides an introduction to basic nutrition concepts and recommendations that are referred to throughout this text. It also provides an overview of the involvement of nutrition and exercise in general health and their associations with major disease processes such as heart disease and osteoporosis. The ensuing chapters discuss basic concepts in exercise and fitness followed by an in-depth examination of the relationships between nutrition and exercise.

For the most part, diet recommendations for the general population also apply to more active populations and athletes. However, the association between nutrition and sport and fitness involves additional considerations as well as controversies. This book enables the reader to identify and understand the basis of these relationships and make sound conclusions on such topics.

■ BASIC NUTRITION CONCEPTS

Where does one begin in the quest to understand nutrition and the influence of diet on physical activity and general health? Simply stated, **nutrition** is the science pertaining to the *nourishment* of the human body, and the nourishment is provided by the foods that we choose to eat. More specifically, numerous factors within food nourish our body. That is where we will begin, by identifying those factors.

Essential and Nonessential Nutrients

The food and beverages we eat and drink provide nourishing substances commonly referred to as **nutrients**. These substances provide some benefit to the body, and each can be classified as either an essential or a nonessential nutrient. Both classes of nutrients are involved in normal or beneficial operations. The **essential nutrients** are substances that are absolutely indispensable but cannot be made in the body, either at all or in adequate amounts in a predictable or reliable manner. If these substances are not provided in the diet, normal operations become hindered and signs of deficiency develop. There are approximately 42 essential nutrients, which fall into several categories (Table 1-1). Water is one essential nutrient. **Vitamins** are defined as essential organic nutrients that do not provide energy. Several **minerals**, which are natural inorganic substances, are essential. Several **amino acids**, nitrogen-containing molecules that can serve as building blocks for protein, are also essential. Two **fatty acids**, organic molecules that are a component of fat, complete the list of essential nutrients.

The **nonessential nutrients** are also important components of normal function but do not have to be part of the diet. The list of nonessential nutrients is much longer and includes carnitine, pyruvate, lipoic acid, several amino acids, and numerous fatty acids. Many of these substances are made within cells as they perform normal operations. Accordingly, a diet lacking them should have no adverse effects on normal body functions.

NUTRACEUTICAL SUBSTANCES.   The list of nutrients also includes nutraceutical substances (Tables 1-2a, b). **Nutraceutical** substances may help prevent and treat certain chronic diseases such as heart disease and cancer and include nonessential nutrients such as *carotenoids, flavonoids,* and *fibers.* Even some essential nutrients such as vitamins C and E are considered nutraceuticals. Their nutraceutical roles in preventing some chronic diseases are in addition to their essentiality, which is based on their ability to prevent deficiency disorders associated with normal operations.

Nutraceutical substances function in several beneficial ways. Some provide antioxidant protection whereas others act as anti-inflammatory agents or positively influence detoxification systems (Table 1-3). Most nutraceutical substances are not made in the human body and are obtained by eating endowed natural and commercially developed foods. These foods are often referred to as **functional foods**,

| Table 1-1 | **Essential Nutrients for Humans** |

Energy Nutrients	Vitamins	Minerals	Other
Protein,[a] carbohydrates,[b] fat	Vitamins A, D, E, K, B_6, B_{12}, and C, thiamin, folate, biotin, niacin, pantothenic acid	Calcium, zinc, copper, sodium, potassium, iron, phosphorus, magnesium, chromium, chloride, molybdenum, fluoride, selenium, manganese, iodide, chromium	Water

[a]Certain amino acids involved in protein production (Chapter 5).
[b]Essential to prevent ketosis (Chapter 4).

Table 1-2a Examples of Nutraceutical Substances Grouped by Natural Food Source

Plant		Animal	Microbial
β-Glucan	Allicin	Conjugated linoleic acid (CLA)	*Saccharomyces boulardii*
Ascorbic acid	*d*-Limonene	Eicosapentaenoic acid (EPA)	(yeast)
γ-Tocotrienol	Genestein	Docosahexaenoic acid (DHA)	*Bifidobacterium bifidum*
Quercetin	Lycopene	Sphingolipids	*Bifidobacterium longum*
Luteolin	Hemicellulose	Choline	*Bifidobacterium infantis*
Cellulose	Lignin	Lecithin	*Lactobacillus acidophilus*
Lutein	Capsaicin	Calcium	(LC1)
Gallic acid	Geraniol	Ubiquinone (coenzyme Q_{10})	*Lactobacillus acidophilus*
Perillyl alcohol	β-Ionone	Selenium	(NCFB 1748)
Indole-3-carbinol	α-Tocopherol	Zinc	*Streptococcus salvarius*
	β-Carotene		subspecies *thermophilus*
Pectin	Nordihydro capsaicin		
Daidzein			
Glutathione	Selenium		
Potassium	Zeaxanthin		

Note: This table includes accepted and purported nutraceutical substances.

Source: Wildman REC, *The Handbook of Nutraceuticals and Functional Foods*, Boca Raton, FL: CRC Press, 2001.

Table 1-2b

Examples of Foods with High Nutraceutical Content

Nutraceutical Substance/Family	Foods of Remarkably High Content
Allyl sulfur compounds	Onions, garlic
Isoflavones	Soybeans and other legumes
Quercetin	Onion, red grapes, citrus fruits, broccoli, Italian yellow squash
Capsaicinoids	Pepper fruit
Eicosapentaenoic acid (EPA) and docosahexaenoic acid (DHA)	Fish oils
Lycopene	Tomatoes and tomato products
Isothiocyanates	Cruciferous vegetables
β-Glucan	Oat bran
Conjugated linoleic acid (CLA)	Beef and dairy
Resveratrol	Grapes (skin), red wine
β-Carotene	Citrus fruits, carrots, squash, pumpkin
Carnosol	Rosemary
Catechins	Teas, berries
Adenosine	Garlic, onion
Indoles	Cabbage, broccoli, cauliflower, kale, brussels sprouts
Curcumin	Turmeric
Ellagic acid	Grapes, strawberries, raspberries, walnuts
Anthocyanins	Red wine
3-*n*-butyl phthalide	Celery
Cellulose	Most plants (component of cell walls)

Note: The substances listed in this table include accepted and purported nutraceutical substances.

Source: Wildman REC, *The Handbook of Nutraceuticals and Functional Foods*, Boca Raton, FL: CRC Press, 2001.

Key Point

Nutrients nourish the body by supporting normal operations. Essential nutrients are required in the diet and nonessential nutrients are not.

which are recognized as helping prevent or treat disease processes and possibly enhancing normal physiological function. As an example, orange juice is viewed as a functional food because it contains vitamin C, folate, and other health-promoting substances. Functional foods can also include those items that are designed to optimize and enhance physical performance. For example, certain sport drinks (such as Gatorade and PowerAde) may be considered a functional food because their contents, namely fluid, carbohydrates, and electrolytes, may enhance physical performance.

Nutrition Recommendations: General Population Versus Athletes

Are nutrition recommendations for highly active people and competitive athletes different from the general population? It certainly might seem so with all the sport nutrition bars and drinks and supplements on the market. The truth is that many of the recommendations for the general public also apply to more active populations. For instance, recommendations for optimal athletic performance emphasize carbohydrates as the predominant source of energy in the diet, to be derived from foods such as whole grain products, fruits, vegetables, legumes, and low-fat dairy products. The same recommendation is made for the general population. In addition, although the quantity of protein required by athletes is higher, its percentage contribution to total energy in the diet is comparable to that for the general population. Furthermore, the required dietary level of many vitamins and minerals is

Table 1-3   Examples of Nutraceuticals Grouped by Mechanism of Action

Anticancer	Positive Influence on Blood Lipids	Antioxidant Activity	Anti-inflammatory	Osteogenic or Bone Protective
Capsaicin	β-Glucan	CLA	Linolenic acid	CLA
Genestein	γ-Tocotrienol	Ascorbic acid	EPA	Soy protein
Daidzein	δ-Tocotrienol	β-Carotene	DHA	Genestein
α-Tocotrienol	Monounsaturated fats (MUFAs)	Polyphenolics	Capsaicin	Daidzein
γ-Tocotrienol	Quercetin	Tocopherols	Quercetin	Calcium
CLA	ω-3 poly-unsaturated fats (PUFAs)	Tocotrienols	Curcumin	
Lactobacillus acidophilus	Resveratrol	Indole-3-carbinol		
Sphingolipids	Tannins	α-Tocopherol		
Limonene	β-Sitosterol	Ellagic acid		
Diallyl sulfide	Saponins	Lycopene		
Ajoene		Lutein		
α-Tocopherol		Glutathione		
Enterolactone		Hydroxytyrosol		
Glycyrrhizin		Luteolin		
Equol		Oleuropein		
Curcumin		Catechins		
Ellagic acid		Gingerol		
Lutein		Chlorogenic acid		
Carnosol		Tannins		
Lactobacillus bulgaricus				

Note: This table includes accepted and purported nutraceutical substances.

Source: Wildman REC, *The Handbook of Nutraceuticals and Functional Foods*, Boca Raton, FL: CRC Press, 2001.

similar to that for less active individuals, reflecting other processes that determine body status of these nutrients (such as absorption efficiency and decreased excretion). Water and electrolyte requirements are clearly different between athletes and the general population, but this is because of a need to recover sweat losses rather than an enhanced requirement.

Although the general recommendations for athletes and the general population are similar, during certain periods athletes may consume a specially designed or restrictive diet. This is often the case when they prepare for competition. For instance, wrestlers may restrict their carbohydrate and fluid intake prior to a match in order to reduce their body mass. Bodybuilders and fitness competitors may also restrict carbohydrate and water consumption for several days prior to competition to reduce the water content of muscle tissue, thereby increasing muscular definition. Some endurance cyclists and runners may refrain from whole grains, legumes, vegetables, and some fruits for several hours or a couple of days prior to competition in an attempt to decrease the potential for food intolerance and to lower total intestinal contents and perhaps reduce body weight. As they see it, a reduction of even a pound of body mass as intestinal content would

Key Point

Dietary recommendations for athletes during training are similar to the general population.

reduce workload at a given intensity during endurance competition. However, these practices are associated with competitions and are usually different from general nutritional recommendations during training.

■ BASIC NUTRITION CONCEPTS *IN REVIEW*

- Essential nutrients are required components of the diet and include water, two fatty acids, several amino acids, vitamins, and certain minerals.
- Nutraceuticals are substances found in foods that can help prevent and treat diseases (such as heart diseases and cancer) and include carotenoids, fibers, and flavonoids.
- Nutritional recommendations for the general population are similar to those for athletes.

RESOURCES FOR NUTRITION INFORMATION

We are bombarded with nutrition information. Although some of this information is credible and helpful, other information may be suspicious and linked to the marketing of products or services. Where can people go to obtain trustworthy information about nutrition? Regardless of athletic level, most people benefit from eating a diet that conforms to general recommendations made by federal institutions and respected private and professional organizations. For instance, government agencies within the United States, United Kingdom, Canada, Germany, and Australia have developed recommendations for nutritional intake. Recommendations made by professional organizations such as the American Dietetic Association (ADA), Dietitians of Canada (DC), Dietitians Association of Australia (DAA), and American College of Sports Medicine (ACSM) are designed to assist their professional members in educating the people they work with, and to provide guidance to the population at large. Private organizations also help shape the recommendations for nutritional intake and behavior. Some private organizations such as the American Heart Association (AHA), National Osteoporosis Foundation (NOF), and the American Cancer Society (ACS) are devoted to providing education designed to help prevent and treat a specific disease. Because the diseases they address are among the major causes of death, their nutritional recommendations are supportive of overall health.

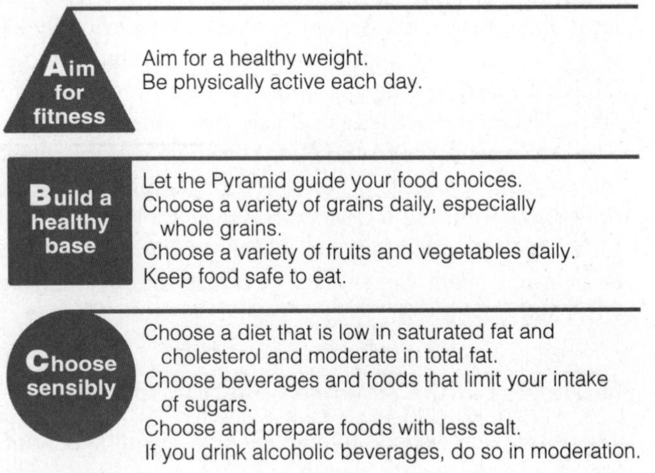

Aim for fitness
- Aim for a healthy weight.
- Be physically active each day.

Build a healthy base
- Let the Pyramid guide your food choices.
- Choose a variety of grains daily, especially whole grains.
- Choose a variety of fruits and vegetables daily.
- Keep food safe to eat.

Choose sensibly
- Choose a diet that is low in saturated fat and cholesterol and moderate in total fat.
- Choose beverages and foods that limit your intake of sugars.
- Choose and prepare foods with less salt.
- If you drink alcoholic beverages, do so in moderation.

Figure 1-1
Dietary guidelines for Americans: the ABCs of good health. [*Note*: These guidelines are intended for adults and healthy children ages 2 and older. *Source*: U.S. Department of Agriculture and U.S. Department of Health and Human Services, *Nutrition and Your Health: Dietary Guidelines for Americans*, Home and Garden Bulletin no. 232 (Washington, D.C.: 2000).]

Key Point

Departments within the governments of the United States, the United Kingdom, Canada, Australia, and Germany provide information for the public on nutrition and diet planning. Private organizations such as the American Heart Association and American Cancer Society also provide dietary information that can help shape nutrition habits.

Recommendations made by these organizations often state that individuals should "eat a balanced diet." What does it mean to eat a balanced and healthy diet? What kind of assistance is available to help people understand nutritional balance and plan their diet? Several resources aid in planning a healthy diet as well as identifying requirements for specific essential nutrients. The United States Department of Agriculture (USDA) and Department of Health and Human Services (DHHS) have developed the *Dietary Guidelines for Americans* and the *Food Guide Pyramid*. The Food and Nutrition Board has developed the Recommended Dietary Allowances (RDA; Appendix A). Other countries have also produced similar resources. For instance, the Canadian government has developed *Canada's Guidelines for Healthy Eating* and *Canada's Food Guide to Healthy Eating*. Australia has *The Australian Guide to Healthy Eating*. These and other educational resources are available on the Internet (Appendix C lists some useful web addresses).

Dietary Guidelines for Americans

The *Dietary Guidelines for Americans* (Figure 1-1) are designed to provide advice for healthy Americans over the age of 2 regarding food choices and physical activity. They promote health and aim to reduce the risk of diseases such as heart disease, certain types of cancer, diabetes, stroke, and osteoporosis. A healthy diet can also reduce major risk factors for chronic disease such as obesity, high blood pressure, and high blood cholesterol. The Dietary Guidelines recognize that food choices, lifestyle, environment, and family history all affect an individual's well-being. The 10 guidelines are sectioned into three compartments forming the ABCs (aim, build, choose) for good health.

AIM FOR FITNESS
- Aim for a healthy weight.
- Be physically active each day.

Following these two guidelines helps keep an individual healthy and fit. Healthy eating and regular physical activity enable people of all ages to work productively, enjoy life, and feel their best. They also help children grow, develop, and do well in school.

Key Point

The Dietary Guidelines provide general dietary and lifestyle recommendations to promote a healthy lifestyle and help prevent disease.

BUILD A HEALTHY BASE
- Let the Food Guide Pyramid structure food choices.
- Choose a variety of grains daily, especially whole grains.
- Choose a variety of fruits and vegetables daily.
- Keep food safe to eat.

Following these guidelines assists an individual in building a base for healthy eating. The Food Guide Pyramid can provide the guidance needed for a person to derive the nutrients the body needs each day. The guidelines also recommend that grains, fruits, and vegetables serve as the foundation of meals. This forms a base for good nutrition and good health and may reduce a person's risk of certain chronic diseases. Flexibility in food choices is encouraged by these guidelines, as well as taking precautions to make sure food choices are safe to eat.

CHOOSE SENSIBLY
- Choose a diet that is low in saturated fat and cholesterol and moderate in total fat.
- Choose beverages and foods to moderate your intake of sugars.
- Choose and prepare foods with less salt.
- If you drink alcoholic beverages, do so in moderation.

These guidelines help an individual make sensible choices that promote health and reduce his or her risk of certain chronic diseases. All foods can be enjoyed as part of a healthy diet as long as one does not overindulge in fat (especially saturated fat), sugars, salt, and alcohol.

Food Guide Pyramid

The Dietary Guidelines for Americans maps out general concepts and recommendations for diet and general fitness. However, it does not allow for meal planning and points to the Food Guide Pyramid (Figure 1-2) to provide diet planning. The Food Guide Pyramid illustrates the importance of balance among food groups in a daily eating pattern. Most of the daily servings of food should be selected from the food groups that are the largest in the picture and closest to the base of the pyramid. More specifically, the Food Guide Pyramid encourages people to:

- Choose most foods from the grain products group (6–11 servings), the vegetable group (3–5 servings), and the fruit group (2–4 servings).

- Eat moderate amounts of foods from the milk group (2–3 servings) and the meat and beans group (2–3 servings).
- Choose sparingly foods that provide few nutrients and are high in fat and sugars.

A range of servings is given for each food group. The smaller number is for people who consume on average 1600 kilocalories (kcal) a day, such as many sedentary women. The larger number is for those who consume about 2800 kcal a day, such as more active men. Individuals need to estimate their energy needs based on equations such as those discussed later in this chapter.

Recommended Dietary Allowances and Dietary Reference Intake

The *Recommended Dietary Allowances* (RDAs) date back to the 1940s and have endured a few revisions. They were designed to provide guidelines for the consumption of specific essential nutrients for different gender and age groups and for reproductive conditions (such as pregnancy and lactation). The RDAs were formulated based on available research information to prevent deficiency and promote adequate status of essential nutrients such as vitamins and minerals. The RDA levels were intended to be adequate for nearly all (97–98%) of the general population.

Recently the RDAs have been expanded into the new *Dietary Reference Intake* (DRI) levels (Table 1-4). The DRIs include the RDAs as well as the *Estimated Average Requirements (EARs)*, *Adequate Intakes (AIs)*, and *Tolerable Upper Intake Limits (ULs)* (Figure 1-3). The EAR is the level of a nutrient that would be adequate for roughly half of a gender and age population. The UL is intended to establish a safety ceiling for nutrient intake for individuals who consume a nutrient in levels that exceed the RDA. The AIs are similar to the RDAs in their design to meet the minimum needs of various populations. However, they differ from the RDA because they apply to nutrients for which the available research did not allow for an RDA to be established. Appendix A provides the most current RDA and DRI levels.

Guidelines of the American Heart Association

The AHA has also developed dietary recommendations for the general population. These recommendations

Key Point

The RDAs have recently been expanded into the Dietary Reference Intakes (DRIs) that include Estimated Average Requirement (EAR), Upper Limit (UL), and Adequate Intake (AI) levels.

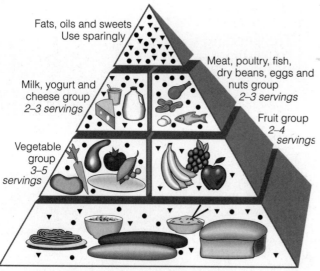

Fats, oils and sweets
Use sparingly

Meat, poultry, fish,
dry beans, eggs and
nuts group
2–3 servings

Milk, yogurt and
cheese group
2–3 servings

Fruit group
2–4 servings

Vegetable
group
3–5 servings

Bread, cereal, rice and pasta group
6–11 servings

KEY
● Fat (naturally occurring and added)
▼ Sugars (added)
These symbols show fats, oils and added sugars in foods.

Serving Sizes for the Food Guide Pyramid

Grain Products Group (bread, cereal, rice, and pasta)
1 slice of bread
1 oz of ready-to-eat cereal
1/2 cup of cooked cereal, rice, or pasta

Vegetable Group
1 cup of raw leafy vegetables
1/2 cup of other vegetables—cooked or chopped raw
3/4 cup of vegetable juice

Fruit Group
1 medium apple, banana, orange
1/2 cup of chopped, cooked, or canned fruit
3/4 cup of fruit juice

Milk Group (milk, yogurt, and cheese)
1 cup of milk or yogurt
1 1/2 oz of natural cheese
2 oz of processed cheese

Meat and Beans Group (meat, poultry, fish, dry beans, eggs, and nuts)
2–3 oz of cooked lean meat, poultry, or fish
1/2 cup of cooked dry beans or 1 egg counts as 1 oz of lean meat
2 tbs of peanut butter or 1/3 cup of nuts count as 1 oz of meat

Figure 1-2

Food Guide Pyramid: A guide to daily food choices. The breadth of the base shows that grains (breads, cereals, rice, and pasta) deserve most emphasis in the diet. The tip is smallest: use fats, oils, and sweets sparingly. Some foods fit into more than one group. Dry beans, peas, and lentils can be counted as servings in either the meat and beans group or vegetable group. These "crossover" foods can be counted as servings from either one or the other group, but not both. Serving sizes are based on both suggested and usually consumed portions necessary to achieve adequate nutrient intake. They differ from serving sizes on the Nutrition Facts label, which reflect portions usually consumed. (*Source*: http://www.nal.usda.gov/fnic/dga/dguide95.html)

Table 1-4

Definitions of Dietary Reference Intakes (DRIs)

Recommended Dietary Allowances (RDAs)	The RDA is the average daily dietary intake level that is sufficient to meet the nutrient requirements of nearly all healthy individuals within an age and gender group. RDAs are provided for essential nutrients for which there is a significant body of knowledge about human needs. RDAs are set at two standard deviations above the EAR for a nutrient.
Estimated Average Requirement (EAR)	The EAR is a nutrient intake value that is estimated to meet the requirement of half the healthy individuals in an age and gender group. The EAR is used to set the RDAs.
Adequate Intake (AI)	The AI level is based on observed or experimentally determined approximations of nutrient intake by a group (or groups) of healthy people. AIs are similar to RDAs and are set when RDAs cannot be established because research knowledge is incomplete.
Upper Limit (UL) (Tolerable Upper Intake Level)	The UL is the highest level of a daily nutrient intake that is likely to pose no threat of adverse health effects to almost all individuals within a population. As intake levels increase above the UL, the risk of adverse effects increases.

largely focus on creating a healthier blood lipid profile and secondarily on achieving a healthier body weight. These recommendations are available on the AHA website, and an abridged listing is presented in Table 1-5. The primary focus of the AHA is the prevention of heart disease, and its recommendations apply to the general population over the age of 2. Heart disease continues to be the major cause of death in the United States (Table 1-6) and a major cause of death in many other countries such as Australia, France, Germany, and the United Kingdom. The recommendations made by the AHA continue to be among the most influential and also help shape recommendations made by many other organizations around the world.

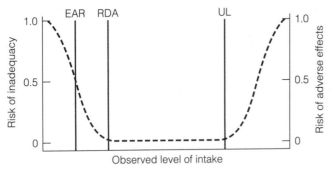

Figure 1-3
Dietary Reference Intake levels compared. The Estimated Average Requirement (EAR) is the intake at which the risk of inadequacy is 0.5 (50%) to an individual. The Recommended Dietary Allowance (RDA) is the intake at which the risk of inadequacy is very small—only 0.02 to 0.03 (2–3%). The Adequate Intake (AI) does not bear a consistent relationship to the EAR or the RDA because it is set without being able to estimate the average requirement. It is assumed that the AI is at or above the RDA if one could be calculated. At intakes between the RDA and the Tolerable Upper Intake Level (UL), the risks of inadequacy and of excess are both close to 0. At intakes above the UL, the risk of adverse effect may increase. (*Source*: DRI: Dietary Reference Intakes for Calcium, Phosphorus, Magnesium, Vitamin D, and Fluoride. Standing Committee on the Scientific Evaluation of Dietary Reference Intakes. Food and Nutrition Board. Institute of Medicine, National Academy Press, 1999.)

Table 1-5

Nutrition Guidelines Provided by the American Heart Association

- Total diet fat intake should not exceed 30% of total energy intake.
- Saturated fatty acid intake should not exceed 8–10% of total energy intake.
- Polyunsaturated fatty acid intake should not exceed 10% of total energy intake.
- Monounsaturated fatty acids can contribute up to 15% of total energy intake.
- Cholesterol intake should be <300 mg per day.
- Sodium intake should be <2400 mg per day (~1¼ teaspoons salt).
- Carbohydrate intake should contribute 55–60% or more of energy intake with emphasis on complex carbohydrates.
- Total energy should be adjusted to achieve and maintain a healthy body weight.

Source: American Heart Association web page (www.americanheart.org).

Key Point

Dietary recommendations made by the American Heart Association apply to the general population because heart disease continues to be the major cause of death in economically prosperous countries such as the United States.

Table 1-6

Ten Leading Causes of Death in the United States

Causes	Percentage of Total Deaths
1. **Heart disease**	31.4
2. **Cancers**	23.3
3. **Strokes**	6.9
4. Chronic obstructive lung disease	4.7
5. Accidents	4.1
6. Pneumonia and influenza	3.7
7. **Diabetes mellitus**	2.7
8. Suicide	1.3
9. Kidney disease	1.1
10. Chronic liver disease	1.1

Note: The diseases in bold type have relationships with diet.

5-a-Day for Better Health Program

The *5-a-Day for Better Health Program* promotes the consumption of five or more servings of fruits and vegetables every day for better health (Figure 1-4). This educational program was established in 1991 as a joint effort of the National Cancer Institute (NCI) in the U.S. Department of Health and Human Services and the Produce for Better Health Foundation, a nonprofit consumer education foundation representing the fruit and vegetable industry. The general concept of this program is based on epidemiological research and clinical trials that suggest that the consumption of at least five servings of fruits and vegetables daily reduces the risk of cancer and promotes general health. Currently, cancer is either the leading or second leading cause of mortality in countries such as the United States, United Kingdom, Australia, Germany, and Canada.

Key Point

The 5-a-Day for Better Health Program promotes the consumption of at least five servings of fruit and vegetables daily to prevent cancer and promote general health.

Food Labeling Information

Food labels also provide information to assist people in planning a healthier diet. Food labels in the United States include the *Nutrition Facts* (Figure 1-5), which were developed by the USDA. The Nutrition Facts include the following information:

- A listing of ingredients in descending order by weight
- Serving size

Figure 1-4
Promotion of better eating habits. (*Source*: www.5aday.gov)

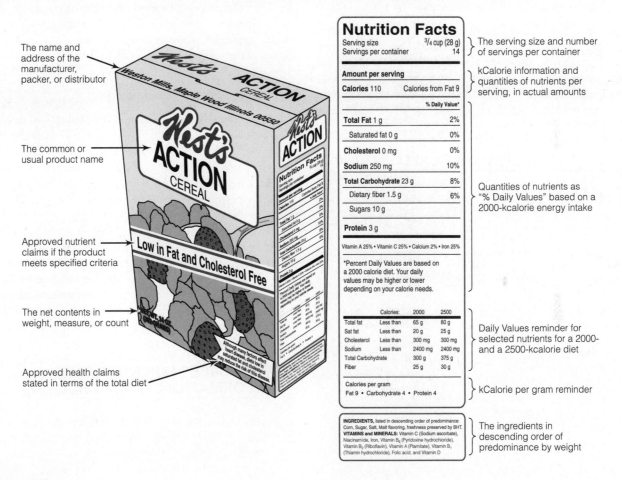

Figure 1-5
Example of a food label.

- Servings per container
- Amount of the following per serving: total energy, total protein, energy contributed by fat, total fat, saturated fat, cholesterol, total carbohydrate, sugar, dietary fiber, vitamin A, vitamin C, calcium, iron, sodium

DAILY VALUES. Because many individuals try to plan their intake of specific nutrients, the Nutrition Facts also include the *Daily Value* for each nutrient. The Daily Value

uses reference nutrition standards to describe how a single serving of a food item relates to the standards. The Daily Values are based on current nutritional recommendation standards, which include the following guidelines:

- A maximum of 30% total energy from fat, or <65 grams (g) total
- A maximum of 10% total energy from saturated fat, or < 20 g

Key Point

Daily Values provided on Nutrition Facts food labels are based on RDA levels as well as current nutrition standards for health.

- A minimum of 60% total energy from carbohydrate
- 10% of total energy from protein
- 10 g of fiber per 1000 kcal
- A maximum of 300 mg of cholesterol
- A maximum of 2400 mg of sodium

The Daily Values for other nutrients such as vitamins A and C, thiamin, riboflavin, niacin, calcium, and iron, are based on RDAs. However, the Daily Value standards are not specific for gender and age, and therefore one quantity applies to all people. The Daily Value is expressed as a percentage and is based on a 2000- or 2500-kcal intake

Table 1-7   Guidelines for Food Label Claims

Food Label Claim	Guidelines for the Food Label Claim (per serving)
Fat free	Must have less than 0.5 g per serving
Saturated fat free	Must contain less than 0.5 g per serving
Cholesterol free	Must contain less than 2 mg per serving
Sugar free	Must contain less than 0.5 g per serving
Sodium free	Must contain less than 5 mg per serving
Calorie free	Must contain less than 5 kcal per serving
Low fat	Must contain no more than 3 g of fat per serving
Low sodium	Must contain less than 40 mg per serving
Low calories	Must contain less than 40 kcal per serving
Low cholesterol	Must contain less than 20 mg per serving
High or good source	One serving must contain at least 20% or more of the recommendation for that nutrient
Reduced, less, or fewer	Must contain at least 25% less of a nutrient, per serving, as compared to the same nutrient in a reference food
More or added	Must contain at least 10% more of the DV for a nutrient as compared to a reference food
Light or lite	The food have at least 50% less fat than a similar, unmodified food which in its unmodified form contains more than 50% of its calories from fat
Lean (meat, fish, poultry)	Must contain less than 10 g of fat, 4 g of saturated fat, and 95 mg of cholesterol per 100 g of the food
Extra lean	Must contain less than 5 g of fat, 2 g of saturated fat, and 95 mg of cholesterol per 100 g of the food

Key Point

The Exchange Lists, developed to help individuals with diabetes mellitus plan their diets, can assist the general population in weight management as well as help structure the diet of athletes.

that approximates recommended energy intake for most people. Therefore, a food providing 250 kcal per serving is listed as 13% or 10% DV for a 2000- or 2500-kcal diet intake, respectively. Beyond the Nutrition Facts, food manufacturers must also follow federal guidelines for other statements they choose to make on a food label. Some of the statements are listed in Table 1-7.

Exchange Lists for Meal Planning

Many people use the Exchange Lists (Table 1-8) to plan a diet consisting of a specific energy level and containing a desirable level of carbohydrate and protein. This system was originally developed to assist individuals with diabetes mellitus in diet planning. However, it has served as the basis for many popular diet programs, and many athletes also use this system for diet planning.

In the Exchange Lists for Meal Planning, foods are listed in three groups (carbohydrate, meat and meat substitutes, and fats) based on their content of energy nutrients. The carbohydrate group has five subgroups (starch, fruit, milk, other carbohydrates, and vegetables), the meat and meat substitutes group contains four subgroups (very lean, lean, medium fat, and high fat), and the fat group has

■ RESOURCES FOR NUTRITION INFORMATION *IN REVIEW*

- Several federal, professional, and private organizations have developed nutritional guidelines, recommendations, and diet planning tools to educate people and help them plan a healthy diet.
- The Dietary Guidelines for Americans provides general recommendations for diet and activity, and the Food Guide Pyramid provides a framework for diet planning.
- The RDAs have been expanded and are now a component of the DRIs, which also include the EAR, AI, and UL levels.
- The 5-a-Day for Better Health Program promotes the consumption of at least five servings of fruits and vegetables daily to decrease the risk of cancer and other diseases.
- Other tools such as the Exchange Lists for meal planning can assist people in diet planning.

| Table 1-8 | **The Exchange List for Meal Planning** |

Food Group/List	Carbohydrate	Protein	Fat	Calories (kcal)
Carbohydrate				
Starch	15	3	1 or less	80
Fruit	15	—	—	60
Milk				
Skim milk	12	8	0–3	90
Low-fat milk	12	8	5	120
Whole milk	12	8	8	150
Other carbohydrates	15	Varies	Varies	Varies
Vegetables	5	2	—	25
Meat and Meat Substitutes				
Very lean	7	—	0–1	35
Lean	7	—	3	55
Medium-fat	7	—	5	75
High-fat	7	—	8	100
Fat	—	—	5	45

Note: A complete listing of foods for each group can be obtained at the website of the American Diabetes Association (www.diabetes.org).

no subgroups. The number of servings for each group is determined, and individuals may exchange items within each group to provide variety and better compliance. Refer to www.diabetes.org for sample diet plans using the Exchange List system for various energy levels, which can be used by the general population in weight management as well as by more active populations and competitive athletes.

■ DIET ASSESSMENT TOOLS AND TECHNIQUES

Many people try to eat a diet that promotes greater health and well-being. How can they be sure of that? Are there ways for people to analyze their diet to validate their efforts and to see what modifications might be needed? An individual's diet composition can be assessed using a variety of tools to understand the types of foods eaten as well as the nutrient composition of the diet and its compliance with current health recommendations. An accurate nutrition assessment also provides valuable information to assist in planning a diet that optimizes performance.

The level and composition of dietary energy and nutrient intake is fairly easy to assess. Such nutritional assessment is commonly practiced by nutritionists working with people in the general population as well as active people and competitive athletes. An accurate diet assessment begins with a *survey* of the food and beverages consumed. Diets can be assessed using nutrition software with up-to-date food databases.

24-Hour Diet Recall

Nutritionists can perform a *24-hour diet recall*, in which individuals verbalize or write down everything they ate and drank over the previous 24 hours (Figure 1-6). The accuracy of this method depends on a person's ability to recall all the food and beverages consumed as well as the quantity and any special preparation. Nutritionists can use food models and pictures to increase the accuracy of the recall. The food and beverages recorded can then be analyzed by hand using nutrition composition tables (Appendix B) or assessed by nutrition assessment software.

Food Diary

Athletes and other individuals can also log the foods and beverages they consume over an established period such as 3, 5, or 7 days. The log is referred to as a *food diary*, and it is the individual's responsibility to precisely record the type, quantity, and preparation style of foods consumed. The accuracy of a food diary increases with appropriate instruction from a trained nutritionist. This is especially true for instruction in recording accurate estimates of food weights and volumes. As with the 24-hour diet recall, the food and beverages logged in the food diary can then be assessed by hand or by utilizing computer software.

24–Hour Diet Recall

Name: _____   Date: _____

Age: _____

Gender: _____

Please list all food and beverages you consumed over the past 24 hours. Please describe the quantity (or estimated size) and any preparation of the food/beverage as well as the manufacturer or restaurant of purchase/consumption.

Food or Beverage Item	Time of Day and Location	Amount	Preparation/Manufacturer or Restaurant

Figure 1-6
Example of a 24-hour diet recall form.

Key Point

Food diaries are logs of food, beverages, and supplements consumed by a person during a specified period (for example, 3 days or a week). The information collected can then be assessed for nutrient composition.

The advantage a food diary has over 24-hour recall is that it provides an assessment of an individual's diet over a longer period. This allows nutritional intake to be averaged and provides a more accurate assessment of the diet and its ability to meet nutrient intake requirements and comply with recommendations. Furthermore, a food diary can accommodate an atypical day of eating (such as a holiday or traveling day) in an assessment of a person's diet.

Food Frequency Questionnaire

Food frequency questionnaires (FFQs) are different from a diet recall or food diary because they ask an individual to indicate the frequency of consumption of specific foods during specific periods. For example, for a question about orange juice consumption the participant indicates how often a cup of orange juice was consumed (daily, weekly, monthly, or never). Information derived from an FFQ provides more insight into the intake levels of different nutrients or nutrient classes over longer periods than does a diet recall or a short-duration food diary. FFQs can have special application to the nutrition assessment of athletic populations during the off-season and the competitive season.

Computer Software for Nutrition Assessment

Diet information collected in the nutrition survey instruments mentioned above can be assessed for nutrient composition and compliance with health recommendations by computer software. Software, such as that provided with this book, can provide baseline information regarding an individual's nutritional intake and allow planning for diet restructuring if needed. Typically, software applications provide estimations of the total energy level and a breakdown of the contributions made by carbohydrate, protein, and fat. In addition, estimations of water, specific carbohydrates, amino acids, fatty acids, vitamins, and minerals are provided. The software can also show ratios important to health, such as ratios of sodium to potassium and of omega-6 to omega-3 polyunsaturated fatty acids.

■ DIET ASSESSMENT TOOLS AND TECHNIQUES *IN REVIEW*

- Tools that describe diet intake and patterns and are used to assess nutritional intake include a 24-hour recall, a food diary, and a food frequency questionnaire.
- Diet information gathered with these tools can be assessed for nutrient composition and appropriateness for general health and physical performance using computer software.

■ CURRENT HEALTH STATUS AND PROMOTION

What does it mean to be healthy? Are people simply healthy or unhealthy, or can there be degrees of health? **Health** is actually a general term and may describe the absence of disease or disorders. It can also be used to address more specific physiological and psychological states such as mental health or cardiovascular health. Health is better considered as a continuum rather than in opposing categories such as "good health" and "bad health." People are said to be in good general health in the absence of infection or illness and when their risk factors for diseases, such as heart disease, are minimal.

Like health, a person's level of **fitness** can be considered on a scale that considers factors such as cardiorespiratory endurance, muscular strength, muscular endurance, body composition, and flexibility. Without question, two of the most significant factors influencing health and fitness are nutrition intake and physical activity.

Health Promotion

Diet and exercise have been topics of concern for a long time. In fact, references to these topics date back to the ancient civilizations of Syria, Egypt, Macedonia, Arabia, Mesopotamia and Persia, India, and China. The interest and discussions continued into modern times, and in the late 1970s and early 1980s several major health organizations took a broader approach to address the health concerns of Americans and used the term "health promotion" to stage the battle. The previous and narrow approach of controlling disease did not lend itself to all the benefits provided by diet, exercise, and other lifestyle changes. Health promotion incorporates the science and art of helping people manage their health status and prevent disease.

In 1979 the surgeon general reported on health promotion and disease prevention, laying the groundwork for health initiatives in the 1980s. The report, *Promoting*

Key Point

Health can be considered as a continuum in which a healthier person is free of illness and presents low risk factors for disease.

Health/Preventing Disease: Objectives for the Nation, published in 1980, listed 226 health objectives for 1990 packaged into 15 areas, two of them related to nutrition and exercise.

Research into and promotion of healthy lifestyles continued through the 1980s, and at the end of that decade the Public Health Service set out several objectives to be met by the year 2000. These were published in 1990 in *Healthy People 2000: National Health Promotion and Disease Prevention Objectives*. In 2000 *Healthy People 2010* came out. The agenda and several strategies for achieving its goals are outlined in Table 1-9. These reports maintain

Table 1-9

Healthy People 2010 Objectives to Increase Physical Activity

- Reduce to 20% the proportion of adults who engage in no leisure-time physical activity.
- Increase to 30% the proportion of adults who engage regularly, preferably daily, in moderate physical activity for at least 30 minutes per day.
- Increase to 30% the proportion of adults who engage in vigorous physical activity that promotes the development and maintenance of cardiorespiratory fitness 3 or more days per week for 20 or more minutes per occasion.
- Increase to 30% the proportion of adults who perform physical activities that enhance and maintain muscular strength and endurance.
- Increase to 43% the proportion of adults who perform physical activities that enhance and maintain flexibility.
- Increase to 35% the proportion of adolescents who engage in moderate physical activity for at least 30 minutes on 5 or more of the previous 7 days.
- Increase to 85% the proportion of adolescents who engage in vigorous physical activity that promotes cardiorespiratory fitness 3 or more days per week for 20 or more minutes per occasion.
- Increase the proportion of the nation's public and private schools that require daily physical education for all students (25% for middle schools and 5% for high schools).
- Increase to 50% the proportion of adolescents who participate in daily school physical education.
- Increase to 50% the proportion of adolescents who spend at least half their school physical education time in physical activity.
- Increase to 75% the proportion of adolescents who view television 2 or fewer hours on a school day.
- Increase access to physical activity spaces and facilities outside of school hours.
- Increase the proportion of trips made by walking and bicycling.
- Increase to 75% the proportion of worksites offering employer-sponsored physical activity and fitness programs.

Source: U.S. Department of Health and Human Services, Public Health Service. *Healthy People 2010: National Health Promotion and Disease Prevention Objectives* (2000).

that people between the ages of 25 and 65 can control most of the major causes of death through lifestyle changes.

Physical Activity Estimates for American Adults

How many people today are getting enough exercise? What is the current trend—are more people exercising regularly today in comparison to years gone by? In the early 1990s the National Center for Health Statistics collected data using the *National Health Interview Survey of Health Promotion and Disease Prevention*, and the Centers for Disease Control and Prevention (CDC) collected data using the *Behavioral Risk Factor Surveillance System* (BHRFSS). From these surveys the CDC concluded that despite all the efforts, information, and promotions, the activity status of the United States was disappointingly low. An estimated 74% of the American public failed to meet recommendations for physical activity. Only 10–15% of Americans were estimated to exercise regularly at or above a moderate intensity level for 20 minutes or longer, three or more times per week, as recommended for cardiovascular benefit. More current estimates state that only approximately 22% of the population get enough exercise, 54% spend inadequate time in physical activity, and 24% are completely sedentary.

Certain economic and social factors may be related to the level of exercise. The CDC collects data through its Behavioral Risk Factor Surveillance System and determined that both education level and income have been associated with exercise patterns in the United States (Figure 1-7). Racial and geographic differences in exercise patterns have also been determined in the United States (Table 1-10). For instance, in a previous survey, people living in California, Oregon, Washington, New Mexico, Colorado, Utah, and Florida were found to be the most active. People in Idaho, Montana, North and South Dakota, Minnesota, and Texas were moderately

Key Point

Roughly three-fourths of the adult population in the United States does not engage in recommended levels of regular physical activity.

active. Least active were people in Oklahoma, Tennessee, Louisiana, Georgia, Indiana, Michigan, Illinois, Maryland, Virginia, and North and South Carolina.

Current data also indicate that males are slightly more active than females and that the elderly tend to exercise less than other groups, relative to appropriate exercise levels. The most common forms of exercise remain walking, jogging, bicycling, and gardening/yard work.

Physical Activity Estimates for Children

Similar data have been collected on the youth of America and the results are surprisingly poor. Several major surveys in the past included the President's Council on Physical Fitness in 1980, Sports School Population Fitness Survey of 1985, Second National Children and Youth Fitness Study of 1987, and the Youth Fitness Behavior Surveillance System of 1991. These studies collectively

Table 1-10   **Race and Physical Activity**

Race	Physically Active (%)	Not Active (%)
White	75.2	24.8
Black	66.8	33.2
Hispanic	68	32
Other	72.5	27.5

Source: CDC National Statistics, 2000.

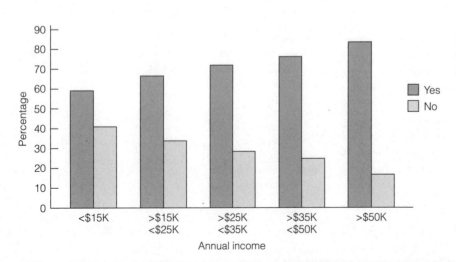

Figure 1-7
Income level and physical activity. (*Source:* Adapted from CDC National Statistics, 2000.)

determined that half of the students surveyed were not able to run a mile in under 10 minutes, flexibility among boys was poor, physical education classes were not taken, body composition (body fat %) had worsened, 50% of the physical activities in schools were not adequate for improving fitness levels, 40% of the students had poor upper body muscular fitness, and physical education programs did not focus on fitness. In the Fall 2000 report to the president, the secretary of health and human services and the secretary of education stated:

> Our nation's young people are, in large measure, inactive, unfit, and increasingly overweight. Physical inactivity threatens to reverse the decades-long progress in reducing deaths from cardiovascular diseases. Ultimately this could have a devastating impact on our national health care budget.

This is an alarming statement, and strategies for addressing these problems have been formulated. See the CDC web page (Appendix C) for the complete report, strategies, and specific objectives.

Nutritional Status and Health

Technological and social changes including TV, video games, computers, and more leisure time resulting from the overall success of our economy seem to have changed the physical activity status of the U.S. population for the worse. How have they affected the nutritional status of our population? At the outset of the 20th century many nutrition concerns were related to nutrient deficiencies, including iron deficiency anemia as well as goiter and rickets, which result from iodine and vitamin D deficiencies, respectively. In the last several decades the occurrence of many of these disorders in developed countries such as the United States has been significantly reduced and/or eradicated by education efforts, product enrichment, fortification, and supplementation. As the 21st century begins, many primary nutritional problems stem from nutritional excesses or imbalances. These include obesity and elevated blood lipid levels.

In the United States, nutrition information has been collected for several decades using survey instruments such as the *National Health and Nutrition Examination Survey* (NHANES). The general population of the United States appears to be increasing in body fat and continues to eat a poor diet relative to recommendations by health agencies and institutions.

BODY WEIGHT AND OBESITY. In 1998 more than 50% (98 million) of the adult population was estimated to be overweight, and current estimates state that an estimated 120 million adults in the United States are overweight or obese (69 million are overweight, 51 million obese). Currently, 61% of American adults, age 20 years and over, are overweight, and 26% are obese.

These numbers have been escalating: the estimated percentage of obese adults increased from 14.5% in 1974 to 22.5% in 1994 and now to 26%. The percentage of obese children has risen as well; in fact the incidence of obesity in this population is among the fastest growing. Chapter 7 discusses obesity further, as it is a major health concern and strongly linked to heart disease, stroke, type 2 diabetes mellitus, some cancers, and a variety of other disorders.

DIET COMPOSITION. As discussed above, many public and private health agencies recommend that people eat a variety of foods that include at least five servings of fruits and vegetables per day and a low level of saturated fat. However, survey data in the United States suggests that more than three-fourths of the adult population fail to eat five or more servings of fruits and vegetables daily. Geography seems to play a role in this. For instance, the states of Louisiana, Georgia, Ohio, West Virginia, Mississippi, Arizona, New Mexico, Kentucky, Iowa, Missouri, South Dakota, and Nebraska have the highest percentage of people failing to meet this recommendation.

■ CURRENT HEALTH STATUS AND PROMOTION *IN REVIEW*

- Only about 20% of adults in the United States engage in regular physical activity.
- More than half the adult population in the United States is overweight, and the percentage of obese adults has steadily increased from 14.5% to 26% from the 1970s to the 2000s.
- Obese children are one of the fastest growing populations and are likely to become obese adults.
- Less than one-fourth of the U.S. adult population eats five or more servings of fruits and vegetables daily.

■ MAJOR DISEASES AND RISK FACTORS

How important are diet and exercise in the prevention of disease? Without question, regular physical activity can reduce the risk for several diseases and risk factors for many others, especially when it is complemented by a balanced diet. However, active people are not resistant to disease and, like the population at large, must be aware of the major risk factors for diseases such as heart disease, osteoporosis, and cancers. This means that all people must be aware of disease **risk factors**, elements that contribute to a disease or process, and how influential they are.

Disease literally means "lack of ease" and describes simple infections resulting in an immune system response and more life-threatening aberrations such as heart

Key Point

Regular physical activity can reduce an individual's risk for diseases such as heart disease and osteoporosis but does not eliminate the risk. People at all activity levels should be aware of risk factors for diseases.

disease, stroke, cancer, and osteoporosis. This section focuses on aspects related to the major diseases (heart disease, cancer, osteoporosis) associated with **mortality** (death) and **morbidity** (the rate of illness) in the United States and other economically developed countries. Obesity and diabetes mellitus are discussed in later chapters.

Genetic Influences on Health

It is often said that some people are "born to be healthier" than others. This implies that genetic endowments, or **heredity**, are involved in determining a person's present and future health. This notion is derived from the fact that "family history" is one of the strongest uncontrollable risk factors for certain diseases. For instance, it seems that members of some families are prone to heart attacks and members of other families to certain forms of cancer.

Certain genes, segments within chromosomes that instruct a cell how to make particular proteins, are linked to several forms of cancer, heart disease, and stroke. These genes are passed on when a person is conceived and may be passed on to his or her children as well. For an individual whose family history suggests a genetic predisposition to develop a certain disease it is important to understand and minimize the other risk factors for that disease.

Genetic links to certain diseases result from alterations, or **mutations**, in normal genes so that they support the development of the disease. Genetic predisposition does not mean that a person inherited an extra or a lesser number of genes, but rather alterations in normal genes. Furthermore, a normal gene can be mutated during a person's life, for example by exposure to radiation or toxins from tobacco smoke. Altered genes can result in the production of proteins with abnormal structure and/or properties, which in turn may support a physiological imbalance contributing to the development of a given disease.

Acute Disease (Infection) and Exercise

Although much discussion of the benefits of exercise focuses on its protective influences against *chronic* degenerative

Key Point

Family history or heredity is one of the most important factors that place a person at greater risk for diseases such as heart disease and cancer.

Key Point

Regular cardiorespiratory exercise may improve immune function and reduce the potential for microbial infection and illness.

diseases (such as heart disease and osteoporosis), regular exercise is also believed to provide some protection against *acute* infections such as the common cold. However, a highly intense and exhaustive bout of exercise or highly strenuous training is claimed to have the opposite effect, rendering one more prone to infection immediately following the exercise bout.

Despite numerous research investigations in this area, it is still difficult to confidently conclude that regular exercise decreases one's risk of infection. Besides its direct physical effects, exercise contributes to psychological and behavioral aspects of life, which have a complicated relationship to decreased incidence of infection and disease. The level of training may be an important factor in predicting the effect of exercise training on sickness. For instance, cardiorespiratory training that increases aerobic potential (as measured by VO_2max, the maximal capacity for oxygen consumption) has been reported to improve immune system potential. However, training that does not result in changes in VO_2max may not evoke positive immune benefit. Therefore, training at a level that allows for cardiorespiratory adaptation may be an important component.

Chronic Diseases

It is often said that a person "catches a cold." This is an acute scenario in which disease symptoms are related to viral infection. In contrast, the major diseases that affect health and lead to death, such as heart disease, cancers, and osteoporosis, are processes that develop over time. Therefore they should be viewed as chronic disorders. Misleadingly, the end result of these processes is often viewed as the disease itself. For instance, heart attacks are much more common as a person ages, which may lead people to falsely believe they do not need to be concerned about heart disease when they are younger. In reality, the process leading up to that heart attack may have started decades earlier.

Key Point

Diseases such as heart disease and osteoporosis are chronic disorders whose processes develop over years and decades.

PROGRESSION OF CHRONIC DISEASES.   Heart disease, cancers, and osteoporosis often start earlier in life than many people realize. For instance, the beginnings of heart disease have been noticed in the arteries of children. For osteoporosis, bones slowly begin to lose structural material during early adulthood, which renders them more susceptible to fractures later in life. The accumulation of mutations leading to cancer can begin in childhood, adolescence, or early adulthood upon initial exposure to cancer-promoting substances or practices (such as smoking and sun exposure).

So, although the critical stage of heart disease, osteoporosis, or cancer is usually associated with older individuals, the initiation probably occurred decades earlier. Once the disease is initiated, slowing its progression is the most important factor in preventing its life-threatening stage (heart attack, hip fracture, cancer). With this in mind, people with a family history or who present risk factors of a disease must manage the "controllable" aspects of their diet and lifestyle to help prevent a rapid development of the disease.

Heart Disease and Stroke

Atherosclerosis is a disease process in which lipid material (largely cholesterol), fibrous proteins, minerals, and other substances accumulate in the walls of arteries. As shown in Figures 1-8 and 1-9, the accumulation of this material can enlarge the wall of the artery and narrow the area in which the blood flows. Atherosclerosis occurs in smaller arteries throughout the body, but the blood vessels in the heart and the brain tend to be the most prone to it. Atherosclerosis of arteries serving heart tissue is one form of **heart disease**, which is any pathological condition of the heart. When blood flow becomes critically compromised, the tissue downstream can die, resulting in myocardial infarction (MI, called "heart attack") if it is in the heart and cerebrovascular accident (CVA, called "stroke") if it is in the brain.

Atherosclerosis is often described as a chronic inflammatory response brought on by tissue injury. Injury to the lining of the arterial wall allows **low-density lipoproteins (LDLs)**, the major carrier of cholesterol in the blood, to freely enter the vessel wall. **Monocytes**, a type of white blood cell, also enter the damaged wall and are transformed into **macrophages**. Macrophages generally function by engulfing foreign matter and tissue debris in a housekeeping effort. LDLs that have been oxidized by **free radicals**, molecules that are highly reactive because they have an unpaired electron, are engulfed by macrophages. Macrophages continue to gorge themselves and eventually burst, spilling their cholesterol content and contributing

Key Point

The major killers of people in developed countries result from disease processes that develop over years or decades. Therefore, prevention often refers to slowing the progression of the disease process.

to the growing atherosclerotic plaque. In addition, smooth muscle cells migrate from the central region of the artery and also engulf oxidized LDLs as well as depositing fibrous proteins. Calcium and other minerals form complexes and adhere to proteins, contributing to the increasing hardness of the plaque.

It appears that an initial injury to the wall of an artery begins the atherosclerotic process. The initial injury may be associated with high blood pressure, cigarette smoking, or elevated blood lipids. The timing of the initial injury may vary from person to person; however, signs of atherosclerosis have become a common observation in children 10–12 years old. Because initiation can occur early in life, the prevention focus is on slowing the progression of atherosclerosis. The rate of progression is strongly influenced by the same risk factors that can cause injury: high blood pressure, elevated LDL-cholesterol, and smoking.

RISK FACTORS FOR HEART DISEASE. Aging and family history of an early heart attack are strong risk factors for heart disease and stroke. However, since these factors are deemed uncontrollable, it is more prudent to attempt to manage the controllable risk factors. These risk factors include both dietary and behavior factors. For instance, a high-fat diet and obesity are associated with elevated blood cholesterol levels. Also, chronic high blood pressure increases the physical trauma to arterial walls, and smoking increases the level of oxidized LDL. Exercise is one of the most significant prevention tools because it increases the vascularity and strength of the heart and the level of cholesterol associated with high-density lipoproteins (HDLs) instead of LDLs. Although an exact mechanism remains elusive, higher levels of HDL-cholesterol are strongly associated with decreased risk of heart disease.

Sadly, a greater number of children are presenting risk factors associated with heart disease. It is not uncommon for children today to exhibit excessive body weight and obesity as well as high blood pressure and hypercholesterolemia. The American Heart Association (AHA) estimates that 2.1 million adolescents between the age of 12 and 17 are cigarette smokers and at least 9 million American children under the age of 5 live with at least one smoker exposing them to secondhand smoke. Alarmingly, the AHA estimates that 3000 American young people become smokers each day.

Key Point

Atherosclerosis is a disease process in which lipid and other materials accumulate in the walls of arteries, progressively narrowing the vessel and restricting blood flow to downstream tissue.

EXERCISE AND HEART DISEASE RISK. Lack of physical activity is recognized as an independent risk factor for heart disease (Table 1-11). Although scientists have yet to find a

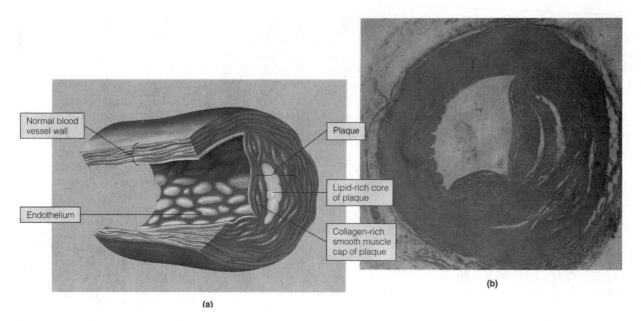

Figure 1-8
Atherosclerotic Plaque (a) Schematic representation of the components of a plaque. (b) Photomicrograph of a severe atherosclerotic plaque in a coronary vessel.

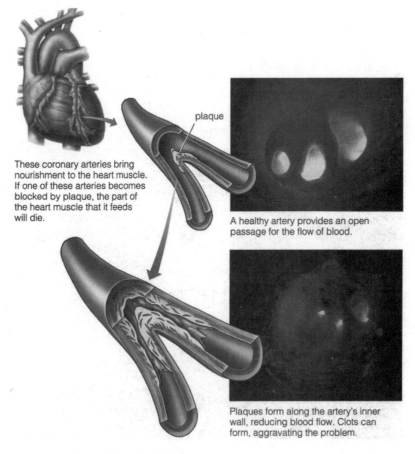

These coronary arteries bring nourishment to the heart muscle. If one of these arteries becomes blocked by plaque, the part of the heart muscle that it feeds will die.

plaque

A healthy artery provides an open passage for the flow of blood.

Plaques form along the artery's inner wall, reducing blood flow. Clots can form, aggravating the problem.

Figure 1-9
The formation of plaques in atherosclerosis

Table 1-11	**Risk Factors for Heart Disease**

Factor	Risk Threshold or Condition
Age	Men over 45 years old Women over 55 years old
Family history	Sudden death or myocardial infarction before 55 (father) or 65 (mother)
Cigarette smoking*	Risk increases with level of smoking or exposure
Hypertension*	Blood pressure >140/90 mm Hg
Dyslipidemia*	Total serum cholesterol >200 mg/dl and/or HDL-cholesterol <35 mg/dl
Diabetes mellitus*	Type 1 DM at >30 years old, type 1 DM longer than 15 years, or type 2 DM longer than 35 years
Sedentary lifestyle*	No regular exercise or active recreational pursuits (physical inactivity)
Obesity*	% Fat: >25% in men, >33% in women Body mass index (weight in kilograms divided by square of height in meters; kg/m^2): >30 for men and women

*Denotes modifiable risk factors

Source: Adapted and modified from American College of Sports Medicine, *ACSM's Guidelines for Exercise Testing and Prescription,* 5th ed. Baltimore: Williams & Wilkins, 1995. Table 1-2, p. 18.

direct relationship between regular exercise and stroke, they acknowledge that regular exercise reduces associated conditions such as atherosclerosis. Furthermore, less active and less fit people have a 30–50% greater risk of developing high blood pressure, another risk factor for heart disease.

Current estimates suggest that up to 250,000 deaths per year in the United States, approximately 12% of total deaths, may be attributable to a lack of regular exercise. It seems that the relative risk of heart disease for an inactive person is 1.5–2.4 times greater than an active person's risk (Table 1-11). This means that lack of regular exercise may be just as dangerous as other controllable risk factors, namely hypercholesterolemia, high blood pressure, and cigarette smoking!

Inactivity is also a growing concern related to children. Children spend an average of 17 hours a week watching TV in addition to the time they spend on video and computer games. It is clear that inactive children are more likely to become inactive adults. To address the health concerns of children and the likelihood of developing major health problems, the AHA has outlined several preventive strategies:

- Engage in regular physical activity.
- Consume a diet low in fat and cholesterol after the age of 2.
- Refrain from the use of tobacco, primarily cigarette smoking.

Key Point

Children in the United States are increasing their chances for developing heart disease by becoming overweight, smoking, and avoiding regular physical activity.

- Maintain appropriate weight levels for a given height.
- Receive regular physical examinations by a physician.

Cancer

Each year more than a half million people in the United States alone die of cancer. This translates to more than 1500 people a day. Cancer follows heart disease on the list of leading causes of death in the United States; one of every four deaths is the result of cancer. The National Institutes of Health (NIH) estimates the overall costs related to the treatment for cancer to exceed $180 billion in the United States.

Cancer is a group of related diseases in which cells lose their ability to govern their own rate of reproduction. As these cells reproduce, a mass of abnormal tissue develops that can invade neighboring tissue and potentially send cancerous cells throughout the body. In order for a cell to lose control it must sustain several mutations in genes that code for proteins involved in cell reproduction. These genes include the *proto-oncogenes* and *tumor suppressor genes*.

It can take decades for an individual to accumulate the necessary mutations to produce a tumor cell. Therefore, cancer development is often viewed as a process just like heart disease. However, its early progression is at the cellular rather than the tissue level. Once a cell accumulates the required number of mutations so that it reproduces uncontrollably, the disease process becomes more like heart disease or osteoporosis in the involvement of a significant amount of tissue. The growing mass of tissue can remain localized or spread throughout the body (metastasize).

One or a combination of three primary methods is utilized to treat tumors and cancer. These methods are surgical removal, radiation, and chemotherapy (treatment with chemicals). Other supportive modalities include hormone treatment, immunotherapy, laser therapy, cryosurgery, and electrodesiccation. The last two methods involve tissue destruction with cold and heat.

Risk of cancer can be defined in a couple of ways. For instance, "lifetime risk" is the probability that a person will develop some type of cancer during his or her lifetime. For American males and females the lifetime risk is roughly one in two and one in three, respectively.

Risk can also be defined as the probability of developing a certain type of cancer relative to exposure to or engagement in a particular risk factor. Some examples of relative risk:

- Male smokers are about 20 times more likely to develop lung cancer than a nonsmoker.

- Women who have a first-degree (mother, sister) family history of breast cancer have twice the risk of developing breast cancer compared to women who do not have a first-degree history of it.
- The incidence of prostate cancer increases with age; 70% of all diagnoses occur in men over 65 years old.

DIET, EXERCISE, AND CANCER RISK.   Diet and physical activity represent the two most significant controllable factors in preventing cancer for nonsmokers, according to the American Cancer Society (ACS). The ACS says that about one-third of the cancer deaths occurring each year are attributable to diet and physical activity–related factors such as obesity. The ACS has updated its *Nutrition and Physical Activity Guidelines*, and they are outlined in Table 1-12. These guidelines also include an explicit *Recommendation for Community Action* to facilitate healthy food choices and opportunities for physical activities in schools, worksites, and communities.

Osteoporosis

Osteoporosis is a condition in which the bones have become porous and weak. It is a major health issue in the United States, threatening as many as 44 million Americans.

| Table 1-12 |

Nutrition Recommendations by the American Cancer Society

Eat a variety of healthful foods, with an emphasis on plant sources.
- Eat five or more servings of a variety of vegetables and fruits each day.
- Choose whole grains in preference to processed (refined) grains and sugars.
- Limit consumption of red meats, especially those high in fat and processed.
- Choose foods that maintain a healthful weight.

Adopt a physically active lifestyle.
- Adults: engage in at least moderate activity for 30 minutes or more on 5 or more days of the week; 45 minutes or more of moderate to vigorous activity on 5 or more days per week may further enhance reductions in the risk of breast and colon cancer.
- Children and adolescents: engage in at least 60 minutes per day of moderate-to-vigorous physical activity at least 5 days per week.

Maintain a healthful weight throughout life.
- Balance caloric intake with physical activity.
- Lose weight if currently overweight or obese.

If you drink alcoholic beverages, limit consumption.

Source: The American Cancer Society web page (www.cancer.org).

Key Point

Cancer is a disease that results from the accumulation of mutations that code for proteins controlling cell turnover.

Ten million are estimated to have the disease and 34 million are estimated to have low bone mass. The complications associated with osteoporosis, such as hip fracture, make this disorder the 12th leading cause of death in the United States. More importantly, it is viewed as one of the most preventable causes of morbidity and mortality. Each year about 260,000 American women suffer hip fractures as a result of osteoporosis. An overview of current and projected estimates of the incidence of low bone mass and osteoporosis is presented in Table 1-13.

The major uncontrollable risk factors for osteoporosis are (1) age, as bones become less dense; (2) gender, because women have less bone tissue and also potentially experience more bone loss after menopause; and (3) race. Caucasian and Asian women are more susceptible than African American and Hispanic women. The major controllable risk factors include diet, level of physical activity, and possibly smoking.

Bone is composed of both an organic matrix, which accounts for about 30% of bone material, and calcium salt mineral deposits that make up the remaining 70%. Roughly 90–95% of the bone matrix is attributable to collagen fibers, which provide incredible tensile strength. The remaining portion of the matrix is a homogeneous medium called *ground substance*. Ground substance is made up of extracellular fluid plus proteoglycans. As discussed in Chapter 5, proteoglycans are composed of proteins plus glycosaminoglycan (GAG) molecules such as *chondroitin sulfate*. *Hyaluronic acid* is another GAG found in bone matrix but is not bound to proteins.

OSTEOPOROSIS PROGRESSION AND PREVENTION A net development of bone material typically continues through the first three decades of life. Just prior to age 30 most people begin to lose bone material and continue to do so throughout the remainder of their lives. This renders bone more susceptible to compression (gravity) and fracture due to falls or contact. Although the focus has been on loss of minerals (especially calcium), osteoporosis results from a general reduction of bone material.

Because the treatment of osteoporosis is difficult and limited, prevention is extremely important. Prevention should focus on two phases:

- Maximize bone development during the first few decades of life and achieve an optimal peak bone mass (PBM).
- Minimize the rate of bone material loss after PBM is achieved. Prevention hinges on physical activity and a well-balanced diet.

Important diet factors include sufficient intake of energy and protein as well as calcium, vitamins C, D, and K,

Table 1-13

State of Low Bone Mass and Osteoporosis in American Adults over 50

Osteoporosis and Low Bone Mass—Females and Males
- Osteoporosis and low bone mass are currently estimated to be a major public health threat for 55% of people aged 50 and older in the United States, almost 44 million people.
- By the year 2010, an estimated 52 million people in this age category will be affected, and if current trends continue, the figure will climb to over 61 million by 2020.
- In 2002, it is estimated that more than 10 million people already have osteoporosis. Approximately 80% of them are women. This figure will rise to almost 12 million people by 2010 and to approximately 14 million by 2020 if additional efforts are not made to stem this disease, which may be largely prevented with lifestyle considerations, and treatment when appropriate.

Osteoporosis and Low Bone Mass—Females
- The number of women age 50 and older who have osteoporosis or are at risk for developing it will increase from almost 30 million in 2002 to over 35 million in 2010 and to approximately 41 million in 2020.
- Of these women, approximately 8 million are estimated to have osteoporosis in 2002. This figure is expected to rise to over 9 million by 2010 and to well over 10 million by 2020.
- Women with low bone mass are estimated at almost 22 million in 2002, almost 26 million in 2010, and over 30 million in 2020.
- Non-Hispanic white women are disproportionately afflicted with this disease, but the number of women of other races and ethnic groups is also significant.

Osteoporosis and Low Bone Mass—Males
- Men with osteoporosis and low bone mass total over 14 million in 2002. This figure is expected to increase to over 17 million in 2010 and to well over 20 million in 2020.
- Almost 12 million men are now estimated to have low bone mass. By 2010, this figure is expected to climb to over 14 million and to more than 17 million men by 2020.
- The prevalence of osteoporosis is estimated at over 2 million men in 2002, almost 3 million in 2010, and well over 3 million men in 2020.
- The number of men of all races and ethnic groups who are affected is significant.

Source: Adapted from National Osteoporosis website (www.nof.org)

copper, iron, zinc, and boron. The role of these nutrients in bone metabolism is discussed in later chapters.

Physical activity applies mechanical stress to bone tissue and stimulates optimal bone matrix development. Several research efforts have demonstrated that more active people have greater bone mass than their less active peers. The relationships between nutrition, physical activity, and bone density and integrity are discussed further in later chapters.

Key Point

Peak bone mass is typically achieved around the age of 30. After that, minerals and protein are gradually lost from bone, rendering it more susceptible to fracture.

■ MAJOR DISEASES AND RISK FACTORS *IN REVIEW*

- Heart disease, stroke, osteoporosis, and cancer should be viewed as the end phase of disease processes (such as atherosclerosis) that were most likely initiated many years prior.
- Heart attacks and strokes are caused by atherosclerotic buildup in the walls of small arteries of the heart and brain.
- The controllable risk factors for heart disease include hypertension, elevated LDL-cholesterol, smoking, and lower levels of physical activity.
- Cancer results from the accumulation of mutations in genes that code for proteins that govern cell reproduction. When this occurs, a cell and its offspring reproduce without control.
- Just before age 30, bones begin to lose minerals and protein, which can compromise bone integrity. Physical activity and a balanced diet can maximize bone mass prior to this phase and also slow the rate of loss after it begins.

■ CONCLUSIONS

Diet and physical activity are recognized worldwide to be among the most important factors influencing general health. Many federal, professional, and private organizations provide nutrition and behavioral information for the general population to assist in nutrition awareness, diet, and lifestyle planning. Individuals can obtain more personal assistance from professionals utilizing diet survey tools and analyzing the data using computer programs.

Current estimates of physical activity and diet in the United States indicate that less than one-fourth of the adult population receives an appropriate level of exercise for health improvement. Meanwhile, the diet of Americans continues to be a concern as less than one-fourth eat the recommended level of fruits and vegetables, and the percentage of overweight and obese children and adults continues to climb. Among the major causes of mortality and morbidity in economically developed countries are heart disease, cancer, and osteoporosis. These are chronic disorders that develop over years or decades. Diet and exercise are clearly involved in reducing the risk of these diseases and slowing their rate of development and need to be a major focus for health promotion.

 **IN FOCUS** OBESITY: A GROWING PROBLEM

Obesity is the nutrition epidemic of modern times in most developed countries, where food is readily and excessively available and daily activities are less laborious. As body composition is a component of fitness, it is easy to deduce that as a person's body fat percentage increases, the level of fitness decreases. In turn, decreased levels of fitness are associated with a variety of risk factors for disease, as described below. This feature presents an overview of one of the most important issues related to human health and fitness. Though not discussed here, many of the concerns associated with obesity in the general population also apply to athletes who can be accurately classified as obese. This might include some heavyweight wrestlers (for example, sumo wrestlers) and some football linemen.

Simply stated, overweight and obesity are caused by a chronic energy imbalance in which energy consumption exceeds expenditure. For each individual, body weight is the result of a combination of genetic, metabolic, behavioral, environmental, cultural, and socioeconomic influences. Behavioral and environmental factors are large contributors to overweight and obesity and provide the greatest opportunity for actions and interventions designed for prevention and treatment. In the United States alone the following facts have been presented as components of *The Surgeon General's Call to Action to Prevent and Decrease Overweight and Obesity* issued in 2001.

General Facts

The level of obese adults continues to increase globally, most significantly in developed countries. In the United States 61% of adults were overweight or obese (BMI > 25) in 1999. Also in 1999, 13% of children aged 6 to 11 years and 14% of adolescents aged 12 to 19 years were overweight. This prevalence has nearly tripled for adolescents in the past two decades. Both males and females of all ages, races, and ethnic groups have demonstrated an increase in overweight and obesity, and approximately 300,000 deaths each year in the United States are associated with obesity. Even adults with a moderate weight excess of 10–20 lb are at greater risk of death. People who are overweight and obese are more prone to heart disease, certain types of cancer, type 2 diabetes, stroke, arthritis, breathing problems, and psychological disorders, such as depression. Alarmingly, obese individuals (BMI > 30) have a 50–100% increased risk of premature death from all causes, compared to individuals with a healthy

weight. As expected, the economic cost of obesity is staggering: about $117 billion for the United States in 2000.

Physical Activity Levels

It is recommended that Americans accumulate at least 30 minutes (for adults) or 60 minutes (for children) of moderate physical activity most days of the week. However, less than one-third of adults engage in the recommended amounts of physical activity. More activity may be needed to prevent weight gain, to lose weight, or to maintain weight loss. Many people live sedentary lives; in fact, 40% of adults in the United States do not participate in any leisure-time physical activity. Among adolescents, 43% watch more than 2 hours of television each day. Physical activity is important in preventing and treating overweight and obesity and is extremely helpful in maintaining weight loss, especially when combined with healthy eating.

Relationship of Overweight and Obesity to Major Diseases

Obesity increases the likelihood (risk) of many diseases such as heart disease, cancer, diabetes mellitus, and arthritis. As an example, the incidence of heart disease is increased in people who are overweight or obese (BMI > 25), as promoted by higher blood pressure and by hypercholesterolemia, hypertriglyceridemia, and decreased HDL levels. The risk of developing type 2 diabetes mellitus is twice as high for people with a weight gain of 11–18 lb compared to individuals who have not gained weight. About 9 out of 10 people diagnosed with type 2 diabetes mellitus in the United States are overweight or obese.

Overweight and obesity are also associated with an increased risk for some types of cancer including endometrial (cancer of the lining of the uterus), colon, gallbladder, prostate, kidney, and breast (postmenopause). Women who experience a weight gain greater than 20 lb from age 18 to midlife have double the risk of developing postmenopausal breast cancer compared to women whose weight remains stable.

Arthritis is more common in overweight and obese people. For instance, for every 2-lb increase in weight, the risk of developing arthritis is increased by 9–13%. Further emphasizing this relationship, symptoms of arthritis tend to improve with weight loss. Sleep apnea (interrupted breathing while sleeping) is more common in obese people, who are also more prone to

OBESITY: A GROWING PROBLEM (CONTINUED)

developing asthma. Meanwhile, the risk of gallbladder disease, incontinence, increased surgical risk, and depression increase along with excessive weight gain.

Overweight women are also more prone to complications during pregnancy (such as high blood pressure and gestational diabetes) and birthing. Infants born to women who are obese during pregnancy are more likely to have high birthweight and, therefore, may face a higher rate of cesarean section delivery and low blood sugar (which can be associated with brain damage and seizures). Furthermore, infants born to mothers who were obese during pregnancy are at greater risk of birth defects, particularly neural tube defects such as spina bifida.

Overweight and Obesity in Children and Adolescents

At the end of the 20th century, 15% of American children and adolescents were overweight or obese (Figure A). This is up from 3–5% in the 1960s, so the incidence of these physical states has tripled in just a few decades. The risk of disease associated with excessive body weight is not limited to adults. It is now clear that the risk factors for heart disease, such as high cholesterol and high blood pressure, occur with increased frequency in overweight children and adolescents compared to those with a healthy weight. Furthermore, type 2 diabetes mellitus in children and adolescence are closely linked to the level of excessive body weight. As for adults, BMI can be used as a predictive tool for obesity in children. However, because BMI decreases during the preschool years and increases thereafter, BMI as a predictor of body weight status and obesity must account for different ages or phases of growth. Therefore, BMI-for-age charts are used, and overweight and obesity is based relative to other children (percentiles). See pages 24–25 for BMI-for-age

gender-specific charts, which present a series of curved lines indicating specific percentiles. Health care professionals use the following established percentile cutoff points to identify underweight and overweight in children based on the following ranges:

- Underweight: BMI-for-age < 5th percentile
- At risk of overweight: BMI-for-age 85th to < 95th percentile
- Overweight: BMI-for-age ≥ 95th percentile

As for adults, lack of physical activity, unhealthy eating patterns, or a combination of the two causes most excessive weight gain during the younger years. Some of the decreased physical activity is related to the popularity of television, computer, and video games (43% of adolescents watch more than 2 hours of television each day). Often, excessive body weight in children is viewed as a "phase" that the child or teen "will grow out of." However, overweight adolescents have a 70% chance of becoming overweight or obese adults, and the risk increases to 80% if at least one parent is overweight or obese. Clearly, the major thrust of educational programs regarding overweight and obesity should focus on prevention, particularly prevention during the younger years.

Prevention and Treatment of Overweight and Obesity

For most overweight and obese individuals, reducing body mass and especially fat mass has proved to be very difficult. In fact, most people in the general population who have participated in weight loss programs and reduced their body weight tend to regain the weight. Weight loss for athletes is different than for the general population because the exercise component probably exists already. It is difficult to say whether weight control and loss of weight after gain is easier for athletes than for the general population.

For all groups, it is clear that the best course of action related to excessive body weight and fat mass is prevention of initial gains. This means that educational and practical efforts to retard the development of obesity should start with younger people. School programs emphasizing healthier eating practices and physical activity are necessary. Regular physical activity is extremely helpful for the prevention of overweight and obesity.

Once the weight gain exists, both physical activity and nutrition manipulation must be considered as the vital components in weight loss, and the focus should be on body fat reduction. Regular physical activity and diet control must be continued to maintain weight loss.

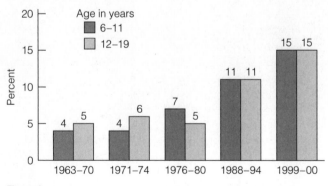

Figure A
NOTES: Excludes pregnant women starting with 1971–74. Pregnancy status not available for 1963–65 and 1966–70. Data for 1963–65 are for children 6–11 years of age; data for 1966–70 are for adolescents 12–17 years of age, not 12–19 years.

OBESITY: A GROWING PROBLEM (CONTINUED)

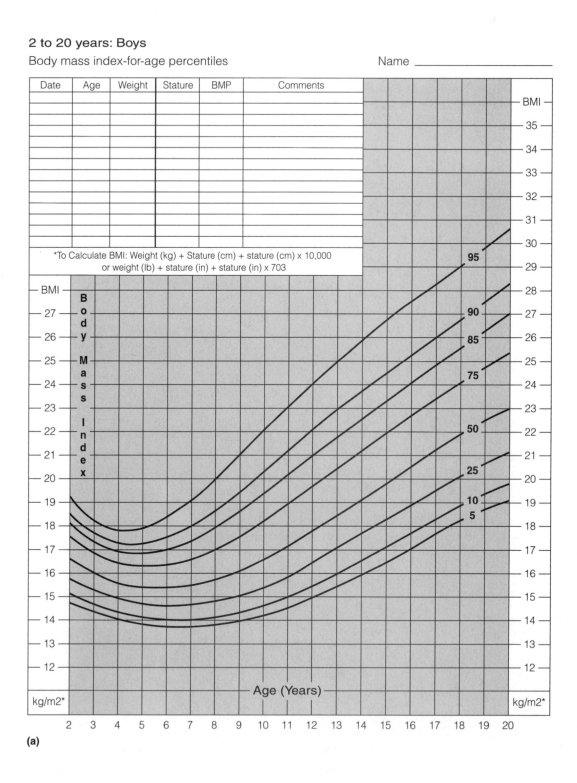

2 to 20 years: Boys
Body mass index-for-age percentiles   Name _____

*To Calculate BMI: Weight (kg) + Stature (cm) + stature (cm) x 10,000
or weight (lb) + stature (in) + stature (in) x 703

(a)

Weight loss itself is not the goal. Achieving and maintaining a healthier body weight and composition conducive to general health and fitness is the real goal. Effective weight control and physical activity help prevent heart disease, help control cholesterol levels and diabetes, reduce bone loss associated with advancing age, lower the risk of certain cancers, and can reduce anxiety and depression.

OBESITY: A GROWING PROBLEM (CONTINUED)

2 to 20 years: Girls
Body mass index-for-age percentiles   Name _____

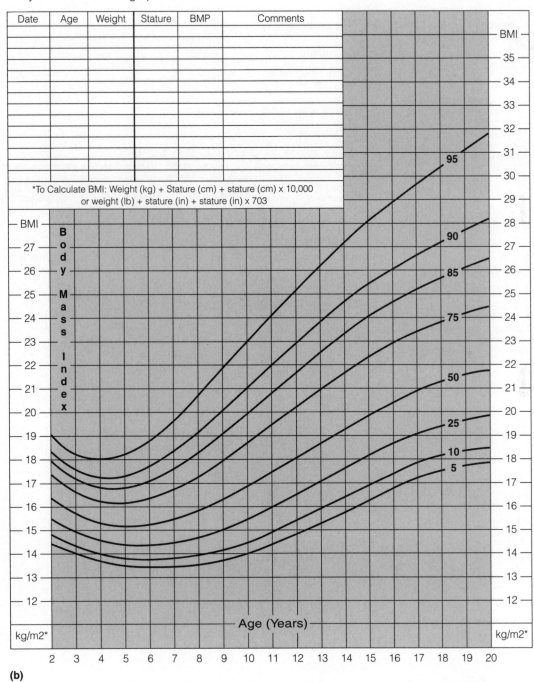

(b)

STUDY QUESTIONS

1. What are the major differences between the Dietary Guidelines for Americans and the Food Guide Pyramid?
2. What are the differences between the RDA, UL, EAR, and AI levels?
3. How were the Daily Values established for Food Facts on labels?
4. What percentage of the American population is getting adequate amounts of physical activity?
5. What is the physical activity status of America's youth?
6. How have estimates of physical activity among adults and children changed over the past few decades in the United States?
7. Are the general dietary recommendations different for highly active individuals and competitive athletes in comparison to the general population?
8. What are the major risk factors for heart disease and osteoporosis?
9. How do diet and exercise influence the development of chronic diseases such as heart disease and osteoporosis?
10. Why are diseases such as atherosclerosis, cancer, and osteoporosis considered chronic diseases?

SUGGESTED READINGS AND WEBSITES

Keast D, Morton AR. Long-term exercise and immune functions. In: *Exercise and Disease* (Watson RR, Eisinger M, eds.), Boca Raton, FL: CRC Press, 1992.

McArdle WD, Katch FI, Katch VL. *Exercise Physiology: Energy, Nutrition, and Human Performance*, 5th ed., Baltimore: Lippincott Williams & Wilkins, 2001.

Pate RR, Pratt M, Blair SN, et al. Physical activity and public health: a recommendation from the Centers for Disease Control and Prevention and the American College of Sports Medicine. *Journal of the American Medical Association* 273:402–407, 1995.

Wildman REC. Classifying nutraceuticals. In: *Handbook of Nutraceuticals and Functional Foods* (Wildman REC, ed.), Boca Raton, FL: CRC Press, 2001.

American Cancer Society. www.cancer.org, 2002.

American Heart Association. www.americanheart.com, 2002.

Centers for Disease Control and Prevention (CDC). National Center for Chronic Disease Prevention and Health Promotion. http://apps.nccd.cdc.gov, 2002.

National Osteoporosis Foundation. www.nof.com, 2002.

ANATOMY AND PHYSIOLOGY OF EXERCISE

Tom Stewart/Corbis

Personal Snapshot

Nancy is a college sophomore majoring in nutrition. She competed in high school athletics, continues to play sports recreationally, and now has an interest in sport nutrition. She would like to become a personal trainer who specializes in exercise nutrition to help the general population become healthier and to help athletes optimize their performance. To accomplish this goal, Nancy needs a strong understanding of muscle tissue and the organ systems related to and supportive of physical activity. But where does she begin?

Chapter Objectives

- Introduce basic concepts in muscle anatomy and physiology.

- Discuss muscle mechanics and their application to physical activity.

- Describe energy systems with special attention to skeletal muscle tissue and specific muscle cell types.

- Discuss the aspects of the cardiovascular system most applicable to physical activity.

- Review basic concepts related to hormones and describe adaptations associated with regular exercise training.

- Discuss organ systems supportive of physical activity.

The science of exercise and performance nutrition is more popular today than ever before. Many resources provide the surface information required for making general recommendations to more active individuals. These are essentially the "what to do" of sport nutrition: for optimal performance eat a certain amount of protein, carbohydrate, and fat; drink a specific amount of water and replace electrolytes lost in sweat; and so on. However, these general recommendations lack fine details and the "whys" of exercise nutrition. Not having a thorough understanding of the relationship between nutrients and physical activity makes it difficult to adapt general recommendations to specific individuals. A sound understanding of the physiological aspects of physical activity empowers individuals to shape nutrition recommendations for themselves and others.

Where does one begin in order to become a capable exercise nutritionist? What scientific foundations are needed to firmly grasp fundamental concepts in exercise nutrition? The quest should begin with an understanding of skeletal muscle and the many organs that play a role in physical activity and exercise. Among the most supportive organ systems for activity are the cardiovascular, pulmonary, endocrine, urinary, and integumentary systems. In accordance, this chapter begins with a review of basic concepts of muscle anatomy and physiology and provides an overview of the cardiovascular system and the hormones relevant to exercise. This overview serves as the foundation for applied exercise nutrition concepts in later chapters as well as sport-specific recommendations for diet and training and the development of sport foods, beverages, and supplements.

■ MUSCLE BASICS

For a lean individual, skeletal muscle is the predominant tissue in the human body, accounting for roughly 40% of body weight. The greater one's muscle mass, the larger the role it plays as a metabolic tissue in the body. Although people exhibit a broad range of shapes and sizes, athletes generally have greater muscle mass percentages than the general population or inactive individuals. This is logical because skeletal muscle contraction serves as the basis of movement and physical activity. But how is muscle tissue designed to provide this function? Are there differences between genders and ages in the design and function of skeletal muscle? This section provides an overview of skeletal muscle design and function and describes differences between genders, age groups, and elite athletes of different sports.

Key Point

Skeletal muscle is generally the predominant tissue of the body for leaner individuals, especially athletes.

Skeletal Muscle Design

The human body contains more than 430 skeletal muscles (Figure 2-1), which are among the most special and interesting tissues in the human body. Each muscle is made up of hundreds of muscle cells. These cells are long and thin, which is why they are commonly referred to as **muscle fibers**. In fact, their diameter may be only 100 micrometers (μm), roughly the thickness of a human hair, yet they can be as long as 750,000 μm, or $2\frac{1}{2}$ ft. As shown in Figure 2-2, each fiber has a connective tissue covering called an *endomysium*. Independent fibers are bundled together, forming *fasciculi* (or fascicles) that are covered by connective tissue called *perimysium*. All of the perimysium is then covered by the *epimysium* that encapsulates each muscle.

From the epimysium, *fascia* and **tendons** are formed to attach the muscle to the connective sheath of bone called the periosteum. So tendons attach skeletal muscle to bone, as shown in Figure 2-2. Muscles generally have two attachment sites. The *proximal* attachment, also known as the *origin*, is generally the site that serves as the stable portion and the *distal* attachment, also known as the *insertion*, is the moving portion.

Because all the connective tissues supporting the muscle and wrapping the bone are continuous with the fasciae and tendon, one fiber could exert tension on the bone, although minimally. These tissues make up the *series-elastic components* (endomysium, perimysium, epimysium, fascia, tendon, periosteum) and act similar to a rubber band grasped to lift a stack of paper it is wrapped around. As muscle contracts it generates an internal tension on the series-elastic components. When the components are taut they create an external tension and cause bones to move, just as the rubber band stretches to the point where the force created is greater than the weight of the paper. However, because the elastic components of the body are separate, less than 100% of the internal tension is transferred to external tension. The series-elastic components also serve to return (retract) the muscle tissue back to its resting length after contractions have ceased.

Muscle Fiber Types

Is skeletal muscle tissue simply skeletal muscle? More specifically, are all muscle fibers the same? Muscle fibers have classically been divided into two primary categories—Type I and Type II. However, many researchers find it more appropriate to consider muscle

Key Point

Individual muscle cells are long and thin and are referred to as fibers. Some fibers may run the entire length of the muscle.

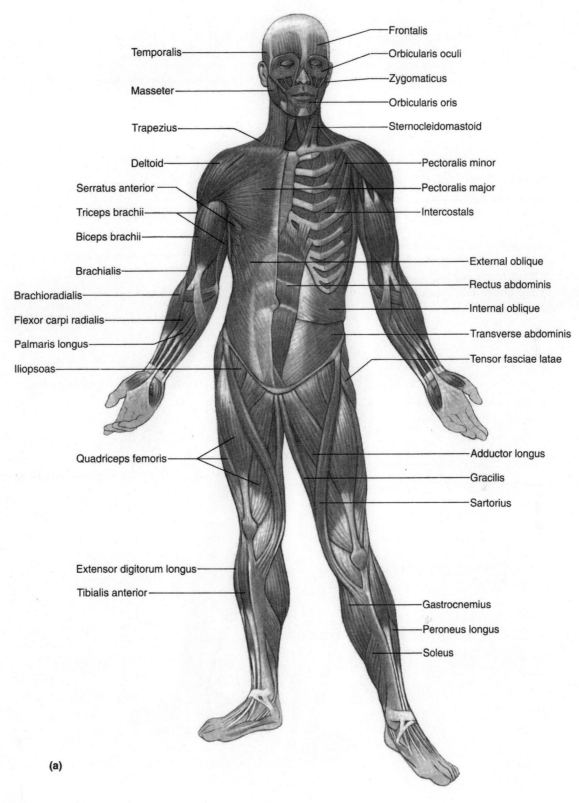

Figure 2-1
Major superficial muscles. (a) Anterior view. From Wingered/*The Human Body*

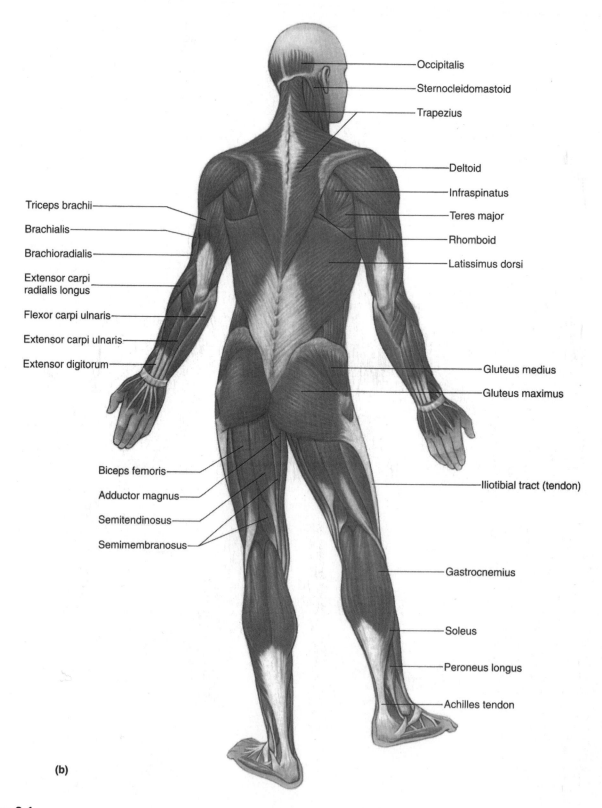

Occipitalis

Sternocleidomastoid

Trapezius

Deltoid

Infraspinatus

Teres major

Rhomboid

Latissimus dorsi

Triceps brachii

Brachialis

Brachioradialis

Extensor carpi
radialis longus

Flexor carpi ulnaris

Extensor carpi ulnaris

Extensor digitorum

Gluteus medius

Gluteus maximus

Biceps femoris

Adductor magnus

Semitendinosus

Semimembranosus

Iliotibial tract (tendon)

Gastrocnemius

Soleus

Peroneus longus

Achilles tendon

(b)

Figure 2-1
Major superficial muscles. (b) Posterior view. From Wingered/*The Human Body*

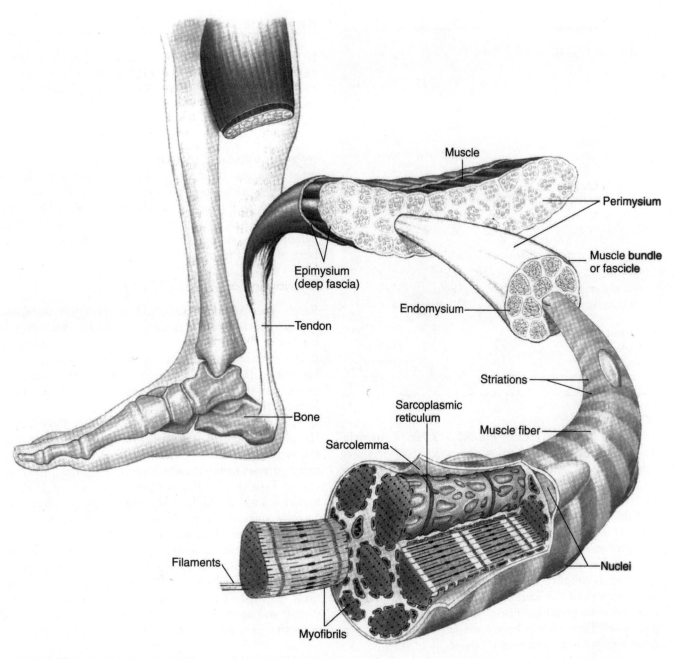

Figure 2-2
Structure of a muscle. A muscle is supported by several layers of connective tissue, which are made visible in this cutaway view. The muscle itself is composed of smaller units, called *muscle bundles* (or *fasciculi*). Each muscle bundle is composed of numerous muscle fibers, each of which contains a number of cylindrical subunits known as *myofibrils*. Myofibrils are made up of protein subunits called *filaments*.

fibers on a continuum, based on metabolic and contractile properties. This is because variations within the Type II category result in subtypes—Type IIa, IIb, and IIc. In addition, all muscle fibers adapt, to some extent, depending on the type of training or stress received. Many of the characteristics of Type I fibers relative to Type II fibers are presented in Table 2-1.

Muscle fibers of the same type are broken into functional groups called motor units, discussed in more detail below. The fibers within a motor unit contract and relax collectively. Without such organization, controlling velocities and forces would be quite difficult. The proportion of fiber types for each muscle is generally based on the function of that muscle (postural control, locomotion, or other).

Table 2-1

Characteristics of Muscle Fiber Types

Characteristic	Type I	Type IIa	Type IIb
Functional Aspects			
Twitch rate	Slow	Fast	Fast
Force production rate	Slow	Fast	Fast
Fatigue resistance	High	Moderate	Low
Structural Aspects			
Mitochondrial density	High	Moderate	Low
Capillary density	High	Moderate	Low
Myoglobin content	High	Moderate	Low
Energy Systems			
Phosphocreatine stores	Low	High	High
Glycogen stores	Low	High	High
Triglyceride stores	High	Moderate	Low
Enzyme System Aspects			
Myosin-ATPase activity	Low	High	High
Glycolytic activity	Low	High	High
Oxidative enzyme activity	High	High	Low

TYPE I (SLOW-TWITCH) FIBERS. **Type I muscle fibers** are often referred to as *slow-twitch (ST)* or slow-oxidative fibers. These fibers are relatively small in diameter and take approximately 110 milliseconds to produce a **twitch** (one cycle of contraction and relaxation). As shown in Table 2-1, Type I muscle fibers have more mitochondria and myoglobin and are associated with more capillaries than fast-twitch fibers. These facets suit Type I muscle fibers for **aerobic**, or oxygen-dependent, energy metabolism and also allow them to be relatively fatigue resistant. Also, because of the increased capillary density, myoglobin, and mitochondria, muscle tissues with a higher proportion of Type I fibers have a deeper red color and are often referred to as "red muscle" tissue.

TYPE II (FAST-TWITCH) FIBERS. Type II muscle fibers are often referred to as *fast-twitch (FT)* muscle fibers. Type II muscle fibers tend to be larger in diameter than Type I fibers and can produce a twitch twice as fast (50 milliseconds). The difference in the rate of tension development between Type I and Type II muscle fibers is largely attributed to the production of the enzyme myosin-ATPase in different forms, or **isozymes**. Type II fibers produce a more rapid form of myosin-ATPase than do Type I muscle fibers.

The fast-twitch fibers are also more efficient in **anaerobic**, or oxygen-independent, energy metabolism and rich in the enzymes needed to break down glycogen for anaerobic glycolysis. In order to serve their fast and forceful contractions, these fibers are equipped with a highly developed *sarcoplasmic reticulum* that delivers and

Key Point

Muscle fibers can generally be classified as either fast twitch or slow twitch. Slow-twitch muscle fibers are more aerobic and fatigue resistant than fast-twitch muscle fibers.

reabsorbs calcium (Ca^{2+}) rapidly. Within the fast-twitch category a few basic subcategories named Type IIa or FOG (fast-oxidative-glycolytic), Type IIb or FG (fast-glycolytic), and Type IIc or unclassified exist. The Type IIa fibers are also called intermediate fibers because they demonstrate some aerobic and fatigue-resistant characteristics of Type I fibers.

Distribution of Muscle Fiber Types

Rarely does an individual excel at seemingly unrelated sports, such as distance running and sprint swimming or ice hockey. This suggests that muscle fiber distribution differs from one person to the next. Is there a genetic predisposition for a certain type of muscle tissue? Furthermore, can the distribution of muscle fiber types differ between the genders and age groups?

On average, most adults have about 50% Type I, 25% IIa, 21% IIb, and 4% IIc fibers in their muscle mass. The relative concentration varies with age, gender, and primary purpose of the muscle itself. For example, the content of Type IIc cells in a newborn infant may be 10 times higher than in an adult. Also, some Type IIa fibers may actually convert to Type IIb during early childhood.

Postural muscles tend to contain more Type I fibers. For instance, the *soleus* (lower calf) and *tibialis anterior* (shin) contain mostly Type I fibers. Muscles not associated with postural control demonstrate greater proportions of Type II fibers. For example, the *rectus femoris* (thigh) and *triceps brachii* (upper arm) muscles tend to be more than two-thirds Type II muscle fibers on average.

GENETIC PREDISPOSITION FOR MUSCLE FIBER TYPE DISTRIBUTION. Genetic factors seem to be the primary variable in determining the proportion of Type I and Type II muscle fibers within muscle tissue. Some athletes are simply blessed with more of the muscle fiber type that is conducive to their sport. Or perhaps more accurately, the muscle fiber proportions determine the sports in which athletes excel. The successful endurance athlete typically has a higher proportion of Type I muscle fibers and the sprinter a higher proportion of Type II muscle fibers. See Table 2-2 for the fiber proportion estimates relative to various athletes.

The genetic predisposition to have a greater proportion of Type I or II muscle fibers, along with an inclination for skill development, raises an interesting question. Did the athlete choose the sport or did the sport choose

Table 2-2

Approximate Fast-Twitch Fiber (Type II) Distribution Among Athletes

Subjects	Male	Female
Nonathletes	54%	49%
Cross-country skiers	36%	41%
Cyclists	41%	49%
Javelin throwers	50%	58%
Runners, 800 meter	52%	39%
Shot putters	62%	49%
Discus throwers	62%	49%
Sprinters	63%	73%

Source: Wilmore JH, Costill DL. *Physiology of Sport and Exercise*, Champaign IL: Human Kinetics, 1999.

the athlete? Certainly, individuals are more likely to enjoy and train at a sport that they seem to excel at from the beginning.

GENDER DIFFERENCES IN MUSCLE FIBER TYPE DISTRIBUTION. Genetic factors for muscle fibers are often evaluated within gender groups; however, not much is known about differences between the genders. For instance, female endurance athletes tend to have higher levels of Type I fibers in muscles related to their sport than their male counterparts. Researchers have attempted to address this. One report summarizing the limited data said that

sedentary males may have larger Type II fibers than women in selected muscles and that the proportion of fiber types found in different types of athletes may vary between genders.[1] In a previous study researchers found that the female middle distance runners examined had a similar proportion of Type I muscle fiber in *gastrocnemius* muscle in their calf relative to their male counterparts.[2] Therefore, until a large study of both genders is conducted it is difficult to state that consistent differences or similarities among the genders exist.

Motor Units

How is the complex chore of coordinating muscle contraction accomplished? The origin of movement begins with an initial impulse developed in the **motor cortex**, the region of the brain responsible for voluntary movement. Nerve cells or motor neurons then activate skeletal muscle cells. The junction between the nerve cell and muscle cell is called the **neuromuscular junction** (Figure 2-3). Each nerve fiber may activate (innervate) a varying number of muscle fibers. The motor neuron and all the muscle fibers it innervates compose a **motor unit**. Motor units can contain a large number of muscle fibers, 1000 or more, for gross motor control. Examples of this occur in the *gastrocnemius* and *tibialis anterior* muscles of the legs. Other motor units contain relatively few fibers, perhaps 15–25, for fine motor control (such as eye movements). All the muscle fibers in a motor unit are of the same type and contract together, providing coordinated and controlled movements. Motor units operate in an "all or none" fashion, and each fiber of the motor unit contracts maximally or not at all.

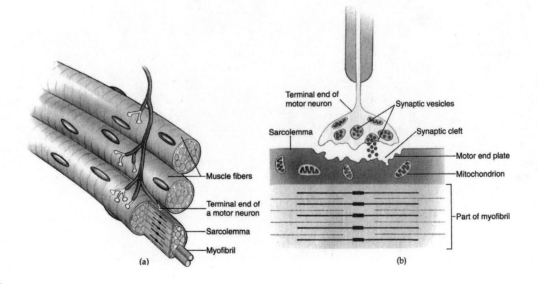

Figure 2-3

The neuromuscular junction. (a) The terminal end of a motor neuron contracts the sarcolemma of a skeletal muscle fiber. (b) In this cutaway close-up, we can see that the contact between neuron and muscle is separated by a narrow space, called the *synaptic cleft*. Notice the neurotransmitters (represented by dots) diffusing across the synaptic cleft.

Key Point

Motor units consist of a motor neuron (nerve cell) and all of the muscle fibers it innervates.

ACETYLCHOLINE. **Acetylcholine** (Figure 2-4) is the neurotransmitter released by motor neurons into the synaptic gap at neuromuscular junctions. Acetylcholine then elicits the "firing" of an action potential, which radiates uniformly outward along the plasma membrane or **sarcolemma** of the muscle fiber. Acetylcholine initiates the action potential by evoking the opening of acetylcholine-gated protein channels. The opening of specific channels allows sodium to traverse the sarcolemma from the

extracellular fluid and depolarize the membrane to threshold, initiating an **action potential**. The action potential is carried toward the center of the muscle fiber by **transverse tubules (T-tubules)**, invaginations of the sarcolemma that reach the sarcoplasmic reticulum. Transverse tubules and the sarcoplasmic reticulum are shown in Figures 2-2 and 2-5, and the involvement of the sarcoplasmic reticulum in initiating contraction is discussed below.

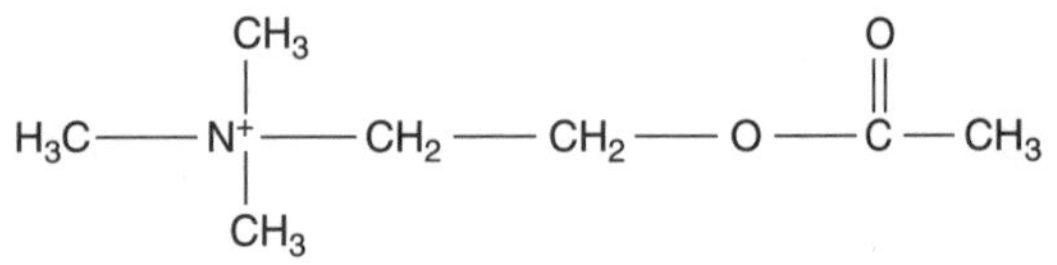

ORDER OF RECRUITMENT.   Motor units contain only one type of muscle fiber, as mentioned above. Furthermore, motor units are recruited in a specific manner, and the primary determinant of the number of motor units recruited is the perceived need for force generation. Motor units containing Type I muscle fibers are recruited prior to motor units containing Type II fibers. This makes sense because during lower intensity exercise plenty of O_2 is available for Type I fibers to perform efficiently and without fatigue. As intensity increases the Type II fibers are recruited, first IIa and then IIb. Figure 2-6 displays a simplified order of recruitment based on the required force for an exercise. It is important to note that the recruitment of Type II fibers during higher intensity activity is in addition to the recruitment of Type I fibers.

$$H_3C - \overset{\overset{\displaystyle CH_3}{|}}{\underset{\underset{\displaystyle CH_3}{|}}{N^+}} - CH_2 - CH_2 - O - \overset{\overset{\displaystyle O}{\|}}{C} - CH_3$$

Figure 2-4
Acetylcholine.

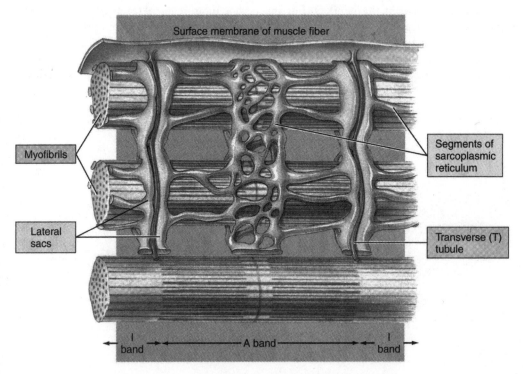

Figure 2-5
The T-tubules and sarcoplasmic reticulum in relationship to the myofibrils. The transverse (T) tubules are membranous perpendicular extensions of the surface membrane that dip deep into the muscle fiber at the junctions between the A and I bands of the myofibrils. The sarcoplasmic reticulum is a fine membranous network that runs longitudinally and surrounds each myofibril, with separate segments encircling each A band and I band. The ends of each segment are expanded to form lateral sacs that lie next to the adjacent T-tubules.

Key Point

Motor units are recruited in a specific order; Type I muscle fibers are recruited first.

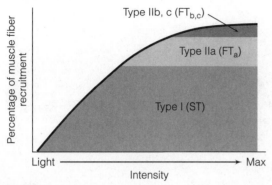

Figure 2-6
Order of motor unit recruitment. Motor units containing Type I fibers are always recruited first and then motor units containing Type II muscle fibers. (ST, Slow twitch, FT, Fast twitch)

Muscle Contraction

To understand how muscle contracts one must look at the composition of each muscle fiber. As shown in Figure 2-7, the cytoplasm of a muscle fiber is occupied by **myofibrils**. The threadlike (1-μm-thick) myofibrils consist of bundles of two types of protein myofilaments, actin and myosin. The thick **myosin** filaments are approximately 16 nm in diameter, and contain stalks of up to 200 myosin molecules. The thin **actin** filaments are approximately 6 nm in diameter and consist of two strands arranged in a double helix (Figures 2-7 and 2-8). Located in the grooves of the actin filament are 40–60 molecules of a protein called **tropomyosin** (Figure 2-8). Additionally, each filament of actin and tropomyosin has a **troponin** protein complex attached to it at regular intervals (Figure 2-8).

SARCOMERES.   The smallest contractile unit of a fiber is called a **sarcomere** (2.6 μm long) and these are repeated the entire length of the muscle fiber. A typical myofibril contains approximately 450 myosin filaments around the middle of each sarcomere and 900 actin filaments at either end of each sarcomere. As displayed in Figures 2-9 and 2-10, each sarcomere is organized into sections. The I band is a section of thin filaments only and is bisected by a dark narrow line called the Z line, where thin filaments attach to one another. The Z line is also the end of a sarcomere. Therefore a sarcomere extends from one Z line to an adjacent Z line. The A bands consist of dark myosin filaments arranged in parallel. The darkest portion of the A band is where the thin and thick

Key Point

Actin and myosin are the structural filaments of muscle contraction. Tropomyosin and the troponin complex function to regulate actin and myosin interaction.

filaments overlap. Within the A band is a lighter region, called the H zone, where the thin filaments do not cross. The M line is a region within the H zone where the thick filaments attach to each other. When a muscle contracts, the Z lines are pulled closer together. As shown in Figure 2-10, this event minimizes H zones visually due to the actin sliding over the myosin toward the center of the sarcomere. The I bands are also reduced due to the pulling of the Z lines toward one another.

SARCOMERE CONTRACTION.   Myosin filaments have protruding globular heads, which bind to actin to form a **cross bridge**, as demonstrated in Figure 2-8. The **sliding filament theory** states that sarcomere contraction occurs when myosin binds to actin filaments and pulls them toward the center of the sarcomere. Thus the thin filaments seem to slide along the myosin filaments (Figure 2-10). This action shortens the width of a sarcomere. The simultaneous contraction of adjacent sarcomeres along the length of the myofibril and among adjacent myofibrils leads to the shortening of the entire muscle fiber. This is done rapidly and repeatedly to accomplish a measurable movement. The role of tropomyosin in the process of sarcomere contraction is to block the active sites between the actin and myosin when the muscle fiber is relaxed. When troponin binds with tropomyosin, the myosin and actin can bind. Therefore, tropomyosin and troponin serve as regulatory proteins and control contraction and relaxation states.

SARCOPLASMIC RETICULUM.   Surrounding each myofibril is an intricate system of tubules termed the **sarcoplasmic reticulum**. The sarcoplasmic reticulum is actually a highly specialized version of the endoplasmic reticulum and is unique to muscle. This organelle terminates in the region around the Z lines and is in association with T-tubules (Figures 2-2 and 2-5). The primary function of the sarcoplasmic reticulum is to store Ca^{2+} between periods of muscle fiber stimulation. When action potentials are carried down the lengths of the T-tubules and subsequently stimulate the sarcoplasmic reticulum, Ca^{2+} is released in large quantities. Calcium then diffuses throughout that local region, binds with troponin, and evokes sarcomere contraction. The binding of Ca^{2+} to troponin results in a conformational change in the protein structure. Troponin then pulls on tropomyosin, exposing the myosin binding sites on actin molecules. The elaborate network of T-tubules and sarcoplasmic reticulum ensure that all aspects of the muscle fiber are exposed to Ca^{2+} at precisely the same moment. This allows for a coordinated contraction of the entire muscle fiber.

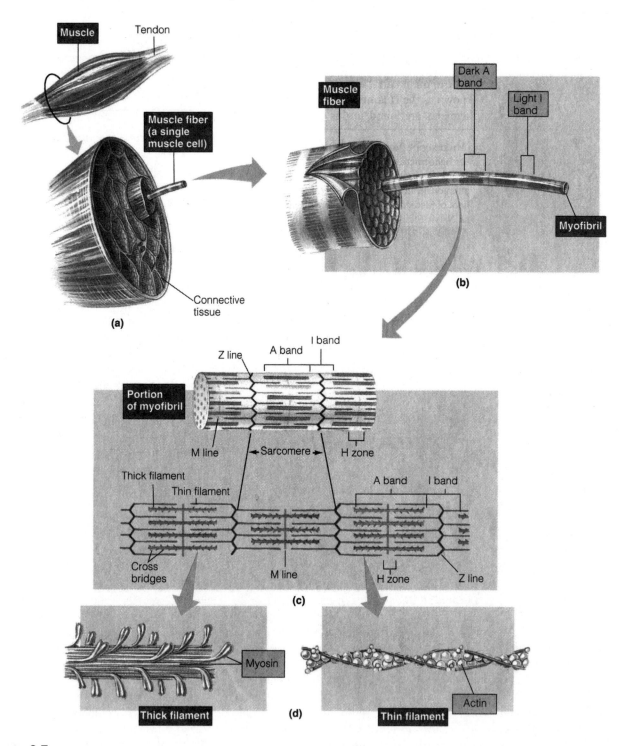

Figure 2-7
Levels of organization in a skeletal muscle. (a) Enlargement of a cross section of a whole muscle. (b) Enlargement of a myofibril within a muscle fiber. (c) Sarcomere components of a myofibril. (d) Protein components of thick and thin filaments.

Key Point

The sarcoplasmic reticulum is a specialized version of the endoplasmic reticulum in muscle fibers and serves as a calcium storage organelle.

Muscle Actions

There are three basic types of muscle contractions, perhaps better termed *muscle actions*. The first type is called **concentric** and occurs when the muscular tension developed internally (cross-bridging) is greater than the resistance. The result is that the muscle contracts while

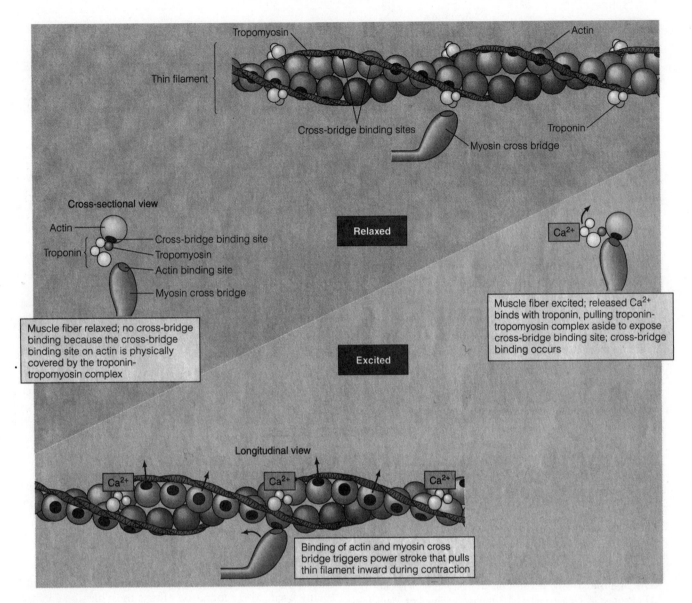

Figure 2-8
Role of calcium in turning on cross bridges.

shortening. The second type is termed **eccentric**, in which the muscle contracts while lengthening. For this muscle action, the tension created internally is less than the resistance, so a lengthening of the muscle occurs. The third type, called **isometric**, is a muscular contraction without any movement because the resistance and tension provided are equal to one another.

An example of the different types of actions can be observed during a bench press. From the arms-extended starting position, the barbell is slowly lowered to the chest (resistance > muscular tension), causing the *triceps brachii* of the upper arms and *pectoralis major* of the chest to be lengthened (eccentric). During the upward phase the barbell is moved away from the chest (resistance < muscular tension) as the *triceps brachii* and *pectoralis major* are shortened (concentric). If there were points in this movement where the barbell was stagnant (such as a pause above the chest) the muscle action would be isometric (resistance = muscular tension).

SKELETAL MUSCLE ENERGY SYSTEMS

Skeletal muscle activity requires a tremendous amount of energy. This includes the energy necessary to perform the physical activity as well as to recover once activity ceases. Adaptive processes such as enhanced protein synthesis and tissue repair further increase energy needs of skeletal muscle. How does muscle generate such high levels of energy? What does it use for fuel, and what are the conditions that allow skeletal muscle to generate energy most efficiently?

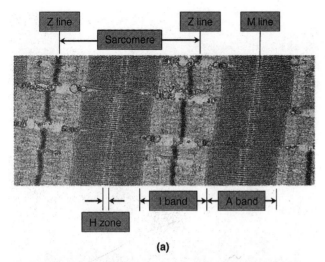

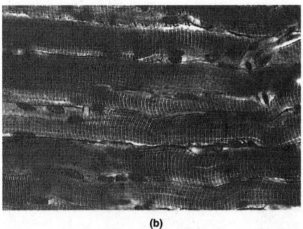

Figure 2-9
Light-microscope view of skeletal muscle components. (a) High-power light-microscope view of a myofibril. (b) Low-power light-microscope view of skeletal muscle fibers. Note striated appearance.
[*Source*: Reprinted with permission from Sydney Schochet Jr., M.D., Professor, Department of Pathology, School of Medicine, West Virginia University: *Diagnostic Pathology of Skeletal Muscle and Nerve* (Stamford, Connecticut: Appleton & Lange, 1986), Figure 1-13.]

Energy for Skeletal Muscle Contraction

What energy molecule is responsible for powering muscle contraction? Furthermore, what do muscle fibers use to generate this energy source? The energy required for sarcomere cross-bridging is supplied by the splitting of *adenosine triphosphate* (*ATP*) into *adenosine diphosphate* (*ADP*) and a free phosphate (PO_4). As shown in Figure 2-11, ATP "energizes" the myosin head by transforming it to the "cocked" position. In the presence of Ca^{2+}, myosin can then form a cross bridge with actin and pull it toward the center of the sarcomere in what has become known as the "power stroke." Next, an additional molecule of ATP is needed on the myosin cross-bridge head to allow it to

■ MUSCLE BASICS *IN REVIEW*

- Skeletal muscle accounts for approximately 40% of an adult's body weight.
- Skeletal muscle is composed of bundled muscle fibers wrapped in connective tissue layers.
- Motor units contain a motor neuron and the skeletal muscle fibers it innervates.
- Muscle motor units are recruited in a specific order: Type I muscle fibers are recruited first, and as intensity increases Type II muscle fibers are recruited in support.
- Muscle contraction reflects a summation of the contraction of sarcomeres in activated muscle fibers.
- Isometric, eccentric, and concentric muscle actions are determined by the relationship between internal tension generated and external force applied to a muscle.

detach from the actin binding site and also allow it to re-cock. This action must be repeated numerous times to sustain the contraction process and to produce movement. Calcium must be continuously available to bind to the troponin, which then binds to tropomyosin, allowing the myosin and actin coupling to occur repeatedly. Without Ca^{2+} the coupling does not occur and the muscle remains in, or reverts back to, a state of relaxation.

Carbohydrates, fatty acids, and amino acids are used by muscle fibers to produce **high-energy phosphate molecules**, most importantly ATP (Figure 2-12), to directly or indirectly provide fuel to the muscles. As mentioned above, ATP powers sarcomere contraction and Ca^{2+} pumping in skeletal muscle and is responsible for powering most other cell operations. ATP is constantly being used, so regeneration efforts must also be constant. The metabolic pathways responsible for generating ATP are discussed in greater detail in Chapters 4–6 and include *glycolysis, fatty acid oxidation*, and the oxidation of certain amino acids (Figure 2-13). Some of these pathways are dependent on O_2; others are not.

HIGH-ENERGY PHOSPHATE MOLECULES. Structurally, ATP consists of an adenine base, linked to a ribose molecule, which itself is linked to three phosphates in series (Figure 2-12). The energy derived from ATP comes from the enzymatic splitting of the high-energy bonds between the phosphate groups. In addition to ATP, muscle fibers produce *guanosine triphosphate* (*GTP*) and *creatine*

Key Point

Adenosine triphosphate is the energy molecule directly responsible for powering muscle contraction.

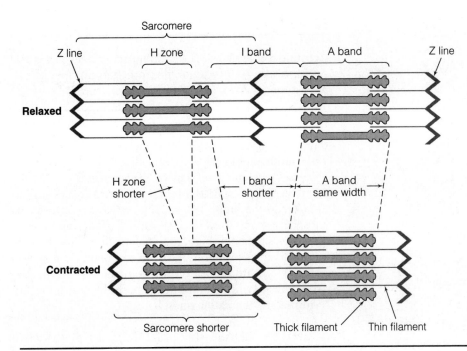

Figure 2-10
Changes in banding pattern during shortening. During muscle contraction, each sarcomere shortens as the thin filaments slide closer together between the thick filaments so that the Z lines are pulled closer together. The width of the A bands does not change as a muscle fiber shortens, but the I bands and H zones become shorter.

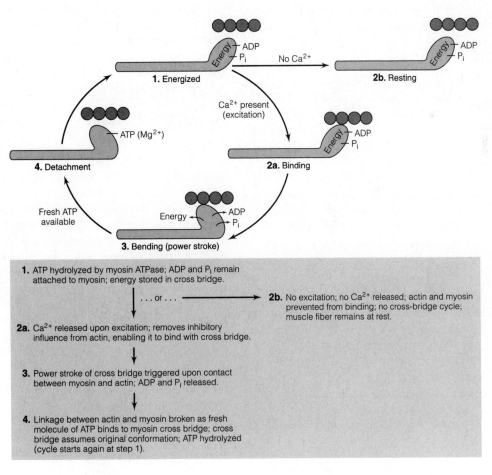

1. ATP hydrolyzed by myosin ATPase; ADP and P_i remain attached to myosin; energy stored in cross bridge.

... or ...

2b. No excitation; no Ca^{2+} released; actin and myosin prevented from binding; no cross-bridge cycle; muscle fiber remains at rest.

2a. Ca^{2+} released upon excitation; removes inhibitory influence from actin, enabling it to bind with cross bridge.

3. Power stroke of cross bridge triggered upon contact between myosin and actin; ADP and P_i released.

4. Linkage between actin and myosin broken as fresh molecule of ATP binds to myosin cross bridge; cross bridge assumes original conformation; ATP hydrolyzed (cycle starts again at step 1).

Figure 2-11
Cross-bridge cycle.

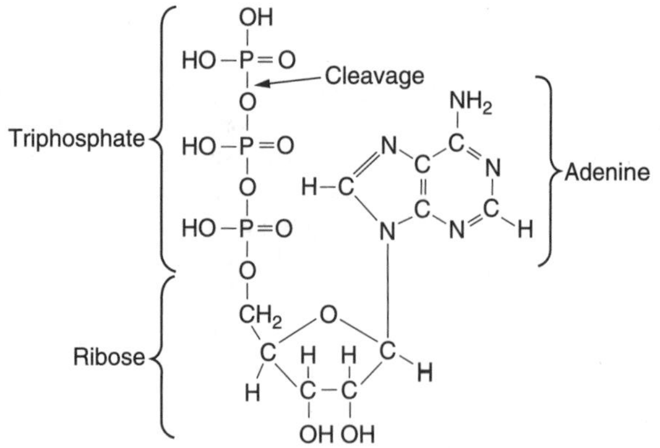

Figure 2-12
Structure of adenosine triphosphate (ATP). ATP is the most important high-energy phosphate molecule and powers muscle contraction. Energy is released from ATP when phosphate bonds are cleaved by ATPase enzymes such as myosin = ATPase in skeletal muscles.

phosphate (CP). CP is particularly important to muscle fiber contraction during strenuous efforts. Collectively, ATP, GTP, and CP are referred to as the high-energy phosphate molecules.

Anaerobic ATP Generation

Muscle fibers have only a small pool of ATP, and during high-intensity exercise this pool is quickly depleted (Figure 2-14). At the same time, muscle fibers attempt to regenerate ATP to sustain activity. Some of the systems involved in ATP regeneration require oxygen and other systems can function without oxygen. Anaerobic mechanisms generate ATP without a need for oxygen and they are of fundamental importance during higher intensity, shorter duration physical efforts. Anaerobic mechanisms of ATP generation are shown in Figure 2-15 and include

- Creatine phosphate (CP)
- Glycolysis to lactic acid
- Adenylyl kinase

ATP AND CREATINE PHOSPHATE. As mentioned, muscle fibers contain a small pool of ATP, which can be complemented by a similarly small amount of CP. See Table 2-3 for the concentration of ATP, CP, and glycogen in the different muscle fiber types. During very intense exercise (such as sprinting or weight training), creatine phosphate serves as the primary mechanism to regenerate ATP. However, because its presence is limited,

Key Point

Muscle fibers are endowed with both aerobic and anaerobic mechanisms for ATP generation.

Key Point

Creatine phosphate is found in muscle fibers and is an important ATP regenerating mechanism during high-intensity exercise.

this process can endure for only a few seconds (Figure 2-14). To continue physical activity, other ATP regenerating processes must increase to meet ATP demands.

Creatine phosphate is found in certain tissues such as muscle and the brain. It is readily available to transfer its phosphate group to ADP to re-form ATP during times of need. As shown in Figure 2-16, creatine phosphate in muscle fibers is produced in the mitochondria and is used in the cytosol in association with sarcomeres. *Creatine kinase* is the enzyme responsible for both the utilization and re-formation of creatine phosphate.

ADENYLYL KINASE (MYOKINASE). As presented in Figure 2-17, when ATP is utilized, either ADP is formed along with phosphate (PO_4), or *adenosine monophosphate* (*AMP*) and *pyrophosphate* (*PP$_i$*) are formed. The first reaction is more prevalent. ATP, ADP, and AMP are interconvertible by the *adenylyl kinase* reaction (Figure 2-17b). Adenylyl kinase is also referred to as *myokinase* in muscle and can be a small yet significant ATP-generating mechanism during intense exercise.

Aerobic ATP Formation

Aerobic mechanisms for generating ATP are those that are oxygen dependent and involve the **electron transport chain** and **oxidative phosphorylation**. The electron transport chain is a series of carrier proteins in mitochondria that pass electrons (removed in oxidation) along their length to generate the necessary energy for the attachment of phosphate to ADP (phosphorylation), thereby forming ATP. Aerobic systems utilize key reactions in metabolic pathways in which electrons are removed (oxidation) from *reactants* and are passed to the electron transport chain in the mitochondria of muscle fibers. Quite simply, muscle fibers that contain more mitochondria and more of the enzymes involved in these reactions possess greater potential for aerobic energy metabolism. Aerobic systems include

- Pyruvate metabolism from glycolysis
- β-Oxidation of fatty acids
- Oxidation of amino acids

Key Point

Aerobic energy systems involve oxidation reactions and are dependent on oxygen and the electron transport chain in mitochondria.

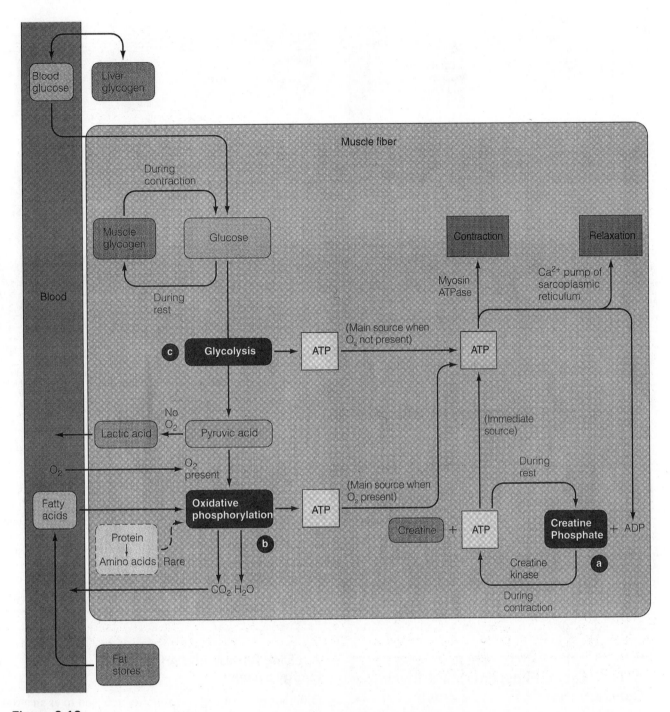

Figure 2-13

Metabolic pathways producing ATP utilized during muscle contraction and relaxation. During muscle contraction, ATP is split by myosin ATPase to power cross-bridge stroking; during relaxation, ATP is needed to run the Ca^{2+} pump that transports Ca^{2+} back into the sarcoplasmic reticulum's lateral sacs. The metabolic pathways that supply the ATP needed to accomplish contraction and relaxation are (a) transfer of a high-energy phosphate from creatine phosphate to ADP (the immediate source); (b) oxidative phosphorylation (the main source when O_2 is present), fueled by glucose derived from muscle glycogen stores or by glucose and fatty acids delivered by the blood; and (c) glycolysis (the main source when O_2 is not present). The end product of glycolysis, pyruvic acid, is converted into lactic acid when the lack of O_2 prevents pyruvic acid from being further processed by the oxidative-phosphorylation pathway.

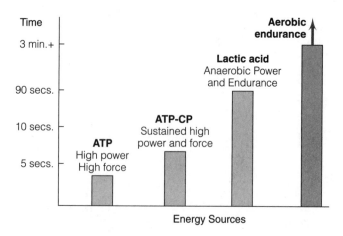

Figure 2-14
Primary energy systems in muscle and characteristics based on the intensity of the exercise and the potential time of sustained activity.

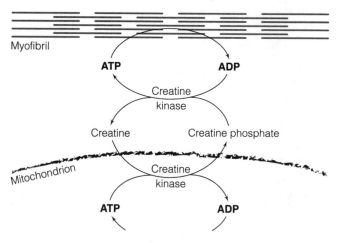

Figure 2-16
Utilization and regeneration of creatine phosphate.

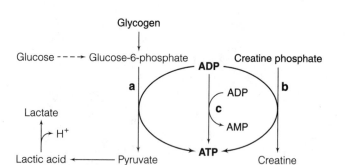

Figure 2-15
Major anaerobic mechanism's of ATP generation in muscle. (a) Anaerobic glycolysis. (b) Creatine phosphate via creatine kinase. (c) Adenylyl kinase (myokinase).

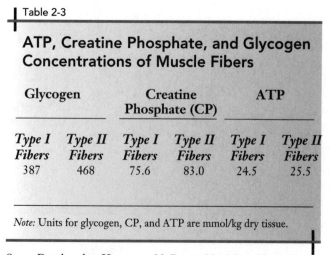

Figure 2-17
The splitting of ATP and regeneration by myokinase. (a) The splitting of ATP yields either ADP and PO_4 or AMP and PP_i. (b) Myokinase in muscle tissue is able to regenerate ATP from ADP and serves as a small yet important ATP regenerating mechanism. [ATP, adenosine triphosphate; ADP, adenosine diphosphate; AMP, adenosine monophosphate; PO_4, phosphate; PP_i, pyrophosphate or diphosphate (PO_4–PO_4)]

Table 2-3

ATP, Creatine Phosphate, and Glycogen Concentrations of Muscle Fibers

Glycogen		Creatine Phosphate (CP)		ATP	
Type I Fibers	*Type II Fibers*	*Type I Fibers*	*Type II Fibers*	*Type I Fibers*	*Type II Fibers*
387	468	75.6	83.0	24.5	25.5

Note: Units for glycogen, CP, and ATP are mmol/kg dry tissue.

Source: Data based on Hargreaves M. *Exercise Metabolism*, Champaign IL: Human Kinetics, 1996.

Applying Aerobic and Anaerobic Energy Systems

Different types of exercise are often referred to as anaerobic or aerobic. For instance, weight lifting is often called an anaerobic effort, and jogging on a treadmill an aerobic activity. Does this mean that ATP regeneration occurs only by aerobic or anaerobic mechanisms, depending on the type of activity? Not necessarily: during exercise, ATP is generated by both aerobic and anaerobic mechanisms. However, if oxygen is available, the intensity of the exercise is the primary factor that determines the ratio of aerobic to anaerobic ATP production. It is important to keep in mind that Type I muscle fibers are always recruited first (see Figure 2-6) and that at least some oxygen is available in muscle tissue. So even if an exercise effort is "all-out" (such as sprinting) and results in fatigue within a few

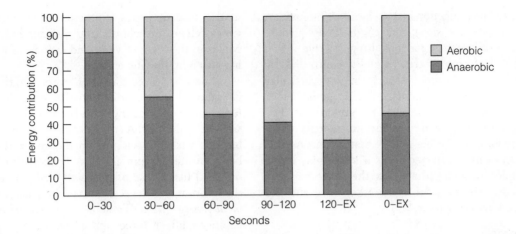

Figure 2-18
Relative contributions made by aerobic and anaerobic energy systems during a 3-minute cycling bout to exhaustion (Ex).
Source: Adapted from Bangsbo et al. *Journal of Physiology* (*London*), 42:539–559, 1990

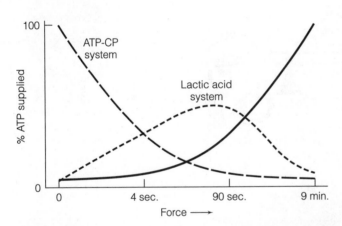

Figure 2-19
Relationship of energy systems during exercise relative to power output and time. Theoretical energy system contribution over time, shown as percentage of maximum ATP production.

seconds, some ATP is still generated via aerobic mechanisms. However, because the contribution made by aerobic mechanisms is greatly overshadowed by anaerobic mechanisms, these types of efforts are referred to as anaerobic. Figure 2-18 demonstrates the contributions made by both anaerobic and aerobic ATP-regenerating mechanisms during 3 minutes of sprint cycling. On the other hand, during exercise such as running a 10-km race, anaerobic mechanisms can make a significant contribution very early but wane as oxygen delivery to muscle is increased and aerobic activities are enhanced. Figure 2-19 displays the relationship between the force requirement and time involved in an exercise and the contribution of aerobic and anaerobic energy systems.

■ SKELETAL MUSCLE ENERGY SYSTEMS *IN REVIEW*

- ATP is the high-energy phosphate molecule responsible for powering muscle contraction.
- Both anaerobic and aerobic mechanisms can regenerate ATP, and the intensity of an effort is the primary determinant of their relative contribution to ATP production.
- Creatine phosphate, anaerobic glycolysis, and myokinase are the principal anaerobic systems, as they can generate ATP directly in muscle fibers.
- β-Oxidation and the breakdown of pyruvate and amino acids are the aerobic systems because they require the electron transport chain for ATP generation.

■ PHYSICAL ASPECTS OF MOVEMENT

An understanding of the mechanics of human movement is needed in order to analyze physical activity and prescribe exercise. The field of biomechanics uses mechanical principles and applies them to biological systems to better understand their function. The frame of the human body consists of the skeletal system; the working force moving the frame is the muscular system. Combining these two systems creates what is termed the musculoskeletal system, which contains bones, joints, connective tissues, and the musculature.

The configuration of these components lends itself to a variety of movement patterns (for example, multiaxial and uniaxial patterns). The various types of joints have inherent mechanical factors that affect their force production capabilities and general function. Beginning with the tendons, which are an extension of one or more muscles, factors associated with function include location, orientation (proximal/distal insertion sites), and length. Consequently, the muscles controlling these tendons also play a role.

Musculoskeletal joints are generally controlled by several muscles. The "prime mover" for any particular joint action is called the **agonist**. Each agonist has a muscle that opposes its action, called the **antagonist**. Together these muscles provide specific movements while stabilizing the joint. Any other muscle that assists or stabilizes the movement is referred to as a **synergist**. For a simple example, in flexing the elbow the *biceps brachii* is the agonist, the *triceps brachii* is the antagonist, and the *brachialis* functions as a synergist.

Lever Systems

The musculoskeletal system acts like the components of a lever system. The levers are the skeletal bones, the fulcrums are the joints, and the effort points are the tendon insertion sites. The effort comes from the forces generated by the contracting muscle, and the resistance comes in the form of work and exercise (for example, lifting a weight, moving a limb).

There are three classes of levers, and the human body is endowed with all of them (Figure 2-20). A first-class lever is that of a seesaw, with the fulcrum between the resistance and effort. A second-class lever is that of a wheelbarrow, with the fulcrum on the end and the resistance between the fulcrum and the effort. A third-class lever is oriented like a pair of tweezers, with the effort between the fulcrum and the resistance. The equation for *mechanical advantage* (MA) (Table 2-4) quantifies how certain orientations affect force and velocity. The primary lever orientation in the human musculoskeletal system is third class; thus humans are built more for speed than strength.

Depending on the lengths of the resistance and effort arms, the mechanical advantage (MA) may be in favor of

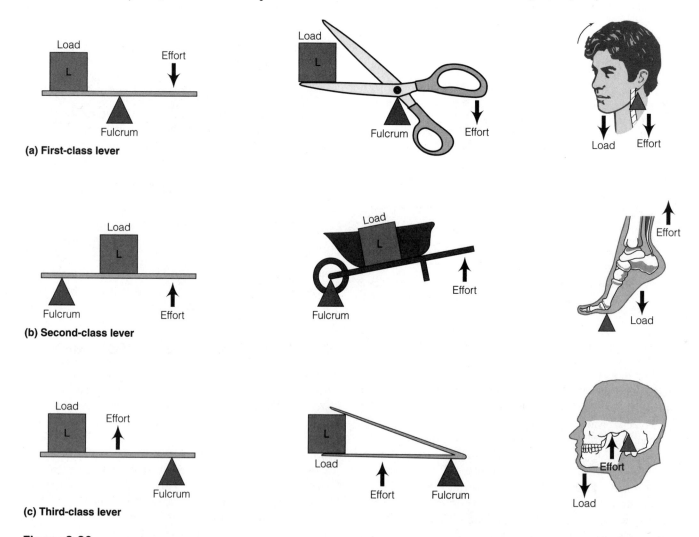

Figure 2-20

The three classes of levers and examples in human anatomy. Authorities still debate whether the ankle is a first-class or second-class lever. *Source*: From Human Anatomy and Physiology, 2nd Edition by Saladin. © McGraw-Hill Companies, Inc.

low strength and high velocity (MA < 1.0) or high strength and low velocity (MA > 1.0). Figure 2-21 depicts the nature of MA during elbow flexion. Less muscular force is needed to accomplish a movement when the length of the force arm is the greatest (increased MA). It should be noted that many muscles act on a joint in a variety of orientations, so each muscle involved makes a different force contribution at different joint angles.

Strength, Force, Work, Power, and Range of Motion

Many terms related to training come up in "gym talk." The use and misuse of some of these terms can be confusing, especially *strength, force, work, power,* and *range of motion.* These terms and variables are interrelated and are quickly discussed here and further developed in Chapter 3. Also, Tables 2-4 and 2-5 outline and define general mechanical concepts relative to human movement.

FORCE.   **Force** can be defined as the capacity to do work or cause physical change. Force is often measured by the amount of resistance it supplies. The most common units used in relation to strength training are pounds and kilograms. For example, if a person is bench pressing a barbell of 250 lb, then 250 lb of force is being produced.

Key Point

The joints of the human body are primarily oriented as third-class levers.

Table 2-4

Mechanical Properties and Equations

Arm or lever: rigid body that rotates about an axis or fulcrum (bones)
Axis of rotation, or fulcrum: the pivot point or axis of a lever
Effort arm, or force arm (Fa): perpendicular distance from the line of action of the force to the axis
Resistance arm (Ra): perpendicular distance from axis to point of resistance
Force (*F*): capacity to do work or cause physical change; energy, strength
Torque (*T*): a force causing rotation

$$T = F \times \text{Fa} \quad \text{or} \quad T = \text{resistance} \times \text{Ra}$$

Mechanical advantage (MA): the ratio of the force arm to the resistance arm

$$\text{MA} = \text{Fa/Ra}$$

Table 2-5

Strength, Work, and Power

Strength: the amount of force (*F*) that can be generated by a muscle or muscle groups
Work (*W*): product of force (*F*) and displacement (*D*)

$$W = F \times D \quad \text{or} \quad W = F \times \text{angular displacement}$$

Power (*P*): the amount of work performed per unit of time

$$P = W/T$$

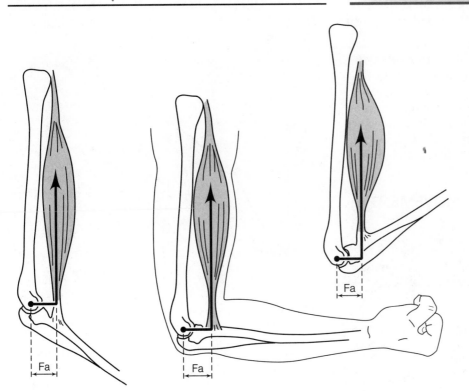

Figure 2-21
Mechanical advantage during elbow flexion. During elbow flexion with the biceps muscle, the perpendicular distance from the joint axis of rotation to the tendon's line of action varies throughout the range of joint motion. When the force arm (Fa) is shorter, there is less mechanical advantage.
[*Source*: Adapted from Bangsbo et al., *Journal of Physiology* (*London*) 42:539–559, 1990.]

Key Point

Strength is the amount of force being generated.

STRENGTH. Muscular **strength** is the maximum amount of force that can be generated. It is one repetition of maximum (1 RM), the largest amount of weight that can be successfully lifted for a given exercise. It should be noted that strength is velocity specific and is affected by the type of muscle action (concentric, eccentric, isometric).

WORK. **Work** is a product of the force produced and the distance an object moves (work = force × distance). For example, for a person bench-pressing a barbell of 200 lb the distance of 1 ft (chest to an arms-extended position), the amount of work completed would be 200 foot-pounds (ft lb). A taller person usually performs more work per repetition than a shorter person because of the greater distance over which force must be applied. Hence, most successful weight lifters are shorter in stature.

POWER. **Power** is the amount of work performed in a specified amount of time (power = work/time). If the exercise in our example above took 2 seconds, the power would be 200 ft lb/2 s = 100 ft lb/s. If a person performed the exercise in just 1 second the power output would be 200 ft lb/s. So the quicker performer has more power. In the sport of powerlifting the amount of time used to execute the lifts is not considered, just the total amount of weight lifted. A better name for this sport might be weight lifting. Regardless of the misuse of terminology, power is a critical component in many athletic performances, especially the major team sports (including football, baseball, hockey, volleyball, and basketball).

RANGE OF MOTION. **Range of motion (ROM)** is the total distance of segment displacement. It is the flexibility or pattern of movement at a joint. Performing an exercise or lift through a full range of motion requires more work than a partial movement because the segments are displaced further. Performing more work requires more energy and potentially increases muscle size and strength throughout that range of motion.

■ PHYSICAL ASPECTS OF MOVEMENT *IN REVIEW*

- The prime mover for a motion is called the agonist and the opposing muscle is called the antagonist. Synergists assist with and help stabilize the joint.
- Muscular strength is the maximum amount of force that can be generated by a muscle or muscle groups. It is the one repetition of maximum.
- Work is the product of force applied over a given distance.
- Power is the amount of work in a given period.

Key Point

Power is generally the most critical physical component of team sports.

■ FACTORS IN FORCE GENERATION

What are the major factors involved in force generation? Can regular physical activity influence these factors? Several biomechanical, anatomical, and physiological parameters are relevant to the force production capabilities of a muscle. The two most significant factors influencing the production of force are the number of motor units recruited and the frequency of their stimulation.

Neural Influences

As discussed previously, muscle fibers associated with a particular motor unit demonstrate an all-or-nothing response to neural signals. Therefore, when the stimulus provided by a motor neuron is strong enough to stimulate contraction, the result is maximal contraction. This is like a light switch: if it is switched on, the light comes on with maximal intensity. In other words, no gradations or "dimmer switches" are associated with the activity of a motor unit. In order to generate varying magnitudes of force the neural signals need only activate the appropriate number of motor units (like turning on more lightbulbs in a room). Physical training enhances this neural control and better synchronizes these motor unit activation patterns. These are often the mechanisms for improved force generation in early phases of training (Figure 2-22).

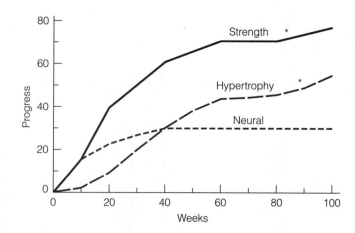

Figure 2-22
Development of strength during resistance training is the result of both neurological and muscular adaptations. The asterisk (*) indicates a potential increase in strength and muscle mass (hypertrophy) above a plateau when an anabolic agent is used.

Key Point

Motor units are activated in an all-or-nothing manner, so recruiting appropriate numbers of motor units determines the magnitude of force generation.

TWITCH AND SUMMATION.   The frequency of stimulation also affects force production, as each neural signal stimulates the entire motor unit to twitch. If the twitches overlap (relaxation was incomplete), the net force production is greater than that of a single twitch. This phenomenon is called **summation** (Figure 2-23).

A related factor in force production is the availability of Ca^{2+}. Calcium affects the number and frequency of active cross bridges within each muscle fiber. The rate of Ca^{2+} release is rapid but not instantaneous, so a muscle requires some time before reaching maximal force production. Also, fast-twitch and slow-twitch muscle fibers vary in their nerve conduction and cross-bridging rates.

Fiber Type and Adaptation

Are there differences in force generation between muscle fiber types? Also, can muscle adapt to generate more force, and if so, are there differences in the degree of adaptation that can occur in Type I versus Type II muscle fibers? As discussed above, Type II muscle fibers have the capacity to generate greater force within a given period than Type I muscle fibers. However, Type II muscle fibers are not recruited until the intensity of the activity is sufficiently high. Therefore, higher intensity training produces greater improvements in force generation because a greater number of motor units, and a greater proportion of Type II motor units, are recruited.

Although the distribution of muscle fiber types is genetically determined, training can result in significant adaptation within the specific fiber types. The type of training stress applied partly determines the degree of adaptation in contractile and physical aspects of the muscle tissue and the fibers themselves. For instance, studies have related the ratio of **cross-sectional areas** (muscle or muscle group diameter estimated by circumference measurements) of Type II to Type I muscle tissue to training intensity for sprinters, jumpers,[3] and weightlifters.[4,5] The high intensity of the training associated with these sports allows adaptation to occur in both muscle fiber types; the change is more significant in the Type II muscle fibers. Thus, strength and power training results in an enhancement of force and power generating capabilities, which are Type II characteristics, within muscle fibers.[6–10] The relatively lower intensity of endurance training tends to result in the development of enhanced aerobic energy

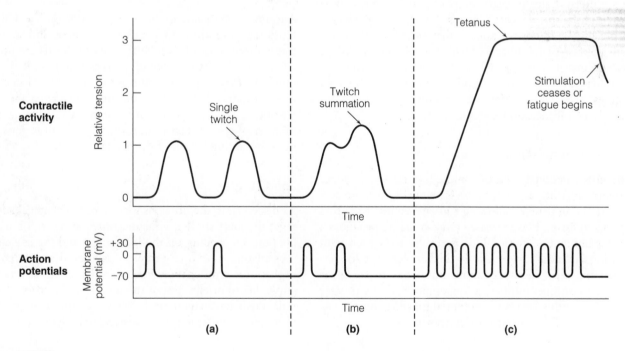

Figure 2-23

Summation and tetanus. (a) If a muscle fiber is restimulated after it has completely relaxed, the second twitch is the same magnitude as the first twitch. (b) If a muscle fiber is restimulated before it has completely relaxed, the second twitch is added on to the first twitch, resolving in summation. (c) If a muscle fiber is stimulated so rapidly that it does not have an opportunity to relax at all between stimuli, a maximal sustained contraction known as tetanus occurs.

capabilities, which are Type I characteristics. This is largely due to the recruitment pattern of motor units, in which Type I muscle fibers are recruited first. If endurance intensity is increased, more Type IIa muscle fibers are recruited in support of Type I fibers. Type IIa fibers have the potential to become more aerobic with appropriate training stimuli.

MUSCLE HYPERTROPHY.   One of the most important adaptations resulting from strength training is the increase in muscle fiber diameter or **hypertrophy**. Muscle hypertrophy leads to an increase in the total cross-sectional area and thus increases the capability to produce force (Figure 2-22). This hypertrophy is due to both the size and number of myofibrils of the Type II and Type I muscle fibers regardless of gender.[11-13]

As mentioned, the fiber types can vary in the level of hypertrophy in response to resistance training. Typically, the hypertrophy experienced by Type II fibers is greater than Type I fibers. Therefore, individuals with a larger proportion of Type II muscle fibers generally respond to resistance training with a higher degree of hypertrophy. Endurance training is generally less intense than weight training and does not recruit the fast-twitch motor units. Thus, one should not expect to gain significant amounts of muscle mass from training aerobically. For example, in comparing long-distance runners (low-intensity exercise) to sprinters (high-intensity exercise), the larger muscle masses of the sprinters are obvious. This is most likely due to the higher intensity training that recruits and conditions the fast-twitch fibers, which have a greater ability to increase in size. Genetic factors also play a significant role as individuals seem to respond to training to different degrees, even when differences in muscle fiber type are taken into consideration.

Cross-Sectional Area

The cross-sectional area of a muscle is related to force production. With a greater cross-sectional area more sarcomeres lie in parallel, allowing a greater number of cross bridges to form. This increased cross-bridging produces a greater amount of force. This is similar to a rope with additional strands that make it thicker, but not longer. When the sarcomeres are more numerous in series or stacked upon one another, a longer fiber results (longer rope). When this occurs the potential for shortening is enhanced if all of the sarcomeres shorten at once.

Key Point

Muscle fiber adaptations are specific to the type of training. Higher intensity training can result in increased force generation and hypertrophy.

Key Point

Certain types of training invoke an increase in skeletal muscle cross-sectional area (hypertrophy), which can lead to greater levels of strength and power.

MUSCLE ORIENTATION.   The various muscle orientations (fusiform, parallel, convergent, unipennate, bipennate, multipennate, circular) are illustrated in Figure 2-24. Clear differences can be observed in the cross-sectional areas of the muscles representing these orientations. For example, the convergent orientation of the *pectoralis major* allows for a greater cross-sectional area versus the fusiform-oriented *biceps brachii*. These fiber orientations affect both the strength and shortening characteristics of the muscle. Perhaps this is why successful throwing athletes (baseball pitchers, javelin throwers) are typically taller and leaner (longer muscles) in structure. Throwing and high-velocity movements require a greater change in muscle length to accommodate the related range of motion. In contrast, athletes associated with strength movements (powerlifters) tend to be bulkier and shorter, which provides a larger cross-sectional area of muscle tissue.

Length-Tension Relationship

Another component of force production is the length-tension relationship. As a muscle lengthens and shortens there are positions in which the myosin heads are unable to bind to the actin, preventing cross-bridging. A smaller number of cross bridges reduces the amount of force that can be produced. Therefore, maximal force from a length-tension relationship occurs at the point where the greatest number of cross bridges can form. This point is generally near the resting length of the muscle.

Joint Angle

As discussed in the previous section on mechanical advantage, the joint angle is a factor contributing to force production. Depending on the joint angle and the associated musculature, the force produced is altered as the lengths of the force arms change. Each joint has a unique strength curve due to orientation of the tendinous insertions. Also, individuals have anatomical differences that may produce differences in strength curves. Muscle length, type of

Key Point

Many muscles act on a joint simultaneously. The resulting force output for any particular angle is the summation of each muscle's contribution of force at that angle.

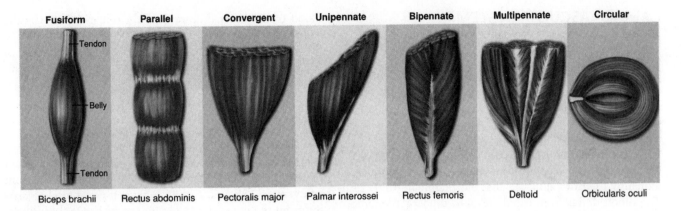

Fusiform	Parallel	Convergent	Unipennate	Bipennate	Multipennate	Circular
Biceps brachii	Rectus abdominis	Pectoralis major	Palmar interossei	Rectus femoris	Deltoid	Orbicularis oculi

Figure 2-24
Classification of muscles according to the arrangement of their fascicles.

action, and velocity must also be considered in relation to force production.

Type of Muscle Action

Eccentric muscle actions can produce higher forces than concentric muscle actions. The force capabilities are typically 120–160% greater in eccentric movements versus concentric.[14–17] Many people have experienced this when performing weight training exercises. The "easy" part of the lift (for example, the downward phase of a squat or bench press) is the eccentric phase and the "hard" part (the upward phase of those exercises) is the concentric phase.

Velocity

Another issue related to force production is the velocity of the movement. With concentric muscle actions, the amount of force produced decreases as velocity increases. It is easier to apply force to a slow-moving object such as a slowly moving medicine ball versus a medicine ball moving rapidly. This is due to the cross-bridging rates of muscle tissue. The more time allotted for cross-bridging and shortening of the sarcomeres, the greater the force generation.

Conversely, with eccentric muscle actions, the amount of force produced increases as velocity increases. Remember that eccentric actions occur when the external force is greater than the internal force; they involve contractions while lengthening. Therefore, eccentric muscle actions are basically attempting to slow or resist the movement. The cross bridges and sarcomeres are already lengthened and opposing the stretch. This is mechanically more efficient

than concentric actions and thus contributes to increased force production as velocity increases.

Body Mass

Is an individual's body weight important in determining strength? If so, should this be a consideration for competitive athletes? Clearly the mass of an individual's body (particularly lean body mass) also plays a role in determining strength and should be a consideration for athletes involved in time-oriented sports (runners, cyclists, swimmers). The relationship between body mass and strength is demonstrated in Newton's second law:

$$\text{Force} = \text{mass} \times \text{acceleration} \ (F = M \times A)$$

If an athlete can maintain acceleration values while increasing mass, more force can be applied. This is also thought of in terms of the ratio of strength to body weight. Algebraically reordering the formula as $A = F/M$ (acceleration = force/mass) clearly states how strength (force) and body weight affect acceleration. Smaller athletes such as gymnasts generally have higher strength to body weight ratios than larger athletes. To optimize force or acceleration an athlete needs to experiment with his or her strength and mass components.

■ FACTORS IN FORCE GENERATION IN REVIEW

- Force generation by muscle is influenced by several factors:
  Frequency of neural stimulation
  Fiber type
  Cross-sectional area of muscle
  Length-tension relationship
  Joint angle
  Type of muscle action
  Velocity of contraction
  Mass of segment

Key Point

The "easy part" of a weight lifting exercise is the eccentric portion and the "hard part" the concentric portion.

CARDIOVASCULAR SYSTEM

Perhaps the most important system that supports exercise is the cardiovascular system. This system delivers O_2, fuel, and hormones to working muscle tissue while removing carbon dioxide (CO_2), lactate, and heat. The adult heart serves to pump blood through about 100,000 miles of blood vessels to all regions of the human body. Blood leaves the heart through the "great arteries," namely the aorta and pulmonary trunk, which feed into the arteries, which themselves feed into the smaller arterioles and subsequently the tiny capillaries that thoroughly infiltrate tissue. Blood drains from capillaries into the larger venules, which drain into the larger veins that ultimately return

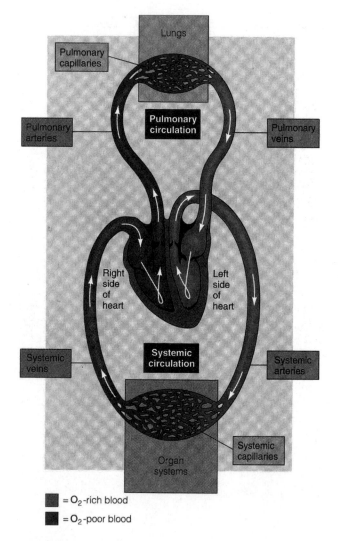

Figure 2-25
Pulmonary and systemic circulation in relation to the heart. The circulatory system consists of two separate vascular loops: the pulmonary circulation, which carries blood between the heart and lungs, and the systemic circulation, which carries blood between the heart and organ systems.

blood to the heart (Figure 2-25). Capillaries serve as the actual sites of exchange of substances between cells and the blood. This section provides a brief overview of some of the key components and operations of the cardiovascular system.

The Heart

The heart consists of four chambers (two atria and two ventricles) that can be divided into a left and a right half (Figure 2-26). The left half, consisting of the left atrium and ventricle, serves to receive oxygen-rich blood returning from the lungs and to pump it to all tissue throughout the body. The right half of the heart, consisting of the right atrium and ventricle, serves to receive oxygen-poor blood returning from tissue throughout the body and to pump it to the lungs. The heart functions as a relay station for moving blood throughout the body in two loops.

The heart is composed primarily of cardiac muscle cells, which are similar to skeletal muscle cells yet retain certain fundamental differences. Although many of the events involved in contraction of cardiac muscle are the same as for skeletal muscle, one obvious difference is that the heart muscle is not attached to bone. Furthermore, the heart does not require an external stimulus (from the brain) to initiate contraction. The stimulus invoking excitability in the heart comes from its own specialized pacemaker region called the *sinoatrial node* or simply the SA node. The heart may beat in excess of 2 billion times throughout human life.

Blood

Blood consists of two main parts, the *formed elements* and the plasma. **Erythrocytes**, more commonly called red blood cells (RBCs), are the most abundant components of the formed elements and function primarily to transport oxygen. **Hematocrit** is the percentage of the blood volume attributable to red blood cells. The hematocrit of a typical adult may be 40–45%, as shown in Figure 2-27. Leukocytes and platelets collectively make up about 1% of blood. **Leukocytes**, or white blood cells (WBCs), are the principal components of the immune system and provide a line of defense against bacteria, viruses, and other intruders. **Platelets** participate in blood clotting.

Plasma accounts for the remaining 55% of the blood. About 92% of the plasma is water. The remaining 8% includes more than 100 different dissolved or suspended

Key Point

The heart serves to pump oxygenated blood to working tissues and to pump oxygen-poor blood to the lungs in one major and continuous loop.

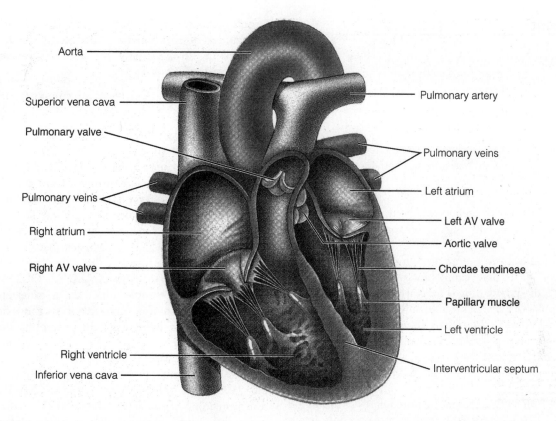

Figure 2-26
The heart. Longitudinal section of the heart, depicting the location of the four heart valves and chambers. (AV, atrioventricular)

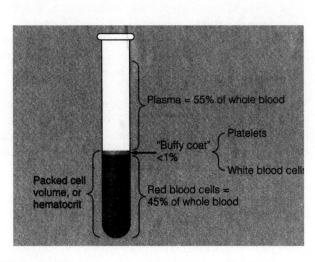

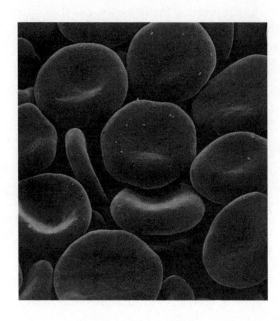

Figure 2-27
Hematocrit and real blood cells (RBCs).

Key Point

Erythrocytes (red blood cells) are packed with hemoglobin, a protein that binds with O_2 molecules for efficient O_2 transport to tissue.

substances such as nutrients, gases, electrolytes, hormones, and proteins such as albumin and clotting factors.

ERYTHROCYTES.   Red blood cells have the responsibility of transporting oxygen throughout the human body. These cells are basically bags of hemoglobin, which constitutes 33% of the mass of an RBC and 95% of its protein. **Hemoglobin** is a large protein molecule containing four atoms of iron that readily bind with oxygen and carry it through the circulation. There are roughly 42–52 million red blood cells per cubic millimeter of blood. A healthy red blood cell contains about 250 million hemoglobin molecules, each with the capacity to bind four O_2 molecules. Therefore, each red blood cell can transport approximately 1 billion molecules of O_2.

CIRCULATORY LOOPS.   When the heart pumps, blood is propelled from the right ventricle into the pulmonary arteries for transport to the lungs. Upon reaching the lungs and the pulmonary capillaries, CO_2 exits the blood and enters the lungs. It is subsequently removed during exhalation. Simultaneously, O_2 enters the blood from the lungs and binds with hemoglobin in red blood cells. The oxygen-containing blood leaves the lungs and travels back to the heart. When the heart contracts, blood is pumped from the left ventricle into the aorta and then into the arteries, arterioles, and finally tiny capillaries in tissue. Blood that has been utilized by tissues drains into small venules, which drain into larger veins and subsequently the inferior and superior vena cava. The venous blood is then pumped to the lungs in order to unload CO_2 and load O_2.

Cardiac Output

How much blood is moved through the circulatory loops every minute? Does this amount increase during physical activity to provide more oxygen-rich blood to working muscle? The volume of blood pumped in each contraction

Key Point

The systemic circulatory loop delivers blood to all tissue throughout the body, while the pulmonary circulatory loop moves blood through the lungs to exchange O_2 and CO_2 with the air.

from the heart and directed toward either the lungs or the remainder of the body is called **stroke volume (SV)**. Multiplying stroke volume by heart rate determines **cardiac output (CO or Q)**.

Cardiac output (CO) = stroke volume (ml)
× heart rate (beats/min)

Cardiac output is the volume of blood pumped out of the heart, either to the lungs or toward body tissue, in 1 minute. It is approximately 5–6 L/min for a resting adult. It should not matter which of the two destinations (lungs or arterial system) one considers, as the two outputs occur simultaneously and have similar stroke volumes. During exercise both heart rate and stroke volume increase, to a point, which increases cardiac output.

TISSUE DISTRIBUTION OF CARDIAC OUTPUT.   During heavy exercise some athletes can experience an increase in cardiac output up to five to six times higher than resting levels to help meet the metabolic demands of working skeletal muscle. Under resting and comfortable environmental conditions about 13% of the left ventricular cardiac output goes to the brain, 4% to the heart, 20–25% to the kidneys, 10% to the skin, and the rest to the remaining tissue in the body, such as the digestive tract, liver, and pancreas. However, during heavy exercise a greater proportion of this cardiac output is routed to working skeletal muscle by decreasing blood flow to less active tissues. Blood is also routed more extensively to the skin for heat removal. After a big meal, and several hours thereafter, a greater proportion of cardiac output is routed to the digestive tract to help metabolize the food.

BLOOD PRESSURE.   Pressure is the force exerted against a given surface area. With regard to the cardiovascular system, the pressure associated with a volume of blood serves as the driving force of circulation. The relatively high pressures generated in the ventricles eject blood out of the heart and through the arterial system. Blood pressure is measured in millimeters of mercury (mm Hg). It is typically measured in a large artery, such as the brachial artery in the arm, and is expressed as systolic pressure over diastolic pressure. For instance, when blood pressure is measured at 120/80 ("120 over 80") the pressure exerted by systemic arterial blood is 120 mm Hg during left ventricular contraction and 80 mm Hg when the left ventricle is relaxing between beats.

Key Point

During exercise blood is shunted from other organs and redistributed to working muscle tissue to meet demand for O_2 and to unload CO_2. Blood is also routed more extensively to the skin for heat removal.

Pulmonary Exchange of Gases

How do we efficiently bring O_2 into the blood for circulation throughout the body? Does this same system allow us to get rid of excessive CO_2? While the left ventricle pumps blood into the systemic circulation, serving all the tissue in the body, the right ventricle pumps blood into the pulmonary circulation with only one destination, the lungs. The pressure generated by the right ventricle is about five times lower than the left ventricle. Thus, the pulmonary circulation is a lower pressure system relative to the systemic circulation. The principal reason for the lower pressure is that pulmonary circulation is much shorter in length and therefore a lower driving pressure is needed. The pulmonary circulation contains only about 9% of our total blood volume or 450 ml. Of this blood, approximately 70 ml is in the capillaries and the remainder is in the pulmonary arteries and veins.

THE LUNGS. Blood surges through the pulmonary trunk and then divides into the left and right pulmonary arteries, which serve their respective lungs. The lungs are among the largest organs in the body and serve as the exchange site between the blood and inhaled air. The elaborate branching of the pulmonary air passages to reach the terminal sites of exchange, called alveoli, allows an exchange surface area in the alveoli approximately 40 times greater than an adult's body surface area (Figure 2-28). Inhaled air moves through the trachea and then branches into the left and right bronchi serving each lung. The bronchi divide into smaller bronchi that continue to branch into numerous bronchioles, which themselves branch into smaller terminal bronchioles and eventually alveoli.

ALVEOLI/CAPILLARY EXCHANGE. Each lung contains 100–150 million alveoli, each covered in a meshlike

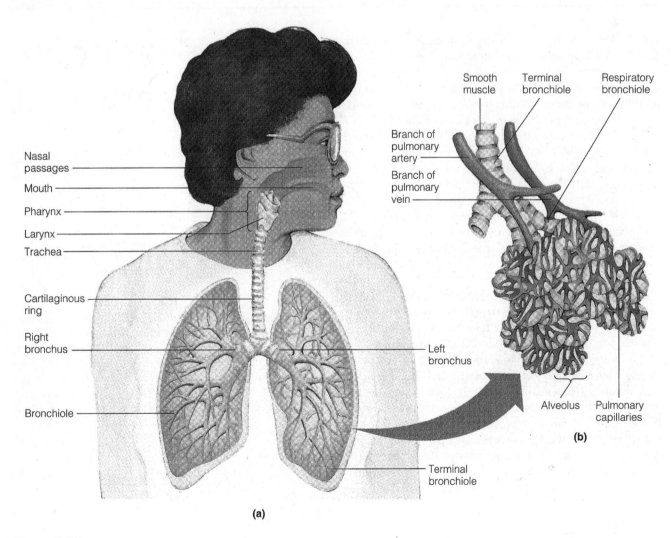

(a)

(b)

Figure 2-28
Alveoli and gas exchange between the lungs and blood.

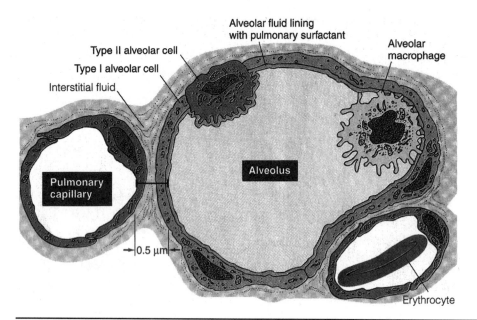

Figure 2-29

Gas exchange between the lung alveolus and capillaries. A schematic representation of a detailed electron microscope view of an alveolus and surrounding capillaries. A single layer of flattened Type I alveolar cells forms the alveolar walls. Type II alveolar cells embedded within the alveolar wall secrete pulmonary surfactant. Wandering alveolar macrophages are found within the alveolar lumen. (The size of the cells and respiratory membrane is exaggerated compared to the size of the alveolar and pulmonary capillary lumens. The diameter of an alveolus is actually about 600 times larger than the intervening space between air and blood.)

capillary blanket. In fact, the capillary density is so thick that 40% of lung volume consists of blood vessels. The interface between the alveoli and capillaries is the site for gas exchange (Figure 2-29). Oxygen enters the blood from the alveoli and associates with hemoglobin in red blood cells. Free oxygen does not easily dissolve into the blood, so hemoglobin is needed to transport it. As mentioned earlier, every red blood cell can carry about 1 billion molecules of oxygen, and adults typically have 4–6 billion circulating red blood cells. Stated in other units, 100 ml of blood can carry as much as 20 ml of oxygen. Because the ability of the blood to carry oxygen is limited by the number of red blood cells and hemoglobin content, some athletes may engage in a practice called "blood doping" that increases red blood cell numbers. This is further discussed in Chapter 13.

OXYGEN SATURATION OF HEMOGLOBIN.
The saturation of hemoglobin with oxygen is about 98% under most normal situations. In the lung alveoli, oxygen exerts a partial pressure of about 104 mm Hg. In accordance with the hemoglobin/oxygen dissociation curve (Figure 2-30) hemoglobin becomes nearly saturated by normal breathing. What happens during exercise? Does increased oxygen removal by muscle and other tissue, as well as changes in pH and body temperature, affect oxygen loading in the lungs? Reduced pH and increased temperature push the curve to the right. In working muscle these factors increase oxygen unloading from hemoglobin, which supplies the increased need for oxygen. Meanwhile the lungs

Key Point

The saturation of hemoglobin with oxygen is nearly complete during typical environmental conditions.

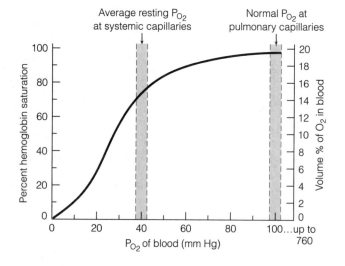

Figure 2-30

Oxygen-hemoglobin (O_2-Hb) dissociation (saturation) curve. The percent hemoglobin saturation (the scale on the left side of the graph) depends on the P_{O_2} of the blood. The relationship between these two variables is depicted by an S-shaped curve with a plateau region between a blood P_{O_2} of 60 and 100 mm Hg and a steep portion between 0 and 60 mm Hg. Another way of expressing the effect of blood P_{O_2} on the amount of O_2 bound with hemoglobin is the volume percent of O_2 in the blood (ml of O_2 bound with hemoglobin in each 100 ml of blood). That relationship is represented by the scale on the right side of the graph.

generally have a higher pH and a cooler temperature, which promote oxygen loading onto hemoglobin.

CARBON MONOXIDE.
It should be mentioned that hemoglobin has a far greater affinity for carbon monoxide (CO) than for oxygen. This means that hemoglobin binds

more CO than O_2 if both are present in the lungs. Therefore, environments where the CO content of the air is greater are not the best environments for training. Athletes should train in areas away from busy streets and industry.

Hormonal Systems

Physical activity involves a complex coordination of physical and metabolic efforts. The motor cortex and cerebellum of the brain are involved in initiating and coordinating muscle activity, and hormones govern most of the metabolic

efforts. Hormones are produced by endocrine glands (Figure 2-31), such as the pituitary gland, parathyroid gland, thyroid gland, hypothalamus, pancreas, stomach, small intestine, adrenal glands, placenta, and gonads (ovaries and testicles). They are largely protein or protein-based (such as glycoproteins), amino acid–based, or cholesterol-derived steroid molecules. Examples of protein hormones include insulin, growth hormone, glucagon, and antidiuretic hormone (ADH). Examples of hormones made from the amino acid tyrosine are epinephrine (adrenaline) and thyroid hormone (T_3/T_4). Steroid hormones are made from cholesterol and include testosterone,

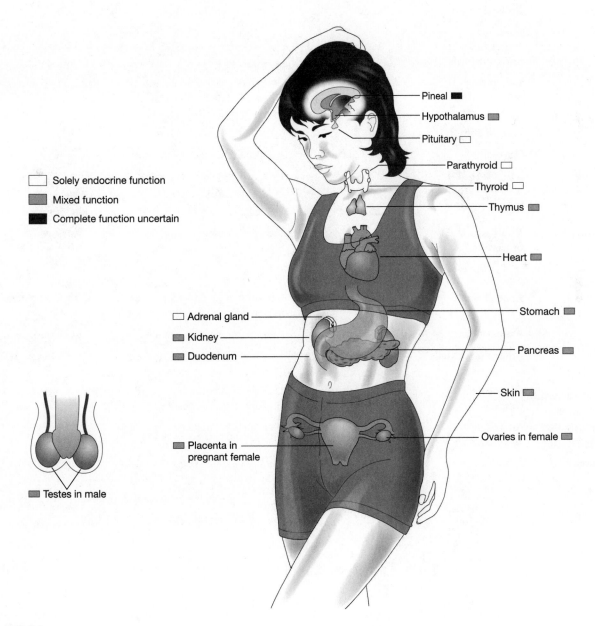

Figure 2-31
The endocrine system.

| Table 2-6 | **Target Organs and Function of Major Hormones** |

Endocrine Gland	Hormone	Target Tissue	Major Functions
Pituitary Anterior lobe	Growth hormone (GH)	Most cells	Promotes development and enlargement of all body tissues, promotes protein synthesis, increases the mobilization of fats for energy, decreases carbohydrate utilization
	Thyroid-stimulating hormone (TSH)	Thyroid	Controls triiodothyronine, thyroxin production by thyroid gland
	Adrenocorticotropin hormone (ACTH)	Adrenal cortex	Controls cortex secretion
	Prolactin	Breasts	Promotes breast development, milk production
	Follicle-stimulating hormone (FSH)	Ovaries, testes	Follicle growth, secretion of estrogen, sperm development
	Luitenizing hormone (LH)	Ovaries, testes	Estrogen, progesterone, testosterone production; release of ovum; follicle rupture
Posterior lobe	Antidiuretic hormone (ADH)	Kidneys	Water excretion, elevation of blood pressure
	Oxytocin	Uterus, breasts	Contraction of uterine muscles, milk secretion
Thyroid	Thyroxine (T_4) Triiodothyronine (T_3)	All cells	Increases metabolism, contractility of heart
	Calcitonin	Bones	Decreases calcium concentration in the blood
Parathyroid	Parathormone or parathyroid hormone (PTH)	Bones, intestines	Increases calcium in blood through its influence on bone, intestine, kidneys
Adrenal Medulla	Epinephrine	Most cells	Glycogen breakdown; increases blood flow to skeletal muscle increases heart rate and O_2 consumption
	Norepeinephrine	Most cells	Constricts arterioles, venules, which elevates blood pressure
Cortex	Mineralocorticoids (aldosterone)	Kidneys	Increase Na^+ retention and K^+ excretion
	Glucocorticoids (cortisol)	Most cells	Control metabolism of carbohydrate, fat and protein; anti-inflammatory response
	Androgens and estrogens	Ovaries, breasts, testes	Promote development of sex characteristics testes
Pancreas	Insulin	All cells	Controls blood glucose, increases glucose uptake and fat synthesis
	Glucagon	All cells	Increases blood glucose, stimulates the breakdown of protein, fat
	Somatostatin	Islets of Langerhans, GI tract	Depresses secretion of insulin, glucagon
Gonads Testes	Testosterone	Sex organs	Promotes development of male sex organs and characteristics, muscle growth
Ovaries	Estrogen	Sex organs	Promotes development of female sex organs and characteristics, increases fat storage, regulates menstruation
Kidneys	Renin	Adrenal cortex	Assists in blood pressure control
	Erythropoietin	Bone marrow	Stimulates erythrocyte production

estrogens, cortisol, progesterone, and aldosterone. Table 2-6 presents some of the major hormones relative to exercise and health and their target tissues and actions.

RECEPTORS AND SECOND MESSENGER SYSTEMS. Hormones are released into circulation and interact with specific receptor complexes in one or more tissues. Only cells that have a specific receptor for a given hormone respond to that hormone. Some cell receptors are located on the plasma membrane and are typically part of a multi-molecule complex that affects events within the cell (Figure 2-32). For instance, the binding of the pancreatic hormone glucagon to its receptor on the plasma membrane of certain liver cells results in an increase in cyclic AMP (cAMP) levels within those cells. Because cAMP is then responsible for initiating the cellular events for which glucagon signaled, cAMP is considered a second messenger. Other second messengers include Ca^{2+}, cGMP, inositol triphosphate, and diacylglycerol. Epinephrine is another hormone that binds with a receptor on the plasma membrane. Like glucagon, it results in an increase in cAMP levels in liver cells, and it also signals muscle fibers. Other hormones, such as thyroid hormone and steroid-based hormones, interact with nuclear receptors. These hormones tend to influence gene expression.

■ CARDIOVASCULAR SYSTEM *IN REVIEW*

- The cardiovascular system is a delivery system for oxygen, nutrients, hormones, and other substances to tissues and also provides a means for removing waste products such as CO_2 and heat from cells.
- Cardiac output is the product of heart rate and stroke volume.
- Exercise results in redistribution of blood to active muscle and the skin.
- Hormones are chemical messengers that include insulin, glucagon, epinephrine, cortisol, growth hormone, and aldosterone.

Conclusions

Skeletal muscle contraction allows for physical activity and is composed of different fiber types (I, IIa, IIb, IIc). The distribution of these fiber types within specific muscle can be influenced by genetic predisposition, training, and age. Type I muscle fibers tend to be more aerobic and fatigue resistant but generate force at a slower rate. Type II muscle fibers are able to generate force more quickly than Type I muscle fibers but are more anaerobic

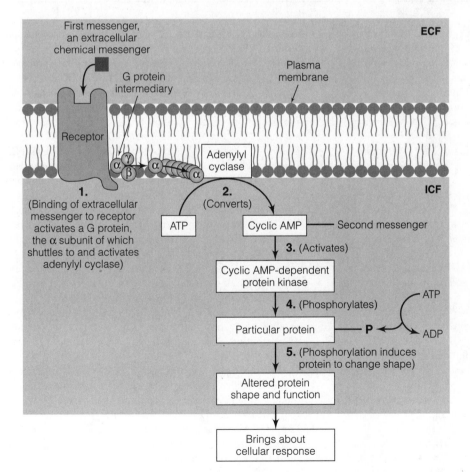

Figure 2-32
Activation of the cyclic AMP second messenger system by an extracellular messenger. (1) Binding of an extracellular chemical messenger, the *first messenger* to a surface membrane receptor activates by means of a G protein intermediary the membrane-bound enzyme adenylyl cyclase, which in turn (2) converts intracellular ATP into cyclic AMP. (3) Cyclic AMP acts as an intracellular *second messenger*, triggering the desired cellular response by activating cAMP-dependent protein kinase, which in turn (4) phosphorylates and thereby (5) modifies a particular intracellular protein. The altered protein then accomplishes the cellular response dictated by the extracellular messenger.

and more prone to fatigue. Sarcomeres, composed of overlapping protein filaments, are the smallest contractile unit within muscle fibers. ATP generated by both aerobic and anaerobic mechanisms is the principal fuel source that powers sarcomere contraction. Muscle fibers can use carbohydrate, fatty acids, and certain amino acids to generate ATP. Skeletal muscle fibers adapt to physical stress. This is seen in improved endurance, strength gains, and muscle hypertrophy, depending on the type of training and genetic potential. In support of skeletal muscle efforts, the cardiovascular system delivers oxygen and nutrients to working muscle and removes waste such as lactate and heat. Also, hormones regulate metabolic events within muscle fibers as well as in other important tissues.

STUDY QUESTIONS

1. How do the series-elastic components function?
2. What are the two major muscle fiber types and what characteristics do each possess?
3. What is the makeup of a motor unit and how does one function?
4. What is the significance of the specific order of motor unit recruitment?
5. What are the types of muscle actions? How do they apply to a squat exercise?
6. What are the mechanisms for anaerobic and aerobic ATP generation?
7. List and discuss the factors influencing force generation.
8. What is the relationship between the intensity of exercise training and the expected adaptation?
9. What is cardiac output and how is it altered during physical activity?
10. What are the major hormones associated with physical activity and how does exercise training influence their role during exercise and at rest?

REFERENCES

1. Chorneyko K, Bourgeois J. Gender differences in skeletal muscle histology and ultrastructure. In: *Gender Differences in Metabolism: Practical and Nutritional Implications* (Tarnopolsky ML, ed.), Boca Raton, FL: CRC Press, 1999.
2. Costill DL, Daniels J, Evans W, Fink W, Krahenbuhl G, Saltin B. Skeletal muscle enzymes and fiber composition in male and female track athletes. *Journal of Applied Physiology* 40(2):149–158, 1976.
3. Edgerton RV. Neuromuscular adaptation to power and endurance work. *Canadian Journal of Applied Sport Science* 1:49–58, 1976.
4. Edstrom L, Ekblom B. Differences in sizes of red and white muscle fibers in vastus lateralis of quadriceps femoris of normal individuals and athletes: relation to physical performance. *Scandinavian Journal of Clinical Laboratory Investigation* 30:175–181, 1972.
5. Prince FP, Hikida RS, Hagerman FC. Human fiber types in power lifters, distance runners, and untrained subjects. *Pflugers Archives* 363:19–26, 1976.
6. Hakkinen K, Komi PV. Alterations of mechanical characteristics of human skeletal muscle during strength training. *European Journal of Applied Physiology* 50:161–172, 1983.
7. Howald H. Training induced morphological and functional changes in skeletal muscle. *International Journal of Sports Medicine* 3:1–12, 1982.
8. Jansson EB, Sjodin, Tesch P. Changes in muscle fiber type distribution in man after physical training. *Acta Physiologica Scandinavica* 104:235–237, 1978.
9. Thorstensson A, Sjodin B, Tesch P, Karlsson J. Actomyosin ATPase, myokinase, CPK, and LDH in human fast and slow twitch muscle fibers. *Acta Physiologica Scandinavica* 99:225–229, 1977.
10. Howley ET, Franks BD. *Health Fitness Instructors Handbook*, 3rd ed., Champaign, IL: Human Kinetics, 1997.
11. Kannus P, Jozsa L, Renstrom P, Jarvinen M, Dvist M, Lehto M, et al. The effects of training, immobilization, and remobilization on musculoskeletal tissue. *Scandinavian Journal of Medicine and Science in Sports* 2:100–118, 1992.
12. Staron RS, Leonardi MJ, Karapondo DL, Malicky ES, Falkel JE, Hagerman FC, Kikada RS. Strength and skeletal muscle adaptations in heavy-resistance trained women after detraining and retraining. *Journal of Applied Physiology* 70:631–640, 1991.
13. Staron RS, Mallicky ES, Leonardi MJ, Falkel JE, Haferman F, Dudley GA. Muscle hypertrophy and fast fiber type conversions in heavy resistance trained women. *European Journal of Applied Physiology* 60:71–79, 1989.
14. Hortobagy T, Katch FI. Eccentric and concentric torque-velocity relationships during arm flexion and extension. *European Journal of Applied Physiology* 60:395–401, 1990.
15. Dudley GA, Tesch PA, Miller BJ, Buchanon P. Importance of eccentric actions in performance adaptations to resistance training. *Aviation, Space, and Environmental Medicine*, 62:543–550, 1991.
16. Hather BM, Tesch PA, Buchanan P, Dudley, GA. Influence of eccentric actions on skeletal muscle adaptations to resistance training. *Acta Physiologica Scandinavica* 143:177–185, 1991.
17. Hortobagyi T, Hill JP, Houmard JA, Fraser DD, Lambert NJ, Israel, RG. Adaptive responses to muscle lengthening and shortening in humans. *Journal of Applied Physiology*, 80:765–772, 1996.

3

BASIC CONCEPTS IN EXERCISE AND FITNESS

Personal Snapshot

Ryan McVay/PhotoDisc/Getty Images

Kim, a 20-year-old college junior, has become interested in strength training and fitness. For the past year she has been lifting weights in the student fitness facility and running around the track and campus grounds. She has tried to figure out her workout routine in the gym by observing others. She has also been reading magazines such as *Runner's World* and *Running* to try to improve her running workouts. However, she is not sure if she is doing the right things to maximize the benefits of her workouts. Kim wonders where she could find basic concepts of training in order to better structure her workouts. She decides to meet with a personal trainer at the local health club to get some initial information.

Chapter Objectives

- Define the terms exercise and fitness.

- Discuss general concepts related to physical training and exercise.

- Introduce the concept of diminishing return.

- Define and discuss the components of fitness.

- Apply the concepts of training (FITT) to the components of fitness.

- Present and discuss the basic recommendations for physical activity.

- Outline and discuss the exercise routine.

Exercise has become a recognizable part of our society, one with numerous benefits when practiced regularly. What exactly is "regular exercise" and how does someone plan to do it? People seem to exercise in a variety of ways. Walkers or runners head to the park, cyclists adventure along paths and roads, and others head to the gym to get their workouts in.

The terms *exercise* and *fitness* are often used in the same discussion, suggesting that they are related concepts. What exactly is fitness and how would regular exercise help someone to achieve it? People commonly say that they just want to be fit, and fitness is part of the title of popular magazines and books. However, many people would struggle to provide an accurate definition of fitness.

The health club and fitness equipment industries try to make exercising as convenient and comfortable as possible. Consequently, there have been several technological advances in fitness equipment. Ideas about exercise, fitness, and health are also being refined and are often aired in the media. This chapter introduces and discusses some formal concepts used in exercise and training to serve as a basis for the principles applied in later chapters. Understanding these concepts will also assist in implementing safe and effective exercise training programs.

■ EXERCISE AND TRAINING CONCEPTS

Exercise is synonymous with physical activity and often refers to a formalized type of physical activity. The more specific term *exercise training* involves consistent physical activity that is initiated to maintain or improve health, fitness, and athletic performance. Such exercise is generally geared to maintain or improve certain physical attributes such as cardiorespiratory endurance or muscular strength. However, most people use the term *exercise* more generally, not distinguishing between mowing the lawn or gardening and cycling or running.

Recognition of the general concepts related to exercise creates a framework for understanding how the body adapts to physical stress. Regular exercise can result in numerous benefits, perhaps foremost a more positive self-image, increased vigor, and improvement in body composition. Also, depending on the nature of the exercise, one might expect improvements in risk factors for chronic diseases, such as reduced blood pressure and lipid levels, as presented in Table 3-1 (discussed further in Chapters 1 and 6). Despite the numerous benefits less than 20% of the general population in the United States, United Kingdom, Canada, and Australia exercise regularly at a level that would allow these benefits to be realized.[1]

The FITT Principle

Discussions of exercise concepts often begin with the *FITT* principle. FITT is simply an acronym for frequency,

---

Table 3-1

Benefits of Regular Physical Activity

Benefits of Regular Physical Activity
↓ Resting heart rate
↑ Cardiac output
↑ Stroke volume
↓ Blood pressure
↑ Maximal oxygen uptake (VO_2max)
↑ HDL (high-density lipoprotein) and ↓ triglycerides
↓ Body fat (adiposity)
↓ All causes of mortality, morbidity, especially coronary artery disease
↓ Risk of certain cancers
↓ Risk of type 2 diabetes mellitus
↓ Risk of osteoporosis

Other Postulated Benefits
↓ Stress
↑ Productivity
↑ Libido (sex drive)
↓ Anxiety/depression
↑ Lean body mass (LBM)

* *Source*: Adapted from American College of Sports Medicine. *ACSM's Guidelines for Exercise Testing and Prescription*, 5th ed., Baltimore: Williams & Wilkins, 1995.

Table 3-2	The FITT Principle

Frequency: how often the activity is performed
Intensity: the level of difficulty at which the activity is performed
Time: the duration of the activity
Type: the specific mode of activity

---

intensity, time, and type, the acute training variables (Table 3-2). Through changes and manipulations of these components, the physiological adaptations and performance characteristics desired can be addressed.

F: FREQUENCY. **Frequency** is how often a particular activity is performed (for example, the number of times per week). This can be used in both general and specific terms. For example, a person might strength train 5 days/week

Key Point

Frequency, intensity, time, and type (FITT) are the major components of exercise training.

but train the hamstring muscles on only one of those days. Also, a participant may engage in various forms of aerobic activity daily but jog only occasionally. Therefore, it is more correct, and informative, to discuss the frequency of specific activities.

How frequently an activity is performed is primarily based on the goals of the program as well as the specific intensity, duration, and type of physical activity. Depending on the physiological adaptations desired, more or less frequent participation may be required. For example, if a person wished to improve athletic performance in the 100-m dash, the activity would most likely need to be performed 4–7 days/week. If instead some basic improvements in health and fitness such as increased cardiorespiratory endurance were desired, 3–5 days/week of jogging might suffice. Typical frequencies for fitness-related endurance training are 3–5 bouts per week, and for strength training 2–4 times per week. Strength training and cardiorespiratory endurance are discussed in detail in Chapters 12 and 13, respectively.

I: INTENSITY. **Intensity** is the level of difficulty or "how hard" one performs an activity. Intensity determines the metabolic demand of exercise. As the level of intensity rises, so does the metabolic demand to perform that activity. For resistance exercises, intensity level is generally expressed in terms of **repetitions of maximum (RM)**. A 1 RM is the maximum amount of weight that can be lifted one time, and an 8 RM is the maximum amount of weight that can be lifted successfully eight times.

Several methods have been used to determine the intensity levels for endurance or aerobic activities. Exercise physiologists have found several variables that either directly or indirectly measure intensity level. These include such formal measures as the volume of oxygen consumed **(VO_2)** and maximal oxygen consumption in a given time period **(VO_2max)**. Other formal measures include heart rate (HR); **heart rate reserve (HRR)**, which is the difference between maximum and resting heart rate; and the **respiratory exchange ratio (RER)**, the ratio of carbon dioxide expired to oxygen consumed at the level of the lungs. More general measures include the **rating of perceived exertion (RPE)**, a subjective assessment by the performer, and the talk test. These are discussed further in Chapter 13.

T: TIME. **Time** is the duration of the activity or "how long" it is performed. For weight training or resistance

Key Point

Intensity refers to the level of exertion during exercise and determines the metabolic demand of the exercise or activity.

exercises, it is more common to use the number of weeks or months rather than the time per training session or exercise. For example, a program may consist of 4–6 weeks of moderate-intensity exercises (8 repetitions of maximum) and then move to 3–4 weeks of high-intensity exercise (4 repetitions of maximum). See the section Periodization below for more information about time.

For endurance activities, time is simply the duration or how long the activity is performed. Using this in conjunction with intensity provides the most accurate training information. For example, a runner may have trained for 1 hour, but within that run there was a low-intensity warm-up of 10 minutes and a 10-minute warm-down. Therefore, only 40 minutes of running may have been at "race pace."

The amount of time spent training is determined by the goals of the program as well as the adaptations desired. Achieving a certain amount of work, expending a certain level of energy (kilocalories), or covering a specified distance often determine training goals. In addressing these goals, the duration of the activity is, at least in part, determined by the intensity level at which the activity is performed. Generally speaking, the greater the intensity level, the shorter the duration. For example, if the goal is to expend 300 kcal, the goal would be met sooner by exercising at an intensity level of 90% of VO_2max versus 70% of VO_2max.

Key Point

The time or duration of the activity is generally dependent on intensity level. As the intensity level increases, the duration of the activity decreases.

T: TYPE. **Type** is the specific modality or exercise performed. There are a variety of exercises for both strength and endurance training and each elicits specific adaptations. Therefore, the specific type of exercise must be considered in order to meet desired adaptations. In other words, the physiological responses and adaptations to exercise are specific to the frequency, intensity, time, and type of exercise encountered. The closer the training program mimics the activity or goals desired, the greater the improvements from training. For example, if a person wishes to improve his or her marathon running, a training program using running as the primary mode of exercise provides the best results. Furthermore, the intensity or running speed for this example needs to be near "race pace" for the best adaptations to occur. Studies have shown that elite endurance athletes show similar VO_2 values regardless of the activity used. However, their peak VO_2 values (VO_2max) are observed in their particular sport because of their adaptations to specific requirements (such as muscular endurance, muscular strength, energy demands, endurance capacity, range of motion) of that activity.

Key Point

Physiological responses and adaptations to exercise are specific to the frequency, intensity, time, and type of exercise.

Endurance training differs from strength training in one major respect. Because all forms of endurance training are dependent on the heart and its delivery of blood, each type of endurance training improves the function of the heart. Consequently, some basic cardiorespiratory improvements can be elicited by endurance training regardless of the type of activity. Therefore, to improve general fitness levels and cardiorespiratory function, any form of endurance training can be effective as long as threshold intensity levels have been reached.

The Law of Diminishing Return

Why does it seem like improvements come so quickly early in the training but wane as the length of the effort is prolonged? Furthermore, why do elite athletes have to practice and train so frequently? The concept of "diminishing return" helps answer these questions (Figure 3-1).

Once initial thresholds have been met relative to frequency, intensity, and time, the return or response to increasing levels of these decreases. For example, if frequency increases from 3 days/week to 4 days/week, the proportional improvement for the added day of training is less. The slope of the line (rise/run) is beginning to flatten. Therefore, the return on the added day of training is

diminished compared to the return previously. This does not mean that improvements cease with increased efforts; it simply means that the proportional return for additional training is less. This is also true for the intensity and duration training variables.

To further explain this concept think about a person who moves from a sedentary lifestyle to one of athletic performance. The potential for improvement of the sedentary person is initially high and then diminishes as improvements are made. Consequently, the level of benefit from a given amount of training begins to diminish as improvement increases. In other words, it gradually takes more effort to keep improving. Therefore, one can expect to see the greatest improvements early in the training process and less as time goes on. For example, an elite marathon runner may have to train for years in order to improve running times significantly, whereas a beginning runner can improve running times easily in just a few weeks or months of training.

Another factor to consider when addressing the training variables is the risk of orthopedic injury. As the frequency, intensity, and time spent training increase, so do the risks for injury. For example, the incidence of injury begins to rise rapidly after 30 minutes of exercise and rises exponentially thereafter. This rise in injury rate also holds true for increasing intensity and frequency levels. Therefore, by maintaining moderate levels of these components, one can accomplish adequate amounts of work while minimizing the risk of injury.

Key Point

The *law of diminishing return* states that in the upper ranges of the frequency, intensity, and time variables of training, the magnitude of improvements decreases.

Overload and Adaptation

Once an exercise program is initiated, is it necessary to keep increasing the intensity and duration for improvements to continue? For the body to improve its function and adapt, some type of stimulus must be present. The stimulus received must be something that is not normally encountered; this is termed **overload**. It can be something completely new or simply a change in frequency, intensity, or duration of the activity. For example, if a person typically uses the elevator to get to the fifth floor of an office building, using the stairs would be a training stimulus not normally encountered and would thus create overload. If the person continues to use the stairs on a regular basis, this becomes the norm and no further overload is present. If, however, the person climbs more stairs or climbs them more quickly, a new type of overload is created and further adaptation can occur. Examples of strength training

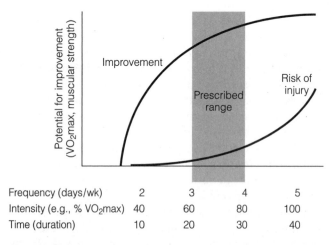

Frequency (days/wk)	2	3	4	5
Intensity (e.g., % VO$_2$max)	40	60	80	100
Time (duration)	10	20	30	40

Figure 3-1
Concept of diminishing return. Increasing the frequency, intensity, and time of exercise has greater proportional benefits in the low and middle ranges than in the upper range. The risk of injury also increases faster in the upper range.

overload are changes in the frequency (training an extra day per week), intensity (using heavier weights), or type (performing different exercises for a given muscle group).

GRADUAL PROGRESSIVE OVERLOAD. Gradual progressive overload (GPO) is a systematic series of gradual progressions resulting from modifications to the frequency, intensity, time, or type of exercise. Results are generally linear in nature. For example, suppose an exercise prescription requires one set of 8–10 repetitions for a given strength training exercise. Once the performer can accomplish 10 repetitions at a given weight, the load should be increased in order to continue the overloading process. A GPO example for an endurance activity could simply be a gradual increase in the distance covered or the time spent exercising.

OVERLOAD-SPECIFIC ADAPTATION. Adaptations are the physiological changes and adjustments resulting from overload. The extent of these adaptations is dependent on the frequency, intensity, time, and type of the activity. In other words, the adaptations that occur are specific to the stimulus received. For example, a weightlifter performing the bench press using two sets of 20 repetitions at an intensity of 60% of 1 RM will develop more muscular endurance than muscular strength because of the relatively low intensity level. A cyclist or runner who trains at low-intensity levels for extended periods will be better adapted to long-enduring cycling or running. To phrase it differently, to excel in a particular sport or activity, performing that specific activity is warranted. Furthermore, participation in the activity needs to be at or near the performance intensity levels desired because one cannot become a great sprinter without sprinting. What this means is that the specific muscle groups used, the range of motion, speed of contraction, type of contraction, and the magnitude of force produced all play a role in specific types of adaptations. To attain the best adaptations for a particular goal, one must simulate these components as closely as possible during training.

Training Peaks, Plateaus, and Staleness

Why does it seem that after a certain fitness level is reached, further improvements are very difficult? The staleness concept refers to a physical state in which no further improvements are realized from training, a state often

Key Point

For the body to improve and adapt, a physical stimulus not normally encountered (overload) must be initiated. The adaptations will be specific to the frequency, intensity, time, and type of overload presented.

called a *peak* or *plateau*. To overcome or avoid staleness it is necessary to manipulate the training variables (FITT) in order to maintain some level of overload. For example, a plateau in bench press strength may come after 3 months of training. In order to create a new stimulus (overload), one may add another set of bench presses or perform a supplementary exercise such as incline presses, which may help overcome this plateau. When athletes approach their physical best, plateaus and peaks become more common. Conversely, improvements come relatively easily for the novice exerciser and few peaks or plateaus are encountered for the first few months of training. Besides experiencing physical peaks and plateaus, athletes often get mentally bored with their training regimens and feel unmotivated. To address both the physiological and mental staleness it is common to manipulate the training variables (FITT) regularly to maintain the physical overload as well as to keep the athlete motivated and challenged.

Key Point

Peaks and plateaus in performance can be overcome by manipulating the training variables (FITT) to create new overload.

Training Volume and Adaptive Response

It is often troublesome to figure out how many sets and repetitions of a given exercise for a muscle group are needed for a given adaptation to occur. To address this issue the amount of work needs to be considered. The concept of **training volume** accounts for the total amount of work performed, also known as "training stress." Higher volumes generally elicit greater adaptations as long as the work is specific for the outcomes desired.

To calculate volume for weight training, the number of sets is multiplied by the number of repetitions and the amount of weight lifted per repetition. For example, if a person performs 3 sets of 8 reps with 200 lb in a bench press exercise, the volume would equal 4800 lb ($3 \times 8 \times 200$ lb). Technically, volume should be in units of work (foot pounds, watts, joules) because work considers the distance (range of motion) over which the force is applied (work = force $\times$ distance). Therefore, if the bench press exercise requires a force to move the barbell 3 ft per repetition, this would be included in the calculation. However, for practical purposes this is generally ignored. The total number of repetitions alone is sometimes used as an estimate of training volume.

For endurance activities, the total amount of work is often given as the distance covered or is estimated by time spent performing the activity. For example, the number of miles run or biked in a given week could be used to determine training volume.

Key Point

Training volume is the total amount of training stress received. For strength training it can be calculated by multiplying sets, reps, and amount of weight lifted per rep. For endurance activities it can be estimated by distance covered or duration of activity.

Training volume is an important consideration when analyzing total training stress. The training stress encountered by the body determines the type and level of physiological adaptations (such as muscle hypertrophy, VO₂max). Training volume must also be evaluated relative to the specific intensity levels of the performed activities. The optimum volume level for any particular purpose is still debated and may be dependent on individual characteristics. Despite the debate, general parameters have been determined and followed by a majority of strength and conditioning coaches for various activities.

Cross Training

Does using multiple exercises enhance adaptations? How will training for multiple activities affect a particular sport? "Cross training" became popular as year-round training became essential to remain competitive in many sports. In fact, most athletes train the entire year although their sports have competitive or official seasons lasting only 4–6 months. The cross training concept is based on utilizing other forms of exercise to elicit improvements for a particular activity. For example, to improve sprinting speed an athlete might utilize training drills in a swimming pool or use resistive training exercises. The idea is to have a "transfer of training effect." For cross training to be most effective, several components need to be addressed. The magnitude of forces and the rate of force development, acceleration, velocity, and range of motion all need to be similar in order for optimal transfers of training to occur. Aside from performance enhancement, cross training is also used to decrease the incidence of overuse injuries and provide variety in the training regimen.

In the fitness realm, the term *cross training* has been used more generally to describe a program with a variety of exercises and activities. Using a treadmill, stationary bike, and swimming to improve cardiorespiratory function has often been considered cross training. Also, a combination of strength and endurance training for general improvements in fitness has been termed cross training among fitness enthusiasts. Despite variations in the term's use, cross training is a valuable training concept utilized by many from both the athletic and fitness areas to improve performance and health while providing variety in their programs.

Periodization

Systematically creating and changing overload generally provides the best results, but how is this done? The term **periodization** refers to systematic or planned variation in the training regimen. This variation elicits desired physiological outcomes (such as hypertrophy) during specific phases of the training program. Periodization provides variety and continues to stimulate and overload the body while allowing for adequate amounts of rest. Most periodized programs revolve around the sports cycle (preseason, in-season) and are planned on a yearly basis. Within the yearly cycle (macrocycle) are several major units constituting months or weeks (mesocycle) and days (microcycle). Each of these cycles should have a purpose and elicit a desired overload or adaptation. Unlike gradual progressive overload, the progression may not be linear because separate performance characteristics are addressed. A periodized program for a team sport athlete is often organized into the following phases:

Hypertrophy → Strength → Power
→ Peaking → Active rest

Periodization also applies to endurance activities. For example, for a runner to successfully compete in a marathon several factors must be addressed. First, the distance of the race must be considered, and second, the rate at which the distance is covered. These two factors determine the performance level of the athlete. For the best performance the runner needs to run at the highest speed while covering the required distance.

Key Point

Periodization provides systematic variation to the training program, allowing for continued overload as well as variety to decrease training boredom.

Using a periodization scheme, the runner concentrates on either of the two components, distance or speed, during specific training days or weeks. This accomplishes overload for the specific characteristic and associated energy systems. Periodization also provides variety to the runner while minimizing orthopedic injuries. The training regimens for other endurance sports are similar, with lighter intensity over long distances on some days and higher intensity over shorter distances on others. For triathletes the training regimen becomes more burdensome because three specific activities must be trained for simultaneously. Despite the number of activities, some type of systematic variation is used to enhance performance, minimize the risk of injury, and help alleviate training boredom.

Interval Training

Is changing the intensity level during an exercise bout beneficial? *Interval training* is the use of varying intensities throughout an exercise bout and is primarily associated with endurance activities. A baseline level of intensity is chosen (such as 50% VO_2max or 65%HR_{max}), and then periodic intervals of higher intensity activities are incorporated. The concept is to use the baseline zone as the active recovery zone for the high-intensity bursts. During these bursts the exerciser increases intensity and works through the anaerobic threshold (see Chapter 13) to the peak zone. When the peak zone is reached the exerciser remains at that intensity level for about 30 seconds. Then the user decreases intensity and returns to the baseline zone to recover. Using this method of training accomplishes overload and also stimulates the anaerobic energy systems. This method of training elicits higher energy expenditures than most steady-state workouts and thus expends more calories.

The levels of intensity are often broken down into specific zones. For example, zone one is the baseline level and could have an intensity level of 50% VO_2max (65% HR_{max}). Zone two could be the anaerobic or lactate threshold level and have intensities of 65–75% VO_2max (75–85% HR_{max}). Zone three for this example could be the peak zone, in which the training intensity approaches maximum levels (say, >85% VO_2max or >90% HR_{max}). Currently, most commercial fitness equipment (such as stationary bikes, treadmills, and steppers) is equipped with various interval programs preprogrammed into the computer displays.

Rest

Does the amount of rest between sets and workouts affect training results? If so, how much rest is needed? The amount of rest between sets (same exercise or muscle group) within a training session as well as between workouts affects adaptation. The rest periods vary depending on the goals of the training program (for example, high-intensity endurance, power, strength, endurance, hypertrophy). High-intensity and power activities require longer periods of rest to regenerate creatine phosphate levels and reestablish intracellular pH. Less rest time is needed for moderate-intensity exercises, and the least rest is needed for endurance activities. Rest is discussed in more detail relative to strength training in Chapter 12 and endurance training in Chapter 13.

For rest periods between strength training workouts, the general guideline is one day off, or a total of 48 hours. This issue is still controversial; the need for rest may depend on the intensity and/or duration of the activity. The recommendation of 48 hours of rest is for any specific muscle group. Therefore, strength training can take place on consecutive days if the same muscle group is not trained on both days. Training routines that break the body into specific regions (such as biceps/back, chest/triceps, legs) are called "split routines" and used by more advanced and experienced lifters.

The guidelines for rest associated with endurance training are less clear. The amount of rest is generally determined by the intensity and duration of the previous workout and by the goals of the program. For basic health improvements, endurance exercise is generally performed 3–5 days a week with a day of rest between most of them. But in training geared for athletic competition, rest is generally less frequent and more dependent on the sport or activity.

■ EXERCISE AND TRAINING CONCEPTS *IN REVIEW*

- Improvements from training come more quickly early in the program and wane as the potential for improvement decreases.
- For a training stimulus or overload to be most effective it needs to closely relate to the outcome desired; because adaptations are specific to the overload encountered.
- Cross training is a useful tool for creating overload while decreasing training boredom and incidence of overuse injuries.
- Periodization is planned variation in the FITT elements and can enhance performance levels, protect against overtraining, and decrease training boredom.
- Adequate amounts of rest are necessary between workouts in order to avoid overtraining and elicit the best responses to training. Rest intervals between sets of an exercise are also important in determining the overload received and the subsequent adaptations.

■ COMPONENTS OF FITNESS

What is fitness and what physical attributes contribute to a certain level of fitness? How are these physical attributes improved, and are they interrelated? **Fitness** may be described as a physical condition that enables a person to appear, feel, and function at a given level. Fitness is composed of five physical variables (Table 3-3). They consist of muscular strength, muscular endurance, cardiorespiratory endurance, flexibility, and body composition. Fitness and

Table 3-3   **Components of Fitness**

Cardiorespiratory endurance
Muscular strength
Muscular endurance
Flexibility
Body composition

exercise programs generally address these components directly or indirectly depending on the goals and aspirations of the participant. These components are briefly defined and discussed below.

Muscular Strength

Muscular strength is the maximum amount of force that can be generated by a muscle or muscle groups. It is one repetition of maximum (1 RM). Muscular force is important to daily life skills, prevention of injury, and sports performance. To improve muscular strength, exercises are generally performed at fairly high intensity levels (for example, 1–6 repetitions of maximum). See Chapter 12 for a more in-depth discussion of strength training relative to muscular strength.

Muscular Endurance

Muscular endurance is the ability to generate submaximal force for a specified amount of time or a set number of repetitions. Performing a higher number of repetitions (for example, >15) generally addresses this component. However, muscular endurance does not necessarily have to be associated with low-intensity levels. For example, one can train for high-intensity muscular endurance by utilizing heavy weights (5 repetitions of maximum) with short rest intervals. Running "wind sprints" (sprints with minimal rest time) is another example of high-intensity muscular endurance training.

Muscular strength and muscular endurance can be thought of as separate components but are often combined and termed "muscular fitness." Most activities and exercise contain both a muscular strength and a muscular endurance component, so combining them to address general health and fitness is appropriate. Strength training exercises have been shown to improve lean body mass, improve function in daily life skills, decrease the number of falls for elderly people, improve self-esteem, and help decrease the incidence of injury.

Key Point

Improving muscular strength can enhance the performance of daily life skills, help reduce the number of falls for elderly people, and decrease the incidence of injury.

For general fitness, strength training exercises should be performed in a smooth, controlled manner throughout the full range of motion. Isometric exercises have been shown to improve strength, but dynamic (concentric/eccentric) exercises better simulate real-world activities and have shown superior results for improvements in muscular strength. Breathing should remain normal with inhalations occurring during the eccentric portion of the exercise and exhalation occurring during the concentric phase. In other words, inhale on the easy part and exhale on the hard part.

Many tests assess muscular strength and endurance. Some commonly used tests are 1 RM bench presses, 1 RM leg presses, push-ups to fatigue, and timed abdominal crunches. Most tests have normative data to evaluate results in relation to age and gender. However, the relative improvements (individual results pre- and posttraining) are much more important than the absolute normative data when considering improvements for general health and fitness.

Key Point

Muscular fitness is a combination of muscular strength and muscular endurance, and most general strength training programs blend those components.

EXERCISE FREQUENCY FOR MUSCULAR FITNESS. The recommended frequency of weight or resistance training is 2–3 days/week. Greater gains may occur with increased frequency, but for the adult fitness population the relative improvements gained from higher frequencies are not warranted. Also, each muscle group may respond differently to the frequency of training, but 2–3 days/week seems to be appropriate for most muscle groups. Training less than 2 days a week may allow for a detraining effect to occur.

EXERCISE INTENSITY FOR MUSCULAR FITNESS. The recommended intensity of resistance exercises for general fitness is 8–12 RM. As discussed in Chapter 2, the number of repetitions helps determine the force outputs needed and consequently the fiber types and number of motor units recruited. Using 8–12 repetitions of maximum elicits an efficient use of both fiber types and thus trains for a combination of muscular strength and endurance. Also, using an 8–12 RM scheme means the total volume of work is large enough to elicit appropriate neural, morphological, and metabolic adaptations.

EXERCISE TIME FOR MUSCULAR FITNESS. The duration of strength training depends on the intensity and total volume of work demanded. The recommendation is for a total body training volume of 8–10 exercises performed in 1–3 sets for 8–12 repetitions. The duration to accomplish this volume ranges from 30 to 60 minutes

Key Point

Strength training sets of 8–12 RM is recommended for general conditioning.

Table 3-4

Sample Strength Training Routine for General Health and Fitness

1. Lat pulldown
2. Leg press
3. Chest press
4. Leg curl
5. Seated row
6. Leg extension
7. Shoulder press
8. Abdominal crunch
9. Back extension

Table 3-5

Recommendations for Muscular Fitness Training

Frequency   2–3 days/week
Intensity   1–3 sets should include 8–12 repetitions of maximum
Type   8–10 exercises focusing on the major muscle groups

depending on the specific exercises chosen and the particular type of equipment used. Time is also used in relation to periodization schemes or other training cycles and accounts for a certain number of months, weeks, or days. This is discussed further in Chapter 12.

EXERCISE TYPE FOR MUSCULAR FITNESS.   The specific types of strength training exercises for improving muscular fitness also need careful consideration. It is recommended that 8–10 exercises be performed that address all the major muscle groups of the body. To accomplish this, many of the exercises must be multijoint or compound in nature. In other words, the exercises incorporate multiple joints as well as many muscle groups simultaneously. Examples of these include the squat, bench press, and pull-ups. Multijoint exercises decrease training time and assists in the development of coordinated functional movement patterns. Table 3-4 outlines a sample routine for general health and fitness. Note that the order of the exercises should be considered: it starts with exercises performed from the larger muscle groups using multiple joints and goes on to the smaller muscle groups using single joints.

The specific modality used (such as free weight or machine) for each of the selected exercises is also relevant to adaptations. Strength training machines generally provide ease of use, structured movement patterns, time efficiency, and various pieces designed for specific muscle groups. This makes the exercise prescription simple and easy to follow. Also, the weight increments are easily adjustable, which allows for easy administration of progressive overload.

Traditional forms of calisthenics (such as push-ups) and other body weight movements (such as lunges) can also be valuable tools to enhance fitness levels. Therefore, as mentioned earlier, the use of a health club is not mandatory for improvements in muscular fitness. Simply using the basic concepts of exercise and convenient modes of training can elicit many of the fitness goals desired. Table 3-5 summarizes the recommendations for muscular fitness.

Cardiorespiratory Endurance

Cardiorespiratory endurance is the ability to sustain physical activity for an extended period. Sustained activity utilizes the aerobic energy system as the primary source of fuel and therefore is also called aerobic, endurance, or cardiovascular exercise. Running, cycling, swimming, stair climbing, and cross-country skiing are among the modes of training. These activities generally utilize large muscle groups in a continuous rhythmic fashion for 15 minutes or more.

Cardiorespiratory endurance is the number one component of fitness because it has the most dramatic impact on the body as a whole. It conditions the heart, lungs, blood vessels, and skeletal muscles and helps maintain healthy energy expenditures. Cardiorespiratory endurance assists in weight management, decreases the risk of heart disease, and can improve self-esteem, reduce stress, increase mental acuity, improve sleeping patterns, and increase libido. Table 3-6 outlines the recommendations for cardiorespiratory endurance exercise.

EXERCISE FREQUENCY FOR CARDIORESPIRATORY ENDURANCE.   The frequency for cardiorespiratory endurance exercise is dependent on the goals of the participant. However, the American College of Sports Medicine (ACSM) recommends 3–5 days/week and preferably some activity daily.[2] Greater improvements (increased VO_2) can be elicited with frequencies above 5 days/week, but as the frequency increases, the magnitude of improvement decreases and the chances of orthopedic injuries increase. Frequencies above 5 days/week have been shown to have little or no further improvement in VO_2max.[3–6] Frequencies below 2 days/week have been shown to have little effect on improving VO_2max.[6–9] Therefore, a frequency range of 3–5 days/week seems most effective for improving VO_2max.

Key Point

Cardiorespiratory endurance is the most important component of fitness!

Table 3-6

Recommendations for Cardiorespiratory Health and Fitness Training

Frequency   3–5 days/week
Intensity   65–90% HR_{max}, 50–85% VO_2max
Time   20–60 minutes, bouts of 10+ minutes totaling > 30 minutes
Type   Continuous, rhythmic movement using large muscle groups

Table 3-7	**Target Heart Rates for Age**	
Age	**Target Heart Rate Zone (50–75% HR_{max})**	**Average Maximum Heart Rate**
20 years	100–150 bpm	200
25 years	98–146 bpm	195
30 years	95–142 bpm	190
35 years	93–138 bpm	185
40 years	90–135 bpm	180
45 years	88–131 bpm	175
50 years	85–127 bpm	170
55 years	83–123 bpm	165
60 years	80–120 bpm	160
65 years	78–116 bpm	155
70 years	75–113 bpm	150

Note: bpm, beats per minute.

EXERCISE INTENSITY FOR CARDIORESPIRATORY ENDURANCE.   The intensity of exercise is also important for improvements in cardiovascular health. Intensity is based on the metabolic demand of the activity. For lower intensity exercise the energy is primarily derived from aerobic processes, whereas higher intensity exercise derives its energy from anaerobic sources. Cardiorespiratory endurance activity derives most of its energy from aerobic sources because the low- to moderate-intensity levels allow ATP to be produced aerobically. The ACSM has recommended intensity levels for percentages of VO_2max [also called VO_2R (for reserve)], heart rate reserve (HRR), maximum heart rate (HR_{max}), and rating of perceived exertion (RPE).[2] The "talk test" is a practical way to evaluate intensity level. When one is engaged in endurance activity, talking should be relatively easy and not labored. Higher intensity levels have benefits; however, for general health and fitness the recommendation is for activity to occur in the low to moderate range of intensity. The recommendations for intensity levels are as follows:

Deconditioned Population
- 40–49% VO_2max
- 40–49% HRR
- 55–64% HR_{max}
- 10–12 RPE

General Population
- 50–80% VO_2max
- 50–80% HRR
- 65–90% HR_{max}
- 13–19 RPE

The listed intensity levels are suggested for improvement of VO_2max. Physical activity at lower intensity levels may have several health benefits but these have not yet been clearly defined. For the deconditioned population, engaging in physical activity at the lower intensity levels in the first column elicits improvements in VO_2max. As fitness levels increase, the intensity level of exercise must also gradually increase in order for improvements to continue. Table 3-7 outlines the target heart rate zones for

Key Point

The recommended intensity level for general cardiorespiratory conditioning is just below the point at which conversation becomes labored.

training at various ages. For a more complete discussion of intensity and cardiorespiratory endurance see Chapter 13.

EXERCISE TIME FOR CARDIORESPIRATORY ENDURANCE.   The time or duration of activity for general conditioning has a recommended range of 20–60 minutes. The combined efforts of several shorter bouts of activity (such as 10 minutes) have also been shown to increase fitness levels. The 60-minute maximum is recommended due to a lack of improvements in VO_2max beyond this period as well as the increased incidence of orthopedic injuries. Duration is based in part on the intensity of the exercise. At lower intensity levels the duration of exercise should be greater to accomplish a similar amount of work (energy expenditure), and conversely for shorter durations a higher intensity should generally be used. The combination of intensity and duration determine the training volume or total amount of work performed. The total work determines the degree of physiological adaptations.

EXERCISE TYPE FOR CARDIORESPIRATORY ENDURANCE.   The type of activity recommended for cardiorespiratory endurance should be continuous and rhythmic while incorporating the use of large muscle groups. It should also minimize high-impact movements, as they may cause unnecessary orthopedic injuries. Although each modality of endurance exercise uses specific muscle groups, generates different forces, utilizes various ranges of motion, and so on, they all elicit cardiovascular benefits. The heart and lungs must work to deliver oxygen-rich blood regardless of the specific muscle tissue to which it is being delivered. Therefore, a variety of exercises train the cardiovascular system. However, for improvements in a specific activity, that particular mode of exercise must be incorporated extensively into the training regimen. In other words, if improvements in running are desired, running must be the primary activity.

The mode of exercise should also minimize orthopedic injuries. Activities such as jumping rope or other high-impact movements may not be appropriate for extended periods. This is especially true for unfit or overweight individuals. The mode of activity should be enjoyable, because adherence to any form of exercise is highly dependent on this factor. Chapter 13 discusses endurance training in detail.

Flexibility

Flexibility is the range of motion for a given joint. Addressing this component includes stretching the various muscles and connective tissues associated with a joint. Improving flexibility is generally thought to enhance joint function and muscular function, reduce the chance of injury, and optimize athletic performance. Flexibility is addressed through four main types of stretching that include static, proprioceptive neuromuscular facilitation (PNF), dynamic, and ballistic.

STATIC STRETCHING. Static stretching exercises are the most commonly used type and consist of relaxing a muscle and then elongating it to a point of mild discomfort. Stretches are generally held at this point of mild discomfort for 10–30 seconds. A number of different body and joint positions are used to stretch specific muscles or many muscle groups simultaneously. It is recommended that a participant consult a coach, trainer, or therapist for proper techniques. A thorough warm-up of the musculature is recommended before any stretching activity. A warm-up increases body temperature and circulation, which help decrease the viscosity of the muscle and improve the plasticity of the various tissues. This is similar to stretching a cold rubber band versus one at room temperature. How to warm up is discussed later in the chapter.

Key Point

Static stretches are the most commonly used, and each stretch is generally held for 10–30 seconds.

PROPRIOCEPTIVE NEUROMUSCULAR FACILITATION. Proprioceptive neuromuscular facilitation (PNF) stretching has proved superior to other forms of stretching, although static stretches remain the most commonly used. PNF stretching involves alternating isometric muscle contraction and passive stretching through defined movements. In a passive stretch, an external force (a partner or machine) is used to create the stretch. Active stretching occurs from forces created by the individual. Several techniques have been added to PNF stretching movements

and include contract/relax, hold/relax, and active/assisted movements. Many of these stretches can be performed without the assistance of a trainer or therapist, but the best results generally occur when assistance is provided.

DYNAMIC STRETCHING. Dynamic stretching uses momentum to create the forces necessary for elongating the muscle tissue. The movements used generally mimic or make up the motion patterns to be encountered in the sport or other exercises that follow. These include such basic movements as arm swings, torso twists, and knee lifts. All movements should be performed under control and kept within the normal range of motion. Dynamic stretches are often combined with the warm-up portion prior to an exercise program; together these components provide what is often termed a "dynamic warm-up." Dynamic warm-ups are functional and may help save time for people on tight schedules.

BALLISTIC STRETCHING. Ballistic stretches are similar to dynamic stretches but differ in one major area—control. Ballistic stretches are not performed under control and are often explosive in nature. They can result in movements beyond the normal range of motion and thus cause muscle tissue damage. Furthermore, the eccentric muscle action necessary to control these ballistic movements may increase delayed onset of muscle soreness (DOMS, discussed in Chapter 12). Ballistic stretches can be useful for improving flexibility but are generally discouraged.

EXERCISE FREQUENCY, INTENSITY, AND TIME FOR FLEXIBILITY. The frequency of stretching exercises should be daily, especially before and after exercise. It is recommended that workouts focus on flexibility training 2–3 days/week. The intensity of stretching should be to a point where mild discomfort is felt. Each static stretch should last 10–30 seconds. The 10-second minimum allows the "stretch reflex" to be overcome and permits elongation of the series-elastic components. The type of stretching exercises may be a combination of static and modified PNF stretches. However, one should understand proper PNF methods before engaging in these types of stretches. (See the accompanying Sport in Practice feature for a comparison of static and PNF stretches.) Table 3-8 summarizes recommendations for flexibility training.

Table 3-8

Recommendations for Flexibility Training

Frequency	At least 2–3 days/week; preferably daily
Intensity	Stretch to a point of mild discomfort
Time	Stretches should be held for 10–30 seconds
Type	Static and modified PNF stretches

Comparison of PNF and Static Stretching

Here is a simple way to compare and contrast two stretching techniques, PNF and static, for the hamstring muscles. Follow the protocol below for an introduction to PNF stretching. Be sure to warm up before attempting these stretches. PNF stretching is used on the hamstring muscle group in the left leg, and static stretching on the right leg hamstring muscles. During the process, compare how the two techniques feel, and note the results of the separate techniques. Is one more efficient than the other? Is one more time consuming?

1. Sit on the ground with legs spread apart to a position of mild discomfort; both knees should be fully extended and the entire leg should be contacting the floor.
2. For PNF stretching on the left leg: Contract the hamstrings for a few seconds by pressing the entire leg against the ground. Relax momentarily and then stretch the hamstring muscles by moving the chest toward the extended left knee for a couple of seconds.
3. From this mildly stretched point, repeat the hastring contraction against the floor for a few seconds and then stretch again to a point of mild discomfort.

4. Continue the series of contractions and stretches as the chest moves closer to the knee. Stop the series of contractions and stretches when the position becomes uncomfortable. Note your head, chest, and back position relative to your extended left leg.
5. Now perform static stretching techniques on the right leg. Gently push the chest and head toward and over the right knee. When a point of mild discomfort is reached, hold for 10–15 seconds. Relax for about 10 seconds and then repeat the process.
6. Use 3–4 repetitions of static stretches on the extended right leg. After the final stretch note the head, chest, and back position relative to the extended right leg.

How do the results of the separate techniques compare? Was there a difference in the way the techniques felt?

During your next stretching session try PNF on the right leg and static stretches on the left. Compare the results once again. Did the stretching technique make a difference? Does one leg seem to have better flexibility than the other?

Each individual has unique flexibility characteristics, to which any flexibility program should cater.

For the best results, stretching exercises should be performed after a warm-up. A warm-up increases body temperature and circulates fluids, allowing the elastic tissues to be stretched without undue trauma. Warm-ups are discussed in more detail later in this chapter.

STRETCHING BY OLDER INDIVIDUALS. Flexibility is a major consideration for the aging population. As people age they generally begin to lose elasticity in many of the body's tissues. This includes the series-elastic components of the musculoskeletal system, particularly the tendons. By including stretching exercises in a complete training regimen, one can maintain and usually improve flexibility of these tissues.

Body Composition

Body composition is the final component of fitness and refers to the relative contributions of different substances and tissues to the total mass of the body. Typically, body composition is assessed to determine body fat percentage and **lean body mass (LBM)**, which consists of essential fat plus the **fat-free mass (FFM)**—all nonfat mass including muscle, bone, skin, and organs.

Ways of assessing body composition range from the simple to the highly technological. Simple procedures include skinfold thickness measurements. More complex are measures such as densitometry by underwater weighing (UWW) and air displacement chambers, and bioelectrical impedance (BIA) sensors. Methods that use complex instrumentation include dual-energy X-ray absorptiometry (DEXA), computerized axial tomography (CAT), sonography, and magnetic resonance imaging (MRI).

The United States has seen an increasing number of people classified as mildly and clinically obese. Unfortunately, obesity has been linked with several cancers and diseases including America's number one killer, heart disease. As people age the trend is generally to accumulate body fat and lose lean body mass. Much of this change is likely due to the decreasing levels of physical activity during aging; however, some genetic and nutritional factors must also be considered. In order to alter the proportions of body fat and fat-free mass, energy expenditure must exceed energy intake. Also, the level of physical activity geared toward muscular development affects these tissue proportions.

To improve body composition it is recommended that a physically active lifestyle consisting of both muscular fitness exercise and aerobic endurance be incorporated with a reduced but healthy energy intake. Body composition is discussed in greater detail as it relates to diet, health, and performance in Chapter 7.

EXERCISE AND BODY COMPOSITION. Many people engage in regular exercise as a mechanism for body weight and composition management. They particularly use aerobic activities, which generally increase fat mobilization and oxidation. Many state that they exercise simply to avoid restricting their energy intake.

Unfortunately, research studies have not found that exercise-only programs for obese people result in significant improvements in body composition. These programs are often unsuccessful because changes in other health factors accompany them, including increased sleeping and resting periods, which decrease energy expenditure, and increased nutritional intake. For an exercise program to be successful, a broader approach to both energy intake and expenditure is needed to elicit proper body composition changes.

Despite the lack of body composition improvements from exercise programs alone, many health benefits still arise from them. These include increased VO_2max, improved lipid levels, improved psychological state, and decreased obesity-related diseases such as diabetes mellitus, heart disease, cancer, and hypertension. It is strongly recommended that obese individuals seek professional medical attention in order to address specific health risk factors as well as to develop long-term treatment plans.

When exercise is prescribed for the obese population, several precautions should be considered. Obese individuals often have lower heat tolerance levels and have movement restrictions, musculoskeletal pain and discomfort, difficulty breathing, and low self-esteem. They also have higher risks for cardiovascular disease, diabetes mellitus, and hypertension. Therefore it is important to properly screen for risk factors for these diseases. For the exercise program, low- to moderate-intensity levels of light and non-weight-bearing exercises (cycling, swimming, walking) are suggested. Such activities minimize orthopedic concerns while allowing the participant to engage in activity for extended periods. These activities also provide positive psychological benefits such as a feeling of personal success and accomplishment. Using an exercise environment that is comfortable and supportive also improves enjoyment levels and subsequent adherence to the exercise program.

■ COMPONENTS OF FITNESS *IN REVIEW*

- The components of fitness are the physical characteristics that determine fitness: muscular strength, muscular endurance, cardiorespiratory endurance, flexibility, and body composition.
- By using the training variables—frequency, intensity, time, and type—to address the components of fitness, positive changes in health can be attained.
- Cardiorespiratory endurance is the most important fitness component.
- Muscular strength and muscular endurance can be addressed specifically but are generally combined and trained together as muscular fitness.
- Body composition alterations are best attained using a global health perspective with a focus on healthy nutritional intake and moderate energy expenditures.

■ COMPONENTS OF A BASIC EXERCISE PRESCRIPTION

The concepts of training have been discussed along with the components of fitness, but how does all this fit together in formulating an exercise routine? Also, what is the typical structure of a "day at the gym"? The basic exercise routine or regimen consists of six items (Table 3-9) that provide a safe and effective protocol for improving and maintaining health and fitness. Prior to engaging in a regular exercise program it is suggested that participants receive a physical examination by a physician. The exercise program should generally follow the order of the components listed in Table 3-9 and discussed below.

Warm-Up

The first item of the exercise prescription is the *warm-up*. This should always be performed prior to any other exercise because it prepares the body for work or exercise. Any light aerobic activity (cycling, walking, light jogging, calisthenics) can be used to warm up, but for best results, performing the actual activity or something similar is recommended. In other words, if jogging is the activity for the day then a light jog or fast walk would probably be the best warm-up activity. During the warm-up the body should gradually increase in temperature so that it begins to feel warm and perspire lightly. For most people the warm-up should take only 5–10 minutes. If the weather is particularly cold or the participant is in the senior population a longer warm-up may be necessary. A warm-up has the following physiological benefits:

- Enhances enzymatic activity in working muscle tissue (which speeds up energy-producing chemical reactions)
- Reduces viscosity of the muscle tissue for improved function and decreased chance for injury

Table 3-9	**The Basic Exercise Prescription**
1. Warm-up	Perform 5–10 minutes of light aerobic activity, increase body temperature, prepare body for work
2. Stretch	Perform static or dynamic stretches for the muscle groups and types of movement patterns to be executed
3. Strength training	Perform 8–10 exercises addressing all the major muscle groups
4. Cardiorespiratory	Perform 30–60 minutes of continuous rhythmic movement
5. Warm-down	Gradually decrease exercise intensity level until heart rate drops to 100 beats per minute or lower
6. Stretch	Perform static stretches for all major muscle groups, especially the muscle groups used

Key Point

A warm-up elevates the body temperature and reduces the viscosity of muscle and connective tissues, preparing the body for activity.

- Increases blood flow to working muscle tissue, which increases temperature and the delivery of nutrients and fuel
- Enhances power output and mechanical efficiency of muscle tissue
- Enhances neural signal delivery for more coordinated movements
- Increases breakdown of oxyhemoglobin and increases the release of oxygen from myoglobin to enhance O_2 delivery
- Decreases viscosity of soft connective tissues, which reduces the chance for injury

As a general rule, the more intense the activity the more time should be spent on the warm-up. This is especially true for dynamic movements observed in many sports because the soft connective tissues (such as cartilage) need to be warmed and ready for the forces encountered during these activities. Most active people would agree that if a proper warm-up is used prior to the activity, performance is improved and the level of soreness postexercise is generally reduced.

Stretching

The next item in the exercise routine is stretching. Stretching should follow a warm-up in order to allow proper warming mechanisms to increase muscle and soft connective tissue temperature and decrease viscosity. These tissues can then be stretched without injury to prepare them for the movement patterns (range of motion) to be encountered. The stretches used should be primarily geared toward the specific muscles and joints that will be used, but it seems advantageous to stretch most of the musculature. Both static and dynamic stretches are recommended.

DYNAMIC WARM-UP. A *dynamic warm-up* and stretching session can be used to combine the first two components of a warm-up and stretch. For example, if a jogging session is to be undertaken, a dynamic warm-up could include walking, walking with exaggerated ranges of motion (such as high knees, hip extension, butt kicks, arm swings, heel raises), and then light jogging with these same types

Key Point

To save time the warm-up and stretch can be combined into a dynamic warm-up.

of movements. The benefits of a dynamic warm-up are that it may address the targeted joints and muscles better than static stretches, and the body does not cool down as it might during static stretches. A dynamic warm-up also saves time by combining the warm-up and stretching, which can be especially beneficial for those with limited time to exercise.

Strength Training

The third item in the exercise routine is the strength training portion. The ACSM strength training recommendation is 2–3 days/week using 8–10 exercises, for 1–3 sets of 8–12 repetitions that utilize most of the body's musculature.[2] Depending on the goals and organization of the training program, strength training may or may not be performed on the same day as the cardiorespiratory training. If the two are to be performed on the same day it is generally recommended that strength training be performed first. There are three reasons for this, and the choice may depend on the goals of the program. First, strength training is generally more intense than the endurance portion, requiring more energy from anaerobic sources that may be more available early in the workout. Second, certain strength exercises (such as squat and shoulder press) require the support and stabilization of the spine. Fatigue of the back, abdomen, and lower extremity muscles prior to these exercises may put the performer at a greater risk for injury. Third, after a vigorous aerobic workout the perspiration produced may allow the body to cool below comfortable levels for the strength training portion.

Cardiorespiratory Endurance Conditioning

Endurance training is the fourth item on the exercise prescription list. As mentioned above, it may or may not occur on the same days as strength training. Cardiorespiratory endurance is the primary component of fitness because of its effect on overall health. The recommended quantity and quality are 3–5 days/week, at 65–90% HR_{max}, for 20–60 minutes, or an accumulated total of 30 minutes daily, as noted in Table 3-6.

It is important to find activities that are enjoyable as well as convenient in order to maintain consistent participation. Exercising with a partner or group of friends can be motivating and enjoyable and is often recommended for adherence purposes; a little motivation from a peer can be quite effective! Scheduling regular days and times ("exercise dates") has also proved effective in increasing adherence to physical activity programs.

Warm-Down

The warm-down, a period in which the intensity level gradually decreases, follows the more active components of the exercise prescription. It is the reverse of a warm-up. It usually follows the aerobic training portion and slowly decreases the heart rate that was elevated from endurance exercise. Gradually decreasing the intensity level allows the pumping of the skeletal muscles to maintain adequate venous return to the heart and hence keep blood from pooling in the extremities. Inadequate venous return can be problematic for the elderly population as well as those with circulatory problems. A warm-down also helps deter muscle stiffness by reducing muscle and blood lactic acid levels and consequently improves recovery time. Finally, following heavy endurance exercise the blood levels of catecholamines increase. Elevated catecholamine levels can adversely affect the heart in certain individuals. Gradually reducing the exercise intensity can reduce the circulating catecholamine levels. A prudent warm-down should get the heart rate to approximately 100 beats per minute or lower. Also, maintaining normal ranges of motion is recommended to capitalize on the skeletal muscle pump for venous return of the blood.

Key Point

Consistent activity participation provides the best overall results. Therefore, it is important to find activities that are enjoyable as well as convenient.

Stretching

The final item in the exercise regimen is again stretching. Performing stretches after a training session helps bring the muscle tissue back to "normal" resting lengths. Physical training elicits contractions of the working muscle tissue and consequently creates tension. This often gives the muscle a feeling of being "pumped" or tight. Stretching allows the muscle tissue to lengthen and thus improves circulation. This reduces recovery time from the activity and potentially decreases muscle soreness.

The exercises performed should address the major musculature of the body and specifically the muscle groups and joints used during the exercise regimen. Stretching exercises should include muscles of the lower back and abdomen, since these muscles are used for virtually all activities. Injuries to the lower back are common because of this heavy use, and maintaining proper flexibility and

strength reduces the incidence of back pain and disorders. Static stretches held for 10–30 seconds for 1–3 repetitions are again the recommended protocol. Stretching after exercise also allows the muscle to remain elongated for an extended period (between activity sessions) and can increase flexibility. Obviously, the more frequent the stretching exercises, the greater the flexibility adaptations.

■ COMPONENTS OF A BASIC EXERCISE PRESCRIPTION *IN REVIEW*

- A proper warm-up prepares the body for physical activity.
- Strength training prior to endurance training may be beneficial due to greater availability of anaerobic fuel sources.
- A warm-down provides adequate venous return to help prevent blood pooling and also allows lactic acid levels to dissipate, potentially decreasing muscle soreness and improving recovery time.

Conclusions

Frequency, intensity, time, and type (FITT) are the training variables associated with physical activity. Applying these variables to exercises for muscular strength, muscular endurance, cardiorespiratory endurance, and flexibility determines how the body will adapt in response to the exercise. The total training volume or training stress of activity relative to a specific intensity is ultimately responsible for the adaptations that occur. As improvements in physical performance are realized the magnitude of further enhancements and the time required for them to occur changes. This is due to the decrease in potential for improvement, and means that peaks and plateaus unfortunately become more common. To reduce the incidence of these stagnant points and improve motivation, constant manipulation of FITT through periodization can be incorporated to maintain adequate overload. The results from all the training as well as nutritional intake ultimately affect physical function, body composition, and athletic performance.

 IN FOCUS   HISTORY AND CURRENT STATUS OF PHYSICAL ACTIVITY AND FITNESS

The Beginning

Long ago activity came in the form of daily work such as hunting, food gathering, and shelter building. Paleolithic researchers have noted that it was common for our ancestors to walk as much as 20 miles to trade their goods and visit family and friends in neighboring villages. Today, this long walk may seem like a laborious chore. However, this trek was often performed after a couple of days of hard work and was considered a form of relaxation and celebration.

Throughout much of recorded history, physical activity has been promoted for improved health, function,

HISTORY AND CURRENT STATUS OF PHYSICAL ACTIVITY AND FITNESS (CONTINUED)

and longevity. With their ideals of physical prowess, athleticism, and health, the ancient Greeks may have had the greatest influence on Western culture's view of physical activity. The Greeks viewed superior athletic achievement as signifying both spiritual and physical strength, almost to the point of rivaling their gods. In the Olympic Games the victors were exalted by the Greeks as men with extraordinary mortal strength.

Some of the early Greek physicians correlated a healthy diet, physical activity, and rest as important contributors to overall well-being. These components were often referred to as the "laws of health." Herodicus in 480 B.C. used what was called "gymnastic medicine" as a means to study physical movement. He is often considered the inspiration for the early emphasis on physical activity.

In Modern Times

During the 1800s some of these Greek philosophies gradually made their way to America. The interest was mostly attributable to transition from the rural life to one of urbanization and industrialization. Farm work was replaced by repetitive movement in the form of factory work, which often lasted 12–16 hours a day. The health of Americans became a concern and the "hygiene" movement soon followed. Under the influence of this movement, physical education was taught in schools and included the early Greek concepts of health. Lectures on diet and regimen included such items as bathing, fresh air, dress, diseases of the skin, and stomach disorders. These items and physical training were regular topics in the *Boston Medical and Surgical Journal*. This wave of reform paved the way for both social and health advocates that soon followed. Dr. Oliver Wendell Holmes, Catherine Beecher, Dr. Dioclesian Lewis, Sylvester Graham, and Dr. William Alcott all played significant roles in the promotion of health and activity. The movement took shape with the opening of gymnasiums and organizations such as the YMCA and YWCA. Their goals were similar to those of today; to promote a proper diet, participation in physical activity, avoidance of alcohol and drugs, and plenty of rest. Coinciding with this movement, colleges and universities began to hire medical doctors (including William Anderson, Dudley Allen Sargent, and Edward Hitchcock) to lecture on health-related topics. Colleges began to form organized athletic competitions, giving birth to intercollegiate athletics. As the popularity of sport grew, colleges and universities began to hire physical educators who taught and promoted sports for improved fitness, character, etiquette, and health.

During the early 1900s many children participated in sports programs, but the benefits of these programs came into question in 1943, when the United States military rejected over 3 million registrants for physical and mental reasons. This spurred a more regimented approach to physical education in place of the emphasis on sports competition and play. It also inspired Thomas Cureton to open and operate the Physical Fitness Research Laboratory at the University of Illinois. The Kraus-Weber tests of minimum muscular fitness of school children were developed and applied. Preliminary results indicated that 58% of American children failed the tests in comparison to only 9% of European children. President Eisenhower was disappointed with these findings and in 1956 formed the President's Council on Youth Fitness and a President's Citizens Advisory Committee on the Fitness of American Youth.

The automobile and other technological advances, sped up by World War II and spread by postwar prosperity, allowed Americans to become increasingly sedentary. Health care costs and the incidence of heart disease quickly rose, obesity became more common, and the general health of America was faltering. Then in 1968, Dr. Kenneth Cooper of the Air Force published his book *Aerobics* and later a second edition, *The New Aerobics*. He challenged Americans to control obesity, heart disease, and rising health care costs by engaging in regular physical activity. He suggested continuous activity of large muscle groups such as jogging, cycling, and swimming, which he termed "aerobic" activity. Formalized research on fitness and components of endurance were also being conducted at places such as the Harvard Fatigue Laboratory.

The early 1970s saw substantial growth in running and other forms of exercise. The birth of health clubs followed in the early 1980s. These clubs offered many activities including racquetball, tennis, aerobic dance, strength training, and swimming. This movement inspired the development of advanced college degrees specializing in corporate fitness, health club administration, fitness management, sport management, health promotion, and others. Also, several private organizations such as the American College of Sports Medicine (ACSM), the National Strength and Conditioning Association (NSCA), and the Aerobic Fitness Association of America (AFAA) now offer specialized training (personal trainer, aerobic instructor, health and fitness instructor) for positions in health club settings.

Private companies and corporations have discovered the benefits of providing health facilities for their employees. It is now common for companies to have on-site fitness centers for use by their employees. Hospitals have joined the health movement by opening and operating health clubs and fitness centers. University and college recreation programs have seen tremendous growth in the 1990s and early 2000s, primarily due to

HISTORY AND CURRENT STATUS OF PHYSICAL ACTIVITY AND FITNESS (CONTINUED)

the fitness movement. Many institutions of higher learning are building and operating large multimillion-dollar recreational facilities to serve students, faculty, staff, and often the local community.

Today

Modern research confirms the link between fitness and health. Physical activity protects against the development of heart disease, reduces cardiovascular mortality, reduces and modifies associated risk factors (such as hypertension, blood lipid levels, insulin resistance, and obesity), improves exercise tolerance, improves functional capacity, and improves psychological well-being and quality of life.

STUDY QUESTIONS

1. List and clearly define the components of fitness. Which one has the most dramatic effect on overall health?
2. List and define the training variables.
3. Using the training variables and components of fitness, describe the basic exercise prescription for general health.
4. How does training volume affect adaptations?
5. Why do improvements come so easily for beginning exercisers?
6. For maintaining a certain level of fitness, which of the training variables is most important?
7. What are the benefits of the warm-up?
8. What is the stretch reflex and how does PNF stretching work?

REFERENCES

1. Pate RR, Pratt M, Blair SN, Haskell WL, Macera CA, Bouchard C, et al. Physical activity and public health: a recommendation from the Centers for Disease Control and Prevention and the American College of Sports Medicine. *Journal of the American Medical Association* 273:402–407, 1995.
2. American College of Sports Medicine. Position stand on the recommended quantity and quality of exercise for developing and maintaining cardiorespiratory and muscular fitness, and flexibility in healthy adults. *Medicine Science Sports and Exercise* 30:975, 1998.
3. Hickson RC, Rosenkoetter MA. Reduced training frequencies and maintenance of increased aerobic power. *Medicine and Science in Sports & Exercise* 13:13–16, 1981.
4. Hickson RC, Knakis C, Davis JR, Moore AM, Rich S. Reduced training duration effects on aerobic power, endurance, and cardiac growth. *Journal of Applied Physiology* 53:225–229, 1982.
5. Hickson RC, Foster C, Pollock ML, Galassi TM, Rich S. Reduced training intensities and loss of aerobic power, endurance, and cardiac growth. *Journal of Applied Physiology* 58:492–499, 1985.
6. Pollock ML. The quantification of endurance training programs. In: *Exercise and Sport Sciences Reviews* (Wilmore JH, ed.), New York: Academic Press 155–188, 1973.
7. Gettman LR, Pollock ML, Durstine JL, Ward A, Ayres J, Linnerud AC. Physiological responses of men to 1, 3, and 5 day per week training programs. *Research Quarterly* 47:638–646, 1976.
8. Wenger HA, Bell GJ. The interactions of intensity, frequency, and duration of exercise training in altering cardiorespiratory fitness. *Sports Medicine* 3:346–356, 1986.
9. Wilmore JH, Costill DL. *Physiology of Sport and Exercise*, Champaign, IL: Human Kinetics, 1994.

CHAPTER

4

CARBOHYDRATES AND EXERCISE

Chapter Objectives

- Introduce basic carbohydrate concepts including type and food sources.

- Discuss the digestion and absorption of carbohydrates.

- Provide an introduction to the key regulatory factors involved in carbohydrate metabolism at rest and during exercise.

- Discuss the relationship between carbohydrate metabolism and exercise intensity and duration, and apply it to different sports.

- Discuss the involvement of carbohydrate in fatigue and how to maximize carbohydrate storage.

- Provide recommendations for carbohydrate consumption before, during, and after training and competition.

Personal Snapshot

Jose Luis Pelaez Inc./Corbis

Tom is a college senior and a competitive triathlete. He shares an apartment with John, a fellow senior who plays on the ice hockey team. Tom and John often discuss training concepts, nutrition, and performance limitations related to their particular sports. They shop for food and eat several meals together, and both of them strive to acquire 55–60% of their diet energy from carbohydrate. In doing so they eat a lot of pastas, breads, fruits, vegetables, and juices as well as low-fat dairy foods. Tom and John also use the same sport drink during training and competition and often pack the same foods when they travel for away games or triathlons. How can two seemingly unrelated sports be so dependent on carbohydrates for optimal competitive performance? Also, why would carbohydrate stores be as important in preventing fatigue in a sport that involves continuous effort (triathlon) as a sport where more time seems to be spend on the bench (ice hockey) than in play?

Pasta dinners are common the evening before a marathon and between stages of multiday cycling tours like the Tour de France. All types of athletes drink carbohydrate-containing sport drinks during training and competition. What makes carbohydrate so special to athletes? Why is it that many athletes pay particular attention to this energy nutrient during training or competition as well as before and after?

Carbohydrates serve as an important fuel resource for physical activity. In the past, researchers focused on endurance sports in studying the relationship between carbohydrate status and metabolism and performance. Today, however, other sports such as weightlifting, soccer, football, and ice hockey are receiving their fair share of the attention. It is now clear that carbohydrate is the most important energy-yielding nutrient for optimizing performance in most sports with greater work output. Additionally, depletion of carbohydrate stores is involved in determining performance limitations (fatigue) in these sports. Therefore, carbohydrate consumption before, during, and after exercise training should be a foremost concern for athletes.

This chapter provides a basic overview of carbohydrate types, food sources, and digestion and absorption. In addition, the metabolism of carbohydrate in different tissues during different *nutrition states* (fed or fasting) are discussed, as these concepts have direct application to energy metabolism during exercise. The influence of exercise intensity and duration are also discussed, as well as

recommendations for carbohydrate intake for athletes. Nutrition in Practice highlights in Chapters 7 and 14 address designing a diet for performance and eating on the road, respectively. These nutritional highlights are based on providing carbohydrate foods to optimize performance.

■ CARBOHYDRATE BASICS

What makes a carbohydrate a carbohydrate, and are there many different kinds of carbohydrates? Indeed, there are numerous types of carbohydrates; thus they are a family of molecules. All carbohydrates share a common molecular foundation in that they are based on monosaccharide molecules. Although many carbohydrate types occur in nature, our discussion includes only the carbohydrates found in greater amounts in the diet and important to the body at rest and during exercise.

Monosaccharides

Monosaccharides are the smallest carbohydrate molecules and include **glucose, fructose**, and **galactose** (Figures 4-1 and 4-2). Both glucose and fructose are naturally found to a varying degree in plant-based foods. For instance, fructose is concentrated in honey and many fruits, contributing much of the sweetness of these foods. Relatively little free galactose is found in foods.

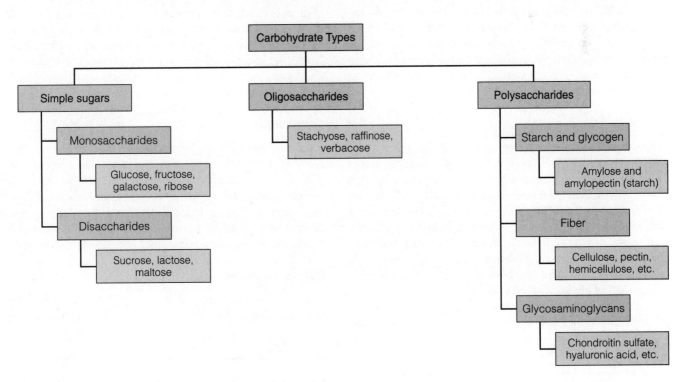

Figure 4-1
Organizational model for major carbohydrate types with examples of each type.

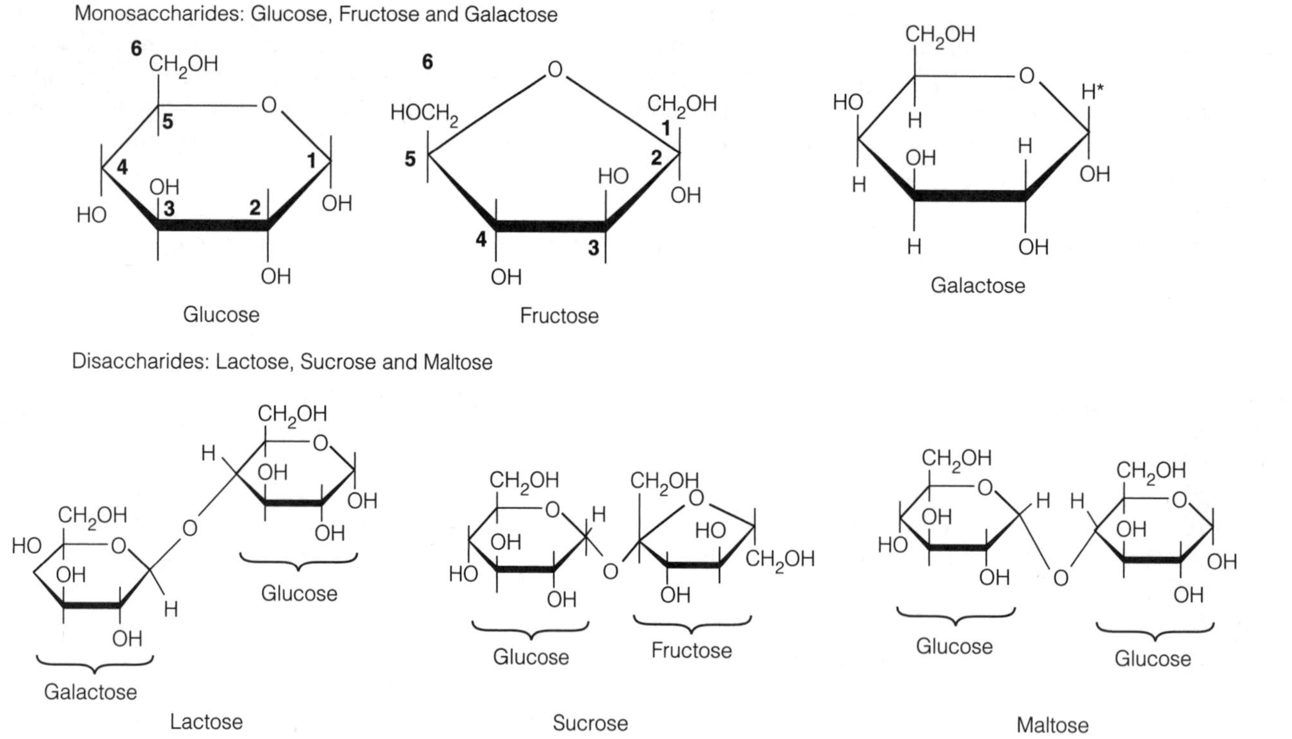

Monosaccharides: Glucose, Fructose and Galactose

Disaccharides: Lactose, Sucrose and Maltose

Figure 4-2
Some monosaccharide and disaccharide chemical structures

In addition to what is provided in the diet, certain tissues in the body can make select monosaccharides. For instance, the liver and the kidneys can make glucose via the conversion of lactate, glycerol, and certain amino acids. In addition, some cells can make galactose, and nearly all cells can produce yet another monosaccharide, **ribose**.

Disaccharides

Disaccharides are two monosaccharides chemically bonded together. As shown in Figure 4-2, glucose serves as half of the disaccharides lactose and sucrose and both halves of maltose. **Sucrose** is glucose linked to fructose and is derived from the sugar cane plant. Sucrose is often called table sugar. **Lactose** is glucose linked to galactose and is the principal carbohydrate found in milk and dairy foods (such as cheese). **Maltose** is found in germinating seeds and malted alcoholic beverages and as mentioned is composed of glucose linked to glucose.

Lactose, sucrose, and maltose, along with the monosaccharides mentioned above, are referred to as **simple sugars**.

Key Point

Monosaccharides are the smallest carbohydrate form and include glucose, galactose, fructose, and ribose.

This is because they are relatively small in comparison to the larger carbohydrate types discussed next. In addition, these carbohydrates elicit a sweet taste when consumed. Table 4-1 displays the simple sugars along with several prominent noncarbohydrate sweeteners and their sweetness relative to sucrose.

Oligosaccharides

Oligosaccharides are composed of 3–10 monosaccharides and include stachyose, raffinose, and verbacose. These oligosaccharides can be found in most legumes, such as peas, green beans, lima beans, pinto beans, black-eyed peas, garbanzo beans (chickpeas), lentils, and soybeans. As with fiber, human enzymes do not efficiently digest some oligosaccharides, including those mentioned.

Polysaccharides

Monosaccharides serve as building blocks not only for the disaccharides and oligosaccharides but for larger

Key Point

Larger carbohydrate molecules are based on links of monosaccharides and include disaccharides, oligosaccharides, and polysaccharides.

| Table 4-1 |

Sweetness of Sugars and Alternatives

Sugar or Sweetener	Sweetness Relative to Sucrose	Common Dietary Sources
Sugars		
Lactose	0.2	Dairy foods
Maltose	0.4	Sprouted seeds
Glucose	0.7	Corn syrup, most carbohydrate sources
Sucrose	1.0	Table sugar, recipe sweetener
Fructose	1.7	Fruit, honey, soft drinks (HFCS)
Sugar Alcohols		
Sorbitol	0.6	Dietetic candies, sugarless gum
Mannitol	0.7	Dietetic candies
Xylitol	0.9	Sugarless gum
Artificial Sweeteners		
Aspartame (Nutrasweet)	200	Diet soft drinks and fruit drinks, tabletop sweetener
Acesulfame-K (acesulfame-potassium)	200	Sugarless gum, diet drink mixes, tabletop sweetener, gelatin and pudding
Saccharin	500 (200–700)	Diet soft drinks, tabletop sweeteners
Sucralose	600	Diet soft drinks, baked goods, sugarless gums

carbohydrates as well. **Polysaccharides** consist of numerous monosaccharides chemically linked together. The most significant polysaccharides to humans are starch and fiber, found in plant foods, and glycogen and glycosaminoglycans, located in muscle and other human tissue.

STARCH   Glucose is the building block of plant **starch**. Straight chains of starch are referred to as **amylose**, and branched chains of amylose are referred to as **amylopectin** (Figure 4-3). The links between glucose monomers in amylose and the branch points of amylopectin are α1-4 and 1-6 glycosidic bonds. Amylopectin appears to branch every 24–30 glucose units.

Key Point

High-fructose corn syrup contains small, branching chains of glucose (maltodextrin) and fructose and is a popular sweetening agent in sport foods and drinks.

Cereal grains such as wheat, rice, and oats and grain products such as flour are good sources of starch, as are potatoes and several fruits and vegetables. Food manufacturers process another form of starch, cornstarch, by partial digestion to make **maltodextrin**, which is small, branching links of glucose. **High-fructose corn syrup** (HFCS) contains partially digested cornstarch and fructose and is a popular commercial sweetener.

GLYCOGEN   **Glycogen**, made up of branching chains of glucose, is produced and stored in the liver and skeletal muscle. About 6–8% of the weight of the liver and 1% of the weight of skeletal muscle is attributed to glycogen. For an adult male, this results in roughly 100 g of glycogen in the liver and 300 g in skeletal muscle. Although glycogen is more concentrated in the liver, the greater total mass of skeletal muscle allows it to have more glycogen. As shown in Figure 4-3, glycogen contains many more branch points than amylopectin. This results in numerous free ends, which in turn allows for a more rapid rate of breakdown and recovery.

Key Point

Glycogen is highly branched, which allows it to be rapidly broken down during periods of need.

FIBER   **Fiber** is a group of structural polymers made by plants that are poorly digested by the digestive enzymes produced by humans. Fiber includes the polysaccharides cellulose, hemicellulose, pectin, algal polysaccharides, and mucilages. Lignin is also considered a fiber although it is constructed from phenolic molecules, not monosaccharides. The bonds between the monosaccharides in fibers are β1-4 and β1-6 glycosidic bonds, which greatly hinder their digestibility by human enzymes.

GLYCOSAMINOGLYCANS   **Glycosaminoglycans (GAGs)**, also referred to as *mucopolysaccharides*, are unique unbranched polysaccharides. There are seven different kinds of GAGs, including **hyaluronic acid** and **chondroitin sulfate**. Hyaluronic acid and chondroitin sulfate are found in joints, where they bind water to maintain fluid volume within the joint and keep cartilage moist and spongy.

The GAGs are composed of repeating disaccharide units in which at least one of the monosaccharides is an amino sugar, namely **glucosamine** or galactosamine. As shown in Figure 4-4, these amino sugars are monosaccharides with an attached amine group. Recently, attention has been focused on glucosamine and GAGs in relation to joint health and recovery from sport-related joint injuries. This is discussed in greater detail in the Chapter 10 feature In Focus: Sport Injury and Nutrition Supplementation.

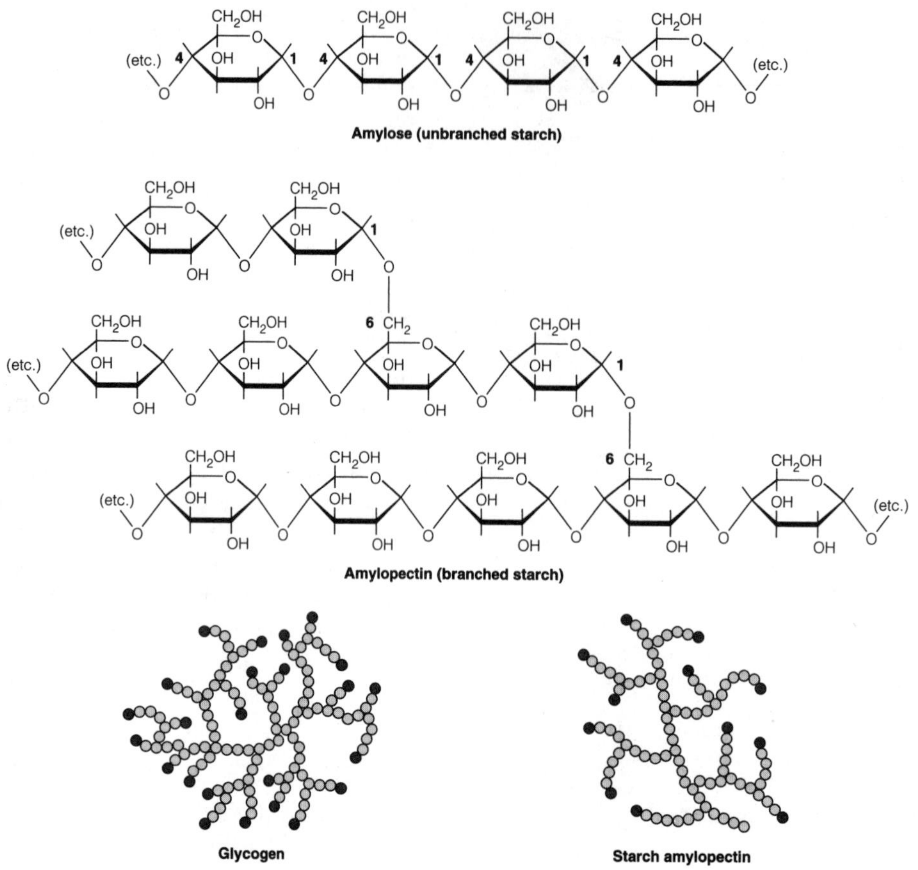

Figure 4-3
Straight chains of starch are called amylose and branched starch is amylopectin. Notice that glycogen contains more branch points than amylopectin.

■ CARBOHYDRATE BASICS *IN REVIEW*

- Monosaccharides are the simplest carbohydrate forms and include glucose, galactose, fructose, and ribose.
- Disaccharides are composed of two monosaccharides chemically bonded together and include sucrose and lactose.
- Oligosaccharides contain 3–10 monosaccharides and are most concentrated in legumes.
- The polysaccharides starch and glycogen are glucose storage polymers for plants and animals, respectively.
- Fibers are structural polymers made by plants and are generally indigestible by humans.
- Glycosaminoglycans (GAGs) are structural polysaccharides found throughout the human body and are especially concentrated in joints.

■ GENERAL CARBOHYDRATE INTAKE AND RECOMMENDATIONS

Carbohydrate is a powerful source of energy, providing 4 kcal/g. For most economically developed societies, carbohydrate is the greatest contributor of energy to the diet. For instance, in the United States approximately 50% of energy in the diet is from carbohydrate. Stated another way, the daily U.S. diet averages roughly 4–5 g of carbohydrate per kilogram of body weight. Although the high carbohydrate consumption in the United States is suggestive of a diet abundant in fruits, vegetables, and whole grain products, this is generally not the case. So what are the major sources of carbohydrate in the diet?

Food Sources of Carbohydrates

Starch provides about half of the carbohydrates in the U.S. diet, or about one-fourth of the total energy. Most of

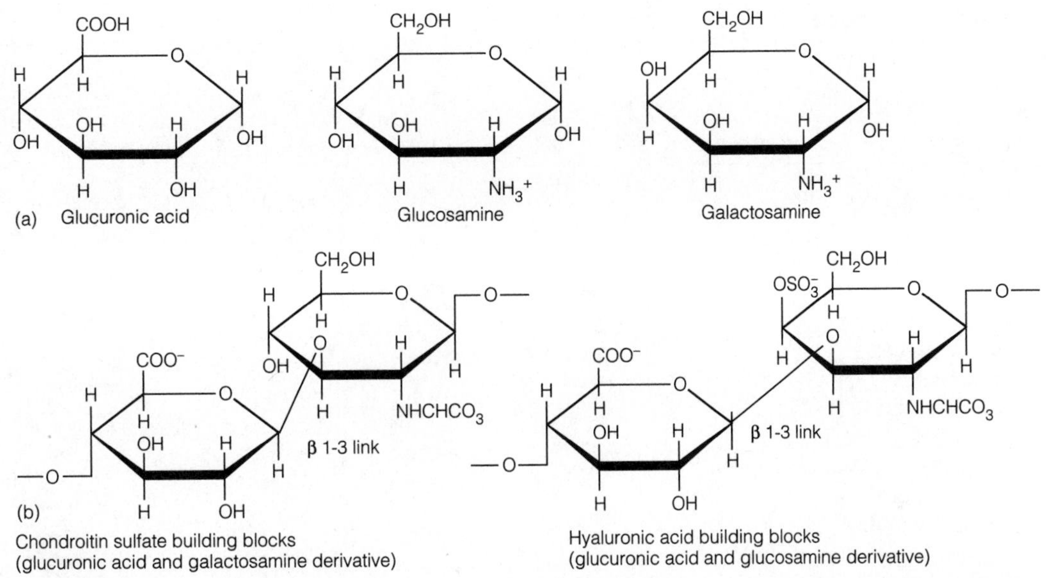

Figure 4-4
Glycosaminoglycan (GAG) building blocks. (a) Glucosamine, galactosamine, and glucoronic acid are important components of (b) the disaccharides used to build glycoaminoglycans (GAGs) such as hyaluronic acid and chondroitin sulfate. GAGs function as structural carbohydrates in connective and other human tissues.

the starch in the diet is derived from pastas, breads, potatoes, and rice. Simple sugars collectively account for the other half of the carbohydrate in the diet. About half of the simple sugars (or one-eighth of the total energy) that Americans consume is sucrose. With the exception of beets and cane, sucrose is not found in appreciable amounts in natural foods. It is primarily consumed in manufactured foods such as soda, candies, and desserts. People also add sugar to foods at home and when they are dining out. For example, sucrose is commonly sprinkled onto breakfast cereal and stirred into coffee and tea. Most of the fructose in the U.S. diet is derived from sweeteners rather than fruits. HFCS is found in soda and many flavored drinks (including sport drinks) and manufactured foods. Fructose is also half of sucrose. The carbohydrate content of common foods is presented in Table 4-2.

Recommendations for Carbohydrate Energy

Most nutrition recommendations state that carbohydrate should make the greatest energy contribution to the diet. For instance, the most recent RDAs recommend that carbohydrate provide 45–65% of dietary energy. In addition, it is recommended that "added sugars" contribute no more than 25% to total energy intake. Added sugars include sucrose and HFCS, which are added to foods in the manufacturing process. These foods include candy, soda, and fruit drinks. The recommendation to minimize added

sugars excludes the simple sugars lactose in dairy products and fructose in fruits because these foods are more concentrated in essential nutrients.

The purposes of a high-carbohydrate diet are to provide energy for all cells throughout the body and to promote the maintenance of carbohydrate stores (glycogen). Also, a high-carbohydrate diet based on fruits, vegetables, and whole grain products provides a bounty of phytochemicals. **Phytochemicals** are substances produced by plants and include many nutraceutical substances, which can promote general health by assisting in the prevention and treatment of diseases, as discussed in Chapter 1. Beneficial phytochemicals include fiber, carotenoids (such as lycopene, lutein, and β-carotene), and polyphenolic molecules (such as genestein and quercetin).

MINIMUM RECOMMENDATIONS FOR DIET CARBOHYDRATE   All cells use glucose for energy to some degree. Red blood cells (RBCs) are unique in that they use

Key Point

It is recommended that a diet contain roughly 45–65% of its energy as carbohydrate and that fruits, vegetables, and whole grain products as well as low-fat dairy foods provide carbohydrates.

Table 4-2   **Carbohydrate Content of Select Foods**

Food	Energy (kcal)	Carbohydrate, Fiber (grams)	Carbohydrate, as % of Total Food Energy
Whole milk (1 c)	150	11, 0	30%
2% Milk (1 c)	121	12, 0	40%
Nonfat (skim) milk (1 c)	85	12, 0	56%
Soy milk (1 c)	81	4, 3	20%
Pepsi Cola (12 oz)	150	41, 0	100%
Gatorade (1 c)	60	15, 0	100%
Swiss cheese (1 oz)	105	1, 0	4%
American cheese (1 oz)	105	<1, 0	4%
Egg, whole (1 medium)	74	1, 0	5%
Banana (1, 8–9″ w/o peel)	109	28, 3	98%
Orange juice (1 c)	112	26, 1	98%
Peach (1 w/o pit)	42	11, 2	100%
Raisins (1 c not pressed down)	435	100, 6	90%
Cauliflower, raw (1/2 c)	12	3, 1	95%
Broccoli, raw (1/2 c)	22	4, 2	73%
Corn, frozen kernels (1/2 c)	66	16, 2	96%
Potato, baked w/skin (1 med)	220	51, 5	93%
Fig bars (4)	223	45, 3	81%
Hamburger bun (1)	123	22, 1	72%
PowerBar (1)	230	45, 1	78%
MET-RX Protein Plus	290	15, 1	21%
Pancakes (4″ diameter, plain)	86	11, 1	51%
Cheerios (1 c)	84	17, 2	81%
Cracklin' Oat Bran	252	48, 8	76%
Raisin bran, Kelloggs (1 c)	200	47, 8	94%
Chicken fajita (1)	405	50, 4	50%
Macaroni & cheese, homemade (1 c)	461	28, 1	24%
Bagel, plain (1)	195	38, 2	78%
Bagel, oat bran (1)	181	38, 3	84%
Bread, white (1 slice of 16/loaf)	120	21, 1	70%
Bread, whole wheat (1 slice of 18/loaf)	70	13, 2	43%
Rice, white (1 c cooked)	205	45, 1	88%
Rice, brown (1 c cooked)	216	45, 4	83%
Popcorn , microwave, low fat (1 c)	25	4, 1	64%
Bean burrito (1)	255	33, 3	52%
Almonds, dry roasted (1/2 c)	405	17, 9	17%
Peanut butter (2 tbs)	190	6, 2	13%
M&Ms, plain (10 pc)	34	5, <1	59%
Fudge, chocolate (1 pc, 17 g)	65	13, 0	80%
Angel food cake (1/12 cake)	72	16, <1	89%
Corn chips (1 c)	140	15, <1	43%
Vanilla wafers (10)	176	29, 1	66%
Apple pie (1/6 pie)	396	57, 3	58%

only glucose. Under normal conditions, the brain uses primarily glucose. Normally, RBCs and the brain of an adult man may use, respectively, 45–50 and 100–125 g of glucose daily. However, during periods of prolonged fasting or decreased carbohydrate availability, the brain adapts to use more ketone bodies for energy.

Setting minimum recommendations for carbohydrate intake starts with the glucose needs of the obligate glucose users (RBCs and brain). Other tissues such as the digestive tract, kidneys, and smooth muscle also use glucose as a primary fuel source. Eating less carbohydrate than is used by tissue as a whole requires the use of glycogen stores and the conversion of other molecules (such as amino acids) to glucose.

It is recommended that adults consume at least 50–100 g of carbohydrate daily to prevent ketosis. **Ketosis** is a metabolic state in which the production of ketone bodies by the liver exceeds the amount that can be metabolized by other tissue (such as muscle and brain). This situation occurs during fasting and also when carbohydrate intake is restricted, as in some popular diet practices. Whether or not ketogenic diets are detrimental to health is a matter of continuous debate among scientists and practitioners. Very low carbohydrate diets are compromised in their phytochemical content, and strict compliance is difficult for many people. Ketogenic diets and their potential influence on health and athletic performance are discussed in Chapter 7.

Recommendations for Fiber

Fiber is not an essential nutrient like vitamin C or iron. However, the health-promoting benefits of consuming fiber are recognized. Consuming recommended levels of fiber can reduce a person's risk of heart disease and maybe colon cancer and possibly assists some people in weight control. In addition, populations that consume lower amounts of fiber are much more prone to **diverticulosis** than populations that eat more fiber. The current RDA for fiber for people 50 years of age or younger is 38 g for men and 25 g for women. For adults over the age of 50, the RDA for fiber is 30 g for men and 21 g for women. However, Americans eat approximately half of the recommended amounts.

Carbohydrate Digestion

How are carbohydrates digested and what happens to those carbohydrates that cannot be digested? Carbohydrate digestion involves several enzymes that liberate monosaccharides for absorption. In brief, saliva and digestive juices from the pancreas contain **amylase**, which begins to split the $\alpha 1$-4 bonds in amylose. Meanwhile, isomaltase is produced by the small intestine to break the branch points ($\alpha 1$-4 bonds) of starch. The small intestine also produces disaccharidases (**sucrase**, **maltase**, and **lactase**) to digest sucrose, maltose, and lactose.

INDIGESTIBLE CARBOHYDRATE Oligosaccharides and fibers are generally resistant to human digestive enzymes. Also, the efficiency of lactose digestion varies among individuals. Undigested carbohydrates become available to bacteria in the latter parts of the digestive tract. The metabolism of these carbohydrates by bacteria can result in the production of the gases H_2, CO_2, and CH_4, which cause bloating, cramping, and flatulence. In addition, these carbohydrates can bind water and possibly cause diarrhea. In the case of undigested lactose these effects are collectively referred to as **lactose intolerance**. On the other hand, as mentioned above, fibers can help prevent diverticulosis, heart disease, and colon cancer.

Key Point

Indigestible carbohydrates such as fibers and lactose are subject to metabolism by bacteria, leading to gas production and potentially causing bloating and cramping.

Carbohydrate Absorption

Absorption of monosaccharides begins with their movement through cells known as **enterocytes** that line the wall of the small intestine (Figure 4-5). As the monosaccharides exit those cells they enter capillaries. The capillaries in the wall of the small intestine drain into the hepatic portal vein. This vein drains not only the small intestine but also the greater digestive tract and the pancreas. This means that the liver has first access to the absorbed monosaccharides and to insulin released by the pancreas. Thus the liver is a primary processing organ for absorbed monosaccharides and will use it for energy and to make glycogen, primarily.

The absorption of glucose and galactose requires sodium (Na^+) and energy (ATP), which indicates their absorption is by **active transport**. The absorption of fructose does not require energy and occurs by **facilitated diffusion**. Thus fructose diffuses back and forth across the lining of the small intestine, leading to a much slower absorption rate in comparison to glucose and galactose.

Glycemic Index

Glycemic index has become an important concept in general nutrition as well as having application to physical performance. Absorbed glucose that is not taken up by the liver circulates throughout the body, increasing blood glucose levels. The degree and duration of blood glucose elevation after eating a food is its **glycemic response**. Glycemic response is measured as the area under the glycemic curve, as displayed in Figure 4-6. The **glycemic index** of a food is the comparison of its glycemic response to a food standard with the same amount (50 g) of carbohydrate (Figure 4-7). Glucose or white bread are used most often as the standard, and Table 4-3 provides the glycemic index of several foods. A more comprehensive list of glycemic index values for various foods has recently been published.[1]

The glycemic index of a food is influenced by several factors. One factor is the type of carbohydrate. As discussed above, only half of the monosaccharide units in lactose and sucrose are glucose, whereas all the monosaccharides in starch are glucose. This suggests that "starchy" foods such as a baked potato might have a higher glycemic index than milk, dairy foods, and "sugary" foods such as candies. Also, the level of protein, fat, and fiber in a food can lower the glycemic index of a food by slowing the rate

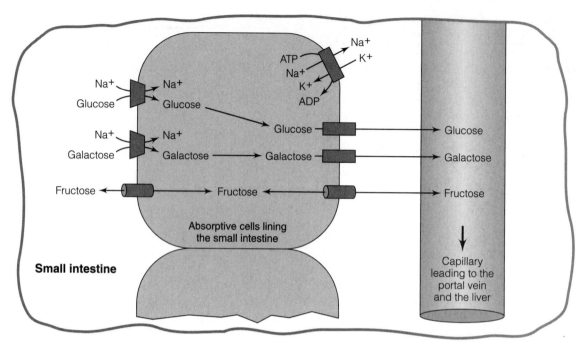

Figure 4-5
Absorption of monosaccharides. The absorption of glucose and galactose require sodium (Na⁺) and energy (ATP) in a carrier-mediated fashion. The absorption of fructose does not require energy and occurs by facilitated diffusion. Monosaccharides travel to the liver via the hepatic portal vein.

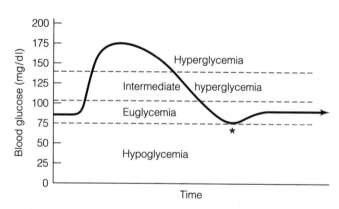

Figure 4-6
Typical glycemic response of a carbohydrate-containing food. *Slight hypoglycemic dip as euglycemia is reestablished.

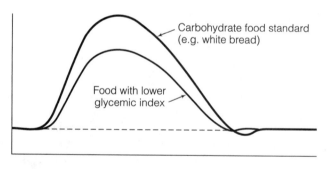

Figure 4-7
Glycemic response of two foods having the same amount of carbohydrate. The glycemic index of the food with the lower response is determined by measuring the area under its curve relative to the area under the standard food's curve.

of digestion and absorption of monosaccharides.[2,3] If monosaccharides are absorbed more slowly, the liver has more opportunity to process them rather than letting them reach general circulation. This helps explain why whole wheat bread can have a lower glycemic index than white bread.

Foods with a higher glycemic index may be undesirable food choices for people with chronic hyperglycemia (as in diabetes mellitus). First, food with a higher glycemic index can worsen a hyperglycemic state. Second, further elevation of circulating glucose could lead to an increase in the level of circulating insulin (**hyperinsulinemia**). For many hyperglycemic people, insulin may already be circulating at normal or elevated levels relative to the blood glucose concentration. Chronic hyperinsulinemia is associated with elevated blood lipids (hypercholesterolemia and hypertriglyceridemia), blood pressure, and body fat.

Table 4-3

Foods Grouped by Relative Glycemic Index[a]

Higher Glycemic Index (>85)	Glucose, sucrose, maple syrup, corn syrup, honey, bagel, candy, corn flakes, carrots, crackers, molasses, potatoes, raisins, sport drinks with simple carbohydrates (Gatorade, Powerade), sport drinks with carbohydrate polymers (GatorLode)
Medium Glycemic Index (60–85)	All-bran cereal, banana, grapes, oatmeal, orange juice, pasta, rice, whole-grain rye bread, yams, corn, baked beans, potato chips
Lower Glycemic Index (<60)	Fructose, apple, applesauce, Cheerios, kidney beans, navy beans, chickpeas, lentils, dates, figs, peaches, plums, ice cream, milk, yogurt, tomato soup

[a]GI listed are relative to glucose.

Source: Adapted from Williams MH. *Nutrition for Health Fitness & Sport*, 5th ed., WCB McGraw Hill, p. 102, 1999.

Key Point

The type of carbohydrate and other factors (such as fiber, protein, fat) in a meal determine a food's glycemic index.

■ GENERAL CARBOHYDRATE INTAKE AND RECOMMENDATIONS *IN REVIEW*

- Carbohydrates are the predominant fuel source for most societies and are largely consumed in fruits and vegetables, dairy products, and prepared foods containing flour, sucrose, or HFCS.
- General recommendations are for carbohydrate to account for 45–65% of total energy.
- A daily intake of at least 50–100 g of carbohydrate is needed to prevent ketosis.
- Carbohydrate is absorbed as monosaccharides, and the liver is a major processing organ.
- Bacteria can metabolize undigested carbohydrates (such as fiber, oligosaccharides, and in some cases lactose) and produce uncomfortable gases and possibly diarrhea.

■ CARBOHYDRATE METABOLISM

If glucose is a primary energy source for all cells in the body, how does it get in and out of cells? Once it is in cells, how is carbohydrate metabolized and what regulatory factors control its breakdown or storage? Furthermore, since glucose is an absolutely necessary fuel to RBCs and important to the brain, how does the body prevent blood glucose levels from falling too low? This section provides an overview of the major metabolic pathways involved in carbohydrate metabolism. Also see In Focus: Nutrition States and Energy Metabolism at the end of this chapter for an overview of general metabolic factors associated with eating (fed state) and periods in between meals (fasting state and starvation).

Glucose must continuously circulate in the blood in order to serve as a fuel resource for cells. An adult male in a fasting state normally has a blood glucose concentration of 70–110 mg/100 ml. If his total blood volume were $5\frac{1}{2}$L this would translate to approximately 5 g of glucose in circulation. Because the level of glucose in the blood can increase after ingesting carbohydrate or decrease during fasting, hormones are employed to control blood glucose levels. The most significant of these hormones are insulin, glucagon, epinephrine, and cortisol.

Insulin promotes reductions in blood glucose during **hyperglycemia**, which is a higher level of blood glucose normally brought on by a meal or other carbohydrate intake. Glucagon, epinephrine, and cortisol attempt to raise blood glucose levels to prevent **hypoglycemia**, a too-low level of blood glucose, during fasting and times of increased glucose demand (such as exercise or stress). These hormones control blood glucose by regulating metabolic pathways involved in glycogen production and breakdown, glucose utilization for energy, and glucose production from other energy molecules. The term **euglycemia** is used to indicate the achievement (via hormone regulation) and maintenance of a fasting blood glucose level despite changing nutrition and metabolic states.

Glucose Transport Proteins

The movement of glucose across cell membranes requires transporters. **Glucose transport (GluT)** proteins are found on the plasma membrane of every cell and the membrane of at least one type of organelle. GluT1 is found on the plasma membrane of all cells and functions optimally during fasting. Neurons of the brain also contain GluT3 to ensure that their high glucose requirement is continuously met. **Hepatocytes** (a major type of liver cell) contain several glucose transport proteins, namely GluT1, GluT2, and GluT7. These transporters provide the basis for the dynamic flux of glucose in and out of these cells during the different metabolic states.

The most recognizable of the GluT proteins is GluT4, which is found in pools inside fat cells (adipocytes)

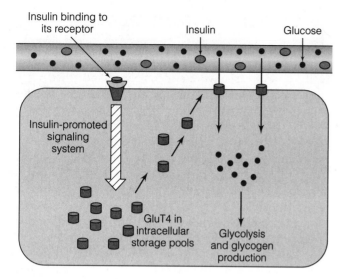

Figure 4-8
The binding of insulin with its receptor on the plasma membrane of skeletal and cardiac muscle cells and adipocytes causes the movement of GluT4 to the plasma membrane. This allows for an exceptional influx of glucose during hyperglycemia.

and skeletal and cardiac muscle cells. During hyperglycemia, GluT4 proteins relocate to the plasma membrane of these cells and increase glucose uptake for storage and/or use as fuel. Insulin promotes this relocation, as shown in Figure 4-8. Exercise also promotes the relocation of GluT4 in working skeletal muscle fibers, as described below. Skeletal muscle also contains GluT5, which is responsible for uptake of fructose when it is available.

Key Point

Glucose crosses cell membranes by way of glucose transport proteins such as GluT4.

Glycolysis

Glycolysis provides the backbone for carbohydrate metabolism. As displayed in the top portion of Figure 4-9, glucose is split into two molecules of pyruvate, and in the process two ATP molecules are generated. Glycolysis does not require O_2 to generate ATP, so it is an anaerobic energy pathway. Insulin promotes glycolysis by increasing the activity of key enzymes (Figure 4-10). Metabolic factors such as ATP content and the levels of some intermediates of glycolysis and the Krebs cycle also influence key steps in glycolysis. All these regulatory efforts help match the flux of carbohydrates through glycolysis with the ATP demands of the cell and vary with the metabolic state of the body (such as fed, fasting, exercising).

GLYCOLYSIS REGULATION   The key enzymes in glycolysis, along with factors that influence their activity, are shown in Figure 4-10. The first step of glycolysis involves the attachment of phosphate to glucose, producing **glucose-6-phosphate**. This is accomplished by *hexokinase* in all cells. Hepatocytes also contain *glucokinase*, which performs the same task. Hexokinase and glucokinase function optimally at different blood glucose levels. Hexokinase functions optimally at a lower blood glucose level, the normal fasting level. This ensures that cells meet their minimal glucose needs at all times. When blood glucose levels are elevated, glucokinase functions maximally. This allows the liver to metabolize larger amounts of glucose during hyperglycemia, such as after eating. Insulin increases the activity of glucokinase.

The enzyme *phosphofructokinase (PFK)* regulates another major step in glycolysis. PFK activity is generally enhanced by AMP, ADP, and insulin, and inhibited by higher ATP levels, glucagon, and citrate, a Krebs cycle intermediate. The influence of these factors depends on the type of cell and the metabolic state of the body. As explained below, during higher intensity exercise the levels of AMP and ADP increase in muscle fibers, which increases the activity of PFK. The enzyme pyruvate kinase regulates the last step of glycolysis, which generates pyruvate. The activity of pyruvate kinase is increased by insulin and decreased by glucagon.

FRUCTOSE AND GALACTOSE METABOLISM   The liver is the primary site of fructose and galactose metabolism. Fructose can also be metabolized by skeletal muscle. Both galactose and fructose are converted to intermediates of glycolysis (Figure 4-9) and can be used for glycogen storage, ATP production, or fat production (liver) depending on the nutritional state in those tissues. The metabolism of these monosaccharides (especially fructose) may slow glucose utilization by inhibiting downstream reactions in glycolysis. This could influence glucose tolerance to some degree.

Fate of Pyruvate

If pyruvate is the product of glycolysis, what is its fate? Pyruvate sits at a metabolic crossroads, as it can be converted to several other molecules. As shown in Figure 4-11, pyruvate can be converted to lactic acid or alanine or enter the mitochondria and be converted to either oxaloacetate (OAA) or acetyl coenzyme A (CoA). The fate of pyruvate largely depends on the type of cell, its metabolic state, and

Key Point

Glycolysis is the backbone of carbohydrate metabolism and is enhanced during a fed state and decreased in fasting conditions.

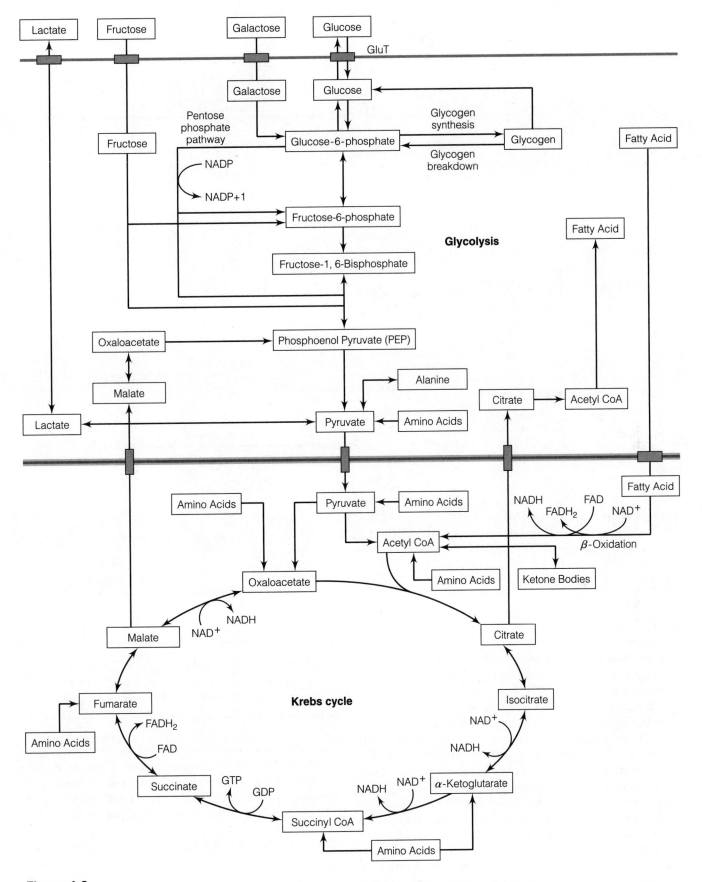

Figure 4-9
General overview of key steps in glycolysis and integrated metabolic pathways. (Some bidirectional steps in glycolysis are not shown.)

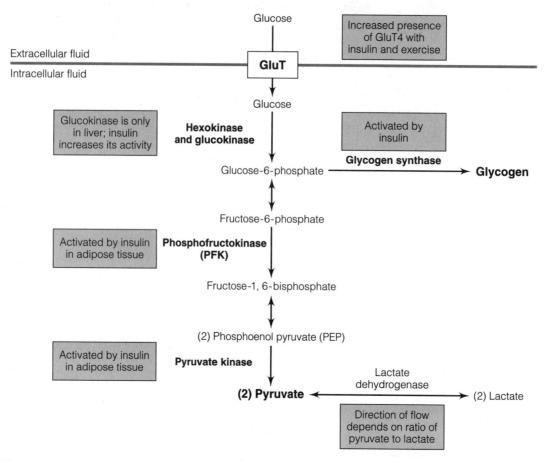

Figure 4-10
Key regulated steps in glycolysis and formation of glycogen and lactate. The activities of glycolysis, glycogen formation, and lactate formation are regulated by hormones and by substrate availability and metabolic factors.

conditions within the cell. In RBCs the only possible fate of pyruvate is conversion to lactic acid by *lactate dehydrogenase*. In muscle cells the degree of lactic acid formation depends on the O_2 availability and mitochondrial content. As discussed in Chapter 2, Type I muscle fibers have greater O_2 availability and more mitochondria and thus produce less lactic acid in comparison to Type II muscle fibers.

Pyruvate can also serve as an acceptor of an amine group from an amino acid; such a transfer is called a **transaminase** reaction, after the enzyme involved. During fasting and endurance exercise, muscle protein is broken down to increase the availability of amino acids to help meet energy requirements. The amine group of certain amino acids is passed to pyruvate, thereby forming alanine. As discussed in more detail in the next chapter, the formation of alanine is an important mechanism for exporting nitrogen out of skeletal muscle during fasting and endurance exercise.

Pyruvate can also enter mitochondria and be converted to acetyl CoA (Figure 4-9). Acetyl CoA is considered the entry molecule of the Krebs cycle. This

Key Point

Depending on the metabolic state and type of cell, pyruvate can be converted to lactic acid (lactate), acetyl CoA, oxaloacetate (OAA), or the amino acid alanine.

conversion occurs in most cells, and *pyruvate dehydrogenase* catalyzes this highly complex reaction. Pyruvate can also be converted to oxaloacetate (OAA), a Krebs cycle intermediate, if conditions dictate. OAA combines with acetyl CoA to form citrate in the first step of the Krebs cycle.

Glycogen Synthesis and Storage

Stored glycogen in muscle and liver serves as an extremely valuable energy resource during fasting and exercise. In the liver, glycogen is found in hepatocytes, and the entire organ may attribute 4% of its mass to this carbohydrate storage after an overnight fast and 8% a couple of hours after a carbohydrate-rich meal. Depending on the nutrition state,

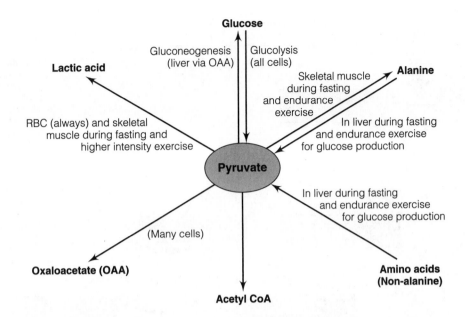

Figure 4-11

Fate of pyuruvate. Pyuruvate sits at the crossroads of several metabolic pathways. Depending on the cell type and metabolic state, pyruvate can be used for fuel in mitochondria (via conversion to acetyl CoA), converted to lactate for exportation from a cell (as in RBCs or skeletal muscle), or used to make glucose (in the liver) or the amino acid alanine (in skeletal muscle).

liver glycogen stores may be 60–120 g, which is the energy equivalent of 240–480 kcal. In muscle, glycogen is found in all muscle fibers, and skeletal muscle mass is typically about 1% glycogen, which can be raised to 2% with exercise training (see below). Adults tend to range between 200 and 500 g of total skeletal muscle glycogen. The broad range is largely attributed to differences in skeletal muscle mass, activity level, and distribution in muscle fiber types.

Glycogen levels tend to be 10–25% higher in Type II versus Type I muscle fibers.[4–7] Resting glycogen levels in regularly active but not highly trained individuals may be approximately 360–400 and 440–480 mmol/kg of dry muscle in Type I and Type II fibers, respectively.[4,6,8,9] Interestingly, skeletal muscle glycogen stores may be lower in children. On the basis of muscle biopsies, researchers have estimated that children may have 50–60% of the muscle glycogen concentration that adults have.[10,11]

Insulin promotes the formation of glycogen in muscle and the liver by increasing the activity of *glycogen synthase* and decreasing the activity of *glycogen phosphorylase*. The higher glycogen concentration in the liver is the result of increased glucose entrance into hepatocytes during hyperglycemia. In support, glucokinase generates large amounts of glucose-6-phosphate, which are used to build glycogen.

Glycogen Breakdown

Glycogen stores are broken down during fasting and exercise, and phosphorylase is the key regulatory enzyme (Figure 4-12). Plasma membranes of skeletal muscle fibers

Key Point

Insulin promotes glycogen synthesis and inhibits glycogen breakdown.

as well liver cells (hepatocytes) contain receptors for epinephrine and norepinephrine. Liver cells also contain glucagon receptors, but skeletal muscle cells do not. This means that epinephrine and norepinephrine stimulate glycogen breakdown in skeletal muscle cells, whereas glucagon, epinephrine, and norepinephrine increase glycogen breakdown in the liver. The binding of glucagon, epinephrine, or norepinephrine to receptors evokes a second messenger system involving *cyclic AMP (cAMP)*, as shown in Figure 4-12. One action of cAMP is to promote the activation of *phosphorylase kinase*, which then activates phosphorylase (*phosphorylase a*). Insulin does the reverse, converting phosphorylase to its inactive form (*phosphorylase b*).

Calcium (Ca^{2+}), AMP and inosine monophosphate (IMP) (via ATP), and phosphate (via creatine phosphate) can increase phosphorylase activity, either directly or indirectly. This is more significant in skeletal muscle cells than in liver cells because of the increased presence of Ca^{2+} in contracting skeletal muscle fibers. During higher intensity exercise, contracting skeletal muscle fibers also have increased levels of AMP and of phosphate from creatine phosphate.

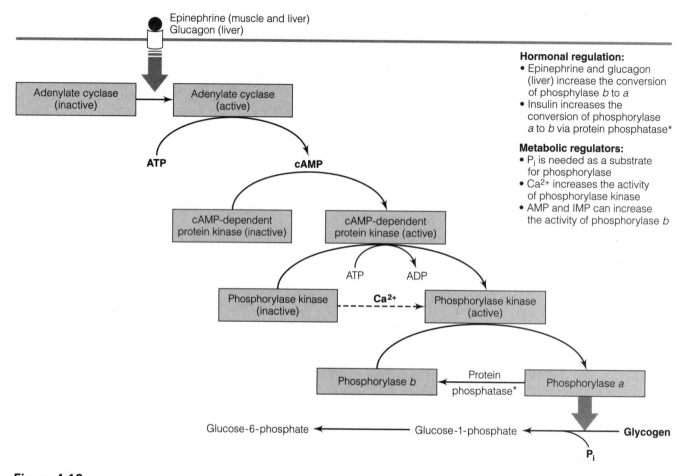

Figure 4-12

Activation of phosphorylase in muscle and the liver. Phosphorylase activity is increased by the binding of epinephrine to receptors in skeletal muscle and liver cells (hepatocytes) and glucagon in the liver. The activation involves a cAMP cascade of steps. Phosphate (P_i), Ca^{2+}, AMP, and IMP all can increase the activity of phosphorylase, either directly or by increasing the activity of phosphorylase kinase.

As shown in Figure 4-13, glycogen phosphorylase releases glucose units from glycogen largely in the form of glucose-1-phosphate, which is then converted to glucose-6-phosphate. Also, a *debranching enzyme* frees glucose at the branch points of glycogen. In liver cells, glucose-6-phosphate is transported to the endoplasmic reticulum, which houses *glucose-6-phosphatase*. Glucose-6-phosphatase detaches the phosphate, and GluT7 then transports the free glucose out of the endoplasmic reticulum. GluT2 allows glucose to leave that liver cell and enter the circulation. Skeletal muscle fibers lack glucose-6-phosphatase and GluT2, so they do not export glycogen-derived glucose. Thus, whereas liver glycogen provides a means to prevent hypoglycemia, the primary purpose of skeletal muscle glycogen is to serve as energy for those cells during physical activity.

Gluconeogenesis

During fasting and exercise, the liver is able to produce glucose and release it into the circulation to help prevent

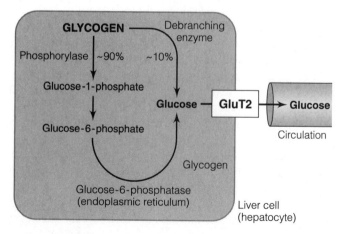

Figure 4-13

Fate of glycogen in the liver. Liver glycogen breakdown is a primary mechanism involved in preventing hypoglycemia during fasting (and exercise).

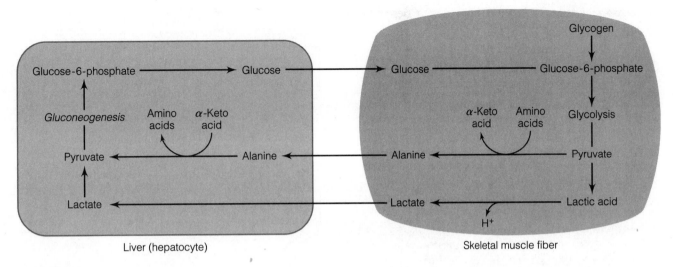

Figure 4-14
Lactate is a primary source of glucose production during endurance exercise. Overall alanine is the predominant amino acid used to produce glucose. The lactate or Cori cycle refers to the conversion of muscle-derived lactate to glucose in the liver, which can circulate back to muscle. The alanine cycle refers to the conversion of muscle-derived alanine to glucose, which can circulate to muscle and be metabolized back to pyruvate and then alanine.

Key Point

Glycogen breakdown is promoted by epinephrine and norepinephrine (muscle and liver) and glucagon (liver).

hypoglycemia. The combined metabolic pathways are referred to as gluconeogenesis, and lactate, glycerol, and certain amino acids are the primary molecules used to make glucose. The hormonal promoters are primarily glucagon, epinephrine, and cortisol. Lactate is derived from RBCs as well as skeletal muscle during fasting periods. However, during higher intensity exercise, more lactate is produced by muscle tissue, especially Type II fibers. Because the glucose derived from lactate originally generated in muscle can circulate back to muscle, as demonstrated in Figure 4-14, the process is often referred to as the lactate cycle. In addition to lactate, glycerol is available to the liver during fasting and endurance exercise. Glycerol is the backbone of stored triglyceride (fat) molecules and is released into circulation when fat stores are broken down (see Chapter 6). Thus adipose tissue is the principal provider of glycerol for gluconeogenesis.

During fasting and endurance exercise, skeletal muscle releases amino acids into the blood. The most important gluconeogenic amino acid is alanine. As shown in Figure 4-14, during fasting and higher intensity endurance exercise alanine is released into the circulation at a level greater than alanine concentration in skeletal muscle protein. This indicates that some alanine is produced using pyruvate as the amine group acceptor. As discussed in Chapter 5, the branched-chain amino acids (BCAAs) and aspartic acid, asparagine, and glutamic acid are the major donors of amine groups as they are used for energy in muscle cells.

Key Point

Amino acids, glycerol, and lactate are used to make glucose in the liver to prevent hypoglycemia during fasting and exercise.

Pentose Phosphate Pathway

The pentose phosphate pathway occurs in the liver (hepatocytes), muscle, kidneys, RBCs, adipose tissue, adrenal cortex, and lactating mammary tissue and is vital to several operations within these tissues. As shown in Figure 4-9, glucose-6-phosphate is the starting molecule of the pentose phosphate pathway. The products of this pathway include intermediates of glycolysis as well as ribose-5-phosphate, which can be used to produce nucleotides. Nucleotides are needed for the production of high-energy phosphates (ATP and GTP), nucleic acids (DNA and RNA), and other molecules.

The pentose phosphate pathway also produces NADPH, the reduced form of NADP⁺. NADPH provides "reducing power" for fatty acid production in the liver and adipose tissue as well as cholesterol in most cells. NADPH is particularly important in red blood cells and probably skeletal muscle cells to regenerate reduced glutathione. As discussed in later chapters, the peptide

glutathione is essential in antioxidation activities and may be particularly important during exercise.

■ CARBOHYDRATE METABOLISM *IN REVIEW*

- Glycolysis is the backbone of carbohydrate metabolism.
- Absorbed glucose in excess of immediate carbohydrate needs for energy is primarily stored in muscle and the liver as glycogen.
- Glycogen provides a buffer against hyperglycemia and hypoglycemia.
- Insulin promotes glucose uptake into muscle and fat tissue for use as fuel and energy storage.
- Glucagon, epinephrine, and cortisol prevent hypoglycemia by increasing glycogen breakdown and promoting glucose formation in the liver so that glucose can be released into circulation.

■ CARBOHYDRATE METABOLISM DURING EXERCISE

Carbohydrate is an important exercise fuel. Furthermore, both muscle glycogen and blood glucose appear to be independently important to performance. During exercise the utilization of carbohydrate by working skeletal muscle increases.[12–15] Samples of muscle tissue (**biopsies**) taken before and after exercise clearly show that muscle glycogen content is reduced after exercise.[13,14] Also, the difference between the level of glucose in the arteries supplying skeletal muscles and the level in the veins that drain them is greater when the muscles are engaged in physical activity.[16,17] Therefore an increased glucose uptake by working muscle complements glycogen breakdown in providing fuel during exercise. Carbohydrate also must be taken seriously as a potential performance-limiting factor. Both hypoglycemia and muscle glycogen depletion have proved to be independently involved in reducing athletic performance and promoting fatigue.[18–20]

As displayed in Figure 4-15, numerous factors influence carbohydrate metabolism during exercise. Among the most significant are the intensity and duration of exercise and the degree of training. Additionally, the nutritional or metabolic state and the level of carbohydrate storage (glycogen) are important factors. This section provides an overview of these and other important factors.

Exercise Intensity and Carbohydrate Utilization by Skeletal Muscle

Carbohydrate is used by skeletal muscle for energy during exercise regardless of the intensity level. However, as the intensity of an exercise bout increases, so too does the reliance on carbohydrate as an energy source. This is

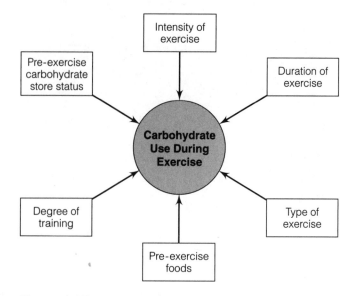

Figure 4-15
Carbohydrate use during about of exercise is influenced by numerous factors.

demonstrated by an elevation in the respiratory exchange ratio (RER) and the level of lactic acid in the blood during higher intensity exercise.[12,21] At roughly 50–65% VO$_2$max, fatty acid utilization tends to peak as a percentage of total energy expended.[2] The range accounts for differences in people including training level and genetics. As exercise intensity increases beyond this level the reliance on carbohydrate, especially muscle glycogen, must increase. This was clearly demonstrated in a study involving young males who performed three endurance exercise bouts at three different intensity levels. As shown in Figure 4-16a–c, after cycling for 30 minutes at 25% VO$_2$max, the most significant fuel source for working muscle was free fatty acids (FFAs) derived from the blood.[22] Meanwhile, plasma glucose was the principal provider of carbohydrate to working muscle. As Figure 4-16a shows, as the exercise intensity increased to 65% VO$_2$max, the contributions made by carbohydrate and fat were approximately equal. At that level, triglyceride stores within the muscle tissue provided a much larger, though still secondary, share of fatty acid fuel. When the exercise bout was performed at 85% VO$_2$max, carbohydrate provided about two-thirds of the energy for working muscle, most of it supplied by glycogen.[22] As exercise intensity climbs above

Key Point

Exercise intensity and duration as well as level of training and nutrition state are among the most significant factors that influence carbohydrate metabolism during exercise.

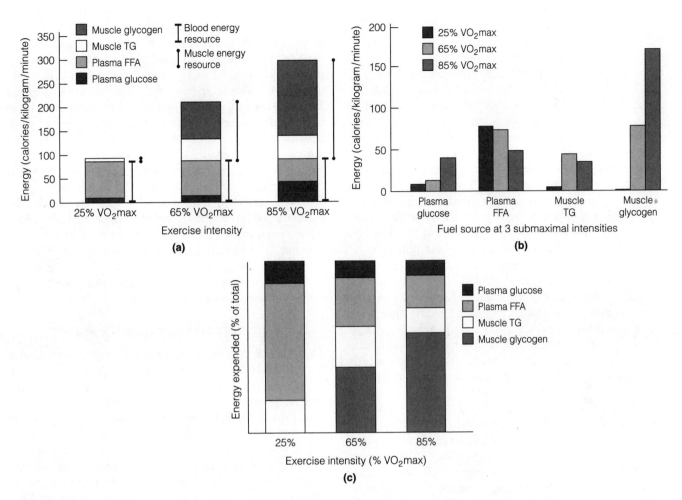

Figure 4-16

Contribution of different fuel sources to working skeletal muscle in cycling exercise (fasting state) at three submaximal intensities in 30-minute bouts. (a) Energy expended and contribution of carbohydrate and fat sources of energy. Note that at 65% VO$_2$max the contribution made by muscle fuel resources begins to exceed blood resources, and that the absolute contribution made by blood resources is consistent despite changes in intensity. (b) Expenditure of each fuel source at the three intensities. (c) Contribution of the fuel sources as a percentage of the total energy expended at each intensity. (1 calorie = $\frac{1}{1000}$ of a kilocalorie. TG, triglyceride; FFA, fatty acid.) (*Source:* Based on data from Romijn, JA, Coyle EG, Sidossis LS, Gastaldelli A, Horowitz JF, Endert E, Wolfe RR. Regulation of endogenous fat and carbohydrate metabolism in relation to exercise intensity and duration. *American Journal of Physiology* 265:E380–E391, 1993.)

100% VO$_2$max (supramaximal) the reliance on carbohydrate is even greater.[23] The ATP requirement during this sprint type of exercise requires a very strong contribution by anaerobic glycolysis. This is reflected in an increased production of lactic acid, which becomes a performance-limiting factor in activities such as sprinting and resistance training.

One important consideration in conducting research that addresses fuel sources during exercise is nutritional state. For instance, the participants in the study discussed above were in a fasting state and did not receive any food or sport drink during the exercise test. However, if carbohydrate were provided prior to or during the exercise bout, the utilization of carbohydrate as a fuel source would be greater, even at the lower intensities.

Obligate Carbohydrate Utilization and the Lactate Threshold

Muscle tissue becomes dependent on carbohydrate as the principal fuel source at higher submaximal exercise intensities and greater. This dependence is primarily based on O$_2$ availability, the recruitment order of muscle fibers, and the level of circulating epinephrine. At lower and moderate submaximal exercise intensities, O$_2$ availability is able to meet demands for aerobic energy metabolism. Fatty acids tend to make a greater fuel contribution as Type I muscle fiber recruitment predominates and the level of circulating epinephrine is relatively low. Then, as exercise intensity is increased, O$_2$ availability becomes insufficient for aerobic metabolism to meet all energy demands. In support, Type II muscle fibers are recruited and the level of circulating epinephrine is increased. As discussed below, epinephrine strongly stimulates the breakdown of glycogen stores, thereby increasing carbohydrate availability for glycolysis within contracting muscle fibers.

The concept of "obligate" carbohydrate use infers a point or threshold in the exercise intensity scale at which additional energy requirements would be met by carbohydrate. However, at what exercise intensity muscle tissue becomes obligated to use carbohydrate is still a matter of debate. One of the most obvious indicators of obligate carbohydrate use is an elevation in blood lactate level. The intensity level at which there is a significant rise in blood lactate levels is often referred to as the *lactate threshold*. The lactate threshold is also referred to as the *anaerobic threshold*, as it is indicative of increased anaerobic glycolysis. The lactate threshold tends to be between 50 and 60% VO_2max in untrained individuals and 70 to 80% VO_2max in highly trained endurance athletes. Lactate is also produced at exercise intensities below the lactate threshold, but because inactive muscle fibers and the heart are able to efficiently metabolize it there is little rise in blood lactate levels. The lactate threshold reflects an intensity at which there is a significant mismatch between lactate production and metabolism.

It is possible that obligate carbohydrate use occurs at an intensity level lower than the lactate threshold as a means of increasing the rate of ATP generation. Increased glycolysis would generate more pyruvate, which in turn could be converted to oxaloacetate (OAA). As demonstrated in Figure 4-9, OAA then would be available to combine with acetyl CoA to form citrate. Acetyl CoA would be derived from fatty acid oxidation as well as from pyruvate generated via glycolysis. This could increase the availability of entry molecules for the Krebs cycle.

Physiological Basis for Carbohydrate Preference at Higher Intensities

Carbohydrate becomes an increasingly important fuel source when exercise is performed at and above the lactate threshold, for several reasons. When exercise is performed at very high intensities (as in sprinting and weightlifting), the extraordinary ATP demand dictates a heavy dependence on anaerobic glycolysis, which supports other anaerobic systems, namely creatine kinase and adenylate kinase. Anaerobic generation of ATP may allow for power outputs that are two to four times greater than those produced at 100% VO_2max. However, this power output can only be sustained for brief periods.

Key Point

As exercise intensity increases, the reliance on carbohydrate increases in working muscle.

The oxidation of carbohydrate may also increase at higher exercise intensities as a matter of efficient O_2 metabolism. Glucose can yield about 10% more energy than fat per volume of O_2 when metabolized aerobically (oxidized). More specifically, glucose can provide 5.10 kcal/L of O_2 in comparison to 4.62 kcal/L of O_2 for fat.[24] Also, as exercise intensity increases, more carbohydrate and less FFAs are available to contracting skeletal muscle fibers. There are several reasons for this. First, epinephrine stimulates the breakdown of muscle glycogen stores, making glucose and glucose-6-phosphate available for glycolysis. Second, during higher intensity exercise the accumulation of lactate in the circulation may slow FFA release from adipose tissue.[24] Third, as more blood is distributed to working skeletal muscle during higher intensity exercise, blood flow through adipose tissue is reduced. Combined, these factors can decrease the availability of FFAs to working skeletal muscle.

Exercise Intensity and Glycogen Breakdown

There is a strong relationship between exercise intensity and the rate of muscle glycogen breakdown. Figure 4-17 displays the rate of degradation of muscle glycogen at three exercise intensities (30%, 60%, and 120% VO_2max).[15,25] This experimental design allowed researchers to estimate the rate of glycogen breakdown during 2 hours of exercise at low and moderate submaximal intensities and during a supramaximal effort that was much briefer due to early exhaustion. Researchers also estimated the rate of glycogen breakdown at 0.7, 1.4, and 3.4 mmol of glycogen per kilogram of muscle per minute at endurance exercise intensities approximating 50%, 75%, and 100% VO_2max.[26] For extreme supramaximal exercise efforts such as weightlifting or sprinting, glycogen is a particularly important fuel resource. For example, 6 seconds of all-out cycling reduced the glycogen content of *vastus lateralis* muscle in the thigh

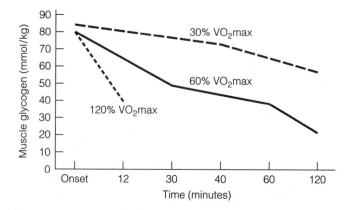

Figure 4-17

Muscle Glycogen depletion in leg muscle (vastus lateralis) during cycling exercise at three different intensities (*Source:* Adapted from references 15 and 68.)

by 14%.[27] A 30-second cycling sprint reduced the glycogen content of the involved leg muscle by as much as 27%.[28]

During lower intensity exercise, glycogen may be reduced only in Type I muscle fibers, as Type II muscle fibers largely remain unrecruited. But as the level of exercise intensity increases and Type II muscle fibers are recruited, they too experience reductions in glycogen content. Because Type II muscle fibers have greater capabilities in anaerobic energy systems, they demonstrate a more rapid depletion of muscle glycogen during maximal and supramaximal exercise than Type I fibers.

GENDER AND AGE AND GLYCOGEN BREAKDOWN

Are there differences in the rate of glycogen breakdown at different ages and between males and females? It does seem that women may have a lower rate of muscle glycogen breakdown than men when participating in endurance activity at the same intensity level.[29] This concept is supported by a consistent finding that women have lower RER measurements when performing the same exercise as men.[29] In children the rate of glycogen breakdown is similar to that in adults when performing the same exercise.[11] This occurs despite lower muscle glycogen concentrations in the children.[11]

Exercise Intensity and Glucose Uptake

The entry of glucose into working muscle fibers is relative to the intensity of an exercise bout.[2,15] For example, when a person is not active (is at rest), skeletal muscle can account for about 15–20% of the total glucose utilized in the body.[15] During exercise this percentage increases as skeletal muscle takes up more glucose from the blood to help fuel muscle contraction. For example, during a bout of cycling at 55–60% VO$_2$max, leg muscles alone are reported to account for as much as 80–85% of glucose utilized by the body.[30] It is logical to assume that this percentage would increase at even higher exercise intensities.[15]

Figure 4-18 displays the difference in glucose uptake by active skeletal muscle at 30% and 60% VO$_2$max. As exercise intensity increases, there is an increase in plasma glucose oxidation (see Figure 4-16b). The energy contribution made by plasma FFA progressively decreases as exercise intensity increases and is compensated by an increased utilization of plasma glucose. This allows the

Key Point

Skeletal muscle is more reliant on carbohydrate at higher exercise intensities due to a decreased O$_2$ availability relative to energy demands, increased glucose availability, and the progressive recruitment of Type II muscle fibers.

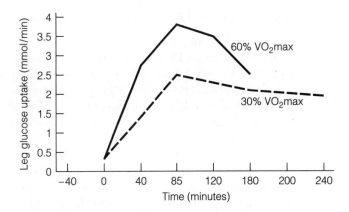

Figure 4-18

Glucose uptake by active skeletal muscle. Glucose uptake into skeletal muscle increases relative to exercise intensity.

energy contribution (grams oxidized per minute) from circulating fuel resources to remain somewhat constant at different exercise intensities (Figure 4-16a). However, as exercise intensity increases, the reliance on muscle fuel sources increases as a percentage of total fuel utilized (Figure 4-16c).

Liver Glucose Output

The intensity and duration of an exercise bout are the most important factors that determine the source and quantity of glucose released by the liver. For example, during the first 30 minutes of moderate to heavy exercise most of the glucose output from the liver is derived from glycogen stores and not gluconeogenesis.[17,31] Even after 1 hour of exercise at a light submaximal intensity the contribution made by gluconeogenesis to glucose release by the liver is still less than 15% of total glucose output.[31]

The intensity of the exercise should also influence the substrates utilized for glucose production, via gluconeogenesis, in the liver. When exercise is performed at an intensity level below the lactate threshold, glycerol is the primary molecule used for gluconeogenesis. As exercise intensity approaches and exceeds the lactate threshold, more lactate becomes available to produce glucose in the liver (Figure 4-14).

An individual's nutrition state also influences how much glucose the liver releases during exercise and what it uses to make that glucose. If an individual has fasted or nearly fasted for a significant period (a day or two), it would be expected that most to practically all of the glucose the liver releases during a bout of exercise is derived from gluconeogenesis.[32] In contrast, if an individual consumes a carbohydrate-containing food or beverage prior to or during exercise, such glucose production and release by the liver would not be necessary.

Exercise Duration and Carbohydrate Utilization

How does exercise duration influence carbohydrate metabolism during exercise? There are several factors to consider, including the time point within the exercise bout and the intensity of the effort at that time point. Furthermore, an individual's level of training and nutritional state influence carbohydrate metabolism during an exercise bout. During prolonged exercise there is an increasing reliance on plasma glucose as a carbohydrate resource.[22,33] This is demonstrated in Figure 4-19, which shows the results of a study involving men who cycled at a moderate submaximal intensity (about 65% VO$_2$max) for 2 hours.[22] During the ride there was a steady increase in the utilization of plasma glucose for fuel that coincided in a predictable manner with a decreased use of glycogen.[34] An increased reliance on plasma glucose for fuel necessitates the release of glucose from the liver to prevent hypoglycemia. It is expected that the maintenance of plasma glucose primarily occurs by liver glycogenolysis and gluconeogenesis.[2,15]

Resistance exercise and sprinting are not considered endurance activity; however, these types of exercises are typically repeated and therefore duration is a consideration. For example, would the details of carbohydrate metabolism be the same during the third or tenth sprint or weight training set as they were during the first? Despite the short nature of supramaximal exercise bouts, glycogen breakdown is significant and subsequent efforts are influenced. For instance, researchers determined that two 30-second cycling sprints to fatigue resulted in a 47% reduction in muscle glycogen content.[35] Likewise, multiexercise resistance workouts to fatigue have been noted to reduce muscle glycogen content by up to 25–40% in the active muscle.[36-38] As the muscle glycogen level is reduced, its contribution to energy expenditure decreases during repeated sprint bouts. This was demonstrated in a study involving men who performed three 30-second maximal cycling sprints separated by 4 minutes of rest.[39] As shown in Figure 4-20, glycogen made a progressively lower contribution to energy expenditure during the later sprints, which led to a decreased work output. Similar to repeated sprinting (as in track or ice hockey), it is common for weight trainers to experience reduced strength and work output during repeated sets that focus on a muscle group (such as bench press or squat). It is logical to think that the decreased performance experienced by weight trainers in subsequent sets is related to reductions in glycogen content.

Carbohydrate Depletion and Fatigue

A strong relationship exists between the depletion of muscle glycogen stores and/or hypoglycemia and the onset of fatigue during endurance exercise bouts. This is especially true when the exercise intensity is in the range of 60–85% VO$_2$max.[40] For highly trained individuals, exercise at this intensity would endure for more than 2–3 hours. This time frame is typical for sports such as endurance cycling (such as tour stages and cycling centuries), marathons (42 kilometers or 26.2 miles), triathlons, and Nordic skiing.

It is unlikely that individuals engaged in lower intensity exercise (<50% VO$_2$max) would exhaust their muscle glycogen stores and thus become fatigued before finishing an exercise bout.[26] This does not mean that muscle glycogen levels are not reduced during lower intensity exercise. They are reduced in active muscle fibers; however, a steady state is achieved in which glycogen breakdown is

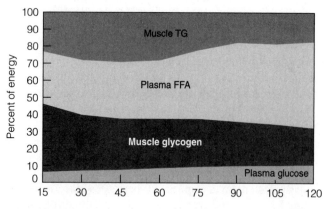

Figure 4-19

Expected fuel sources (% of energy expended) during a 4-hour cycling bout at 65% VO$_2$max in pre-exercise fasted young males. (*Source:* Based on data reported from reference 22.)

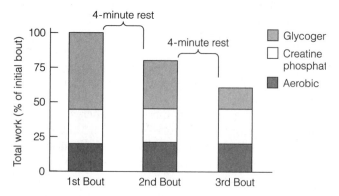

Figure 4-20

Contribution toward ATP regeneration during three 30-minute bouts of maximal cycle sprinting with 4 minutes of rest between bouts. The contributions made by aerobic efforts and creatine phosphate were consistent in the three bouts. The contribution made by glycogen was progressively lower during the second and third bouts, which accounted for the reduction in work output during those bouts. (*Source:* Adapted from reference 39)

balanced by the minimal glycogen synthesis efforts occurring at the same time.[41,42] Athletes exercising at higher submaximal intensities ($\geq$90% VO$_2$max) would also end an exercise bout prior to exhausting their muscle glycogen stores.[2] Here the fatigue would be related to other factors such as lactic acid production.

PERIPHERAL VERSUS CENTRAL FATIGUE RELATED TO CARBOHYDRATE AVAILABILITY   When endurance exercise continues for more than an hour, muscle glycogen content wanes and makes a progressively lower contribution to energy expenditure. Reduced glycogen utilization is compensated by increased plasma glucose utilization; however, the compensation is incomplete. Muscle performance fades soon after muscle glycogen reaches its lowest level. Reduced muscle performance is referred to as **peripheral fatigue** (muscular fatigue) and is based on an inability to meet ATP demands to sustain a higher level of performance when carbohydrate availability is reduced.

Exercise is often halted at this point of peripheral fatigue. Or, an athlete might continue, but at a reduced performance level. For instance, a distance runner who experiences peripheral fatigue prior to completing a marathon may record a final mile time that is slower than at the 10- or 20-mile point. Some athletes can experience sluggishness and reduced performance when they train to near or complete exhaustion several days in a row. It is important to remember that the recovery of muscle glycogen stores can take a day after reaching a nadir. Therefore, subsequent training bouts initiated prior to recovering glycogen stores can progressively lessen muscle glycogen stores. In this scenario, chronic overtraining leads to premature fatigue.

The maintenance of blood glucose is critical as an energy supply not only for working muscle but for the central nervous system as well. For instance, as blood glucose levels approach 45 mg/100 ml one may experience lightheadedness, lethargy, and nausea **(neuroglucopenia)**.[40] These symptoms are associated with decreased performance and are hallmarks of **central fatigue**. Less than half of individuals performing endurance exercise at 60–70% VO$_2$max for $2\frac{1}{2}$–$3\frac{1}{2}$ hours would experience hypoglycemia. However, there is convincing evidence that carbohydrate intake during endurance exercise of more than 2 hours can increase glucose utilization and improve performance by delaying fatigue.[43] Although much of this effect is probably attributable to providing fuel for muscle, it may also be related to subtle factors associated with central fatigue. Extreme central fatigue is often referred to as "hitting the wall" or **bonking**, truly an experience not soon forgotten and often shared among the endurance faithful. The term "bonk" is derived from a British military slang meaning to bomb. It is also the root of bonkers, which of course means to go mad or out of one's mind.

Key Point

Fatigue is often related to depletion of muscle glycogen or to hypoglycemia. Muscle glycogen depletion results in a reduction in muscle performance (peripheral fatigue), whereas hypoglycemia can result in central fatigue as glucose supply to the brain is compromised.

■ CARBOHYDRATE METABOLISM DURING EXERCISE *IN REVIEW*

- Carbohydrate becomes an increasingly greater contributor of energy for contracting muscle at intensities greater than 50–60% VO$_2$max.
- Fatigue related to muscle carbohydrate depletion and hypoglycemia is mostly experienced at exercise intensities between 60 and 85% VO$_2$max.
- Decreased glycogen availability can decrease performance during repeated supramaximal efforts such as weight training and sprinting.

■ REGULATION OF CARBOHYDRATE METABOLISM DURING EXERCISE

The availability of glucose to working muscle cells depends on the rates of glycogen breakdown and glucose uptake. How are these two processes regulated during exercise to allow for optimal carbohydrate availability? Both are influenced hormonally. Furthermore, skeletal muscle phosphorylase and key enzymes of glycolysis are influenced by metabolic factors associated with muscle fiber activity. Combined, these hormonal and metabolic factors attempt to match carbohydrate availability to the ATP demands of a particular exercise.

Glucose Uptake into Skeletal Muscle

As discussed above, glucose uptake into skeletal muscle is greatly enhanced during exercise. Because glucose is transported into muscle cells via GluT, an increased uptake is highly suggestive of enhanced GluT activity. GluT4 is the glucose transport protein responsible for increased glucose uptake into muscle fibers when stimulated by insulin during hyperglycemia. Thus the notion that glucose uptake is enhanced by exercise, via GluT4, seems to contrast with what would be expected, because insulin release is inhibited as the intensity of an exercise bout increases. More specifically, insulin secretion from the pancreas is inhibited when the level of exertion climbs above 50% VO$_2$max.[44] The sympathetic nervous system is partly responsible for the inhibition. In addition, circulating epinephrine and to a lesser degree, norepinephrine and growth hormone dampen insulin release during exercise.[12]

Because insulin release is inhibited during exercise, it is possible for blood glucose to become slightly elevated during early endurance exercise.[44] Reduced insulin levels during exercise should have a profound impact on energy metabolism in skeletal muscle as well as other tissue. First, a reduced insulin presence allows for a greater rate of muscle glycogen breakdown. Second, it allows for a greater rate of fatty acid release from adipose tissue, which would be available to muscle and other tissue (see Chapter 6). Additionally, other tissue such as the heart and inactive muscle tissue may actually begin to use less glucose during exercise.[12] Combined, these events would increase the availability of blood glucose to working muscle.

INSULIN-INDEPENDENT GLUCOSE UPTAKE Exercise results in the movement of GluT4 from its internal pool to the plasma membrane of muscle fibers.[15,45–49] The fact that this occurs even when insulin release is inhibited during higher intensity exercise suggests that exercise-related factors have an independent effect on GluT4 relocation and/or enhance the potency of the remaining insulin. There is scientific support for the notion that insulin and muscle fiber activity evoke GluT4 migration by different mechanisms.[50,51] First, insulin and muscle fiber contraction appear to have an additive effect. Second, researchers have reported that insulin and exercise may actually signal different intracellular pools of GluT4.[45,52,53] However, since exercise increases the sensitivity of muscle to insulin, it would seem that there is at least some overlap between mechanisms.[54]

The exact intracellular mechanisms leading to exercise-induced relocation of GluT4 remain somewhat elusive. Researchers have determined that increased intracellular calcium, as well as hypoxia, can increase glucose transport into muscle tissue.[55,56] Both of these conditions are associated with exercise. The potential role of epinephrine is unclear so far.[15,57–59]

GENDER CONSIDERATIONS AND GLUCOSE UPTAKE
Are there differences in exercise-related glucose uptake into skeletal muscle between males and females? The fluctuation of hormones during the menstrual cycle in females can affect glucose uptake into muscle and other tissue such as the central nervous system. It stands to reason then that altered glucose uptake into muscle and the brain could in turn influence performance.[60] Although it seems

Key Point

Increased glucose uptake into working skeletal muscle fibers can occur without insulin; however, insulin enhances the uptake.

that both estrogen (17-β estradiol) and progesterone may enhance insulin release from the pancreas, only estrogen enhances insulin-promoted glucose uptake into tissue such as muscle. Progesterone may actually inhibit insulin-promoted glucose uptake. In addition, estrogen enhances glucose uptake across the blood-brain barrier, which is not insulin-mediated and is believed to primarily occur via GluT1.[60] Therefore, estrogen may allow for optimal performance by increasing glucose uptake into muscle and the brain during exercise and thus decrease the potential for premature peripheral and central fatigue. With this in mind, exercise trial protocols involving females need to consider whether the participants are menstruating as well as the stage of the menstrual cycle at the time of testing.

Glycogen Breakdown in Active Muscle Tissue

The regulation of glycogen breakdown involves various hormones (such as epinephrine and insulin) and intracellular factors [such as Ca^{2+} and phosphate (P_i)], as demonstrated in Figure 4-12. In several studies investigating the rate of muscle glycogen breakdown, researchers have determined that higher levels of pre-exercise glycogen resulted in a greater rate of breakdown.[15,61–65] This was demonstrated in a study involving young males who were fed either a high-carbohydrate diet (>80% of energy) or a low-carbohydrate diet (10–25% of energy).[66] Their muscle glycogen levels were determined prior to and after 40 minutes of exercise at 65% VO_2max. As shown in Figure 4-21, the rate of glycogen depletion was significantly

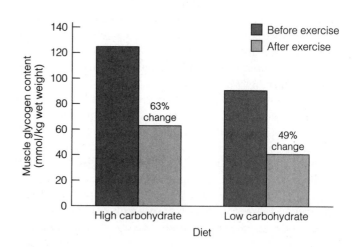

Figure 4-21
Pre-exercise muscle glycogen content can increase the rate of breakdown before and after exercise. Young males fed a high-carbohydrate diet (>80% of energy from carbohydrate) had higher rate of glycogen utilization compared to a similar group fed a low-carbohydrate diet (10–25% of energy). Despite the higher glycogen utilization, the muscle glycogen content in the first group was still higher after cycling for 40 minutes at 65–70% VO_2max. (*Source:* Adapted from references 15 and 66.)

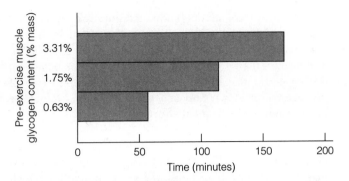

Figure 4-22
Cycling time to exhaution when beginning exercise at different glycogen contents (% of mass). Cycling time exhaustion is extended when athletes begin endurance exercise with greater muscle glycogen content. (*Source:* Adapted from reference 67.)

greater after consuming the high-carbohydrate diet, which led to higher pre-exercise glycogen levels. The increased rate of glycogen breakdown is partly due to increased substrate availability for phosphorylase.

Even though the rate of glycogen breakdown is greater in muscle with a higher pre-exercise glycogen content, the higher initial content should still allow an individual to have more glycogen at various time points of endurance exercise. As shown in Figure 4-22, this should allow a longer duration of endurance exercise. In fact, the positive effect of higher pre-exercise muscle glycogen on endurance performance (such as distance running and cycling) is generally accepted.[15,67] For sports involving intermittent supramaximal efforts separated by periods of rest (weight training, soccer, football), having greater muscle glycogen stores may enhance performance as well.

The rate of glycogen breakdown is greatest during the earliest stages of exercise (Figure 4-17). However, the rate of glycogen breakdown slows as exercise is continued, despite a continuance of the same intensity.[15,25] It may be that the rate of glycogen breakdown slows due to a decrease in substrate availability.[15,61,68] That is, as the content of glycogen is reduced there are fewer free ends on the glycogen structure. However, a similar pattern of rapid and then slowing glycogen breakdown may actually occur regardless of the glycogen level at the onset of exercise.[68]

HORMONAL AND METABOLIC INFLUENCES ON GLYCOGEN BREAKDOWN Epinephrine is the most potent hormone that promotes glycogen breakdown in muscle fibers. By binding to receptors on the plasma membrane (sarcolemma), epinephrine activates phosphorylase kinase, which in turn activates phosphorylase, as shown in Figure 4-12. An increase in the level of Ca^{2+} within muscle fibers can also activate phosphorylase kinase.[69] This means that even though phosphorylase kinase may be in its inactive form, its activity can be immediately

increased by several hundredfold via the binding of Ca^{2+}. Also, Ca^{2+} is needed to maximally activate this enzyme. Thus, Ca^{2+} appears to be an important intracellular factor in the immediate and maximal glycogen breakdown response at the onset of exercise.

Phosphate has been suggested as an activator of phosphorylase.[15,69] Perhaps the most important explanation is that P_i serves as a substrate for phosphorylase, as shown in Figure 4-12. As glycogen is broken down, glucose-1-phosphate is generated. In active muscle fibers, the breakdown of creatine phosphate (CP) and ATP would increase the availability of P_i to phosphorylase.[15] In fact, P_i (and Ca^{2+}) may be important in explaining why glycogen breakdown is not significant in inactive muscle during exercise, despite elevated epinephrine and decreased insulin levels. Because energy requirements are elevated in muscle fibers during higher intensity exercise, the concentration of AMP and IMP would also increase. Both AMP and IMP appear to directly activate glycogen phosphorylase.[69]

■ **REGULATION OF CARBOHYDRATE METABOLISM DURING EXERCISE IN REVIEW**

- GluT4 protein is responsible for the increased glucose uptake during exercise.
- GluT4 relocates from intracellular pools to the plasma membrane of active muscle fibers in a seemingly insulin-independent manner during exercise.
- Metabolic factors associated with muscle contraction may make insulin more potent during exercise.
- Glycogen breakdown is enhanced hormonally by epinephrine and metabolically by Ca^{2+}, P_i, AMP, and IMP.
- Creatine phosphate and ATP may be important providers of P_i for glycogen breakdown.

■ **TRAINING ADAPTIONS IN CARBOHYDRATE METABOLISM**

Without question, carbohydrate is the most important fuel source for muscle during supramaximal efforts. Furthermore, for well-hydrated individuals, carbohydrate is the most significant performance-limiting nutrient during endurance efforts performed at 60–85% VO_2max. Thus, it would be beneficial for muscle systems to adapt to improve carbohydrate storage and aspects of carbohydrate metabolism during higher submaximal and supramaximal exercise. This section provides an overview of several aspects of exercise and carbohydrate metabolism. Chapters 6 and 7 include discussions of how a low-carbohydrate diet increases the utilization of fat at rest and during exercise.

Adaptations Influencing Carbohydrate Utilization

One of the most consistent and impressive changes associated with endurance training is a reduction in respiratory exchange ratio (RER). This is evident when measuring whole-body RER, as well as when samples of the arterial and venous blood of an exercising arm or leg are obtained and the gases assessed. A reduced RER suggests an adaptation in the processes that allow muscle to oxidize fatty acids during exercise.[71] As would be expected, the reduced RER associated with endurance training is complemented by an increase in the anaerobic threshold. As mentioned above, untrained individuals can begin to accumulate lactate at approximately 50% VO_2max. In contrast, endurance-trained individuals may not begin to accumulate lactate until exercise intensity climbs to 70–80% VO_2max.[24]

The ability to utilize more fat for fuel during exercise is a cumulative reflection of metabolic adaptations in trained muscle fibers as well as improved aspects of oxygen delivery. Endurance training increases the concentration of mitochondria in muscle fibers as well as the oxidative capacity of that tissue (see Chapter 13). Endurance training also increases the content of fat within muscle fibers, thereby putting the fatty acids at the site of utilization. Increased vascularization of muscle and enhancements in cardiovascular performance do their part to increase O_2 delivery to active muscle fibers. All totaled, these adaptations increase the ability of muscle to oxidize fatty acids during submaximal exercise, thereby decreasing carbohydrate utilization. Thus, it would seem that reduced carbohydrate utilization during endurance activity is the result of adaptations in fat oxidation, rather than direct changes in carbohydrate metabolism.

Adaptations in Muscle Glycogen Content

Glycogen becomes an increasingly important fuel source as exercise intensity is increased. Thus, an enhancement of the capacity of muscle to store glycogen could improve performance in sports in which performance would be

Key Point

Endurance training increases the oxidative capabilities of muscle tissue, which in turn increases the use of fat and decreases the dependence on carbohydrate during exercise.

limited by glycogen depletion. An increase in skeletal muscle glycogen storage occurs as a result of both submaximal and supramaximal exercise training. In addition, a reduced rate of glycogen breakdown would be beneficial during endurance exercise, whereas an increased rate of glycogen breakdown and anaerobic glycolysis might be beneficial in supramaximal efforts.

EXERCISE-INDUCED ADAPTATIONS IN MUSCLE GLYCOGEN CONTENT   Training can increase glycogen stores in both fiber types; the key to predicting which fiber types experience the adaptation is the intensity of the training. Moderate submaximal intensity training (such as 60% VO_2max) leads to an increase in the glycogen content in muscle, primarily in Type I muscle fibers. This is because they were the fibers recruited during training. However, as the intensity level of training bouts is increased and more Type IIa and IIb fibers are recruited, the muscle glycogen content increases in these muscle fiber types. For instance, 4–5 months of weight training or 8 weeks of intense cycling training can increase glycogen in muscle tissue by as much as 36%.[71,72]

EXERCISE-INDUCED ADAPTATIONS IN MUSCLE GLYCOGEN BREAKDOWN   A decreased rate of glycogen breakdown during submaximal intensity exercise tends to complement the increased glycogen storage capability in trained muscle fibers. For instance, when trained and untrained individuals exercise at the same intensity (say 75–80% VO_2max), the rate of muscle glycogen utilization is similar.[73] However, because the work output is greater for the trained individuals, their rate of glycogen use is actually lower (glycogen utilized per unit of work output). One method used to test the notion that training decreases the rate of muscle glycogen breakdown is to exercise train one leg and compare it to the other, untrained leg.[74,75] Yet some researchers are not convinced that this is a true direct training effect. They contend that the slower rate of glycogen breakdown is an indirect event and that the increased ability to use fatty acids is the real adaptation.[75a]

In contrast, strength and sprint training may enhance the rate of muscle glycogen breakdown. For instance, researchers have determined that the activity of phosphorylase as well as phosphofructokinase and lactate dehydrogenase may increase 10–25% following 30-second sprint training.[76] In theory this could lead to greater glycolytic activity and allow muscle to develop greater tension and for a longer period. In addition, this type of adaptation could be important in recovery operations between repeated sprints.

Key Point

Exercise training leads to an increase in muscle glycogen content.

Adaptations in Liver Glycogen

Maintenance of an appropriate blood glucose level depends on the rate of utilization by muscle and other tissue and the replenishment of glucose. If carbohydrate is not ingested, the supply of glucose at rest and during exercise depends on glycogen breakdown and gluconeogenesis in the liver. Knowing this, it would be important to discern whether exercise training leads to adaptations in liver glycogen breakdown and gluconeogenesis. As important as this question is, direct study of the impact of exercise training on liver glycogen metabolism in people is extremely limited. It is not common for liver samples to be obtained during exercise studies. This means that researchers mostly rely on the results of animal studies or indirect estimations in human trials to surmise what is really happening.

Based on the available information from research, it seems that when individuals go from an untrained to a trained state, the rate of breakdown of liver glycogen is slowed and matches a reduced glucose production during exercise.[31] These metabolic events appear to be related to the decreased glucose uptake into trained versus untrained skeletal muscle fibers during exercise as they use more fatty acids for fuel. Therefore, whether there is a direct training effect on glycogen breakdown or gluconeogenic mechanisms in the liver or the reduced glucose output is an indirect effect of other factors (such as hormones) remains debatable. For instance, reduced glucose uptake into muscle during exercise would allow blood glucose to be more stable during exercise. This would reduce the presence and influence of glucagon and to a lesser extent epinephrine on glycogen breakdown and glucose production during exercise.

■ TRAINING ADAPTATIONS IN CARBOHYDRATE METABOLISM *IN REVIEW*

- Endurance training allows muscle to use more fat and thus decrease its dependence on carbohydrate.
- Endurance training increases the oxidative (aerobic) capacity of muscle fibers, and the lactate threshold is increased.
- The maximal level of glycogen within skeletal muscle increases as a result of both endurance and resistance training.
- The rate of glycogen breakdown during exercise is potentially slowed as a result of endurance training and increased as a result of supramaximal training.
- Muscle fibers trained at supramaximal intensities experience adaptations that allow for greater anaerobic glycolysis.

■ CARBOHYDRATE BEFORE, DURING, AND AFTER EXERCISE

Many of the studies discussed above involved fasted participants. Studies such as these provide basic information on exercise metabolism; however, some prudence is required when interpreting the results. For instance, the participants would begin the exercise bout in a fasting state, which from a hormonal standpoint is supportive of fatty acid oxidation. Also, fasting overnight or for more than 8 hours can result in reductions in glycogen stores in the liver and to some degree in skeletal muscle. The reduction in carbohydrate stores could then influence the contribution made by carbohydrate to energy expenditure during exercise.

In the "real world," exercise more often occurs in the late morning, afternoon, or early evening. These are when most school and college athletic teams practice and when games are played. Late afternoon to early evening is also when many people visit health clubs to train. Surely then, most individuals would have eaten within a few hours before exercise and some would have eaten just before exercise. Even when collegiate swimmers arise at 6 A.M. for swim practice, they tend to eat or drink something, such as a bagel or soda, first.

Ingesting carbohydrate before exercise can improve performance. However, the timing and quantity of carbohydrate needed to maximize performance has been debated and investigated. In addition, the type of carbohydrate has become a consideration for some people in light of glycemic index and gastrointestinal comfort. Carbohydrate should also be the foundation of food or beverages consumed after a training session or competition. Efficient recovery of muscle and liver glycogen stores prepares an athlete for the next training session or competition. This section provides an overview as well as recommendations for consuming carbohydrate before, during, and after exercise. These recommendations, along with general recommendations for daily carbohydrate consumption, are presented in Table 4-4.

Carbohydrate Before Exercise

It is generally believed that a well-tolerated carbohydrate-containing food and/or fluid consumed prior to exercise (minutes to hours) can improve performance. What are the reasons for this? One reason is a maximization of glycogen stores at the onset of exercise. Another reason is

Key Point

Although many exercise studies use "fasting" participants, people who train in the late morning and later in the day are more likely be in a "fed" state at the onset of exercise.

| Table 4-4 | **Carbohydrate Recommendations for Athletes** |

Experimentation
- Athletes need to experiment with different protocols for providing carbohydrate in the days leading to exercise and before, during, and after exercise.
- Experimentation needs to occur during training to allow for optimal performance during competition.

General Carbohydrate Intake
- Eat 50–70% of total energy as carbohydrate with a variety of natural carbohydrate food sources including fruits and vegetables and their juices as well as whole grain products and low-fat dairy food.
- A general recommendation for athletes expressed in grams would be 6–10 g of carbohydrate per kilogram of body weight depending on gender and type of sport.

Carbohydrate 3–4 Hours Prior to Exercise
- 200–350 g of carbohydrate 3–4 hours prior to exercise for an adult maximizes glycogen stores at the onset of exercise and thus enhances performance. Body size, or more appropriately glycogen storage potential, accounts for the broad range.

Carbohydrate 30–60 Minutes Before Exercise
- Ingest 50–75 g of carbohydrate 30–60 minutes prior to exercise.

Carbohydrate Immediately Before Exercise
- If carbohydrate is to be consumed immediately prior to exercise (<5 minutes), 50 g (less for women) can enhance performance without overloading the stomach at the onset of exercise.
- The ergogenic potential of carbohydrate eaten immediately prior to exercise is similar to carbohydrate ingested during exercise.

Carbohydrate During Exercise
- An endurance athlete should strive to ingest 30–60 g of carbohydrate per hour of performance to maintain blood glucose levels.
- Drinking 600–1200 ml of a 6–8% carbohydrate sport drink per hour provides the recommended level.
- Glucose, sucrose, or maltodextrin are probably about the same with regard to enhancing performance; however, fructose may be a detriment by causing intestinal discomfort.

Carbohydrate After Exercise
- Ingest carbohydrate as soon as possible after completion of a glycogen-depleting exercise session.
- Ingest 1.5 g of carbohydrate per kilogram of body weight during the first 30 minutes and every 2 hours for at least 4–6 hours thereafter.
- Ingesting carbohydrate late in an endurance bout may be beneficial if recovery time to next bout is limited (<1 day).

that it would allow an individual to begin exercise in a hyperglycemic state, which might decrease the dependence on muscle glycogen during exercise. However, several factors need to be considered including the timing, amount, and type of carbohydrate as well as the impact of the carbohydrate on hormonal factors and the metabolism of other energy nutrients. One of the most important considerations is athlete individuality. Athletes need to experiment with different protocols during training to maximize potential benefits during competition. Recommendations are often provided in ranges or based on body weight to allow for individuality. For instance, two men having the same body weight but different body composition might have different glycogen storage capacities as well as glycemic indices and insulin responses. Therefore personal experimentation is necessary.

CARBOHYDRATE 3–6 HOURS PRIOR TO EXERCISE
It is fairly well established that exercise performance can

be improved by eating a carbohydrate-rich meal 3–4 hours prior to prolonged exercise versus fasting during that same period.[2,77] Prior to an endurance training or competition, the maximization of skeletal muscle and liver glycogen stores, via carbohydrate consumption, is extremely important. For instance, researchers reported that an initial muscle glycogen content of 3.31 g/100 g of wet muscle allowed their participants to tolerate a standard endurance workload for 167 minutes prior to fatiguing (Figure 4-22).[67] When the participants' initial muscle glycogen content was reduced to 1.75 or 0.63 g/100 g of wet muscle, they could perform for only 114 and 57 minutes, respectively, at that same workload prior to fatigue.

During the night (of sleep) prior to the exercise, significant reductions can be experienced in the liver glycogen stores. Although muscle glycogen stores are not reduced as dramatically overnight, there is still a potential to "top off" their levels. Eating carbohydrate 3–4 hours prior to exercise raises blood glucose and insulin levels, which drive glycogen recovery in muscle and the liver. A secondary effect of a well-timed pre-exercise meal may be the alleviation of hunger, which may develop just before or at the onset of exercise. For some individuals, hunger can be uncomfortable and distracting to performance. Many individuals claim that they do not perform well on an "empty stomach."

The effects of a carbohydrate-containing meal may last 4–8 hours depending on the size and composition of the meal. For instance, the complete absorption and metabolism of 100 g of glucose requires approximately 4 hours. The length of time depends on volume and content of food. In addition, the influence of insulin lingers for a while, even after blood glucose returns to a fasting level. Therefore, exercise would begin in a fed state or a postabsorptive or early fasting state (for more about this see In Focus: Nutrition States and Energy Metabolism on page 110). In either case the glycogen stores have been maximized relative to the carbohydrate level and the influences of insulin would be waning. This might be the case for an individual who ate a large breakfast and trained in the early afternoon.

As presented in Table 4-4, a general recommendation for carbohydrate intake 3–4 hours prior to exercise for an adult is 200–350 g. This would be appropriate to maximize glycogen stores at the onset of exercise and thus enhance performance. Many athletes choose foods that they have tolerated well in the past and that have minimal indigestible material (such as fiber). This meal should be lower in fat to allow for an optimal rate of emptying from the stomach and should provide fluids to optimize hydration status. Typically this is a major meal (breakfast, lunch, or dinner), depending on the time of training. For instance, an endurance cyclist's breakfast containing six medium pancakes, $\frac{1}{4}$ cup syrup, 12 oz of orange juice, and two pieces of toast with jelly would provide approximately 200 g of carbohydrate. However, not everyone tolerates a large meal so close to exercise. Therefore athletes must experiment. In addition, these recommendations are more applicable to athletes engaged in heavy training. Recreational athletes, such as an individual who is going to run 4–6 miles at a pace of 8 minutes/mile does not have to be so attentive to these recommendations. Sport drinks and bars are convenient and generally lower in volume and are popular choices for individuals seeking a lower volume of food and carbohydrate. The feature Nutrition in Practice: Designing an Optimal Training Diet for Sport Performance in Chapter 7 provides a closer look at carbohydrate consumption in meals in relation to training and competition.

Key Point

Carbohydrate ingested 3–6 hours prior to exercise can improve performance.

CARBOHYDRATE 30–60 MINUTES PRIOR TO EXERCISE How would drinking a cola or sport drink or eating a bagel within in an hour of exercise affect performance? Blood insulin levels peak roughly 30–60 minutes after ingesting carbohydrate. This led to a long-standing notion that consuming a carbohydrate-containing food or drink within an hour prior to the onset of exercise might negatively influence performance. It was reasoned that elevated insulin levels at the onset of exercise would inhibit the mobilization of FFAs from adipose tissue, making working skeletal muscle more dependent on carbohydrate. Also, insulin would increase the uptake of glucose into muscle while decreasing the release of glucose into circulation by the liver. This could result in hypoglycemia and decreased performance.[78–81] For instance, the results of one study suggested that 75 g of glucose provided to individuals 30–45 minutes prior to cycling at 80% VO$_2$max resulted in an earlier fatigue than when the individuals received water.[80]

However, several studies support the notion that consuming a carbohydrate resource within the hour prior to endurance exercise can improve performance.[2,77,79] For instance, when cyclists were provided with a carbohydrate supplement containing 1.0 g/kg of body weight 45 minutes prior to riding at 73% VO$_2$max, they were able to perform for 13 minutes longer than when carbohydrate was not provided (placebo).[82] In another study, trained cyclists were provided with either a carbohydrate solution or placebo 60 minutes prior to cycling at 70% VO$_2$max, after which they performed a 45-minute higher intensity cycling time-trial.[83] Here carbohydrate ingestion ($\geq$1.1 g/kg of body weight) allowed for a 12% improvement in performance. As expected, the level of circulating insulin was elevated and the FFA level was lower at the onset of exercise for the cyclists who received the carbohydrate solution. Also as expected, more carbohydrate was utilized as fuel when the cyclists ingested the carbohydrate solution versus the placebo. However, other research efforts have suggested that eating a carbohydrate-containing food 30–45 minutes prior to an endurance exercise trial neither increased performance measurements nor caused them to decline.[82,84] In these studies both glucose and insulin levels were elevated in the blood at the onset of exercise.

The type and intensity of exercise at the onset may be important factors. For instance, sports requiring greater muscle involvement (such as Nordic skiing and running) and performed at higher submaximal intensities at the onset of exercise might have a greater potential for decreased performance. The dip in blood glucose that may be experienced after ingesting carbohydrate 30–60 minutes prior to exercise seems to reach its low point at approximately 15 minutes after the onset of exercise.[70] For some individuals the transient hypoglycemia can be enough to result in performance reduction, but for others it is not. Certainly athletes need to experiment with carbohydrate consumption within an hour of exercise during training periods. This assists an athlete in determining which protocol allows them to maximize their training sessions. It can also assist them in predicting what might happen during competition.

CARBOHYDRATE IMMEDIATELY PRIOR TO EXERCISE

Quite often carbohydrate is consumed just prior to exercise (≤5 minutes), such as consuming sport drink just prior to the start of a race. For some athletes the consumption of carbohydrate during a race is problematic due to difficulties experienced with drinking while moving. This might be the case in a 10- to 15-kilometer race for runners who cannot afford to slow their running motion to more efficiently ingest a sport drink. Therefore ingesting 30–50 g of carbohydrate a few minutes prior to the onset of a race would be easier than trying to consume recommended levels during the event.

It is generally believed that consuming carbohydrate just prior to exercise can result in improvements in performance. For instance, researchers provided cyclists either a solution containing 45 g of glucose polymer, a carbohydrate-containing bar, or a placebo 5 minutes prior to the exercise trial.[85] When the carbohydrate was ingested prior to exercise the cyclists were able to generate a 10% greater average of work output after cycling for 45 minutes at 80% of their VO_2max.

It is logical to think that any improvements in performance related to ingesting carbohydrate immediately before exercise would basically be the same as ingesting carbohydrate early in the session. For instance, researchers provided one group of participants with 30 g of carbohydrate (10% carbohydrate solution) just prior to high-intensity cycling for 1 hour. They gave a second group the same amount of carbohydrate split into portions consumed just prior to and every 15 minutes during the exercise bout. The researchers compared the results of both carbohydrate protocol trials to another trial in which carbohydrate was not provided at all (placebo).[86] Carbohydrate ingestion prior to but not during cycling resulted in less reduction of power output late in the cycling trial compared to the placebo group. Approximately the same performance benefits occurred when the carbohydrate solution was ingested during the cycling bout. This supports the notion that during higher intensity exercise, which would last a shorter time (<1 hour), performance can be enhanced by ingesting carbohydrate just minutes prior to the onset of exercise. However, added benefits are not likely to result from consuming additional carbohydrate during the exercise.[2,87]

Glycogen Loading (Muscle Glycogen Supercompensation)

Glycogen loading is also known as *carbohydrate loading, carbo loading,* and *muscle glycogen supercompensation.* This practice provides a means of maximizing muscle (and liver) glycogen stores at the onset of exercise and is most often used to enhance endurance performance of more than 2 hours. As muscular fatigue is closely associated with depletion of muscle glycogen stores, maximizing these stores could allow an individual to perform at a moderate to higher submaximal intensity level (60–85% VO_2max) longer before exhaustion. Glycogen loading protocols can increase muscle glycogen stores by 50–60% in comparison to simply consuming a mixed diet.

Glycogen loading can enhance the performance of athletes in sports such as marathons, triathlons, ultramarathons, ultra-endurance events, Nordic skiing, and long-distance swimming or cycling. In addition, it has been suggested that late performance in intermittent high-intensity sports such as soccer and football may be improved by glycogen loading. However, not all athletes benefit from glycogen loading. In theory, sports involving jumping and sprinting and short-distance efforts (10 km or less) might be hindered by the extra mass of carbohydrate and associated water. Researchers are currently working on this issue and athletes need to experiment during training periods. The feature Nutrition in Practice: How to Glycogen Load provides instructions for training and diet in the days leading up to competition.

CLASSIC METHOD OF GLYCOGEN LOADING

The classic method of glycogen loading involves depleting muscle tissue of its glycogen over a 3-day period after a bout of exhaustive exercise by eating a very low carbohydrate diet (<5% of energy). As shown in Figure 4-23, this period is then followed by exhaustive exercise and three days of consuming a very high carbohydrate diet (>90% of energy). However, many individuals experience hypoglycemia, irritability, and fatigue during the first half of the protocol and demonstrate poor compliance during the latter half of the protocol. Furthermore, exercise to exhaustion a few days prior to competition can place an athlete at higher risk of injury. For these reasons, alternative carbohydrate loading methods have emerged.

MODERN METHOD OF GLYCOGEN LOADING

More modern approaches to glycogen loading appear to

NUTRITION IN PRACTICE

How to Glycogen Load

Many athletes, such as distance runners and cyclists, attempt to maximize their muscle glycogen content prior to competition. More modern methods of carbohydrate supercompensation or glycogen loading do not involve the rigid carbohydrate-restricting practice of the earlier models. One popular method, outlined in Table A, lasts 7 days; the 7th day is the day of competition. The exercise intensity is maintained at 70–75% VO2max during the

6 days prior to an event while the duration of training is reduced. At the same time, carbohydrate intake is increased from 50% to 70% of energy intake or at least 4–10 g of carbohydrate per kilogram of body weight.

Athletes need to experiment with glycogen loading protocols during training periods to find what works best for them. It is important to consume appropriate levels of water during glycogen loading.

Table A	**A 7-Day Training and Diet Protocol for Glycogen Loading**	
Time Prior to Competition	**Duration and Intensity of Training**	**Dietary Carbohydrate as % of Energy (or g/kg of body weight)**
6th day	90 minutes at ~70–75% VO$_2$max	50% of energy (4–5 g/kg)
5th day	40 minutes at ~70–75% VO$_2$max	50% of energy (4–5 g/kg)
4th day	40 minutes at ~70–75% VO$_2$max	50% of energy (4–5 g/kg)
3rd day	20 minutes at ~70–75% VO$_2$max Rest muscle while not training	70% of energy (10 g/kg) Hydrate
2nd day	20 minutes at ~70–75% VO$_2$max Rest muscle while not training	70% of energy (10 g/kg) Hydrate
1st day (day before)	Rest muscle as much as possible	70% of energy (10 g/kg) Hydrate copiously
Competition	Rest muscle prior to competition	Eat carbohydrate-based meal >2–3 hours prior if possible; ingest carbohydrate 15–30 minutes prior if desired; hydrate appropriately

be very tolerable and equally as effective in maximizing glycogen stores. As displayed in Figure 4-23, one popular method entails a tapering down of exercise duration while maintaining the intensity levels at 75% VO$_2$max during the 6 days prior to a competition. During this time the contribution made by carbohydrate to the energy in the diet increases. Here the energy level must also be appropriate to balance expenditure.

The duration of the exercise bouts are 90 minutes for day 1, 40 minutes for days 2 and 3, and 20 minutes for days 4 and 5. Day 6 is a day of muscular rest followed by competition on day 7. The carbohydrate contribution to the diet is set at 50% of energy or 4 g/kg of body weight for days 1–3 of the loading phase. During days 4–6 it is set at 70% of energy or at least 10 g/kg of body weight. Thus carbohydrate levels may be increased from approximately 290–300 g to more than 550–600 g during the last 3 days. Consumption of such a high carbohydrate load usually is

in the form of pastas, rice, and breads. However, synthetic low-residue, energy-dense, maltodextrin-rich drinks are often used to partially substitute for some of the food volume. Carbohydrate consumption culminates in a pre-event meal 3–4 hours prior to competition that contains approximately 250–300 g of carbohydrate.

Key Point

Glycogen loading programs manipulate exercise training and diet in an attempt to maximize muscle glycogen stores at the onset of competition.

Carbohydrate Type Before Exercise

Foods contain different types of carbohydrates. Natural carbohydrate-containing foods mostly contain glucose,

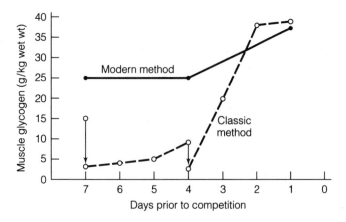

Figure 4-23
Two glycogen loading protocols. The graph compares muscle glycogen content during the classic and the mordern approach to glycogen loading. Exhaustive exercise (*) was a component of the classic method of glycogen loading. (*Source:* Adapted from reference 111.)

fructose, lactose, and starch as well as indigestible fibers and oligosaccharides. Commercially manufactured foods contain some of these carbohydrates plus sucrose and high-fructose corn syrup. The type of carbohydrate ingested prior to exercise requires consideration along with the quantity.

The absorption of the monosaccharide fructose occurs more slowly and results in a less dramatic increase in glucose (lower glycemic index) and corresponding insulin response. Therefore it has been theorized that fructose may provide a performance advantage over glucose and sucrose when consumed prior to exercise. For example, cyclists who ingested 75 g of glucose 45 minutes prior to exercise had a threefold higher plasma glucose level and a 2.5-fold higher insulin level in comparison to those who ingested the same amount of fructose.[79] However, after cycling for 30 minutes at 75% VO_2max, the glucose ingestion resulted in lower plasma glucose and insulin levels versus fructose. The results of this study suggest that fructose consumption during exercise may be more beneficial than glucose by maintaining a more favorable level of blood glucose later in the exercise session and possibly reducing the potential for fatigue. However, this remains to be proven. For instance, researchers reported that the consumption of fructose versus glucose did not improve cycling performance by extending the period before exhaustion.[89] Other research efforts have also failed to demonstrate that fructose consumption during exercise is more beneficial than glucose alone.[90]

There may be reason for some concern about using fructose as an exclusive carbohydrate resource consumed just prior to or during exercise. Some research findings suggest that the consumption of fructose in an amount greater than 50 g can result in cramping and diarrhea due

Key Point

Athletes need to experiment with different types of carbohydrates consumed prior to training or competition to gauge their effect on performance.

to incomplete absorption.[91] Therefore, because fructose has not been shown to be more beneficial than other carbohydrate types, and there seems to be the potential for gastrointestinal discomfort, it is not advocated by most exercise nutritionists as a stand-alone carbohydrate source.

GLYCEMIC INDEX   How important to physical performance is the glycemic index of food consumed prior to exercise? Certainly carbohydrate ingested ≤1 hour prior to exercise would increase the level of glucose and insulin in the blood.[92] It stands to reason, then, that the inhibitory influence of insulin on glycogen breakdown and liver glucose production would conflict with exercise metabolic needs and negatively influence performance. Therefore, it has been suggested that foods having a lower glycemic index may be more appropriate than higher glycemic index foods for consumption just prior to exercise. To try to prove or disprove this notion, researchers provided individuals with lower glycemic index foods (rice or spaghetti) or higher glycemic index foods (glucose, potatoes, or bread) 1 hour prior to cycling 60% at VO_2max. Eating the lower glycemic index foods did result in lower insulin levels and a lower RER and higher plasma glucose levels after 30 minutes of exercise.[93] In a similar study, researchers fed trained cyclists either lentils (lower glycemic index) or potatoes or glucose (higher glycemic index) 1 hour before cycling till exhaustion.[94] After eating the lentils the cyclists had lower insulin and higher plasma glucose levels 30 minutes prior to and right before exercise. During exercise, the cyclists receiving the lentils demonstrated higher levels of circulating FFAs and lower carbohydrate oxidation (lower RER) during exercise. Also, the lower glycemic index (lentils) meal resulted in an average time till exhaustion that was 9 minutes longer than when the cyclists received glucose and 20 minutes longer than when the cyclists received potatoes. Other research efforts have suggested that foods having different glycemic indices can influence blood glucose and insulin levels at the onset and during exercise and can influence RER during exercise.[95,96]

Presently it is not clear whether glycemic index strongly influences athletic performance.[96] However, some clarity may be added by considering exercise intensity.[2] For instance, performance could theoretically be improved when consuming a carbohydrate source with a lower glycemic index prior to prolonged moderate intensity activity. A greater availability of plasma FFAs and the exogenous glucose would serve as fuel for muscle fibers at this intensity. With this in mind, many sport nutritionists recommend consuming a lower glycemic index food as a precompetition carbohydrate source. However, when the level of intensity exceeds the anaerobic threshold the advantage may be nullified.[97]

INDIGESTIBLE CARBOHYDRATE FOODS   Foods containing fiber and oligosaccharide can increase the content within the digestive tract. This results in two potential concerns. First, bacterial metabolism of these carbohydrates produces gas, which can cause bloating and abdominal pain. For this reason some individuals do not eat fiber- and oligosaccharide-containing foods during the hours prior to exercise. Second, undigested carbohydrate increases the mass of the stool in the colon. This is a cumulative effect of the presence of undigested carbohydrate, the water it attracts, and the proliferation of bacteria. Whether an increased mass is detrimental to performance remains to be resolved. However, some endurance athletes refrain from eating whole grain products, fruits, and vegetables for a day or so prior to competition. The extra weight has been suggested to be a performance consideration, as it would increase an athlete's workload. The extra mass associated with indigestible carbohydrate foods might also be a consideration for athletes such as wrestlers and boxers who compete in weight classes.

Key Point

Some athletes may refrain from eating foods containing indigestible carbohydrates such as fiber and oligosaccharides prior to training and competition.

Carbohydrate During Exercise

The consumption of carbohydrate during exercise is common to many sports. Carbohydrate-containing sport drinks are certainly seen during endurance sports (in cycling bottles and on support tables during marathons) as well as on the sidelines of football games and on the ice hockey bench. The provided carbohydrate increases the availability of carbohydrate to working muscle fibers, which can have a positive influence on endurance performance. Evidence is also mounting that ingesting carbohydrate may improve performance in intermittently higher intensity sports such as football, ice hockey, and soccer.

The type of carbohydrate provided during exercise can increase its availability to working muscle tissue, making it more available for oxidation. For instance, glucose, maltose, sucrose, amylopectin, and maltodextrins are oxidized at higher rates than fructose, amylose, and galactose.[98] Thus, the former are more common than the latter in commercially available carbohydrate-based sport drinks and related semisolid and solid products. However, when

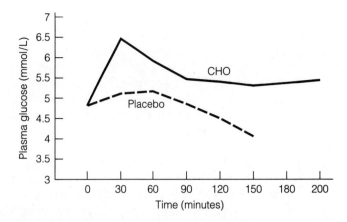

Figure 4-24
Glucose ingestion during endurance exercise can extend performance time till fatigue. The asterisk (*) indicates time point of fatigue. (CHO, carbohydrate)

these products contain a mixture of carbohydrate types to adjust their osmolality and taste, the importance of individual types is lessened. This means that someone using sport drinks and bars does not need to pay strict attention to the effect of individual carbohydrate types. The specific effects of different carbohydrate types are addressed below, and more information on the carbohydrate content of commercial sport drinks and bars is provided in Chapter 11.

ENDURANCE SPORTS   During lower intensity exercise (30% VO_2max) the ingestion of carbohydrate increases both insulin and glucose levels.[2] This results in a twofold increase in skeletal muscle glucose uptake and utilization and a decreased use of plasma FFA as an energy source.[99] However, carbohydrate ingestion during moderate-intensity (50–75% VO_2max) exercise does not result in such dramatic increases in blood glucose and insulin levels. This may be explained mostly by an increased rate of blood glucose utilization in conjunction with a blunted insulin response relative to the exercise intensity.

Carbohydrate ingestion during prolonged endurance exercise can enhance performance (Figure 4-24).[100,101] This is one of the founding concepts of the sport drink industry. It is logical to assume that if more glucose is made available to working muscle tissue, it could make a greater contribution to the total carbohydrate utilization. This process would decrease the dependence on muscle glycogen and thus slow the rate at which it is utilized. Simply said, ingesting carbohydrate during exercise might "spare" muscle glycogen. However, it is not clear whether this is indeed the case.

Researchers have determined that when blood glucose levels are maintained at three times higher than a typical fasting blood glucose level, muscle glycogen utilization during intense exercise might be reduced by 20–25%.[102] However, to accomplish this level, the researchers needed to infuse glucose directly into circulation. Achieving this level of blood glucose via carbohydrate ingestion is not practical. Furthermore, when blood glucose levels were maintained at about two times fasting level, via infusion, there was no effect on the rate of glycogen utilization.[103] Other research studies have also suggested that ingesting carbohydrate during submaximal exercise does not affect the rate of muscle glycogen breakdown.[101,104] These studies were performed using cycling as the mode of exercise. It should be mentioned that the effects of ingesting carbohydrate on muscle glycogen breakdown in runners might be different. Studies involving treadmill running have suggested that carbohydrate ingestion may actually slow the rate of glycogen breakdown, especially in Type I fibers.[105,106] This discrepancy between the findings on glycogen sparing with carbohydrate consumption during exercise remains to be solved, but clearly it strengthens the case for focusing on the maximization of pre-exercise glycogen stores.

Carbohydrate ingestion during endurance exercise can extend performance time prior to fatigue.[101] Since the rate of muscle glycogen breakdown may or may not be affected, researchers have concluded that the most significant ergogenic impact of carbohydrate ingestion during endurance exercise is to maintain blood glucose levels (euglycemia) and to provide carbohydrate to muscle when glycogen levels are nearing depletion later in the bout.[15,103,107] Carbohydrate ingestion during exercise does seem to slow liver glycogen breakdown.[98]

Glucose uptake and oxidation may peak around 1.0–1.2 g/min late in exercise.[70,92] Even when higher amounts of glucose are consumed during exercise, glucose oxidation still plateaus.[92,108] An endurance athlete should strive to ingest 30–60 g of carbohydrate per hour of performance to maintain blood glucose levels and optimize glucose uptake and oxidation. Drinking 600–1200 ml of a 6–8% carbohydrate sport drink per hour provides the recommended level.

Glucose, sucrose, or maltodextrin are probably about the same with regard to enhancing performance; however, fructose may be a detriment by causing intestinal discomfort. Athletes need to be a little cautious not to overconsume carbohydrate by using a more concentrated sport drink or food or by drinking too much of a sport drink. Sport drinks and foods of higher carbohydrate content may slow water absorption and potentially lead to stomach and intestinal

Key Point

Ingesting carbohydrate during most sports with longer duration (sustained and intermittent efforts) may improve performance.

discomfort. In addition, even though glucose is absorbed in a sodium-dependent system, extra sodium is not necessary to maximize carbohydrate absorption.[109] There seems to be plenty of sodium in digestive secretions to assist glucose absorption.

HIGHER INTENSITY SPORTS   The potential ergogenic effects of carbohydrate ingestion during sports with intermittent bouts of extremely high intensity (such as weight training and sprinting) remain to be resolved but show promise. For instance, muscle (*vastus lateralis*) glycogen levels were measured prior to and after 4 hours of lower intensity exercise (50% VO$_2$max) with intermittent high-intensity bouts of exercise at 100% VO$_2$max for 30 seconds each.[110] Throughout the session, participants received a carbohydrate/energy source (43 g sucrose, 9 g fat, and 3 g protein) or a placebo. At the end of the exercise trial the individuals receiving the carbohydrate-containing drink had approximately 22% of their pre-exercise muscle glycogen while the placebo group had only 12%. That is, those receiving carbohydrate used 11% less glycogen during the 4-hour exercise trial. It was speculated that in between the high-intensity bouts, the ingested carbohydrate and resulting hyperglycemia could have led to some glycogen synthesis, thereby reducing the net glycogen utilization. Experimental designs such as these may provide some insight into muscle glycogen metabolism and the influence of sport drinks during sports such as soccer, ice hockey, and lacrosse.

Carbohydrate After Exercise

Eating after a training session is extremely important to maximize the beneficial physiological effects of the exercise through repair and adaptation operations. In addition, the postexercise meal provides fuel resources to begin to recover energy stores utilized during the exercise. These provide two primary reasons for a postexercise meal.

Endurance athletes training till exhaustion or near exhaustion on successive days are likely to require a carbohydrate contribution of at least 60% of their total energy intake.[111] When an exercise bout depletes muscle glycogen stores, it can take up to 24 hours to replenish these stores. The rate of muscle glycogen recovery may be roughly 7–8% for the first 1–2 hours and then slow to about 5% per hour thereafter. Therefore an endurance athlete training or competing to near or complete exhaustion of glycogen stores might not achieve complete recovery of muscle glycogen at the onset of exercise on the subsequent day. This recovery rate also provides one reason why heavy resistance training sessions focusing on specific muscle groups should be separated by a couple of days.

The rate of glycogen recovery is more rapid after complete or near-complete exhaustion. This is partly explained by glycogen synthase. Simply stated, the greater the glycogen depletion, the higher the activity of glycogen

Key Point

Appropriate carbohydrate consumption after exercise is necessary for maximal recovery of muscle glycogen stores.

synthase in response to postexercise carbohydrate intake. This also helps explain why the rate of glycogen recovery is greater in the first hour or so after an exhaustive exercise bout. Furthermore, glucose transport and the sensitivity of muscle to insulin remain increased during this time. This allows for more glucose to be available in recovering muscle at the same time that glycogen synthesis is more potent.

It is important for athletes to consume a carbohydrate-containing meal as quickly as possible after endurance efforts. Higher glycemic index foods may allow for a more rapid glycogen recovery versus lower glycemic index foods. Also, several sport drinks are available with a higher carbohydrate content than recommended during exercise. High-carbohydrate loading and recovery drinks that may be 20% carbohydrate by weight are marketed for rapid muscle glycogen restoration as well as part of carbohydrate loading regimens. Because it would substitute for other carbohydrate foods, this product is fortified with B-complex and C vitamins that naturally occur in these foods and are important for carbohydrate metabolism. Chapter 7 provides greater detail about designing a postexercise meal as a component of an athlete's diet.

LEVEL OF DIETARY CARBOHYDRATE   There may be an upper level for carbohydrate intake with regard to maximizing glycogen recovery. A carbohydrate intake greater than 500–600 g might not make an additional contribution to the recovery of glycogen stores for typical endurance athletes involved in heavy training. Also, lower levels of dietary carbohydrate might lead to lesser glycogen stores. Therefore, it is important for endurance athletes to gauge their carbohydrate intake as both a percentage of total energy and grams of carbohydrate provided. For instance, a runner eating 3500 kcal with 65% of the energy derived from carbohydrate would get about 568 g of carbohydrate a day. However, if only 50% of total energy is provided by carbohydrate, 438 g would have been provided. For higher energy expenditure, a lesser percentage of energy can be provided as carbohydrate and still achieve an appropriate carbohydrate quantity. For instance, a runner who requires 4000 kcal a day might be able to recover glycogen stores when only 50% of the energy is derived from carbohydrate.

Based on body weight, a general recommendation for carbohydrate intake is 1.5 g/kg of body weight during the first 30 minutes and again every 2–3 hours thereafter, which may allow for maximal glycogen recovery during the hours after exhaustive exercise.[112,113] However, many factors individualize requirements, including the absolute quantity of muscle glycogen depletion during exercise based on skeletal muscle mass differences and type of exercise.

POSTEXERCISE PROTEIN AND GLYCOGEN RECOVERY   It has been speculated that the addition of protein might increase the efficiency of postexercise glycogen recovery in response to diet carbohydrate. This notion is supported by some research. For instance, researchers depleted leg muscle glycogen stores of athletes by having them cycle for 2.5 hours.[114] They were then provided either carbohydrate alone or carbohydrate plus protein to promote glycogen recovery. After 4 hours, muscle glycogen recovery was greater with the protein supplement to carbohydrate. However, when a relatively high amount of carbohydrate was provided after glycogen depletion exercise, additional protein did not further enhance glycogen recovery.[115]

CARBOHYDRATE AND PROTEIN TURNOVER   As discussed in Chapter 5, net protein balance reflects an algebraic balance between degradation and synthesis. Although muscle protein balance may be positive for as much as a day after a training session, within the first few hours the balance may actually be zero or even negative. This suggests that early after hard exercise training, muscle protein degradation can equal or exceed protein synthesis. In an effort to minimize postexercise protein degradation it is important to eat some carbohydrate along with protein (essential amino acids) immediately following exercise. A meal containing higher glycemic index foods may also be beneficial to evoke a greater insulin response, which is an important anabolic factor.

■ CARBOHYDRATE BEFORE, DURING, AND AFTER EXERCISE *IN REVIEW*

- It is common for individuals to eat or drink carbohydrate-containing foods before and during exercise for practical and ergogenic reasons.
- Carbohydrate intake 4–6 hours prior to endurance exercise can enhance endurance performance.
- Carbohydrate intake 30–60 minutes prior to exercise can result in hyperglycemia and hyperinsulemia at the onset of exercise and hypoglycemia early in exercise. The effect on performance may be individualized and influenced by the level of intensity.
- Carbohydrate ingested just before exercise has the same effect as carbohydrate ingested during exercise.
- Carbohydrate ingested during endurance exercise can prolong endurance performance by providing a carbohydrate resource later in exercise.
- Carbohydrate should be ingested immediately after completion of exhaustive exercise and every couple of hours thereafter to promote efficient recovery.

Conclusions

Plants and many prepared foods as well as milk and dairy products provide dietary carbohydrates. Glucose is the most significant absorbed carbohydrate type and is a primary fuel for all cells. The level of circulating glucose influences the level of the metabolic hormones, namely insulin and glucagon. At moderate submaximal intensity and above, insulin release is blunted and epinephrine is elevated, which allow for a breakdown of muscle glycogen. Circulating epinephrine and intracellular substances (Ca^{2+}, P_i, AMP, IMP) increase glycogen breakdown in active muscle fibers. Depletion of muscle glycogen is associated with muscular fatigue, and hypoglycemia is associated with central fatigue.

Higher carbohydrate diets are recommended for all athletes. In addition, glycogen loading protocols can help maximize muscle glycogen stores at the onset of endurance competition. Providing a carbohydrate source during endurance exercise can extend performance time before exhaustion. Postexercise carbohydrate is needed to recover glycogen stores and promote a faster return to a positive protein balance.

 **IN FOCUS**   EATING ENOUGH CARBOHYDRATE

Eating a high-carbohydrate diet is important to optimize performance in many sports. This is especially true of endurance sports and probably true for weight-training athletes. Estimating how much carbohydrate a diet contains is fairly easy. One method is to estimate the amount of carbohydrate needed (Table A) and then use the Exchange List System in the Appendix (with the information in Tables A and B here) to plan for carbohydrate consumption. This promotes the consumption of a variety of carbohydrate sources including fruits and vegetables.

Two other ways to calculate the carbohydrate component of the diet are to consult a registered dietitian or to use food analysis software. A registered dietitian can be found locally in the telephone book or by using the "Find a Dietitian" service offered by the American Dietetic Association (www.eatright.org). A program has also been made available by the USDA (www.usda. gov/cnpp/). In either case a food diary will be needed (see Chapter 1). A food diary is a log of all the foods consumed during the days to be assessed. Meticulous effort is needed in recording the type of food and quantity of food eaten along with any preparation or cooking considerations.

Table A

A Carbohydrate Contributions of Food Groups to Provide 50–70% of Total Dietary Energy

	Total Diet Energy (kcal)				
	2000	*2500*	*3000*	*3500*	*4000*
Energy from carbohydrate (50–70%)	1000–1400	1250–1750	1500–2100	1750–2450	2000–2800
Total grams of carbohydrate	250–350	310–440	375–525	438–612	500–700
Distribution of Carbohydrate					
Starch group	10–14	12–16	14–18	16–20	18–22
Fruit group	3–4	4–6	5–7	6–8	7–9
Milk group	2–3	3–4	4–5	5–6	6–7
Vegetable group	3–4	4–5	5–6	6–7	7–8
Other carbohydrate[a]	2–3	2–3	3–4	3–4	4–5

[a]Carbohydrate may be derived from sport bars and drinks utilized during training and/or recovery.

EATING ENOUGH CARBOHYDRATE (CONTINUED)

| Table B | Approximate Carbohydrate Content of Grouped Foods |

Carbohydrate Food Group	Grams of Carbohydrate	Foods	Approximate Energy (kcal)
Starch (foods with higher starch content)	15	1 slice bread	80
		¾ c ready-to-eat cereal	
		½ c cooked pasta	
		½ c cooked beans	
		½ c cooked corn, peas, or yams	
		1 small (3 oz) potato	
		½ bagel, English muffin, or bun	
		1 tortilla, waffle, roll, taco, or matzo	80
Fruit	15	1 small banana, nectarine, apple, or orange	60
		½ grapefruit, pear, or papaya	
		17 small grapes	
		½ cantaloupe or 1 c cubed cantaloupe	
		2 tbs raisins	
		½ c orange, apple, or grapefruit juice	
		1½ c dried figs	60
Milk	12	1 c milk	48
		¾ c nonfat or low-fat yogurt	
		⅓ c nonfat dry milk	48
Vegetables	5	½ c cooked carrots, greens, green beans, brussels sprouts, beets, broccoli, cauliflower, or spinach	
		1 c raw carrots, radishes, or salad greens	
		1 large tomato	25
Other carbohydrates	15	2 small cookies	Varies
		1 small brownie or piece of cake	
		5 vanilla wafers	
		1 granola bar soda (6 oz)	
		sport drink (8 oz)	Varies

Note: Red meats, chicken, fish, eggs, plant oils, and animal fats are carbohydrate void.

IN FOCUS   NUTRITION STATES AND ENERGY METABOLISM

Athletes and active people alike need to understand the basic concepts of **nutrition state** as it determines the net impact on energy stores. In general, energy stores are constantly in flux between net gain and loss, based on the timing and composition of a person's last meal, and hormones provide most of the regulation. Nutrition state strongly influences energy metabolism during physical activity as well as the recovery of energy nutrient stores after exercise. In addition, nutrition state is a foremost consideration for manipulation of body weight and composition, as discussed in Chapter 7.

The human body is often said to be in either a *fed* or a *fasting* state. However, the initiation and cessation of these states is more analogous to a "dimmer switch," which results in an overlap in these opposing activities, rather than a simple on or off mechanism. For instance, the effects of fasting persist for approximately a half-hour after eating a meal. Likewise, the transition to a fasting state occurs while the operations

NUTRITION STATES AND ENERGY METABOLISM (CONTINUED)

of a fed state slowly wane. Therefore contrasting metabolic pathways can operate simultaneously within certain tissues, especially the liver, skeletal and cardiac muscle, and adipose tissue. Although at first glance this may seem futile, it has several advantages, such as aggressive glycogen recovery early in a fed state. This feature addresses phases of nutrition states and how they might influence exercise energy metabolism and vice versa.

Fed State

The fed state begins with the ingestion of food items. The liver removes most of the absorbed galactose and fructose as well as a significant portion of the glucose from the hepatic portal vein. However, much of the glucose enters the general circulation and becomes available to other tissue. Depending on the type and size of a

meal and other factors, the concentration of blood glucose can easily increase to more than 140 mg/100 ml.

As displayed in Figure A, insulin is released from the pancreas relative to the elevated level of circulating glucose. Insulin promotes the uptake of glucose into skeletal muscle fibers and fat cells (adipocytes) by increasing the presence of GluT4 on their plasma membranes. Because skeletal muscle and fat tissue combined may account for more than half of the total body mass, the net effect is a fairly rapid return to the fasting level of blood glucose. Increased GluT4 uptake of glucose by the heart also makes a contribution in reducing blood glucose levels.

The liver (hepatocytes) takes up more glucose per gram of tissue than any other tissue. This reflects the action of GluT2 and glucokinase and increased processing of glucose in these cells (glycogen synthesis and glycolysis).

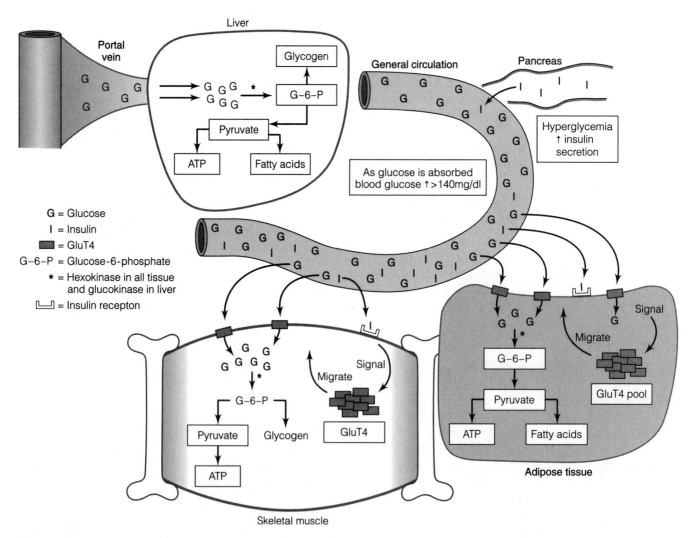

Figure A
Glucose absorption and metabolism.

NUTRITION STATES AND ENERGY METABOLISM (CONTINUED)

Early Fed State

During the early fed state, which is the first half hour of a fed state, the effects of the previous fast persist. Some of the newly arriving energy nutrients can actually be metabolized as if the tissue were still in the fasting state. Although this may seem inefficient, it allows for an expedient recovery of liver and muscle glycogen stores. This is because during the fast, fatty acids served as the primary fuel source in these tissues and are still being used in the early fed state, but to a lessening degree. When glucose enters these cells, the levels of ATP and citrate, a Krebs cycle intermediate, are high, and both of them inhibit phosphofructokinase (PFK) in muscle and liver cells (Figure B). This allows more glucose to be directed toward glycogen production. In addition, pyruvate dehydrogenase may be partly inhibited, which may increase the production of lactate in skeletal muscle in a very early fed state (Figure B). Some of this lactate can circulate to the liver and serve as a precursor for gluconeogenesis and glycogen synthesis. It is therefore possible for some of the absorbed glucose to become part of liver glycogen after going through glycolysis in skeletal muscle.

Engaged Fed State

After a half hour or so, the hormonal and enzymatic activities associated with fasting have been minimized and tissue becomes fully engaged in a fed state. At this time, fatty acid oxidation is minimized in the liver and skeletal muscle and the associated inhibition of glycolysis and pyruvate dehydrogenase is removed. More glucose can now flux through glycolysis to pyruvate. Pyruvate can then be further metabolized in mitochondria. Here again the ATP and citrate levels produced by pyruvate utilization slow PFK activity to a degree that allows continued (but slower) glycogen production.

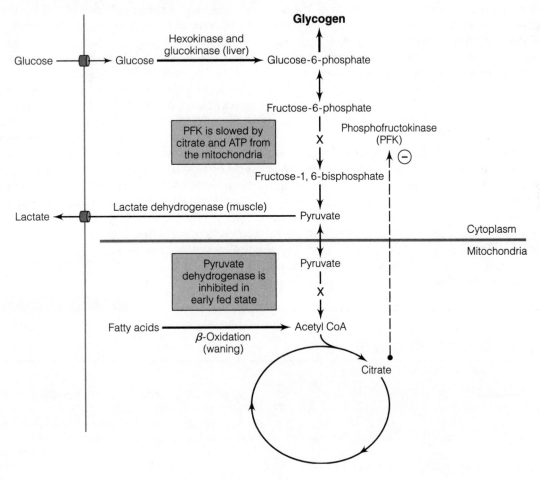

Figure B

Glycogen formation in an early fed state. Inhibition of glycolysis and pyruvate dehydrogenase allow for greater glycogen formation in muscle and the liver.

NUTRITION STATES AND ENERGY METABOLISM (CONTINUED)

In skeletal muscle the increased glucose uptake in a fed state slows as glycogen stores are recovered. Once glycogen stores are maximal, glucose uptake tends to reflect immediate energy needs of the muscle fiber. Therefore, a person would have a greater potential for glucose uptake in a fed state if he or she recently completed a bout of exercise, producing a greater need for glycogen recovery. In addition, exercise training can enhance the glycogen storage potential of muscle. These two factors help us understand some of the differences between active and sedentary people in **glucose tolerance** (the ability to appropriately reestablish euglycemia following meal-induced hyperglycemia).

Fasting State

The conversion from a fed state to a fasting state is not immediate. Insulin levels decrease as circulating glucose levels return to a fasting level. However, the effect of insulin on metabolic pathways can linger during the early transition to a fasting state. The later transition to an adaptive fasting state (starvation) involving marked gluconeogenesis and ketogenesis is also not immediate. This transition period is largely attributable to the short-lived involvement of liver glycogen in maintaining blood glucose levels.

Postabsorptive Fasting State

The postabsorptive or early fasting phase refers to the period when carbohydrate is no longer being absorbed and blood glucose has returned to a fasting level. For a brief time, glucose remains the predominant fuel of all tissue. This explains why blood glucose levels may actually dip below normal fasting levels for several minutes (Figure 4-6).

Glucagon is released from the pancreas as blood glucose levels decline. Glucagon promotes immediate glycogen breakdown in the liver and begins to increase the formation of glucose (gluconeogenesis). For the first few hours glycogen provides most of the glucose that is released into the blood by the liver. However, as time goes on, more and more of the glucose released into the blood is derived from gluconeogenesis.

As the fast continues, the levels of cortisol and epinephrine also increase. Both hormones support glycogen breakdown in the liver. Epinephrine and cortisol also promote the breakdown of glycogen and protein, respectively, in skeletal muscle cells. The glycogen-derived glucose-6-phosphate and glucose, along with some of the protein-derived amino acids, help fuel skeletal muscle, and other amino acids can circulate to the liver for glucose production.

Adaptive Fasting State (Starvation)

Liver glycogen stores are limited and can be nearly exhausted within the first day of fasting. Once this occurs, all glucose released into the blood by the liver is derived from gluconeogenesis. Amino acids from skeletal muscle are the greatest contributors to gluconeogenic efforts, followed by lactate and glycerol. Meanwhile, fatty acids derived from adipose tissue are the primary fuel source for the liver and muscle during fasting. The acetyl CoA derived from fatty acid oxidation becomes excessive and promotes the formation of ketone bodies in the liver. Ketone bodies can then leave the liver and serve as an alternative fuel source for muscle and other tissue. During this time the brain also utilizes ketone bodies for fuel. In fact, roughly half of the brain's energy needs can be met by ketone bodies after a few days of an adaptive fast. This is a survival mechanism to decrease the need for glucose and thus slow down the loss of body protein.

STUDY QUESTIONS

1. Describe the different types of carbohydrates.
2. How are starch and glycogen similar and how are they different?
3. What are the differences between digestible and indigestible carbohydrate types?
4. Describe the processes involved in carbohydrate absorption.
5. What are GluTs, and which GluT is responsive to insulin and exercise?
6. What tissues in the body have a significant amount of glycogen?
7. How do the muscle fiber types differ in their glycogen content?
8. What are the major regulatory factors involved in glycogen breakdown at rest and during exercise?
9. What is fatigue, and how is carbohydrate involved in fatigue during physical activity?
10. What are recommendations for carbohydrate before and during exercise?
11. Beyond carbohydrate, what other energy nutrient may be important for efficient glycogen recovery?

REFERENCES

1. Foster-Powell K, Holt SH, Brand-Miller JC. International table of glycemic index and glycemic load values: 2002. *American Journal of Clinical Nutrition* 76(1):5–56, 2002.

2. Wilkinson JG, Liebman M. Carbohydrate metabolism in sport and exercise. In: *Nutrition in Exercise and Sport*, 3rd ed. (Wolinsky I, ed.), Boca Raton, FL: CRC Press, 1998.

3. Hollenbeck CB, Coulston AM. The clinical utility of the glycemic index and its application to mixed meals. *Canadian Journal of Physiology and Pharmacology* 69:100–107, 1991.

4. Greenhaff PL, Soderlund K, Re J-M, Hultman E. Energy metabolism in single human muscle fibers during intermittent contraction with occluded circulation. *Journal of Physiology* 460:443–453, 1993.

5. Soderlund K, Greenhaff PL, Hultman E. Energy metabolism in type I and type II human muscle fibers during short term electrical stimulation at different frequencies. *Acta Physiologica Scandinavica* 144:15–22, 1992.

6. Spiret LL. Anaerobic metabolism during high-intensity exercise. In: *Exercise Metabolism* (Hargreaves M, ed.), Champaign, IL: Human Kinetics, 1994.

7. Greenhaff PL, Ren, J-M, Soderlund K, Hultman E. Energy metabolism in single human muscle fibers during contraction without and with epinephrine infusion. *American Journal of Physiology* 260:E713–718, 1991.

8. Soderlund K, Hultman E. ATP content in single fiber from human skeletal muscle after electrical sktimulation and during recovery. *Acta Physiologica Scandinavica* 139:459–466, 1990.

9. Tesch P, Thorsson A, Fujitsuka N. Creatine phosphate in fiber types of skeletal muscle before and after exhaustive exercise. *Journal of Applied Physiology* 66:1756–1759, 1989.

10. Eriksson BO, Karlsson J, Saltin B. Muscle metabolites during exercise in pubertal boys. *Acta Paediatrica Scandinavica Supplement* 217:154–157, 1971.

11. Eriksson BO, Gollnick PD, Saltin B. Muscle metabolism and enzyme activities after training in boys 11–13 years old. *Acta Physiologica Scandinavica* 87(4):485–497, 1973.

12. Brooks GA. Physical activity and carbohydrate metabolism. In: *Exercise, Fitness, and Health: A Consensus of Current Knowledge* (Bouchard C, Shephard RJ, Stephens T, Sutton JR, McPherson BD, eds.), Boston, MA: Blackwell Science, 1990.

13. Bergstromn J, Hultman E. A study of the glycogen metabolism during exercise in man. *Scandinavian Journal of Clinical and Laboratory Investigation* 19:218–228, 1967.

14. Hermansen L, Hultman E, Saltin B. Muscle glycogen during prolonged severe exercise. *Acta Physiologica Scandinavica* 71:129–139, 1967.

15. Hargreaves M. Skeletal muscle carbohydrate metabolism during exercise. In: *Exercise Metabolism* (Hargreaves M, ed.), Champaign, IL: Human Kinetics, 1995.

16. Sanders CA, Levinson GE, Abelman WH, Freinkel N. Effect of exercise on the peripheriperal utilization of glucose. *New England Journal of Medicine*. 271:220–225, 1964.

17. Wahren J, Idstrom JP, Ahlborg G, Jorfeldt J. Glucose metabolism during leg exercise in man. *Journal of Clinical Investigation* 50:2715–2725, 1971.

18. Constantin-Teodosiu D, Cederblad G, Hultman E. PDC activity and acetyl group acclimation in skeletal muscle during prolonged exercise. *Journal of Applied Physiology* 73:2403–2407, 1992.

19. Coyle EF, Coggan AR, Hemmert MK, Ivy JL. Muscle glycogen utilization during prolonged strenuous exercise when fed carbohydrate. *Journal of Applied Physiology* 61:165–172, 1986.

20. Sahlin K, Katz A, Broberg S. Tricarboxylic acid cycle intermediates in human muscle during prolonged exercise. *American Journal of Physiology* 259:C834–841, 1990.

21. Hultman E. *Physiological role of muscle glycogen in man, with special reference to exercise. Physiology of muscular exercise.* Monograph 15. New York: American Heart Association, 1967:I99–I112. Circulation RC 681 A105, 1967.

22. Romijn JA, Coyle EF, Sidossis LS, Gastaldelli A, Horowitz JF, Endert E, Wolfe RR. Regulation of endogenous fat and carbohydrate metabolism in relation to exercise intensity and duration. *American Journal of Physiology* 265:E380–E391, 1993.

23. Rankin J. Dietary carbohydrate and performance of brief, intense exercise. GSSI *Sports Science Exchange* 13(4):1–4, 2000.

24. Coleman EJ. Carbohydrate—The master fuel. In: *Nutrition for Sport and Exercise*, 2nd ed. (Berning JR, Steen SN, eds.), Gaithersburg, MD: Aspen, 1998.

25. Gollnick PD, Piehl K, Saltin B. Selective glycogen depletion pattern in human muscle fibers after exercise of varying intensity and at varying pedalling rates. *Journal of Physiology* 241:45–57, 1974.

26. Saltin B, Gollnick PD. Fuel for muscular exercise: role of carbohydrate. In: *Exercise, Nutrition and Energy Metabolism* (Horton ES, Terjung RL, eds.), New York: Macmillan, 1988.

27. Gaitanos GC, Williams C, Boobis LH, Brooks S. Human muscle metabolism during intermittent maximal exercise. *Journal of Applied Physiology* 75:712–719, 1993.

28. Esbjornsson-Liljedahl M, Sundberg B, Norman B, Jansson E. Metabolic response in type I and type II muscle fibers during 30-s cycling sprint in men and women. *Journal of Applied Physiology* 87:1326–1332.

29. Tarnopolsky MA. Gender differences in metabolism; nutrition and supplements. *Journal of Science and Medicine in Sport* 3(3):287–298, 2000.

30. Kjaer M, Kiens B, Hargreaves M, Richter EA. Influence of active muscle mass on glucose homeostasis during exercise in humans. *Journal of Applied Physiology* 71:552–557, 1991.

31. Kjaer M. Hepatic fuel metabolism during exercise. In: *Exercise Metabolism* (Hargreaves M, ed.), Champaign, IL: Human Kinetics, 1995.

32. Bjorkman O, Eriksson LS. Splanchnic glucose metabolism during leg exercise in 60-hour fasted human subjects. *American Journal of Physiology* 245:E443–448, 1983.

33. Romijn JA, Coyle EF, Sidossis LS, Rosenblatt J, Wolfe RR. Substrate metabolism during different exercise intensities in endurance-trained women. *Journal of Applied Physiology* 88(5):1707–1714, 2000.

34. Coggan AR. Plasma glucose metabolism during exercise in humans. *Sports Medicine* 11:102–110, 1991.

35. Hargreaves M, Finn JP, Withers JA, Halbert JA, Scroop GC, MacKay M, Snow RJ, Carey MF. Effect of muscle glycogen availability on maximal exercise performance. *European Journal of Applied Physiology* 75:188–192, 1997.

36. MacDougall J, Ray S, McCartney N, Sale D, Lee P, Garner S. Substrate utilization during weightlifting. *Medicine and Science in Sports and Exercise* 20:S66, 1988.

37. Roberts RA, Pearson DR, Costill DL, Fink WJ, Pascoe DD, Benedict MA, Lambert CP, Zachweija JJ. Muscle glycogenolysis during differing intensities of weight-resistance exercise. *Journal of Applied Physiology* 70(4):1700–1706, 1991.

38. Tesch PA, Poultz-Snyder LL, Ystrom L, Castro MJ, Dudley GA. Skeletal muscle glycogen loss evoked by resistance training. *Journal of Strength & Conditioning Research* 12:67–73, 1998.

39. Spriet LL, Lindinger MI, McKelvie RS, Heigenhauser GJ, Jones NL. Muscle glycogenolysis and H$^+$ concentration during maximal intermittent cycling. *Journal of Applied Physiology* 66(1):8–13, 1989.

40. Sherman WM, Wimer GS, Insufficient dietary carbohydrate during training: does it impair athletic performance? *International Journal of Sport Nutrition* 1:28–36, 1991.

41. Price TB, Rothman DL, Avison MJ, Buonamico P, Shulman RG. 13C-NMR measurements of muscle glycogen during low intensity exercise. *Journal of Applied Physiology* 70, 1836–1842, 1991.

42. Price TB, Taylor R, Mason GF, Rothman DL, Shulman GI, Shulman RG. Turnover of human muscle glycogen with low-intensity exercise. *Medicine and Science in Sport and Exercise* 26:983–989, 1991.

43. Coyle EF, Hagberg JM, Hurley BF, Martin WH, Ehsani AA, Hooloszy JO. Carbohydrate feeding during prolonged strenuous exercise can delay fatigue. *Journal of Applied Physiology* 55:230–240, 1980.

44. Brooks GA, Butterfield GE, Wolfe RR, Groves BM, Mazzaeo RS, Sutton JR, Worfel EE, Reeves JT. Increased dependence on blood glucose after acclimatization to 4,300 m. *Journal of Applied Physiology* 70:919–927, 1991.

45. Ploug T, Wojtaszewski J, Kristiansen S, Hespel P, Galbo H, Richter EA. Glucose transport and transporters in muscle giant vesicles: differential effects of insulin and contractions. *American Journal of Physiology* 264:E270–278, 1993.

46. Douen AG, Ramlal T, Cartee GD, Klip A. Exercise modulates the insulin-induced translocation of glucose transporters in rat skeletal muscle. *FEBS Letters* 261:256–260, 1990.

47. Douen AG, Ramlal T, Klip A, Young DA, Cartee GD, Holloszy JO. Exercise-induced increase in glucose transporters in plasma membranes of rat skeletal muscle. *Endocrinology* 124:449–454, 1989.

48. Goodyear LJ, Hirshman MF, King PA, Horton ED, Thompson CM, Horton ES. Skeletal muscle plasma membrane glucose transport and glucose receptors after exercise. *Journal of Applied Physiology* 68:193–198, 1990.

49. Goodyear LJ, King PA, Hirshman MF, Thompson CM, Horton ED, Horton ES. Contractile activity increases plasma membrane glucose transporters in the absence of insulin. *American Journal of Physiology* 258:E667–672, 1990.

50. Constable SH, Favier RJ, Cartee GD, Young DA, Holloszy JO. Muscle glucose transport: interactions of in vitro contractions, insulin and exercise. *Journal of Applied Physiology* 64:2329–2332, 1988.

51. Nesher R, Karl IE, Kipnis DM. Dissociation of effects of insulin and contraction on glucose transport in rat epitrochleris muscle. *American Journal of Physiology* 249:C226–232, 1985.

52. Douen AG, Ramlal T, Rastogi S, Bilan PJ, Cartee GD, Vranic M, Hollszy JO, Klip A. Exercise induces recruitment of the "insulin-responsive glucose transporter": evidence for distinct intracellular insulin- and exercise-recruitable transporter pools in skeletal muscle. *Journal of Biological Chemistry* 265:13427–13430, 1990.

53. Ploug T, Galbo H, Ohkuwa T, Tranum-Jensen J, Vinten J. Kinetics of glucose transport in rat skeletal muscle membrane vesicles: effects of insulin and contractions. *American Journal of Physiology* 262:E700–711, 1992.

54. Wasserman DH, Geer RJ, Rice DE, Bracy D, Flakoll PJ, Brown LL, Hill JO, Abumrad NN. Interactions of exercise and insulin action in humans. *American Journal of Physiology* 260:E37–45, 1991.

55. Hollszy JO, Constable SH, Young DA. Activation of glucose transport in muscle by exercise. *Diabetes/Metabolism Reviews* 1:409–423, 1986.

56. Young JH, Gulve EA, Hollszy JO. Calcium stimulates glucose transport in skeletal muscle by a pathway independent of contraction. *American Journal of Physiology* 260:C555–561, 1991.

57. Jansson E, Hjemdahl P, Kaijser L. Epinephrine-induced changes in muscle carbohydrate metabolism during exercise in male subjects. *Journal of Applied Physiology* 60:1466–1470, 1986.

58. Hartling OJ, Trap-Jensen JP. Stimulation of β-adrenoreceptors in the exercising human forearm. *Clinical Physiology* 2:363–371, 1982.

59. Raz I, Katz A, Spencer MK. Epinephrine inhibits insulin-mediated glycogenesis but enhances glycolysis in human skeletal muscle. *American Journal of Physiology* 260:E430–435, 1991.

60. Ruby BC. Gender differences in carbohydrate metabolism: rest, exercise, and post exercise, In: *Gender differences in metabolism* (Tarnopolsky M, ed.), Boca Raton, FL: CRC Press, 1999.

61. Hespel P, Ricter EA. Mechanism linking glycogen concentration and glycogenolytic rate in perfused contracting rat skeletal muscle. *Biochemistry Journal* 284:777–780, 1992.

62. Galbo H, Holst JJ, Christensen NJ. The effect of different diets and of insulin on the hormonal response to prolonged exercise. *Acta Physiologica Scandinavica* 107:19–32, 1979.

63. Gollnick PD, Pernow B, Essen B, Jansson E, Saltin B. Availability of glycogen and plasma FFA for substrate utilization in leg muscle of man during exercise. *Clinical Physiology* 1:27–42, 1981.

64. Gollnick PD, Piehl K, Saubert CW, Armstrong RB, Saltin B. Diet, exercise and glycogen in human muscle fibers. *Journal of Applied Physiology* 33:421–425, 1972.

65. Hespel P, Richter EA. Glucose uptake and transport in contracting, perfused rat muscle with different pre-contraction glycogen concentrations. *Journal of Physiology* 427:347–359, 1990.

66. Hargreaves M, McConell G, Proietto J. Influence of muscle glycogen on glycogenolysis and glucose uptake during exercise. *Journal of Applied Physiology* 78:288–292, 1995.

67. Bergstrom J, Hermansen L, Hultman E, Saltin B. Diet, muscle glycogen and physical performance. *Acta Physiologica Scandinavica* 71:140–150, 1967.

68. Entman ML, Keslensky SS, Chu A, Van Winkle WB. The sarcoplasmic reticulum-glycogenolytic complex in mammalian fast-twitch skeletal muscle. *Journal of Biological Chemistry* 255:6245, 1980.

69. Mayes PA. Metabolism of glycogen. In: *Harper's Biochemistry* 24th ed. (Murray RK, Granner DK, Mayes PA, Rodwell VW, eds.), Stamford, CT: Appleton and Lange, 1996.

70. Omitted in proof.

71. MacDougall JD, Ward GR, Sale DG, Sutton JR. Biochemical adaptation of human skeletal muscle to heavy resistance training and immobilization. *Journal of Applied Physiology* 43:700–703, 1977.

72. Boobis LH, Williams C, Wooton SA. Influence of sprint training on muscle metabolsim during brief maximal exercise in man. *Journal of Physiology London* 342:36P–37P, 1983 (Abstract).

73. Hermansen L, Hultman E, Saltin B. Muscle glycogen during prolonged severe exercise. *Acta Physiologica Scandinavica* 71(2):129–139, 1967.

74. Henriksson J. Training induced adaptation of skeletal muscle and metabolism during submaximal exercise. *Journal of Physiology* 270(3):661–675, 1977.

75. Saltin B, Nazar K, Costill DL, Stein E, Jansson E, Essen B, Gollnick D. The nature of the training response; peripheral and central adaptations of one-legged exercise. *Acta Physiologica Scandinavica* 96(3):289–305, 1976.

75a. Coggan AR, Williams BD. Metabolic adaptations to endurance training: substrate metabolism during exercise. In: *Exercise Metabolism* (Hargreaves M, ed.), Champaign, IL: Human Kinetics, 1995.

76. Wilmore JH, Costill DL. Metabolic adaptations to training. In: *Physiology of Exercise and Sport*, Champaign, IL: Human Kinetics, 1994.

77. Coyle EF. Substrate utilization during exercise in active people. *American Journal of Clinical Nutrition* 61(suppl.): 968S–973S, 1995.

78. Costill DL, Coyle E, Dalsky G, Evans W, Fink W, Hoopes D. Effects of elevated plasma FFA and insulin on muscle glycogen usage during exercise. *Journal of Applied Physiology* 43:695–699, 1977.

79. Koivisto VA, Karvonen S, Nikkila EA. Carbohydrate ingestion before exercise: comparison of glucose, fructose, and sweet placebo. *Journal of Applied Physiology* 51:783–790, 1981.

80. Foster C, Costill DL, Fink WJ. Effects of pre-exercise feeding on endurance performance. *Medicine and Science in Sports and Exercise* 11:1–13, 1979.

81. Keller K, Schwarzkopf R. Pre-exercise snacks may decrease exercise performance. *Physiology of Sports Medicine* 12:89, 1984.

82. Gleeson M, Maughan RJ, Greenhaff PL. Comparison of the effects of pre-exercise feedings of glucose, glycerol, and placebo on endurance and fuel homeostasis in man. *European Journal of Applied Physiology* 55:645–653, 1986.

83. Sherman WM, Peden MC, Wright DA. Carbohydrate feeding 1 hour prior to exercise improves cycling performance. *American Journal of Clinical Nutrition* 54:866–870, 1991.

84. Devlin JT, Calles-Escandon J, Horton ES. Effects of preexercise snack feeding on endurance cycle exercise. *Journal of Applied Physiology* 60(3):980–985, 1986.

85. Neufer PD, Costill DL, Flynn MG, Kirwan JP, Mitchell JP, Houmard J. Improvements in exercise performance: effects of carbohydrate feeding and diet. *Journal of Applied Physiology* 62:983–988, 1987.

86. Anantaraman R, Carmines A, Gaesser GA, Weltman A. Effects of carbohydrate supplementation on performance during 1 hour of high-intensity exercise. *International Journal of Sports Medicine* 16:461–465, 1995.

87. Gisolfi CV, Duchman SM. Guidelines for optimal replacement beverages for different athletic events. *Medicine and Science in Sport and Exercise* 24:679–687, 1992.

88. Omitted in proof.

89. Hargreaves M, Costill DL, Fink WJ, King DS, Fielding RA. Effect of pre-exercise carbohydrate feedings on endurance cycling performance. *Medicine and Science in Sports and Exercise* 19:33–40, 1987.

90. Brundle S, Thayer R, Taylor AW. Comparison of fructose and glucose ingestion before and during endurance cycling to exhaustion. *Journal of Sports Medicine and Physical Fitness* 40(4):343–349, 2000.

91. Ravich WJ, Bayless TM, Thomas M. Fructose: incomplete intestinal absorption in humans. *Gastroenterology* 84:26–32, 1983.

92. Hargreaves M. Metabolic response to carbohydrate ingestion: effects on exercise performance. In: *Perspectives in Exercise Science and Sport Medicine*, vol. 12, *The Metabolic Basis of Performance in Exercise and Sport* (Lamb DR, Murray R, eds.), Carmel, IN: Cooper, 1999.

93. Guezennec CY, Satabin P, Duforez F, Koziet J, Antoine JM. The role of type and structure of complex carbohydrates response to physical exercise. *International Journal of Sports Medicine* 14:224–230, 1993.

94. Thomas DE, Brotherhood JR, Brand JC. Carbohydrate feeding before exercise: effect of glycemic index. *International Journal of Sports Medicine* 12(2):180–186, 1991.

95. Jarvis JK, Pearson D, Oliner CM, Schoeller DA. The effect of food matrix on carbohydrate utilization during moderate exercise. *Medicine and Science in Sports and Exercise* 24: 320–326, 1992.

96. Paul GL, Rokusek JT, Dykstra GL, Boileau RA, Layman DK. Oat, wheat or corn cereal ingestion before exercise alters metabolism in humans. *Journal of Nutrition* 126(5):1372–1381, 1994.

97. Anderson M, Bergman EA, Nethery VM. Pre-exercise meal affects ride time to fatigue in trained cyclists. *Journal of the American Dietetic Association* 94:1152–1156, 1994.

98. Jeukendrup AE, Jentjens R. Oxidation of carbohydrate feedings during prolonged exercise: current thoughts, guidelines and directions for future research. *Sports Medicine* 29(6):407–24, 2000.

99. Coggan AR, Coyle EF. Carbohydrate ingestion during prolonged exercise: effects on metabolism and performance. *Exercise and Sport Science Reviews* 19:1–25, 1991.

100. Coggan AR, Coyle EF. Reversal of fatigue during prolonged exercise by carbohydrate infusion or ingestion. *Journal of Applied Physiology* 63:2388–2395, 1987.

101. Coyle EF, Coggan AR, Hemmert MK, Ivy JL. Muscle glycogen utilization during prolonged strenuous exercise when fed carbohydrate. *Journal of Applied Physiology* 61:165–172, 1986.

102. Bergstrom J, Hultman E. A study of the glycogen metabolism during exercise in man. *Scandinavian Journal of Clinical and Laboratory Investigation* 19:218–228, 1967.

103. Coyle EF, Hamilton MT, Gonzalez-Alonza J, Mountain SJ, Ivy JL. Carbohydrate metabolism during intense exercise when hyperglycemic. *Journal of Applied Physiology* 70:834–840, 1991.

104. Hargreaves M, Briggs CA. Effect of carbohydrate ingestion on exercise metabolism. *Journal of Applied Physiology* 65:1553–1555, 1988.

105. Tsintzas OK, Williams C, Boobis L, Greenhaff P. Carbohydrate ingestion and glycogen utilization in different muscle fibre types in man. *Journal of Physiology* 489:242–250, 1995.

106. Tsintzas OK, Williams C, Boobis L, Greenhaff P. Carbohydrate ingestion and single muscle fibre glycogen metabolism during prolonged running in men. *Journal of Applied Physiology* 81:801–809, 1996.

107. Coyle EF. Carbohydrate supplementation during exercise. *Journal of Nutrition* 122:788–794, 1992.

108. Wagenmakers AJM, Brouns F, Saris WHM, Halliday D. Oxidation rates of orally ingested carbohydrates during prolonged exercise in men. *Journal of Applied Physiology* 75:2774–2780, 1993.

109. Hargreaves M, Costill D, Burke L, McConell G, Febbraio M. Influence of sodium on glucose bioavailability during exercise. *Medicine and Science in Sports and Exercise* 26(3):365–368, 1994.

110. Hargreaves M, Costill DL, Coggan A, Fink WJ, Nishibata I. Effect of carbohydrate feedings on muscle glycogen utilization and exercise performance. *Medicine and Science of Sport and Exercise* 16:219–222, 1984.

111. Sherman W. Carbohydrates, muscle glycogen and muscle glycogen supercompensation. In: *Ergogenic Aids in Sports* (Williams MH, ed.), Champaign, IL: Human Kinetics, 1983.

112. Friedman JE, Neufer PD, Dohm GL. Regulation of glycogen resynthesis following exercise. *Sports Medicine* 11: 232–240, 1991.

113. Ivy JL. Carbohydrate supplementation for rapid muscle glycogen storage in the hours immediately after exercise. In: *The Theory and Practice of Athletic Nutrition: Bridging the Gap* (Grandjean AC, Storlie J, eds.), Report of the Ross Symposium, pp. 47–65, Columbus OH: Ross Laboratories, 1989.

114. Ivy JL, Goforth HW, Damon BM, McCauley TR, Parsons EC, Price TB. Early postexercise muscle glycogen recovery is enhanced with a carbohydrate-protein supplement. *Journal of Applied Physiology* 93(4):1337–1344, 2002.

115. Jentjens RL, van Loon LJ, Mann CH, Wagenmakers AJ, Jeukendrup AE. Addition of protein and amino acids to carbohydrates does not enhance postexercise muscle glycogen synthesis. *Journal of Applied Physiology* 91(2):839–846, 2001.

AMINO ACIDS, PROTEIN, AND EXERCISE

Personal Snapshot

Susan is a competitive distance runner who will train for and compete in several marathons during the upcoming year. Her diet consists mostly of carbohydrate-rich foods such as pasta, breads, fruits, and juices, as she is well aware of the importance of carbohydrate to her running performance. However, Susan is not sure how much protein she needs while maintaining such a high level of training and competition. She has heard that bodybuilders and football players require more protein than the average person, but because her physique is unlike their very muscular one, she wonders if the higher requirements apply to her. Beyond muscle mass development, are there other reasons why endurance athletes, such as Susan, should pay special attention to protein intake and body protein status? To answer some of her questions Susan seeks the assistance of a registered dietitian to help plan her diet and determine whether she needs more protein than recommended for the general population.

Chapter Objectives

- Introduce basic concepts related to protein and amino acids including food sources, digestion, and absorption.

- Discuss the metabolism of amino acids in different tissues and the influence of nutrition state on those processes.

- Describe how exercise influences the metabolism of protein and amino acids.

- Present a recommendation for protein intake for active people.

While carbohydrate has been the energy nutrient most commonly associated with endurance exercise, protein is most associated with weight training. In fact, protein has been the nutrient of primary interest to the weight training and bodybuilding community for decades. Very little has occurred over the past few decades of practice and research to lower protein from its exalted height. Why is this? Is it because these sports are associated with greater protein mass, and once water is accounted for, muscle tissue is largely protein?

People have long held protein in high esteem with regard to athletic performance. The interest seems as old as organized athletic competition itself. For instance, the ancient Greeks and the Romans had a strong interest in the relationship between protein and athletic performance. They believed that their athletes and warriors could achieve greater strength and endurance by eating the protein-rich flesh of animals that possessed such qualities. This early idea has endured, as suggested by recent polls of high school and college athletes and coaches who believe that performance can be improved by a high-protein diet and that protein intake is the most important nutritional factor in increasing muscle mass.

Today, sport nutritionists know that in comparison to the general population, individuals who train for strength and power (such as bodybuilders and football players) have elevated protein requirements that are related to the level of resistance training. Endurance athletes (such as distance runners and cyclists, and triathletes) also have elevated protein needs related to the level of training. For people in resistance training, protein requirements are increased to allow for an enhancement and maintenance of lean body mass (LBM). For endurance athletes much of the increased need can be attributed to an increased oxidation of amino acids. However, for both types of athletes, the positive influence of dietary protein probably has a ceiling. That is, once enough protein is provided in the diet to maximize and compensate for the physical training, additional protein will not provide further enhancement.

This chapter provides an overview of basic concepts involving amino acids and protein, including their basic structure and roles. Food sources and the digestion and absorption of protein and amino acids are also discussed. With that groundwork laid, the chapter then details the influence of exercise on protein metabolism as well as the protein requirements of physically active individuals. Many of the common theories, practices, and controversies are presented, including what is unknown about their validity and efficacy.

■ AMINO ACIDS AND PROTEIN BASICS

Protein is a major structural and functional component of the human body. What are proteins, how are they constructed, and what do they look like? Are all amino acids

used to make proteins? What makes amino acids different from other energy molecules? This section provides an overview of amino acids and protein that is needed to understand applied concepts of protein metabolism in response to physical activity discussed in later sections.

Amino Acids

Proteins are composed of amino acids. So, the most obvious role of amino acids is to serve as "building blocks" for proteins. However, some amino acids have independent functions or are used to make other important molecules such as neurotransmitters and hormones. How are the various amino acids similar and how are they different?

As shown in Figure 5-1, all amino acids contain a centralized or alpha (α) carbon, and thus are referred to as α-amino acids. Attached to the α carbon are an amine group, a carboxyl group, and a hydrogen atom. Also attached to the α carbon is a *side group*, which differs in size, structure, and charge. Having different side groups makes one amino acid unique from another. For instance, phenylalanine contains a hydrocarbon ring, which makes it the largest amino acid. The side group of glycine is merely a hydrogen atom, making it the smallest amino acid. Table 5-1 and Figure 5-2 present the amino acids found in protein, and Table 5-1 lists a few other amino acids important to the body but not employed to make proteins.

AMINO ACID STEREOISOMERS (D AND L). Each amino acid has two **stereoisomer** forms, or mirror-image forms (Figure 5-3). For the most part, levo (L) isomers are

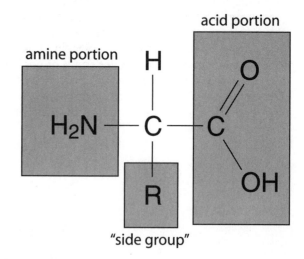

Figure 5-1

Basic architecture of an amino acid. Bonded to the central carbon (α carbon) are amine and carboxylic acid groups, a side chain (R group), and a hydrogen (H) atom.

Table 5-1   Amino Acids Found in the Human Body

Essential Amino Acids Found in Protein	Nonessential Amino Acids Found in Protein	Nonessential Amino Acids Not Found in Protein
Tryptophan	Glycine	Ornithine
Valine	Aspartic acid	Taurine
Threonine	Asparagine	γ-Aminobutyric acid (GABA)
Isoleucine	Proline	Citrulline
Lysine	Glutamine	Cystine
Leucine	Glutamic acid	
Phenylalanine	Arginine	
Methionine	Cysteine	
Histidine	Tyrosine	
Arginine[a]	Serine	
	Alanine	
	Hydoxyproline[b]	
	Hydroxylysine[b]	
	3-Methyl histidine[b]	

[a]Considered by some to be essential during growth spurts.
[b]Formed in posttranslational modification.

amino acid forms found naturally, whereas most dextro (D) forms are synthesized in laboratories. Thus, the amino acids found in foods and the body are L-α-amino acids, referred to more simply as L-amino acids. Quite often the isomeric L form is indicated in amino acid supplements and sport foods. For instance, L-arginine and L-ornithine are two fairly popular nutrition supplements.

Key Point

Amino acids have a consistent structure and include a nitrogen-containing amine group. For most amino acids, the L stereoisomer form is found in nature and the D form is synthetic.

Protein Formation and Posttranslational Modification

How are amino acids used to make proteins? Also, how many amino acids can be used to make proteins? Only 20 of the amino acids can be linked together by **peptide bonds** to build protein, in a biological process referred to as **translation** (Figure 5-4). Translation is a precise process whereby amino acids are linked together based on the instructions in genes. Each gene contains a DNA sequence that specifies a particular protein, which has unique properties and functions.

Some amino acids are modified after a protein is constructed. Because the modification occurs after translation, the processing is called a **posttranslational modification**

(Table 5-1). Some of the amino acids that can undergo posttranslational modification are proline, lysine, serine, and histidine. However, these are specific events and do not occur in all proteins. For example, collagen is rich in *hydroxylysine* and *hydroxyproline*, and muscle contractile proteins, primarily actin, contain *3-methyl-histidine* (3-MH). However, most other proteins do not contain these modified amino acids.

Amino acids that have been posttranslationally modified provide researchers with an interesting research tool. When a protein containing these amino acids is broken down, these amino acids cannot be reused to make proteins. These amino acids subsequently diffuse from tissue, circulate, and are excreted in the urine. Therefore, measurement of these amino acids in the blood and urine can be used to estimate the rate of breakdown of the protein in which they were found. As examples, blood and urinary levels of 3-MH provide some insight into the rate of muscle protein breakdown, as described later in the chapter. Also, hydroxylysine and hydroxyproline can be used to estimate the breakdown of collagen, as from the skeletal muscular system, because collagen is particularly dense in these tissues.

Key Point

Amino acids are linked together by peptide bonds during protein translation. The instructions for linking amino acids in a specific sequence are coded in DNA.

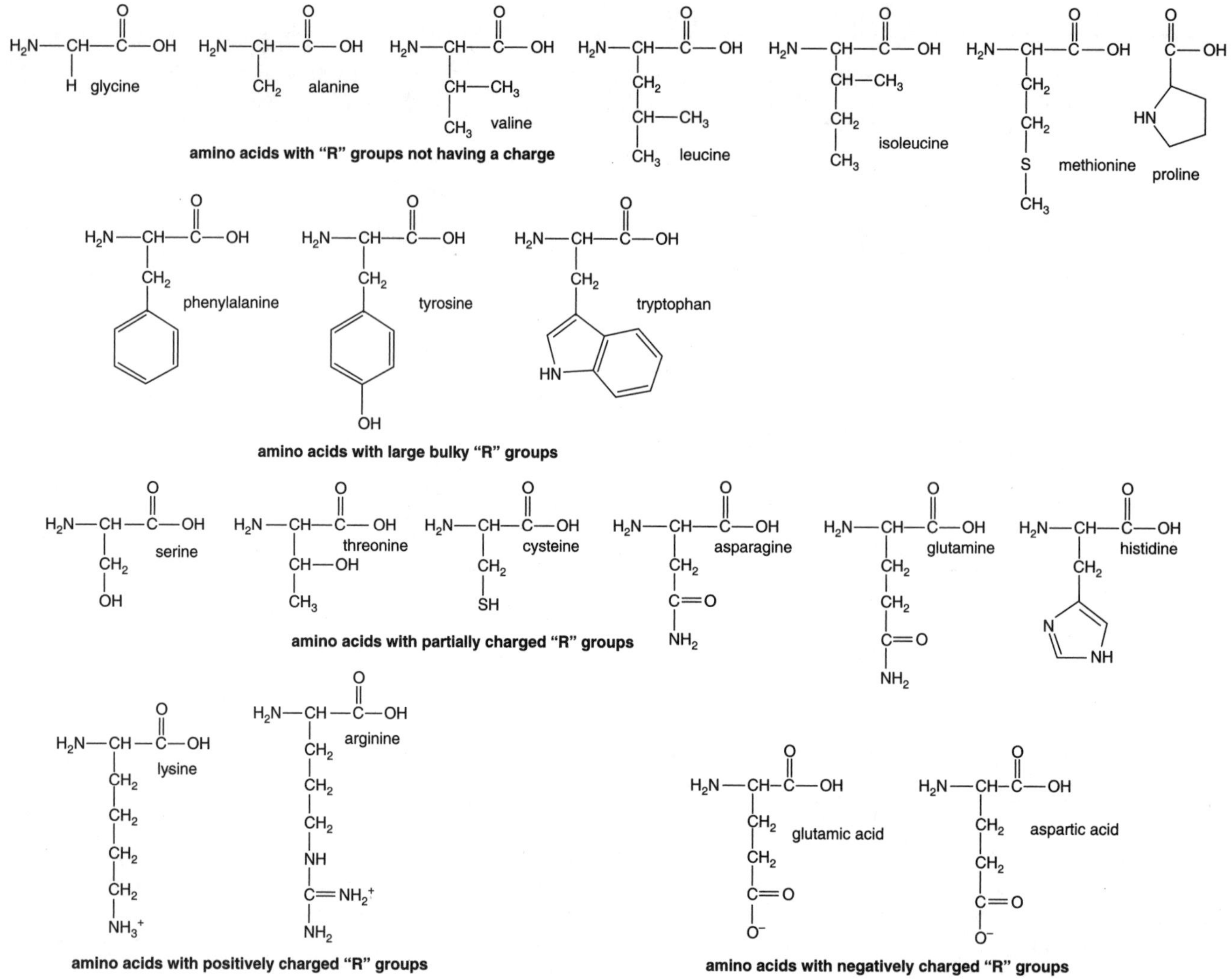

Figure 5-2

Basic amino acid design and the 20 amino acids used to make proteins.

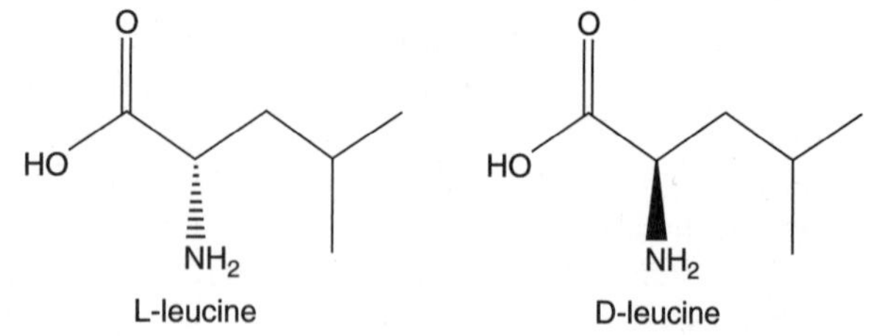

Figure 5-3

D and L stereoisomer forms of leucine. Even though these structures look identical they cannot be superimposed. For instance, the amine group (NH₂) of two isomers of leucine is oriented on opposite sides of the molecule (the dashed wedge is a bond extending away from the viewer; the solid wedge extends forward).

Protein Structure

The sequence of amino acids is responsible for determining the final three-dimensional structure of a protein (Figure 5-5). A protein contorts until it achieves its most stable three-dimensional design. For some proteins the final design is fairly linear, while others take on a seemingly disorganized balled or globbed-up design. The latter type, **globular proteins**, led to the coining of the term *globulin* (such as α-globulin, γ-globulin, hemoglobin). The former type, **fibrous proteins** or **protein fibers**, are more elongated or fibrous. Many other proteins are not easily defined as being either globular or fibrous.

The final and functional form of many proteins is that of a multiprotein complex. Proteins that are used to form a functional protein complex are often referred to as **subunits**. For instance, collagen fibers in connective tissue

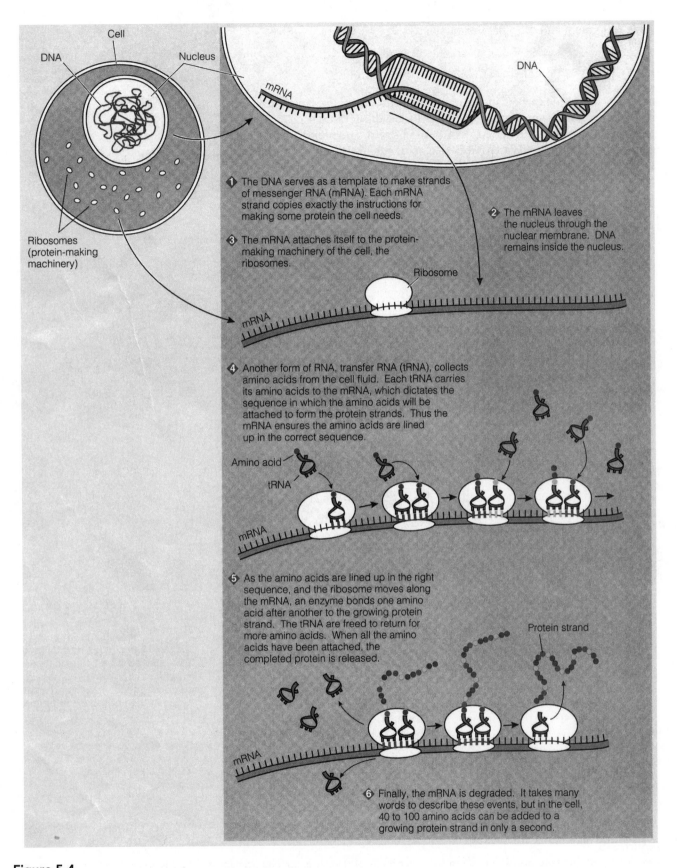

Figure 5-4

Protein synthesis. (1) *Transcription*. The DNA segment, or gene, specifying one polypeptide uncoils, and one strand acts as a template for the synthesis of a complementary mRNA molecule. **(2)** *Translation begins*. Messenger RNA from the nucleus attaches to a small ribosomal subunit in the cytoplasm. **(3)** Transfer RNA transports amino acids to the mRNA strand and recognizes the mRNA codon calling for its amino acid by its ability to base pair with the codon (via its anticodon). The ribosome then assembles and translation begins. **(4)** The ribosome moves along the mRNA strand as each codon is read sequentially. **(5)** As each amino acid is bound to the next by a peptide bond, its tRNA is released. The polypeptide chain is released when the termination (stop) codon is read.

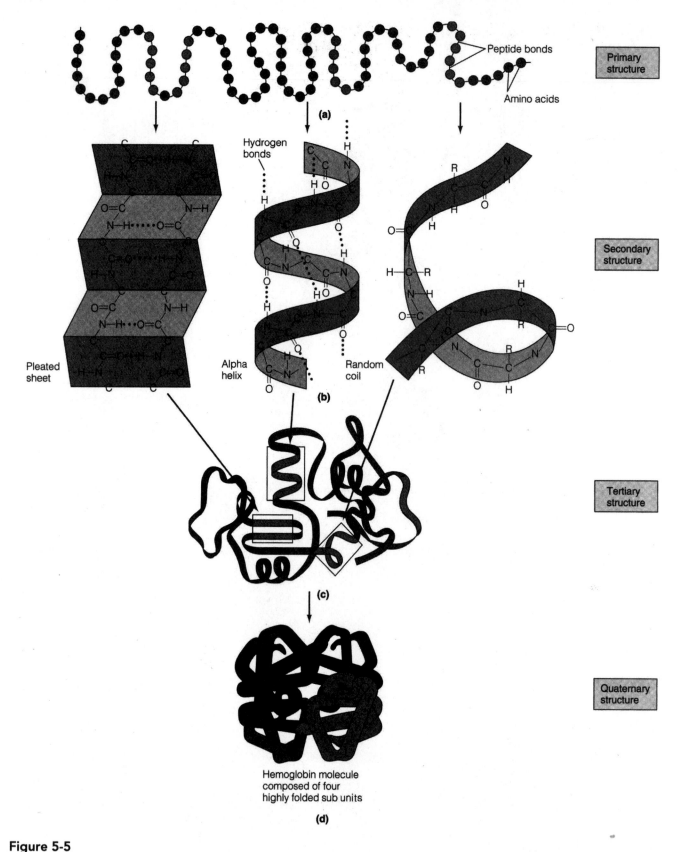

Figure 5-5

Levels of protein structure. Proteins can have four levels of structure. (a) The primary structure is a particular sequence of amino acids bonded in a chain. (b) At the secondary level, hydrogen bonding occurs between various amino acids within the chain, causing the chain to assume a particular shape. The most common secondary protein structure in the body is the alpha helix. (c) The tertiary structure is formed by the folding of the secondary structure into a functional three-dimensional configuration. (d) Some proteins form a fourth level of structure composed of several polypeptides, a exemplified by hemoglobin.

Key Point

Proteins can have fibrous or globular appearances, and many final proteins are actually complexes of a few or numerous subunits.

are actually made up of hundreds of smaller collagen proteins. Likewise, actin (the thin filament in the muscle contractile structure) contains hundreds of G-actin globular subunits (see Figure 2-8). Myosin (the thick contractile filament) and hemoglobin in red blood cells (Figure 5-5) contain six and four protein subunits, respectively.

■ AMINO ACIDS AND PROTEIN BASICS *IN REVIEW*

- All amino acids share a basic design and each has a unique side group.
- Twenty amino acids are linked together by peptide bonds in various sequences to build proteins.
- Some amino acids can be modified after a protein is made, in a process called posttranslational modification.
- The final three-dimensional design of a protein is based on the sequence of amino acids and determines its function.
- Many proteins function as complexes of smaller protein subunits.

■ ROLES OF PROTEIN AND AMINO ACIDS

Protein is a major component of the human body, contributing about 14–16% to the total mass of a lean adult. Therefore a 154-lb (70-kg) man might contain approximately 22 lb (11 kg) of protein. Skeletal muscle accounts for about 62% of body protein, and skin and blood each account for approximately 15%. Furthermore, once water is removed from the body, roughly half of the average "dry weight" of cells and tissue is protein. However, this is an average; at the two extremes are muscle and adipose tissue. Protein accounts for more than 80% of the dry weight of skeletal muscle and less than 10% of the dry weight of adipose tissue.

Body protein is constantly broken down **(catabolism)** and synthesized **(anabolism)**. For a typical adult, approximately 200–400 g of protein are broken down daily in tissue throughout the body. During that same day the same level of protein is made in order to meet the homeostatic needs of cells and tissue. It is important to realize that protein breakdown and synthesis are not the same system operating in opposite directions. Instead, these processes

function simultaneously and the dynamic balance between protein breakdown and synthesis is called **protein turnover**.

Some of the most significant roles for proteins in the body are to serve as structural components (as in connective tissue), contractile filaments and complexes (myosin, actin, tropomyosin, and troponin), antibodies for immune responses, blood clotting factors, transporters in circulation and within cell membranes, neurotransmitters, hormones, and enzymes (Figure 5-6). Also, protein is an energy resource, although protein is usually not viewed from an energy perspective as carbohydrate and fat are. Protein, more specifically amino acids, provides approximately 4 kcal/g.

Key Point

Protein turnover reflects both protein synthesis and breakdown, which employ different mechanisms. Thus, *net* protein balance is the algebraic sum of these two opposing events.

Enzymes Are Proteins

Enzymes are proteins that function to catalyze biochemical reactions. Enzymes function by interacting with the reacting substances of a chemical reaction and lowering the energy requirements for the chemical reaction to occur. In general, enzyme names end in *-ase*, which follows the type of reaction they catalyze. For instance, a *dehydrogenase* (such as pyruvate dehydrogenase) tends to facilitate the loss of hydrogen (and an electron) from a reactant. A *transferase* facilitates the transfer of a group from one molecule to another (such as alanine transferase).

SKELETAL MUSCLE ENZYMES.   Skeletal muscle fibers contain several unique enzymes that are extremely important to contraction. For instance, myosin has an enzyme component that hydrolyzes ATP to power muscle sarcomere contraction, as described in Chapter 2. Another fairly unique enzyme in muscle fibers is creatine kinase, which catalyzes the formation of creatine phosphate and the transfer of its phosphate to ADP to form ATP. Additionally, muscle fibers contain ATPase that liberates the

Key Point

Proteins serve as enzymes, antibodies, for immune responses, transporters in the blood and within cell membranes, neurotransmitters, hormones, blood clotting factors, contractile filaments and complexes (myosin, actin, tropomyosin, and troponin) and structural components (as in connective tissue).

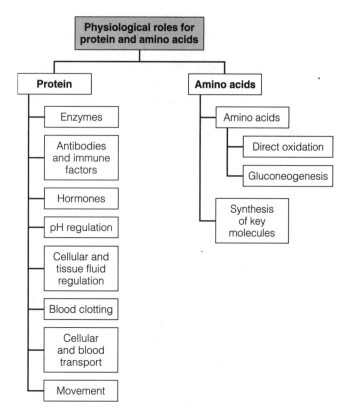

Figure 5-6
Physiological roles of protein and amino acids.

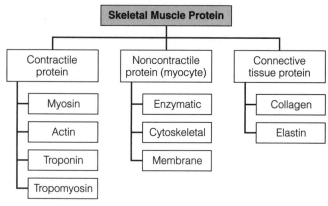

Figure 5-7
Major types of proteins in skeletal muscle.

energy needed to power the pumping of calcium (Ca^{2+}) into the sarcoplasmic reticulum and across the plasma membrane (sarcolemma).

Contractile Proteins

Contractile proteins, such as actin, myosin, tropomyosin, and troponin, collectively are responsible for about two-thirds of total muscle protein mass.[1] Thus, the remaining protein, which is noncontractile in function, contributes about one-third of the protein mass of skeletal muscle. These proteins include connective tissue, enzymes, receptors, and transporters (Figure 5-7).

Myosin is one of the most abundant proteins in the body, accounting for as much as 55% of the protein mass of skeletal muscle. Myosin contains three pairs of proteins: one pair of heavy chains and two different pairs of light chains. The heavy chains alone are about 1800 amino acids long and rank among the longest proteins in humans. Actin begins as a smaller, globular protein, G-actin, which is *polymerized* to produce a fibrous structure called F-actin (see Figure 2-8).

The fibrous protein tropomyosin associates with F-actin in a lengthwise manner (see Figure 2-8). This protein serves to cover binding sites for myosin on G-actin

when a muscle fiber is not stimulated. Troponin is not a single protein but a complex of three spherical proteins. These proteins are associated with one another while one of the troponin proteins is associated with tropomyosin. When calcium (Ca^{2+}) floods into a stimulated skeletal muscle fiber, it strongly interacts with the troponin complex. The Ca^{2+}-troponin interaction results in a conformational change in the troponin complex. This serves to pull tropomyosin away from the myosin-binding site on actin. The power stroke can then follow. See Chapter 2 for a more detailed explanation of cross-bridging and sarcomere contraction.

Protein Transporters

In red blood cells (RBCs), hemoglobin transports O_2. A typical adult male might have just under a pound ($<\frac{1}{2}$ kg) of this protein in his blood. The plasma portion of the blood also contains transport proteins. The liver makes these molecules as well as most other plasma proteins. The total protein content of plasma for an adult is 7.0–7.5 g/100 ml.

Perhaps the most versatile and by far the most abundant plasma protein is **albumin**, which contains 610 amino acids. The level of albumin in the blood of an adult usually is 3.5–5.0 g/100 ml. Albumin transports many substances including free fatty acids (FFAs), bile acids, and several minerals. As discussed in Chapter 6, the transport of FFAs to working muscle fibers during exercise is of

Key Point

Hemoglobin in RBCs and albumin in the plasma carry O_2 and fatty acids, respectively, in circulation and are vital to muscle energy metabolism at rest and during exercise.

fundamental importance during submaximal endurance exercise. Other transport proteins found in the plasma include transferrin, ceruloplasmin, and vitamin D binding protein (DBP). Protein-containing lipoproteins are major transporters of lipids and lipid-soluble substances (such as fat-soluble vitamins).

Connective Tissue Proteins

Connective tissue protein structures such as *collagen* and *elastin* provide strength and elasticity properties to tissue. Collagen, the most abundant protein in the body, is a fibrous protein found in tendons, bones, cartilage, skin, the cornea, and in between cells (interstitial spaces). Elastin is different from collagen in that collagen provides tensile strength and elastin provides recoil or elasticity properties to tissue. Elastin is an especially important component of vascular tissue such as the aorta and other arteries as well as arterioles. Elastin allows for expansion of a vessel under pressure and recoil as the pressure wanes. In contrast, the protein α-*keratin* in hair, nails, and skin provides much of the structural integrity associated with these tissues.

Protein Hormones

Many of the hormones circulating in the blood are proteins or amino acid derivatives. For example, insulin and glucagon are proteins produced and released by the pancreas relative to blood glucose levels. The hypothalamus, one of the most active endocrine organs, produces nine protein hormones, including thyrotropin-releasing hormone (TRH) and growth hormone—releasing hormone (GHRH). These hormones help regulate the levels of thyroid hormone and growth hormone, respectively. The adrenal glands produce epinephrine and norepinephrine, both derived from the amino acid tyrosine. The anterior pituitary gland produces eight protein hormones, including thyroid-stimulating hormone (TSH) and growth hormone (GH). One of the two hormones produced by the posterior pituitary gland is a protein, antidiuretic hormone (ADH), which is fundamentally important in regulating blood pressure, as discussed in Chapter 8.

Key Point

Many hormones are proteins or are derived from the amino acid tyrosine.

Water Balance and pH Regulation

Proteins are also important in regulating body water balance and distribution. As most proteins have limited ability to freely diffuse through a capillary wall or cell membranes, they exert *osmotic* pressure in either the intracellular or extracellular fluid, or across organelle membranes. Plasma proteins, especially albumin, are important factors in establishing water balance between tissue and the blood by providing osmotic properties of the plasma. Along with *hydrostatic* pressure, osmotic pressure determines the flux of water between the blood and tissue. This is explored further in Chapter 8.

Proteins also help regulate the natural pH or acid-base balance. Proteins act as natural buffers because their amino acids provide both acidic and basic groups. The carboxyl group of amino acids helps neutralize excess bases, and the amine groups help neutralize excess acid. The natural buffering capacity of amino acids and some proteins (such as carbonic anhydrase) are critical for cellular and extracellular environments, metabolic functions, and biochemical reactions and help manage some of the excessive H^+ produced during higher intensity exercise.

Key Point

Plasma protein, in particular albumin, helps maintain blood water content (volume) that is quite important during exercise.

Antibodies, Transport Proteins, and Other Protein Roles

Antibodies are proteins, and insufficient protein intake over time can lead to a decreased resistance to infections. Proteins also act as membrane transporters, such as GluT4 and Na^+/K^+-ATPase, and as receptors for signaling molecules like hormones and neurotransmitters. Additionally, there are many intracellular binding proteins, such as metallothionein and ferritin. Metallothionein binds many minor minerals, and ferritin is an iron storage protein. Both of these proteins are discussed in Chapter 10 and have application to exercise. The visual pigment *opsin*, found in the retina of the eye, is a key protein in vision.

Amino Acids as Energy Molecules

Cells have a small pool of free amino acids. These, combined with the free amino acids in the circulation, make up the **amino acid pool** of the body. The pool within a cell reflects the balance between protein synthesis and breakdown within that cell as well as free amino acid exchange with the blood. These amino acids can serve as direct or indirect energy resources for some cells, such as muscle and liver cells. Once the nitrogen is removed from an amino acid, the **carbon skeleton** can be oxidized directly within certain cells or converted to another energy molecule. Because certain amino acids can be used to make glucose and/or ketone bodies in the liver during exercise or fasting, they are often referred to as **glucogenic**

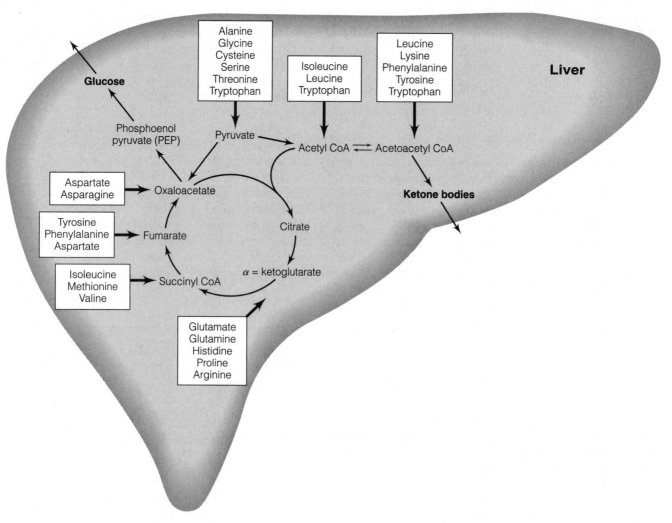

Figure 5-8
Conversion of amino acids to energy molecules. The carbon skeletons of some amino acids can be used to make either glucose and ketone bodies or both in the liver during fasting and endurance activity.

or **ketogenic**, respectively. For instance, phenylalanine, tyrosine, leucine, and isoleucine are broken down to the ketone body *acetoacetate*; thus they are deemed ketogenic. Other amino acids are broken down to pyruvate, oxaloacetate (OAA), α-ketoglurate, succinate, and fumarate, which can be used in gluconeogenesis to produce glucose and are thus glucogenic. Figure 5-8 shows the conversion of amino acids to either glucose or ketone bodies.

Key Point

The plasma and the intracellular fluid contain a small amount of free amino acids called the amino acid pool. In fasting and during certain types of exercise these amino acids are used for energy or are converted to glucose and ketone bodies.

Amino Acid–Derived Molecules

Certain amino acids are also used to make important biochemical substances. Tyrosine is one of the most physiologically versatile amino acids. For instance, tyrosine is used by the thyroid gland to make thyroid hormone (T_3/T_4) and by other tissues to make dopamine, epinephrine, and norepinephrine. The sulfur-containing amino acids (cysteine and methionine) give rise to the nonessential amino acid taurine. Although taurine is not used as a building block for protein, it is used for retinal processes related to vision, in membrane stability, as a component of bile acids, and possibly as a neurotransmitter. Tryptophan is used to make **serotonin**, a critical neurotransmitter. Even the simplest amino acid, glycine, is involved in the formation of creatine as well as the **porphyrin** ring of hemoglobin. Ornithine is needed to properly dispose of nitrogenous waste molecules (ammonia) via the urea cycle.

Table 5-2   Amino Acid–Derived Substances

Substance	Amino Acids Utilized in the Synthesis
Choline	Serine
Niacin	Tryptophan
Glutathione	Cysteine, glycine, glutamic acid
Serotonin	Tryptophan
Carnitine	Lysine, methionine
Carnosine	Histidine, alanine
Creatine	Arginine, glycine, methionine
Pyrimidines	Aspartate, glutamine
Purines	Aspartate, glutamine, glycine
Epinephrine, norepinephrine, dopamine, thyroid hormone	Phenylalanine or tyrosine

Table 5-2 summarizes some of the molecules that are derived from amino acids.

Amino Acids and Single-Carbon Donation

The amino acid methionine is an important component of single-carbon unit transfer mechanisms that are used to produce certain molecules. For instance, methionine can be converted to **S-adenosyl methionine (SAM or SAMe)**, as shown in Figure 5-9. SAM is then the principal methyl donor used to make molecules such as carnitine, creatine, epinephrine, and purines, all of which have application to sport performance. Once the methyl group is removed from SAM the resulting molecule is *S*-adenosyl homocysteine (SAH), which can be converted to homocysteine. Homocysteine can be converted back to methionine to restart the process; however, the assistance of vitamin B_{12}, 5-methyl THF (folate), and betaine are required. Recently attention has been focused on the relationship between homocysteinemia and the risk of heart disease. Additionally, SAM is used as a nutrition supplement in the treatment of sport injury to connective tissue and joint inflammation. This is discussed in the Chapter 10 feature In Focus: Sport Injury and Nutrition Supplementation.

■ ROLES OF PROTEIN AND AMINO ACIDS *IN REVIEW*

- Proteins have numerous roles in the human body including functioning as enzymes, contractile proteins, transport proteins, blood clotting factors, and hormones.
- Certain proteins can help buffer intracellular and extracellular fluids, and plasma proteins, especially albumin, are important in maintaining blood volume.
- Some amino acids can be used to produce hormones (including thyroid hormone and catecholamines) and neurotransmitters (such as serotonin).
- Amino acid carbon skeletons can be oxidized for energy in certain cells or converted to glucose or ketone bodies in the liver.
- Methionine is important in single-carbon transfer reactions that are needed to build certain molecules.

■ AMINO ACID METABOLISM

The amino acid pool in the plasma reflects the dynamic exchange of these nutrients between cells. How are amino acids exchanged between tissue, and do skeletal muscle and the liver have special roles in the exchange? Furthermore, how are amino acids processed inside of cells, and how is the nitrogen component of amino acids handled? This section provides an overview of the basic processing systems for amino acids, including the handling of the nitrogen in amino acids.

Transamination and Deamination of Amino Acids

As mentioned, the nitrogen-containing amine group of amino acids can be removed and passed to another molecule to create a nonessential amino acid (Figure 5-10). This process is called **transamination**. Transamination

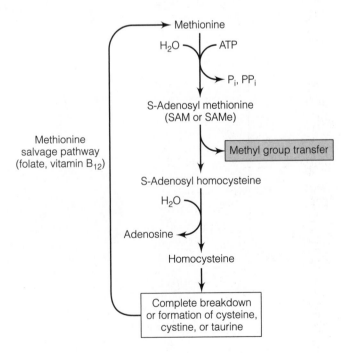

Figure 5-9
SAM is needed to form carnitine, creatine, epinephrine, purines.

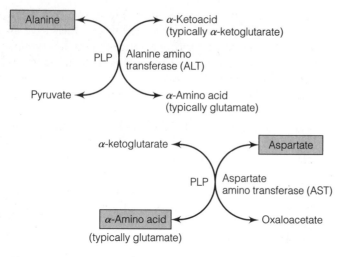

Figure 5-10

Transamination reactions. Transfer of an amino group from alanine (top) and aspartate (bottom) to from other amino acids from their respective α-keto acids. (PLP, pyridoxal S′-phosphate)

reactions involve an *amino transferase* or *transaminase* enzyme, and the acceptor of the amine group is an α-keto acid, such as pyruvate or α-ketoglutarate. Amino transferases require the assistance of vitamin B_6 (pyridoxal phosphate). Pyruvate and α-ketoglutarate are the two

α-keto acids most often used by cells to accept the amine group. In the process, pyruvate becomes alanine and α-ketoglutarate becomes glutamate; thus two nonessential amino acids are formed. Alanine transaminase (ALT) and aspartate transaminase (AST) are present in most tissues and their actions are demonstrated in Figure 5-10. The presence of tissue-specific isomer forms of these enzymes (such as cardiac and hepatic forms) in the blood is a diagnostic indicator of heart or liver pathology. Likewise, transaminase enzymes from skeletal muscle might indicate muscle tissue damage.

In addition, an amine group can be enzymatically "plucked" from an amino acid in a process called **deamination**. Deamination reactions liberate an amine group in the form of ammonia (NH_4^+), which can then be used to form urea, the principal molecule used to excrete nitrogen. Glutaminase, asparaginase, and the amino acid oxidases all produce ammonia. However, glutamate dehydrogenase generates the most ammonia and is important in the liver in generating ammonia for urea production. The release of NH_4^+ and its incorporation into urea formation is shown in Figure 5-11.

Key Point

Transamination and deamination reactions allow for the removal of the amine group from amino acids.

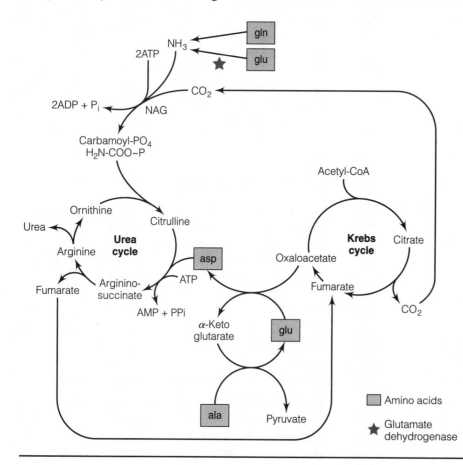

Figure 5-11

Interrelationships of amino acids and the urea and Krebs cycles in the liver. Ala, alanine; asp, aspartate; gln, glutamine; glu, glutamate.

□ Amino acids

★ Glutamate dehydrogenase

Table 5-3

Major Forms of Nitrogen Excreted in the Urine

Urinary Nitrogen Form	Expected Daily Loss (g/24 h)	% of Total Nitrogen
Urea	30	85–86%
Ammonia (NH_4^+)	<1	2–3%
Creatinine	1–2	4–5%
Uric acid	≤1	2–3%

Urea Production and Nitrogen Disposal

The removal of the nitrogen derived from amino acids is critical for survival. If the blood ammonia levels are allowed to increase, a toxicity syndrome can result that especially affects the brain. As presented in Table 5-3, typically more than 80% of the nitrogen in the urine is in the form of urea. Other forms of nitrogen include free ammonia, creatinine, and uric acid. Several enzymes produce free ammonia, the most productive of which is glutamate dehydrogenase. When amino acids are broken down in the liver, much of their nitrogen is passed (transaminated) to α-ketoglutarate to form glutamate. As mentioned above, glutamate can then be deaminated, producing free ammonia that is available to become part of urea, as shown in Figure 5-11.

Urea production occurs largely in the liver and to a lesser degree in the kidneys. The urea cycle is composed of five principal reactions that convert ammonia, CO_2, and the α-amino nitrogen of aspartate into urea. Two of the reactions involved in the urea cycle occur within mitochondria, and the remaining reactions occur within the cytoplasm. The entire process is considered a cycle because aspartate is both an initial reactant and a final product (Figure 5-11). Because the buildup of ammonia has been theorized to be involved in muscular fatigue, aspartate has been discussed as a potential performance-enhancing aid (see Chapter 11).

Key Point

Urea is formed in the liver and is the primary means of nitrogen disposal from the body.

Amino Acid Exchange Between Tissues

Amino acids are in a continuous flux between incorporation into proteins and the free amino acid pool in cells and circulation. The content of the amino acids in the pool is relatively small, representing only about 0.5–1.0% of all the amino acids in the body. Because of the relatively large amount of skeletal muscle, about three-quarters of the free amino acid pool in the body is found within skeletal muscle tissue.[2] The amino acids not in the pool (≥99%) are largely incorporated into protein structures. For instance, an adult male weighing about 154 lb (70 kg) might contain roughly 22 lb (11 kg) of protein. Approximately 14 lb (7 kg) of that would be found in skeletal muscle, which might contain only 120 g of free amino acids. Only about 5 g of amino acids would be present in his circulation.

Because it is the predominant tissue in the body and thus has a large free amino acid pool, it makes sense that skeletal muscle is fundamentally involved in the coordinated exchange and metabolism of amino acids. Furthermore, the breakdown of certain amino acids in skeletal muscle fibers during exercise might serve to maintain levels of Krebs cycle intermediates within those cells for energy production, as shown in Figure 5-11. As nitrogen is removed from amino acids it must ultimately reach the liver or kidneys for urea formation and excretion. As carriers of amine groups, certain amino acids, namely alanine and glutamine, are extremely important in nitrogen excretion, as also shown in Figure 5-11.

Amino Acid Oxidation in Liver and Skeletal Muscle

The liver is the primary organ for amino acid metabolism and oxidation. It is able to oxidize most amino acids. In contrast, skeletal muscle is able to completely oxidize six amino acids (Figure 5-12). These are the three branched chain amino acids (BCAAs), namely leucine, isoleucine, and valine, as well as glutamate, aspartate, and asparagine. Actually, skeletal muscle is more efficient at metabolizing BCAAs than the liver, and as discussed below, the BCAAs have a special role in regulating muscle protein synthesis, making them particularly important to skeletal muscle. Alanine can also be oxidized by skeletal muscle. However, during *catabolic* situations (such as fasting and endurance exercise) there is a tendency toward the formation of alanine and its release from skeletal muscle (see Figure 4-14).

Amino Acid Release from Skeletal Muscle

Physiological *stress*, such as starvation, trauma, or sepsis, results in an increased release of amino acids by skeletal muscle into circulation.[3,4] However, not all of the amino acids are released in proportion to muscle protein concentration. For instance, BCAAs are released into circulation in lower proportion than their presence in muscle protein, and alanine and glutamine in higher proportion than their presence. This would be explained if BCAAs are broken

Key Point

Skeletal muscle has the ability to break down six amino acids for energy purposes, namely BCAAs, glutamate, aspartate, and asparagine.

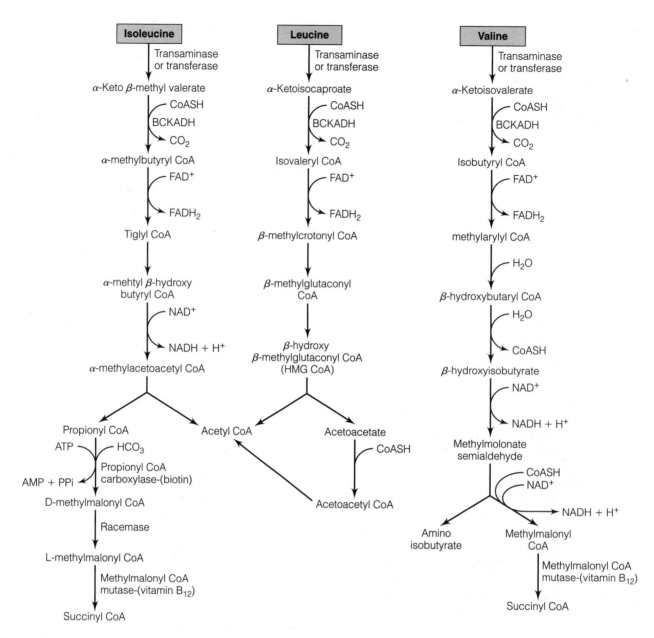

Figure 5-12
Metabolism of branched chain amino acids. BCKADH, branched chain α-keto acid dehydrogenase, requiring thiamin as TPP, niacin as NADH, Mg^{2+}, and CoA from pantothenate.

down at the same time alanine and glutamine are produced. This is indeed the case. In fact, much of the alanine and glutamine produced is attributable to the breakdown of BCAAs and the transfer of their amine group to pyruvate and glutamate.[1,5] As discussed below, increased release of alanine and glutamine occurs in a similar manner during prolonged higher submaximal intensity exercise.[6] This means that exercise of this type could be considered a metabolic stress.

Glutamine and alanine account for approximately 15% of the amino acids in muscle protein; however, they can account for as much as 50–80% of the amino acids released from muscle during such stress. In contrast, the BCAAs, which constitute about 19% of muscle protein, are released from skeletal muscle in lower amounts or not at all during physiological stress. The same is true for glutamate, which accounts for 7% of muscle protein, and aspartate (aspartic acid) and asparagine, which combined account for 9% of muscle protein.[7] Most other skeletal muscle amino acids are released into circulation in quantities relative to their content in muscle proteins,[7] which suggests they undergo little if any metabolism.

Key Point

During fasting (and prolonged exercise), skeletal muscle releases alanine and glutamine into circulation at levels much higher than that found in muscle proteins.

Skeletal Muscle Utilization of Branched Chain Amino Acids

As mentioned above, skeletal muscle can metabolize six amino acids and use the carbon skeletons for energy (oxidation). α-Ketoglutarate appears to be crucial in the transfer of the α-amine group from the carbon skeletons. The first step in the metabolism of BCAAs is the transfer of their amine group to α-ketoglutarate, forming glutamate (Figure 5-10). The same enzyme system catalyzes this reaction for all three BCAAs; however, the liver contains lower levels of this enzyme than skeletal muscle. Thus, skeletal muscle is the primary site of BCAA metabolism, especially during exercise.

The resulting carbon skeletons of the BCAA aminotransferase reaction are referred to as *branched chain α-ketoacids* (BCAKA) or branched chain oxo-acids (BCOA). As shown in Figure 5-12, the BCAKA are then metabolized by *branched chain α-ketoacid dehydrogenase* (BCKADH). Both the liver and skeletal muscle produce this enzyme, and it is found in mitochondria. So after they are formed in the cytosol, the BCAKA must enter the mitochondria to be further metabolized. It should be mentioned that some of the BCAKA formed in skeletal muscle fibers can diffuse from those cells and circulate to the liver for oxidation as well. Once BCKADH metabolizes the α-ketoacids, the derived carbon skeletons are further processed to either acetyl CoA or the Krebs cycle intermediate succinyl CoA (Figures 5-8 and 5-12). BCKADH activity is regulated by the metabolic state of cells in a manner similar to pyruvate dehydrogenase (see Chapter 4).

■ AMINO ACID METABOLISM *IN REVIEW*

- Alanine and glutamine are formed in muscle during the breakdown of BCAAs and are released from skeletal muscle in relatively high levels during fasting and prolonged exercise.
- Transamination reactions transfer an amine group from an amino acid to an α-keto acid to form a nonessential amino acid.
- Deaminases remove the amine group from an amino acid, forming ammonia (NH_4^+).
- Urea is the primary means of excreting nitrogen from the body.
- BCAAs are broken down to acetyl CoA and succinyl CoA.

■ DIETARY PROTEIN AND DIGESTION AND ABSORPTION

As protein is vital to life, all organisms including those used as food contain protein. However, the protein content varies among different life forms. Once protein is eaten, significant digestion must occur to break it down to amino acids, which are the absorbed form of dietary protein. This means that efficient protein digestion must deal with complex three-dimensional structures and protein complexes as well as the peptide bonds of a myriad of potential amino acid linkages. This section provides an overview of diet protein level and food sources as well as discussion of amino acid *essentiality*. It also discusses the processes involved in digestion and absorption, along with initial processing of absorbed amino acids. Protein dietary requirements for the general population and using nitrogen balance to estimate protein needs are also covered.

Dietary Essential and Nonessential Amino Acids

Human tissues are unable to make certain amino acids, either at all or in adequate amounts to meet the needs for maintenance and health. These amino acids have long been referred to as dietarily *essential*. These **essential amino acids** are tryptophan, valine, threonine, isoleucine, leucine, lysine, phenylalanine, methionine, and histidine (Table 5-1). Arginine has often been considered conditionally an *essential* or *semi-essential* amino acid, as it might become essential to the diet during periods of rapid growth. The remaining 10 amino acids are deemed *nonessential* amino acids, as they are produced in adequate quantities in tissue under most normal circumstances. For instance, two of the nonessential amino acids, namely cysteine and tyrosine, are made from the essential amino acids methionine (with help from serine) and phenylalanine, respectively. All other amino acids are derived from intermediates of energy pathways or other nonessential amino acids. For instance, alanine is readily derived from pyruvate. Glutamine is derived from glutamate, which itself is a nonessential amino acid as it can be made from the Krebs cycle intermediate molecule α-ketoglutarate.

More modern nutrition terminology uses *indispensable* and *dispensable* in place of essential and nonessential, respectively. Also, as shown in Figure 5-13, the dispensable amino acids can be separated into two subclasses, the dispensable (absolute) and the *conditionally indispensable*. The amino acids deemed conditionally indispensable are dispensable during normal physiological situations; however, they can become indispensable if problems involving their synthesis occur. This could be the case in premature newborns, which often experience a decreased ability to produce arginine. Also, illnesses involving the small intestine can decrease the production of conditionally indispensable amino acids, thereby making them indispensable components in the diet. The influence of exercise training on

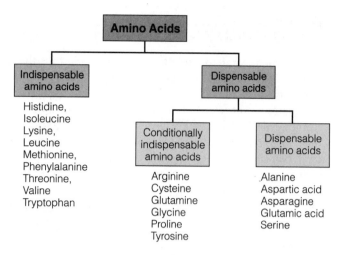

Figure 5-13
The dispensable and indispensable amino acids. Note that the dispensable amino acids are subdivided into the dispensable and conditionally indispensable amino acids. The conditionally indispensable amino acids are those that might not efficiently be synthesized during certain problematic physiological situations.

the metabolism of specific conditionally indispensable amino acids will be researched more thoroughly in the years to come.

Key Point

Nine amino acids are dietarily essential, or indispensable, at all stages of life because they cannot be made in cells, or cannot be made in sufficient quantity to match cellular needs.

Food Protein

Protein is derived from all natural foods. In general, foods of animal origin have greater protein content than plants and plant-derived foods (Table 5-4). This is largely due to the skeletal muscle of animals. The protein content of a diet designed to optimize athletic performance is presented in Chapter 7 in the feature Nutrition in Practice: Designing an Optimal Diet for Sport Performance.

MEAT PROTEINS.   Animal flesh (meat) is largely skeletal muscle and adipose tissue. Skeletal muscle is roughly 20–25% protein, and most of what remains is water. The primary proteins found in meats generally are the same as those found in human skeletal muscle, namely myosin, actin, troponin, tropomyosin, and connective tissue proteins. Meats can be tenderized by breaking up the connective network using meat tenderizers (**collagenases**) or physical efforts.

Table 5-4

Approximate Protein Content of Various Foods

Food	Protein (g)
Beef (3 oz)	28
Pork (3 oz)	28
Cod, poached (32 oz)	21
Oysters (32 oz)	17
Milk (1 c)	8
Cheddar cheese (1 oz)	7
Egg (1 large)	6
Peanut butter (1 tbs)	8
Potato (1)	3
Bread (1 slice)	2
Banana (1 med)	1
Carrots, sliced (2 c)	1
Apple (1)	2
Sugar, oil	0

Although meats are a rich source of protein, the contribution of protein to the total energy obtained from meat depends on the amount of adipose tissue in the meat. Chicken tends to store more fat beneath the skin (subcutaneously) whereas cattle have adipose tissue *marbled* throughout the skeletal muscle. Thus, removing the skin of a chicken breast or thigh would greatly lower the fat content and increase the protein content (as a percentage of total energy). Meanwhile, beef needs to be carefully trimmed of visible fat, and different cuts of beef (such as rib versus flank) vary in the extent of marbling. Choosing leaner cuts of beef and trimming away visible fat increases the percentage of energy derived from protein.

FISH PROTEIN.   Fish proteins are very similar to mammalian meat proteins. However, the skeletal muscle of fish is organized with shorter fibers and arranged between connective tissue sheets. Microscopically, the striated appearance of fish myofibrils is similar to mammals and contains the same proteins. Fish is typically considered a lean protein source as it contains relatively little fat. For instance, water-packed tuna derives more than 80% of its energy from protein. This is true for many fish. However, check the food label, as some fish are fatty or perhaps canned in oils.

MILK PROTEIN.   Milk proteins include **casein** and serum proteins. Casein proteins account for roughly 78%

Key Point

The protein in meats and fish are similar to human skeletal muscle proteins.

of milk's nitrogen-based mass and the serum proteins for about 17%. The remaining nitrogen compounds found in milk are mostly amino acids and peptides. Casein is a family of spherical phosphoproteins, and the serum proteins, or **whey** proteins, consist largely of β-lactoglobulin, lactalbumin, immunoglobulins, and other albumins. β-Lactoglobulin is the most abundant whey protein and is rich in lysine, leucine, glutamic acid, and aspartic acid. The immunoglobulins in milk are largely immunoglobulin M (IgM), IgA, and IgG.

Key Point

Milk proteins are the caseins and whey proteins; meats are largely contractile proteins and connective tissue. Whey is a popular protein supplement.

EGG PROTEIN. Egg whites contain at least eight proteins, including ovalbumin, conalbumin, ovomucoid, avidin, flavoprotein-apoprotein, "proteinase inhibitor," ovomucin, and globulins. These proteins collectively account for about 11% of liquid egg white. Egg yolk proteins are found in the form of emulsifying lipoproteins due to the high lipid content of the yolk. Because egg yolk is fat rich, egg white has a higher protein content (% of total energy) than a whole egg or the yolk.

WHEAT PROTEIN. Wheat contains some proteins that are unique and allow for bread making. There are four main protein fractions in wheat proteins: albumin, globulin, gliadin, and glutelin. When the latter two proteins, as part of flour, are mixed with water, they form gluten. Gluten's elastic property allows for a structural network that holds together other bread components such as starch and air, giving bread structure. Gluten proteins are relatively high in glutamine and relatively low in lysine, methionine, and tryptophan.

SOY PROTEIN. Soy proteins are a good source of all essential amino acids except methionine and tryptophan. Soy proteins do not include gliadin or glutelin, so soy flour is not typically used in breads. However, special additives can be used to enhance the volume of breads made with soy flour. For the most part, soy proteins are complex multiprotein globulins. Isolated soy protein (ISP) is used commercially in drinks mixed with fruit and water, coffee whiteners, liquid whipped toppings, and sour cream dressings. Soy proteins are often found associated with

Key Point

Egg white contains protein and very little fat, making it a protein-rich food source.

Key Point

Wheat and many other cereal grains contain proteins that allow bread to expand in volume.

isoflavone molecules such as *genestein* and *daidzein* that might have health-promoting properties and are considered *nutraceuticals* (see Chapter 1).

Protein Consumption

Adults in the United States derive nearly two-thirds of their dietary protein from animal sources. This is among the highest levels in the world. Many African and Asian societies derive only about one-fifth of their protein from animals and get the remainder from plant sources. American adults tend to eat about 100–125 g of protein daily, with women at the lower end and men toward the higher end of this range. The protein intake of athletes can vary tremendously depending on their sport and eating philosophies, as discussed below and in Chapter 14.

Protein Digestion

Protein digestion is very complex and is summarized in Figure 5-14. In addition to dietary protein, some protein (such as enzymes) is found in digestive juices (saliva, stomach juice, and pancreatic and intestinal secretions) as well as in the cells that are sloughed from the lining of the tract. In fact, about 100–125 g of protein per day might enter the digestive tract of an adult in addition to what is consumed in the diet.

The objective of protein digestion is to dismantle proteins, regardless of source, and liberate the amino acids for absorption. Protein-digesting enzymes or **proteases** split peptide bonds (such as carboxypeptidase A and B). Because of the numerous different amino acid links, several different proteases are needed.

STOMACH. Protein digestion begins in the stomach. Secretion of a juice rich in hydrochloric acid (HCl) serves to create an extremely acidic environment (pH 1.5–2.5) within the stomach. This is important to protein digestion for two reasons. First, the low pH activates the first protein-digesting enzyme, *pepsin* (Figure 5-14). Second, the acidity of the stomach juice straightens out complex protein structures. Because proteins in nature have a specific three-dimensional design, straightening proteins out is referred to as **denaturing** them. The denaturing of proteins increases the efficiency of pepsin and subsequent protein-digesting enzymes.

PANCREATIC PROTEASES. Pepsin is able to digest larger protein chains into much shorter fragments and some amino acids. As the stomach empties the protein fragments into the small intestine, a battery of protein-digesting

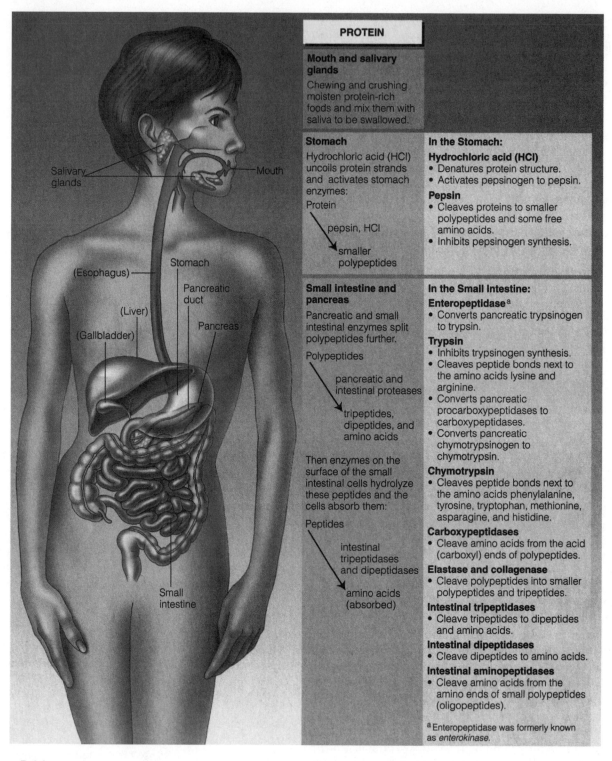

PROTEIN

Mouth and salivary glands

Chewing and crushing moisten protein-rich foods and mix them with saliva to be swallowed.

Stomach

Hydrochloric acid (HCl) uncoils protein strands and activates stomach enzymes:

Protein

↓ pepsin, HCl

smaller polypeptides

In the Stomach:

Hydrochloric acid (HCl)
• Denatures protein structure.
• Activates pepsinogen to pepsin.

Pepsin
• Cleaves proteins to smaller polypeptides and some free amino acids.
• Inhibits pepsinogen synthesis.

Small intestine and pancreas

Pancreatic and small intestinal enzymes split polypeptides further.

Polypeptides

↓ pancreatic and intestinal proteases

tripeptides, dipeptides, and amino acids

Then enzymes on the surface of the small intestinal cells hydrolyze these peptides and the cells absorb them:

Peptides

↓ intestinal tripeptidases and dipeptidases

amino acids (absorbed)

In the Small Intestine:

Enteropeptidase [a]
• Converts pancreatic trypsinogen to trypsin.

Trypsin
• Inhibits trypsinogen synthesis.
• Cleaves peptide bonds next to the amino acids lysine and arginine.
• Converts pancreatic procarboxypeptidases to carboxypeptidases.
• Converts pancreatic chymotrypsinogen to chymotrypsin.

Chymotrypsin
• Cleaves peptide bonds next to the amino acids phenylalanine, tyrosine, tryptophan, methionine, asparagine, and histidine.

Carboxypeptidases
• Cleave amino acids from the acid (carboxyl) ends of polypeptides.

Elastase and collagenase
• Cleave polypeptides into smaller polypeptides and tripeptides.

Intestinal tripeptidases
• Cleave tripeptides to dipeptides and amino acids.

Intestinal dipeptidases
• Cleave dipeptides to amino acids.

Intestinal aminopeptidases
• Cleave amino acids from the amino ends of small polypeptides (oligopeptides).

[a] Enteropeptidase was formerly known as *enterokinase*.

Salivary glands — Mouth — Stomach — (Esophagus) — Pancreatic duct — (Liver) — Pancreas — (Gallbladder) — Small intestine

Figure 5-14
Protein digestion in the gastrointestinal tract.

enzymes produced by the pancreas and small intestine continue the assault. These include *trypsin, chymotrypsin, elastase,* and *carboxypeptidases A* and *B.* Collagenase and *elastase* help digest connective tissue, making animal flesh proteins in muscle cells more available for digestion.

INTESTINAL PROTEASES. The potency of pancreatic proteases is very respectable, as they nearly complete protein digestion. However, their activity is fairly short, as they are subject to digestion themselves. The products of these proteases are mostly free amino acids and short

Key Point

Proteins in foods vary in size, shape, and amino acid composition, making protein digestion complicated.

peptides, most of which are just two or three amino acids linked together (dipeptides and tripeptides). Final digestion of the short peptides occurs via enzymes on the plasma membrane of the cells lining the small intestine (enterocytes) or within those cells. This means that not only are amino acids taken up by the cells lining the small intestine but so are a significant quantity of dipeptides, tripeptides, and maybe slightly longer peptides.

Amino Acid Absorption

The absorption of amino acids requires energy and involves amino acid transporters. Furthermore, these carriers require sodium to operate effectively. Table 5-5 presents the four principal kinds of amino acid transporters. Because there are more than 20 amino acids requiring uptake, some competition can occur. This raises questions as to whether supplementation of individual amino acids might decrease the absorption of other amino acids utilizing the same transporter.

The coupling of amino acid absorption with sodium (Na^+) is similar to that of glucose and galactose, described in the previous chapter. Na^+/K^+-ATPase maintains very low Na^+ concentration in the absorptive cells of the small intestine (Figure 5-15). Amino acids in the small intestine bind to a transporter on the plasma membrane and, along with Na^+, are brought into the cell as Na^+ moves down its concentration gradient. Thus the energy requirement for

Key Point

Amino acids are the absorbed form of protein.

Table 5-5	**Amino Acid Transport Systems**

1. **Neutral amino acids**: tyrosine, tryptophan and phenylalanine, alanine, serine, threonine, valine, leucine, isoleucine, glycine, methionine, histidine, glutamine, asparagine, cysteine, proline, hydroxyproline
2. **Basic amino acids**: lysine, arginine, ornithine, cystine
3. **Dicarboxylic acids**: glutamic and aspartic acids
4. **Imino acids**: proline, hydroxyproline, glycine, proline, hydroxyproline, taurine, D-alanine, and γ-aminobutyric acid (GABA)

Note: Some amino acids can use two transport systems.

Source: Adapted from reference 1.

amino acid absorption is to maintain the low level of Na^+ within that cell.

Rate of Absorption of Amino Acids

The absorption rate of amino acids from a food or supplement can influence the rate of availability of amino acids to the liver and other tissue. This, in turn, can influence how the amino acids are metabolized: whether they are oxidized, used to make glucose/glycogen, or become involved in protein synthesis (see In Focus: Nutrition States and Energy Metabolism in Chapter 4). If the rate of absorption of amino acids is higher for a food or supplement, it is referred to as a *fast* protein; if the amino acids are absorbed at a lower rate, the protein is referred to as a *slow* protein.

INTACT PROTEIN VERSUS FREE AMINO ACIDS.   It seems reasonable to assume that the time from ingestion to absorption would be greater for intact proteins than for free amino acids because of the steps involved in digesting protein to amino acids. This would mean that free amino acids would raise plasma amino acids more quickly and significantly than amino acids from intact protein; however, the effect would wane more quickly.

DIFFERENT PROTEIN SOURCES.   Different protein isolates can have different rates of amino acid absorption. Much of the difference in absorption is related to the protein's complexity or other characteristics that influence the time needed for complete digestion and absorption. For instance, when comparing the rate of amino acid availability of casein and whey protein, whey is a fast protein and casein is a slow protein source.[8] Whey protein is more water soluble and empties from the stomach more rapidly than casein, which tends to curdle.

As discussed below, the concept of fast and slow proteins might be an important consideration for maximizing gains in body protein following certain types of training. For instance, because the effects associated with whey and casein proteins differ, both having their desirable attributes, one might be best served by combining these sources. However, as proteins are combined and mixed with other nutrients, especially carbohydrate and fat, the differences in absorption rate are reduced.

MIXED MEALS AND AMINO ACID ABSORPTION.   When amino acids or intact protein are consumed as a component of a meal or food containing other energy

Key Point

The rate of amino acid absorption from free amino acid sources (such as supplements) might be a little quicker than amino acids from intact protein of a similar amino acid content.

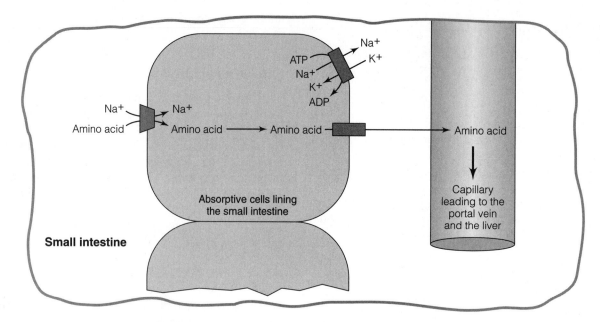

Figure 5-15
Amino acids are absorbed by cells of the small intestine wall. The absorption requires Na⁺ and ATP. Some amino acids are used by these cells and others travel via the hepatic portal vein to the liver, where they may be used or sent into the general circulation.

nutrients, the rate of amino acid absorption becomes equalized. For instance, it would be expected that the amino acids of two sport bars containing either intact protein or free amino acids, or two sport bars containing whey or casein, would be absorbed at a similar rate if the sport bars contained a similar level of carbohydrate and fat.

Intestinal and Liver Metabolism of Absorbed Amino Acids

The cells lining the small intestine retain a significant portion of the amino acids being absorbed. As the portal blood surges through the liver, yet more amino acids are removed. Combined, these tissues might take up and metabolize a sizable amount of the absorbed amino acids during the "first pass" of the amino acids through the bloodstream. More specifically, it has been estimated that the small intestine and liver extract 20–90% of absorbed amino acids.[9] The reason for the great range is primarily attributed to the preferential uptake of different amino acids. For instance, only about 20% of absorbed leucine might be extracted while more than 50% of the tryptophan is removed.[10,11] In addition, the liver is particularly aggressive at removing glutamate (glutamic acid) during the first pass. The amino acids extracted by the intestine and liver can be used to make proteins. They can also be broken down for energy or used by the liver to make glucose and glycogen in the early fed state (see Chapter 4). As a result, urea is generated and released into circulation for excretion by the kidneys.

Key Point

The small intestine and the liver get "first shot" at absorbed amino acids and are very involved in their metabolism and presence in the general circulation.

Protein Quantity

It is important not to confuse protein *quality* with protein *quantity*; however, both are important to protein nutrition. Quite simply, protein quantity refers to the amount of protein in the diet. Measuring the nitrogen content in food assesses protein quantity, as most of the nitrogen is found in the amino acids that make up protein. Although certain other molecules contain nitrogen, such as nucleic acids and some vitamins, their contribution to the nitrogen content of a food is relatively small. In general, protein is 16% nitrogen. So, multiply the grams of nitrogen by 6.25 to estimate protein content, and multiply grams of protein by 0.16 to estimate grams of nitrogen.

Example of estimating grams of protein from grams of nitrogen:

$$12 \text{ g of nitrogen} \times 6.25 = 75 \text{ g of protein}$$

Example of estimating grams of nitrogen from grams of protein:

$$80 \text{ g of protein} \times 0.16 = 12.8 \text{ g of nitrogen}$$

Protein Quality

Protein quality is different from quantity in that it implies that comparisons have been made among proteins. Quite simply, protein quality refers to how a food protein compares to the needs of human cells. If the protein is digestible and the available amino acids, particularly the essential amino acids, are proportionate with the needs of cells, then the food protein is deemed to have higher quality. However, if a food protein fails to provide all the essential amino acids in levels proportionate to cell needs, it is said to be "limited" at the level of the least available essential amino acid. In addition, protein synthesis in tissue would be limited to the level of availability of that amino acid; hence the term **limiting amino acid**.

Food proteins are often classified based on **biological value**, which in turn is based on the amino acid composition of the food protein relative to human needs.

- *High biological value (complete protein)*: the protein source has all nine essential amino acids in proportion to human cell needs and supports growth and maintenance. This is normally protein of animal origin.
- *Low biological value (incomplete protein)*: the protein source contains all nine essential amino acids but not in proportion to human cell needs. Plant proteins often fall into this category. Corn provides an example because it is limited in the amino acid lysine.

There are several methods available to assess protein quality. In addition to the essential amino acid content, protein quality accounts for digestibility of a food and food protein. Quite simply, how much benefit would be derived from a food protein containing all of the essential amino acids in levels that are proportionate to cell needs, if the protein is too complex to be digested?

Key Point

Protein quality varies from food to food and is determined by how well a food's essential amino acid composition matches up with the needs of human cells.

Protein Requirements for the General Population

Protein is needed as a component of the diet to replace the amino acids that are broken down for energy or metabolized in an irreversible fashion. For a typical adult, protein losses for energy purposes are approximately 20–30 g daily. Most of the protein (amino acid) loss is attributable to energy processes. However, small amounts of amino acids are irreversibly metabolized in the creation of neurotransmitters, hormones, and other molecules such as

Key Point

The current RDA for protein is 0.8 g/kg of body weight for adults, and an acceptable range of protein intake is 10–35% of energy.

carnitine, creatine, and bases (purines and pyrimidines) for nucleotides and nucleic acid synthesis (see Table 5-2). Also, posttranslational modification of amino acids for a specific protein type renders them unusable in future protein synthesis, after that protein is broken down.

As a component of the newest Dietary Reference Intakes (DRIs), the Recommended Dietary Allowance (RDA) for protein for American adults is 0.8 g/kg of body weight. If the diet contains a variety of foods, this level of protein intake should provide all essential amino acids in more than adequate amounts along with a rich complement of nonessential amino acids. The DRIs also list the acceptable range of protein intake as 10–35% of energy. Appendix A presents the current recommendations for protein and amino acids for specific genders and ages.

Nitrogen Balance Estimation of Protein Needs

To estimate protein needs, researchers and clinicians can take advantage of the fact that protein contains nitrogen. A person is said to be in **nitrogen balance (N_{bal})** when the amount of nitrogen lost from the body over a period matches the nitrogen content of the diet. Said another way, nitrogen balance is the state in which protein is neither lost nor gained from body tissue as a whole. If more nitrogen is lost from the body than is consumed in the diet, a person is said to be in *negative* nitrogen balance. Negative nitrogen balance ($N_{in} < N_{out}$) occurs during starvation and other stress situations such as burn injury and other traumas. *Positive* nitrogen balance occurs when there is a building of tissues, such as during growth and tissue repair. In this situation more nitrogen is retained than excreted ($N_{in} > N_{out}$). This is the goal of weight trainers and bodybuilders attempting to build mass.

To increase the accuracy of N_{bal} assessment, the nitrogen content of the diet as well as the amount of nitrogen lost in the urine, feces, and sweat must be precisely assessed. As described above, the nitrogen in feces is derived not only from unabsorbed protein in food but also from digestive secretions, sloughed-off cells, and bacteria. Even skin, hair, and fingernail nitrogen is accounted for during meticulous evaluation; however, estimation standards are available. Table 5-6 provides typical quantities of nitrogen loss by various routes throughout a day.

$$\text{Nitrogen balance} = \text{diet N intake} - (\text{urine N} + \text{fecal N} + \text{skin N})$$

Table 5-6

Estimated Daily Loss of Nitrogen by Various Routes

Route	Daily Nitrogen Loss (mg/kg of body weight)
Urine	37
Feces	12
Cutaneous	3
Minor routes (nasal secretions, expired breath, seminal fluid)	2
Total	**54**

■ DIETARY PROTEIN AND DIGESTION AND ABSORPTION *IN REVIEW*

- Food proteins contain a variety of protein types.
- Efficient protein digestion must straighten out complex three-dimensional proteins and split the peptide bonds between numerous combinations of amino acids.
- Amino acids are absorbed via enterocytes and enter the portal vein. Much of the absorbed amino acids can be retained by the enterocytes and liver during this "first pass."
- The rate of absorption of specific proteins can vary due to differences associated with their digestion.
- Nitrogen balance can be used to estimate changes in muscle protein content but it really reflects whole body protein status.

■ AMINO ACID AND PROTEIN METABOLISM DURING EXERCISE

This section presents one of the most fascinating and at times controversial areas of exercise nutrition. Metabolism of amino acid and protein during exercise, as well as during recovery and after adaptation, is more complex than for the other energy nutrients. Unlike carbohydrates, for which the focus is mostly on glucose and glycogen, there are numerous amino acids to be considered. Unlike stored molecules of carbohydrate (glycogen) and fat, which are largely energy storage structures, there are not inert amino acid molecules for energy storage. As mentioned, at least 99% of amino acids in the tissues at any time are linked together to provide structure and function to cells and tissue.

Amino acids also differ from carbohydrates and fat in that they are not completely metabolized to just CO_2 and H_2O. Nitrogen must be removed from amino acids, transported, and eventually excreted.

With these concepts in mind, some of the biggest questions with regard to exercise and the metabolism of protein and amino acids are as follows:

- How does a bout of exercise and its type, intensity, and duration influence protein turnover (synthesis and breakdown)?
- Are amino acids an important energy source and do they serve other functions during exercise?
- How is nitrogen metabolized during exercise?
- How might factors involved in carbohydrate and fat metabolism influence protein and amino acid metabolism during exercise?
- How does training influence any of the functions noted in the previous questions?
- What are the needs for specific amino acids and protein for individuals involved in sport training?
- Is the timing of protein and carbohydrate ingestion relative to exercise important?

General Influence of Exercise on Energy Metabolism and Protein Turnover

Exercise has several hormonal and metabolic similarities to fasting in that there is a net breakdown of energy nutrient stores. Yet, there are several striking differences between fasting and exercise. First and foremost, exercise involves a dramatic increase in energy expenditure. In fact, total energy expenditure during exercise can increase several times above the resting state. Another major difference between fasting and exercise is the greater focus on skeletal muscle as a fuel resource during activity. For instance, during supramaximal exercise (as in sprinting) nearly all the energy utilized to power the exercise activities is derived from within the working muscle fibers themselves. During exercise performed at a moderate to higher submaximal intensity, only about half the energy utilized is derived from within muscle. The difference in the management of fuel storage resources during fasting and exercise suggests different activity levels of regulatory hormones during these metabolic states. For instance, epinephrine is more important during exercise, and glucagon is more important during fasting. This makes sense, as skeletal muscle fibers do not have glucagon receptors. Cortisol, however, is an important hormone in both fasting and exercise.

Exercise and Hormonal Influences on Protein Metabolism

In muscle tissue, protein turnover largely reflects the synthesis and breakdown of contractile proteins, connective tissue, metabolic support proteins (such as myoglobin and proteins in electron transport complexes), ion pumps (Ca^{2+}, Na^+/K^+), transport proteins, and enzymes. If exercise does influence protein turnover, how does it do so, and what factors are responsible for promoting such activities? One of the most important ways that exercise influences protein turnover is by altering the level of key hormones. Among the most important messengers are insulin, cortisol, growth hormone, insulin-like growth factor-1 (IGF-1), and testosterone.

Hormones affect muscle protein turnover by influencing protein synthesis or breakdown, or both. Hormones that are independently associated with a net protein synthesis are deemed *anabolic* and those that result in net breakdown are *catabolic*. However, these hormones need to be considered conjointly since they are always present to varying degrees in circulation. Therefore their relative concentrations can exert **antagonistic** and counterbalancing influences, or be additive and even **synergistic**, having a greater effect together than the sum of their separate effects.

Although a single exercise bout can modify the circulating levels of hormones that regulate protein turnover, the effects of chronic exercise training must also be considered. For instance, a single exercise bout can increase the circulating level of growth hormone and testosterone. However, this response to an exercise bout tends to be less dramatic when the same workout is performed regularly. Therefore, the training stimulus must be increased to maintain a consistent exercise-induced rise in growth hormone and testosterone.

Key Point

Muscle protein turnover largely reflects synthesis and breakdown of contractile and connective tissue proteins, enzymes, ion pumps, transporters, and metabolic support proteins.

INSULIN.   Insulin release is strongly stimulated by elevated blood glucose and less potently by an elevation of amino acids. Insulin is believed to promote the uptake of circulating amino acids into certain cells, particularly skeletal muscle fibers. In skeletal muscle, insulin promotes a net protein synthesis by positively influencing synthesis and negatively influencing breakdown. However, the influence of insulin might have a greater influence on limiting protein breakdown.[11–13]

As discussed in Chapter 4, insulin release can be blunted during exercise when the intensity level and duration surpass certain thresholds. This would clearly be the case for endurance exercise of at least a moderate submaximal intensity. Even when carbohydrate is provided during exercise, insulin release is still blunted versus what would be expected. The decreased influence of insulin during exercise allows for greater glycogen and fat breakdown. Also, this hormonal modification might allow for a net protein breakdown during certain sports such as endurance cycling and running. This would make more

Key Point

Much of the anabolic influence of growth hormone on skeletal muscle protein turnover is mediated through IGF-1.

amino acids available to active skeletal muscle and for gluconeogenesis in the liver.

CORTISOL.   Cortisol is released by the adrenal glands during periods of physiological or nutritional stress. As a stress hormone, cortisol promotes a general increase in energy nutrient availability. This might include amino acids derived from skeletal muscle protein, involving a net protein breakdown and amino acid release from muscle. It is currently believed that cortisol has a greater influence on muscle protein breakdown than on protein synthesis.

Like insulin, cortisol seems to vary in level during exercise in relationship to the intensity and duration of the activity. This means that an intensity/duration threshold may be associated with cortisol release. Strenuous submaximal endurance exercise can increase the level of circulating cortisol. So does weight training if it involves multiple sets of strenuous effort.

GROWTH HORMONE.   Growth hormone levels are known to increase during exercise and remain elevated for a corresponding time afterward. However, some researchers have determined that if an exercise bout is more extreme (such as running a marathon or triathlon), the postexercise response might be blunted.[14] Additionally, some differences in the growth hormone response to exercise seem to occur as the result of aging.[15]

Increased levels of amino acids in circulation after a protein-rich meal also raise growth hormone levels. Growth hormone is associated with increased amino acid uptake into muscle cells and protein synthesis. Although growth hormone might have an independent influence on protein synthesis, some if not the majority of its action might be mediated through the release of other factors, especially insulin-like growth factor 1.

INSULIN-LIKE GROWTH FACTOR 1.   It is often overlooked that the anabolic effects of growth hormone are actually mediated through other growth factors. Insulin-like growth factor 1 (IGF-1, also called somatomedin C) is perhaps the most significant of the factors.[12] Even if growth hormone levels are normal, growth can be inhibited if the level of IGF-1 is low. This is an important concept, especially when considering methods purported to raise growth hormone levels, such as supplementation of ornithine, arginine, and lysine, as discussed in Chapter 11.

TESTOSTERONE.   Testosterone and testosterone analogs (such as synthetic anabolic steroids, discussed in Chapter 11)

Key Point

Several hormones influence one or both sides of protein turnover; they include insulin, cortisol, growth hormone, and testosterone.

are well known as promoters of protein synthesis in muscle.[16,17] Whether or not they invoke increased amino acid uptake into muscle is still unclear; however, they do appear to enhance the utilization of amino acids from the intracellular pool for protein synthesis. Therefore, even though uptake might not be increased, amino acid release from muscle cells is lessened, leading to net protein accretion.

Testosterone levels are known to increase in response to resistance exercise and might increase during endurance exercise. However, researchers have noted that higher training volumes for endurance athletes might decrease resting testosterone levels.[18,19] This reflects either an increased efficiency in protein metabolism or having crossed an overtraining threshold.

Many researchers suggest that testosterone as an anabolic factor should be measured in concert with cortisol as a counterbalancing catabolic factor.[20] Furthermore, they suggest that the testosterone-to-cortisol ratio (T:C) is more predictive of net protein metabolism than testosterone or cortisol independently.[21] However, this ratio might have greater application to male athletes rather than females.

Experimental Investigations of Protein Metabolism

As for other nutrients, modern knowledge pertaining to protein metabolism during and after exercise is the result of combined research information derived from studies involving humans as well as animals. Much of the animal research has involved rodents and to a lesser degree dogs. The choice of experimental subject is often a matter of control of investigative variables and practical issues of the experimental design.

EXERCISE AND PROTEIN RESEARCH IN HUMANS. Using human subjects has the advantage of direct application of the exercise research information to athletes and other active people. In addition, research studies can be performed with specific populations of people (such as untrained/trained, male/female, and youth/adult). However, human research is at times limited in the breadth of information that it can provide. For instance, to assess protein metabolism in muscle tissue a biopsy is often required, and a more comprehensive evaluation of muscle protein turnover requires a greater sampling of tissue. The influence of exercise on protein metabolism in other tissue

offers similar if not greater challenges. Certainly the liver is important in amino acid metabolism during exercise. However, liver biopsies typically are not taken during exercise studies. Other tissues subject to exercise-induced adaptations (such as bone, lung, and digestive tract) also cannot be assessed directly. Thus, the effects of exercise on various tissues require some speculation based on indirect assessment techniques (such as dual energy X-ray absorptiometry, or DXA) and metabolites and other markers of physiological processes (see below).

EXERCISE AND PROTEIN RESEARCH IN ANIMALS. Research designs that utilize animals offer several advantages over human research in that more tissue is available to researchers, allowing for a more comprehensive evaluation. In addition, research studies involving animal subjects allow for greater control of the different variables that influence protein metabolism. The diet and nonexercise behavior can be controlled, which might decrease some of the variability of the derived information.

Several drawbacks are associated with exercise research involving animals. For instance, the direct application of the information derived from animal research can be problematic since some biochemical and physiological differences exist. With regard to protein metabolism, major considerations include differences in endocrinology and the metabolism of some amino acids (such as BCAAs). In addition, animals might not voluntarily perform desired exercises. Thus, to induce an exercise bout some type of involuntary motivational application may be involved that could induce the element of stress. Yet, information from research on animals often provides tremendous support and direction to human research.

Tools to Assess Protein Metabolism and Exercise

The assessment of protein and amino acid metabolism in muscle and other tissue can be made using direct and/or indirect methods. Direct measurement involves biopsy of muscle or other tissue and includes molecular biology techniques to study changes in the events of protein synthesis. Indirect assessment is more common and can involve the collection of urine and blood and measuring the factors within those fluids that are associated with protein turnover.

ISOTOPES. One way to measure how a nutrient is used in the body is to have subjects ingest a form that incorporates an isotope whose presence can be measured later. All atoms of a given element contain the same number of protons. However, atoms of a given element can contain varying levels of neutrons. Atoms of an element with differing numbers of neutrons are called **isotopes**.

If the neutrons of an isotope slowly break down, releasing radioactive energy, it is referred to as a **radioactive**

Key Point

Labeled amino acids contain isotopes that can be used to assess amino acid metabolism.

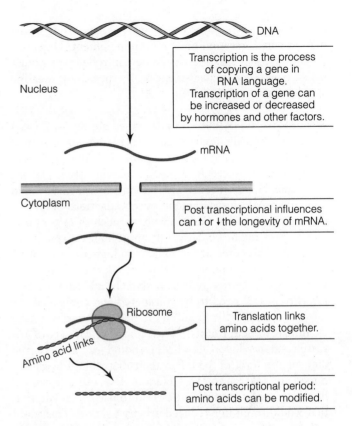

Figure 5-16

The stages involved in the formation of a protein include transcription, posttranscription, translation, and posttranslation.

isotope or *radioisotope*. The emitted energy is in the form of alpha, beta, or gamma rays, which can be measured with equipment such as a *Geiger counter* or a *gamma counter*. Radioisotopes that can be used for protein assessment include ^{14}C, ^{15}N, ^{17}O, and ^{3}H, and amino acids containing these isotopes are sometimes referred to as *radio-labeled*.

Nonradioactive or **stable isotopes** contain neutrons that are not degrading. The greatest advantage of using stable isotopes is that it avoids concern about the harmful effects of radioactivity. The presence of stable isotopes can be assessed by mass spectrometry. The stable carbon and nitrogen isotopes ^{13}C and ^{15}N are often used to form tracers of amino acid metabolism.

Amino acids containing isotopes are often referred to as being *labeled* because they can be identified and quantified in the laboratory. The labeled amino acids are often called *tracers*, as they can be tracked through tissue to provide information on the metabolism of the labeled amino acids. Labeled leucine, lysine, and phenylalanine are among the most commonly used tracers in muscle protein metabolism. Leucine is a BCAA, so labeled leucine can be used to estimate BCAA metabolism and oxidation during exercise. Phenylalanine and lysine are not oxidized in muscle tissue and can be used to estimate other aspects of protein turnover, such as incorporation into protein to reflect protein synthesis.

MOLECULAR BIOLOGY ASSESSMENT. If a sample of muscle tissue (biopsy) is available to a researcher, it can be assessed for changes in the content of messenger RNA (mRNA) and proteins. For instance, researchers can measure the level of mRNA and proteins using research protocols called **Northern blotting** and **Western blotting**, respectively. These research tools allow researchers to understand the impact of exercise and other metabolic states on the transcription of a gene (mRNA production) and to evaluate the translation of the mRNA to a protein.

The rate of protein synthesis can be influenced at several levels. These include transcriptional, posttranscriptional, translational, and posttranslational events (Figures 5-4 and 5-16). The levels of mRNA and muscle proteins reflect a balance between synthetic and degradative operations. For instance, mRNA levels for actin can be altered by factors involved in its production (transcription) as well as its stability or longevity once it is formed.

NITROGEN BALANCE. Nitrogen balance is perhaps one of the most recognizable tools used to assess body protein metabolism. As described above, nitrogen balance provides indirect evidence of body protein metabolism. Because more than half of the nitrogen in the body is in skeletal muscle, nitrogen balance can be used to assess protein metabolism in skeletal muscle. This notion is believed to be fairly accurate when the only experimental variable is the exercise stimulus. However, nitrogen balance should not be used as the sole assessment tool for muscle protein turnover. This is especially true when energy and protein levels are not rigorously controlled. For instance, nitrogen balance can remain negative if an insufficient level of energy is consumed, even when protein intake is high. In addition, as protein intake increases, nitrogen excretion and thus the balanced level increase as well. This might lead some researchers to overestimate protein needs.

CREATININE. Creatinine is a metabolite of creatine found primarily in skeletal and cardiac muscle and the brain. Free creatine in muscle is a reflection of the balance between the utilization of creatine phosphate for ATP production in the cytosol and the synthesis of creatine phosphate in mitochondria (see Figure 2-16). In the process, a predictable fraction of the free creatine is irreversibly converted to creatinine. Creatinine then diffuses out of muscle cells and circulates to the kidneys. The level of creatinine in the urine can be used as supporting information when assessing muscle protein content. However, dietary levels of creatine should be controlled when creatinine is used as an assessment tool, especially if creatine supplementation is involved.

3-METHYL HISTIDINE. As described above, 3-MH is a posttranslationally modified amino acid found primarily in actin filaments in muscle and certain other tissues. 3-MH typically is measured in extracted muscle tissue, plasma, and urine; however, some caution must be exercised when applying 3-MH measurements in investigations of muscle protein degradation. For instance, as much as 25% of urinary 3-MH might be derived from nonmuscle sources at rest,[22] and this percentage can be even higher in rodents.[23] These nonmuscle pools of 3-MH are derived from areas such as the digestive tract and skin. In addition, dietary levels of 3-MH need to be considered and controlled. As most of the 3-MH is found in the muscle, it is also found in animal muscle (meats). Therefore, participants of research studies involving 3-MH should eat a meat-free diet, or the level of 3-MH in their diet should be determined and taken into consideration. Also, exercise can influence kidney function and/or lead to partial dehydration that would result in a reduced urinary volume. This could influence the output of 3-MH for a time.

The ratio of 3-MH to creatinine is often measured in exercise studies. As both 3-MH and creatinine are related to lean body mass, using the ratio of 3-MH to creatinine helps normalize the data in groups of individuals with varying lean body masses. For instance, enhanced protein breakdown and release of 3-MH should increase the ratio of 3-MH to creatinine in the urine.

BLOOD UREA NITROGEN. Blood urea nitrogen (BUN) can also provide supporting information regarding muscle protein metabolism. BUN largely reflects the urea made by the liver destined for urinary excretion. Therefore, BUN and urinary urea should correlate closely in healthy people. An elevated BUN level suggests an increased degree of protein breakdown and amino acid use for fuel. However, BUN can increase when higher protein diets are consumed. Therefore, the level of dietary protein must be controlled in studies using BUN as an assessment tool.

HYDROXYLATED AMINO ACIDS. Skeletal muscle contains collagen protein as part of its vast connective tissue support system. During the remodeling of connective

Key Point

3-MH is a commonly used estimator of muscle protein breakdown. However, some assumptions must be taken into consideration when using 3-MH.

tissue in skeletal muscle, as stimulated by exercise training, hydroxylated amino acids such as hydroxylysine and hydroxyproline are released from skeletal muscle. However, these markers of collagen turnover cannot be used exclusively to assess muscle connective tissue metabolism since exercise also influences collagen remodeling in bone and other tissues.

■ AMINO ACID AND PROTEIN METABOLISM DURING EXERCISE *IN REVIEW*

- Protein turnover in skeletal muscle involves contractile and connective tissue protein, enzymes, ion pumps and transporters, and metabolic proteins (such as myoglobin).
- Several hormones influence skeletal muscle protein turnover, including testosterone, growth hormone, insulin, IGF-1, and cortisol.
- Knowledge regarding the influence of exercise on protein metabolism is derived from research involving both humans and animals.
- Several estimators of protein breakdown, synthesis, and oxidation are available to researchers, including 3-MH, radioactive and stable isotopes, and molecular biology tools.

■ RESISTANCE EXERCISE AND PROTEIN TURNOVER

As described above, protein synthesis and its counterpart, protein breakdown, occur simultaneously. Therefore, a net gain or loss of muscle protein reflects an imbalance between these opposing efforts and it is important to understand how different types of exercise influence both sides of protein turnover. In turn, this would assist coaches and athletes in designing training programs and diets to maximize the desired outcomes of exercise training.

For resistance training the goal is to create an imbalance between muscle protein synthesis and breakdown that allows for a chronic net positive protein balance. Many of the research tools mentioned above (labeled amino acids, 3-MH, N-balance) can be used to study the influence of different training and nutritional regimens on protein accruement. Researchers continue to fill in the gaps in knowledge and a more complete picture should be developed during the coming years. This section describes how resistance exercise can influence protein turnover and also provides insight into the potential for nutrition intake to support or negate desirable outcomes of training.

Resistance Exercise and Protein Synthesis

As would be expected, protein synthesis is enhanced for an extended period after the completion of a bout of strenuous

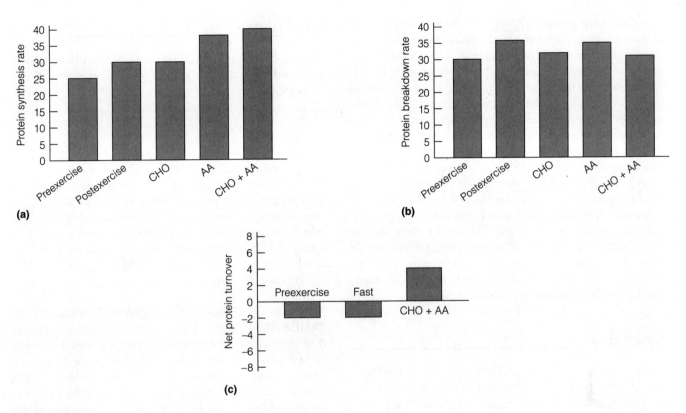

Figure 5-17

Expected influence of a bout of weight training on (a) muscle protein synthesis, (b) protein breakdown, and (c) net protein turnover 3–4 hours after training and while fasting (postexercise) or when carbohydrate (CHO), essential amino acids (AA), or both (CHO + AA) were provided immediately after training. Pre-exercise was a fasting state, and units are arbitrary.

resistance training. For instance, researchers reported that muscle protein synthesis was increased for as long as 24 hours after a resistance training session.[24] However, the rate of protein synthesis during strenuous resistance training might be unchanged or even slightly depressed. This suggests that protein synthesis operations do not increase dramatically until after the exercise stress has ceased (Figure 5-17a). The most outstanding factors that influence the degree of postexercise protein synthesis enhancement include the intensity or workload of the training session, the level of training of the individual, and the nutritional intake in association with the training session.

TYPE OF PROTEINS AFFECTED BY RESISTANCE TRAINING.   At this time researchers can only speculate based on limited research information as to which body proteins respond to resistance training. However, it would seem logical that the majority of postexercise increases in protein synthesis would be related to contractile proteins such as actin and myosin. In addition, it seems likely that these adaptations would occur across all fiber types, but to varying degrees. Some enhancement in protein synthesis would also be expected for noncontractile proteins. For instance, the level of some enzymes involved in anaerobic metabolism (for example, creatine kinase) should increase

in response to resistance training (see Chapter 4). Proteins required for the remodeling and strengthening of connective tissue should also have increased synthesis.[9]

LEVEL OF RESISTANCE TRAINING.   The level of resistance training is also an important consideration. For instance, researchers determined that protein synthesis remained elevated for 2 days after an isolated bout of resistance exercise in previously untrained individuals.[25] However, the impact of a similar training session seems to be lower and wanes earlier when individuals have trained regularly.[26] Therefore, in order for a trained individual to experience the same effect on protein synthesis, the training workload would have to be increased. Said another way, the trained subjects would need to work harder to continue to make significant gains. This concept of the progression of training workload might initiate the quest for unnatural means for enhancing muscle mass in some individuals (see Chapter 11).

BASAL PROTEIN SYNTHESIS AND RESISTANCE TRAINING.   Without doubt, appropriate resistance training results in muscle hypertrophy for most people. Also, it would be hard to argue that muscle hypertrophy is unrelated to protein accrual in the trained muscle tissue.

However, what is easily arguable is whether regular resistance training results in enhanced protein synthesis and net positive *basal* protein turnover at rest. The primary problem in confidently concluding that regular training enhances basal protein synthesis lies in the vague timeline of when the influence of the last training session ends. As the influence of a training session on protein synthesis can last a day or more, measurements taken within 48 hours after the most recent training session must be questioned. Yet, if a researcher waits too long after the last training session (say a week), it is difficult to know when the *detraining* of muscle would begin. Perhaps this question can be addressed with experimental designs that include progressively longer periods between training sessions. This might allow for the identification of a plateau in protein synthesis rate that is higher or the same as pretraining measurements. This type of experimental design would also be helpful in determining the timeline for the onset of detraining.

Key Point

Protein synthesis is increased after resistance training and is influenced by the intensity (workload) of the resistance exercise session and level of training.

Resistance Exercise and Protein Breakdown

Resistance exercise does not seem to result in significant changes in protein breakdown during a training session. Some researchers have speculated that since a typical resistance training session actually involves more time resting than exercising, the workload is similar to sustained lower submaximal intensities, which are not associated with protein breakdown.[27] On the other hand, resistance training seems to enhance the potential for muscle protein breakdown for several hours after completing a strenuous training session (Figure 5-17b).[9,12] In one study, researchers determined that muscle protein breakdown was elevated by 30% 3 hours after the exercise session and was reduced to 18% above resting levels after 24 hours.[25] The level of muscle protein breakdown had returned to pre-exercise levels (baseline) at 48 hours. Yet, estimations of the timeline and magnitude of the breakdown processes have varied in research studies owing to differences in participants and experimental designs.[25,28]

TYPE OF RESISTANCE TRAINING AND PROTEIN BREAKDOWN.   One important influence on protein breakdown resulting from resistance exercise is the type of muscle action. *Eccentric* muscle actions, which are contractions while lengthening, have been studied more than their *concentric* counterparts. Logically, eccentric muscle actions would result in more muscle damage or injury to muscle tissue, and allow for greater repair and recovery operations. This notion is supported by the ability of eccentric

Key Point

Resistance training enhances both protein breakdown and synthesis rates for several hours after completion of exercises.

loadings to result in the activation of certain proteases in muscle, such as *calpain*.[29] It is possible that eccentric contraction–induced injury to the plasma membrane and sarcoplasmic reticulum membrane enhances calcium presence in muscle fibers and calcium-associated activation of calpain and other proteases. Furthermore, researchers have reported that the magnitude of muscle damage and delayed onset muscular soreness (DOMS) is greater after eccentric than after concentric muscle actions.[30]

TYPE OF PROTEIN BREAKDOWN DURING RESISTANCE TRAINING.   Early research findings suggested that there is a preferential pattern for muscle protein breakdown in response to resistance training. For instance, researchers reported that urinary levels of 3-MH were not altered by upper and lower body weightlifting for the subsequent 24-hour period.[31,32] This suggests that enhanced breakdown of contractile protein does not occur as the result of an isolated bout of resistance training.[33] However, it does seem that 3-MH excretion in the urine can increase with more time and/or training sessions.[25,34] Therefore, the enhanced protein breakdown resulting from regular resistance training seems to affect both contractile and noncontractile proteins.

Resistance Training and Amino Acid Oxidation

Most research in the area of protein metabolism and resistance training has focused on protein synthesis and the enhancement of muscle mass. Meanwhile, very little research has attempted to answer the question of whether an isolated bout of resistance training or chronic training enhances protein oxidation in muscle or other tissues. One group of researchers attempted to shed light on this topic by infusing leucine during a resistance training workout.[35] The training session lasted an hour and consisted of three sets of nine separate exercises, each performed at 70% of 1 RM. The results of this study suggest that an isolated bout of resistance training did not enhance leucine oxidation in muscle or other tissue. This means that the increased protein need of people engaged in resistance training, by and large, is attributable to the accruement and maintenance of a greater mass of skeletal muscle protein.

Influence of Food on Protein Turnover After Resistance Training

Resistance training is believed to enhance both protein synthesis and breakdown in the hours following a workout

Key Point

Resistance training does not increase amino acid oxidation.

(Figure 5-17). However, the degree of change in either side of protein turnover is modified by the consumption (or lack of consumption) of carbohydrate and/or protein. For instance, when essential amino acids are provided after a training bout, the degree of muscle protein synthesis is enhanced.[9,12,36] When carbohydrate is provided with the amino acids there might be, at best, a slight increase in protein synthesis above the essential amino acids alone. When carbohydrate is provided by itself, protein breakdown is lessened but the effect on protein synthesis is minimal.

It appears that the consumption of a posttraining food source of carbohydrate and protein, especially essential amino acids, is necessary to maximize muscle protein enhancement in response to resistance training (see below for recommendations). Although most of the research to date has involved males, this response is believed to have application to both genders. Age, however, might be an influential factor because in elderly individuals amino acid ingestion may provide less enhancement of protein synthesis following resistance training.[9] Much more research is needed in this area, and hormone levels require serious consideration when making comparisons between different age groups.

■ RESISTANCE EXERCISE AND PROTEIN TURNOVER *IN REVIEW*

- Protein synthesis might be unchanged or reduced during resistance exercise and can increase for several hours after the training is complete, depending on the intensity and level of training.
- Protein breakdown might not occur during typical resistance training but may increase for several hours after training.
- Amino acid oxidation is probably not enhanced during resistance exercise.
- Consumption of a carbohydrate- and protein-containing meal immediately following resistance training is important to minimize protein breakdown and maximize protein synthesis in muscle.

Key Point

Eating carbohydrate and protein immediately after resistance training decreases muscle protein breakdown and enhances protein synthesis, leading to a net positive protein balance.

■ ENDURANCE EXERCISE AND PROTEIN TURNOVER

Endurance exercise is performed at significantly lower intensity levels than resistance exercise. However, the much longer duration of exercise creates a sustained physiological stress on muscle and other tissue. In response, the hormonal milieu can promote a net negative protein balance during endurance exercise. While muscle protein synthesis is unchanged or decreased, muscle protein breakdown can increase. In accordance, the pool of free amino acids in muscle fibers can increase and some amino acids can be used as fuel and others released into the circulation. This section describes how endurance exercise influences protein turnover and also provides insight into the potential for nutrition intake to support or negate desirable outcomes of training.

Endurance Exercise and Muscle Protein Synthesis

Endurance exercise can decrease muscle (and liver) protein synthesis in an intensity- and duration-related manner.[37–39] To be more specific, shorter duration endurance exercise at a higher submaximal intensity (>70% VO_2max) or longer duration endurance at a more moderate intensity (50–65% VO_2max) can decrease the rate of muscle protein synthesis during that bout. For instance, researchers determined that during $3\frac{3}{4}$ hours of treadmill exercise at 50% VO_2max, whole body protein synthesis had decreased by 14%.[40] The researchers conducting this investigation also measured protein synthesis during recovery and determined that protein synthesis increased by 22% above pre-exercise levels. The findings of another study also support the notion that the rate of muscle protein synthesis increases for several hours following endurance exercise.[41]

The recovery and enhancement of muscle protein synthesis following endurance exercise levels might not be immediate.[42] Furthermore, the postexercise increase in protein synthesis might be dampened when an endurance exercise bout is more extreme. Therefore, endurance sessions at higher submaximal intensity for longer periods might negate some or all of the potentially positive influence on protein synthesis. Not only will the more extreme endurance effort lessen the rise in protein synthesis following exercise, it also shortens the duration of the enhancement. In addition, the degree of enhancement of protein synthesis might lessen with chronic training.[12] Thus, in accordance with the law of diminishing return discussed in Chapter 3, this allows for greater gains in adaptation and performance earlier in a training program and subsequent performance plateaus after months of invariable training.

TYPES OF PROTEINS SYNTHESIZED IN RESPONSE TO ENDURANCE EXERCISE.   Protein synthesis can be enhanced for several hours following endurance training

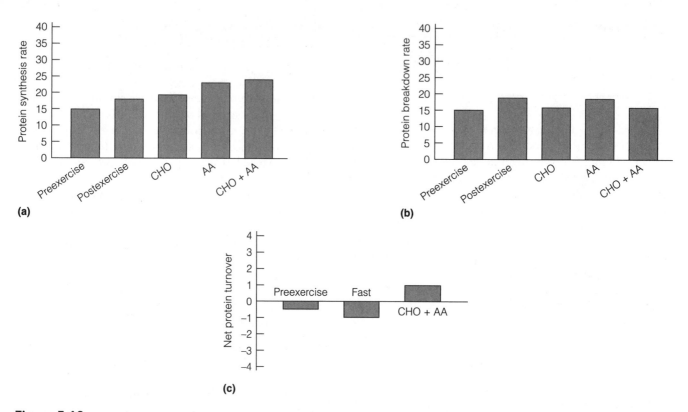

Figure 5-18
Expected influence of a bout of endurance training on (a) muscle protein synthesis, (b) protein breakdown, and (c) net protein turnover 3–4 hours after training and while fasting (posttraining) or when carbohydrate (CHO), essential amino acids (AA), or both carbohydrate and essential amino acids (CHO + AA) were provided immediately after training. Preexercise was a fasting state, and units are arbitrary.

(Figure 5-18a). The intensity level of endurance training leads one to expect that most of the protein synthesis would be in relation to aerobic energy metabolism. For instance, endurance training can enhance the content of myoglobin, mitochondria, and β-oxidative enzyme in muscle. In addition, enhanced production of lipoprotein lipase (LPL), antioxidant enzymes, and fatty acid binding proteins (FABPs) might contribute to increased protein content. Some enhancement of proteins unrelated to aerobic energy metabolism is also likely. These enhancements would account for at least some of the gains in strength and power noted in some studies involving endurance training.

PRECISION OF ESTIMATES OF PROTEIN SYNTHESIS RATES. Estimations of the degree of suppression of protein synthesis can vary based on the measurement tools used in exercise studies. This concept was demonstrated when researchers used two different labeled amino acids to estimate protein synthesis rates during 105 minutes of cycling at 30% VO₂max.[43,44] They found that when they used labeled leucine and lysine, the reduction in protein synthesis was estimated at 48% and 17%, respectively.

Therefore, both amino acids revealed a reduction in protein synthesis; however, the magnitude of the reduction was estimated to be nearly three times greater when using the labeled leucine. Information such as this confirms that although general protein requirements increase as the result of strenuous endurance exercise, the metabolism and requirements for specific amino acid vary.[33]

Endurance Exercise and Protein Breakdown

Protein breakdown can occur in liver and skeletal muscle during moderate and higher intensity endurance exercise (Figure 5-18b). For instance, researchers estimated that during 3¾ hours of treadmill exercise at 50% VO₂max, the rate

Key Point

Endurance exercise can decrease the rate of protein synthesis in skeletal muscle in an intensity- and duration-related manner and increase the rate of protein synthesis for several hours after exercise.

of protein degradation had increased from rest by 54%.[40] In this study the researchers also estimated protein breakdown during recovery and found that the rate of breakdown returned to the pre-exercise level soon after the completion of exercise. The potential for enhanced protein breakdown during endurance efforts is supported by research findings indicating that the level of leucine (a BCAA) oxidized during a bout of prolonged endurance exercise is about 25 times greater than the content of free leucine in the amino acid pool in muscle, liver, and plasma.[38] Therefore, protein breakdown would have had to occur to make more leucine available for oxidation.

Several factors require further research in regard to protein breakdown. For instance, the influences of pre-exercise muscle glycogen status and carbohydrate consumption during endurance exercise remain to be firmly determined. Also, if the ingestion of carbohydrate were able to lessen muscle protein breakdown during and after exercise, this might positively influence performance adaptations of chronic training. This would be another positive influence of consuming a sport drink during endurance efforts.

PREFERENTIAL PROTEIN BREAKDOWN DURING ENDURANCE EXERCISE. A hierarchy of protein breakdown during endurance exercise may exist. Some research information suggests that noncontractile proteins are preferentially broken down during endurance exercise and the rate of contractile protein degradation might be suppressed.[37] However, there are some discrepancies in the evidence. For instance, when an individual performs the same endurance exercise over several days, more contractile protein is broken down.[33] Therefore the level of training can be an important consideration. Researchers are currently attempting to add clarity to this area of exercise nutrition.

Key Point

Protein breakdown can increase during endurance exercise efforts.

Amino Acids as Fuel During Endurance Exercise

Although carbohydrate and fat provide the majority of the energy during endurance exercise, one cannot overlook amino acids as a third potential energy source. Researchers have reported that 3–18% of the energy expended during longer bouts of submaximal exercise is attributable to amino acids.[37] Much of the energy derived from amino acids during endurance exercise occurs via direct oxidation in active skeletal muscle fibers, and the remainder is mostly attributable to glucose created via gluconeogenesis in the liver.

Several factors influence the level of contribution made by amino acids to exercise fuel. Perhaps the most significant factor is amino acid availability in the intracellular amino acid pool. This means that muscle protein breakdown and even amino acid uptake from circulation can be important factors. When considering protein breakdown, intensity and duration of the exercise are important factors. Elevations in cortisol levels might lead to muscle protein breakdown. Meanwhile, the blunting of insulin release when exercise intensity climbs above 50% VO_2max might render muscle protein more vulnerable to breakdown.

ENDURANCE TRAINING ADAPTATIONS AND AMINO ACID OXIDATION. The contribution made by amino acids to energy expenditure probably decreases as a result of the metabolic adaptations promoted by regular endurance training.[45] These adaptations would involve changes in the metabolism of amino acids and/or other energy nutrients. For instance, increased fatty acid oxidative capabilities result in reductions in glucose utilization at a constant workload after several weeks of training. This might also decrease the need for amino acids during endurance exercise at the same level of intensity. In addition, an enhancement of muscle glycogen stores in trained muscle fibers might decrease the need for amino acid as fuel. However, more research is needed in this area to prove that regular endurance training promotes the partial sparing of muscle protein.

Influence of Food on Protein Turnover After Resistance Training

Researchers have noted that the consumption of a carbohydrate source or complete nutrition supplement immediately following exercise could abate most, if not all, of the reduced postexercise protein synthesis (Figure 5-18a).[46] For instance, a 26–30% reduction of protein synthesis rate was measured 1 hour after treadmill running (2 hours), when food was omitted postexercise. However, the consumption of carbohydrate after exercise lessened the reduction of protein synthesis rate to only 10% of pre-exercise levels. Interestingly, the consumption of a complete nutritional supplement resulted in a protein synthesis rate that was 8% greater than nonexercise levels. These results suggest that the rate of protein synthesis after exercise is strongly and positively influenced by meal-induced factors such as elevations in insulin and availability of amino acids. Perhaps the influences of a postexercise carbohydrate and protein meal are similar to what has been observed after resistance exercise.[12] See below for recommendations for posttraining meals.

■ ENDURANCE EXERCISE AND PROTEIN TURNOVER *IN REVIEW*

- Protein synthesis might be unchanged or reduced during endurance exercise depending on the intensity of the bout and the level of training.
- Protein breakdown can occur during higher intensity, longer duration endurance exercise.
- Amino acid oxidation can occur during prolonged moderate to higher intensity submaximal endurance exercise.
- Ingestion of a meal that includes carbohydrate and amino acids immediately following endurance training is probably important to minimize protein breakdown and maximize protein synthesis in muscle.

■ PROTEIN INTAKE AND REQUIREMENTS OF ATHLETES

The need for dietary protein for athletes is obvious, but how much protein is still a matter of debate. A few decades ago it seemed that there were two schools of thought. Some researchers felt that protein requirements were increased substantially during periods of strenuous training. This view has endured for decades and was surely shared by the strength and power athletes of the time, though not necessarily by endurance athletes. Many other researchers were not convinced that athletes had protein requirements beyond those of the general population. They noted that protein requirements are expressed as grams of protein per kilogram of body weight, so more muscular and heavier athletes (such as football players and bodybuilders) already have higher calculated recommendations. And, even if protein requirements were slightly higher, the RDA was already "padded" (two standard deviations above the average) to provide any additional protein requirements. In addition, it has long been believed that high-protein diets are detrimental to human health. As discussed in the feature In Focus: High-Protein Diets and Health Concerns at the end of the chapter, the strong statements made greatly outweighed the available research of the day.

Protein Intake of Strength and Power Athletes

The protein consumption of various strength and power athletes has been reported by researchers in terms of grams of protein per day, percentage of energy consumed, grams per kilogram of body weight, or a combination of two or all three measures. For instance, the average protein intake of female bodybuilders in seven studies was 103 ± 42 g daily or 27.9% ± 9% of total energy.[47] Male body builders averaged 195 ± 38 g of protein, or 28 ± 10% of their total energy intake, over twelve studies.[47] Male

Key Point

Protein intake levels are expressed in g/day, g/kg of body weight, or % of energy consumed.

weightlifters and wrestlers averaged 17 ± 3% of their energy as protein over six combined studies, and football players averaged 156 ± 44 g of protein or 17 ± 3% of total energy over three studies. When protein intake was expressed relative to body weight, the football players averaged 1.7 g/kg.[47]

Several variables could influence the level of protein consumed by a strength training athlete, including sport position, age, and timing. For instance, researchers determined that the defensive backs in their study had lower protein intakes than offensive and defensive linemen.[48] Another study reported the average protein intake of junior high school, senior high school, and college football players as 14%, 16%, and 22%, respectively.[49,50] Other factors such as income and availability of protein supplements and high-protein sport bars and shakes can also influence protein intake. Some athletes might fail to consume adequate levels of protein because they associate many good protein sources with a higher fat content. This might be the case with whole milk and certain cuts of animal meat. See the accompanying feature, Nutrition in Practice: How to Increase Protein Intake Without Increasing Fat Intake.

Protein Requirements of Strength Training Athletes

Clearly, the protein needs for athletes engaged in strenuous strength training are increased. For these athletes the enhanced protein need is related to the accumulation and maintenance of a greater than typical muscle mass. Much of the information supportive of an enhanced protein need is derived from nitrogen balance studies. Several studies have demonstrated that recommendations for the general public such as the American RDA (protein intake of 0.8 g/kg of body weight) would put most serious resistance training athletes in negative nitrogen balance.[51,52]

The results of earlier studies suggested that protein intake of more than 2 g/kg of body weight daily is necessary to maintain nitrogen balance during high-intensity training, such as top weight lifters preparing for international-level competition.[53,54] However, more recent studies suggest a lower protein requirement to achieve nitrogen balance during strength training. For instance, one study involved strength trained and sedentary individuals who were fed diets with a protein content of either 0.86, 1.4, or 2.4 g/kg of body weight.[55] The researchers conducting this study reported that the trained participants required an average protein intake of 1.41 g/kg

How to Increase Protein Intake Without Increasing Fat Intake

Consuming adequate protein should be a foremost consideration for both strength/power and endurance athletes. However, some athletes find it difficult to eat the recommended level of protein. In particular, they avoid consuming some good sources of protein because of their fat content or their source. The following are several ways to enhance protein levels in the diet without adding a lot of fat.

- Mix tuna with chopped pickles to produce a tuna salad without mayonnaise.
- Use three or four egg whites to each whole egg when making eggs, omelets, and breakfast burritos.
- Mix nonfat dry milk in a blender with skim milk, nonfat frozen yogurt, and fruit.

- Grill or bake fish seasoned with lemon and spices.
- Grill chicken (add barbecue sauce or season as desired) to use as a topping for homemade pizza.
- Use nonfat cottage cheese as a topping for baked potatoes.

In addition, several sport foods and supplements are designed to enhance protein intake without providing significant fat. For instance, a 72-g packet of MET-Rx powder contains 37 g of protein, which accounts for roughly 57% of the energy in the powder. The Protein Plus Bar, also manufactured by MET-Rx, contains 32 g of protein. The Myoplex Plus Deluxe Bar by EAS contains 24 g of protein and the Promax Bar by SportPharma contains 20 g of protein.

---

of body weight to achieve nitrogen balance, whereas the sedentary group required only 0.69 g/kg. Based on these statistical averages (means), finding a level two standard deviations above the average (as used to determine RDA) resulted in recommendations of 0.89 and 1.76 g/kg of body weight for sedentary and trained individuals, respectively. As the protein balance for the sedentary group approximated the RDA for protein in Canada and the United States, the credibility of the study was easily supported.

Thus it seems clear that protein requirements are increased as a result of higher intensity resistance training. Most of the studies discussed above suggest that the protein requirement for nitrogen balance in strength and power athletes is about 100% higher than the RDA in the U.S., or perhaps higher. Many researchers feel that 1.4–1.75 g of protein is adequate to achieve nitrogen balance in most strength-training athletes. Given the high-energy intakes required by these athletes, the protein requirement to achieve nitrogen balance is below 20% of total energy intake.

GREATER INTAKES OF PROTEIN FOR POSITIVE NITROGEN BALANCE. Many studies have been performed in an attempt to determine protein requirements for strength and power athletes. In those studies nitrogen balance was determined and used as the basis for recommendations. However, a question such as this is often posed: "Does one need additional protein during periods of protein accretion such as the growth phase of bodybuilding or a lineman trying to enhance muscle mass during the off-season?" One might use the information from the study discussed above involving three levels of protein provided to training weight lifters.[55] In that study, labeled

leucine was used to determine the protein level threshold above which additional amino acids were oxidized. The researchers determined that when the level of protein was provided at 0.86 and 1.4 g/kg of body weight, there was not a change in leucine oxidation. However, when the rate of leucine oxidation was compared between the participants receiving 2.4 versus 1.4 g/kg of body weight, the higher protein intake was associated with greater leucine oxidation, indicative of excess. Thus, when protein intake achieves nitrogen balance, extra protein might not allow for greater benefit than that afforded by the training stimulus.

Many of the studies to date have estimated protein needs for nitrogen balance while the training protocol remained constant for several weeks. The results of some of these studies suggest that the level of dietary protein needed to achieve nitrogen balance is highest at the beginning of an exercise program and is reduced as the program continues and is unvaried.[56] Therefore, as the training leads to neuromuscular adaptations, the overload stimulus is lessened. Yet, if the training workload were to be progressively increased, enhanced protein requirements might be sustained. The findings of earlier studies have suggested that protein needs were relative to training intensity and degree of overload.[53,54]

PROTEIN CYCLING. It might be that when a person eats a higher protein diet, the intestines and the liver take up and metabolize more of the absorbed amino acids. While this idea is largely attributable to studies performed in rats,[11,56] it is logical that some or much of the same effect would occur in humans. This operation serves as a buffer to limit the level of circulating amino acids after a high protein intake. In accordance, some strength and power athletes practice dietary protein cycling in an attempt to maximize the quantity of amino acids that get past the intestines and liver and into the general circulation and thus to muscle. For instance, they cycle their protein intake between a lower intake (such as 0.8–1.0 g/kg) and a higher intake (such as 1.7 g/kg). The switch between the levels would occur every 3 days. This protocol is based on the notion that it takes several days for the liver and intestines to adapt their amino acid retention during a higher protein intake. Thus protein levels are switched back to the lower level before adaptation can occur during the higher intake. Whether or not this has an impact on the level of amino acids available to skeletal muscle and protein turnover remains to be studied.

Key Point

Protein recommendations for strength and power athletes is 1.4–1.75 g/kg of body weight to achieve nitrogen balance.

Protein Intake of Endurance Athletes

Nutrition surveys of endurance athletes have revealed that the protein intake among male triathletes was 15% of total energy.[57] Similarly, the information obtained from surveys of female competitive distance runners estimated their protein intakes to be around 13–15% of total energy.[58-60] Shortly prior to competition, the contribution of protein to their total energy intake might decrease as the runners focus more on carbohydrate.[61] Vegetarianism seems to be more prevalent among endurance athletes, especially distance runners, than among strength and power athletes. Protein and other nutritional considerations of a vegetarian diet are addressed in Chapter 15.

Key Point

The average protein intake (g/day) of strength and power athletes tends to be a little higher than that of endurance athletes.

Protein Requirements for Endurance Training

Unlike resistance training, endurance training can increase the oxidation of amino acids in working skeletal muscle and the liver. As mentioned above, skeletal muscle cells have the ability to directly use six amino acids for energy purposes, and the remainder can be released from muscle during exercise. Estimates of amino acid contributions to the total energy expended during endurance efforts have varied with the sport, intensity level, and

duration of the exercise bout. Estimates range from 3–18% as discussed above.

It appears that the protein requirements for endurance athletes are similar yet slightly less than, those of strength training athletes. A protein intake of 1.2—1.4 g/kg of body weight should suffice for most competitive endurance athletes. The requirement varies from athlete to athlete as well as day to day depending on changes in training intensity and duration and other factors. In any case, when an endurance athlete consumes enough energy for weight maintenance, providing 15–20% of the energy as protein should suffice.

Key Point

Protein recommendations for competitive endurance athletes are 1.4–1.75 g/kg of body weight, similar to bodybuilders and other weight trainers.

Total Energy and Carbohydrate Intake and Protein Turnover

Consuming an appropriate level of energy to at least maintain body weight is crucial to maintaining muscle protein mass. In addition, to maximally benefit from resistance training by increasing muscle mass without reducing other energy stores would require an energy intake in excess of that needed for weight maintenance. Thus, as many athletes restrict their energy intake to reduce weight and body fat, muscle protein can be compromised in the process. Inadequate energy intake can also lead to incomplete recovery of muscle protein in endurance athletes and compromise metabolic adaptation processes.

In light of the recent information on the importance of insulin in reducing muscle protein turnover, the consumption of lower carbohydrate diets by athletes during training seems potentially detrimental to performance. In addition, during periods when very little carbohydrate is consumed, more protein would be necessary in the diet, as amino acids are used for gluconeogenesis. It does appear that when carbohydrate intake is significantly reduced, muscle and certain other tissues (such as the liver) adapt to use more fat as a fuel source at rest and during exercise. How that might influence the reliance on amino acids as gluconeogenic substrates, especially during exercise, is still unknown. Some athletes approaching competition (such as wrestlers and bodybuilders) switch to a very low carbohydrate diet in conjunction with other efforts;

Key Point

Appropriate levels of energy and carbohydrate are necessary to dedicate more protein to repair, recovery, and adaptation processes associated with training.

however, these practices are not generally recommended, as discussed in Chapter 14.

Timing of Protein Consumption and Postexercise Protein Turnover

One important factor during postexercise recovery of protein synthesis is the timing and composition of the next meal. The consumption of amino acids, especially essential amino acids, allows for an influx into muscle cells and promotes protein synthesis. Meanwhile, carbohydrate provides fuel and also increases the uptake of amino acids into muscle, thus supporting protein synthesis and decreasing protein breakdown. The effects on net protein turnover of consuming both carbohydrate and protein appear to be additive.[62]

At this time one can only speculate as to the ideal composition of the posttraining/postcompetition meal. A minimum recommendation of 0.5 g of protein per kilogram of body weight would seem to be sufficient. This level of protein would complement 1.5 g of carbohydrate per kilogram of body weight recommended after a strenuous training bout or competition, as discussed in Chapter 4. Using the recommendations, the meal would provide at least 135 g of carbohydrate and 45 g of protein for a 200-lb (90-kg) weight trainer and at least 35 g of protein and 105 g of carbohydrate for a 160-lb (73-kg) endurance athlete. These general recommendations would provide about 560–720 kcal for these athletes. This might be about one-fifth or one-sixth of the athlete's daily energy needs for weight maintenance. In addition, these recommendations assume that these athletes will eat again within 3–4 hours.

Key Point

Athletes should strive to consume at least 1 g of carbohydrate and 0.5 g of protein per kilogram of body weight as soon as possible after completing a strenuous training session.

ORDER OF CARBOHYDRATE AND PROTEIN IN POSTTRAINING FEEDING. It has been clearly established that consuming carbohydrate and protein immediately following resistance or endurance exercise can have a positive influence on both sides of the protein turnover coin. Yet, there might be reason to ponder whether the carbohydrate should precede the amino acid source. The notion is that the absorption of amino acids would not occur until insulin levels are elevated and the metabolic state of the liver and skeletal muscle is shifting to a "fed" state. This might reduce the immediate breakdown and oxidation of absorbed amino acids. However, the trade-off is that the stimulating effect of amino acids on protein

synthesis would be postponed following the exercise bout. Therefore, it is still recommended that the postexercise meal contain both carbohydrate and protein.

CONSUMING CARBOHYDRATE AND PROTEIN DURING EXERCISE. It is also easy to speculate that consuming a sport drink during a training session might not only benefit immediate performance but also support better recovery and adaptation. This is because at the close of a training session insulin levels would already be increased. The result should be decreased oxidation of amino acids ingested in the postexercise meal. Also, if a significant carbohydrate source is consumed during a training session, the order of posttraining carbohydrate and protein discussed above might become a moot point.

If a weight trainer spends extended periods of time in the gym and works multiple muscle groups, timing of carbohydrate and protein intake becomes a consideration as well. For instance, if a weight trainer performs 12 sets of exercises that focus on the chest, then moves on to triceps and then legs, significant time can pass between the completion of the chest exercise and the close of the training session. It would seem that in this situation the weight trainer should consider consuming a carbohydrate and protein resource (such as a sport bar) prior to or during the training session.

Number of Meals and Energy Ratios to Support Desirable Muscle Protein Turnover

It seems logical that a strength or endurance athlete would be better served by a diet separating food and energy nutrients into numerous smaller meals over the waking hours. This would theoretically allow for a more consistent influence of insulin over the course of the day and potentially minimize catabolic periods during extended stretches between meals. In addition, a more consistent availability of amino acids, especially the essential amino acids, might have a more desirable influence on protein synthesis.

Practical recommendations for athletes would be to determine their energy requirements and then split that energy into the number of meals for that day. For instance, if a 170-lb (77-kg) male runner has energy requirements of 3200 kcal to maintain body weight, five to seven meals would consist of roughly 460–640 kcal each. If 17% of his energy intake is to be protein, about 136 g of protein would be distributed throughout the day. If 40 g of that protein is dedicated to the posttraining meal then the remaining meals would contain 15–20 g of protein each. Many athletes

Key Point

Many sport nutritionists and coaches recommend that athletes partition their protein into several smaller meals over the course of the day for a more consistent availability of amino acids.

also dedicate more energy to the first meal of the day (breakfast) and the posttraining meal. This is demonstrated in Chapter 7 in the feature Nutrition in Practice: Designing an Optimal Training Diet for Sport Performance.

Use of Protein as a Nutrition Supplement for Athletes

Protein supplements are extremely popular although the vehicle of delivery has changed to some degree over the past decades. In past years, protein was largely supplemented in the form of powders used to make drinks and shakes. Today, high-protein sport bars and ready to drink (RTD) sources are available. Quite often protein shakes are used as meal replacements, as described in detail in Chapter 11. Meanwhile, some coaches and athletes believe that protein supplements in addition to what seems to be adequate dietary protein enhances muscle protein development. Researchers have attempted to determine if protein (or amino acid) supplements could have a positive influence on muscle protein turnover. For instance, one study was conducted to assess whether supplemental protein enhanced muscle and strength gains in young males who participated in a weight training program.[63] Half of the participants received a protein supplement that raised their protein intake to approximately 2.62 g/kg of body weight. Another group of males received a supplement in which carbohydrate was exchanged for protein. For this group, protein consumption was 1.35 g/kg of body weight daily. After $3\frac{1}{2}$ weeks of training, the level of protein necessary for nitrogen balance was 1.4–1.5 g/kg of body weight, regardless of the diet level of protein. The additional protein provided by a protein supplement did not lead to an enhanced gain in muscle mass or strength. The results of this study do not lend support to the use of protein supplementation as a means of inducing increased muscle mass development when protein requirements are already being achieved by food.

WHEY PROTEIN. Whey protein has become one of the most popular types of protein used in powders and high-protein sport bars and shakes. For instance, recent surveys of competitive male and female bodybuilders revealed that more than half of these athletes used whey protein supplements during precompetition preparation.[64,65] Whey protein is a complete protein source from milk and is easily digestible. Whey is also rich in BCAAs, which are touted to have an anabolic effect (see Chapter 11). Outside of some basic speculation, whey protein has yet to be proven more effective in positively influencing muscle protein turnover than other proteins. One study involving whey protein supplementation did demonstrate that whey supplementation (1.2 g/kg of body weight daily) resulted in a higher level of lean body mass in comparison to supplementation of carbohydrate of the same energy value.[66] However, the results of this study are difficult to apply because the protein and carbohydrate intake was different between the groups.

Table 5-7

Popular Amino Acid Supplements and Purported Effects

Amino Acid	Purported Ergogenic Effect
Glutamine	Promote positive protein balance and muscle mass enhancement
Arginine, ornithine, lysine	Increase release of growth hormone
Aspartic acid	Decrease presence of ammonia during sustained higher intensity exercise and decrease fatigue
Leucine, isoleucine, valine (BCAA)	Decrease central fatigue and perhaps promote protein synthesis in muscle

Individual or Combined Amino Acid Supplements

Based on their established physiological roles, as well as unproven notions, individual amino acid supplements are used by many athletes in an attempt to increase strength and/or muscle mass. Table 5-7 presents a list of theoretical performance benefits of amino acid supplementation. On the list are three amino acids, namely ornithine, arginine, and lysine, that are purported to increase the release of growth hormone from the pituitary gland. Many athletes use glutamine supplements in an attempt to reduce protein degradation after extreme exercise. Other athletes experiment with BCAA in an attempt to decrease central fatigue and possibly increase protein synthesis. The potential effects of these and other amino acid supplements remain somewhat controversial and are discussed in Chapter 11.

■ PROTEIN INTAKE AND REQUIREMENTS OF ATHLETES *IN REVIEW*

- Protein recommendations for both endurance and strength/power athletes are 1.2–1.4 g/kg of body weight a day respectively or 15–20% of energy needed for weight maintenance.
- Consuming a posttraining meal containing both carbohydrate and protein maximizes the positive effects of the exercise on muscle protein turnover; however, the levels and ratios are unclear at present.
- Many researchers and coaches recommend partitioning protein into several smaller meals consumed throughout the day.
- Protein supplementation is not necessary if diet protein levels are adequate.

Conclusions

Proteins are composed of nitrogen-endowed amino acids and are the structural and functional basis of cells and tissue. Proteins serve as transporters, hormones, enzymes, clotting factors, contractile and structural apparatus, and so on. Individual amino acids can be used to build important molecules, and a small amount of free amino acids is found in intracellular and extracellular fluids. Food proteins are derived from all natural foods. Complex proteins are denatured by stomach acid and digested to the absorbable amino acids by a battery of proteases in stomach and pancreatic juices and within cells lining the small intestine. Amino acids are exchanged between tissues, and the liver is able to oxidize most amino acids, while skeletal muscle can oxidize six or seven. Protein turnover in muscle and other tissue takes into consideration both protein synthesis and protein breakdown, which are independent operations.

Endurance exercise can result in increased protein breakdown and decreased protein synthesis during the bout. After the endurance training session, protein synthesis and breakdown can be significantly increased for several hours. Resistance exercise training may not significantly influence protein turnover during a training session; however, both protein breakdown and synthesis reduction are enhanced after the training session is complete. For both types of exercise, a posttraining meal containing carbohydrate and protein can minimize protein breakdown and enhance protein synthesis in the hours that follow the completion of exercise. A general recommendation for daily protein for endurance and strength athletes is 1.2–1.4 g/kg of body weight respectively to at least achieve nitrogen balance. Immediately following strenuous weight or endurance training, protein intake of 0.5 g/kg of body weight is recommended. This is complemented by carbohydrate intake of at least 1.5 g/kg of body weight. Protein supplementation has not proved efficacious when recommended diet protein levels are met.

IN FOCUS   HIGH-PROTEIN DIETS AND HEALTH CONCERNS

Strength and power athletes have long advocated higher protein diets to maximize performance in their sports. Nutritional surveys and personal communications suggest that some athletes consume 200–300 g of protein or more daily. This might be three to four times the RDA for a 190-lb (86-kg) adult male weight lifter. Higher protein intakes represent an attempt to maximize anabolic events in muscle tissue as well as to substitute for carbohydrate and/or fat energy. However, nutritionists have long been skeptical that these diets can cause greater gains in muscle mass and strength. Many also believe that a higher protein intake would take a toll on human tissue and cause deterioration of function.

For as much "bad press" as high-protein diets receive, one would expect to find overwhelming research evidence demonstrating that a high-protein diet contributes to an acceleration of disease processes. Surprisingly, however, relatively little direct information is available about the negative effects of a high-protein diet. Part of the problem lies in how to define a high-protein diet. How much protein and what types of proteins constitute a high-protein intake? Is a high-protein diet one that provides more than 150 g daily, or 1.8 g/kg of body weight, or more than 20% of energy intake? Also, different protein types have varying levels of amino acids, which affects protein turnover and other processes in tissue.

Researchers often have a difficult time "teasing out" the true effects of a higher protein intake from the concomitant dietary and behavioral factors that are often present. For instance, people who eat a high-protein diet, which is derived from red meat and eggs, might also be less active, smoke, and eat few fruits and vegetables. In this scenario, researchers must try to account for numerous potentially negative factors in order to assess the independent influence of protein level.

Much of the negative mind-set associated with a higher protein intake is based on research involving individuals with progressive kidney failure. When they are fed a higher protein intake their kidney function deteriorates more rapidly. How does this translate to individuals with healthy functioning kidneys? As noted in a recent review of the subject, there is at this time only a weak case for renal problems resulting from a higher protein diet in healthy adults.[67]

A high-protein diet might be detrimental to health in that many high-protein foods such as red meat, eggs and cheeses are rich in saturated fat and cholesterol and somewhat lower in some health-promoting nutraceuticals compared to higher carbohydrate sources such as fruits and vegetables.[2] Thus a more likely problem with a high intake of protein is that it commonly provides saturated fatty acids and to some extent replaces health-promoting fruits, vegetables, and whole grains. Also, meats are often charred, which leads to the formation of potentially dangerous amine compounds (free radicals).[68]

Early humans and their predecessors more than likely ate a diet rich in protein, obtained from wild

HIGH-PROTEIN DIETS AND HEALTH CONCERNS (CONTINUED)

game and vegetation (roots, vegetables, and so on). It is logical to think that the human genome evolved in conjunction with such a diet. Furthermore, chimpanzees, whose genome is the closest to that of humans, eat a high-protein diet. So perhaps it is not protein quantity so much as the protein source that is a potential detriment to health. For instance, the domestication of animals has drastically changed the amount of fat and fatty acid composition of animal flesh. The feed provided to farm animals leads to animal foods that have more saturated fatty acids that raise blood cholesterol levels as well as more ω-6 polyunsaturated fatty acids (PUFA) and less ω-3 PUFA. Consuming a diet with a higher ω-6:ω-3 PUFA ratio might promote inflammation, vasoconstriction, platelet aggregation, and cell growth. These mechanisms are involved in disease processes such as arthritis, heart disease, and cancers.

Reliable conclusions are impossible without further information. Solid research efforts focusing specifically on protein level and determining potential differences between protein types are needed. Also, it is important for researchers and health care practitioners to identify the difference between a high-protein diet and a nutritionally undesirable diet that is higher in protein. For instance, one could obtain more than 20% of energy from protein while eating only vegetables and soy-based products. How many people would consider that a disease-promoting diet?

STUDY QUESTIONS

1. How is the basic structure of amino acids unique among energy nutrients?
2. Which amino acids are used to make human proteins, and what is posttranslational modification of amino acids?
3. What type of bond links amino acids together, and what are the basic designs of proteins?
4. Name the different physiological roles for protein and amino acids.
5. Which amino acids are used by skeletal muscle for energy?
6. What is the relationship between protein synthesis and breakdown in protein turnover in skeletal muscle and the body?
7. Which hormones are believed to be very influential in protein turnover?
8. How does resistance and endurance training influence both sides of protein turnover?
9. What are recommendations for protein intake for athletes?
10. What influence do nonprotein energy sources (such as carbohydrate) and total energy intake have on protein turnover?

SUGGESTED READINGS AND REFERENCES

Groff JL, Gropper SS. *Advanced Nutrition and Energy Metabolism*, 3rd ed., Atlanta: Wadsworth, 2002.

Brown A. *Understanding Food: Principles and Preparation*, Atlanta: Wadsworth, 2002.

Wildman REC, Medeiros DM. Protein. In: *Advanced Human Nutrition*, Boca Raton, FL: CRC Press, 2000.

1. Haymond WM, Miles JM. Branched-chain amino acids as a major source of alanine nitrogen in man. *Diabetes* 31:86–89, 1982.
2. Smith K, Rennie MJ. Protein turnover and amino acid metabolism in human skeletal muscle. *Clinical Endocrinology* 4:461–498, 1990.
3. Clowes GHA, Randall HT, Cha CJ. Amino acid and energy metabolism in septic and traumatized patients. *Journal of Enteral and Parenteral Nutrition* 4:195–205, 1980.
4. Marliss EB, Aoki TT, Pozefsky AS, Most AS, Cahill GF. Muscle and splanchnic glutamine and glutamic acid metabolism in postabsorptive and starved man. *Journal of Clinical Investigation* 50:814–817, 1971.

5. Darmaun D, Déchelotte P. Role of leucine as a precursor of glutamine α-amino nitrogen in vivo in humans. *American Journal of Physiology* 260:E326–E329, 1991.
6. Van Hall G, Saltin B, van der Vusse GJ, Soderlund K, Wagenmakers AJM. Deamination of amino acids as a source of ammonia production in human skeletal muscle during prolonged exercise. *Journal of Physiology* 489:251–261, 1995.
7. Wagenmakers AJM. Muscle amino acid metabolism at rest and during exercise: Role in human physiology and metabolism. *Exercise and Sport Sciences Reviews* 26:287–314, 1998.
8. Dangin M, Boirie Y, Guillet C, Beaufrere B. Influence of the protein digestion rate on protein turnover in young and elderly subjects. *Journal of Nutrition* 132(10):3228S–3233S, 2002.
9. Rennie MJ, Tipton KD. Protein and amino acid metabolism during and after exercise and the effects of nutrition. *Annual Review of Nutrition* 20:457–483, 2000.

10. Fouillet H, Bos C, Gaudichon C, Tome D. Approaches to quantifying protein metabolism in response to nutrient ingestion. *Journal of Nutrition* 132(10):3208S–3218S, 2002.

11. Morens C, Gaudichon C, Metges CC, Fromentin G, Baglieri A, Even PC, Huneau JF, Tome D. A high-protein meal exceeds anabolic and catabolic capacities in rats adapted to a normal protein diet. *Journal of Nutrition* 130(9):2312–2321, 2000.

12. Tipton KD, Wolfe RR. Exercise, protein metabolism, and muscle growth. *International Journal of Sport Nutrition and Exercise Metabolism* 11(1):109–132, 2001.

13. Kimball SR, Farrell PA, Jefferson LS. Invited review: Role of insulin in translational control of protein synthesis in skeletal muscle by amino acids or exercise. *Journal of Applied Physiology* 93(3):1168–1180, 2002.

14. Banfi G, Marinelli M, Roi GS, Colombini A, Pontillo M, Giacometti M, Wade S. Growth hormone and insulin-like growth factor I in athletes performing a marathon at 4000 m of altitude. *Growth Regulation* 4(2):82–86, 1994.

15. Zaccaria M, Varnier M, Piazza P, Noventa D, Ermolao A. Blunted growth hormone response to maximal exercise in middle-aged versus young subjects and no effect of endurance training. *Journal of Clinical Endocrinology and Metabolism* 84(7):2303–2307, 1999.

16. Sheffield-Moore M. Androgens and the control of skeletal muscle protein synthesis. *Annals of Medicine* 32(3):181–186, 2000.

17. Wolfe R, Ferrando A, Sheffield-Moore M, Urban R. Testosterone and muscle protein metabolism. *Mayo Clinic Proceedings* 75(Suppl):S55–S59; discussion S59–60, 2000.

18. Flynn MG, Pizza FX, Brolinson PG. Hormonal responses to excessive training: influence of cross training. *International Journal of Sports Medicine* 18(3):191–196, 1997.

19. Gulledge TP, Hackney AC. Reproducibility of low resting testosterone concentrations in endurance trained men. *European Journal of Applied Physiology and Occupational Physiology* 73(6):582–583, 1996.

20. Hoogeveen AR, Zonderland ML. Relationships between testosterone, cortisol and performance in professional cyclists. *International Journal of Sports Medicine* 17(6):423–428, 1996.

21. Urhausen A, Gabriel H, Kindermann W. Blood hormones as markers of training stress and overtraining. *Sports Medicine* 20(4):251–276, 1995.

22. Afting EG, Bernhardt W, Janzen RWC, Rothig HJ. Quantitative importance of non-skeletal muscle N-methylhistidine and creatinine in human urine. *Biochemistry Journal* 220:449–452, 1981.

23. Dohm GL, Tapscott EB, Kasperek GJ. Protein degradation during endurance exercise and recovery. *Medicine and Science in Sports and Exercise* 19:S166–S171, 1987.

24. Chelsey A, MacDougall JD, Tarnpolsky MA, Atkinson SA, Smith K. Changes in human muscle protein synthesis after resistance exercise. *Journal of Applied Physiology* 73:1383–1388, 1992.

25. Phillips SM, Tipton KD, Aarsland AA, Wolfe SE, Wolfe RR. Mixed muscle protein synthesis and breakdown following resistance exercise in humans. *American Journal of Physiology* 273:E99, 1997.

26. Phillips SM, Tipton KD, Ferrando A, Wolfe RR. Resistance training reduces the acute exercise-induced increase in muscle protein turnover. *American Journal of Physiology* 276:E118–E124, 1999.

27. Phillips SM. Protein metabolism and exercise: potential sex-based differences. In: *Gender Differences in Metabolism* (Tarnopolsky M, ed.), Boca Raton, FL: CRC Press, 1999.

28. Lowe DA, Warren GL, Ingalls CP, Boorstein DB, Armstrong RB. Muscle function and protein metabolism after initiation of eccentric contraction-induced injury. *Journal of Applied Physiology* 79(4):1260–1270, 1995.

29. Belcastro AN. Skeletal muscle calcium-activated neutral protease (calpain) with exercise. *Journal of Applied Physiology* 74:1381–1386, 1993.

30. Croisier JL, Camus G, Deby-Dupont G, Bertrand F, Lhermerout C, Crielaard JM, Juchmes-Ferir A, Deby C, Albert A, Lamy M. Myocellular enzyme leakage, polymorphonuclear neutrophil activation and delayed onset muscle soreness induced by isokinetic eccentric exercise. *Archives of Physiology and Biochemistry* 104(3):322–329, 1996.

31. Hickson JF, Wolinsky I, Rodriguez GP, Pivarnik JM, Kent MC, Shier NW. Failure of weight training to affect urinary indices of protein metabolism in men. *Medicine and Science in Sports and Exercise* 18:563–567, 1986.

32. Horswill CA, Layman DK, Boileau RA, Williams RT, Askew EW. Biphasic changes in 3-methylhistidine and hydroxyproline following acute weight-training exercise. *International Journal of Sports Medicine* 9:245–250, 1988.

33. Paul GL, Gautsch TA, Layman DK. Amino acid and protein metabolism during exercise and recovery. In: *Nutrition in Exercise and Sport*, Boca Raton, FL: CRC Press, 1998.

34. Pivarnik JM, Hickson JF, Wolinsky I. Urinary 3-methylhistidine excretion increases with reported repeated weight training. *Medicine and Science in Sport and Exercise* 21(3):283–287, 1989.

35. Tarnopolsky MA, Atkinson SA, MacDougall JD, Senor BB, Lemon PWR, Schwartz H. Whole body leucine metabolism during and after resistance exercise in fed humans. *Medicine and Science in Sports and Exercise* 23:326–333, 1991.

36. Wolfe RR. Regulation of muscle protein by amino acids. *Journal of Nutrition* 132(10):3219S–3224S, 2002.

37. Graham TE, Rush JWE, MacLean DA. Skeletal muscle amino acid metabolism and ammonia production during exercise. In: *Exercise Metabolism* (Hargreaves M, ed.), Champaign, IL: Human Kinetics, 1995.

38. Dohm GL. Protein as a fuel for endurance exercise. *Exercise and Sport Science Review* 14:143–173, 1986.

39. Booth FW, Watson PA. Control of adaptations in protein levels in response to exercise. *Federal Proceedings* 44:2293–2300, 1985.

40. Rennie MJ, Edwards RHT, Krywawych S, Davies CTM, Halliday D, Waterlow JC, Millward DJ. Effect of exercise on protein turnover in man. *Clinical Sciences* 61:627–639, 1981.

41. Carraro F, Staurt CA, Hartl WH, Rosenblat J, Wolfe RR. Effect of exercise and recovery on muscle protein synthesis in human subjects. *American Journal of Physiology* 259:E470–E478, 1990.

42. Layman DK, Paul GL, Olken MH. Amino acid metabolism during exercise. In: *Nutrition in Exercise and Sport* (Wolinsky I, Hickson JF, eds.), Boca Raton, FL: CRC Press, 1994.

43. Wolfe RR, Goodenough RD, Wolfe MH, Royle GT, Nadel ER. Isotopic analysis of leucine and urea metabolism in exercising humans. *Journal of Applied Physiology* 52: 458–466, 1982.

44. Wolfe RR, Wolfe MH, Nadel ER, Shaw JHF. Isotopic determination of amino acid-urea interactions in exercise in humans. *Journal of Applied Physiology* 56:221–229, 1984.

45. McKenzie S, Phillips SM, Carter SL, Lowther S, Gibala, MJ, Tarnopolsky MA. Endurance exercise training attenuates leucine oxidation and BCOAD activation during exercise in humans. *American Journal of Physiology* 278(4): E580–E587, 2000.

46. Paul GL, Layman DK. Post-exercise feedings stimulate skeletal muscle protein synthesis and alter plasma amino acids. *FASEB Journal* 9:A746, 1995.

47. Bazzare TL. Nutrition and strength. In: *Nutrition in Exercise and Sport*, 3rd ed. (Wolinsky I, ed.), Boca Raton, FL: CRC Press, 1997.

48. Short S, Short WR. Four-year study of university athletes' dietary intake. *Journal of the American Dietetic Association* 82:632–645, 1983.

49. Hickson J, Duke MA, Risser WL, Johnson CW, Palmer R, Stockton JE. Nutritional intake from food sources in high school football athletes. *Journal of the American Dietetics Association* 87:1656–1659, 1987.

50. Hickson J, Wolinsky K, Pivarnik JM, Neuman EA, Itak JF, Stockton JE. Nutrition profile of football athletes eating from a training table. *Nutrition Research* 7:27–35, 1987.

51. Phillips SM, Atkinson SA, Tarnpolsky MA, MacDougall JD. Gender differences in leucine kinetics and nitrogen balance in endurance athletes. *Journal of Applied Physiology* 75:2134–2141, 1994.

52. Tarnopolsky MA, MacDougall JD, Atkinson SA. Influence of protein intake and training status on nitrogen balance and lean body mass. *Journal of Applied Physiology* 64:187–194, 1988.

53. Celejowa I, Homa M. Food intake, N and energy balance in Polish weightlifters during a training camp. *Nutrition Metabolism* 12:259–266, 1970.

54. Laritcheva JA, Yalovaya NI, Shubinb VI, Smirnov PV. Study of energy expenditure and protein needs of top weight lifters. In: *Nutrition, Physical Fitness, and Health* (Parizkova J, Rogozkin VA, ed.), Baltimore: University Park Press, 1978.

55. Tarnopolsky MA, Atkinson SA, MacDougall JD, Chesley A, Phillips S, Schwarcz HP. Evaluation of protein requirements for trained strength athletes. *Journal of Applied Physiology* 73:1986–1985, 1992.

56. Morens C, Gaudichon C, Fromentin G, Marsset-Baglieri A, Bensaid A, Larue-Achagiotis C, Luengo C, Tome D. Daily delivery of dietary nitrogen to the periphery is stable in rats adapted to increased protein intake. *American Journal of Physiology* 281(4):E826–E836, 2001.

57. Bazarre TL, Marquart LF, Izurieta IM, Jones A. Incidence of poor nutrition status among triathletes and control subjects. *Medicine and Science in Sport and Exercise* 18:590–598, 1986.

58. Manore MM, Besenfelder PD, Wells CL, Carroll SS, Hooker SP. Nutrient intakes and iron status in female long-distance runners during training. *Journal of the American Dietetics Association* 89(2):257–259, 1989.

59. Deuster PA, Kyle SB, Moser PB, Vigersky RA, Singh A, Schoomaker EB. Nutritional survey of highly trained women runners. *American Journal of Clinical Dietetics* 44(6): 954–962, 1986.

60. Baer JT, Taper LJ. Amenorrheic and eumenorrheic adolescent runners: dietary intake and exercise training status. *Journal of the American Dietetics Association* 92(1):89–91, 1992.

61. Peters EM, Goetzsche JM. Dietary practices of South African ultradistance runners. *International Journal of Sport Nutrition* 7(2):80–103, 1997.

62. Miller SL, Tipton KD, Chinkes DL, Wolf SE, Wolfe RR. Independent and combined effects of amino acids and glucose after resistance exercise. *Medicine and Science in Sports and Exercise* 35(3):449–455, 2003.

63. Lemon PW, Tarnopolsky MA, MacDougall JD, Atkinson SA. Protein requirements and muscle mass/strength changes during intensive training in novice body builders. *Journal of Applied Physiology* 73(2):767–775, 1992.

64. Saylor K, Wildman REC, Willard G. Dietary practices, physical assessment and blood chemistry of competitive female bodybuilders. *FASEB Abstract*, 2001.

65. Saylor K, Wildman REC, Willard G. Dietary practices, physical assessment and blood chemistry of competitive male bodybuilders. *Sports Medicine* (submitted).

66. Burke DG, Chilibeck PD, Davidson KS, Candow DG, Farthing J, Smith-Palmer T. The effect of whey protein supplementation with and without creatine monohydrate combined with resistance training on lean tissue mass and muscle strength. *International Journal of Sport Nutrition and Exercise Metabolism* 11(3):349–364, 2001.

67. Millward DJ, Optimal intakes of protein in the human diet. *Proceedings of the Nutrition Society* 58(2):403–413, 1999.

68. Bingham SA. High-meat diets and cancer risk. *Proceedings of the Nutrition Society* 58(2):243–248, 1999.

FAT, CHOLESTEROL, AND EXERCISE

Brian Leatart/Foodpix

Personal Snapshot

Keisha is a competitive short-distance runner. Recently she ran a 5-km race and at the postrace festivities she overheard some other runners talking about how eating a higher fat diet can enhance fat utilization during a run. Keisha wondered if this is something she should consider. On one hand she is very dedicated to her sport, yet on the other hand Keisha is not sure if she would be able to structure her diet to eat a higher level of fat. She decides to make an appointment with a registered dietitian who works with athletes to have her current diet analyzed and to see what foods would be in a high-fat diet.

Chapter Objectives

- Introduce basic concepts related to fat and cholesterol including their structure, food sources, digestion and absorption, and physiological roles.

- Provide an introduction to key regulatory factors involved in fat and cholesterol metabolism at rest and during exercise.

- Discuss the relationship between fat metabolism, exercise intensity and duration, and apply it to different sports.

- Discuss key points related to adapting to a high-fat diet including the sports it might apply to, as well as the limitations of a high-fat diet.

- Provide recommendations for fat consumption before, during, and after training and competition.

Although many people in the general population seem to fear fat, endurance athletes recognize its true importance to performance. For these athletes, fat serves as a principal source of fuel that helps power muscle contraction during endurance activities. In fact, at lower intensities fat can be the primary fuel source for working muscle. Although fat's percentage contribution to the total energy used by muscle during exercise decreases as intensity increases, fat remains a significant source of fuel during an endurance effort. Endurance athletes also seem to understand that regular training can increase fat storage in muscle tissue as well as increase the ability of that tissue to use fat as an energy source during exercise. Furthermore, fat in their diet can help decrease the volume of food they would have to eat in order to achieve energy balance. For elite endurance athletes, fear of accumulating body fat is usually not an issue in light of their tremendous energy expenditure. On the opposite side of the exercise coin, body fat is the bane of bodybuilders, especially as they approach competition. The same is true of several other athletic groups such as gymnasts, sprinters, and ice skaters. Many of these athletes eat a diet that is relatively low in fat. In addition, although their sports are not as endurance oriented as distance cycling or running, their training might include these types of endurance activities in an attempt to maximize their leanness.

This chapter provides an overview of basic concepts related to fat and cholesterol metabolism, both at rest and during different types of exercise. Metabolic adaptations related to fat and other lipids in response to regular exercise are described, and the In Focus feature at the end of the chapter discusses why exercise promotes cardiovascular health and longevity. The next chapter provides a closer look at body fat mass and its relation to different sports.

■ FAT AND CHOLESTEROL

Fat and cholesterol belong to a special class of molecules called **lipids**. The most salient characteristic of lipids is water insolubility. Because of this property, additional efforts are required to efficiently digest and absorb lipid molecules into the body as well as to circulate these substances within the blood. This section explores basic structural concepts related to fat and cholesterol.

Fat

Fat is the common term for **triglycerides** (triacylglycerol) and is the most potent energy nutrient. As shown in Figure 6-1, a triglyceride molecule is a combination of three fatty acids linked to a glycerol molecule backbone. Both the fatty acids and glycerol can be used for energy, which makes triglycerides a complete energy storage molecule. All triglyceride molecules have this general design; however, the fatty acids that are linked to glycerol are quite variable. The specific types of fatty acids attached to a glycerol help determine the physical state of a fat source, as discussed shortly.

If glycerol has one only attached fatty acid, then it is a monoglyceride. Diglycerides have two fatty acids attached to glycerol (Figure 6-1). Although monoglycerides and diglycerides are considered fat, triglycerides are the predominant form of fat found in the body and the diet.

Key Point

Triglyceride (fat) molecules are composed of three fatty acids linked to a glycerol backbone.

Fatty Acids

Like monosaccharides and amino acids, fatty acids come in numerous types, many of which are presented in Table 6-1. Researchers have devised systems for describing the different fatty acids. The most popular system is the omega system, which is based on the Greek alphabet. Omega (ω) is the last letter in that alphabet, and alpha (α) is the first. If a fatty acid is linked to glycerol, the second carbon closest to the link is referred to as the alpha carbon, as shown in Figure 6-2. Said another way, starting

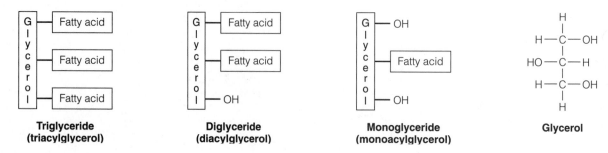

Figure 6-1
Triglyceride, diglyceride, and monoglyceride.

Table 6-1   **Common Fatty Acids**

Saturated Fatty Acids	Nomenclature	Unsaturated Fatty Acid	Nomenclature
Acetic acid	2:0	Palmitoleic acid	16:1 ω-9
Butyric acid	4:0	Oleic acid	18:1 ω-9
Caproic acid	6:0	Linoleic acid	18:2 ω-6
Caprylic acid	8:0	Linolenic acid	18:3 ω-3
Capric acid	10:0	Arachidonic acid	20:4 ω-6
Lauric acid	12:0	Eicosapentaenoic acid	20:5 ω-3
Myristic acid	14:0	Docosahexaenoic acid	22:6 ω-3
Palmitic acid	16:0		
Stearic acid	18:0		
Arachidic acid	20:0		

at the carboxylic acid (COOH) end of the fatty acid, the next carbon is the α carbon. The carbon furthest from the linkage to glycerol is called the ω carbon. As shown in Figure 6-2, this carbon can also be viewed as a methyl group (–CH$_3$). It is important to realize that no matter how many carbons are in a fatty acid chain, the carbon atoms at these locations are addressed in this manner. For instance, in palmitic acid, the 16-carbon fatty acid shown

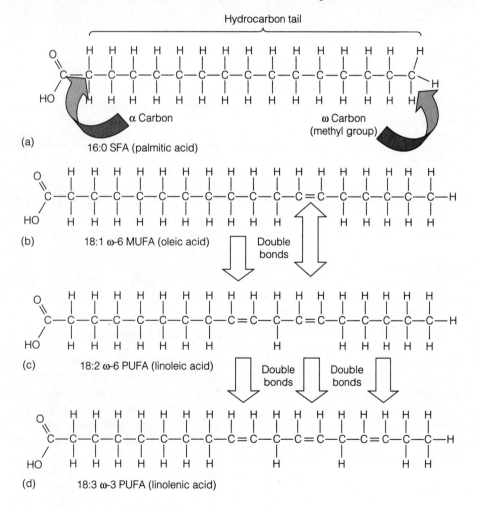

Figure 6-2
(a) Saturated fatty acid (palmitic acid) showing the alpha (α) and omega (ω) carbons, (b) A Monounsaturated fatty acid (MUFA), oleic acid; (c, d) Polyunsaturated fatty acids (PUFAs), linoleic and linolenic acid.

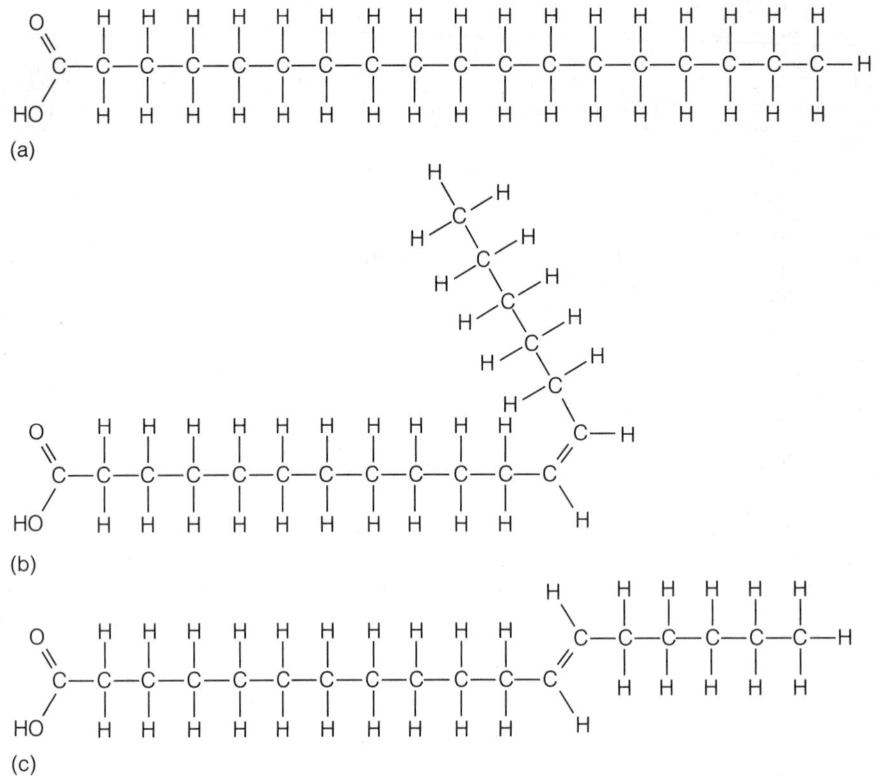

Figure 6-3
(a) A saturated fatty acid (18:0), (b) a *cis* and (c) a *trans* configured monounsaturated fatty acid (18:1 n-6). The *cis* double bond kinks the fatty acid tail, whereas in the *trans* configuration it does not.

in Figure 6-2, the α carbon and the ω carbon are separated by 13 carbon atoms. In the two-carbon fatty acid (acetic acid), the α carbon and the ω carbon are the same carbon.

Fatty acids are in essence a string of carbon atoms linked together. As their molecular design can be long, much like a tail on a dog, they are often referred to as fatty acid "tails" (Figure 6-2). With the exception of the carboxylic acid end of a fatty acid, which is also the point of attachment to glycerol, the remaining tail is all carbon and hydrogen, or "hydrocarbon." In foods, as well as in the body, the length of fatty acids can vary by as much as 20 or so carbons, typically between 2 and 26 carbons. If a fatty acid has four carbons or fewer, it is referred to as a **short chain fatty acid (SCFA)**.[1] If a fatty acid chain has 6–12 carbons, 12–22 carbons, or 24 carbons or more, it is referred as a **medium chain fatty acid**

(MCFA), a **long chain fatty acid (LCFA)**, or a **very long chain fatty acid (VLCFA)**, respectively. Most fatty acids found in nature have an even number of carbons; however, some odd-chain fatty acids exist or are created in metabolic pathways. For example, the breakdown of certain amino acids produces *propionic acid*, which has three carbons.

Fatty Acid Saturation

Fatty acids can differ in their degree of **saturation**. Saturation refers to whether all the carbons between the α and ω carbons are covalently bonded to two atoms of hydrogen. If this is the case, the carbons are said to be saturated with hydrogen and the fatty acid is referred to as a **saturated fatty acid (SFA)** (Figures 6-2 and 6-3). Quite simply, a saturated fatty acid is holding onto the maximal amount of hydrogen, just as a saturated sponge could hold no more water. However, if there are one or more points within a fatty acid where adjacent carbons are linked to only one hydrogen atom each, it is an **unsaturated fatty acid**. Here, not all of the carbons between the α and ω carbons are saturated with hydrogen, sort of like a dry spot on a wet sponge.

Key Point

Fatty acids can vary in length, usually from 2 to 26 carbons, and in their degree of saturation.

As a general biological rule, when two adjacent carbon atoms in a fatty acid are bonded to only one atom of hydrogen each, they must bond to each other twice, forming a *double bond*. If a fatty acid has but one point of unsaturation or double bond, it is referred to as a **monounsaturated fatty acid (MUFA)**. If there is more than one double bond, then it is a **polyunsaturated fatty acid (PUFA)**. Examples of SFAs, MUFAs, and PUFAs are all displayed in Figure 6-2. Although saturation/unsaturation may seem like a minor point, it influences the structure of a fatty acid significantly. Points of unsaturation can "kink" the molecule, as displayed in Figure 6-3. Both saturated and unsaturated fatty acids can be used for energy; unsaturated fatty acids require the assistance of a couple more enzymes.

TRANS FATTY ACIDS.   A closer look at the double bond reveals whether both hydrogen atoms are positioned on the same side or opposite sides of the double bond. If the hydrogen atoms attached to the double-bonded carbons are positioned on the same side of the double bond, it is called a *cis* configuration. This is by far the prevalent way for these hydrogen atoms to be positioned (Figure 6-4). If the hydrogen atoms bonded to the carbon are on opposite sides of the double bond, it is referred to as a *trans* fatty acid. This means that an unsaturated fatty acid can have *cis* and *trans* **isomer** forms as shown in Figure 6-4.

Over the last decade or so, there has been growing interest in the presence of *trans* fatty acids in the diet and their impact on health. Although a *cis* versus *trans* configuration might seem like a minor point in regard to fatty acid design, they impart different physical properties to a fatty acid. *Trans* fatty acids tend to act more like saturated than unsaturated fatty acids. One reason is that *cis* double bonds allow for a kinking of the fatty acid tail, whereas *trans* double bonds do not kink, so that the chain resembles a saturated fatty acid (Figure 6-3). This may affect LDL removal from the blood and heart disease risk as discussed below and at the end of the chapter in the In Focus feature.

Key Point

Triglycerides in oils tend to have relatively higher levels of unsaturated fatty acids than those in more solid fat sources such as butter.

Oil Versus Fat

"Oils" tend to contain more unsaturated fatty acids and "fats" tend to contain more saturated fatty acids. As shown in Figure 6-5, butter (solid) contains roughly two-thirds saturated fatty acids, whereas olive oil (liquid) contains roughly 90% unsaturated fatty acids. The physical state of a triglyceride source is largely determined by the ratio of saturated to unsaturated fatty acids. The kinking of unsaturated fatty acids allows triglycerides to "push out" on adjacent molecules, which keeps the triglyceride molecules from packing together tightly and solidifying.

HYDROGENATION.   As researchers came to recognize the relationship between a diet that is higher in saturated fat and cholesterol and the risk of heart disease, food manufacturers began to offer solidified oils as healthier alternatives. Oils tend to have much less saturated fatty acid and do not contain cholesterol. Food manufacturers began to market solidified oils (as **margarine**) to replace butter at the table and in recipes. These solid and semisolid oils are prepared by adding hydrogen atoms to the unsaturated fatty acids, thereby making them more saturated and solidifying the oil. This process is called **hydrogenation** and involves heating a source of unsaturated fatty acids (that is, oil), which excites fatty acid molecules and allows for molecular changes. Hydrogen gas is applied at a high pressure and some of the double bonds give way, so that hydrogen is added. During hydrogenation some of the PUFAs are converted to MUFAs and some of the MUFAs are converted to SFAs (Table 6-2 and Figure 6-6).

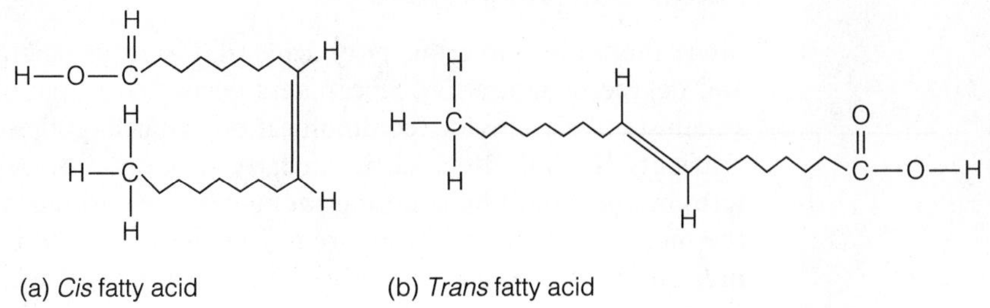

(a) *Cis* fatty acid   (b) *Trans* fatty acid

Figure 6-4

Cis and *trans* fatty acids compared. (a) A *cis* fatty acid has its hydrogens on the same side of the double bond; *cis* molecules fold back into tail-like formation. Most naturally occuring unsaturated fatty acids in foods are *cis*. (b) A *trans* fatty acids has its hydrogens on the opposite sides of the double bond; *trans* molecules are more linear. The *trans* from typically occurs in partially hydrogenated foods when hydrogen atoms shift around some double bonds and change the configuration from *cis* to *trans*.

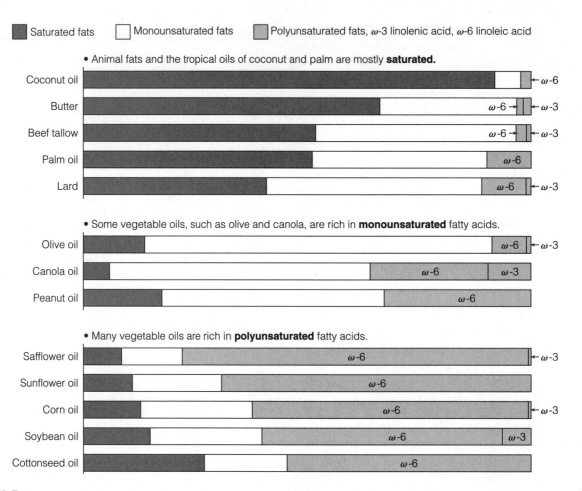

Figure 6-5
Comparison of dietary fats. Most fats are a mixture of saturated, monounsaturated, and polyunsaturated fatty acids.

One of the most popular oils used in hydrogenation processes is soybean oil, which accounts for more than 80% of the oil used to make margarine and roughly two-thirds of that used to make shortening.

Table 6-2

Hydrogenated Corn Oil Margarine Comparison

Triglyceride Source	SFA	MUFA	PUFA	*Trans* Fatty Acids (%)
Butter fat	66%	30%	4%	4%
Corn oil	13%	25%	62%	<1%
Margarine (hydrogenated corn oil)	17%	49%	34%	2%

Note: SFA, saturated fatty acid; MUFA, monounsaturated fatty acid, PUFA, polyunsaturated fatty acid.

Key Point

Hydrogenation adds hydrogen to double bonds and makes an unsaturated fatty acid more saturated and a triglyceride source more solid.

Fatty Acid Nomenclature

Since there are numerous fatty acids, differing in length and degree of saturation, researchers devised a couple of systems for simplifying communication regarding these nutrients. To indicate a particular fatty acid, one first determines its length by counting the number of carbons in the molecule, as shown in Figure 6-7. This is followed by indicating the number of double bonds in the fatty acid. For instance, if a fatty acid is 18 carbons long and contains one double bond, it is indicated as 18:1. Likewise, if a fatty acid is 16 carbons long with no double bonds, it is indicated as 16:0. So the nomenclature states the length first, then the number of double bonds. However, when a fatty acid is unsaturated one more step is needed to indicate the

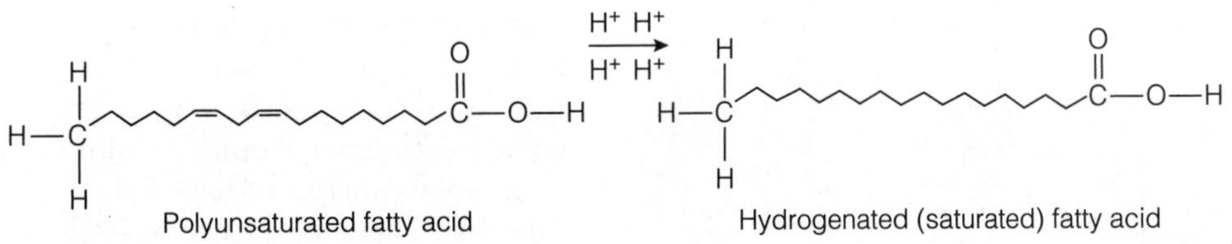

Figure 6-6
Hydrogenation. Double bonds carry a slightly negative charge and readily accept positively charged hydrogen atoms creating a saturated fatty acid.

positioning of the points of unsaturation, or double bond(s).

OMEGA SYSTEM. The omega (ω) system was devised to identify the positions of double bonds. This system counts the number of carbons from the ω end until the first carbon of the first double bond is reached. As shown in Figure 6-7, if the first double bond appears at the 6th carbon in from the ω carbon, than the fatty acid is called an ω-6 fatty acid. Likewise, if the first double bond appears at the 3rd or the 9th carbon the fatty acid would be an ω-3 or ω-9 unsaturated fatty acid, respectively. When dealing with PUFAs only the position of the first double bond has to be indicated, as any subsequent double bonds typically occur after one saturated carbon (see Figure 6-7). For example, 18:3 ω-3 fatty acid has three double bonds positioned at the 3rd, 6th, and 9th carbons in from the omega end. However, there can be exceptions, as in the case of *conjugated linoleic acid (CLA)*. In CLA, the double bonds

are adjacent to one another. CLA is produced by bacteria in the stomach of ruminants such as cows and is absorbed into their body and ultimately assimilated into their flesh (beef) and milk (dairy foods).

Cholesterol

Cholesterol is not an energy nutrient; however, it has structural roles in tissue and serves as a precursor for steroids and vitamin D. As displayed in Figure 6-8, the cholesterol molecule contains four rings (A, B, C, and D). Molecules possessing these four-ring foundations are referred to as **sterols**. Both plants and animals make sterols; however, only animals make cholesterol. Molecules derived from cholesterol that maintain the four-ring structure (such as testosterone) are referred to as **steroids**. Cholesterol is discussed in this chapter because of its close association with fat in many foods and during digestion, absorption, and circulation. In addition, fat and cholesterol

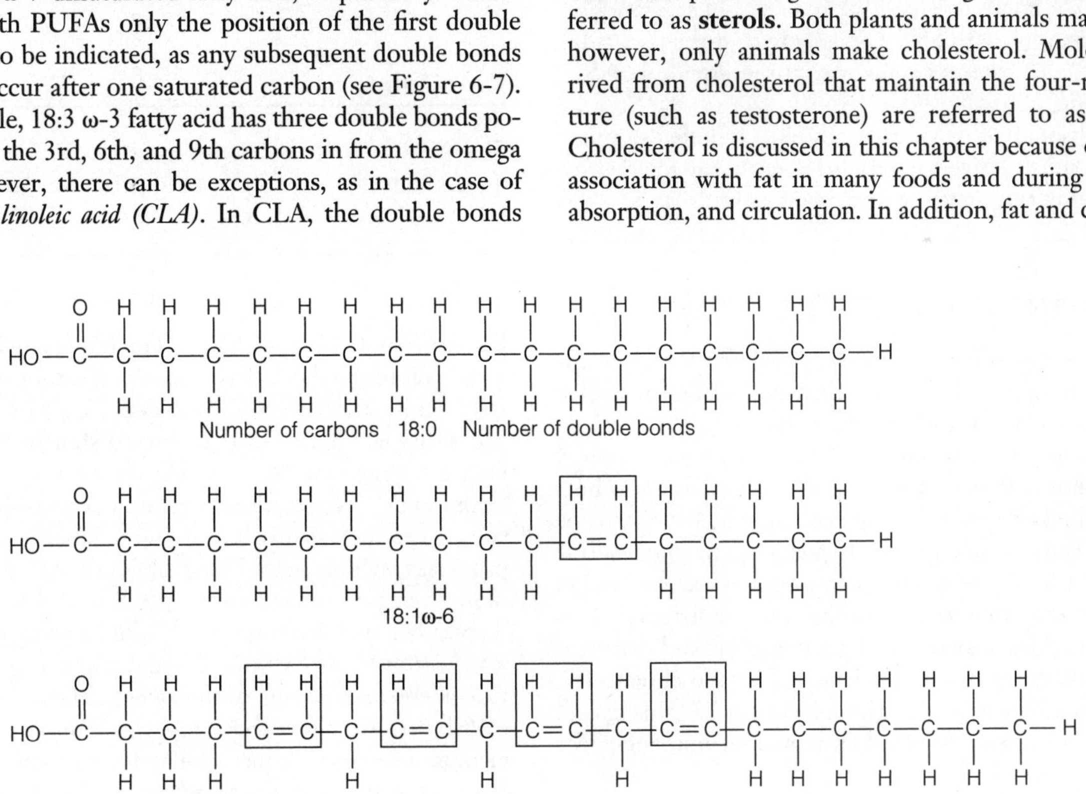

Figure 6-7
Fatty acid nomenclature. Position of the alpha (α) and omega (ω) carbons on fatty acids. The ω system of fatty acid nomenclature involves counting in from the ω end to the position of the first carbon of the first double bond.

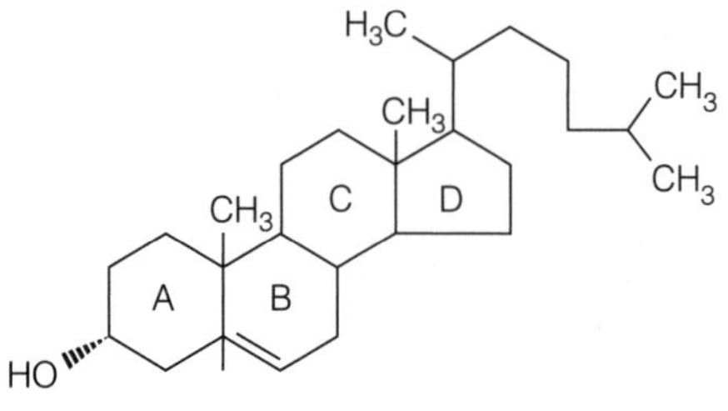

Figure 6-8
Cholesterol.

can be discussed together because of their relationship in estimating heart disease risk, as discussed in the In Focus feature at the end of the chapter.

■ FAT AND CHOLESTEROL *IN REVIEW*

- Lipids are relatively insoluble in water and include most fatty acids, triglycerides, and cholesterol.
- Fatty acids can vary in length and in whether and to what degree they are saturated.
- Double bonds are found at points of unsaturation; *cis*-configured double bonds "kink" the fatty acid molecules whereas *trans*-configured double bonds do not.
- Cholesterol is a sterol produced by animals, and molecules derived from cholesterol that maintain the four-ring structure are called steroids.

■ FAT AND CHOLESTEROL INTAKE

Fat is an energy-dense molecule, providing 9 kcal of energy per gram, which is more than double that provided by an equal mass of carbohydrate and protein. Additionally, gram for gram, fat contains almost 30% more energy than alcohol (7 kcal/g). Cholesterol in foods can also help nourish the body by reducing the amount that has to be made in cells. What foods provide fat and cholesterol? Once these foods are eaten, what processes are involved in effectively digesting and absorbing these nutrients? This section provides answers to these questions and discusses recommendations for dietary levels of fat and cholesterol. In addition, the feature Nutrition in Practice: Athletes and Dietary Fat presents levels of fat intake recommended for more active populations.

General Recommendations for Fat and Cholesterol

Over the past couple of decades, recommendations for fat and cholesterol have been made with special consideration to reducing risk factors for heart disease. Although absolute

American Recommended Dietary Allowance (RDA) levels were not available, recommendations by organizations such as the American Heart Association (AHA) have guided recommendations by professional organizations such as the American Medical Association (AMA) and the American Dietetic Association (ADA). The AHA recommends that less than 30% and 10% of a person's total energy consumption be derived from fat and saturated fat, respectively.

The newest U.S. Dietary Reference Intake publications include Acceptable Macronutrient Distribution Ranges (AMDRs), which include the recommended range for fat intake of 20–35% of the total energy in the diet. This range is in conjunction with a recommended range of carbohydrate intake of 45–65% of total energy in a person's diet. Meanwhile, Adequate Intake (AI) and RDA levels have not been set for many types of fatty acids and cholesterol. Because saturated fatty acids, monounsaturated fatty acids [such as 18:1 ω-9 MUFA (oleic acid)], and cholesterol are made in adequate amounts in the body, they are not dietarily essential. Therefore an AI or RDA has not been set for these nutrients. On the other hand, because dietary deficiency of ω-3 and ω-6 PUFA are known to exist, DRI levels have been set for these essential nutrients.

Key Point

The AHA recommends that less than 30% and 10% of a person's total energy consumption be derived from fat and saturated fat, respectively.

ESSENTIAL FATTY ACIDS. The requirement of at least a little fat in the diet is based on the need for two essential fatty acids, linoleic acid (18:2 ω-6 PUFA) and linolenic acid (18:3 ω-3 PUFA). As discussed shortly, the essential fatty acids (ω-3 and ω-6 PUFA) are used by cells to make eicosanoids. These molecules function as hormonelike substances and are involved in the regulation of numerous processes such as local blood pressure and clotting, hormone release, inflammation, and bronchiole air flow. Typically a diet containing 3–5% of its energy as fat from a variety of foods (such as fish and plant oils) should provide adequate amounts of the essential fatty acids. The oil of fish provide two ω-3 PUFAs that can be used to make eicosanoids in a manner similar to linolenic acid. These fatty acids are **eicosapentaenoic acid (EPA)** (20:5 ω-3) and **docosahexaenoic acid (DHA)** (22:6 ω-3). An AI for linoleic acid has been set at 17 and 12 g/day for adult men and women, respectively, with an AMDR set at 5–10% of total energy. The AI for linolenic acid is 1.6 and 1.1 g/day for adult men and women, respectively, with the AMDR set at 0.6–1.1% of total energy. Not enough information

NUTRITION IN PRACTICE

Athletes and Dietary Fat

Recommendations for dietary fat for athletes and active people have a broader range than recommendations for carbohydrate and protein. For some athletes, such as bodybuilders and wrestlers, the dietary fat level is drastically restricted to promote greater leanness, while for others, such as distance cyclists, fat might be sought out to enhance the energy content of the diet while decreasing the volume of food. Also, some sport nutritionists advocate high-fat diets (>70% of energy) for athletes preparing for protracted lower intensity endurance events, such as ultramarathons. So what is the right amount of fat for an athlete to eat?

Deciding what level of dietary fat is best depends on the athlete and is thus individualized. As discussed in the next chapter, designing the energy proportioning of an athlete's diet begins with assessing that person's needs for carbohydrate and protein. However, athletes must recognize the importance of fat in their diet, as it is a source of essential fatty acids and promotes efficient digestion and absorption of other essential lipid molecules such as the fat-soluble vitamins. The U.S. Food and Nutrition Board states that the Acceptable Macronutrient Distribution Range (AMDR) is 20–35% of total energy, and this could certainly apply to some athletes. However, each athlete must be considered individually based on metabolic needs and diet characteristics. Table A shows the fat intake required to provide 20–35% of total energy at different energy levels. Dietary fat should be derived from a variety of fat sources including nuts, seeds, dairy, vegetable oils, and animal products (such as fish and dairy).

Table A

Daily Fat Intake at Various Energy Levels

Total Diet Energy (kcal)	20–35% Energy from Fat (kcal)	Total Grams of Fat
2000	400–700	44–78
2500	500–875	55–97
3000	600–1050	66–116
3500	700–1225	77–136
4000	800–1400	88–156

is available to set Tolerable Upper Intake Levels (ULs) for total fat, cholesterol, or individual fatty acids. However, it has been generally recommended that the level of saturated fatty acids as well as *trans* fatty acids be minimized as part of a healthy diet.

Key Point

Fat may not be an essential component of the diet as an energy resource; however, at least 3–5% of dietary energy should come from fat to provide essential fatty acids.

CHOLESTEROL.   In the 1970s it seemed that cholesterol-phobia struck the American population and other countries. Cholesterol information was put on food labels, and even fruits were sometimes labeled "a cholesterol-free food." For the general population a cholesterol intake less than 300 mg has long been advocated by organizations such as the AHA and ADA. At present AI, RDA, and UL amounts have not been set for cholesterol. Because it is not an essential nutrient and because of its long history of association with heart disease, it is generally recommended that cholesterol be limited in the diet.

Food Sources of Fat and Cholesterol

The fat content of common foods is presented in Table 6-3. Fat can be derived from animals by eating meats (skeletal muscle and organs) or by consuming mammalian milk and milk-derived foods (dairy). For instance, butter is made from the fat in milk, lard is hog fat, and tallow is the fat of cattle or sheep. The fatty acid profiles of these foods differ somewhat; however, meats tend to contain higher amounts of saturated long chain fatty acids (LCFAs), such as palmitic acid. Milk fat also contains some saturated short chain fatty acids (SCFAs), such as butyric acid. Oil is a natural component of plant tissue (such as leaves and stems) and seeds. Oils can be extracted from seeds or other plant components by *pressing* or other methods of extraction. Common oils include sunflower, safflower, corn, olive, coconut, and palm oil.

ESSENTIAL FATTY ACIDS IN FOODS.   Good sources of linoleic acid are corn oil, soybean oil, cottonseed oil, sunflower oil, and safflower oil. For linolenic acid, good

Table 6-3	**Approximate Fat and Cholesterol Content of Various Foods**

Food (Serving Size)	Fat as % of Mass (Mass per Serving)	Cholesterol as % of Mass (Mass per Serving)
Animal Product		
Beef (4 oz)	32% (9 g)	<1% (101 mg)
Bologna (1 pc)	29% (7 g)	1% (13 mg)
Butter (1 ea)	82% (4 g)	2% (11 mg)
Chicken, white meat (3 oz)	4% (3 g)	<1% (41 mg)
Cheese, cheddar (1 c)	32% (44 g)	1% (139 mg)
Cheese, cottage (4%) (1 c)	4% (10 g)	<1% (31 mg)
Codfish (4 oz)	<1% (7 g)	Trace (35 mg)
Egg, whole (1 ea)	12% (5 g)	4% (213 mg)
Egg, white (1ea)	<1%	Trace (Trace)
Halibut (4 oz)	3% (9 g)	Trace (131 mg)
Hamburger (4 oz)	13% (14 g)	<1% (58 mg)
Lamb chops (2.5 oz)	36% (18 g)	1% (84 mg)
Mackerel (4 oz)	6% (16 g)	Trace (78 mg)
Margarine (1 pat)	82% (3 g)	—
Milk, whole (1 c)	3% (8 g)	<1% (33 mg)
Milk, skim (1 c)	Trace (Trace)	Trace (Trace)
Pork chops (2.5 oz)	21% (8 g)	1% (18 mg)
Pork sausage (1 pc)	46% (6 g)	1% (11 mg)
Salmon (4 oz)	4% (23 g)	Trace (Trace)
Plant Products		
Avocado (1 ea)	13% (27 g)	—
Bread, white (1 pc)	4% (1 g)	<1% (Trace)
Cereals and grains (1 c)	1–2% (1 g)	—
Crackers (4 pc)	1% (2 g)	—
Fruits (1 ea)	<1% (<1 g)	—
Leafy vegetables ($\frac{1}{2}$ c)	<1% (<1 g)	—
Legumes ($\frac{1}{2}$ c)	<1% (<1 g)	—
Margarine (1 tbs)	82% (6 g)	—
Root vegetables ($\frac{1}{2}$ c)	<1% (<1 g)	—

sources are soybean, canola, linseed, and rapeseed oils as well as some leafy vegetables. The fatty tissue of marine mammals (such as whale, seal, and walrus) and the oil derived from cold-water fish (cod, herring, menhaden, salmon, tuna, pollock) and crustaceans such as lobster, crab and shrimp provide EPA and DHA.

TRANS FATTY ACIDS IN FOODS.   *Trans* fatty acids can be found in most fat sources, although their prevalence is very low. However, meat, milk, and dairy fat from ruminant animals may contain 2–8% of their fatty acids as *trans* fatty acids.[1] These fatty acids are actually created by bacteria in the stomach (rumen) of these animals, which include cows, goats, sheep, deer, and camels. The *trans* fatty acids are then absorbed from their gut and become incorporated in their tissue and milk. Additionally, *trans* fatty acids can be created during the hydrogenation of oils (in making margarine and other products). Typically, about half of the *trans* fatty acids in the American diet are derived from animal sources and the remaining half from processed oils, either consumed plain or used in recipes. Cookies, crackers, and other snack foods that utilize hydrogenated vegetable oil may contain up to 9–10% of their fatty acids as *trans* fatty acids. Salad oils may contain 8–17% *trans* fatty acids, and shortening can vary between 14 and 60%.

CHOLESTEROL IN FOODS.   Cholesterol is a key component of animal cell membranes. Mammals also include some cholesterol in the milk they produce. Therefore, eating animal tissue or consuming milk or dairy supplies cholesterol. The cholesterol content varies to some degree among animal food sources, as shown in Table 6-3.

Key Point

Linoleic acid (18:2 ω-6) and linolenic acid (18:3 ω-3) are essential fatty acids. Two ω-3 PUFAs in seafood can substitute for linolenic acid.

Key Point

Food manufacturers attempt to reduce the fat content of a food by using fat substitutes such as Olestra and Simplesse.

Fat Substitutes

Fat tends to impart a smooth texture and tastier quality to many foods. For example, most of ice cream's taste and "mouth-feel" are the result of its rich fat content. However, along with the positive sensory attributes associated with the fat in foods are some potentially negative attributes: fat enhances the energy content of a food, too much so for the health of many people, and a diet rich in fat contradicts health recommendations. Thus food scientists have long searched for fat substitutes, with the goal of identifying and/or manufacturing substances that provide the desirable mouth feel and taste of fat but provide less or no food energy. One of the most prevalent fat substitutes is Olestra, which is several fatty acids attached to a molecule of sucrose (Figure 6-9). The fatty acids allow for much of the desirable sensory qualities of normal food fat. However, since Olestra cannot be digested and absorbed, this substitute comes without an energy expense. Some concerns have been expressed regarding the large-scale use of Olestra in foods. One concern is that it may bind vitamin E and perhaps other lipid nutrients in the digestive tract and decrease their absorption. For this reason Olestra products are fortified with fat-soluble vitamins. It is believed by some researchers that consuming large amounts of Olestra may also result in gastrointestinal disturbances.

Simplesse, another prevalent fat substitute, is a product of milk and egg proteins, mixed and heat treated until fine mistlike protein globules are formed. These protein globules taste and provide a mouth feel similar to fat. This substitute yields a fraction of the energy of fat. The application of Simplesse is limited to cool or cold items, such as cheese, cold desserts, mayonnaise, yogurt, and salad

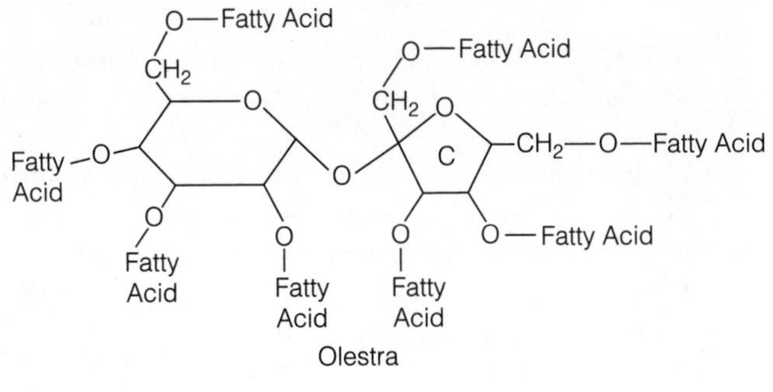

Figure 6-9
Olestra, a synthetic fat substitute that provides no food energy.

dressings. Heat breaks down the fine protein globules, so Simplesse is inappropriate for baked or fried items.

Digestion of Triglycerides and Cholesterol

Efficient fat digestion involves **lipase** enzymes, which break down lipids, and detergent substances. Both the saliva and the digestive juice delivered to the small intestine from the pancreas contain lipases that digest fat. As shown in Figure 6-10, the digestion of fat also requires **bile** to serve as an **emulsifier** or detergent to maximize the efficiency. Bile is a complex solution produced in the liver and delivered to the small intestine through a series of ducts. During periods between meals, about half of the bile can drain into the gallbladder, where it is concentrated and stored. When fat-containing food particles arrive in the small intestine, bile is squeezed out of the gallbladder and travels to the small intestine through the common bile duct, as shown in Figure 6-11. The detergent molecules in bile coat droplets of lipid so that lipid digestive enzymes can efficiently break them down (Figure 6-10).

A fat-digesting enzyme called *lingual lipase* is present in the saliva and begins to cleave fatty acids from glycerol while food is in the mouth. However, this action is short-lived, and most of the digestion of fat is performed by *pancreatic lipase*. Pancreatic lipase detaches two fatty acids from glycerol, which results in a monoglyceride and two fatty acids, as shown in Figure 6-12. The remaining fatty acid might be detached from some of the monoglycerides with the assistance of another enzyme, resulting in glycerol and a fatty acid. Therefore the products of fat digestion are fatty acids, monoglycerides, and glycerol. These substances are now small enough to move into the cells lining the small intestine, along with cholesterol.

Key Point

Fat is digested by lipase enzymes found in saliva and the small intestine.

Absorption of Fat and Cholesterol

Inside the cells lining the wall of the small intestine, triglycerides are re-formed from fatty acids and monoglycerides, as shown in Figure 6-12. These lipids are then incorporated into **chylomicrons**, a form of lipoprotein, that are released from the cell and enter a lymphatic vessel *(lacteal)*. Chylomicrons are too large to immediately enter capillaries in the wall of the small intestine and must enter the body via the lymphatic circulation. Eventually the lymphatic circulation drains into the general circulation via the thoracic ducts. Interestingly, short chain and medium chain fatty acids and glycerol released during digestion can enter capillaries in the wall of the small intestines.

Much of the fat that circulates in chylomicrons is deposited in adipose tissue. In addition, there is opportunity

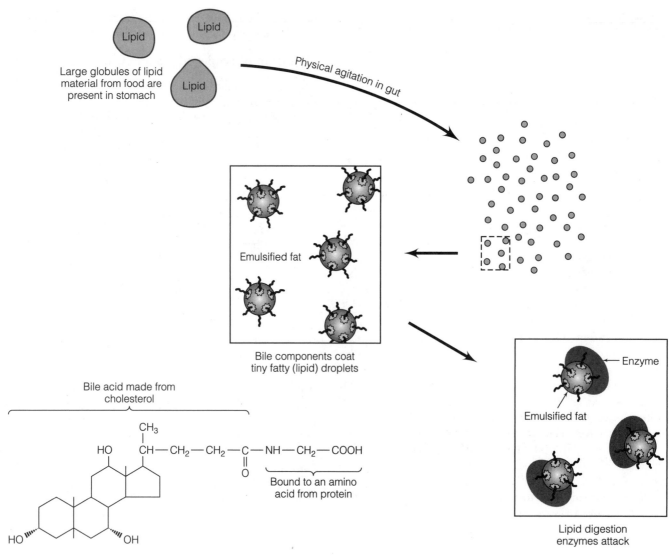

Figure 6-10
Fat digestion.

for other tissue such as skeletal and cardiac muscle to take up some of the fat. This becomes more significant when people train and enhance their skeletal muscle mass (resistance training) and/or when they increase the fat storage potential of existing skeletal muscle (cardiorespiratory training). As shown in Figure 6-13, as fat is removed from chylomicrons they become smaller and are taken out of the blood by the liver. Any remaining fat and cholesterol become the property of the liver. Small chylomicrons are referred to as *chylomicron remnants*, and a chylomicron may only circulate for half an hour to an hour before being removed.

Key Point

Dietary fat and cholesterol are absorbed mostly in the form of chylomicrons, which enter the blood by first traveling through the lymphatic circulation.

■ FAT AND CHOLESTEROL INTAKE *IN REVIEW*

- Animal foods provide both fat and cholesterol whereas plant foods can contain sterols other than cholesterol.
- Hydrogenation reduces the number of double bonds in triglycerides, which solidifies a fat source; examples of the results are margarine and shortening.
- Lipases are enzymes present in the saliva and in pancreatic juices; they digest fat molecules.
- In the wall of the small intestine, triglycerides and cholesterol are packaged into chylomicrons, which enter the lymphatic circulation and then the blood.
- Fat substitutes are popular because they provide no fat energy (Olestra) or little fat energy (Simplesse).

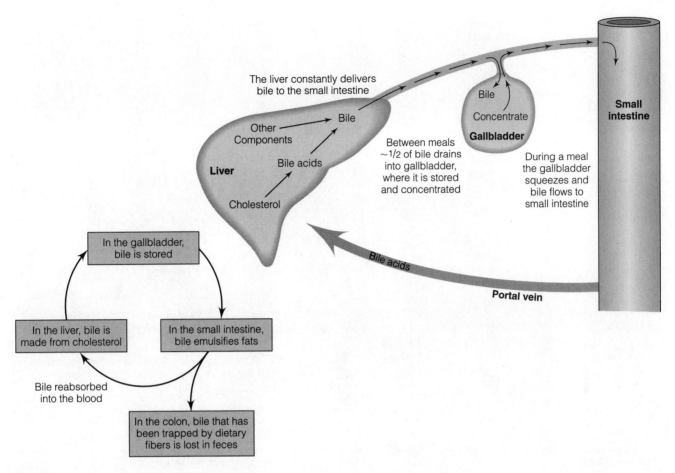

Figure 6-11
Enterohepatic circulation. Most of the bile released into the small intestine is reabsorbed and sent back to the liver to be reused. This cycle is called the **enterohepatic circulation** of bile. Some bile is excreted.

■ PHYSIOLOGICAL ROLES OF FAT AND CHOLESTEROL

When people think of fat they generally think of energy, and more specifically, energy storage (adipose tissue). Yet beyond providing energy, fat and some fatty acids have other important physiological roles. Cholesterol is also an important molecule in the body. This section provides an overview of the major roles of fat, fatty acids, and cholesterol in the body, to serve as the basis for understanding the role of these nutrients in exercise.

Fat as Energy Storage

Fat is an energy source for many types of cells in the body and is the primary means of storing excessive energy. Storing excessive energy as fat rather than as protein or carbohydrate has several significant advantages. First, more than twice the amount of energy can be stored in 1 g of fat as in an equivalent mass of carbohydrate or protein; 9 versus 4 kcal/g, as mentioned previously. Second, stored fat is associated with very little water because of its general water insolubility. This is the opposite of carbohydrate and protein; for instance, each gram of stored carbohydrate (glycogen) is associated with 3 g of water. The net effect of storing excessive energy as fat instead of carbohydrate or protein is that body weight and volume are minimized, as demonstrated mathematically in Table 6-4.

GENDER, AGE, AND BODY FAT.   During childhood the difference in body fat between males and females is less obvious than it is in teenagers and adults. After birth and during infancy and early childhood there are minimal but appreciable differences between the level of body fat in boys and girls. Boys tend to have a little more total body mass than girls as well as a slightly higher percentage of fat-free mass (FFM) and lean body mass (LBM). As discussed in the next chapter, FFM includes all nonfat tissue, whereas LBM is FFM plus essential fat tissue. At the onset of puberty, males experience a burst in the development of FFM. At roughly the same developmental time, females tend to experience a similar growth burst, but it involves more body fat accumulation. At about 10 years of age,

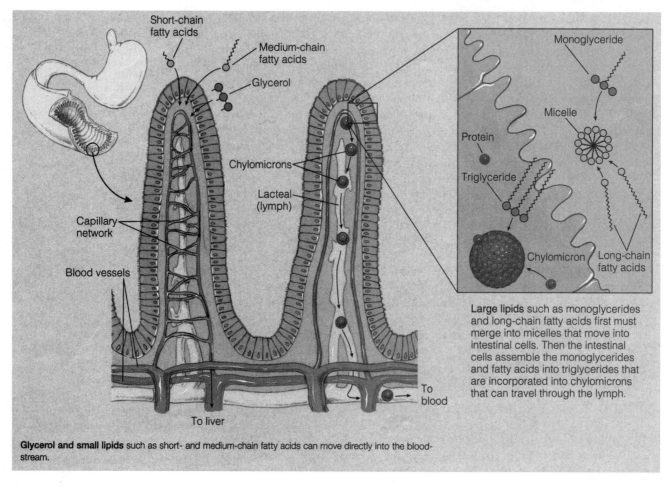

Glycerol and small lipids such as short- and medium-chain fatty acids can move directly into the blood-stream.

Large lipids such as monoglycerides and long-chain fatty acids first must merge into micelles that move into intestinal cells. Then the intestinal cells assemble the monoglycerides and fatty acids into triglycerides that are incorporated into chylomicrons that can travel through the lymph.

Figure 6-12
Absorption of lipids. The end products of fat digestion are mostly monoglycerides, some fatty acids, and very little glycerol. Their absorption differs depending on their size. (In reality, molecules of fatty acid are too small to see without a powerful microscope, while villi are visible to the naked eye.)

boys and girls may have a body fat percentage of about 13% and 19%, respectively. Body fat percentage tends to increase for both genders during adolescence and early adulthood, with a greater increase for females. Upon reaching early adulthood, males might attribute approximately 14–16% of their mass to fat and females 22–24%. Also, females tend to achieve their maximal FFM by approximately age 18, whereas males tend to increase in FFM into their early 20s.[1]

For some very lean individuals, body fat can represent as little as 4–5% of body mass, while in morbidly obese individuals it can account for as much as 66–70% of their mass. Women tend to have more body fat than men. There is a noticeable trend for the level of body fat to increase as people age. Some of this effect is related to a reduction in activity that is common as people age. Other possibilities such as changes in metabolic hormone levels during aging have been suggested.

Adipocytes and Adipose Tissue

Many cell types, especially muscle, contain fat droplets serving as an energy reserve. However, the majority of stored fat is housed within **adipocytes** (Figure 6-14), which are held together by connective tissue in **adipose tissue**. More than 80% of the mass of adipocytes is attributable to stored triglyceride. Adipose tissue is found beneath the skin (subcutaneous stores) and in and around vital organs (visceral stores) and can expand by both *hypertrophy* of existing fat cells and **hyperplasia**, the development of new adipocytes. As discussed in previous chapters, hypertrophy is a general term that can be applied to the growth of cells or tissue. Why adipose tissue expands is obvious: as a result of chronic overconsumption of energy. The next question is whether the tissue expansion reflects adipocyte hypertrophy or hyperplasia or a combination of the two. The short answer is maybe both; however, most adipose expansion is attributable to adipocyte hypertrophy.

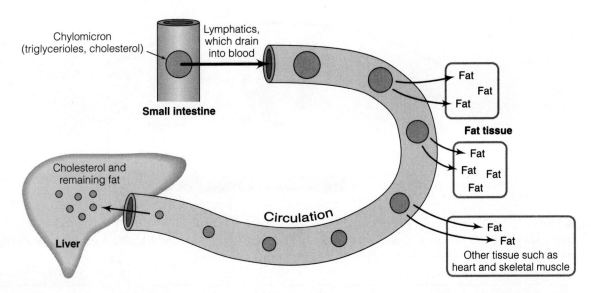

Figure 6-13
Chylomicron circulation and removal. As chylomicrons circulate, their triglyceride cargo is digested by lipoprotein lipase and the result-ing chylomicron remnants are eventually taken up by the liver. Fatty acids move into tissue and are primarily used to make triglycerides.

Table 6-4

Fat Versus Carbohydrate Storage: Hypothetical Example

	Stored Excessive Energy as Fat	Stored Excessive Energy as Carbohydrate
Body fat	15%	0%
Mass of fat	30 lb	0
Mass of associated water	3 lb	0
Body carbohydrate	1%	15.5%
Mass of carbohydrate	2 lb	70 lb
Mass of associated water	6 lb	210 lb
Total body mass	**200 lb**	**~450 lb**

Note: For a 200-lb male whose actual body composition is roughly 15% body fat and 1% carbohydrate, the last column shows the change in body weight if the body could store all excessive energy as carbohydrate. These numbers are hypothetical and are based on the assumptions that fat tissue would be roughly 10% H_2O and that 3 g of water is associated with every gram of glycogen stored. Also, the increased body mass associated with augmentations of body protein and mineral for greater supporting musculature and bone has not been accounted for, nor have other factors such as expanded blood volume and organ sizes.

Once a fat cell has matured it loses its ability to divide into two fat cells. In accordance, as excessive energy is stored in the fat tissue, for the most part individual fat cells are merely getting larger. However, some new fat cells can begin to appear as more and more fat is stored, which means that the potential for hyperplasia must reside in precursor forms of fat cells.[1] Deep within fat deposits are fibroblast-like adipocyte precursor cells, or stem cells, that can reproduce, as shown in Figure 6-15. Numerous factors seem to regulate the reproduction of these cells including tumor necrosis factor-α (TNF-α), which is released by the mature adipocytes as they hypertrophy. Thus, swollen adipocytes signal stem cells to produce new cells.

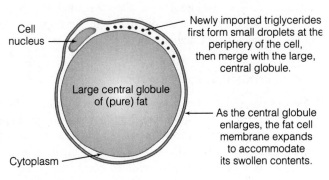

Figure 6-14
An adipocyte. An adipose, or fat, cell seems to expand almost indefinitely. The more fat it stores, the larger it grows.

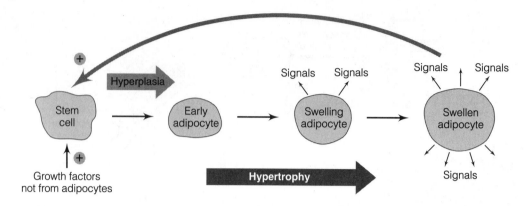

Figure 6-15
Hypertrophy and Hyperplasia of adipose tissue. Fat is accumulated in existing adipocytes, which release growth factors as they swell (hypertrophy) that stimulate production of new adipocytes (hyperplasia).

Key Point

Triglycerides are stored within cells as droplets. Skeletal muscle cells contain several smaller droplets whereas adipocytes contain only one or just a few very large droplet(s).

Adipose Tissue and Energy Status Regulation

Adipocytes have long been considered simply inert containers of energy; however, they are now being recognized as having a more complex involvement in regulation of energy homeostasis and body composition.[1] Adipocytes in adipose tissue are able to gauge energy status and then communicate with the brain via hormones. This concept makes perfect sense. What tissue is better suited than adipose tissue to gauge the status of body energy stores? Leptin and other factors released by adipose tissue may help regulate appetite and energy expenditure. Overexpanded fat cells release a few other substances such as angiotensinogen and TNF-α, as mentioned above. Angiotensinogen is speculated to be involved in the development of obesity-related hypertension, and leptin may be involved in the development of obesity-related type 2 diabetes mellitus.

Tissue Protection and Insulation

Adipose tissue has other roles beyond energy storage. For instance, fat stores in and around internal organs provide some cushioning, and subcutaneous fat provides a layer of padding that protects the underlying skeletal muscle. As they become more lean just prior to competition, bodybuilders refrain from contact activities to reduce the potential for bruising. Visceral and subcutaneous fat also provides insulation. Temperature is important to the optimal functioning of enzyme systems in organs and tissue, and fat serves as a layer of insulation that impedes heat

transfer, especially transfer away from the body (convection). This layer of fat can be particularly important for athletes participating in water or cool-weather sports, as described below.

Phospholipids

Phospholipids may be viewed as structurally similar to triglycerides. However, one fatty acid is substituted with a phosphate, which itself is attached to another substance such as choline or inositol (Figure 6-16). Phospholipids can interact with both water-soluble substances and lipids, so they can be found in cell membranes and lipoprotein shells and are a key emulsifier component of bile. There are several types of phospholipids including phosphatidylinositol, phosphatidylethanolamine, phosphatidylserine, and phosphatidylcholine (lecithin). Researchers are beginning to focus on the phospholipids of the plasma membrane (sarcolemma) and sarcoplasmic reticulum of skeletal muscle cells. Some of the questions that need to be resolved by research is how phospholipids interact with key proteins in those membranes that are important to muscle contraction, such as receptors and Ca^{2+} pumps. Another is how diet and exercise influence the fatty acid composition of these membranes.

PUFAs and Eicosanoid Production

Certain polyunsaturated fatty acids found as part of plasma membrane phospholipids can be used to make

Key Point

Adipocytes function as gauges for energy status because changes in the level of stored triglyceride can result in the release of hormones, some of which can influence appetite and maybe metabolic rate.

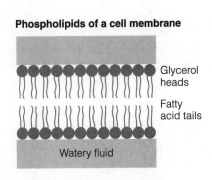

Phospholipids of a cell membrane

Glycerol heads

Fatty acid tails

Watery fluid

From 2 fatty acids

From glycerol

From phosphate

From choline

Figure 6-16

A lecithin. This is one of the lecithins. Other lecithins have different fatty acids at the upper two positions. Notice that a molecule of lecithin is similar to a triglyceride but contains only two fatty acids. The third position is occupied by a phosphate group and a molecule of choline.

Key Point

Eicosanoids are hormonelike regulators and derivatives of ω-3 and ω-6 PUFAs. They help regulate local events such as blood pressure and inflammation.

powerful hormonelike molecules called *eicosanoids*. Eicosanoids include thromboxanes, leukotrienes, and prostaglandins and are involved in the regulation of local events such as blood vessel and bronchiole constriction/ dilation, blood clotting, and inflammatory processes. Eicosanoids are derived from the ω-3 and ω-6 PUFAs, which seem to compete for the same converting enzyme systems (such as *cyclooxygenase* and *lipoxygenase*). The regulation of these enzymes has become of interest to researchers and health practitioners because many of the events regulated by eicosanoids are involved in disease processes such as arthritis and heart disease. Furthermore, as discussed below, many researchers are interested in the influence of exercise on the production of eicosanoids and the effect on blood flow to muscle and other events.

Structural and Functional Role of Cholesterol

Cholesterol can be made in numerous cell types, and under normal situations adequate levels of cholesterol are produced. An adult may make about $\frac{1}{2}$ to 1 g of cholesterol per day; the liver is the most productive tissue in this endeavor, producing more than 10% of all of the cholesterol made.[1,2] The intestines make another 10% of the cholesterol produced. The recognition of a strong relationship between elevations in blood cholesterol levels and heart disease has led to a disdain for cholesterol by the general public. However, cholesterol is an extremely important molecule. It is a vital component of cell membranes and is used in the body to manufacture many vital substances. These substances include bile acids, vitamin D, testosterone, estrogens, dehydroepiandrosterone (DHEA), aldosterone, progesterone, and cortisol.

■ PHYSIOLOGICAL ROLES OF FAT AND CHOLESTEROL *IN REVIEW*

- Fat serves as a potent energy molecule and is stored in cells as droplets; most is stored in adipocytes.
- Adipocytes gauge energy storage status and send chemical messengers (hormones) to the brain to help regulate appetite.
- Phospholipids are found in cell membranes, lipoprotein shells, and bile.
- Essential fatty acids are used to make eicosanoids, which help regulate local events.
- Cholesterol serves as the precursor for vitamin D and steroid molecules such as testosterone, estrogens, and aldosterone.

■ FAT METABOLISM FOR ENERGY

Both fatty acids and glycerol components from triglyceride molecules can be used for fuel. The fatty acids are oxidized in several types of tissue including the liver and muscle. What steps are needed for fat to be used as fuel? Furthermore, what conditions promote fat utilization for energy? This section provides an overview of how fat is used for fuel in the human body and more specifically in skeletal muscle and the liver.

Fat Mobilization from Adipose Tissue

Fat in adipose tissue can be broken down when it is needed as an energy source, namely during fasting and during activity. For instance, more than half the energy used by working muscle at lower intensities (<25% VO_2max) is derived from free fatty acids (FFAs) in the blood. The mobilization of fat from adipose tissue is promoted by the increased presence of hormones associated with exercise and fasting. These are glucagon, epinephrine (and norepinephrine), and cortisol. Also, hormones such as adrenocorticotropic hormone (ACTH), thyroid-stimulating hormone (TSH), growth hormone, and α- and β-melanocyte-stimulating hormones (MSH) can also increase fat mobilization.[3]

Key Point

The levels of insulin relative to the levels of the catabolic hormones (glucagon, catecholamines, and cortisol) dictate whether fatty acids are released from adipose tissue.

RELEASE OF FATTY ACIDS FROM ADIPOSE TISSUE. The circulating level of FFAs derived from adipose tissue triglycerides depends on two factors, as shown in Figures 6–17 and 6–18. The first factor is the rate of **lipolysis**, or liberation of FFAs from triglyceride molecules, as

Key Point

The level of circulating glycerol can be used as an indicator of fat breakdown and mobilization.

promoted by the hormones mentioned above, relative to the rate of reattachment (re-esterification).[4] The second factor involves the association of FFAs with the protein **albumin**, the principal FFA carrier in the blood. The most important enzyme involved in the release of fatty acids from adipose tissue is **hormone-sensitive lipase (HSL)**. Two other enzymes, called *monoacylglycerol lipase (MGL)* and *diacylglycerol lipase (DGL)* are also involved in fat mobilization. Researchers tend to focus on HSL because it is the rate-limiting factor in fat mobilization. HSL cleaves the outer fatty acids, those at positions 1 and 3 in the triglyceride. MGL can then cleave the final middle fatty acid (position 2).[4] As fatty acids are cleaved from glycerol by HSL they can be reattached by *acyltransferase* enzymes. The activity of these enzymes appears to increase when circulating insulin levels are higher. Therefore, the primary factor that dictates the availability of FFAs in adipocytes is the relative levels of hormones that increase HSL activity and of insulin, which increases the activity of counterbalancing reattachment enzymes.

The rate of lipolysis can be estimated either by measuring the activity of HSL or the release of glycerol from adipose tissue. The rate of appearance (R_a) of glycerol in the blood is often used when assessing fat mobilization. Once all three fatty acids are removed from glycerol, free

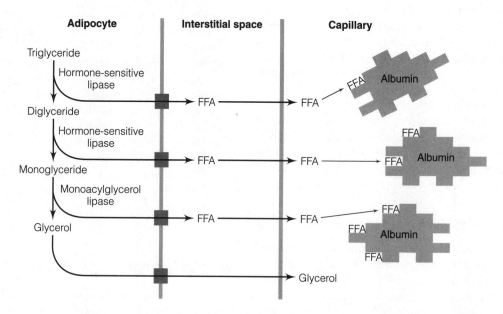

Figure 6-17

The breakdown of triglycerides (lipolysis) in adipose tissue. Hormone-sensitive lipase is activated by hormones including epinephrine. The released fatty acids can be transported out of adipocytes and associated with albumin in circulation.

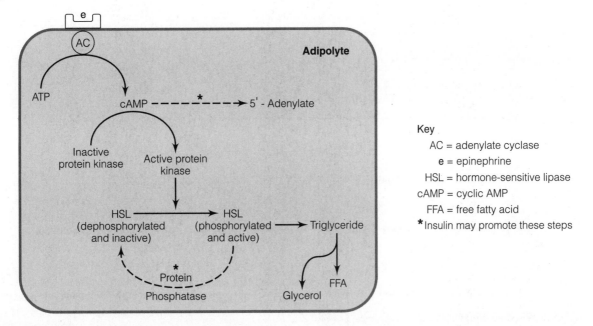

Figure 6-18
Cellular signals for and against lipolysis. During Exercise, hormone-sensitive lipase is primarily activated by epinephrine. cAMP serves as the second messenger.

glycerol can also diffuse into the blood. Adipose tissue fails to produce significant amounts of an enzyme called *glycerol kinase*, which would allow the reattachment of a fatty acid to glycerol.[4]

Fatty Acid Influx into Muscle Fibers

As the blood level of FFAs increases so does FFA utilization in muscle during rest and exercise.[4,5] In order for FFAs circulating in the blood to be used for fuel they must dissociate from albumin and cross over the plasma membrane of cells, as shown at the top of Figure 6-19. The classical view of FFA entry into muscle cells is that they simply diffuse through the plasma membrane (sarcolemma). This would mean that as the concentration of FFAs increases during fasting and exercise so does the diffusion gradient, resulting in greater influx and utilization. Yet many researchers have questioned this concept, stating that during situations in which fatty acid use increases significantly (fasting and exercise), the remarkably high fatty acid influx could not be attributable to simple diffusion alone.[4,6] In addition, more recent research efforts are revealing that specific proteins are likely to be responsible for much of the FFA entry in cells including muscle, especially during fasting and exercise.[7–13] These proteins include a fatty acid translocase (FAT), a fatty acid transport protein (FABP), and a fatty acid binding protein (FABP$_{PM}$) associated with the plasma membrane.[9,10] Future research will focus on understanding how these proteins work together to allow for efficient fatty acid availability in muscle cells and how the proteins are regulated. Questions about how exercise influences the level and function of these proteins will be addressed as well. Some research is already suggesting that a fatty acid transporter is actually brought to the plasma membrane, in a manner similar to glucose transport protein 4 (GluT4, described in Chapter 4), during times of increased requirement.[7]

Key Point

Fatty acid transport proteins increase fatty acid entry into muscle cells during fasting and exercise.

Fat Use as Fuel

Fatty acids are used for fuel by a number of cell types; however, in order to use fatty acids, cells must have mitochondria and an adequate supply of oxygen (O_2). Fatty acids are primarily oxidized in a metabolic pathway referred to as β-oxidation, which generates *acetyl CoA* as well as reduced electron carriers (NADH and FADH$_2$) (Figure 6-19). Acetyl CoA may then enter the Krebs cycle, and the reduced electron carriers transfer the electrons to the electron transport chain. The key enzymes involved in β-oxidation are *acyl-CoA dehydrogenase, enoyl-CoA hydratase, 3-hydroxyacylacyl-CoA dehydrogenase,* and *thiolase*.[1] Researchers can measure changes in the activity level of these enzymes to help demonstrate metabolic adaptation in muscle that results from endurance training.

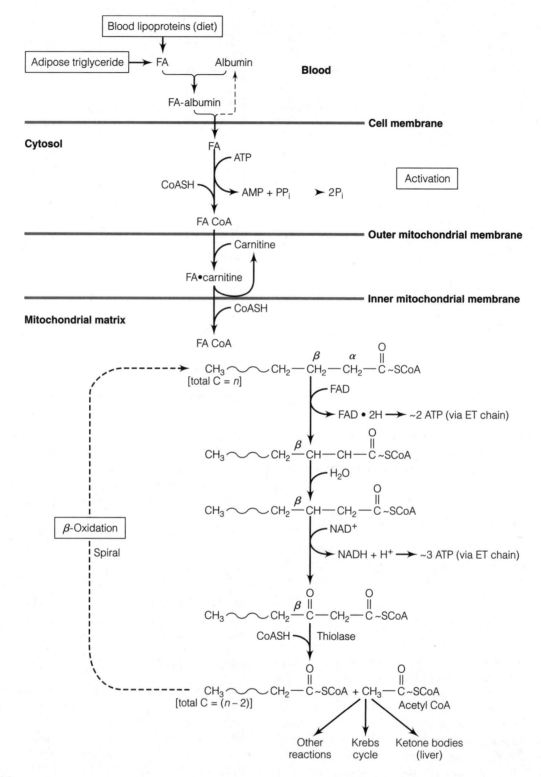

Figure 6-19

Generation of acetyl CoA, NADH, and FADH₂ by β-oxidation. Acetyl CoA can be used to form citrate, the first molecule in the Krebs cycle, and NADH and FADH₂ can pass the electrons to the electron transport (ET) chain. (FA, fatty acid)

Key Point

Fatty acids are broken down in mitochondria in a series of reactions called β-oxidation.

Glycerol can also be used as a fuel source if it enters glycolysis, but perhaps more importantly it can be converted to glucose (gluconeogenesis) by the liver to help maintain blood glucose levels during fasting and prolonged exercise.

Carnitine Transport System for Fatty Acids

Longer chain fatty acids require the assistance of carnitine to cross the inner membrane of mitochondria. As Figure 6-20 shows, an enzyme complex called **carnitine palmitoyl transferase I (CPT-I)** is associated with the outer mitochondrial membrane.[1] "Palmitoyl" refers to palmitic acid (16:0), one of the predominant fatty acids to utilize this system. CPT-I takes an activated fatty acid (acyl CoA) and attaches it to the molecule carnitine. As shown in Figure 6-20, during this reaction coenzyme A (CoA) is removed from the fatty acid, and the fatty acid/carnitine complex (acylcarnitine)

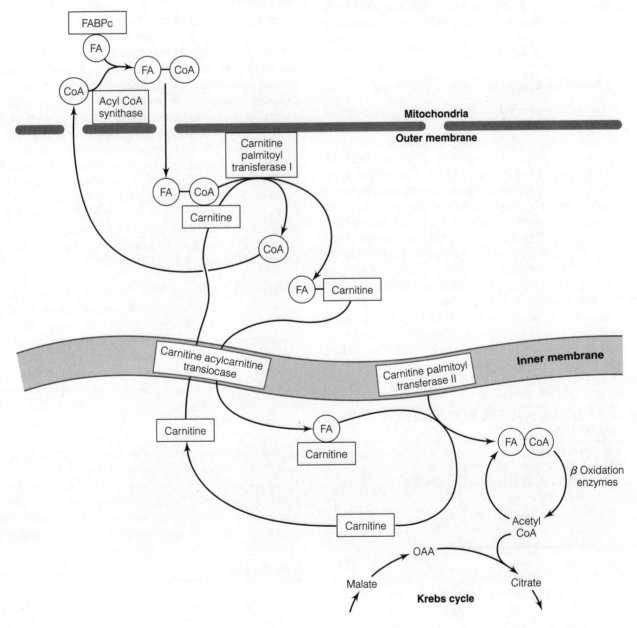

Figure 6-20

Carnitine–fatty acid transport system. Longer chain fatty acids (FA) require the assistance of a carnitine-dependent transport system to enter mitochondria and subsequently engage in β-oxidation.

Key Point

The movement of longer chain fatty acids into mitochondria for aerobic metabolism requires carnitine.

can move across the inner mitochondrial membrane via *carnitine acylcarnitine translocase*. As also shown in Figure 6-20, once acylcarnitine is within the mitochondrial matrix, carnitine is split from the fatty acid and CoA is again attached, reforming the activated fatty acid. Because free carnitine is actually exchanged for acylcarnitine via the carnitine acylcarnitine translocase process, carnitine does not build up in the mitochondrial matrix. Interestingly, shorter chain fatty acids do not require the carnitine transport system to gain access to the mitochondrial matrix.[1]

Fatty Acid Oxidation

Fatty acid oxidation is set up by a couple of events within cells such as muscle fibers. First fatty acids interact with a special protein that helps guide them in subsequent metabolic events, namely fatty acid binding protein (FABP$_c$). Second, coenzyme A (CoA) must be attached to fatty acids to activate it, as mentioned above. During exercise more and more fatty acids will be used as fuel, especially at lower to moderate intensities and in Type I and IIa muscle fibers. Therefore, more and more fatty acids must enter mitochondria in muscle cells, where the enzymes of β-oxidation are housed. As mentioned previously, muscle cells have intracellular pools of triglyceride, which exist as droplets adjacent to mitochondria and are more abundant in Type I and IIa muscle fibers. These droplets, in addition to fatty acids derived from circulation, provide a resource for fatty acids in muscle fibers. As would be expected based on their greater reliance on fatty acid for fuel, Type I muscle fibers are more concentrated with FABP$_c$ than Type II fibers.[14]

■ FAT METABOLISM FOR ENERGY IN REVIEW

- When fat is mobilized from fat cells (adipocytes), free fatty acids diffuse into the blood and associate with albumin.
- Hormone-sensitive lipase (HSL) is the key enzyme involved in fat mobilization. Its activity is increased by glucagon, epinephrine, and cortisol and slowed by insulin.
- Glycerol is also released from fat cells during lypolysis and its rate of appearance (R$_a$) in the blood can be used to gauge the rate of fat breakdown.
- Carnitine is needed to help shuttle longer chain fatty acids into mitochondria so that they can engage in β-oxidation.
- Mitochondria and β-oxidation enzymes are densely found in Type I muscle fibers.

■ LIPOPROTEIN DESIGN AND METABOLISM

Lipid-carrying shuttles are produced by the liver and small intestine to transport triglycerides, cholesterol, and other lipid molecules in circulation. These shuttles are called **lipoproteins** and can be viewed as submarines with lipid cargo. For instance, absorbed fat and cholesterol enter the lymphatic circulation aboard chylomicrons, which eventually enter the blood. This section provides an overview of lipoprotein design as well as metabolism of lipoproteins.

General Lipoprotein Design

With the exception of small amounts of FFAs associated with albumin, lipids are shuttled around in circulation as part of lipoproteins. As shown in Figure 6-21, lipoproteins are composed of a protein-containing shell encasing a core of lipid substances, namely triglycerides and *cholesterol esters*, which are cholesterol linked to a fatty acid. The shell also contains cholesterol and phospholipids such as lecithin. Lipoproteins can be divided into major general classes based on their relative densities, which means the different lipoprotein fractions can be separated by a centrifuge and then quantified in a laboratory. In order of increasing density the primary lipoproteins are chylomicrons, very low density lipoproteins (VLDLs), low-density lipoproteins (LDLs), and high-density lipoproteins (HDLs). It is the lipid-to-protein ratio that dictates their density, as demonstrated in Figure 6-21.

APOLIPOPROTEINS. The proteins found in the lipoprotein shell are called **apolipoproteins**, or more simply apoproteins. Apolipoproteins not only make the lipoprotein more water soluble but also activate metabolizing enzymes and interact with specific receptors in tissue throughout the body. These apolipoprotein services allow a lipoprotein to load and unload some of its lipid cargo or to be removed from the blood and degraded. The naming of apolipoproteins follows an A,B,C,D,E classification with subtypes. Different lipoproteins contain different combinations of apolipoproteins. For instance, circulating chylomicrons contain apolipoproteins C$_{II}$, B-48, E, and A. Apolipoprotein C$_{II}$ is necessary to activate lipoprotein lipase (discussed next) for fat deposition, and apolipoprotein E is necessary for chylomicrons to dock with their receptors on the plasma membranes of liver cells for uptake.[1,3]

Key Point

Lipoproteins circulate fat and cholesterol to tissue throughout the body.

Chylomicrons

Chylomicrons are made by the cells lining the small intestine and transport diet-derived lipid throughout the body

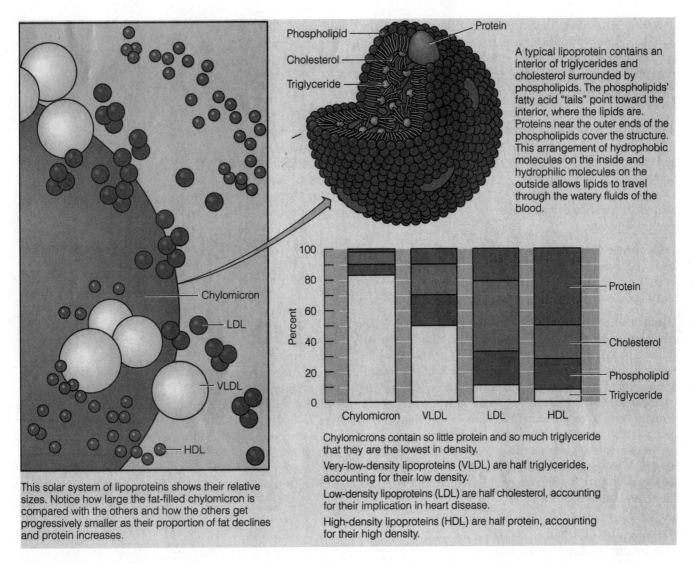

Figure 6-21
Sizes and compositions of the lipoproteins.

(Figures 6–13). The lipid composition of chylomicrons reflects dietary lipid intake, which is typically more than 75 g of fat and less than $\frac{1}{2}$ g of cholesterol daily. Therefore the core of chylomicrons contains mostly fat. One of the most important factors influencing the metabolism of chylomicrons (and VLDL) is the enzyme *lipoprotein lipase (LPL)*. LPL digests triglycerides in circulating chylomicrons to free fatty acids, monoglyceride, and glycerol. The fatty acids then enter the adjacent cells and are either stored as triglyceride molecules or used for energy. LPL is produced in higher amounts by muscle, fat tissue, and lactating mammary glands and in lower amounts in the spleen, lungs, kidneys, aorta, and diaphragm.[1]

Once LPL is exported from a cell it inserts itself into the wall of an adjacent capillary. LPL sticks out of the capillary wall into the blood like a fallen tree extending into a river. As chylomicrons circulate they are lanced by

LPL, which then digests the fat molecules. Apolipoprotein C_{II} in the shell of chylomicrons (and in VLDLs) activates LPL, and the released fatty acids are then able to diffuse into the tissue served by the capillary (for example, adipocytes, muscle cells, mammary tissue). Once most of the fat (≥90%) has been removed, the chylomicron remnant is removed from the blood by the liver and broken down. Any cholesterol and leftover fat (~10%) become the property of the liver, and some may be used to make VLDL.[1] Chylomicron remnants are removed from

Key Point

Lipoprotein lipase digests triglycerides in chylomicrons and VLDLs so that the fatty acids can enter tissue and be used for fuel or stored.

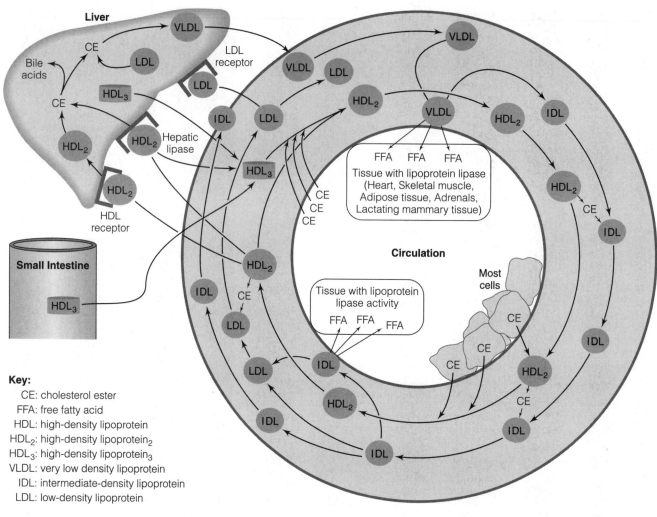

Key:
   CE: cholesterol ester
   FFA: free fatty acid
   HDL: high-density lipoprotein
   HDL$_2$: high-density lipoprotein$_2$
   HDL$_3$: high-density lipoprotein$_3$
   VLDL: very low density lipoprotein
   IDL: intermediate-density lipoprotein
   LDL: low-density lipoprotein

Figure 6-22
Cholesterol circulation in the body.

circulation about 30 minutes to an hour after their predecessor first entered the blood.

Very Low Density and Low-Density Lipoproteins

Not only does the liver receive cholesterol and some fat from chylomicrons, it is also the primary cholesterol- and triglyceride-producing organ in the body. Fat and cholesterol in excess of the needs of liver cells are packaged into **very low density lipoproteins (VLDLs)** and released into circulation. As VLDLs circulate, they unload much of their fat cargo, primarily into fat cells. As shown in Figure 6-22, the lipid-to-protein ratio decreases as VLDLs lose fat, rendering them increasingly dense until they become **low-density lipoproteins (LDLs).** LDLs have two fates: (1) to continue circulating and deposit cholesterol in various tissues; and (2) to interact with the LDL-receptors

on tissue and be removed from the blood and broken down. LDL receptors are produced by many tissues; however, the liver appears to handle about 70% of the task.[3] It should be recognized that as VLDLs are converted to LDLs, there is an intermediate point where they are recognized as intermediate-density lipoproteins (IDLs). Some of the IDLs can also be removed from the blood; however, the majority continues to complete the transformation to LDL. Both VLDLs and chylomicrons can be a FFA source for fat storage in skeletal muscle at rest as well a FFA resource for working muscle, as discussed below.

Key Point

VLDLs are produced in the liver and deliver fat and cholesterol to tissue throughout the body.

Key Point

Lipoproteins transport fat, cholesterol, and other lipid molecules throughout the body and include chylomicrons, VLDLs, LDLs, and HDLs.

High-Density Lipoproteins

High-density lipoproteins (HDLs) are made in the liver and to a lesser extent in the intestines. Their function is to circulate and accumulate cholesterol and return it to the liver. The whole process is interesting because in order for HDLs to return cholesterol to the liver, some of the cholesterol is first passed to circulating LDL and chylomicron remnants, both of which are subject to removal from circulation. When HDLs are made and released they are disc-shaped and often called *discoidal* HDL. As an HDL circulates, its core accumulates cholesterol, some derived from other lipoproteins and other derived from tissue throughout the body. The cholesterol can then be delivered to the liver as the HDL interacts with its receptors or be passed to other lipoproteins prior to their extraction from the blood.

As shown in Figure 6-22, as an HDL accumulates cholesterol in its core, its size and density decrease, transforming it from its highest density form (HDL_3) to a lower density form (HDL_2). At the same time, HDL_2 can be transformed back toward HDL_3 as an enzyme called *hepatic lipase* digests and removes some of its lipid, making HDL more dense again. HDL_2 concentrations are inversely related to the incidence of coronary artery diseases. See the feature In Focus: Nutrition, Exercise, Blood Lipids, and Heart Disease Risk at the end of this chapter for more discussion on this topic.

■ LIPOPROTEIN DESIGN AND METABOLISM *IN REVIEW*

- Lipoproteins are lipid transporters. They contain triglycerides and cholesterol esters in their core, and protein, phospholipids, and cholesterol in their shell.
- Lipoprotein lipase digests triglycerides in VLDLs and chylomicrons. The fatty acids then migrate into adipose tissue as well as skeletal and cardiac muscle and other tissues.
- LDLs are major cholesterol-carrying lipoproteins and are derived from VLDL in circulation.

■ FAT METABOLISM AND EXERCISE

During exercise the fatty acids that help fuel muscle contraction come from within the muscle cells themselves or from circulation. During endurance exercise, the relative contribution made by fat to muscle fuel use decreases with increasing intensity. Fat is an important fuel during recovery periods between efforts of intermittent sports (such as soccer, hockey, and weight training) and after the competition or training bout. In addition, glycerol released by adipose tissue can serve as a resource for gluconeogenesis to help maintain blood glucose levels during exercise. What are the major factors that influence fat use during and after exercise? For starters, several hormones increase fatty acid availability from adipose tissue. At the same time, certain metabolic factors that occur during exercise support fat utilization, including O_2 availability and the presence of β-oxidative enzymes and carnitine. This section provides an overview of the most important factors influencing fat metabolism during and between exercise bouts.

Hormonal Influence on Fat Metabolism During Exercise

The availability of fatty acids to working muscle is important to efficient exercise metabolism. Greater availability of fatty acids during exercise and an increased capacity to use them for fuel in muscle cells can reduce carbohydrate use and maximize the length of an exercise bout. Triglyceride breakdown (lipolysis) in adipose tissue is under hormonal control, and among the most important stimulators of lipolysis during exercise are the catecholamines (epinephrine and norepinephrine).[15] The levels of these hormones increase relative to the intensity of an exercise bout. Cortisol is another important hormone that can stimulate HSL activity and would also be released during higher intensity, longer duration exercise. Glucagon can also increase the activity of HSL if blood glucose levels decrease during exercise. In contrast, insulin is the most potent hormonal inhibitor of lipolysis, as it decreases the activity of HSL.[16] Therefore, the most potent lipolytic hormonal scenario would result from moderate to higher intensity submaximal exercise, when increased levels of catecholamines in circulation occur simultaneously with a blunting of insulin release (see In Focus: Nutrition States and Energy Metabolism in Chapter 4 for more detail on this topic).

The concentration of glucose in the blood can influence the rate of lipolysis by promoting changes in circulating hormones, and it might also have a hormone-independent role.[17] Some evidence for this independent role is derived from studies in which samples of human adipose tissue were bathed in solutions that contained varying levels of glucose.[18] As the level of glucose increased in the solution, there was a corresponding reduction in the rate of fat breakdown. In addition, when a glucose solution was

Key Point

The release of fatty acids into the blood is increased by epinephrine and decreased by insulin.

infused directly into the blood of healthy individuals, there was a reduction in the rate of glycerol appearance in the blood.[19] Certainly it could be argued that the suppression of glycerol appearance may be related to the hyperglycemia-induced insulin release. However, when researchers performed similar studies in which they inhibited the release of insulin during hyperglycemia, the release of fatty acids into circulation still decreased in comparison to a fasting state.[10] This seems to suggest that plasma glucose levels have a direct impact on lipolysis in adipose tissue and that this effect is to some degree independent of insulin levels. This may be a factor during exercise when a carbohydrate-containing sport drink is utilized or an individual eats prior to exercise.

Oxygen Availability and Fat Utilization During Exercise

Unlike glucose, which can undergo some anaerobic metabolism to generate ATP in muscle and other cells, the oxidation of fatty acids is O_2 dependent. Therefore, one of the most important factors for fat utilization by skeletal muscle during exercise is O_2 availability. This begins to explain why there is a decreasing reliance on fat as fuel for skeletal muscle as intensity climbs, thereby creating a mismatch between ATP demand and O_2 availability. Also, as discussed in Chapter 4, the energy efficiency of O_2 utilization (in kilocalories per liter of O_2) is approximately 10% less for fat than for carbohydrate.

Muscle blood flow is increased during exercise via the pumping action of skeletal muscle and is modulated by several neurological and metabolic factors.[20] The contraction of muscle tissue (the "muscle pump") increases blood pressure in muscle, which drives blood through blood vessels as well as into muscle tissue. In addition, muscle contraction increases venous blood pressure, which increases blood flow to the heart, which in turn can be oxygenated in the lungs and eventually pumped back to working muscle. Neurological factors help regulate the degree of blood vessel constriction promoted by the sympathetic nervous system (via norepinephrine) in working muscle. Furthermore, metabolic factors such as nitric oxide during exercise can be released by the blood vessels in muscle, and promote vasodilation.[21,22]

EICOSANOIDS AND MUSCLE BLOOD FLOW DURING EXERCISE. In an attempt to maximize O_2 availability to working muscle tissue during exercise, the production of certain eicosanoids might be regulated.[21,22] Although this area of research is fairly new, and thus conclusions cannot be made at this time, preliminary evidence suggests that increased production of certain eicosanoids (the prostaglandins PGI_2, PGE_2, and PGF_1) promotes vasodilation in muscle tissue during exercise. At the same time the production of another eicosanoid (thromboxane A_2, TXA_2) that would promote the constriction of blood vessels is inhibited in working muscle tissue during exercise. When combined,

these events would help increase blood flow to working muscle fibers, thus providing O_2 as well as energy nutrients while removing CO_2, heat, and excessive lactate.

Key Point

Eicosanoids derived from PUFAs in the walls of blood vessels in muscle may help to maintain maximal blood flow to working skeletal muscle.

Transport of Fatty Acids in Circulation During Exercise

Because FFAs are transported in the blood by albumin, the two most significant factors determining FFA transport in the blood are the rate of blood flow through adipose tissue and the level of albumin in the blood.[1,4,23,24] Simply said, as more FFAs are released into the plasma, more bind to albumin, and the FFA concentration of has been noted to increase as much as 20-fold during endurance exercise.[4]

Each albumin molecule in circulation has the ability to bind 10 fatty acids; however, only three of these sites are recognized as "high-affinity" areas.[25] Said another way, the binding affinity of albumin for FFAs decreases as the molar ratio of FFA/albumin increases. This allows for more FFAs to become dissociated from albumin in muscle blood vessels, thereby increasing the availability of FFAs to working muscle.[6] Furthermore, endurance exercise training has been noted to potentially increase albumin content of the blood, which would increase the level of FFAs in circulation during exercise.[26]

Key Point

O_2 availability, albumin, and eisocanoids are also important regulators of fat utilization during exercise.

Fatty Acid Resources for Muscle During Exercise

Fatty acids are made available to mitochondria within muscle fibers from either inside a muscle fiber or outside those cells. Fatty acid resources outside of a muscle fiber include adipose tissue in muscle tissue or elsewhere in the body (peripheral) as well as circulating lipoproteins. Fatty acids can also be derived from fat stores within muscle fibers.

PERIPHERAL ADIPOSE TISSUE AS A MUSCLE FUEL RESOURCE DURING EXERCISE. Adipose tissue might account for roughly 5–25% of total body mass in most athletes. Therefore vast amounts of triglyceride stored in adipose tissue can serve as a significant resource of fatty acids for muscle fibers during exercise. For instance, the release of FFAs and glycerol into circulation has been noted to increase twofold after 30 minutes of treadmill exercise at 40% VO_2max and fivefold after 4 hours of submaximal treadmill exercise.[27] Furthermore, 30 minutes of moderate intensity exercise lead to an increased glycerol content in adipose tissue (subcutaneous), indicating that lipolysis had increased in that tissue.[28] The availability of this fatty acid resource depends on the degree of hormonal stimulation as well as the availability of albumin and the circulatory flow through adipose tissue. Exercise intensity and duration, as well as nutrition state, are among the most significant factors influencing the hormonal scenario and circulation.[17,27,29]

MUSCLE INTRACELLULAR FAT STORES AS AN ENERGY RESOURCE DURING EXERCISE.

Muscle cells, particularly Type I and IIa fibers, are endowed with droplets of lipid substance, much of which is triglyceride. These droplets are found in proximity, if not directly adjacent, to mitochondria. This intracellular fatty acid resource is important to the fibers in which it is found. Although researchers are still trying to figure out the processes involved, the breakdown of these triglycerides might occur in a manner that is similar to the breakdown of glycogen. This would mean that circulating hormones, such as epinephrine, as well as intracellular metabolic factors (such as Ca^{2+} and AMP) would be involved. Muscle intracellular triglyceride resources are gaining recognition as an important fuel source for muscle fibers during submaximal exercise, especially of a moderate intensity.[30] Furthermore, it is interesting that the energy level found in these stores approximates the energy level of muscle glycogen stores.[31]

LIPOPROTEIN FAT AS A MUSCLE FUEL RESOURCE DURING EXERCISE.

Fatty acids from circulating lipoproteins, primarily chylomicrons and VLDLs, are also available to muscle during exercise. Thus the timing and composition of the most recent meal is a factor in the fat contribution made by lipoproteins during an exercise bout. As discussed above, the release of fatty acids from fat circulating in lipoproteins is dependent on LPL activity, which increases during certain types of exercise and remains elevated for some time during recovery, which allows for recovery of intracellular fat stores.

Key Point

Fatty acids are available to working muscle fibers from adipose tissue found in muscle tissue and other regions of the body, from lipoproteins, and also from intracellular storage droplets.

Exercise Intensity and Fat Utilization

The rate at which FFAs are oxidized in muscle tissue is largely dependent on the intensity of an exercise and its duration.[4] Figure 6-23a displays the contribution of fuel substrates to energy expenditure after 30 minutes of exercise at 25%, 65%, and 85% VO₂max in fasting individuals. At a lower intensity, such as 25% VO₂max shown in Figure 6-23,

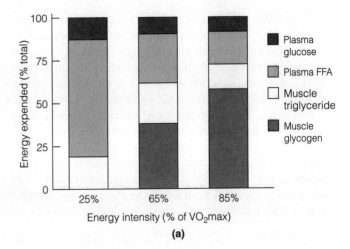

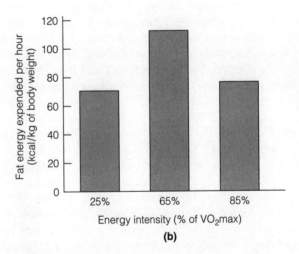

Figure 6-23

Fuel source utilization at three exercise intensities. (a) Expected percentage contributions of fuel sources to muscle energy expenditure after 30 minutes of submaximal exercise. At the lowest intensity (25% VO₂max) the utilization of fat as fuel (%) is highest. (b) However, the quantity of fat used is highest during exercise at moderate intensity (65% VO₂max). (*Source:* Based on reference 23.)

the predominant fuel source is plasma FFAs, followed by muscle fatty acids and plasma glucose.[17] Then, as the intensity increases to 65% VO₂max, the contribution made by plasma FFAs can decrease slightly while the contribution made by muscle fatty acids increases. At this intensity total energy expenditure is roughly split between carbohydrate and fat. This means that there is a decreased reliance on fatty acids as an energy source when expressed as a percentage of total fuel utilized. However, as total energy expenditure has approximately doubled, the amount of fatty acid used is greater than at the lower intensity (Figure 6-23b). As exercise intensity increases further (85% VO₂max) carbohydrate utilization increases and fatty acid utilization decreases [both as a percentage of energy and as an absolute amount (grams)] in comparison to moderate intensity.

For intermittent sports such as soccer and ice hockey, fatty acid oxidation undulates throughout a practice or game. During the bouts of sprinting in these sports, the greater force and power output dictate a greater recruitment of Type II muscle fibers and a greater use of the anaerobic ATP-generating mechanisms, such as creatine phosphate (CP) and anaerobic glycolysis. Although fatty acid oxidation still occurs during sprinting, its contribution to ATP production during that time is reduced substantially, probably to less than 10% of total energy. However, in between sprints (walking, light jogging, bench time) fatty acids make a greater contribution to the recovery of ATP and CP levels and help fuel other recovery operations.

EXERCISE INTENSITY AND MAXIMAL FAT BURNING.
For the general population, many of whom engage in regular exercise to reduce body fat, it is important to understand the difference between the percentage of fat fuel used and the absolute quantity of fat used. Based on the relationship between exercise intensity and fat utilization discussed above, low-intensity exercise uses a greater percentage of fat fuel but maximal fatty acid utilization by quantity appears to occur at a moderate submaximal intensity. Exercising above that intensity level increases the use of carbohydrate and decreases total fat utilization. However, from the available exercise studies comparing 25%, 65%, and 85% VO_2max intensities, it is not known if more fat may have been utilized at 60% or even 50% VO_2max.[17] As 65% VO_2max is probably associated with the anaerobic threshold (lactic acid), it is probably safe to assume that it would approximate the point of maximal aerobic ATP production and fatty acid utilization.

Factors such as level of training and genetic predisposition for muscle fiber type content tend to make intensity level of maximal fat use individualized. Therefore, the "fat-burning zone" often posted on aerobic equipment such as stationary bikes and steppers is presented as a range of associated heart rates. Furthermore, research studies suggest that fat oxidation and FFA and glucose metabolism are somewhat similar in people separated by 50 years when controlling for factors such as body composition.[32] Thus when it comes to maximizing fat utilization during exercise, the level of intensity and degree of training are more important factors than age.

Key Point

Exercise intensity is inversely related to fat utilization expressed as a percentage of total energy used. Meanwhile, moderate intensity levels are associated with the greatest quantity of fat utilization.

Exercise Duration and Fat Metabolism

Fatty acid utilization tends to increase throughout the duration of exercise at lower to moderate submaximal

Key Point

Fatty acid utilization during prolonged exercise tends to increase throughout the duration of the effort.

intensity. The availabilities of glucose, fatty acid, and O_2 at a given time point of the prolonged activity are the most influential factors. During the early stages of prolonged activity, fat contribution to energy expenditure by working skeletal muscle increases slowly. This has led to the commonly stated concept that it takes at least 10–15 minutes or so to achieve the "fat–burning zone" associated with exercise. There are several reasons why the oxidation of fatty acids is somewhat slow to reach a maximal point. First, cardiac output and breathing depth and rate need to increase to provide more oxygenated blood to reach working muscle and lessen early exercise O_2 debt (see Chapter 2). These changes are complemented by local dilation of blood vessels serving muscle cells hard at work. In addition, mobilization of fatty acids from intracellular stores and adipose tissue take a little time to maximize. When O_2 and fatty acid availability are optimized, fatty acid oxidation is also optimized. Then as exercise continues, the contribution made by fat to fueling muscle contraction can slowly increase, especially if carbohydrate is not consumed during the endurance bout. This is because as glycogen stores wane, muscle cells begin to rely more on fatty acid oxidation, as shown in Figure 4–19.

Exercise and the Carnitine Transport System

The flux of fatty acids into muscle cell mitochondria is a principal determinant of fatty acid use during exercise. The carnitine transport system is the primary system for transporting fatty acids into the mitochondria, and this system is under some regulatory control by **malonyl CoA**. Insulin increases the activity of *acetyl CoA carboxylase*, the enzyme that produces malonyl CoA. In turn, malonyl CoA decreases the activity of CPT-I, which results in decreased fatty acid influx and oxidation. Since insulin levels are elevated when blood glucose is high (fed state), this serves to decrease fat use as fuel in cells. In contrast to insulin, epinephrine slows the activity of acetyl CoA carboxylase. It would seem then that during exercise, elevated epinephrine and decreased insulin in the blood would lead to a general inhibition of acetyl CoA carboxylase activity and thus reduced malonyl CoA content in muscle fibers.[33] As Figure 6-24 shows, muscle malonyl CoA levels decrease significantly after 30 minutes of treadmill exercise.[34]

Ketone Body Use as Fuel During Exercise

Ketone bodies are produced in the liver during prolonged fasting, periods of very low carbohydrate intake, and uncontrolled diabetes mellitus. As displayed in Figure 6-25,

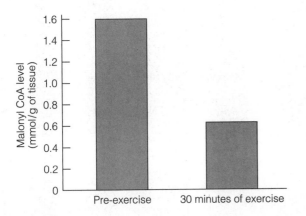

Figure 6-24

Expected decrease in malonyl CoA level after 30 minutes of cardiorespiratory exercise. (*Source:* Based on reference 34.)

Key Point

The level of malonyl CoA in skeletal muscle is a primary regulator of the carnitine transport system that is involved in transporting most fatty acids into mitochondria.

acetyl CoA derived from excessive fatty acid oxidation serves as the preliminary building block for ketone bodies. Ketone bodies are produced as an alternative fuel source for tissue such as skeletal and cardiac muscle and the brain. Having this alternative fuel source decreases the reliance of these tissues on plasma glucose for energy. It has been reported that plasma levels of ketone bodies do rise gradually during endurance exercise; however, their contribution to muscle energy metabolism is minimal.[4,35] Even in ketotic diabetic patients, whose ketone body levels were very high, the contribution of ketone bodies to leg exercise was still below 5%.[36] Therefore, although some ketone bodies can be used as fuel during exercise, their effect is minimal.

Medium Chain Triglycerides as Fuel During Exercise

Medium chain triglycerides (MCTs) contain medium chain fatty acids (MCFAs). When these fatty acids are removed during digestion they can diffuse directly into capillaries in the wall of the small intestine and circulate immediately in the blood. Thus MCFAs generally avoid being incorporated into chylomicrons and the lymphatic

Key Point

Ketone bodies, produced during fasting and restricted carbohydrate intake, can be used as fuel during exercise; however, they do not influence physical performance.

circulation. As the free fatty acids derived from MCTs would be available for working muscle cells, it has been speculated that MCTs provided just before or during exercise might spare muscle glycogen. Research findings do support the notion that MCTs can serve as an energy source during endurance exercise, especially when consumed with carbohydrate; however, there is generally no impact on exercise performance.[37] Furthermore, the potential for gastrointestinal discomfort, including diarrhea, hampers the use of MCTs by many athletes. The use of MCTs to enhance athletic performance is discussed in more detail in Chapter 11.

■ FAT METABOLISM AND EXERCISE IN REVIEW

- Exercise intensity dictates the contribution made by fatty acids to muscle energy expenditure in an inverse manner.
- As exercise is prolonged the contribution made by fatty acids to energy expenditure tends to increase.
- The carnitine transport system is inhibited by the level of malonyl CoA, which is decreased during exercise.
- Ketone bodies produced during prolonged exercise can be used as fuel; however, their contribution is small.
- Medium chain triglycerides (containing medium chain fatty acids) can be absorbed directly into the blood and can serve as a fuel source during exercise; however, performance does not seem to be enhanced by MCT intake.

■ ADAPTATIONS IN FAT METABOLISM

Fat is often viewed as a desirable fuel source during exercise because for every gram of fatty acid used, 2 g of carbohydrate might be spared. This would slow the rate of glycogen loss within muscle fibers and possibly extend performance time prior to fatigue. Using more fat as an energy source during exercise is often considered desirable because it would promote increased leanness of an individual. This may improve appearance, which could be important in sports that involve subjective scoring (such as ice skating, fitness competitions, and bodybuilding). As a result of exercise training more fat can be used as fuel during exercise; this section provides an overview of some of the most significant adaptations that occur in muscle and other tissue in response to exercise training to increase fat utilization.

Increased Fat Use as Fuel Resulting from Exercise Training

Endurance exercise training leads to adaptations in muscle and other tissue that is supportive of greater fat utilization at a given exercise workload.[38] If muscle uses more fat as

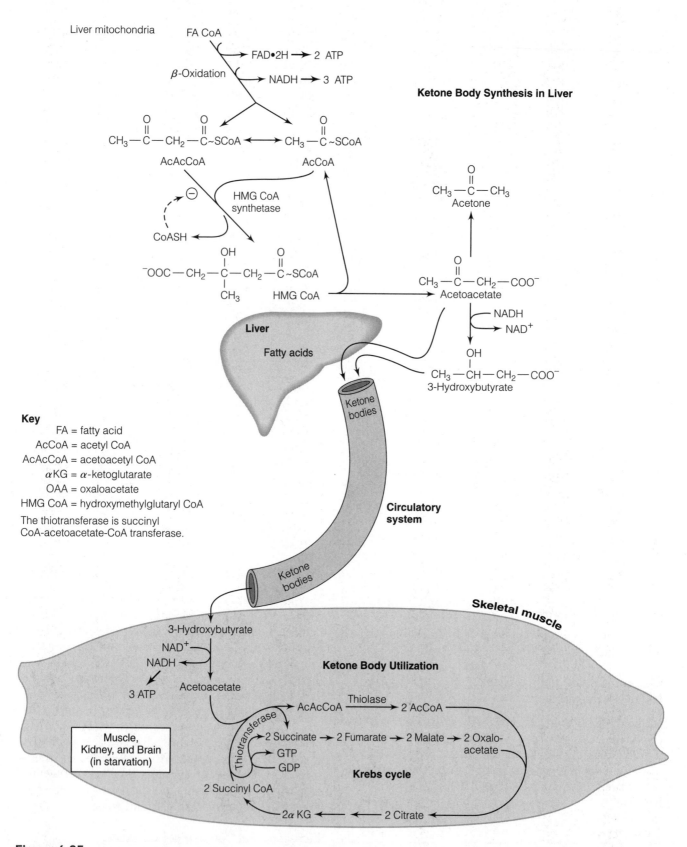

Figure 6-25

Ketone bodies are found in the liver when fatty acid breakdown is excessive during fasting and very low carbohydrate diets. Ketone bodies are used for fuel by muscle to a limited degree during exercise.

fuel during a given level of exercise there is the potential to spare glycogen at that level. For instance, when males engaged in a 12-week endurance exercise program (64% VO$_2$max), their respiratory exchange ratio (RER) was lower during the exercise test in the trained state, indicating a greater reliance on fatty acids.[39] It was estimated that the energy expenditure derived from fat during the exercise test increased from 35% before training to 57% after training, and that the use of muscle glycogen was 41% lower. Other research studies have also shown this to be the case.[40]

Key Point

Endurance exercise training enhances fat utilization in muscle fibers, which can decrease the reliance on glycogen for fuel.

Adaptations in Muscle Fibers That Enhance Fat Use as Fuel

As discussed in Chapter 2, successful endurance athletes tend to have a higher proportion (>60%) of Type I muscle fibers. Because exercise does not seem to cause changes in muscle fiber type—that is, it does not appear to change fibers from Type II to I—the higher proportion of Type I fibers observed in the muscles of these athletes is probably genetic.[41] This also means that the adaptations that occur in muscle serve to make existing muscle fibers better aerobic machines. It is important to realize that only the recruited and thus trained fibers undergo significant adaptation. At the intensity level associated with endurance training, Type I and IIa fibers adapt more than Type IIb fibers. The changes that occur include an enhancement of myoglobin, which is the O$_2$ storage protein in muscle fibers, and of mitochondria and oxidative enzyme activity, which are involved in fat metabolism.[42,43]

Adaptations in the Cardiovascular System That Enhance Fat Use as Fuel

Endurance training results in significant adaptations in the cardiovascular system, including an increased stroke volume and enhanced cardiac output potential. In addition, more blood vessels are developed in the heart as well as in trained muscle tissue.[15] The increase in capillary density occurs in association with both types of muscle fibers, although the adaptation might be most dramatic in association with Type IIb fibers provided that the exercise intensity is high enough. This is because Type IIb fibers are poorly vascularized in comparison to Type I and even Type IIa, so they have the greatest potential for improvement.[44,45] Increased capillarity improves O$_2$ extraction by working tissue, thereby increasing aerobic energy metabolism.[46] In addition, the level of RBCs in circulation may be increased as a result of cardiorespiratory training.

Key Point

The net effect of cellular (muscle fiber) and cardiovascular adaptations to cardiorespiratory training is an enhanced oxidative or aerobic potential during exercise.

Adaptations in Lipoprotein Lipase Activity

As discussed above, LPL is responsible for digesting fat molecules circulating aboard lipoproteins, primarily chylomicrons and VLDLs. LPL activity becomes elevated and remains elevated for several hours in response to a bout of endurance exercise. The increase in LPL activity postexercise is important to replenish intramuscular triglyceride content. The enhanced LPL activity is reflected in lower plasma triglyceride levels during that same time. Therefore, lower plasma triglyceride levels resulting from endurance training may partly reflect an enhanced LPL activity in skeletal muscle in trained individuals in comparison to sedentary people.

An increase in LPL activity could be the result of increased quantity of LPL and/or increased activity of the existing LPL. Some evidence suggests that increased quantity is responsible for at least some of the increased LPL activity.[47,48] For instance, a 3-month low-intensity training program increased the production of mRNA for skeletal muscle LPL.[48] Furthermore, the quantity of LPL probably stays elevated for several hours, maybe 8 hours or more, but should return to pre-exercise levels within 24 hours.

Adaptations to a High-Fat Diet and Exercise

Recently the notion of eating a high-fat diet (>65% of energy) to enhance fat oxidation during endurance sports has resurfaced. This concept has been known as "fat loading." The idea is that if an athlete eats a high-fat, low-carbohydrate diet, muscle cells will adapt to use more fat as fuel during rest and exercise. This in turn would reduce the reliance on carbohydrate and perhaps prolong endurance efforts that might be limited by muscle glycogen depletion. However, it is not exactly clear what a high-fat, low-carbohydrate diet means. For instance, if fat is to provide 65% of total energy and protein intake is adequate (15–20% of energy), carbohydrate would be limited to 15–20% of energy. Thus a distance runner requiring 3800 kcal daily would ingest 140–190 g of carbohydrate and roughly 275 g of fat. Several important considerations

Key Point

A bout of endurance exercise can increase the activity of lipoprotein lipase, which helps replenish muscle fiber fat stores after exercise and lower circulating triglyceride levels.

are associated with this protocol, including the length of time the diet is used, the individual's tolerance for the diet, and the nature of the sport (its intensity and duration). In addition, long-term issues, such as the influence on general health, must be considered.

As carbohydrate is replaced in the diet with fat energy, muscle tissue responds by using more fat for energy. This means that the combination of low carbohydrate and high fat may enhance the levels of β-oxidation enzymes so that fat oxidation would be increased both at rest and during endurance exercise.[49] The adaptation occurs over several days, and high-fat diets probably would need to be adhered to for about a week to maximize this potential.[50] The adaptation can be tracked by measuring the respiratory exchange ratio (RER) during rest and exercise. (The Chapter 13 In Focus discusses these measures.) In addition, the increased ability to use fat as a fuel source continues for a couple of days even when an individual returns to a higher carbohydrate diet. Some researchers feel that this concept can be applied to certain types of sustained physical activity lasting several hours. For instance, what if athletes follow a high-fat diet for at least a week and then switch to a high-carbohydrate diet for a day prior to endurance competition? Would they be able to use more fat during the competition and also start with greater glycogen stores in comparison to just eating a high-fat diet?

The type of sport an athlete competes in is perhaps the most important consideration. For athletes who compete in higher submaximal intensity sports (such as 5-km runners) and intermittent supramaximal intensity sports (such as ice hockey and soccer), a high-fat diet is probably not worthwhile. The intensity level of these sports is generally at or above the anaerobic threshold and carbohydrate is the principal fuel. In addition, substituting carbohydrate with fat energy in the diet can reduce glycogen stores in these athletes. For sports that apply a lower to moderate intensity over a longer period (such as ultramarathons), a high-fat diet might be worth considering. These sports involve continuous performance for many hours (often >6 hours). These athletes may experiment with a high-fat diet during a training simulation of an upcoming event.

Adhering to a high-fat diet for extended periods (such as weeks and months) can be problematic for some athletes and also have long-term health considerations. Also, athletes using natural foods may find it difficult to consume the proportion of fat required for a high-fat diet. They might need to increase their consumption of manufactured

Key Point

Adaptation to a high-fat diet (fat loading) for at least 1 week prior to an ultra-endurance activity has become popular again although conclusive research into its efficacy is lacking.

foods such as lower carbohydrate sport bars and shakes and use low-carbohydrate flours and baking mixes to produce pastas, breads, and other baked goods.

Exercise Adaptation in Blood Lipid Levels

Many types of exercise appear to positively influence blood lipid profiles. Positive changes in lipid profiles related to exercise include reductions in total and LDL-cholesterol levels, raised HDL-cholesterol levels, and reduced triglyceride levels. Certainly positive changes are more associated with endurance training such as running, cycling, and stairstepping than with weight training sports such as Olympic lifting and power lifting. Most studies involving endurance exercise have indicated that aerobic training increases HDL-cholesterol, perhaps more so within the HDL_2 fraction, and decreases serum triglyceride levels.[51] The influence of aerobic exercise on total cholesterol levels can require some scrutiny. For instance, if total cholesterol levels remain constant, this may actually reflect that decreases in VLDL- and LDL-cholesterol have been balanced by increases in HDL-cholesterol levels. Because VLDLs are the major carrier of triglycerides in a fasting state, reductions in VLDL-cholesterol levels should decline hand-in-hand with reductions in serum triglyceride. As training would increase LPL activity, this may account for some of the reductions in triglyceride levels, especially if the blood draw occurred within or close to 24 hours after a training session or competition. For instance, plasma triglyceride levels have been observed to be reduced for 1–2 days after a 70-km cross-country ski race.[15]

Key Point

Regular cardiorespiratory exercise can increase HDL-cholesterol levels and possibly decrease total and LDL-cholesterol levels, thereby decreasing one's risk of heart disease.

Exercise workload might be one of the most important factors that predict how blood lipid levels would respond to exercise training. For instance, researchers assessed the information derived from 10 endurance studies involving men and women and determined that for each 10 km/week run or jogged, the HDL-cholesterol levels were roughly 3 mg/100 ml of blood higher for both males and females.[52] Furthermore, the HDL_2-cholesterol levels were determined to be 1.5 mg and 2.7 mg/100 ml of blood higher for the men and women, respectively. Researchers have suggested that an "exercise dosage" of 13–20 km/week of jogging for 4–6 months is required for significant benefit.[53]

The research regarding weight training and potential changes in blood lipids has been difficult to interpret due to uncontrolled factors such as changes in diet and body composition and the use of anabolic agents. For instance, the use of anabolic agents by individuals involved in resistance training can actually increase LDL-cholesterol while decreasing HDL-cholesterol levels.[54–56]

■ ADAPTATIONS IN FAT METABOLISM *IN REVIEW*

- Muscle tissue adapts in response to cardiorespiratory training by increasing its oxidative capabilities, which increases fat and decreases glycogen use during an exercise bout.
- Cardiovascular adaptations to cardiorespiratory training include enhanced stroke volume and cardiac output potential and increased RBC formation and vascularization of muscle tissue.
- High-fat, low-carbohydrate diets can enhance the fat-metabolizing capabilities of muscle and may have application in low- to moderate-intensity sports with long durations.
- HDL-cholesterol tends to increase and total and LDL-cholesterol may decrease in response to regular cardiorespiratory training.

Conclusions

Triglycerides (fat) and cholesterol belong to a class of molecules referred to as lipids whose most salient feature is water insolubility. Lipoproteins transport fat and cholesterol. Fat is primarily stored in adipose tissue and secondarily within cells such as muscle. The minimum requirement for fat in the diet is about 3–5% of total energy to provide the essential fatty acids (linoleic and linolenic acids), which are precursors for eicosanoids. Fatty acids are a fuel resource for working muscle and are derived from intracellular stores and from circulation. Meanwhile, glycerol derived from stored or circulating triglycerides can serve as a gluconeogenic resource to help prevent hypoglycemia during exercise. In response to cardiorespiratory training, adaptations within muscle fibers and the cardiovascular system allow muscle to have enhanced oxidative capabilities and use more fatty acids for fuel at a given exercise intensity. Adaptations in blood lipids in response to exercise training may decrease an individual's risk of heart disease. The level of body fat must also be considered by athletes as a factor in aesthetic characteristics and/or performance measures that affect sport success.

 **IN FOCUS**   NUTRITION, EXERCISE, BLOOD LIPIDS, AND HEART DISEASE RISK

Heart disease continues to be a foremost health concern in developed countries. Although inactivity is a known risk factor, regular exercise does not eliminate the need for concern. Therefore, both active and inactive people should understand the risk factors for heart disease and engage in dietary practices and other behaviors that minimize their risk. This In Focus discusses the role of exercise and diet in minimizing a person's heart disease risk.

The level of total cholesterol and lipoprotein cholesterol fractions remains one of the most important "controllable" risk factors for heart disease. Reductions in total and LDL-cholesterol levels and elevations in HDL-cholesterol levels can reduce an athlete's risk of heart disease. Recently, the American College of Cardiology categorized LDL-cholesterol, HDL-cholesterol, and triglycerides based on the predicted success of interventions to reduce the risk of heart disease.[57] Category I risk factors are those for which interventions have been classified as "proven" to reduce the risk of heart disease. Category II and III are risk factors for which intervention is "likely to" and "might" reduce the incidence of cardiac events, respectively. LDL-cholesterol, HDL-cholesterol, and blood triglycerides have been placed into Categories I, II, and III, respectively.

Total blood cholesterol refers to the sum of the cholesterol in all of the lipoproteins circulating in the blood at the time it was drawn. Because blood is typically drawn after an overnight fast, there should not be chylomicrons in the blood. As mentioned in this chapter, chylomicrons usually circulate for less than 1 hour before they are removed from the blood in the form of a chylomicron remnant. Thus, depending on the lipid content and size of a meal, chylomicrons should be completely removed from the blood within 6–8 hours after a meal. Thus the sum of VLDL, LDL, and HDL-cholesterol should approximate the level of total cholesterol. See Table A for a typical blood lipid profile.

Higher levels of total blood cholesterol and LDL-cholesterol are strongly associated with heart disease. As most of the cholesterol circulating in humans is carried aboard LDL, total cholesterol and LDL-cholesterol levels are directly related. LDL is also the primary lipoprotein that crosses from circulation into the walls of arteries and propagates *atherosclerosis*. Higher LDL-cholesterol levels imply that there is more circulating LDL, which increases the potential for LDL migration into arterial walls. Atherosclerosis tends to occur in, but is certainly not limited to, smaller arteries of the heart and brain, which can eventually lead to heart attack and stroke, respectively. Increases in HDL-cholesterol

NUTRITION, EXERCISE, BLOOD LIPIDS, AND HEART DISEASE RISK (CONTINUED)

Table A

An Example of a Blood Lipid Profile for a 34-Year-Old Male

Lipid	Result	Normal Range	Level Considered a Risk Factor
Triglycerides (TG)	137 mg/dl	0–210 mg/dl	>150 mg/dl
Total cholesterol (TC)	163 mg/dl	50–200 mg/dl	>200 mg/dl[a]
HDL-cholesterol (HDL-C)	42 mg/dl	30–90 mg/dl	<40 mg/dl
VLDL-cholesterol (VLDL-C)	27 mg/dl	5–40 mg/dl	Not determined
LDL-cholesterol (LDL-C)	94 mg/dl	50–130 mg/dl	>160 mg/dl[b]

[a] TC must be evaluated with respect to HDL-C.
[b] The risk factor level is lower when other risk factors are present.

Table B

Recommended Gram Levels of Fat for Different Energy Levels

Energy Level (kcal/day)	Total Fat: ≤30% (g/day)	Saturated Fat: <10% (g/day)	Saturated Fat: <7% (g/day)
1200	≤40	<13	<9
1500	≤50	<17	<12
1800	≤60	<20	<14
2000	≤67	<22	<16
2200	≤73	<24	<17
2500	≤83	<28	<19
3000	≤100	<33	<23
3500	≤117	<38	<27
4000	≤133	<44	<31

levels, especially HDL_2, are associated with a decreased risk. Cardiorespiratory training can increase HDL-cholesterol levels, and smoking can decrease these levels. Smoking and a diet high in saturated fat can increase total and LDL-cholesterol levels.

One of the most important events involved in the atherosclerotic process is the continued oxidation of LDL. This includes LDL in circulation as well as LDL that has moved from circulation into the wall of an artery at the site of an injury. Researchers now know that exercise with higher O_2 demands produces a corresponding level of free radicals (see In Focus: Antioxidant Supplementation by Athletes in Chapter 9). This raises the question of whether exercise may actually be detrimental with regard to the atherosclerotic process. The answer is probably not. Muscle and other tissues adapt to increase the activity of antioxidant systems in response to training. Thus exercise may increase cardiovascular protection because its enhancement of antioxidant capabilities would persist throughout a 24-hour period, and exercise probably occurs for less than 10% of that time.

A person's diet is believed to have a profound impact on blood lipid levels. In general, diets that contain excessive energy and fat, especially saturated fat, are associated with elevated total and LDL-cholesterol

levels. General dietary recommendations to reduce the risk of heart disease in people over the age of 2 include limiting fat intake to <30% of energy with no more than 10% derived from saturated fat (Table B). If the level of LDL-cholesterol is above goal range, additional therapeutic lifestyle changes are suggested, such as consuming a diet with <7% of calories from saturated fat and <200 mg/day of cholesterol (Table B). Other dietary options include ingesting plant stanols/sterols (≤2 g/day) and/or increasing soluble fiber intake by 10–25 g/day. Also, reduction of body weight and in particular body fat levels should be considered. For less active individuals, increasing physical activity can help metabolize circulating lipoproteins.

Serum triglyceride levels can be used to assess heart disease risk; however, because they are Category III they should not be used as a stand-alone assessment tool. When triglyceride levels are elevated, the HDL-cholesterol level tends to be lower. Because exercise leads to increased metabolism of circulating triglycerides and increases HDL-cholesterol levels, regular exercise tends to produce a potent one-two punch in reducing a person's risk. A level of HDL-cholesterol of less than 40 mg/dl is considered a risk factor for heart disease, and a level greater than 60 mg/dl is considered protective.

Current knowledge indicates that the most significant ways a person can reduce the risk of developing heart disease is to plan for a heart-healthy diet and exercise regularly. These, in combination with not smoking and leading a less stressful life, will go a long way. For more information please see the website of the American Heart Association (www.americanheart.org).

STUDY QUESTIONS

1. What is the basic design of fatty acids and fat molecules?
2. How is fat digested and absorbed into the body?
3. How are fat and cholesterol molecules transported throughout the body in the blood?
4. What are the most important steps in mobilizing fatty acids from adipose tissue?
5. What are the different fat resources available to skeletal muscle during exercise?
6. What organelle is responsible for oxidizing fatty acids and how important is O_2 to the process?
7. How do intensity and duration of exercise influence fat metabolism?
8. How is carnitine involved in fatty acid metabolism?
9. What adaptations might occur that would increase fat utilization during exercise?
10. What are recommendations for fat consumption for the general population as well as athletes?

REFERENCES

1. Wildman REC, Medeiros DM. Lipids. In: *Advanced Human Nutrition*, Boca Raton, FL: CRC Press, 2000.
2. Mayes PA. Cholesterol synthesis, transport and excretion. In: *Harper's Biochemistry*, 24th ed. (Murray RK, Granner DK, Mayes PA, Rodwell VW, eds.), Stamford, CT: Appleton & Lange, 1998.
3. Mayes PA. Lipid transport and storage. In: *Harper's Biochemistry*, 24th ed. (Murray RK, Granner DK, Mayes PA, Rodwell VW, eds.), Stamford, CT: Appleton & Lange, 1998.
4. Turcotte LP, Richter EA, Kiens B. Lipid metabolism during exercise. In: *Exercise Metabolism* (Hargreaves M, ed.), Champaign, IL: Human Kinetics, 1995.
5. Hagenfeldt L, Wahren J, Pernow B, Raf L. Uptake of individual free fatty acids by skeletal muscle and liver in man. *Journal of Clinical Investigation* 51(9):2324–2330, 1972.
6. Potter BJ, Sorrentino D, Berk PD. Mechanisms of cellular uptake of free fatty acids. *Annual Review of Nutrition* 9:253–270, 1989.
7. Glatz JF, Luiken JJ, Bonen A. Involvement of membrane-associated proteins in the acute regulation of cellular fatty acid uptake. *Journal of Molecular Neuroscience* 16(2–3):123–132, 2001.
8. Binnert C, Koistinen HA, Martin G, Andreelli F, Ebeling P, Koivisto VA, et al. Fatty acid transport protein-1 mRNA expression in skeletal muscle and in adipose tissue in humans. *American Journal of Physiology: Endocrinology and Metabolism* 279(5): E1072–1079, 2000.
9. Luiken JJ, Miskovic D, Arumugam Y, Glatz JF, Bonen A. Skeletal muscle fatty acid transport and transporters. *International Journal of Sport Nutrition and Exercise Metabolism* 11 Suppl: S92–96, 2001.
10. Zimmerman AW, Veerkamp JH. New insights into the structure and function of fatty acid-binding proteins. *Cell Molecular Life Science* 59(7):1096–1116, 2002.
11. Abumrad NA, Perkins RC, Park JH, Park CR. Mechanism of long chain fatty acid permeation in the isolated adipocyte. *Journal of Biological Chemistry* 10; 256(17):9183–9191, 1981.
12. Abumrad NA, Perry PR, Whitesell RR. Stimulation by epinephrine of the membrane transport of long chain fatty acid in the adipocyte. *Journal of Biological Chemistry* 260(18): 9969–9971, 1985.
13. Abumrad NA, Harmon CM, Barnela US, Whitesell RR. Insulin antagonism of catecholamine stimulation of fatty acid transport in the adipocyte. Studies on its mechanism of action. *Journal of Biological Chemistry* 15; 263(29):14678–14683, 1988.
14. Miller WC, Hickson RC, Bass NM. Fatty acid binding proteins in the three types of rat skeletal muscle. *Proceedings of the Society of Experimental Biology and Medicine* 189(2):183–188, 1988.
15. Wildman REC, Medeiros DM. Energy metabolism. In: *Advanced Human Nutrition*, Boca Raton, FL: CRC Press, 2000.
16. Hales CN, Luzio JP, Siddle K. Hormonal control of adipose-tissue lipolysis. *Biochemical Society Symposium* 43:97–135, 1978.
17. Romijn JA, Coyle EF, Sidossis LS, Gastaldelli A, Horowitz JF, Endert E, Wolfe RR. Regulation of endogenous fat and carbohydrate metabolism in relation to exercise intensity and duration. *American Journal of Physiology* 265:E380–E391.
18. Arner P, Bolinder J, Ostman J. Glucose stimulation of the antilipolytic effect of insulin in humans. *Science* 3;220(4601): 1057–1059, 1983.
19. Wolfe RR, Peters EJ, Klein S, Holland OB, Rosenblatt J, Gary H, Jr. Effect of short-term fasting on lipolytic responsiveness in normal and obese human subjects. *American Journal of Physiology* 252(2 Pt 1):E189–196, 1987.
20. Laughlin MH, Korzick DH. Vascular smooth muscle: integrator of vasoactive signals during exercise hyperemia. *Medicine and Science in Sports and Exercise* 33(1):81–91, 2001.
21. Hester RL, Choi J. Blood flow control during exercise: role for the venular endothelium? *Exercise and Sport Science Review* 30(4): 147–151, 2002.
22. Boushel R, Langberg H, Gemmer C, Olesen J, Crameri R, Scheede C, et al. Combined inhibition of nitric oxide and prostaglandins reduces human skeletal muscle blood flow during exercise. *Journal of Physiology* 1;543(Pt 2):691–698, 2002.
23. Carlson MG, Snead WL, Hill JO, Nurjhan N, Campbell PJ. Glucose regulation of lipid metabolism in humans. *American Journal of Physiology* 261(6 Pt 1):E815–820, 1991.
24. Bulow J. Regulation of lipid mobilization in exercise. *Canadian Journal of Sport Science* 12(Suppl):117S–119S, 1987.
25. Spector AA, Fletcher JE, Ashbrook JD. Analysis of long-chain free fatty acid binding to bovine serum albumin by determination of stepwise equilibrium constants. *Biochemistry* 17(10):3229–3232, 1971.
26. Koch G, Rocker L. Plasma volume and intravascular protein masses in trained boys and fit young men. *Journal of Applied Physiology* 43(6):1085–1088, 1977.

27. Wolfe RR, Klein S, Carraro F, Weber JM. Role of triglyceride-fatty acid cycle in controlling fat metabolism in humans during and after exercise. *American Journal of Physiology* 258(2 Pt 1):E382–389, 1990.

28. Arner P, Kriegholm E, Engfeldt P, Bolinder J. Adrenergic regulation of lipolysis *in situ* at rest and during exercise. *Journal of Clinical Investigation* 85(3):893–898, 1990.

29. Wahrenberg H, Bolinder J, Arner P. Adrenergic regulation of lipolysis in human fat cells during exercise. *European Journal of Clinical Investigation* 21(5):534–541, 1991.

30. Watt MJ, Heigenhauser GJ, Spriet LL. Intramuscular triacylglycerol utilization in human skeletal muscle during exercise: is there a controversy? *Journal of Applied Physiology* 93(4):1185–1195, 2002.

31. Spriet LL. Regulation of skeletal muscle fat oxidation during exercise in humans. *Medicine and Science in Sports and Exercise* 34(9):1477–1484, 2002.

32. Bonadonna RC, Groop LC, Simonson DC, DeFronzo RA. Free fatty acid and glucose metabolism in human aging: evidence for operation of the Randle cycle. *American Journal of Physiology* 266(3 Pt 1):E501–509, 1994.

33. Odland LM, Heigenhauser GJ, Spriet LL. Effects of high fat provision on muscle PDH activation and malonyl-CoA content in moderate exercise. *Journal of Applied Physiology* 89(6):2352–2358, 2000.

34. Winder WW, Arogyasami J, Barton RJ, Elayan IM, Vehrs PR. Muscle malonyl-CoA decreases during exercise. *Journal of Applied Physiology* 67(6):2230–2233, 1989.

35. Turcotte LP, Richter EA, Kiens B. Increased plasma FFA uptake and oxidation during prolonged exercise in trained vs. untrained humans. *American Journal of Physiology* 262:E791–99, 1992.

36. Wahren J, Hagenfeldt L, Felig P. Glucose and free fatty acid utilization in exercise. Studies in normal and diabetic man. *Israel Journal of Medical Science* 11(6):551–559, 1975.

37. Jeukendrup AE, Saris WH, Schrauwen P, Brouns F, Wagenmakers AJ. Metabolic availability of medium-chain triglycerides coingested with carbohydrates during prolonged exercise. *Journal of Applied Physiology* 79(3):756–762, 1995.

38. Robinson DM, Ogilvie RW, Tullson PC, Terjung RL. Increased peak oxygen consumption of trained muscle requires increased electron flux capacity. *Journal of Applied Physiology* 77:1941–1952, 1994.

39. Hurley BF, Nemeth PM, Martin WH, Hagberg JM, Dalsky GP, Holloszy JO. Muscle triglyceride utilization during exercise: effect of training. *Journal of Applied Physiology* 60(2):562–567, 1986.

40. Martin WH, 3d, Dalsky GP, Hurley BF, Matthews DE, Bier DM, Hagberg JM, et al. Effect of endurance training on plasma free fatty acid turnover and oxidation during exercise. *American Journal of Physiology* 265:E708–714, 1993.

41. Fink WJ, Costill DL, Pollock ML. Submaximal and maximal working capacity of elite distance runners. Part II. Muscle fiber composition and enzyme activities. *Annals of the New York Academy of Sciences* 301:323–327, 1977.

42. Pattengale PK. Augmentation of skeletal muscle myoglobin by a program of treadmill running. *American Journal of Physiology* 213(3): 783–785, 1967.

43. Holloszy JO. Biochemical adaptations in muscle. Effects of exercise on mitochondrial oxygen uptake and respiratory enzyme activity in skeletal muscle. *Journal of Biological Chemistry* 242:2278–2282, 1967.

44. Saltin B, Gollnick PD. Skeletal muscle adaptability: significance for metabolism and performance. In: *Handbook of Physiology*, Sec. 10, *Skeletal Muscle* (Peachey LD, ed.), Baltimore, MD: Williams & Wilkins, pp. 555–631, 1983.

45. Yang HT, Ogilvie RW, Terjung RL. Peripheral adaptations in trained aged rats with femoral artery stenosis. *Circulation Research* 74:235–243, 1994.

46. Saltin B, Nazar K, Costill DL, Stein E, Jansson E, Essen B, Gollnick PD. The nature of the training response: peripheral and central adaptations of one-legged exercise. *Acta Physiologica Scandinavica* 96:289–305, 1976.

47. Seip RL, Semenkovich CF. Skeletal muscle lipoprotein lipase: molecular regulation and physiological effects in relation to exercise. *Exercise and Sport Sciences Review* 26:191–218, 1998.

48. Schrauwen P, van Aggel-Leijssen DP, Hul G, Wagenmakers AJ, Vidal H, Saris WH, van Baak MA. The effect of a 3-month low-intensity endurance training program on fat oxidation and acetyl-CoA carboxylase-2 expression. *Diabetes* 51(7):2220–2226, 2002.

49. Carey AL, Staudacher HM, Cummings NK, Stepto NK, Nikolopoulos V, Burke LM, Hawley JA. Effects of fat adaptation and carbohydrate restoration on prolonged endurance exercise. *Journal of Applied Physiology* 91(1): 115–122, 2001.

50. Helge JW. Adaptation to a fat-rich diet: effects on endurance performance in humans. *Sports Medicine* 30(5): 347–357, 2000.

51. Leon AS, Sanchez OA. Response of blood lipids to exercise training alone or combined with dietary intervention. *Medicine and Science in Sports and Exercise* 33(6 Suppl):S502–515, 2001.

52. Hartung GH, Lally DA, Prins J, Goebert DA. Relation of high-density lipoprotein cholesterol to physical activity levels in men and women. *Medicine, Exercise, Nutrition and Health* 1:293–299, 1992.

53. Murray TD, Squires WG, Hartung GH, Bunger J. Effects of diet and exercise on lipids and lipoproteins. In: *Nutrition in Exercise and Sport*, 3rd ed. (Wolinsky I, ed.), Boca Raton, FL: CRC Press, 1998.

54. Hurley BF, Seals DR, Hagberg JM, Goldberg AC, Ostrove SM, Holloszy JO, et al. High-density-lipoprotein cholesterol in bodybuilders v powerlifters. Negative effects of androgen use. *Journal of the American Medical Association* 252(4):507–513, 1984.

55. Costill DL, Pearson DR, Fink WJ. Anabolic steroid use among athletes: changes in HDL-C levels. *Physiology and Sportsmedicine* 12:112–118, 1984.

56. Webb OL, Laskarzewski PM, Glueck CJ. Severe depression of high-density lipoprotein cholesterol levels in weight lifters and body builders by self-administered exogenous testosterone and anabolic-androgenic steroids. *Metabolism* 33(11): 971–975, 1984.

57. Leon AS, Sanchez OA. Response of blood lipids to exercise training alone or combined with dietary intervention. *Medicine and Science in Sports and Exercise* 33(6 Suppl):S502–515, 2001.

ENERGY BALANCE, BODY WEIGHT AND COMPOSITION FOR SPORT AND FITNESS

CHAPTER

7

Personal Snapshot

John, a college sophomore, is 6'3" and 235 lb. He plays defensive end and thinks that he could improve his play if he gained 20 lb during the off-season. He discusses this with his coaches and team physician, who are not convinced that a change in John's body weight would really improve his performance. Even though a weight gain would seem to provide a competitive advantage over other players currently weighing the same as he does now, what if John's strength is unchanged but his quickness, agility, and stamina become compromised? This might be the case if the majority of his weight gain was body fat. With their help, John decides that his focus should be on increasing his body weight with little change in body composition or an improvement in leaner body tissue, primarily skeletal muscle. Therefore, the weight gain would occur without decreasing his strength and power relative to body weight, and it might increase relative strength and power. To begin, John's body strength and weight and composition are accurately assessed in the exercise physiology laboratory on campus. Next, John will meet with the team strength and conditioning coach and the nutritionist to plan his diet and training programs.

Chapter Objectives

- Describe the difference between body weight and composition and discuss their application to exercise and sport performance.

- Define overweight and obesity and discuss the potential health consequences of excessive body fat.

- Provide an overview of the different techniques commonly available to assess body composition.

- Describe energy balance and the influence of energy nutrients and physical activity on body weight and composition.

- Discuss recommendations for achieving a desired body weight and/or composition for athletic performance and health.

- Discuss different diet programs that are commonly promoted for active individuals.

Body weight and composition are important considerations for athletes as well as people in general. Athletes and coaches must determine the body weight and composition at which the physical performance for an athlete is optimized. Some sports rank athletes in a competition based on body composition. For instance, bodybuilders are judged based on maximal skeletal muscle mass and minimal levels of body fat. Meanwhile, fitness competitors often demonstrate a muscular and very lean physique since body composition is a component of fitness. Even in subjectively evaluated sports such as figure skating and gymnastics where body composition is not a criteria on which athletes are scored, body composition has been noted to possibly influence a score given to an athlete.

What determines body weight and composition? Among the most influential factors are energy balance and the type of training. Energy balance, the balance between energy intake and expenditure, is an important consideration for athletes and nonathletes alike. Nonathletes must balance energy intake and expenditure in order to achieve or maintain a body weight and composition that optimizes health. For the competitive athlete, the consumption of energy nutrients can help fuel physical efforts and promote efficient recovery and adaptation. In addition, athletes participating in sports involving weight classes such as wrestling, boxing, and judo can benefit by understanding basic concepts of energy balance for weight control. This chapter provides an overview of concepts related to body weight and composition and discusses diet strategies for changing weight and composition. In addition, popular diets and diet concepts are discussed.

■ BODY WEIGHT AND COMPOSITION

Body weight and composition are not synonymous although they are often used interchangeably. **Body weight** is the total mass of a person and is expressed in pounds (lb) or kilograms (kg). **Body composition** refers to the relative contributions to a person's mass made by different substances or tissues that make up the body. This section looks more closely at the components of body composition as well as the relationship between body weight and composition and health.

Body Weight and Health

Body weight has long been used as a predictor of health because of its association with body composition, especially for inactive people. Generally speaking, heavier people tend to have greater amounts and percentages of fat. Likewise, inactivity in individuals is a reasonable predictor of body fat levels. As the amount of body fat increases above desirable levels the risk of certain diseases such as heart disease rises in a related manner. With this in mind,

weight standards have been developed to determine body weight ranges that can be used to characterize a population and be used in health promotion (Table 7-1). The concept of a **healthy weight** was developed to demonstrate a body weight specific to gender, height, and frame size in which there is a strong association with good health and longevity. One is said to be overweight when his or her body weight falls above the healthy weight range. When body weight increases above or falls below the healthy weight range, the person is at greater risk of illness. In 1959 the Metropolitan Life Insurance Company developed its height and weight tables, which were modified in 1983 and are still used as references today (Table 7-2). The company presented weight ranges for a given height for adults and used these as the basis for setting premiums for insurance policies. Standards such as healthy weight have been used for estimating obesity as well. For instance, the threshold for estimating obesity from height for the general population is set at >120% of the healthy weight for a gender and height.

Table 7-1 Healthy Weights for Adults

Height[a]	Weight (lb)[a]	
	Midpoint	*Range*
4'10"	105	91–119
4'11"	109	94–124
5'0"	112	97–128
5'1"	116	101–132
5'2"	120	104–137
5'3"	124	107–141
5'4"	128	111–146
5'5"	132	114–150
5'6"	136	118–155
5'7"	140	121–160
5'8"	144	125–164
5'9"	149	129–169
5'10"	153	132–174
5'11"	157	136–179
6'0"	162	140–184
6'1"	166	144–189
6'2"	171	148–195
6'3"	176	152–200
6'4"	180	156–205
6'5"	185	160–211
6'6"	190	164–216

Note: The higher weights in the ranges generally apply to men, who tend to have more muscle and bone; the lower weights more often apply to women, who have less muscle and bone.
[a]Without shoes or clothes

Source: Report of the Dietary Guidelines Advisory committee on the Dietary Guidelines for Americans (Washington, D.C.: Government Printing Office, 1995).

| Table 7-2 | **Metropolitan Life Insurance Height and Weight Tables** |

Men				Women			
Height (inches)	**Small Frame**	**Medium Frame**	**Large Frame**	**Height (inches)**	**Small Frame**	**Medium Frame**	**Large Frame**
62	128–134	131–141	138–150	58	102–111	109–121	118–131
63	130–136	133–143	140–153	59	103–113	111–123	120–134
64	132–138	135–145	142–156	60	104–115	113–126	122–137
65	134–140	137–148	144–160	61	106–118	115–129	125–140
66	136–142	139–151	146–164	62	108–121	118–132	128–143
67	138–145	145–154	149–168	63	111–124	121–135	131–147
68	140–148	145–157	152–172	64	114–127	124–138	134–151
69	142–151	148–160	155–176	65	117–130	127–141	137–155
70	144–154	151–163	158–180	66	120–133	130–144	140–159
71	146–157	154–166	161–184	67	123–136	133–147	143–163
72	149–160	157–170	164–188	68	126–139	136–150	146–167
73	152–164	160–174	168–192	69	129–142	139–153	149–170
74	155–168	164–178	172–197	70	132–145	142–156	152–173
75	158–172	167–182	176–202	71	135–148	145–159	155–176
76	162–176	171–187	181–207	72	138–151	148–162	158–179

Note: Weights at ages 25–59 based on lowest mortality. Weight in pounds according to frame (in indoor clothing weighing 3 lb; shoes with 1-inch heels). There are 12 inches in a foot (for example, 6 feet is 72 inches). To convert inches to centimeters multiply by 2.54 (for example, 72 inches is roughly 183 cm).

BODY MASS INDEX. **Body mass index (BMI)** is a method available to express body size and is calculated in the metric system by dividing weight by the square of height, as shown in Table 7-3. Although different estimates exist, a BMI of 18.5–24.9 is considered normal for a member of the general population. A BMI between 25 and 29.9 is considered overweight, and a BMI of 30 is the lower cutoff for obesity, as shown in Table 7-3 and Figure 7-1. Figure 7-2 presents the disease risk associated with the range of BMI values for adults. As BMI increases above 25 the risk of disease increases, as does the rate of mortality.

For instance, an adult with a BMI of 37 has double the risk of all-cause mortality that a person with a BMI of 23 has. However, when BMI falls below 20 the risk of disease and mortality also increases. Therefore the relationship between BMI and morbidity and mortality is J-shaped. However, many researchers think the risk of a lower BMI is overstated because the lowered BMI for some people was the result of disease and/or related to cigarette smoking. Interestingly, the BMI-mortality curve for women is below that for men at all BMI levels. The curve shown in Figure 7-2 displays the average risk for men and women; the curve would be slightly higher to show the average risk for men and slightly lower to show the average risk for women.

BODY WEIGHT AND HEALTH FOR ACTIVE PEOPLE. For more active people and athletes, body weight alone cannot be used to estimate health risk. Many athletes engaged in strength- and power-oriented sports maintain greater muscle mass and might be characterized as overweight or even obese by general standards, yet they may

| Table 7-3 |

Calculating Body Mass Index (BMI)

BMI can be calculated in metric or U.S. units using the following formulas:

$$\text{BMI} = \frac{(\text{weight in kilograms})}{(\text{height in meters})^2} = \frac{(\text{weight in pounds}/2.2)}{(\text{height in inches}/39.37)^2}$$

18.5   25   30

Under weight   Healthy weight   Overweight   Obesity

Source: Reprinted from the *Journal of the American Dietetic Association,* Vol. 91:843, R. P. Abernathy, "Body Mass Index," © 1991, with permission from the American Dietetic Association.

Key Point

Body weight and BMI are often used to predict overweight and obesity in the general population, but they are not accurate in more active populations.

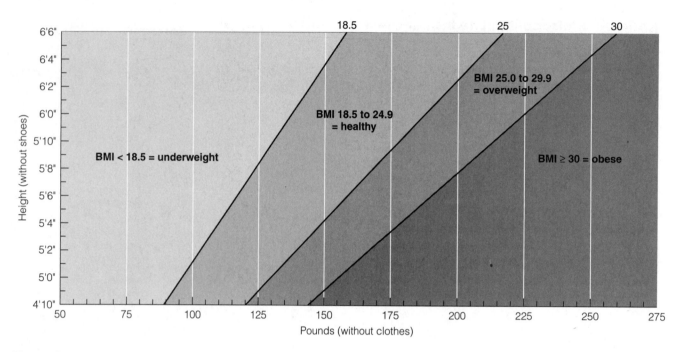

Figure 7-1
Body mass index (BMI) (*Source:* U.S. Department of Agriculture and U.S. Department of Health and Human Services, *Nutrition and Your Health: Dietary Guidelines for Americans*, Washington, DC, p. 7, 2000).

have a relatively low body fat level and therefore a desirable body composition and low health risk. For instance, a 5′10″, 225-lb bodybuilder would have a BMI of 32, in the obese range. However, probably no one would classify him as such. Therefore, body weight and BMI are practical estimators of body size and overweight and obesity in the general population; however, in more active populations their precision is decreased.

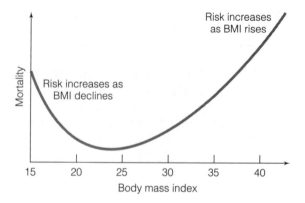

Figure 7-2
This J-shaped curve describes the relationship between body mass index (BMI) and mortality and shows that both underweight and overweight present risks of a premature death.

Components of Body Composition

The human body is composed of many substances, as alluded to in the preceding chapters. Water accounts for about 60% (56–64%) of a typical lean adult's mass. Protein and fat each account for about 14–16% of the total mass of an adult; minerals, 4–6%; and carbohydrate, less than 1%. Table 7-4 presents the major substances contributing to body mass for lean adults. However, grouping certain tissues together, as in Figure 7-3, is more common when assessing body composition. Depending on the technique used to assess body composition, the components identified are most often expressed as mass quantities such as fat mass (FM), lean body mass (LBM), fat-free mass (FFM), and bone mineral mass (BMM), as defined in Table 7-5. In addition, body composition is expressed as percentages of total body mass. These include percent body fat (%BF) or %FM, and lean body mass (%LBM).

Often FFM and LBM are used interchangeably. However, FFM includes only nonfat components of the body whereas LBM includes essential fat depots as well as FFM. **Essential fat** is found associated with the bone marrow, central nervous system, and internal organs. **Nonessential fat** is found within adipose tissue. Women also have essential body fat associated with mammary glands and the pelvic regions. Essential body fat percentages for an adult male and female are approximately 3–5% and 12% of body weight, respectively, as shown in Figure 7-3.

Table 7-4

Theoretical Contributors to Body Weight

Component (Substance)	Lean Man (%)	Lean Woman (%)
Water	62	59
Fat	16	22
Protein	16	14
Minerals	5–6	4–5
Carbohydrate	<1	<1

Because body composition varies among people, two males or females having the same height and weight can have differences in body composition. In addition, two people of different height and weight can have similar body composition. Although changes in body weight and composition are often associated, a person can experience a change in body composition but not weight and vice versa, although the latter is less likely. Therefore the physical assessment of a person for health and performance should include both body weight and composition, especially if the person is involved in physical training.

Key Point

Lean body mass (LBM) includes fat-free mass plus essential body fat.

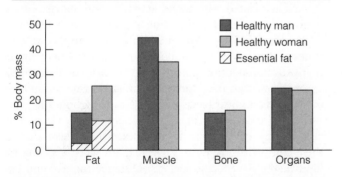

Figure 7-3
Adult male and female body compositions compared. The differences between male and female body compositions become apparent during adolescence. Lean body mass (primarily muscle) increases more in males than in females. Fat assumes a larger percentage of female body composition as essential body fat is deposited in the mammary glands and pelvic region in preparation for childbearing. Both men and women have essential fat associated with the bone marrow, the central nervous system, and the internal organs. (*Source*: Wildman REC, Medeiros DM. *Advanced Human Nutrition*, Boca Raton, FL: CRC Press, pp. 321–323, 2000. Used with permission.)

Table 7-5

Body Composition Compartments

Component	Characteristics of Component
Fat mass (FM)	Mass of body fat (fat is defined as any lipid material that would be soluble and extractable in ether)
% Body fat (%BF)	Percentage of total body mass that is fat mass
Fat-free mass (FFM)	Mass of body substances that are not fat including water, protein, and minerals as found in organs, muscle, bone
Lean body mass (LBM)	Mass of FFM plus essential body fat
Total body water (TBW)	Total of intracellular and extracellular water
Bone mineral mass (BMM)	Mass of mineral content of bone based on estimators of bone density

Body Composition and Health

Most of the attention to body composition and its influence on health has focused on the proportion of body fat. The percentage of body fat can vary from 3–5% of mass in excessively lean adults to as much as 65–70% of the total mass in excessively (morbidly) obese individuals (Figure 7-4). Body fat can influence health in several ways; this is clearly demonstrated as body fat increases over time and medical situations develop and worsen in a relative manner. However, when body fat levels fall too low, the resultant *excessive leanness* is potentially problematic as well.

Fat-free mass is a desirable component of body composition for health. Higher levels of FFM can predict more desirable bone density and skeletal muscle mass. Greater bone density is associated with greater bone integrity, which reduces the current risk of fracture as well as the risk of osteoporosis in the future. Likewise, having more skeletal muscle may be associated with greater strength and decreased risk of physical injury. Increased skeletal muscle mass is related to increased daily energy expenditure, which can reduce the possibility of weight gain and obesity. In addition, having more skeletal muscle mass can improve glucose tolerance and decrease the risk of diabetes mellitus. Conversely, decreases in FFM can result in lower energy expenditure and increased proneness to weight gain, reduced glucose tolerance, and decreased strength and physical capabilities.

OBESITY.   **Obesity** is a state of excessive body fatness. For men, a level of body fat above 25% is considered the lower threshold for obesity. Women are considered obese

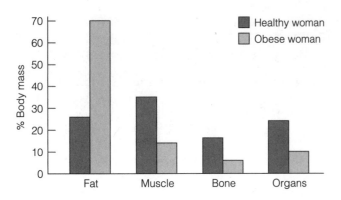
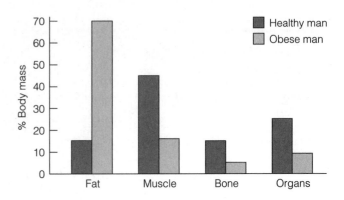

Figure 7-4

Healthy and morbidly obese body compostions compared. Body fat may vary from as little as 5% extremely lean adult to as much as 70% in excessively obese adults. As the percentage of fat increases, the percentage of other body components decreases. *Source:* Adapted with permission from Wildman REC, Medeiros DM. *Advanced Human Nutrition*, Boca Raton, FL: CRC Press, pp. 321–323, 2000.

when their body fat exceeds 33%. Obesity is associated with various risk factors for diseases, such as the heart disease risk factors of high blood pressure, elevated levels of total and LDL-cholesterol and triglycerides, and low levels of HDL-cholesterol. In addition, excessive body mass associated with obesity places increased mechanical stress on joints, and many obese individuals suffer from joint disorders (Table 7-6). See In Focus: Obesity in Chapter 1.

Although a higher level of body fat is not common among athletes, it is associated with some sports. For instance, sumo and heavyweight wrestlers and football linemen tend to have relatively high levels of body fat. Whether this is a health problem during the competing

years remains to be conclusively determined. However, former athletes who maintain the eating pattern of their competitive years but not the level of training and energy expenditure are susceptible to increased adiposity and obesity.

Key Point

Both obesity and excessive leanness are associated with health risks.

EXCESSIVE LEANNESS.   Extremely low levels of body fat are associated with some activities and sports, either transiently or chronically. For instance, some athletes, such as bodybuilders who may only compete a couple of times a year, reduce their body fat by several percentage points as competition approaches. Others, such as fitness competitors who may compete numerous times annually, may maintain lower levels of body fat throughout the year. Athletes participating in certain sports involving weight classes may consciously become excessively lean during their competitive season. This includes wrestlers, boxers, lightweight rowers, and jockeys. Distance runners and cyclists also tend to have very low levels of body fat, usually as a result of the high energy demands of their sport.

Approximately 3–5% for men and 12% for women have been suggested as the minimal levels of body fat compatible with health.[1] Excessive leanness was once assumed to be a primary cause of reduced estrogen production and disruption of menstrual cycle in some female athletes. Over the past decade or so it has become more clear that amenorrhea may be more directly related to a chronic underconsumption of energy. Female athletes and their coaches should be aware of disruptions in menstruation and seek medical assistance to manage the situation. Often an excessively lean body reflects a psychological

Table 7-6

Disorders and Diseases Associated with Obesity

Tissue or Physiological System	Pathology or Disorder
Cardiovascular system	Hypertension Atherosclerosis
Hormonal/metabolic systems	Diabetes (type 2) Hypercholesterolemia Hypertriglyceridemia Gout
Joints and connective tissue	Arthritis of knees and hips Bone spurs
Respiratory systems	Obstructive sleep apnea
Psychological state	Depression Social isolation
General physical state	Decreased mobility Decreased vitality

condition, as discussed in the feature In Focus: Eating Disorders and the Female Athlete Triad in Chapter 15.

■ BODY WEIGHT AND COMPOSITION *IN REVIEW*

- Body weight can be used as a predictor of health in the general population but not in more active people and athletes.
- Obesity and excessive leanness are associated with potential health risks.
- Men and women are considered obese when body fat percentage exceeds 25% and 33%, respectively.
- Men and women are considered excessively lean when their body fat percentage falls below 3–5% and 12%, respectively.
- Obesity and associated health concerns can be a consideration for some athletes during and after their playing years.

■ BODY COMPOSITION ASSESSMENT

Body composition refers to the absolute and relative contributions made by different substances or tissues to total body mass. Certainly, the assessment of body composition provides useful information regarding health as well as physical performance. What tools do researchers and clinicians use to assess body composition, and what advantages and disadvantages are associated with these various tools? This section provides an overview of how body composition is assessed as well as considerations for each system.

Comparison of Tools Used for Body Composition Assessment

Several methods are available to estimate body composition including underwater weighing (UWW), bioelectrical impedance analysis (BIA), and dual-energy X-ray absorptiometry (DXA or DEXA). Some techniques are considered more accurate and are used to validate the accuracy of other methods. These techniques are often referred to as **gold standards** or **criterion methods**. For instance, UWW and DXA are considered gold standards for body composition assessment in adults.

Some of the techniques are able to separate body composition into two components; others can estimate three or four components. For instance, densitometry and BIA separate body composition into FM and FFM, whereas DXA is able to discern three compartments (FM, FFM, and BMM). Some of the body composition assessment techniques involve large, immobile machinery or facilities (including UWW and DXA) and can be used only in a controlled *laboratory* situation. Other body composition

Figure 7-5
Using hydrodensitometry to assess body fat. The difference between a person's weight on land and in water allows for a calculation of body density, from which percentage of body fat is calculated.

assessment tools (such as skinfold and BIA) are mobile, allowing researchers and practitioners to estimate body composition in the *field* (that is, not in a laboratory). Only the more commonly applied body composition assessment tools and techniques are discussed below.

Key Point

Body composition can be assessed by several methods including underwater weighing, (UWW), dual-energy X-ray absorptiometry (DXA), bioelectrical impedance analysis (BIA), and skinfold assessment.

Underwater Weighing

Underwater weighing (UWW) is also referred to as *hydrostatic weighing* or *hydrodensitometry*. UWW assessment of body composition is based on density, which is mass divided by volume. The densities of FM and FFM are approximately 0.9 g/cm³ and 1.1 g/cm³. As shown in Figure 7-5, UWW involves submerging an individual in a tank of water and then determining his or her weight. An individual's weight underwater predicts body volume, and body density is determined by dividing weight on land by volume. Once body density is determined it can be used in an equation specifically developed for age, gender, and race. For instance, an equation developed for young white males appears to be accurate to ±2.7%.[2] However, if that same equation were applied to the general population, the accuracy is decreased to ±4%.[3] Therefore, population-specific equations must be applied as available. Specific equations may reflect variations in FFM density between the genders and among racial groups; equations for various populations of women and African Americans have been developed and await validation.[4,5]

UWW is not as accurate a predictor of body composition in children as it is in adults. This is due to density differences in FFM at different points of development. The density of FFM varies based on the contributions made by water, protein, and minerals, which can change during growth. FFM can be slightly less dense during childhood, which promotes an overestimation of %BF; thus an equation specific to children should be applied.[6]

One of the most important considerations when using UWW is gas in the lungs and the digestive tract. The presence of these gases increases the buoyancy and thus decreases the weight of an individual submerged in the tank. This would lead to an overestimation of body volume and body fat. To minimize the error associated with the digestive tract, individuals should be fasted for at least 6–8 hours and asked to evacuate their colon prior to assessment. Then a standard factor is applied for residual gas in the digestive tract. Lung gases are addressed in one of two ways. One way is to determine an individual's **residual volume (RV)**, which is the volume of air remaining in the lungs after forceful expulsion. The individual is then weighed underwater with only RV in the lungs. The second way of addressing lung gas involves a breathing tube; then an individual's normal breathing volume **(tidal volume)** and RV are taken into account in the final calculations.

Plethysmography

Like UWW, air displacement **plethysmography** applies body density in estimating composition. Air displacement plethysmography uses a sealed chamber of known volume, such as the Bod Pod (by Life Measurement Instruments). As shown in Figure 7-6, an individual sits in the chamber in thin, tight clothing, for example, a bathing suit, and breathes normally. The concept is similar to underwater weighing, but air is displaced instead of water to determine volume. The Bod Pod is considered a precise tool for body composition, and its estimates of %BF are similar to those from UWW.[7,8] In addition, air displacement plethysmography is considered a more comfortable procedure than UWW. Methodological considerations include transient variations in body surface unrelated to changes in body composition, such as abdominal bloating after a large meal.

Key Point

Underwater weighing and plethysmography estimate body composition based on body density.

Dual-Energy X-Ray Absorptiometry

Dual-energy X-ray absorptiometry (DXA or DEXA) allows for a more comprehensive assessment of body

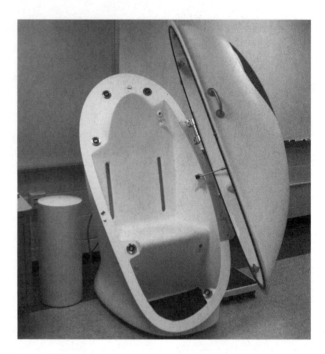

Figure 7-6
Air displacement plethysmography uses a sealed chamber, such as the Bod Pod.

composition because it quantifies three compartments, namely FM, FFM in soft tissue, and BMM. Thus, FFM is separated into bone and nonbone material. DXA also allows for regional body tissue assessment that includes bone density for osteoporosis risk assessment. DXA assessment times are short (10–20 minutes) and comfortable, as an individual lies on the machine table (Figure 7-7). In addition, DXA is generally believed to be safe, as the radiation dose is 800–2000 times lower than in a typical chest X-ray. However, the most significant factor limiting the use of DXA equipment is its size and cost. Therefore, DXA is primarily found in clinical institutions and research facilities.

Key Point

Dual-energy X-ray absorptiometry (DXA) measures FM, soft-tissue FFM, and BMM. The expense of the equipment limits its general use.

Bioelectrical Impedance Analysis

Bioelectrical impedance analysis (BIA) has become one of the most popular tools for estimating body composition. Conventional BIA passes a small electrical current from one body extremity to another (for example, right wrist to right ankle) while a person lies on a non-conducting surface (Figure 7-8). Electrical current is transmitted

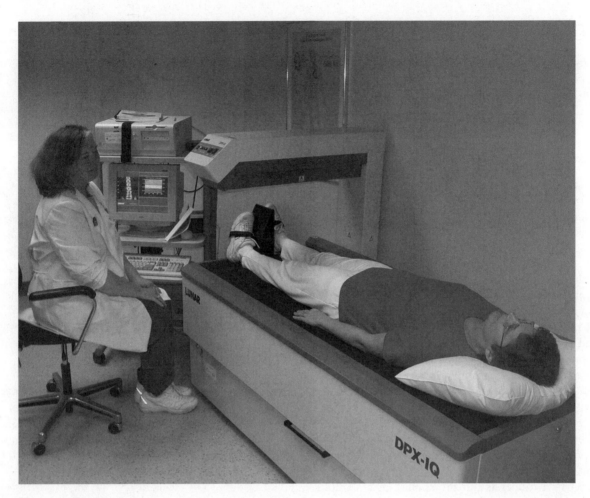

Figure 7-7
Assessment of body composition using Dual-energy X-ray absorptiometry. (*Source*: Courtesy of Mercy Medical Center, Sioux City.)

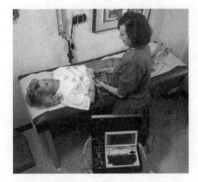

Figure 7-8
Using bioelectrical impedance to assess body fat. A low-intensity electrical current enters an electrode at one part of the body and passes out an electrode at another point. Adipose tissue provides resistance to electrical current, and electrolyte-containing fluids, found primarily in lean body tissues, readily conduct electrical current. Thus percentage of body fat can be estimated from resistance.

through body tissue from one electrode to the other electrode. Impedance to flow of electric current (measured in ohms) is attributable to FM, and this measurement is applied in an equation specific to the age and gender group of the person being assessed.

BIA has the potential to be an accurate ($\pm 5\%$ BF) assessment tool of body composition if the appropriate prediction equation is applied for a person or a group being assessed.[9] Individuals need to be in good hydration status and electrolyte balance, as these factors influence the conduction of electricity. Recently, handheld and foot scale BIA devices have become available to researchers and the general public. The ability of these units to provide precise estimation of body composition remains to be conclusively determined. Therefore researchers should not use them solely to predict body composition. However, they may be helpful in providing rough approximations of changes in an individual's body composition over time.

Key Point

Skinfold Assessment

Body composition can be estimated by measuring the thickness of the layer of adipose tissue beneath the skin; that is, the **subcutaneous** fat. Skinfold thickness measurements are obtained with calipers at several anatomical locations (Figure 7-9). Primary sites of measurement are the triceps, abdomen, subscapular area, thigh, and suprailiac area, as listed in Table 7-7. Secondary sites include the chest, midaxillary, and the medial calf. Three general assumptions associated with skinfold measures are that (1) a direct relationship exists between the quantity of fat deposited just below the skin (subcutaneous fat) and total body fat; (2) the thickness of the skin and subcutaneous adipose tissue has a constant compressibility throughout the body; and (3) the thickness of skin is negligible and a constant fraction of skinfold measurements. Skin thickness tends to vary between 0.5 and 2.0 mm.

The use of skinfold calipers requires training to maximize precision. Several factors must be taken into consideration to improve the accuracy of skinfold assessment. First, when a caliper is initially applied to a skinfold, a brief period is necessary for the tissue to compress prior to the reading of the thickness. After a couple of seconds the tips of the caliper have appropriately compressed the skinfold. However, the reading continues to become smaller as the calipers compress the skinfold and force fluids from the area. Error increases if an accurate measurement is not

| Table 7-7 | **Sites of Skinfold Measurements** |

Location	Measurement Technique
Triceps	Vertical skinfold measurement on the back of the arm at the midpoint between the tip of the shoulder (lateral process of the acromion) and the tip of the elbow (olecranon process of the ulna)
Subscapular	Oblique skinfold measurement made just beneath the tip of the scapula (inferior angle of the scapula bone)
Abdominal	Vertical skinfold measurement made approximately 3 cm to the right and 1 cm below the midpoint of the belly button (umbilicus)
Suprailiac	Slightly oblique skinfold measurement just above the iliac crest at the midaxillary line
Thigh	Anterior vertical fold measured at the midpoint between the superior border of the kneecap (patella) and the inguinal crease
Pectoral (chest)	Skinfold measurement as high as possible, just beneath the nipple, along the line from the anterior axillary fold to the nipple
Medial calf	As subject sits with right leg flexed approximately 90° at the knee, vertical skinfold measurement made at the medial region of maximum calf circumference
Midaxillary	Skinfold measurement at the right midaxillary line at the level of the superior aspect of the xiphoid process

Source: Wildman REC, Medeiros DM. *Advanced Human Nutrition*, Boca Raton, FL: CRC Press, p. 326, 2000.

obtained after 4–5 seconds. Skinfold measurements can be taken on either side of the body, and inexperienced skinfold technicians are encouraged to mark the body at points of measurement. If population-specific equations are applied, skinfold assessment can provide a fairly accurate prediction (±3–4% of body weight).[10]

■ BODY COMPOSITION ASSESSMENT IN REVIEW

- Body composition is the proportion of lean body tissue to fat tissue within the body.
- Body composition can be assessed several ways; the most accurate common techniques are UWW and DXA.
- Plethysmography and UWW estimate body composition based on density.
- Skinfold assessment is the cheapest and most widely used method for estimating body composition.

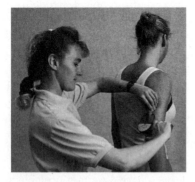

Figure 7-9
Using fatfold measures to assess body fat. The assessor measures subcutaneous fat using a caliper to gauge the thickness of a fold of skin on the back of the arm (triceps), below the shoulder blade (subscapular), and on other areas of the body. These measures are compared with standards and used to estimate total body fat.

■ BODY WEIGHT AND COMPOSITION FOR SPORT PERFORMANCE

Body weight and composition can vary among athletes depending on the sport and possibly personal preference (Figure 7-10). Some sports such as bodybuilding are associated with greater FFM and minimal FM, as body composition is a component of the subjective scoring system. Looking at body composition purely from a physical performance perspective, as the %FFM increases at a given weight, so does athletic performance. The opposite can be said of FM: as %BF increases at a given weight, performance tends to decrease. In addition to aesthetics and performance, the level of body fat must be evaluated for its impact on an athlete's health. Expected body fat levels of competitive athletes are displayed in Figure 7-11 and are discussed further in Chapter 14. This section provides a closer look at the relationship between athletic performance and body weight and composition.

Effect of Body Weight and Composition on Athletic Performance

Body weight and composition should be a consideration for most athletes. Many sports including wrestling, judo, and boxing have body weight classes. In addition, some organized youth sports, such as football, are divided into weight classes. It is up to the athlete and coaches to determine the body weight and composition that will lead to optimal performance. Some of the most important

Figure 7-10
The sport dictates the height and weight of an athlete's body.

| Table 7-8

Determining an Appropriate Body Weight and Composition

Important considerations for determining an appropriate body weight and composition for an athletic individual:

- Determine a body composition and weight that is associated with greater performance for a given sport.
- Determine if the modified body weight and composition would pose a risk of injury or disease.
- Determine the individual's personal body weight and composition history.
- Determine the individual's diet composition and direction of modification if needed.
- Determine realistic goals for any changes in body weight and composition based on current physical state, diet, and attitude toward modification.

Key Point

Percentage of fat-free mass (%FFM) is positively related to physical performance at a given weight.

considerations for body weight and composition for athletes and active people are outlined in Table 7-8. These considerations not only address issues of performance but general physical and psychological health.

Body mass is an important consideration for individuals competing in sports that involve body contact and positioning such as American football, rugby, and basketball. Here a heavier player would seem to be at an advantage over a lighter player. On the other hand, a lighter runner or cyclist would seem to be at an advantage over a heavier runner or cyclist as they would have less *work* to perform. Body mass can also influence physical performance in sports that involve jumping (such as the long jump and volleyball) and climbing.

EFFECT OF BODY WEIGHT AND COMPOSITION ON SPEED. For sports that involve sprinting, additional body mass attributable to body fat can be a performance hindrance. The statement of the physical law relating force, mass, and acceleration

$$\text{Force} = \text{mass} \times \text{acceleration}$$

or

$$\text{Acceleration} = \frac{\text{force}}{\text{mass}}$$

shows that when mass increases, acceleration decreases unless the increased mass is compensated for by increased force generation. This would be the case if the weight

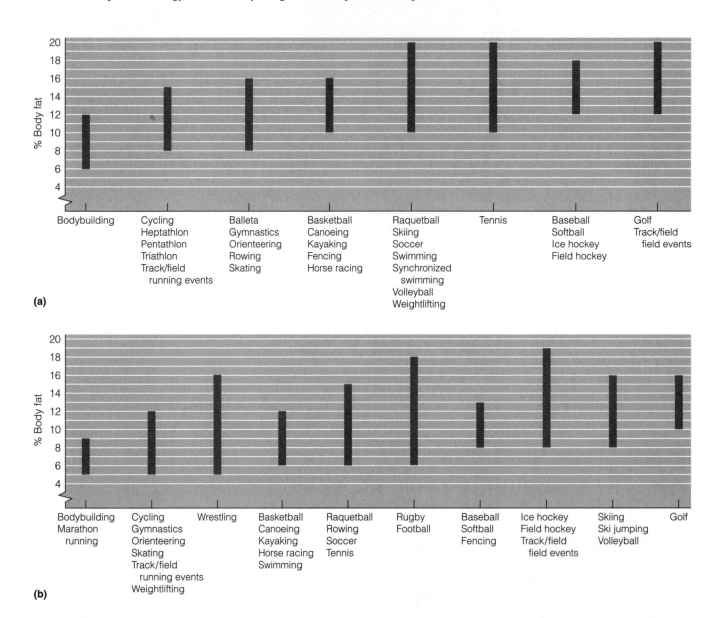

Figure 7-11
Reported ranges of percentage of body fat of (a) female and (b) male athletes from various sports. (*Source*: Adapted with permission from M. Manore and J. Thompson, *Sports Nutrition for Health and Performance*, p. 209, 210. © 2000 by Human Kinetic.)

gain were largely attributable to increased skeletal muscle mass related to sport movement. In fact, force generation can increase, allowing for an enhancement of acceleration rate. Therefore, enhancement of skeletal muscle mass must apply sport-specific training in order to increase performance, as discussed in Chapter 14. However, if the increased mass were disproportionately attributable to adipose tissue, then acceleration would be hindered.

Increased adipose tissue can also increase the drag of an athlete moving through either air or water. This can be an important consideration for sprint cyclists (for example, velodrome cyclists) and runners/hurdlers (for example, in 100-m or 5-km events). Athletes involved in these sports tend to be leaner, perhaps 8–12% body fat.[10]

EFFECT OF BODY COMPOSITION ON ENDURANCE.
The impact of body weight and composition on endurance performance must take into consideration two factors: (1) the potential for body fat to serve as a fuel resource during exercise, and (2) the demands of an increased workload due to additional body mass. While successful endurance athletes tend to have body fat levels lower than those of other athletes, rarely does a successful endurance athlete have excessively low body fat level. Researchers have reported that competitive male distance runners can average roughly 5% and women 14% body fat.[11–13] A survey of female U.S. and Australian National Road Cycling Teams from 1980 to 2000 revealed 7 to 12% body fat.[14]

Extremely low body fat levels in endurance athletes reflect not only the nature of their sport but to some degree the athlete's nutritional state. Excessively lean endurance athletes may not be consuming adequate energy to maintain body fat mass. Endurance athletes often need to eat fat at a level that exceeds general health recommendations. The addition of some energy-dense, fat-laden foods (such as high-fat meats, ice cream, and candy bars) can reduce the food volume necessary to maintain weight. This is discussed further in Chapter 14.

Key Point

Endurance athletes must take into consideration the need for body fat as a fuel resource during exercise as well as the workload associated with additional body weight when deciding on an optimal body composition.

EFFECT OF BODY COMPOSITION ON WATER SPORT PERFORMANCE. Body fat is less dense than water, so it can enhance the buoyancy of athletes engaged in water sports. This may provide some advantage in the water; however, a greater surface area also increases drag, so too much body fat can be problematic for performance by sprint swimmers and water polo players. Researchers have reported that competitive female swimmers have an average of 16–25% body fat, and male swimmers average 9–14%.[15,16] Distance swimmers, such as those who swim the English Channel, tend to have higher body fat levels (22%) to provide thermal benefits as well as buoyancy. The English Channel has water temperatures well below body temperature and extra fat helps the body maintain temperature.

Body Composition and Aesthetic Aspects of Sport

In certain sports, subjective judging involves a visual assessment of body composition. Bodybuilders demonstrate extreme leanness at competition, attributing as little as 4–8% of their body mass to fat.[17–19] This level might be somewhat misleading because bodybuilders assessed just prior to or during competition can be dehydrated. Competitive female bodybuilders tend to have higher levels of body fat than their male counterparts but only by a couple of percentage points (perhaps 6–10%). In addition, body

Key Point

Water athletes can benefit from body fat's buoyancy and insulation against body heat loss in cooler water.

fat levels are important during scoring for fitness competitions. As body fat is a component of *fitness* (see Chapter 3), the more successful competitors tend to be more lean. However, technical equipment is not applied in this sport to assess body composition.

Sports such as figure skating and gymnastics involve subjective scoring. Although scoring is supposed to be limited to the examination of technical skills during a routine or presentation, many athletes feel that body composition may bias judging. In addition, the popularity of an athlete with fans and the marketability of professional athletes may be related to body composition. For instance, highly muscular and very lean professional wrestlers (sport entertainment) might enjoy greater popularity than their counterparts with higher levels of body fat. The lean and muscular appearance of professional tennis players (for example, Serena and Venus Williams) certainly increases their appeal to the public. Meanwhile, the revealing suits and uniforms in sports such as figure skating and sand volleyball allow for greater scrutiny of body composition by the public. In such sports, body fat is said to have "nowhere to hide" and athletes are prone to be more conscious of their appearance. They might also be more prone to develop disordered patterns of eating (see the In Focus feature in Chapter 15).

Key Point

The level of body fat can influence subjective scoring and popularity in some sports as well as physical aspects of performance and sport success.

■ BODY WEIGHT AND COMPOSITION FOR SPORT PERFORMANCE *IN REVIEW*

- Body fat is a consideration for many athletes because of its relationship to performance and/or aesthetics.
- Increased mass attributable to adipose tissue can reduce acceleration of sprint athletes unless force is also enhanced by sport-specific training.
- In some sports, judging involves body composition assessment (as in bodybuilding and fitness competitions). In other sports it may be a factor in judging although it should not be (ice skating, gymnastics).

■ ENERGY BALANCE AND BODY WEIGHT AND COMPOSITION

The level of diet energy relative to energy expenditure is the most important determinant of body weight and an influential factor in determining body composition.

Table 7-9 **Energy Balance Equations**

Energy balance is the relationship between the level of energy intake and expenditure. *Energy intake* is the energy absorbed and maintained by the body. *Energy expended* is the energy used in cellular metabolism or lost from excretory routes (feces, urine, skin). The percentage energy balance is calculated as follows:

$$\text{Energy balance (\%)} = \frac{\text{Energy intake (kcal)}}{\text{Energy expenditure (kcal)}} \times 100\%$$

Energy balance can be negative, positive, or nil. A positive energy balance promotes energy accumulation and weight gain, and a negative energy balance has the opposite effect. The following table shows three examples of the energy balance relationship.

Energy Intake	–	Energy Expenditure	=	Energy Balance (kcal)	Energy Balance (%)	Net Result
~3500	–	~3500	=	0	0	No change in energy status and weight
3650	–	3500	=	+150	104	Energy surplus and potential increase in body weight
3300	–	3500	=	−200	94	Energy deficit and potential body weight reduction

Energy balance occurs when energy intake matches energy expenditure (Table 7-9). In accordance, a person would be in a *positive energy state* when energy intake exceeds expenditure during a specified period. A *negative energy state* would occur when energy expenditure exceeds intake. This section provides an overview of energy balance, methods of assessing energy expenditure, and the major components of energy expenditure.

Energy Intake

Energy intake is one component of the energy balance equation. Energy is brought into the body through the consumption of carbohydrate, protein, and fat and to a lesser degree alcohol. As discussed in previous chapters and shown in Table 7-10, the energy density varies among these substances. Because of the increased energy expenditure associated with exercise, athletes and other active people require more energy than inactive people to achieve energy balance and maintain their body weight.

Several diet assessment tools are available to estimate energy intake. Among the most useful nutrition assessment tools are the 24-hour diet recall and food diaries, which were discussed in Chapter 1. A food frequency questionnaire (FFQ) would be less helpful in estimating a person's energy intake but would provide valuable information regarding the types and sources of energy consumed over longer periods, such as weeks or months.

Table 7-10

Energy Densities of Carbohydrate, Protein, and Fat

Energy Nutrient	Energy Density (kcal/g)	RQ (RER)[a]	Energy Value (kcal/L of O$_2$)[a]	Energy Components or Usable Form
Carbohydrate	4.2	1.0	5.05	Monosaccharides
Protein	4.2	0.80	4.50	Amino acids
Fat	9.5	0.71	4.70	Fatty acids and glycerol
Mixed diet		0.82	4.82	All the above

[a]Respiratory quotient (RQ), respiratory exchange ratio (RER), and energy value in terms of kcal/liter of O$_2$ are discussed further in the feature In Focus: Understanding Respiratory Exchange Ratio (RER) and Respiratory Quotient (RQ) at the end of Chapter 13.

Key Point

Energy balance is the algebraic sum of energy intake and energy expenditure within a period of time.

Energy Expenditure

On the other side of the energy balance equation is energy expenditure. Energy nutrients are broken down in tissue to power muscle contraction and other operations in cells. As a result of these activities, energy is released from the body in the form of heat. Heat energy is measured in **kilocalories (kcal)**, where 1 kcal is the amount of heat required to raise the temperature of 1 kg of H_2O by 1°C; it is the equivalent of 4.184 kilojoules (kJ). The term **metabolism** refers to all energy-releasing processes in human tissue, which can be measured as energy expenditure from the body. **Metabolic rate** describes the amount of energy released in a given unit of time (for example, 1 hour).

Key Point

Metabolism refers to the physiological processes of tissue and is reflected in energy expenditure.

Tools Used to Assess Energy Expenditure

Knowing an individual's energy expenditure provides the basis for establishing energy balance or creating energy imbalance leading to changes in body weight and composition. The amount of energy expended by the human body can be measured directly (direct calorimetry) or estimated indirectly by measuring the amount of gases utilized and produced during energy metabolism (indirect calorimetry).

DIRECT CALORIMETRY. Energy expenditure can be measured by applying **direct calorimetry**. To do so, an individual enters an insulated room or **metabolic chamber** for a specific period (such as 24 hours) and the heat dissipated by his or her body is measured (Figure 7-12).

That heat release warms a layer of water or other fluid surrounding the chamber, and the change in fluid temperature reflects the person's energy expenditure. Although direct calorimetry is very accurate, the cost and complexity of the metabolic chamber prohibits its widespread use. In fact, only a dozen or so direct calorimetry units are operational throughout the world. Researchers have sought less expensive and more practical ways to estimate energy expenditure and have used direct calorimetry to validate the accuracy and precision of these alternatives.

INDIRECT CALORIMETRY. A much more practical method of estimating energy expenditure is **indirect calorimetry**. Most exercise physiology laboratories are equipped with a machine called a **metabolic cart** that measures VO_2 and VCO_2. As shown in Figure 7-13, tubing connects the metabolic cart to a person. The individual can be at rest or engaged in physical activity such as running on a treadmill or cycling in a stationary apparatus. The volume of carbon dioxide produced is divided by

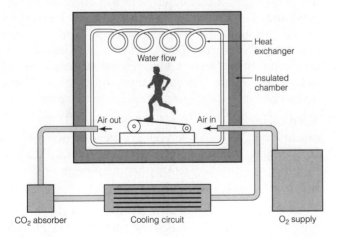

Figure 7-12
A simplified version of the human calorimeter used to measure direct body heat loss (that is, energy expenditure).

Figure 7-13
Indirect calorimetry estimates energy expenditure using a metabolic cart.

the volume of oxygen consumed to yield the respiratory exchange ratio, or RER (that is, RER = VCO_2/VO_2). RER can be used to predict the contribution made by fat and carbohydrate to total energy expenditure, because the oxidations of these energy nutrients are associated with different ratios of VO_2 and CO_2. The RER values of glucose and fat are 1.0 and 0.71, respectively. Therefore, RER is a range, and the closer a person's RER is to the value for either glucose or fat, the greater the contribution made by that fuel source to energy expenditure. For instance, an RER of 0.90 is roughly two-thirds of the range in the direction of glucose utilization; thus approximately two-thirds of the fuel used during the measurement was derived from glucose. However, this does not take into account the contribution made by protein. This method is discussed in more detail in the Chapter 13 feature In Focus: Respiratory Exchange Ratio and Respiratory Quotient.

Also, once RER is calculated for a specific period (such as 1 hour), VO_2 and VCO_2 can be used to estimate energy expenditure during that time. An example of estimating energy expenditure and fuel mixture using RER is also presented in the Chapter 13 In Focus. Estimating energy expenditure and fuel utilization from RER has provided researchers with valuable information regarding the influence of an acute bout of exercise on energy metabolism as well as the effects of regular training. In addition, RER is now used in some weight management programs to better predict energy expenditure for diet prescription.

Key Point

Researchers commonly use indirect calorimetry to estimate energy expenditure and the contribution of carbohydrate and fat to energy expenditure.

DOUBLY LABELED WATER. **Doubly labeled water** (DLW) utilizes water molecules containing the stable isotopes of hydrogen and oxygen: ^2H (deuterium) and ^{18}O. To begin, a known amount of ^2H_2O and $H_2^{18}O$ is ingested by an individual or infused into a vein. Then, energy expenditure is estimated based on the processing of the two forms of water in body water, which indicates the production of CO_2 during energy metabolism. Here is how it works. After ^2H_2O and $H_2^{18}O$ equilibrate throughout body fluid, they are slowly lost from the body. ^2H remains a component of body water and is lost from the body during normal water loss (such as urine and sweat). Meanwhile, some of the ^{18}O from $H_2^{18}O$ becomes part of CO_2 produced by the **carbonic anhydrase** system in the blood. Figure 7-14 provides an overview of this system, which serves as a means of circulating greater amounts of CO_2 from tissue to the lungs. CO_2 production over time can be estimated by the difference in levels of ^2H_2O and $H_2^{18}O$ in

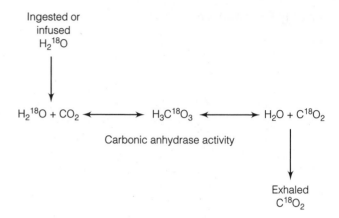

Figure 7-14
Carbonic anhydrase activity results in the incorporate of ^{18}O in CO_2. Doubly labeled water (DLL) can accurately estimate energy expenditure using the difference between levels of ^2H_2O and $H_2^{18}O$ to estimate CO_2 production, which in turn is used to estimate energy expenditure.

saliva and urine. Then VCO_2 can be used to estimate energy expenditure by applying a specific factor.

Key Point

Energy expenditure and estimations of contribution made by carbohydrate and fat can be assessed using indirect calorimetry.

Components of Energy Expenditure

The rate of energy expenditure fluctuates throughout the day. This suggests that different factors affect energy expenditure. For the most part, these factors are skeletal muscle activity, the consumption of nutrients, and changes in the climate. These factors tend to increase energy expenditure above a foundational or **basal** level of energy expenditure. Therefore **total energy expenditure** is separated into several distinguishable components.

Total energy expenditure (TEE) = basal metabolic rate (BMR)
+ thermal effect of activity (TEA)
+ thermal effect of food (TEF)
+ Adaptive thermogenesis (AT)

or

Total energy expenditure (TEE) = resting metabolic rate (RMR)
+ thermal effect of activity (TEA)
+ Adaptive thermogenesis (AT)

BASAL AND RESTING METABOLIC RATE. **Basal metabolism** is the energy expended by processes during a period of nonactive rest (for example, immediately after waking). It is measured in a climate-controlled room (not cold or warm) and at least 12 hours after the consumption of a meal. **Basal metabolic rate (BMR)** is the measurement of basal metabolism during a specific period (such as 1 hour or 1 day). BMR is often used interchangeably with **resting metabolic rate (RMR)**; however, RMR restricts food consumption for only 4 hours prior to assessment and can be measured later in the day. This makes RMR more practical for researchers; however, RMR tends to be about 10% greater than BMR.

The energy expended during BMR or RMR is related to homeostasis and includes the energy expended for cell turnover, resting heart rate and respiration, urine production, and the synthesis of proteins, nucleic acids, and other substances. Typically, 50–75% and 60–75% of TEE is attributable to BMR and RMR, respectively, for the general population. Both BMR and RMR can be estimated mathematically; the BMR calculations are presented in Table 7-11. However, one limitation of these calculations is that they are not very sensitive to body composition and tend to overestimate BMR (RMR) in heavier individuals with a higher %FM and to underestimate BMR (RMR) in athletes with a higher %FFM.

The relative contribution made by different tissue and organs to BMR (RMR) varies. The most metabolically active tissues (measured in kilocalories expended per gram of tissue) are organs such as the heart, kidneys, lungs, brain, and the liver. Collectively these organs account for only about 5% of an adult's body weight, yet the energy they expend might account for 50–60% of BMR. Although resting skeletal muscle is not as metabolically active as these organs it can contribute about 25% to BMR for inactive people and 30–40% in athletes. Most of the basal metabolic operations of skeletal muscle are related to protein turnover. Collectively, the energy expenditure (at rest) of FFM approximates 13–28 kcal/kg per day.[20,21] Adipose tissue has a very low metabolic rate relative to its mass and normally contributes less to BMR (RMR), an estimated 6.5 kcal/kg per day.[21] Variations in FFM have been shown to explain 65–90% of the variation in BMR (RMR) energy expeditures.[22–26]

BMR for males tends to be a little higher than for females since men tend to have a greater skeletal muscle to adipose tissue ratio. This concept is easily supported by the gender differences in O_2 consumption (VO_2). On average, women tend to consume only about 80% as much O_2 (per body mass) as men. However, when males and females are compared related to LBM, BMR is approximately the same.

Meanwhile, BMR per kilogram of body weight is highest during infancy and declines with age. Here again BMR is mostly related to LBM, which tends to be higher during infancy and lower in older individuals. This is because infants tend to have a higher percentage of FFM than adults and are also engaged in very rapid tissue growth.

THERMAL EFFECT OF ACTIVITY (TEA).

For most people, BMR (RMR) is the greatest component of TEE throughout a day, followed by skeletal muscle activity. However, for more active people such as athletes, the thermal effect of activity (TEA) can become the greatest contributor to energy expenditure throughout a day (Figure 7-15). Skeletal muscle activity is very costly from an ATP standpoint, as both muscle contraction and relaxation require ATP. In addition to energy for physical movement (such as walking, talking, running, climbing stairs) TEA

| Table 7-11 |

Three Methods of Estimating BMR

All formulas here use metabolic rates in kilocalories per day, body weight in kilograms, and height in centimeters.

Method 1: Simple application of body mass

$$BMR = BW \times 24 \text{ hours}$$

Method 2: Body weight raised to the power of three-fourths and multiplied by 70

$$BMR = 70 \times BW^{75}$$

Method 3: Harris and Benedict equation for BMR

Men: $BMR = 66 + (13.7 \times BW)$
$+ (5 \times ht) - (6.8 \times age)$

Women: $BMR = 655 + (9.6 \times BW)$
$+ (1.7 \times ht) - (4.7 \times age)$

Calculated estimates of BMR for a 35-year-old man who is 180 lb (82 kg) and 5'11" (180 cm) using the three different equations:

1. 1963 kcal
2. 1904 kcal
3. 1853 kcal

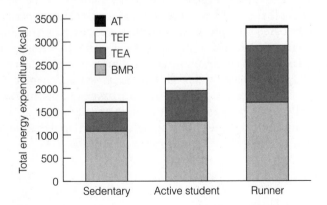

Figure 7-15
Contributions made by BMR, TEA, TEF, and AT toward total energy expenditure for three women of the same age, height, and weight.

Table 7-12   **Estimating Daily Energy Expenditure at Various Activity Levels**

Level of Intensity	Type of Activity	Activity Factor (× BMR)	Energy Expenditure (kcal/kg/day)
Very light	Seated and standing activities, painting trades, driving, laboratory work, typing, sewing, ironing, cooking, playing cards, playing a musical instrument	1.3 (men) 1.3 (women)	31 30
Light	Walking on a level surface at 2.5 to 3 mph, garage work, electrical trades, carpentry, restaurant trades, housecleaning, child care, golf, sailing, table tennis	1.6 (men) 1.5 (women)	38 35
Moderate	Walking 3.5 to 4 mph, weeding and hoeing, carrying a load, cycling, skiing, tennis, dancing	1.7 (men) 1.6 (women)	41 37
Heavy	Walking with a load uphill, tree felling, heavy manual digging, basketball, climbing, football, soccer	2.1 (men) 1.9 (women)	50 44
Exceptional	Training in professional or world-class athletic events	2.4 (men) 2.2 (women)	58 51

Note: The second section of the "How to" above describes how to use this table and explains that the estimate reflects both physical activity *and* basal metabolic activity.

Source: Reprinted with permission from *Dietary Reference Intakes*, © 1989 by the National Academy of Sciences. Courtesy of the National Academies Press, Washington, D.C.

includes the energy expended to maintain position and posture (Table 7-12). For instance, sitting on a stool without back support increases TEA by 3–5% over lounging on a recliner, and standing requires yet more energy.[1]

The energy expense of exercise training can be very large; Table 7-13 provides estimates of energy expenditure based on body weight. Furthermore, exercise training increases not only daily TEA but also BMR, because of recovery and adaptive operations. This results in increased VO_2 for a related time following exercise [see Chapter 13 for an overview of postexercise oxygen consumption (EPOC)]. To estimate TEA an individual can keep an activity log over a 24-hour period and then apply energy-equivalent coefficients, such as those in Table 7-12.

THERMAL EFFECT OF FOOD (TEF). TEF is the increase in energy expenditure associated with the consumption of food. It represents an increase in TEE attributable to the events associated with digestive and absorptive operations and the metabolism and storage of nutrients. In general, TEF can be estimated at 10% (5–15%) of total energy intake during a day. For instance, TEF can be estimated at 250 kcal for an individual who eats a mixed diet containing 2500 kcal over a 24-hour period.

The TEF associated with a particular meal varies depending on the size and composition of meals. Obviously, larger meals tend to have greater TEF than smaller meals. However, meals with a higher proportion of carbohydrate and protein tend to have higher TEF per volume. This is

Key Point

Although BMR can be higher for more active people, its contribution to TEE is lower for an active person than an inactive person.

Key Point

Total energy expenditure reflects the energy released by homeostatic operations, activity, digestion, processing of food, and adaptation to changes in environmental temperature.

| Table 7-13 **Energy Spent on Various Activities** |

The values listed in this table reflect both the energy spent in physical activity *and* the amount used for BMR

Activity	Energy Spent (kcal/lb/min[a])	Energy Spent at Different Body Weights (kcal/min)				
		110 lb	*125 lb*	*150 lb*	*175 lb*	*200 lb*
Aerobic dance (vigorous)	.062	6.8	7.8	9.3	10.9	12.4
Basketball (vigorous, full court)	.097	10.7	12.1	14.6	17.0	19.4
Bicycling						
13 mph	.045	5.0	5.6	6.8	7.9	9.0
15 mph	.049	5.4	6.1	7.4	8.6	9.8
17 mph	.057	6.3	7.1	8.6	10.0	11.4
19 mph	.076	8.4	9.5	11.4	13.3	15.2
21 mph	.090	9.9	11.3	13.5	15.8	18.0
23 mph	.109	12.0	13.6	16.4	19.0	21.8
25 mph	.139	15.3	17.4	20.9	24.3	27.8
Canoeing, flat water, moderate pace	.045	5.0	5.6	6.8	7.9	9.0
Cross-country skiing, 8 mph	.104	11.4	13.0	15.6	18.2	20.8
Golf (carrying clubs)	.045	5.0	5.6	6.8	7.9	9.0
Handball	.078	8.6	9.8	11.7	13.7	15.6
Horseback riding (trot)	.052	5.7	6.5	7.8	9.1	10.4
Rowing (vigorous)	.097	10.7	12.1	14.6	17.0	19.4
Running						
5 mph	.061	6.7	7.6	9.2	10.7	12.2
6 mph	.074	8.1	9.2	11.1	13.0	14.8
7.5 mph	.094	10.3	11.8	14.1	16.4	18.8
9 mph	.103	11.3	12.9	15.5	18.0	20.6
10 mph	.114	12.5	14.3	17.1	20.0	22.9
11 mph	.131	14.4	16.4	19.7	22.9	26.2
Soccer (vigorous)	.097	10.7	12.1	14.6	17.0	19.4
Studying	.011	1.2	1.4	1.7	1.9	2.2
Swimming						
20 yd/min	.032	3.5	4.0	4.8	5.6	6.4
45 yd/min	.058	6.4	7.3	8.7	10.2	11.6
50 yd/min	.070	7.7	8.8	10.5	12.3	14.0
Table tennis (skilled)	.045	5.0	5.6	6.8	7.9	9.0
Tennis (beginner)	.032	3.5	4.0	4.8	5.6	6.4
Walking (brisk pace)						
3.5 mph	.035	3.9	4.4	5.2	6.1	7.0
4.5 mph	.048	5.3	6.0	7.2	8.4	9.6
Wheelchair basketball	.084	9.2	10.5	12.6	14.7	16.8
Wheeling self in wheelchair	.030	3.3	3.8	4.5	5.3	6.0

[a]To calculate kcal spent per minute of activity for a specific body weight, multiply kcal/lb/min by the weight and then multiply that number by the number of minutes spent in the activity. For example, if a person weighing 142 lb spends 30 minutes doing vigorous aerobic dance, $0.062 \times 142 = 8.8$ kcal per minute; 8.8×30 (minutes) $= 264$ total kcal spent.

due to the varying energy expenditure associated with assimilating different energy nutrients into different body energy stores. For instance, the cost for storing glucose as glycogen is 7% of the available energy. The energy cost of converting glucose to fatty acids and storing it as a component of fat is roughly 26%. Storing food fat as body fat costs about 3% of the available energy, and storing protein as body protein costs approximately 24%. Researchers often use these energy conversion estimates to support the notion that a higher carbohydrate and lower fat intake is more conducive to weight control. Quite simply, a person would store less fat if the excess diet energy came in the form of carbohydrate instead of fat.

ADAPTIVE THERMOGENESIS (AT). Energy expenditure can increase and even decrease due to changes in environmental temperature and exposure to radiant energy (such as sunlight). This attempt to manipulate energy expenditure to regulate body temperature is called adaptive thermogenesis. For instance, when **core body temperature** (the temperature in and around vital organs) increases during exercise or in warm climates, blood flow to the skin is increased and sweating is enhanced. This is an attempt to circulate more heat to the skin for heat release via sweating. On the other hand, when body temperature cools, less blood is circulated to the skin to conserve core body heat and shivering is stimulated. Shivering is tiny involuntary muscle contractions that are intended to generate heat. During infancy, the presence of **brown adipose tissue (BAT)** can assist in heat production, as this form of tissue is specifically designed to generate heat.

Exposure to cold and a reduction in core temperature can increase the release of thyroid hormone (T_3/T_4). Thyroid hormone increases the general metabolic rate of many cells, thereby generating heat. However, this response may be more associated with chronic exposure to a cold environment and not necessarily an acute response. Thus it may be more applicable to athletes who regularly train for long periods in a cool environment, either on land or in water.

ENERGY BALANCE AND BODY WEIGHT AND COMPOSITION *IN REVIEW*

- Energy balance is the algebraic sum of energy intake and expenditure.
- Total energy expenditure has four primary components: BMR (RMR), TEA, TEF, and AT.
- For inactive people, BMR is the greatest contributor to total energy expenditure; however, for athletes and other very active people, TEA can become the greatest contributor.
- Vital organs and skeletal muscle are the major contributors to BMR.
- Exercise training increases TEA throughout a day and can increase BMR as well because of repair and recovery operations.

CHANGING BODY WEIGHT AND COMPOSITION

Many athletes and nonathletes attempt to increase or decrease body weight and manipulate their body composition to achieve performance, fitness, and health goals.

Prerequisite to changing body weight and composition is an understanding of the components of energy expenditure, as explained above, and how to appropriately create an energy imbalance that allows for the change. This section provides an overview of some of the most important considerations for manipulating body weight and composition.

Human Energy Stores

Many tissues contain stored energy that enables the body to endure extended periods without energy consumption. As presented in Table 7-4, once water and minerals are accounted for, the majority of the human body is composed of molecules having energy potential. This means that body weight reflects body energy status. For example, a muscular and lean 154-lb (70-kg) runner with body composition measures of 10% body fat, 20% protein, and 1% carbohydrate theoretically contains a little less than 120,000 kcal of stored energy. This allows more than enough energy for individuals to endure fasting periods and to engage in bouts of physical activity with higher energy demands (such as a marathon). The liver, skeletal muscle, and adipose tissue are the primary energy storage tissues, and during periods of fasting or exercise, the energy stored in these cells becomes available to body tissue via the circulation. Hormones govern the storage and release of energy within this tissue as discussed in the feature In Focus: Nutrition States and Energy Metabolism in Chapter 4.

SKELETAL MUSCLE ENERGY STORES. Skeletal muscle cells have limited yet significant stores of fat and carbohydrate. This tissue also has a relatively large store of energy in the form of protein. Muscle tissue is composed of approximately 22% protein, 1–2% carbohydrate (glycogen), and a little less than 1% fat. Thus a 176-lb (80-kg) adult male who is 45% skeletal muscle would have approximately 8000 g (8 kg) of protein, 540 g of glycogen, and 200 g of fat in his skeletal muscle. During exercise these stores are broken down to help fuel the efforts of working muscle. Also, skeletal muscle stores are valuable energy resources during fasting and starvation.

ADIPOSE TISSUE ENERGY STORES. Adipose tissue is perhaps the most obvious energy storage tissue in the human body. In fact, adipose tissue attributes about 86% of its mass to stored fat. Adipose tissue is found throughout the body including the subcutaneous layer, which lies just beneath the skin, and the **visceral** stores in association with the vital organs and within skeletal muscle. In inactive people with higher levels of body fat, more fat is distributed subcutaneously and viscerally. For lean and muscular athletes, particularly those who train aerobically, the level of subcutaneous adipose tissue might underestimate FM because these individuals have enhanced fat stores within muscle fibers.

Energy from stored fat (triglycerides) in adipose tissue is derived from fatty acids and glycerol. During exercise and fasting, fatty acids diffuse out of adipocytes and are transported to tissue such as skeletal muscle and the liver to serve as fuel (see Figure 6-20). Meanwhile, glycerol can circulate to the liver and be converted to glucose, which can then be released into the blood. Triglyceride breakdown in adipose tissue is stimulated by a variety of hormones including glucagon, epinephrine, and cortisol. Therefore, fat breakdown is increased during fasting, exercise, and stress.

LIVER ENERGY STORES. Glycogen accounts for about 6–8% of the mass of the liver, which equates to about 75–100 g of glucose. As discussed in detail in Chapter 4, this storage of carbohydrate (glycogen) is an extremely important energy source during fasting and exercise because it is a source of blood glucose. Liver glycogen stores are rapidly broken down during fasting, primarily due to glucagon stimulation. For instance, liver glycogen stores might be reduced by half during an overnight sleep. Higher intensity exercise and the associated rise in epinephrine also stimulate liver glycogen breakdown.

Key Point

The principal energy stores in the human body are in the liver, adipose tissue, and skeletal muscle.

Body Weight Fluctuation

Scales are found in many homes and locker rooms. People who step on the scale every day or several times a week realize that their body weight can fluctuate in a manner seemingly unrelated to energy intake. These are short-term fluctuations in body mass, oscillating between small losses and gains ($\leq$1–2%) in body weight. Greater changes in body weight over time are more reflective of an energy imbalance.

SHORT-TERM WEIGHT FLUCTUATION. People can experience minor fluctuations in body weight even when energy intake matches energy expenditure. However, in this case body weight changes are more associated with nonenergy content. For instance, changes in hydration status and content in the digestive tract can account for weight fluctuation. Changes in hydration status can result from an imbalance between water intake and loss, especially when a person engages in exercise (which increases sweating and exhalation). In addition, minor changes in muscle and liver glycogen levels can influence the amount of water maintained or lost from the body, as an estimated 3 g of water is associated with each gram of glycogen. Therefore, a person who has 400 g of glycogen also has

Key Point

Short-term fluctuations in body weight are caused more by changes in body water status, whereas long-term fluctuations are more associated with energy imbalance.

1200 g (1.2 kg) or 2.4 lb of associated water. A change in glycogen content from 400 to 300 g would theoretically reduce body water by more than $\frac{1}{2}$ lb ($\frac{1}{4}$ kg).

LONG-TERM WEIGHT GAIN. It is estimated that American adults experience an increase in body weight of roughly $\frac{1}{2}$ lb each year or about 5 lb (2.3 kg) per decade as they age. This weight change is undesirable, as more often than not it reflects a net positive energy balance and accruement of stored energy as fat. This energy imbalance tends to be the result of decreased physical activity and/or increased energy intake. Although most of such weight gain involves an expansion of adipose tissue, approximately 15% of the weight gain in sedentary people would be attributable to FFM,[1] representing an increase in the total content of body water, protein, and minerals. This is explained by the need for materials to expand adipose tissue, the higher volume of skin and blood required, and increased bone density and skeletal muscle mass associated with increased mechanical stress.

RAPID WEIGHT LOSS. Rapid weight loss can result in at least two undesirable situations that can hinder further weight loss and/or management of weight loss. First, drastic energy restriction results in a decrease in circulating thyroid hormone (T_3/T_4).[1] Thyroid hormone is one of the most potent stimulators of metabolism in cells. Therefore, reduced levels of thyroid hormone can decrease an individual's BMR and TEE during the restriction. Also, drastic energy restriction can lead to loss of skeletal muscle mass and a corresponding decrease in energy expenditure. Prolonged fasting or semistarvation can increase circulating levels of cortisol, a powerful catabolic hormone. Elevated cortisol levels are associated with increased skeletal muscle protein breakdown, yielding amino acids, many of which can serve as precursors for gluconeogenesis.

Key Point

Rapid weight loss can be problematic because it may reduce circulating thyroid hormone levels and lean body mass, and both these losses will reduce energy expenditure.

Key Point

Planning for effective weight loss requires an understanding of the starting point and realistic goals.

Planning for Effective Body Weight and Fat Reduction

Effective manipulation of body weight and composition requires practical planning and implementation. Planning for body weight and fat loss must begin with an understanding of the starting point and the desired process of weight loss. For instance, an overweight (overfat) person must realize that his or her body weight reflects the presence of lean tissue as well as fat. For instance, a 225-lb man with 29% body fat must realize that fat accounts for 65 lb (30 kg) of his mass. Second, the goal should be to reduce body fat, which in most situations would reduce body weight. Third, the person needs to understand that humans are not supposed to be "fat free." As discussed, body fat serves several vital roles, and excessive leanness can have physiological consequences. Therefore, the goal is not 0% BF but a lower level of body fat associated with optimal performance and health. For athletes, reductions in body weight and changing body composition are more associated with sport performance than improvements in general health. These individuals need to focus on achieving a desired body composition as much as, if not more than, desired body weight. For instance, a 145-lb female figure skater with 25% body fat should decide on a realistic goal for her body fat level. Proper planning and achievement of long-term energy imbalance favoring loss of body fat is the focus, rather than loss of body weight. See the feature Nutrition in Practice: Determining Optimal Body Weight Based on Desired Body Composition for an example.

Creating an Energy Imbalance

Changes in body energy status reflect the relationship between energy expenditure and consumption, as shown in the equation in Table 7-9. Reductions in body energy status can be accomplished by manipulating either variable on the right side of the equation. Therefore, one can either increase energy expenditure or reduce energy input to create an energy imbalance favoring body fat reduction. Whichever the case, energy balance (%) on the left side of the equation should not be chronically less than 80%; that is, energy intake should usually be at least four-fifths of energy expenditure.[1] In addition, the energy imbalance should not exceed 500–1000 kcal daily, depending on a person's size and energy needs.

RESTRICTING DIET ENERGY. Exercise became a mainstream tool for weight control in the last couple of decades. Before then, weight reduction occurred largely by semistarvation diets and over-the-counter diet aids and prescription drugs. Commercially available meal-replacement powders and shakes became popular in the 1970s and 1980s and are still popular today. **Semistarvation** diets tend to restrict energy intake to about half to three-quarters of what would be needed for energy balance, and people might limit themselves to 1000–1200 kcal/day.

NUTRITION IN PRACTICE   Determining Optimal Body Weight Based on Desired Body Composition

Because body weight and composition are related to physical performance, athletes and people in general should determine their optimal body weight based on body composition. This decreases the attention on an ideal body weight and focuses on body composition. The questions to be answered prior to determining an appropriate body weight are these:

1. What is the current weight and body composition?
2. What is the goal for percentage of body fat that would be appropriate for performance, and is body weight a consideration?
3. Is the body fat goal realistic for the individual?
4. What is a realistic goal for creating an energy imbalance (total change and time allotted)?

Example

Susie is an aspiring triathlete who trains five to six times a week. At 5'8" and 150 lb (68 kg), she was recently assessed by DXA and found to have 22% body fat. She believes that she could enjoy greater performance if she weighed 10–15 lb less while maintaining the same amount of muscle mass. She thinks that 15% body fat is achievable. Her calculations indicate she should strive to reduce her body weight to approximately 138 lb (63 kg). She should now seek out the assistance of a registered dietitian who specializes in working with athletes to help her plan for effective weight loss. Her goal would be to create an energy imbalance that is not less than 80–90% of her current energy needs.

- Current weight   150 lb (68 kg)
- Current body fat   22% or 33 lb (15 kg)
- Current fat-free mass   78% or 117 lb (53 kg)

- Body composition goal   85% FFM and 15% FM
- Body weight goal   117 lb ÷ 0.85 = 138 lb (63 kg)

Diet energy should not be restricted to less than 80% of TEE unless medically supervised.[1] For example, a sedentary woman who expends 2000 kcal daily should not restrict her intake to less than 1600 kcal. However, if that same woman increases her daily activity and expends 2200 kcal, her energy intake should not be less than 1760 kcal. This recommendation for an energy restriction is considered mild and is believed to promote compliance and to minimize the reduction of FFM. Another concern with drastic energy restrictions is what happens after a goal weight has been achieved. Has the individual learned anything that will help with weight maintenance?

Key Point

Body weight/fat reduction requires an energy imbalance that can be produced by decreased energy consumption or increased expenditure.

INCREASING TOTAL ENERGY EXPENDITURE.   For a more sedentary person, increasing the level of physical activity has several benefits. Most obvious are an increased TEA, which in turn increases TEE. Second, exercise training can increase BMR (RMR) due to repair and recovery operations and possibly muscle hypertrophy. As the size of vital organs cannot be changed voluntarily, increasing skeletal muscle mass is the most significant natural way to increase BMR. Third, participation in regular physical activity occupies time that might have been spent eating or being sedentary (such as watching TV or "surfing" the Internet). Finally, exercise and physical accomplishments can improve one's self-image, which can lead to greater dietary and exercise compliance.

RESISTANCE TRAINING AND RMR.   Resistance training is touted as an effective means to increase fat-free mass (FFM) and in turn increase BMR or RMR. For example, it would be expected that a man with 88% FFM (12% BF) would have a higher RMR than another man of the same weight but who has 80% FFM (20% BF). However, the relationship has not proven to be as simple as researchers and health practitioners had hoped, as age and gender can play significant roles in how resistance training ultimately affects an individual's RMR. When looking at middle-aged adult populations (35–65 years old), research studies have suggested that a significant increase in RMR after progressive resistance training programs can be expected.[27–29] For instance, researchers reported a 7.7% RMR increase in 50- to 65-year-old males after a 16-week heavy resistance training program.[29] Similar results were determined in moderately obese women (38 ± 0.9 years), as researchers noted a 44 kcal/day increase in RMR after 20 weeks of high-intensity resistance training.[30] Yet, the same results have not always replicated in younger populations.[31–33] For example, no significant changes in RMR were found in

Key Point

Resistance training can increase BMR (RMR) by increasing LBM.

18- to 35-year-old males after 12 weeks of high-intensity resistance training[31] or in another study involving 12 weeks of resistance in young males.[33] Yet, as discussed later, resistance training can be crucial to maintaining or minimizing the loss of LBM and RMR during weight loss.[34]

With regard to gender, researchers have reported conflicting results as to whether resistance training has the same impact on females as males.[35] For instance, 24 weeks of progressive resistance training resulted in a 7% increase in RMR in males, whereas females experienced no significant changes in RMR despite similar increases in FFM (~1.5 kg).[35] Similarly, other researchers noted that despite significant increases in muscular strength after 6 months of resistance training, total daily energy expenditure was unchanged.[36] However, these studies contrast with previous research that noted significant increases in RMR for females after resistance training.[30]

Planning for Weight Gain

For athletes, weight gain should primarily focus on increasing FFM. Once again, performance is typically enhanced by increased FFM. Meanwhile, performance tends to decrease with increased levels of FM. The rate of weight gain for more dynamic athletes should be limited by rate of change in %BF. For example, if a football linebacker wants to increase his body weight from 215 lb to 240 lb during the off-season, he would need to create a positive energy imbalance. However, the rate of gain should be controlled so that the contribution made by FM is minimized or does not exceed a certain goal. This is demonstrated in the accompanying feature, Nutrition in Practice: Planning to Gain Weight.

■ CHANGING BODY WEIGHT AND COMPOSITION *IN REVIEW*

- Body weight/fat reduction requires the creation of an energy imbalance, which can be achieved by increasing energy expenditure or decreasing consumption.
- Adipose tissue, skeletal muscle, and the liver are the major energy storage tissues of the body.
- Planned weight loss needs to focus on the preservation of FFM, and weight gain should result from increased FFM (such as skeletal muscle).
- Planned weight gain for athletes should focus on limiting the contribution of fat, thereby allowing an individual to improve performance.

NUTRITION IN PRACTICE   **Planning to Gain Weight**

Athletes and active people attempting to increase their body weight should understand not only their starting physical state but also how best to accomplish their goals for body weight and composition. As increased performance is associated with FFM and decreased performance is associated with FM, the primary focus should be on developing greater FFM while FM is reduced, unchanged, or perhaps increases only slightly. Consider the following example.

Example

Tom is a 21-year-old outfielder for a Division I collegiate baseball team. He is 5'11" and 175 lb and currently has 15% body fat. During the off-season he would like to increase his weight in a way that allows for more hitting power but does not want to compromise speed in the outfield and around the bases. Tom's coach and he decide that 185 lb is the goal and that they will plan how to gain 10 lb of FFM. Last year, during the off-season, he trained two or three times a week with weights (5 sets per muscle group) and ran 5 miles five or six times a week. What should he do this off-season?

Planning for Weight Gain

The off-season lasts 4 months prior to spring camp. This allows 17 weeks for the modification in body weight. Tom needs to focus on increasing FFM and plans to gain a little more than $\frac{1}{2}$ lb per week. Therefore, the energy imbalance favoring weight gain must be slight, not to exceed 120% of energy requirements for weight maintenance. At the same time, Tom will need to increase his resistance training volume to provide greater overload stimulus. Meanwhile, he can alter his running program to reduce his training volume (intensity × distance); however, these changes must be considered in his energy expenditure.

Starting Point
- Body weight   175 lb (80 kg)
- Body fat %   15%
- Fat mass   175 lb × 0.15 = 26 lb (12 kg)
- Fat-free mass   175 lb − 26 lb = **149 lb** (68 kg)

Goal
- Body weight   185 lb (84 kg)
- Fat-free mass   **149 lb** + 10 lb = 159 lb (72 kg)
- Fat mass   185 lb − 159 lb = 26 lb (12 kg)
- Body fat %   26 ÷ 185 × 100% = 14%

Currently Tom requires 3400 kcal for weight maintenance. A nutritional assessment using a 3-day food diary estimates that the contributions to total energy intake from carbohydrate, protein, and fat are 60%, 15%, and 25%, respectively. He knows that FFM is a combination of mostly protein, water, and minerals, so he plans to eat an additional 200 kcal/day and increase his protein intake to 20% of energy intake, as shown in Table A. Tom will keep his carbohydrate intake at 60% of energy, and the remaining energy will come by way of fat. He will also drink plenty of water and monitor changes in body weight and composition using BIA and skinfold assessment performed by the strength and conditioning coach.

Table A

Example of a Diet Plan to Gain Weight

Total Energy Intake	Energy Nutrient Contribution		
	Carbohydrate	*Protein*	*Fat*
3400 kcal (current)	60%	15%	25%
	510 g	127 g	94 g
3600 kcal (modified diet)	60%	20%	20%
	540 g	180 g	80 g

■ DIET CONCEPTS

Diet books have been a consistent component on the *New York Times* bestseller list for decades and have included *The Zone, Sugar Busters,* and *Dr. Atkins' Diet Revolution.* These books describe protocols for nutrition planning in an attempt to help people lose weight and/or change their body composition and promote health. In addition, some of these programs claim to be ideal for athletes.

The various diet programs address two major questions in making a weight loss plan: how much should energy intake be restricted, and what energy nutrient ratios should be applied? Current information indicates that overly restrictive energy intakes are problematic, and that different energy nutrient ratios might lead to similar levels in weight loss. At least this seems to be the case when swapping carbohydrate for fat energy in weight loss programs. Changes in body composition and compliance or attrition are among the most important considerations with any weight loss protocol. This section provides an overview of current diet concepts as well as a discussion of popular fat-burning "tricks."

Energy Nutrient Ratios and Nutrient Restriction

Popular diet programs tend to focus on a specific energy nutrient ratio or energy nutrient restriction. For instance, *The Zone* is based on an energy nutrient ratio of 40:30:30, meaning that 40% of daily energy is derived from carbohydrate and 30% each from protein and fat. Meanwhile, *Sugar Busters* and *Dr. Atkins' Diet Revolution* advocate consuming a diet that is very low in carbohydrate, to the point of **ketogenesis** (the formation of ketone bodies in the liver when fatty acids or other ketogenic molecules are used for energy). The energy ratios promoted in these books challenge the recommendations made by organizations such as the American Heart Association, which recommend that >50% of energy be derived from carbohydrate and only about 15% from protein. All three of these energy nutrient ratios have been advocated in both weight loss and maintenance programs.

So, is there an optimal energy nutrient ratio that people and athletes should follow? It is very difficult to say. Each person would have to be assessed individually using several considerations. For instance, what is the activity level and health status of the individual? Certainly athletes should strive to consume a diet containing higher levels of carbohydrate, as discussed in Chapters 4 and 14. But what about more sedentary individuals who are hyperglycemic and obese? Would a lower carbohydrate intake, such as 40–45% of total energy intake or less, be helpful in a weight loss or maintenance effort? Even the newest DRI recommendations for carbohydrate are provided as a range (40–65% of energy), reinforcing the individuality of the issue. Surely these will be key questions to be addressed by researchers in the coming years.

Energy nutrient ratios and restrictions continue to draw a lot of public attention, and researchers are attempting to gain a better understanding. Often the focus of popular systems is on hormone levels and the processes related to fat breakdown. Of the hormones involved, insulin has gotten the most attention because it inhibits fat breakdown and release from adipose tissue as well as fatty acid oxidation in tissue. So, it would be logical to think that a lower carbohydrate intake would allow for greater fat use for fuel, as discussed below. However, there is a flip side that needs to be considered. Insulin plays a role in reducing

Key Point

Lower carbohydrate diets are popular and are purported to improve the fat-burning process by reducing the influence of insulin.

skeletal muscle protein breakdown, and skeletal muscle is an important metabolic tissue. So one important consideration of a very low carbohydrate diet might be the preservation of LBM during weight loss or maintenance.

ENERGY NUTRIENT OXIDATIVE PREFERENCE. When an individual eats a high-carbohydrate diet, he or she will oxidize more carbohydrate and less fat. This suggests a preference of energy nutrient oxidation resulting from energy nutrient consumption. Yet there are a couple of considerations. First, the oxidation of carbohydrate and protein seems to be highly related to their dietary level but fat oxidation is not. Therefore, there may be an *oxidative hierarchy* for the energy nutrients, at least acutely.[37,38] For instance, when a person eats a meal containing 25 g of carbohydrate and 100 g of fat, more carbohydrate would be oxidized initially. Second, when a person shifts from eating a higher carbohydrate diet to a lower carbohydrate diet, the adaptation to greater fat utilization takes several days to become fully engaged. Therefore, diet carbohydrate and amino acids are preferred fuel sources after eating, and adaptation to greater fat utilization takes several days and occurs only when carbohydrate intake is low.

The preference for carbohydrate and amino acids as fuel is due to hormonal influences and storage capabilities. As glucose is absorbed the circulating insulin level is elevated, thereby increasing glycolysis in tissue such as the liver and muscle. In addition, the absorption of amino acids increases both insulin and glucagon, which increases amino acid metabolism. On the other hand, the absorption of fat does not evoke a hormonal response. Thus, the control of fat oxidation is more indirect and inversely related to the metabolism of glucose and amino acids. As discussed in the previous chapter, it takes days for tissue to adapt to use more fat. For instance, researchers reported that it took 7 days for healthy, nonobese young males and females to reach zero fat balance (intake = oxidation), after switching from a low-fat to high-fat diet at an energy level that equaled their expenditure.[38] Therefore, for 6 days the participants would seem to have been in a positive fat balance, or accruing body fat.

ENERGY NUTRIENTS AND SATIETY. Is it possible that different energy nutrient ratios influence energy intake by regulating satiety? Nutrition textbooks often state that a food (or meal) with a higher fat content can induce satiety by eliciting the release of the hormone cholecystokinin (CCK).[39] Yet, populations around the world that eat a higher fat diet tend to be heavier and more obese. Some explanation for this contradiction might be found in the difference in food energy densities.[40] Simply stated, a higher fat diet may result in the consumption of less food volume, but more food energy. So, when considering the amount of energy consumed and not volume, a higher fat diet might actually have a lower satiating power, meaning it allows many calories to be consumed before satiety, and carbohydrates and protein a greater satiating effect.

Higher Carbohydrate Diets

People often ask why the general recommendations for carbohydrate intake are so high relative to the other energy nutrients. Is there scientific proof that a higher carbohydrate intake is healthier? Or were recommendations for carbohydrate intake determined mostly by default? For instance, if an individual is supposed to eat <30% of total energy as fat and about 12–15% as protein, then carbohydrate must make up the difference. Such a default recommendation, not considering alcohol, would be based on the following equation.

Carbohydrate = 100% of total diet energy recommendation − diet energy from fat + protein

RESEARCH EVIDENCE SUPPORTING A HIGHER CARBOHYDRATE DIET.

Epidemiological studies suggest that obesity is more common in populations that eat a diet rich in fat and less common in populations that eat a higher carbohydrate diet.[40,41] Also, population studies suggest that a higher carbohydrate diet and physical activity together are the key to weight gain prevention. For instance, a recent report stated that data from numerous rural areas in China indicated that adult males eat 30% more energy (kcal/kg of body weight) than American men, yet their BMI was 25% lower.[41] In addition, the diets of the Chinese men were low in fat (14.5%) and rich in plant-based foods, which provided 90% of their protein and roughly 33 g of fiber daily. The researchers concluded that a significantly higher level of physical activity accounted for much of the difference in BMI. Somewhat similar observations were made involving a population of women in Sweden.[42]

When energy is restricted, are there advantages to having more carbohydrate in the diet? The short answer is probably. Using semistarvation diets, for example, the energy nutrient ratio influences insulin and blood lipid levels; however, it might not influence the rate of body weight and fat loss. For instance, researchers randomly assigned obese adults to 1000-kcal diets containing an energy nutrient ratio of either 15/32/53 or 45/29/26 (% energy from carbohydrate/protein/fat).[43] After 6 weeks, weight loss was similar between the groups, as were the reductions in total body fat and waist-to-hip ratio. Plasma insulin and triglyceride levels were lowered in the 15% carbohydrate diet, but not in the 45% carbohydrate diet, which might be a consideration for individuals also coping with elevated blood insulin and triglyceride levels.

In another study, obese women were provided either a 25%, 45% or 75% carbohydrate diet for 10 weeks at an energy level of 1200 kcal.[44] Again, the weight and body fat losses were not different between groups. Considering the

Key Point

The energy nutrient ratio of weight-reducing diet plans might not make a difference in weight and fat loss, but a higher carbohydrate and protein intake helps minimize the loss of LBM.

results of these and other studies, it would seem that the energy nutrient ratio of a semistarvation diet might not influence the rate of body fat loss. However, one important consideration is maintenance or minimization of LBM during energy restriction. At this time the results of some, but not all, research studies involving energy restriction suggest that dedicating more of the energy to protein may be an important factor in preserving LBM.[45–47]

Ketogenic Diets (Very Low Carbohydrate Diets)

The application of the ketogenic diet for weight reduction has been discussed for decades. In fact, a review of professional journals yields numerous research papers on the topic dating back 30–40 years.[48] Dr. Atkins first marketed his low-carbohydrate weight reduction program in the 1970s, and statements made in his book were immediately met with criticism.[49,50] Despite the omnipresence of this diet in the media and pop culture, relatively little research exists to evaluate its long-term efficacy and safety. As mentioned above, eating a diet lower in carbohydrate (as a percentage of total energy) is advocated to reduce the metabolic influences of insulin. Insulin inhibits the breakdown of triglycerides in adipose tissue and promotes fat production in the liver and adipose tissue. Logically, the net effect of chronically elevated insulin levels would be more fat production and less use of body fat. But what do we really know about the efficacy and safety of long-term carbohydrate restriction?

KETOGENIC DIETS AND HEALTH RISKS.

Among the most significant concerns surrounding the diet plans that have relatively low carbohydrate and high fat content is the potential impact on risk factors for heart disease. This concern was supported a couple of decades ago when it was reported in an Australian medical journal that two individuals following a ketogenic diet experienced a rise in blood cholesterol levels.[51] Later, researchers reported the results of another study involving lean men following a ketogenic diet containing 35–50 kcal and 1.75 g of protein per kilogram of body weight and <20 g of carbohydrate.[52] The energy level was set to maintain energy balance and weight stability. The researchers noted that the diet resulted in an increase in the average total cholesterol levels from 159 to 208 mg/100 ml of blood while triglycerides fell from 107 to 79 mg/100 ml. The diet was well tolerated and did not result in a pathological change in serum lipid levels. In another study, 12 men switched from their habitual diet (17% protein, 47% carbohydrate, and 32% fat) to a ketogenic diet (30% protein, 8% carbohydrate, and 61% fat) for 6 weeks.[53] The men averaged a 34% and a 33% reduction in fasting insulin and triglyceride levels, respectively, while total cholesterol, LDL-cholesterol, and oxidized LDL remained the same. In addition, it was noted that HDL-cholesterol levels tended to increase with the ketogenic diet, although the effect was not statistically significant.

The results of available research to date do not strongly support the notion that very low carbohydrate, higher fat diets can unfavorably influence blood lipid levels and increase disease risk. Furthermore, reductions in fasting lipid levels reported by some researchers are highly suggestive of increased fat metabolism. However, fat metabolism has upper limits, and data from these energy-restricted studies cannot be easily applied to an adequate energy intake.

Key Point

Ketogenic diets involve a very low carbohydrate intake to the point of excessive ketone body production.

CONSIDERATIONS WITH KETOGENIC DIETS. Despite the popularity of this dietary regimen, many questions remain. Furthermore, researchers must be very thorough in designing weight-loss studies that use ketogenic diets. For instance, weight loss might be more rapid following an energy-restricted diet that contains a very low level of carbohydrate than following a diet with a higher carbohydrate percentage. However, much of the additional weight loss would be attributable to water loss, as glycogen stores would be drastically reduced. Therefore, changes in body composition must also be tracked. Furthermore, as insulin seems to be a potent hormone involved in limiting skeletal muscle protein breakdown, researchers are encouraged to assess protein turnover as well.

The impact of very low carbohydrate diets on exercise performance and adherence to an exercise program must be considered as well. If the carbohydrate restriction leads to decreased performance during strength and endurance training, this can influence energy expenditure and body composition over time. In addition, very low carbohydrate diets reduce the intake of phytochemicals. Phytochemicals are substances found in plants, many of which have health-promoting properties such as bolstering antioxidant efforts, stoking up detoxification enzyme systems, and inhibiting cell growth mechanisms involved in cancer and heart disease.[54]

Popular Fat-Burning "Tricks"

Several "tricks" have been publicized in magazines and books that are supposed to enhance "fat burning." These tend to involve the timing of exercise and/or meals. Although many of these tricks appear to have a scientific rationale, they lack proper investigation. Therefore, each of these notions should be carefully evaluated for individual merit, especially when considering how to achieve an individual's long-term goals. Some of the most common notions include carbohydrate tapering throughout the day, performing aerobic exercise before eating in the morning, and not eating for a couple of hours after aerobic exercise.

CARBOHYDRATE TAPERING. Carbohydrate tapering involves eating meals with higher carbohydrate content in the morning (and after heavy exercise), and then tapering the level of carbohydrate in the remaining meals of that day. The idea is that the higher carbohydrate consumption in the morning and after exercise would promote recovery of glycogen stores that were reduced during the night and during exercise. The reduced levels of carbohydrate in the meals later in the day would mean more fat is burned at night and during sleep. One of the concepts important to this notion is that the glycemic index associated with a meal would be lowest in the morning and after exercise and higher later in the day. This is because the rate of glycogen recovery is most rapid when stores are lower, and glycogen levels might be lowest in the morning after an overnight fast. This would allow for a more aggressive glucose removal from circulation and in turn a lower glycemic index for a food. Carbohydrate tapering is one of the most interesting notions for enhanced fat burning; however, the type, intensity, duration, and timing of exercise and other activities need to be considered. For instance, the glycogen recovery can take several hours to a day, depending on the degree of exhaustion of those stores. Therefore, reducing the level of carbohydrate as the day continues can lead to incomplete recovery of glycogen stores and potential reduction in athletic performance in subsequent exercise bouts.

CARDIOMORNING. Is it better to perform cardiorespiratory exercise first thing in the morning or later in the day? Here the idea is that during sleep, fat burning is maximized and thus cardiorespiratory exercise first thing in the morning (before eating) could amplify fat burning. However, there are several factors to consider when evaluating this fat-burning method. The intensity of exercise is important, as moderate to higher submaximal intensity exercise requires carbohydrate as a foundational fuel source. Since carbohydrate is not consumed before and/or during the morning exercise, the reduced level of glycogen in skeletal muscle could become a performance-limiting factor during moderate to higher submaximal exercise efforts. For fat-burning purposes, the morning exercise might be performed at a lower intensity, which utilizes the highest percentage of fat, as discussed in Chapters 4 and 6. However, the total quantity of fat oxidized would be greater at a moderate submaximal intensity because of greater overall energy expenditure.

AFTERBURN. The "afterburn" involves withholding food for a couple of hours after finishing cardiorespiratory exercise. This notion is based on the effect of exercise on fat storage breakdown and fatty acid oxidation in muscle and the liver. By not eating for a couple of hours after exercise, one could maximize fat use during that time, as fat mobilization and oxidation processes are already prominent. Here again the intensity of the exercise is a factor along with long-term goals. Exercise performed at moderate to higher submaximal intensities can induce a negative protein balance after the exercise bout. Therefore, waiting several hours to eat carbohydrate and protein may lead to increased protein breakdown affecting muscle mass and subsequently BMR (RMR). However, if the intensity is lowered to increase the percentage of fat utilized during exercise, then TEA and adaptations also would be reduced.

IRON THEN CARDIO. The theory here is that weight training at the beginning of an exercise session will decrease muscle glycogen stores and force muscle to use a greater percentage of fat during the cardiorespiratory exercise that follows. However, it would seem that for this to be true the resistance training would have to involve the same muscles used in the cardiorespiratory exercise. For instance, this concept would not apply if a person trained the chest, shoulders, and back and then hopped on the treadmill. In addition, the intensity of the cardiorespiratory exercise may be a factor. Higher submaximal intensity running or cycling may negatively influence the effect of the resistance training on protein turnover in muscle. Therefore it is generally recommended that cardiorespiratory exercise involving higher workloads be separated from resistance training sessions by as much time as possible, especially when the same muscle groups are involved.

■ DIET CONCEPTS *IN REVIEW*

- Recommendations for a higher carbohydrate diet for active people have epidemiological and physiological foundations.
- Popular fat-burning notions are theoretical and require research.
- Popular diet programs often challenge the recommendations for energy nutrient consumption made by organizations such as the American Heart Association and American Dietetic Association.

IN FOCUS   DESIGNING AN OPTIMAL TRAINING DIET FOR SPORT PERFORMANCE

Designing the optimal diet for athletic performance can be challenging. The first step is to determine how much energy is expended in a day and set goals for body weight and composition. Is weight to be maintained, increased, or decreased? Second, set practical goals for the distribution of diet energy into carbohydrate, protein, and fat. This involves considering recommendations on two bases: on the basis of nutrient intake per kilogram of body weight and on the basis of nutrients as a percentage of diet energy. Table A demonstrates the distribution of energy based on these two systems. Third, distribute the energy intake throughout the number of meals and snacks to be consumed. Special attention is given to the first meal of the day, and the time of workout is considered.

Example

Tyrell is a 165-lb (75-kg) runner. He is pleased with his weight but would like to make sure he is eating the right proportions of energy nutrients. He wants to plan to eat five to six times a day. His training runs are usually planned for 5:00 P.M. when he gets home from school.

1. Determine energy needs for weight maintenance. Since Tyrell does not have access to indirect calorimetry, he can estimate his energy requirements based on an energy expenditure equation and confirm the estimate by eating that energy level without experiencing fluctuations in body weight.

   Tyrell has estimated his total energy expenditure (TEE) using the Harris Benedict equation (see Table 7-11) as an estimate of BMR and has also factored in physical activity (thermal effect of activity, TEA), the thermal effect of food (TEF), and adaptive thermogenesis (AT). Assessment of his 3-day diet history provided an estimation of the thermal effect of food (TEF). The energy intake level found by nutritional assessment of his diet was 4287 kcal/day, which closely matched his estimated TEE of 4326 kcal/day. On the basis of this information, Tyrell plans for an intake of 4300 kcal/day, as he is not looking to alter his body weight or composition at this time.

DESIGNING AN OPTIMAL TRAINING DIET FOR SPORT PERFORMANCE (CONTINUED)

Table A   Energy Nutrient Requirements Based on Body Mass and Energy Needs

Energy Nutrient	Requirement Based on Body Mass (nutrient g/kg of body weight)		Requirement Based on % of Total Energy	
	Recommendation	*Applied to Example*	*Recommendation*	*Applied to Example*
Carbohydrate	8–10 g/kg	600–750 g (2400–3000 kcal, 56–70% of energy)	60%	645 g (2580 kcal)
Protein	1.4–1.75 g/kg	105–131 g (420–525 kcal, 10–12% of energy)	15–20%	161–215 g (645–860 kcal)
Fat	Remaining energy	86–164 g (775–1480 kcal, 18–34% of energy)	20–25%	95–120 g (860–1075 kcal)

Table B   Distribution of Energy over Six Meals

Energy Nutrient (Daily Intake)	Breakfast (7:00 A.M.)	Snack (10:30 A.M.)	Lunch (1:00 P.M.)	Snack (3:00 P.M.)	Dinner (6:30 P.M.)	Snack (9:00 P.M.)
Energy (4300 kcal)	1050 kcal	395 kcal	945 kcal	300 kcal	1095 kcal	515 kcal
Carbohydrate (645 g)	150 g	50 g	150 g	75 g	150 g	70 g
Protein (160 g)	45 g	15 g	30 g	<1 g	45 g	25 g
Fat (120 g)	30 g	15 g	25 g	<1 g	35 g	15 g

Note: Training run at 5:00 P.M. (10 km at 7 min/mile pace). Afternoon snack includes 12-oz (355-ml) sport drink with 6% carbohydrate.

2. Determine minimal requirements for carbohydrate and protein. Tyrell has estimated his requirements for carbohydrate, protein, and fat based on his body mass and on estimated energy requirements (Table A). On the basis of these estimates he plans for 60% of his energy to be derived from carbohydrate, 15% from protein, and 25% from fat.
3. Distribute the energy intake over the course of the day. For Tyrell, breakfast and dinner will be larger meals, as they must account for recovery of energy stores depleted during sleep and training, respectively (Table B). Also, since his training run will take place prior to dinner, the snack before will focus more on providing carbohydrate and less on fat, and dinner will have a higher fat content than other meals. The evening snack needs to contain some fat and fiber to slow the absorption of energy nutrients and allow for a longer absorptive period. This might also minimize the influence of catabolic factors such as cortisol during sleep.

Conclusions

Body weight and composition are important components of exercise and sport performance for athletes, as an ideal body weight and composition can optimize performance. Body weight and composition are also important health considerations for everyone because overweight and obesity can have serious health consequences. The intake of energy nutrients and level of physical activity influence body weight and composition. Several techniques have been developed to assess body composition including densitometry, BIA, DXA, and skinfold assessment. Many diet and exercise programs have been advocated to change body weight and composition; however, the efficacy of different programs requires consideration and further research.

STUDY QUESTIONS

1. How are body weight and composition related to health?
2. What are the different systems used to estimate body composition and what are their strengths and weaknesses?
3. What are the four components of energy expenditure?
4. How does exercise influence TEE and what is the potential impact of resistance training on RMR?
5. What are the most important considerations for altering body weight by athletes?
6. How might altering the energy nutrient ratio influence the oxidation of carbohydrate and fat?
7. What are the most important considerations associated with popular "fat-burning tricks"?

REFERENCES

1. Wildman REC, Medeiros DM. Body composition and obesity. In: *Advanced Human Nutrition*, Boca Raton, FL: CRC Press, 2000.
2. Lohman TG. Skinfolds and body density and their relation to body fatness: a review. *Human Biology* 53(2):181–225, 1981.
3. Siri WE. Body composition from fluid spaces and density: analysis of methods. In: *Techniques for Measuring Body Composition* (Brozek J, Hensel A, ed.), Washington DC: National Academy of Sciences National Research Council, 1961.
4. Lohman TG. *Advances in Body Composition Assessment*, Champlaign, IL: Human Kinetics, 1992.
5. Schutte JE, Townsend EJ, Hugg J, Shoup RF, Malina RM, Blomqvist CG. Density of lean body mass is greater in blacks than in whites. *Journal of Applied Physiology* 56(6):1647–1649, 1984.
6. Lohman TG. Applicability of body composition techniques and constants for children and youths. In: *Exercise and Sports Science Reviews*, 14th ed. (Pandolf K, ed.), New York: Macmillan, 1986.
7. McCrory MA, Mole PA, Gomez TD, Dewey KG, Bernauer EM. Body composition by air-displacement plethysmography by using predicted and measured thoracic gas volumes. *Journal of Applied Physiology* 84(4):1475–1479, 1998.
8. Biaggi RR, Vollman MW, Nies MA, Brener CE, Flakoll PJ, Levenhagen DK, Sun M, Karabulut Z, Chen KY. Comparison of air-displacement plethysmography with hydrostatic weighing and bioelectrical impedance analysis for the assessment of body composition in healthy adults. *American Journal of Clinical Nutrition* 69(5):898–903, 1999.
9. National Institutes of Health. *Bioelectrical Impedance Analysis of Body Composition Measurement*. NIH Technological Assessment Statement, Dec. 12–14, pp. 1–35, 1996.
10. Withers RT, Craig NP, Bourdon PC, Norton KI. Relative body fat and anthropometric prediction of body density of male athletes. *European Journal of Applied Physiology and Occupational Physiology* 56(2):191–200, 1987.
11. Pollock ML, Gettman LR, Jackson A, Ayres J, Ward A, Linnerud AC. Body composition of elite class distance runners. *Annals of the New York Academy of Sciences* 301:361–370, 1977.
12. Graves JE, Pollock ML, Sparling PB. Body composition of elite female distance runners. *International Journal of Sports Medicine* 2:96–102, 1987.
13. Wilmore JH, Brown CH, Davis JA. Body physique and composition of the female distance runner. *Annals of the New York Academy of Sciences* 301:764–776, 1977.
14. Martin DT, McLean B, Trewin C, Lee H, Victor J, Hahn AG. Physiological characteristics of nationally competitive female road cyclists and demands of competition. *Sports Medicine* 31(7):469–477, 2001.
15. Melesky BW, Shoup RF, Malina RM. Size, physique and body composition of competitive female swimmers 11 through 20 years of age. *Human Biology* 54:609–617, 1982.
16. Siders WA, Lukaski HC, Bolonchuk WW. Relationships among swimming performance, body composition and somatotype in competitive collegiate swimmers. *Journal of Sports Medicine and Physical Fitness* 33(2):166–171, 1993.
17. Sinning WE, Dolny DG, Little KD, Cunningham LN, Racaniello A, Siconolfi SF, Sholes JL. Validity of "generalized" equations for body composition analysis in male athletes. *Medicine and Science in Sports and Exercise* 17(1):124–130, 1985.
18. Saylor K, Wildman REC, Willard G. Dietary practices, physical assessment and blood chemistry of competitive male bodybuilders. *FASEB Abstract*, 2002.

19. Wildman REC, Saylor K, Willard G. Dietary practices, physical assessment and blood chemistry of competitive female bodybuilders. *FASEB Abstract*, 2001.

20. Garby L, Garrow JS, Jorgensen B, Lammert O, Madsen K, Sorensen P, Webster J. Relation between energy expenditure and body composition in man: specific energy expenditure in vivo of fat and fat-free tissue. *European Journal of Clinical Nutrition* 42(4):301–305, 1988.

21. Illner K, Brinkmann G, Heller M, Bosy-Westphal A, Muller MJ. Metabolically active components of fat free mass and resting energy expenditure in nonobese adults. *American Journal of Physiology: Endocrinology and Metabolism.* 278(2):E308–E315, 2000.

22. Astrup A, Buemann B, Christensen NJ, Madsen J, Gluud C, Bennett P, Svenstrup B. The contribution of body composition, substrates, and hormones to the variability in energy expenditure and substrate utilization in premenopausal women. *Journal of Clinical Endocrinology and Metabolism* 74:279–286, 1992.

23. Cunningham, J.C. (1991). Body composition as a determinant of energy expenditure: a synthetic review and a proposed general prediction equation. *American Journal of Clinical Nutrition* 54:963–969, 1991.

24. Nelson KM. Prediction of resting energy expenditure from fat-free mass and fat mass. *American Journal of Clinical Nutrition* 56(5):848–856, 1992.

25. Nielsen S, Hensrud DD, Romanski S, Levine JA, Burguera B, Jensen MD. Body composition and resting energy expenditure in humans: role of fat, fat-free mass and extracellular fluid. *International Journal of Obesity and Related Metabolic Disorders* 24(9):1153–1157, 2000.

26. Sparti A, DeLany JP, de la Bretonne JA, Sander GE, Bray GA. Relationship between resting metabolic rate and the composition of the fat-free mass. *Metabolism* 46(10):1225–1230, 1997.

27. Campbell WW, Crim MC, Young VR, Evans WJ. Increased energy requirements and changes in body composition with resistance training in older adults. *American Journal of Clinical Nutrition* 60:167–175, 1994.

28. Rall LC, Meydani SN, Kehayias JJ, Dawson-Hughes B, Roubenoff R. The effect of progressive resistance training in rheumatoid arthritis. Increased strength without changes in energy balance or body composition. *Arthritis and Rheumatism* 39(3):415–426, 1996.

29. Pratley R, Nicklas B, Rubin M, Miller J, Smith A, Smith M, Hurley B, Goldberg A. Strength training increases resting metabolic rate and norepinephrine levels in healthy 50- to 65-yr-old men. *Journal of Applied Physiology* 76(1):133–137, 1994.

30. Byrne, HK, Wilmore JH. The effects of a 20-week exercise training program on resting metabolic rate in previously sedentary, moderately obese women. *International Journal of Sport Nutrition and Exercise Metabolism* 11(1):15–31, 2001.

31. Broeder CE, Burrhus KA, Svanevik LS, Wilmore JH. The effects of either high-intensity resistance or endurance training on resting metabolic rate. *American Journal of Clinical Nutrition* 55(4):802–810, 1992.

32. Cullinen K, Caldwell M. Weight training increase fat-free mass and strength in untrained young women. *Journal of the American Dietetics Association* 98(4):414–418.

33. van Etten LM, Westerterp KR, Verstappen FT. Effect of weight-training on energy expenditure and substrate utilization during sleep. *Medicine and Science in Sports and Exercise* 27(2):188–193, 1995.

34. Bryner RW, Ullrich IH, Sauers J, Donley D, Hornsby G, Kolar M, Yeater R. Effects of resistance vs. aerobic training combined with an 800 calorie liquid diet on lean body mass and resting metabolic rate. *Journal of the American College of Nutrition* 18(2):115–121, 1999.

35. Lemmer JT, Ivey FM, Ryan AS, Martel GF, Hurlbut DE, Metter JE, Fozard JL, Fleg JL, Hurley BF. Effect of strength training on resting metabolic rate and physical activity: age and gender comparisons. *Medicine and Science in Sports and Exercise* 33(4):532–541, 2001.

36. Poehlman ET, Denino WF, Beckett T, Kinaman KA, Dionne IJ, Dvorak R. Effects of endurance and resistance training on total daily energy expenditure in young women: a controlled randomized trial. *Journal of Clinical Endocrinology and Metabolism* 87(3):1004–1009, 2002.

37. Tataranni PA, Ravussin E. Effect of fat intake on energy balance. *Annals of the New York Academy of Sciences* 23; 819:37–43, 1997.

38. Schrauwen, P, van Marken Lichtenbelt WD, Saris WH, Westerterp KR. Changes in fat oxidation in response to a high fat diet. *American Journal of Clinical Nutrition* 66(2):276–282, 1997.

39. Wildman REC, Medeiros DM. Lipids. In: *Advanced Human Nutrition*. Boca Raton, FL: CRC Press, 2000.

40. Golay A, Bobbioni E. The role of dietary fat in obesity. *International Journal of Obesity and Related Metabolic Disorders* 21(Suppl. 3):S2–11, 1997.

41. Campbell TC, Chen J. Energy balance: interpretation of data from rural China. *Toxicological Sciences* 52(2 Suppl.):87–94, 1999.

42. Lissner L, Heitmann BL, Bengtsson C. Low-fat diets may prevent weight gain in sedentary women: prospective observations from the population study of women in Gothenburg, Sweden. *Obesity Research* 5(1):43–48, 1997.

43. Golay A, Allaz AF, Morel Y, de Tonnac N, Tankova S, Reaven G. Similar weight loss with low- or high-carbohydrate diets. *American Journal of Clinical Nutrition* 63(2): 174–178, 1996.

44. Alford BB, Blankenship AC, Hagen RD. The effects of variations in carbohydrate, protein, and fat content of the diet upon weight loss, blood values, and nutrient intake of adult obese women. *Journal of the American Dietetic Association* 90(4):534–540, 1990.

45. Piatti PM, Monti F, Fermo I, Baruffaldi L, Nasser R, Santambrogio G, Librenti MC, Galli-Kienle M, Pontiroli AE, Pozza G. Hypocaloric high-protein diet improves glucose oxidation and spares lean body mass: comparison to hypocaloric high-carbohydrate diet. *Metabolism* 43(12):1481–1487, 1994.

46. Layman DK, Boileau RA, Erickson DJ, Painter JE, Shiue H, Sather C, Christou DD. A reduced ratio of dietary carbohydrate to protein improves body composition and blood lipid profiles during weight loss in adult women. *Journal of Nutrition* 133(2):411–417, 2003.

47. Luscombe ND, Clifton PM, Noakes M, Farnsworth E, Wittert G. Effect of a high-protein, energy-restricted diet

on weight loss and energy expenditure after weight stabilization in hyperinsulinemic subjects. *International Journal of Obesity and Related Metabolic Disorders* 27(5):582–590, 2003.

48. Weight reduction: fasting versus a ketogenic diet. *Nutrition Reviews* 24(5):133–134, 1966.

49. Hirschel B. Dr. Atkins' dietetic revolution: a critique. *Schweizerische Medizinische Wochenschrift* 23;107(29):1017–1025, 1977.

50. Forster H. Is the Atkins diet safe in respect to health? *Fortschritte der Medizin* 14;96(34):1697–1702, 1978.

51. Elliot B, Roeser HP, Warrell A, Linton I, Owens P, Gaffney T. Effect of a high energy, low carbohydrate diet on serum levels of lipids and lipoproteins. *Medical Journal of Australia* 7;1(5):237–240, 1981.

52. Phinney SD, Bistrian BR, Wolfe RR, Blackburn GL. The human metabolic response to chronic ketosis without caloric restriction: physical and biochemical adaptation. *Metabolism* 32(8):757–768, 1983.

53. Sharman MJ, Kraemer WJ, Love DM, Avery NG, Gomez AL, Scheett TP, Volek JS. A ketogenic diet favorably affects serum biomarkers for cardiovascular disease in normal-weight men. *Journal of Nutrition* 132(7):1879–1885, 2002.

54. Wildman REC. Nutraceuticals: A brief review of historical and teleological aspects. In: *Nutraceuticals and Functional Foods* (Wildman REC, ed.), Boca Raton, FL: CRC Press, 2001.

WATER, HYDRATION, AND EXERCISE

Personal Snapshot

Steve is a competitive high school tennis player. He has complained to his coach about feeling weak and prone to mistakes later in practices and games. Steve's coach suspects dehydration and has him complete a 24-hour diet recall. He also weighs Steve before and after practice to estimate sweat volume. The assessment of Steve's diet reveals that he consumes about 3 L of water from fluids and foods and drinks about 1 L of water during practice and games in the form of a sport drink. According to his coach's estimates, Steve would need about 5 L of water to balance what he loses throughout the day and during practice as sweat. How important is body water to Steve's optimal performance? What can Steve do to ensure optimal hydration status during practice and competition?

Chapter Objectives

- Provide an overview of the properties of water and its role in thermoregulation and other processes.

- Discuss water balance as the net result of water intake from diet and metabolism and water loss with special attention to sweat.

- Describe the involvement of dehydration in reduced heat tolerance and fatigue.

- Discuss the impact of dehydration on exercise performance.

- Provide recommendations for water consumption during training and competition.

How important is body water to physical performance? Incredibly important! Body water or **hydration** status must be a foremost consideration for active people and competitive athletes. This is because performance can decrease even when the reduction in hydration status, or **dehydration**, seems subtle. For athletes such as distance runners and cyclists, dehydration can lead to reduced endurance and premature fatigue. Dehydration can result in reduced power and strength for athletes participating in sports such as football, ice hockey, soccer, tennis, and weightlifting. So proper hydration before and during training or competition is essential to optimal performance. However, some athletes, such as bodybuilders and fitness competitors, consciously attempt to reduce their body water content for aesthetic reasons, and others, such as wrestlers and rowers, do it to reduce their body weight to compete in a lower weight class. These individuals must consider potential reductions in performance and the negative effects of dehydration on their health.

Water is also the "playing field" for several sports, such as swimming, water polo, ice hockey, skiing, and figure skating. Body temperature regulation is a consideration for training or competing in these sports. For example, does exercise in cool water change how heat is released from the body? In other sports, special clothing may affect body temperature regulation. Ice and snow sports like ice hockey involve insulated uniforms or undergarment materials (such as polypropylene), and in other sports such as downhill ski racing and figure and speed skating the participants are either scantily clad or clad in a sheer material. How do these situations affect heat loss via sweating? This chapter explores the basic properties of water and its role in regulating body temperature and blood pressure. In addition, the relationship between hydration status and physical performance are explored and recommendations for water consumption before, during, and after exercise are presented.

■ BODY WATER BASICS

The human body is mostly water—about 56–64% of the weight of the average adult.[1] But water is not evenly distributed in the body. A fairly lean 175-lb (80-kg) man, who is 60% water, can attribute approximately 105 lb (about 48 kg) of his mass to water. Water is found in all tissues of the body, but its presence varies from tissue to tissue.

Properties of Water

Water is the ideal medium for the body. What properties make it so? First, water is an excellent solvent because it is a **polar** molecule. Although it has no net electrical charge (overall it is neutral), the unequal sharing of electrons within its covalent bonds leads to partial negative and positive charges on the oxygen and hydrogen atoms, respectively (Figure 8-1). This allows water molecules to electrically interact with other water molecules and other partially or fully charged substances such as electrolytes, glucose, amino acids, and proteins.

Second, water has a relatively high **specific heat**, a measure of the calorie energy required to raise a gram of a substance by 1°C. Water has a specific heat of 1.0 compared to 0.83 for the entire body of an adult male; the

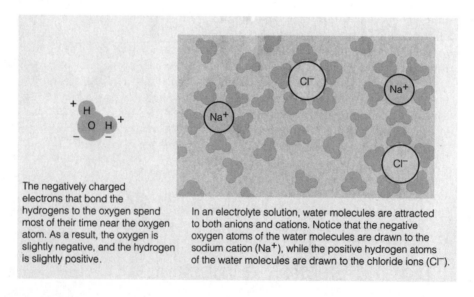

The negatively charged electrons that bond the hydrogens to the oxygen spend most of their time near the oxygen atom. As a result, the oxygen is slightly negative, and the hydrogen is slightly positive.

In an electrolyte solution, water molecules are attracted to both anions and cations. Notice that the negative oxygen atoms of the water molecules are drawn to the sodium cation (Na⁺), while the positive hydrogen atoms of the water molecules are drawn to the chloride ions (Cl⁻).

Figure 8-1
Water dissolves salts and follows electrolytes. The structural arrangement of the two hydrogen atoms and one oxygen atom enables water to dissolve salts. Water's role as a solvent is one of its most valuable characteristics.

lower overall figure is due to the presence of proteins, fat, and other substances. The contribution of water helps slow the rise in body temperature in hot environments and/or during periods of greater heat production such as exercise. Therefore, body water is viewed as a "heat buffer." In addition, the high specific heat of water minimizes the amount of sweat needed to dissipate surplus heat from the body. For instance, it takes approximately 580 kcal of heat to vaporize a liter of sweat water.

Key Point

Body water acts as a "heat buffer" and also allows for a lot of heat to be released from the body in a small amount of sweat.

Distribution of Body Water

How is water distributed throughout the body? About two-thirds of the water in the body is found within cells as the basis of **intracellular fluid.** The remaining one-third exists outside of cells as the basis of **extracellular fluid.** The extracellular fluid includes the fluids between cells, called **interstitial fluid,** and the plasma portion of blood. Figure 8-2 presents the separation of body water into different anatomical *compartments.*

Water moves in and out of cells through protein channels called **aquaporins** in cell membranes. Water is also free to move between the blood and tissue through holes **(fenestrations** and **intercellular clefts)** in capillary walls. Hydrostatic and osmotic pressures govern the movement of water in and out of cells and between tissue and the blood. **Hydrostatic pressure** (such as blood pressure) is a "pushing force." **Osmotic pressure** is a "pulling force" and results in **osmosis,** the movement of water down its concentration gradient through a semipermeable barrier such as a cell membrane or capillary wall. Osmosis

is depicted in Figure 8-3, and the relationship between hydrostatic and osmotic forces in determining water *flux* is demonstrated in Figure 8-4.

PLASMA WATER.   Roughly half of the total blood volume is water. This can be reasoned as follows: If the hematocrit, or red blood cells (RBCs), is 44% of an adult man's total blood volume (about 5 L), and white blood cells are about 1%, this leaves 55% of his blood as plasma. If his plasma is about 90% water, this translates to about 50% of his total blood volume or about $2\frac{1}{2}$ L.

Electrolytes, mainly sodium and chloride, also play a significant role in determining the osmotic properties of the blood. Figure 8-5 presents the concentration of electrolytes and other osmotic factors in body fluids. Another influential factor determining plasma water volume is protein level. The protein content of the plasma is roughly 7 g/100 ml; albumin typically accounts for more than half of total plasma protein (3–5 g/100 ml). Protein exerts a powerful osmotic effect on water, and any changes in plasma protein levels affect the water content of the blood. For instance, if plasma protein levels decrease (as in malnutrition), plasma water volume decreases in response.

Key Point

Body water serves as the basis for the intracellular and extracellular fluids. The extracellular fluid includes interstitial fluid and plasma.

SKELETAL MUSCLE AND ADIPOSE TISSUE.   Roughly 72–73% of the mass of skeletal muscle is water (Figure 8-6). Only 10% or less of the weight of adipose tissue is water.

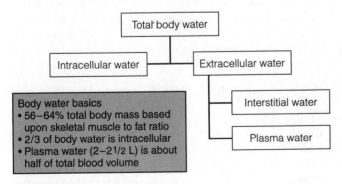

Figure 8-2
Distribution of body water into intracellular and extracellular compartments. The extracellular compartment includes the plasma and interstitial fluids.

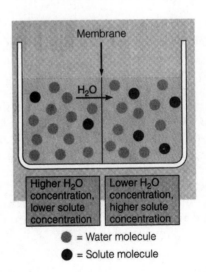

Figure 8-3
Osmosis is the net diffusion of water down its own concentration gradient (to the area of higher solute concentration).

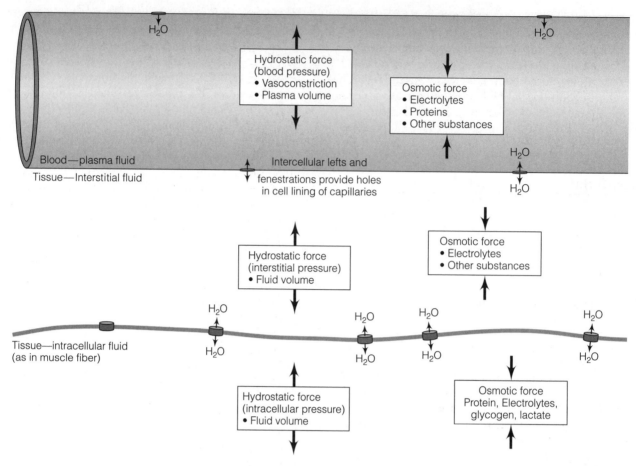

Figure 8-4

Water flux between the plasma and the tissue interstitial and intracellular fluids is governed by hydrostatic pressure (such as blood presure) and osmotic forces of dissolved substances (elecrolytes, protein, and so on).

As skeletal muscle and adipose tissue are the predominating types of body tissues, their relative contribution to total body weight largely determines body water percentage. So a bodybuilder with 8% body fat (92% fat-free mass) has a higher body water percentage than another male who weighs the same but has a body fat level of 20% (80% fat-free mass). Because men tend to have a greater percentage of skeletal muscle and a lower percentage of body fat compared to women, they generally have a greater percentage of body water.

The difference in water content between skeletal muscle and adipose tissue is primarily due to the intracellular content of osmotically active molecules. Muscle fibers are concentrated with protein and also contain glycogen, creatine phosphate, and amino acids. Collectively, these and other substances create a significant osmotic force, which pull water into muscle fibers. For instance, each gram of glycogen associates with about 3 g of water.

Assessment of Body Water

Plasma volume and total body water mass can be estimated in several ways. Dyes such as indocyanine green and RBCs containing the chromium isotope (^{51}Cr) can be infused into the blood to estimate plasma volume. After allowing an appropriate period of time for equilibration throughout the blood, a sample of blood is drawn. Plasma volume is then estimated by comparing the concentration of dye or ^{51}Cr-labeled RBCs in the infused solution to the concentration in the blood sample. Labeled RBCs can also assist in assessing changes in the number of RBCs, as discussed later in the chapter. Here is the equation:

Key Point

The ratio of skeletal muscle to adipose tissue is the primary factor dictating body water percentage and mass.

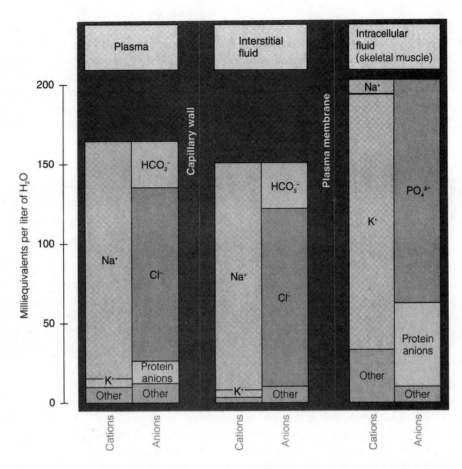

Figure 8-5
Ionic composition of the major body fluid compartments.

$$C_1V_1 = C_2V_2$$

where:

C_1 = concentration of the dye or labeled RBCs in the infused solution

V_1 = volume of the infused solution

C_2 = concentration of the dye or labeled RBCs in the sample of blood

V_2 = volume of the sample of blood

Methods for estimating total body water include dual energy X-ray absorptiometry (DXA or DEXA), bioelectrical impedance assessment (BIA), and labeled water. These techniques are described in Chapter 7 as they are used to assess body composition (DXA and BIA) and to estimate energy expenditure (labeled water). The use of labeled water for estimating total body water can involve different isotope forms of water and involves a different procedure than estimating energy expenditure. Two forms of water that can be used to assess total body water include an isotope of hydrogen. Deuterium (^{2}H) is a stable isotope of hydrogen, meaning it does not emit radioactivity. Tritium (^{3}H) is radioactive, making tritium-labeled water (^{3}H$_2$O) less popular than deuterium-labeled water (^{2}H$_2$O), especially when assessing children and pregnant women.

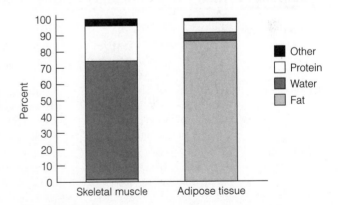

Figure 8-6
Composition difference between skeletal muscle and adipose (fat) tissue. Skeletal muscle is largely water and protein while adipose tissue is mostly fat and very little water, protein, and other material.

Assessment of total body water using labeled water begins after an individual fasts for 8–10 hours. Then a sample of saliva, urine, or blood is collected and participants ingest 10 g of labeled water as part of juice or another fluid. After allowing at least 2 hours for complete absorption and equilibration throughout body water compartments, samples of blood, saliva, and urine are collected and their labeled water content is assessed using laboratory equipment.

■ BODY WATER BASICS *IN REVIEW*

- Water is the predominant substance of the body for most individuals and has several unique qualities such as a relatively high specific heat.
- Hydrostatic and osmotic forces dictate net water fluxes across cell membranes and between tissues and the blood.
- Body water percentage is dictated by the ratio of skeletal muscle to adipose tissue.
- Total body water and plasma water can be estimated using dyes and isotopes.

■ ROLES OF BODY WATER

If water serves as the medium of the body and has unique properties, what physiological roles does it play? Water plays several roles in basic human function and **homeostasis**; the maintenance of nearly constant internal conditions. For instance, water provides the basis for blood pressure and body temperature regulation. In addition, it

is the basis of digestive juices, excretory fluids such as sweat and urine, and the fluid found within joints. Furthermore, the hydration status of cells may have a regulatory influence on the metabolism of energy nutrients. More specifically, dehydration may promote glycogen and protein breakdown in tissue such as skeletal muscle. This could affect exercise energy metabolism and recovery and adaptive processes, as discussed later in this chapter.

Sweating

One of the most obvious physiological roles of water is its involvement in body temperature regulation. Along with **convection, conduction,** and **radiation,** sweating is a primary mechanism for heat loss from the body. As described in Table 8-1, climate and other factors influence the relative contributions made by these mechanisms. Average core body temperature is approximately 98.6°F (37°C), and the maintenance of this temperature is critical to survival. **Hyperthermia,** a rise in core body temperature, leads to changes in enzyme activities and other cellular

Key Point

Sweating is an important mechanism for heat release from the body.

| Table 8-1 | **Primary Mechanisms of Body Heat Transfer** |

Mechanism of Heat Transfer	Principle of Heat Transfer Mechanism	Factors That Increase Heat Loss from the Body by This Process	Factors That Decrease Heat Loss from the Body by This Process
Convection	Transfer of heat to/from the body with the surrounding fluid medium (air or water)	Cool medium such as a cool or cold day, a cool stadium (ice rink), or coolpool water (as in water polo, distance swimming)	Warm days or training or competing in a warm facility (summer football camp, baseball, soccer)
Conduction	Transfer of heat to/from the body by contact with an object or surface	Contact with a surface or object that has a temperature the same as or cooler than skin (baseball bats, racquets, football in colder environment)	Contact with a surface or object that has a temperature the same temperature as or warmer than skin (not significant in most sports)
Radiation	Transfer of energy between the body and other objects via electromagnetic radiation	Sports played in warm climates and on sunny days (baseball, football, soccer, lacrosse, field hockey)	Sports played on cool or cold days or in a cool facility (skating, ice hockey, winter football)
Evaporation (sweat)	Transfer of heat energy from the body to surrounding air by heating sweat water to its vapor point	Warm environments. As environmental temperature increases the effectiveness of the other mechanisms decreases	Cool environments (air and water), primarily due to increased heat loss by the other mechanisms (running in the winter, ice skating)

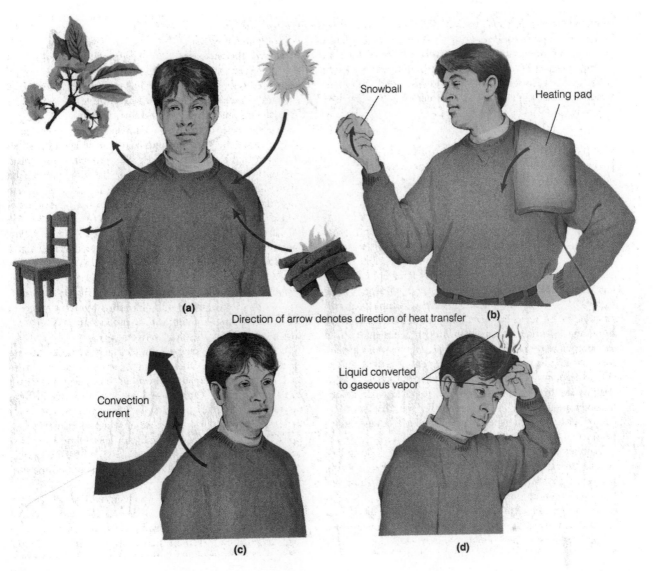

Direction of arrow denotes direction of heat transfer

Figure 8-7
Mechanisms of heat transfer. (a) Radiation—the transfer of heat energy from a warmer object to a cooler one via electromagnetic waves ("heat waves"), which travel through space. (b) Conduction—the transfer of heat from a cooler object that is in direct contact with the warmer one. The heat is transferred through the movement of thermal energy between adjacent molecules. (c) Convection—the transfer of heat energy by air currents. Cool air warmed by the body raises and is replaced by more cool air. This process is enhanced by the forced movement of air across the body sweating.

irregularities that can disrupt homeostasis. During increased metabolic heat generation, such as during exercise, sweat production increases. Furthermore, as increases occur in environmental temperature and/or **humidity** (the level of moisture in the air), the capabilities of the other heat-releasing mechanisms decrease and sweating has to increase to compensate (Figure 8-7).

SWEAT GLANDS.   **Sweat glands** are found in skin tissue covering the entire body. As shown in Figure 8-8, each sweat gland consists of a deep subdermal coiled portion and a duct that extends to the skin surface. These glands are stimulated by the sympathetic nervous system, and the neurotransmitter applied is *acetylcholine*. During rest, sweat gland stimulation is largely under the control of the hypothalamus and dictated by core body temperature. As body core temperature increases on a warm and sunny day, stimulation of sweat glands increases. During exercise, additional stimulation of sweat glands comes from circulating *epinephrine* and *norepinephrine*.

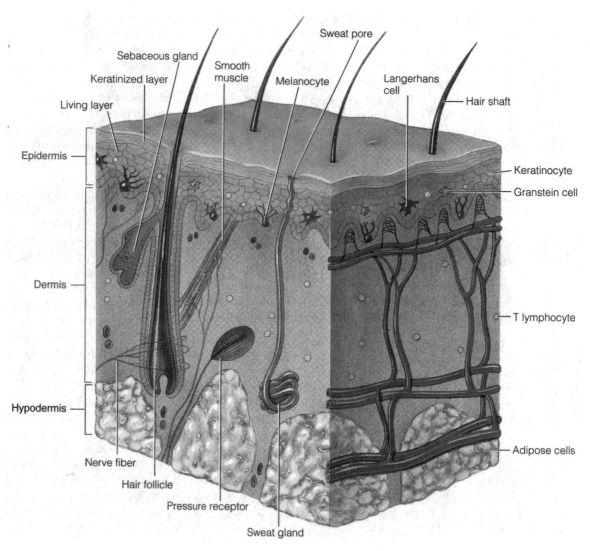

Figure 8-8
Anatomy of the skin. The skin consists of two layers, a keratinized outer epidermis and a richly vascularized inner, connective tissue dermis. Special infoldings of the epidermis form the sweat glands, sebaceous glands, and hair follicles.

SWEAT COMPOSITION. When the coiled subdermal cells are stimulated they secrete a **primary sweat** solution that is similar in electrolyte composition to plasma. However, primary sweat contains relatively small amounts of the other plasma solutes such as protein. Primary sweat flows along the length of the sweat gland tubule toward the skin surface. The rate of flow through the tubule is dictated by the rate of primary sweat production, which itself is dictated by the degree of stimulation.

As the primary sweat flows through the tubule its composition is modified. **Reabsorption** of sodium and chloride by the cells lining the tubule depends on the flow rate. If the stimulation is weak, resulting in a slower flow rate, nearly all of the sodium and chloride are reabsorbed. Aiding in the reabsorption of water is the osmotic force developed by sodium and chloride uptake into tubule cells. As water is reabsorbed, the remaining solutes in sweat become more concentrated. These solutes include lactic acid, urea, and potassium.

If sweat flow through the tubule is rapid, then the reabsorption of sodium, chloride, and water is reduced and more reaches the skin surface. For example, when a copious amount of the primary sweat is produced, only about half of the sodium and chloride along with minimal

Key Point

Sweat rate is a principal determinant of sweat composition. During heavy sweating the concentration of electrolytes is greater than when sweating is milder.

amounts of water are reabsorbed. This results in a final sweat concentration of sodium and chloride that is roughly half that found in the plasma. This is noticeable on a warm, dry day as evaporated sweat water leaves behind sodium and chloride caked on the skin.

Urine

Another obvious role of body water is to serve as the basis of urine. Urine is a composite of water, electrolytes, urea, creatinine, and trace amounts of glucose, amino acids, and proteins. During rest the kidneys receive approximately 20–22% of the cardiac output from the left ventricle, about 1 L of blood per minute. Efficient urine formation is crucial in removing nitrogen waste and excessive amounts of electrolytes and certain other substances from the body. Urine production decreases during exercise and with dehydration. But it will not cease, as waste materials still must be removed.

Millions of structures called *nephrons* in the kidneys process the blood to produce urine. Figure 8-9 shows a schematic diagram of one nephron. The arteriole serving the nephron forms a ball of capillaries called the *glomerulus*. The high capillary pressure (hydrostatic pressure) within the glomerulus forces plasma-derived fluid (filtrate) into *Bowman's capsule*, which encapsulates the glomerulus at the beginning of the nephron. Bowman's capsule then drains into the first segment of the renal tubule system, the *proximal convoluted tubule* (*PCT*) (Figure 8-9). Much like the cells lining the small intestine, the cells of the PCT are designed for efficient absorption and are densely packed with mitochondria to power these operations. Glucose and amino acids are efficiently reabsorbed in the PCT. More than half the electrolytes and about two-thirds of the water are also reabsorbed in the PCT. Reabsorption of the remaining water and sodium in the later segments of the renal tubule system are strongly influenced by hormones that regulate blood pressure.

WATER REABSORPTION. Water reabsorption occurs osmotically throughout the renal tubule system as dictated by the reabsorption of the solutes. Water moves through the spaces (tight junctions) between the cells that line the tubule as well as through aquaporins in the plasma membrane. As fluid moves along the tubule into the ensuing *loop of Henle*, *distal tubule*, and *collecting duct*, the tight junctions become tighter. This decreases the opportunity for water to move between these cells. Therefore in the latter segments of the tubule system, water reabsorption is more reliant on the presence of aquaporins in the plasma membranes. Hormones govern the quantity of aquaporins, as discussed next.

Blood Pressure

Because water is the predominant component of the blood, it is fundamentally involved in establishing blood volume. Blood volume then influences blood pressure. Several systems regulate plasma volume. Antidiuretic hormone and aldosterone are two of the most influential hormones involved.

ANTIDIURETIC HORMONE. **Antidiuretic hormone (ADH),** also known as *vasopressin*, regulates the reabsorption of water in the latter segments of the nephron tubule system. As presented in Figure 8-10, the level of circulating ADH becomes elevated when the solute concentration, or **osmolality,** of the extracellular fluid increases. This, in turn, is sensed by osmoreceptors in the hypothalamus, which stimulate the release of ADH as well as initiating **thirst.** ADH promotes the presence of more aquaporins in the later segments of the tubule system, allowing more water to be reabsorbed. This helps preserve blood volume and pressure. Conversely, when ADH is not present the later segments of the tubule remain relatively impermeable to water. In this situation the final urine becomes more dilute. Thirst, on the other hand, promotes water consumption, which dilutes the extracellular fluid and thereby decreases its omolality.

ALDOSTERONE. **Aldosterone** promotes sodium reabsorption in the later segments of the nephron tubule system. Preservation of plasma sodium levels helps stabilize blood volume by osmotically attracting water. The major stimulus for aldosterone release is complex and involves the **renin-angiotensin system** (Figure 8-11). Briefly, when blood pressure or flow is reduced through the kidneys, renin is released into circulation. Renin is an enzyme that converts the plasma protein *angiotensinogen* to *angiotensin I*. As angiotensin I circulates it is converted to angiotensin II by *angiotensin converting enzymes* (ACEs) in tissue such as the lungs. Angiotensin II then stimulates the release of aldosterone and also promotes **vasoconstriction** (the constriction of blood vessels, which raises blood pressure). Combined, the actions of aldosterone on plasma volume and vasoconstriction increase blood pressure.

Key Point

Urine production decreases during exercise and dehydration.

Key Point

Water is important in maintaining blood volume and thus pressure. Antidiuretic hormone (ADH) and aldosterone are two of the most influential hormones involved in preserving blood water levels.

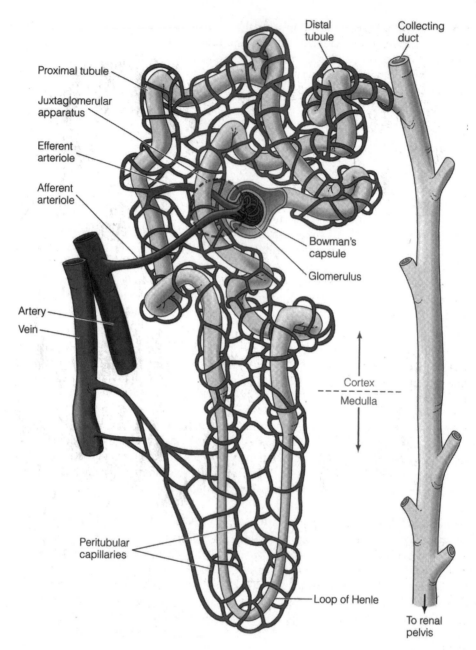

Figure 8-9
A schematic representation of the major components of the nephron.

■ ROLES OF BODY WATER *IN REVIEW*

- Sweating is an important mechanism for removing excessive heat from the body, and sweat composition is dictated by sweat rate.
- Water is a key factor in urine formation to excrete nitrogen and other substances.
- Regulation of plasma water by aldosterone and ADH is the focus of some processes that control blood pressure.

■ WATER BALANCE

If water is the basis for urine and sweat, how much water do we lose by these routes throughout a day? Do other mechanisms also result in water loss? How do we balance water losses to prevent dehydration? People experience significant water flux in and out of the body on a daily basis. For instance, an adult may lose the equivalent of 4% of body mass as water in a typical day, more if involved in heavy physical activity. To prevent dehydration, water intake and generation must balance daily water loss.

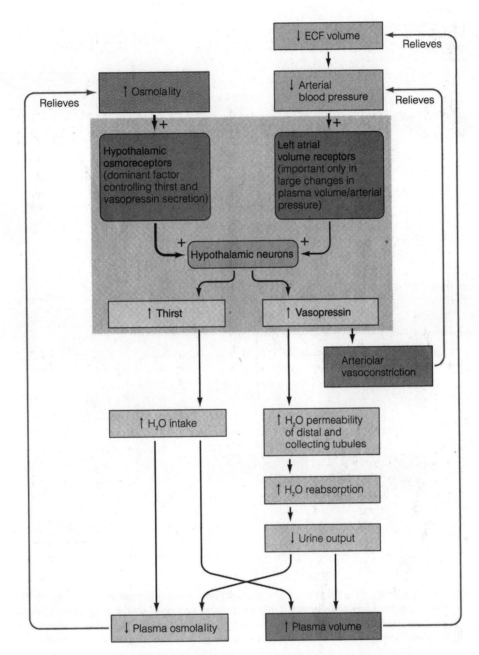

Figure 8-10
Control of increased antidiuretic hormone (vasopressin) secretion and thirst during an H_2O deficit.

Humans do not have cellular stores of water, so when a reduction in the water content of extracellular fluid draws water from intracellular fluid, it will have physiological consequences.

Water Intake and Metabolic Generation

Water loss must be balanced by water provided to the body in order to prevent dehydration. How is water provided to the body? Must we derive all of it from the fluids we drink? Not necessarily. Water is brought into the body by ingesting both fluids and solid foods, and some water is generated within cells during metabolic processes. For a typical adult about two-thirds of the water entering the body is as water and other water-based fluids. The remaining one-third comes by way of the moisture in foods. As displayed in Table 8-2, fruits and vegetables rank among the best food sources of water, as many of these foods are greater than 70% water by mass. The feature In Focus: Water Balance of an Athlete at the end of this chapter presents water balance of a runner and a nonrunner.

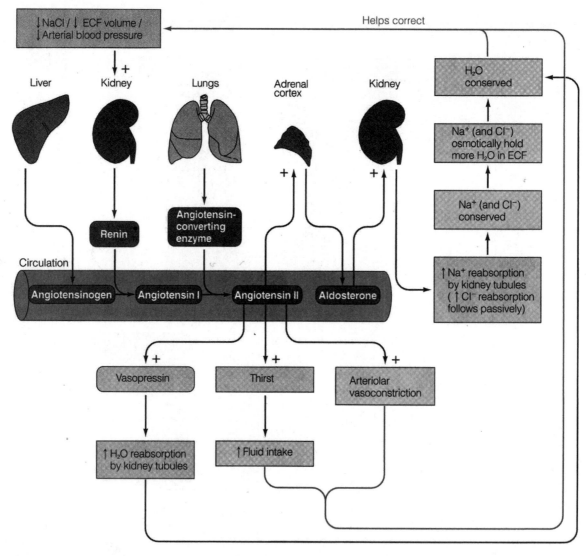

Figure 8-11

Renin-angiotensin-aldosterone system. The kidneys secrete the hormone renin in response to a reducton in NaCl/ECF volume/ arterial blood pressure. Renin activates angiotensinogen, a plasma protein produced by the liver, into angiotensin I. Angiotensin I is converted into angiotensin II by angiotensin-converting enzyme produced in the lungs. Angiotensin II stimulates the adrenal cortex to secrete the hormone aldosterone, which stimulates Na^+ reabsorption by the kidneys. The resultant retention of Na^+ exerts an osmotic effect that holds more H_2O in ECF. Together the conserved Na^+ and H_2O help correct the original stimuli that activated this renin-angiostensin-aldosterone system. Angiostensin II also exerts other effects that help rectify the original stimuli.

Water is a product of the combustion of the various energy nutrients. Combined, these processes generate approximately 300–400 ml of water daily. The amount of water generated metabolically is relative to the amount of energy expended. Therefore as activity increases so does water production in cells. Water production depends not only on how much energy is expended but also on what cells use for fuel. For instance, the complete oxidation of 1 mole of glucose and 1 mole of palmitic acid (a saturated fatty acid) yields 6 and 16 moles of water, respectively. This is demonstrated in the balanced equations below. These equations were also presented in Chapters 4 and 7

with regarding to energy metabolism during exercise and for weight control.

$$\text{Glucose } C_6H_{12}O_6 + 6\,O_2 \rightarrow 6\,CO_2 + 6\,\mathbf{H_2O}$$

$$\text{Palmitic acid } C_{16}H_{32}O_2 + 23\,O_2 \rightarrow 16\,CO_2 + 16\,\mathbf{H_2O}$$

Water Loss

Water loss from the body occurs by several routes. The most significant are sweat, urine, feces, and exhalation of air humidified in the lungs. The total water loss for an adult is typically 2–3 L/day. Water loss by exhalation of

Table 8-2

Water Content of Various Foods[a]

Food	Approximate Water %
Tomato	95
Lettuce	95
Cabbage	92
Beer	90
Orange	87
Apple juice	87
Milk	87
Potato	78
Banana	75
Chicken	70
Bread, white	35
Jam	28
Honey	20
Butter	16
Rice	12
Shortening	0

[a]Approximate % of total weight

water vapor is usually about 200–400 ml/day, but arid environments and increased breathing volume during physical activity increase this amount. Water loss through fecal elimination is usually about 100 ml/day, depending on one's intake of undigested and absorbed materials such as plant fibers. (Plant fibers can bind water and thus increase fecal moisture and volume.)

Water losses in urine can be $1-1\frac{1}{2}$ L/day for an adult—more if a person drinks lots of water and water-based fluids. Conversely, the urinary loss may be lowered if a person drinks less fluid or if water loss by other means increases. Under any circumstances, however, the average adult must excrete 400–600 ml/day of urine each day simply to remove nitrogenous waste, such as urea.

Sweating makes a significant contribution to water loss even for inactive people. In a comfortable environment, sweating is minimal and typically goes unnoticed. Under these circumstances an adult can lose approximately 500 ml of water each day through sweat. As core body temperature increases in warmer environments or during physical activity, sweat production increases and can exceed 2–3 L/hour for some highly trained athletes.

Estimating Water Needs

General recommendations for water consumption are 2–3 L daily for adults to balance losses. Water requirements vary among individuals; estimates of individual water requirements can be derived from energy expenditure. For instance, 1.0 and 1.5 ml of water are recommended per kilocalorie expended for inactive adults and children,

respectively. More active adults may need 1.5 ml/kcal to cover increased water losses due to sweating. Still, these recommendations are fairly general and the sport and climate influence water requirements for individual athletes.

Thirst is the perceived need for water, is a symptom of dehydration, and is often cited as the gauge for water need. The hypothalamus serves as the control center for thirst, and there is a direct relationship between plasma osmolality and the intensity of thirst. This means that an individual must be slightly dehydrated in order for thirst to be evoked. For an average person this may not be a concern, but for a competitive athlete it may be the difference between optimal and decreased performance. In addition, the thirst mechanism may be blunted in vigorous activity and should not be used alone to gauge hydration status by athletes, as discussed below.

Key Point

Thirst is a symptom of dehydration, and because physical performance may be reduced at this point, athletes should not wait for thirst as a signal to drink fluids.

Physiological Impact of Hydration Imbalance

Most people realize that dehydration can be problematic. At what level of dehydration do negative changes begin to occur? Can **overhydration,** the opposite of dehydration, also influence physiological function? As shown in Table 8-3, water loss approximating 1–2% of body weight can result in

Table 8-3

Effects of Body Water Loss on Physiological Performance

% Body Weight Loss as Water	Physiological Effect
1–2%	Thirst, some fatigue and minor reductions in strength
3–4%	Reduction in maximal aerobic power and endurance, increased rate of overheating due to plasma volume reductions, compromised thermoregulation
5–6%	Decreased concentration and focus, headache, increased breathing, reduction in regulation of thermoneutrality, decreased cardiac output, chills, nausea, rapid pulse
7–10%	Dizziness, muscle spasms, poor balance, delirium, exhaustion, collapse, progressive reductions in plasma volume, potential cardiogenic shock

reduced strength and weakness. Further dehydration leads to more significant deficits in strength and physical performance. In addition, breathing and heart rate can increase and an individual becomes more prone to dizziness and muscle cramps. Heat intolerance and exhaustion are both associated with body water losses of 5–7%. As even more body water is lost, blood volume can decrease, rendering an individual at risk for cardiogenic shock, discussed in more detail below. Coma and possibly death may soon follow.

HYDRATION STATUS AND METABOLISM.

Hydration status can influence the energy metabolism within certain cells. When researchers perform experiments to overhydrate the body, the swelling of skeletal muscle fibers seems to induce protein synthesis and might also lead to a general inhibition of glycogenolysis, glycolysis, and protein breakdown and to enhanced fat breakdown and utilization as fuel.[2,3] Inhibition of protein and glycogen breakdown during overhydration can serve as a means of limiting the amount of osmotically active molecules in the intracellular fluid, namely amino acids, glucose-6-phosphate, and glucose. At the same time, cells pump electrolytes out into the extracellular fluid. These efforts attempt to limit cell swelling to prevent and/or limit damage to cellular structures. To study the effects of overhydration, researchers have used chemicals similar to ADH (such as desmopressin), then intravenously infused a **hypotonic** saline solution (a solution with a greater percentage of water than in blood) and had the participants liberally ingest water.[2,3] This may affect the validity of the research, as it raises questions about real-world application.

In the opposite case, dehydration, it has been theorized that a decrease in cell volume promotes the breakdown of glycogen and protein in some cells such as skeletal muscle fibers.[2,3] The amino acids and glucose-6-phosphate liberated by the breakdown of proteins and glycogen would then draw water into the cell in an attempt to restore cell volume. It follows that cellular dehydration can be considered a catabolic influence. Whether the dehydration typical of some sports such as distance running has a negative influence on energy metabolism during exercise and on glycogen recovery and protein turnover during the hours that follow the exercise remains to be determined.

■ WATER BALANCE *IN REVIEW*

- Body water status reflects the algebraic sum of water losses and provision.
- Water loss routes include sweating, breathing, fecal loss, and urine.
- Water is available from fluids and the moisture in foods as well as the water produced when energy nutrients are oxidized to form ATP.
- Cell hydration status may influence the metabolism of carbohydrate and protein.

■ EXERCISE AND BODY WATER DISTRIBUTION

Exercise results in many physiological events that can influence the distribution of water within the body. These include both physical and physiological changes that influence hydrostatic and osmotic pressures. In addition, water loss from the body increases during exercise as the result of increased sweating and breathing volume. This section provides an overview of the most significant events that influence water flux across cell membranes and between tissue and the blood and affect water loss from the body.

Body Water Shifts During Exercise

Can exercise result in shifts in body water between different compartments? If so, what factors are involved? It does seem that total blood volume can decrease during the initial phase of endurance exercise performed at a moderate submaximal intensity or higher.[4–7] Reduction in blood volume reflects a flux of water from the plasma into the interstitial and intracellular fluids in active skeletal muscle. Logically, the net movement of water out of the plasma is related to the amount of skeletal muscle that is active during exercise. Fluid flux from the plasma into muscle tissue also occurs during strength and power exercise bouts (such as weight training). This leads to the transient swelling of muscle tissue referred to as the "pump."

Both hydrostatic and osmotic forces contribute to fluid shifts during exercise. Early in exercise, increased systemic blood pressure and local muscle contractions drive plasma fluid into working muscle tissue via hydrostatic force. The osmolality of interstitial and intracellular fluids within active muscle tissue increases during exercise. This is largely due to increasing concentrations of glucose and glycolysis intermediates, hydrogen ions (H^+), and lactate. Also, heat dissipation by sweating can lead to a continued reduction in blood volume during sustained exercise. Combined, these factors tend to reduce plasma volume early in exercise.

Key Point

Shifts of fluid into skeletal muscle during higher-intensity training (such as weight training) result in the "pump" fondly discussed by many people.

Stabilization of Plasma Volume

If water shifts from the plasma into tissue during exercise, what impact can it have on an individual's performance? Reductions in plasma volume must be stabilized in order

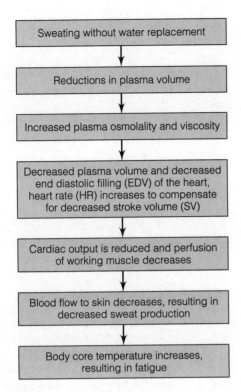

Figure 8-12
Physiological effects of reduced plasma volume.

to avoid detrimental reductions in cardiovascular efficiency. As shown in Figure 8-12, both blood pressure and cardiac output eventually are lowered. Meanwhile, blood viscosity tends to increase, as its solutes are progressively concentrated. This is referred to as **hemoconcentration.** The combination of these factors compromises the delivery of blood to tissue as well as the ability of the body to dissipate heat. Therefore, a stabilization of plasma volume during exercise is crucial to continuation of effectual performance. Partial to full recovery of plasma volume, via fluid consumption, would also reduce cardiac burden at a given exercise level.

Plasma volume tends to decrease during the early phase of exercise until a new balance between the blood and interstitial and intracellular fluids is established. As plasma water is forced into tissue, the osmolality of the blood increases and exerts a counterbalancing pull. Also, as the volume of the interstitial fluid compartment increases, so does the hydrostatic pressure it exerts. This is an impeding force against further fluid flux from the blood. Thus plasma volume stabilizes as a new balance develops between all opposing hydrostatic and osmotic forces.

Hormones that affect water and sodium loss (such as aldosterone and ADH) in the urine are also involved in minimizing the reduction in plasma volume. Because both aldosterone and ADH levels tend to increase relative to

the level of exercise, urine production is reduced. Still, fluid consumption is very important in stabilizing and perhaps promoting the recovery of plasma volume during sustained exercise.

CLIMATE INFLUENCES ON STABILIZATION OF PLASMA VOLUME DURING EXERCISE. Climate is important in influencing the stabilization of plasma volume during exercise. For instance, reduced plasma volume levels tend to stabilize during endurance exercise in temperate or cool climates. This was demonstrated in a study involving female and male runners who ran a marathon in mild conditions (17.5–20.4°C).[5] Figure 8-13 shows the results. At the 6-km point of the marathon, the female and male runners experienced a reduction in plasma volume of 8.5% and 6.5%, respectively, even though their body weight was reduced <1%.[5] However, although the participants lost more weight during the remainder of the race, their plasma volume stabilized and was comparable to the 6-km assessment. Researchers reported similar results with cyclists working at 50% of maximal work rate in a temperate (22°C) and cool environment (14°C).[4] In contrast, when the participants cycled in a hot environment (36°C), their plasma volume did not stabilize, due to the increased degree of sweating. Therefore, stabilization of plasma volume is hindered in a hot environment, when

Key Point

Body water can shift from the plasma into tissue and working muscle fibers during exercise. However, plasma volume is stabilized if sweat water losses are balanced by fluid consumption.

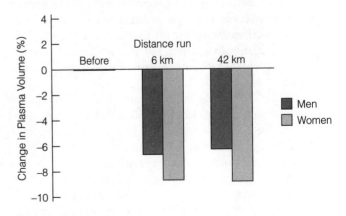

Figure 8-13
Plasma volume decreases in the early phase of an endurance exercise bout and then stabilizes for a marathon run in mild weather (17.5–20.4°C). (*Source*: Based on data from reference 5.)

sweating is heavy, without appropriate compensation by water consumption.

POSTURAL EFFECTS.   Changes in one's posture can influence how water shifts between the different compartments. To demonstrate this, researchers assessed fluid shifts in eight males exposed to three postural positions—lying down, seated, or standing.[7] Compared to the seated position, plasma volume increased when they were lying down and decreased while standing. Therefore the potential effect of posture needs to be considered when assessing body fluid distribution and changes during exercise. The posture for baseline measurements should be appropriate for the type of exercise. For instance, cyclists should be assessed in the cycling position and runners in the standing position at rest prior to an exercise trial.

Adaptations in Blood Volume

If fluid shifts from the plasma are potentially problematic, does exercise training lead to adaptations in blood volume? The answer is not clear and may be conditional. Some studies have shown that endurance training does lead to an expansion of blood volume, regardless of an individual's age.[8,9] The increased blood volume has been noted to be 5–20% above pretraining levels.[10] Furthermore, as intensity increases, less time is needed during training bouts to elicit such a change.[9,11] However, not all research has demonstrated a direct relationship between exercise training and an expansion in blood volume.[12]

Some researchers have proposed that some individual variations in blood volume change may be related to dietary sodium levels during training.[13] That is, individuals who consumed more sodium during a training program tended to experience a greater degree of plasma expansion. This might be because a higher sodium intake allows for a quicker and/or better recovery of the sodium lost in sweat. This in turn would improve rehydration efforts and perhaps allow for expansion of the plasma. Further investigation is needed, but this does emphasize the importance of controlling nutrition factors in exercise studies.

Keep in mind that an expansion of blood volume involves an increase in either plasma volume or red blood cells, or both. If the expansion in blood volume were caused by an increase in plasma volume alone, then a reduction in hematocrit would be expected. Or if there were an increase in the number of RBCs in the absence of changes in plasma volume, then an increase in hematocrit would be expected. Therefore any assessment of changes in blood volume needs to consider all blood components. Using dyes and labeled RBCs, as discussed earlier, could be helpful in identifying changes in both plasma volume and the number of RBCs when assessing changes in blood volume.

■ EXERCISE AND BODY WATER DISTRIBUTION *IN REVIEW*

- Both hydrostatic and osmotic factors lead to a shift in plasma water into working muscle tissue during exercise.
- Plasma volume can decrease and then stabilize during endurance exercise and repeated strength exercises (such as weightlifting).
- Excessive sweating can hinder the stabilization of plasma volume during exercise if sweat volume is not compensated for by fluid consumption.
- Blood volume may expand as a result of endurance training.

■ EXERCISE-INDUCED SWEATING

The breakdown of energy nutrients to fuel muscle contraction generates excessive heat. Depending on the type, intensity, and duration of exercise, energy expenditure can be greatly increased. For example, a runner may expend five times more energy during a run than when resting. Knowing this, how important is sweating in the dissipation of heat during physical activity? In a word: extremely! To dissipate this substantial energy load, sweating becomes the primary mechanism for releasing heat during exercise, at least for sports on land. Increased sympathetic stimulation and circulating epinephrine are largely responsible for stimulating the higher sweat rates during exercise.

Maximal Sweat Rates

Sweat rates vary among active people and athletes. Some of the most important nonhereditary factors that determine maximal sweat rates include age, sport, climate, acclimatization, hydration status, and perhaps body composition. A common estimate of sweat during athletic competition is 1–2 L/hour, but a rate of 2–3 L/hour or more would not be uncommon for others, say for a tennis player during the first hour of competition on a hot day. Also, it is logical that larger individuals would have higher maximal sweat rates during a given exercise task than smaller individuals.

On a hot and sunny day, sweating must compensate for reductions in other mechanisms of heat release. However, higher humidity levels make it more difficult for sweat to evaporate. This decreases the efficiency of sweating as a

Key Point

Athletes can sweat a few liters per hour during competition on a hot day.

mechanism to dissipate heat. Therefore the potential for heat-related complications during exercise on a hot day increases as humidity increases.

SWEATING OF CHILDREN.   Children have a greater ratio of surface area to body mass than adults.[14] For instance, the ratio for an 8-year-old boy may be 40% greater than for a young man. This allows children to rely more on nonsweating mechanisms of heat loss, namely convection, conduction, and radiation in a warmer environment. During exercise children tend to have a lower sweat rate than adults; not until late puberty are they able to match adults in sweating capabilities.[14] Therefore the opportunity for heat intolerance is greater for children participating in sports on a hot day than for adults.

The lower sweat rate of children is due to a lower production of sweat per gland, not because of fewer glands. Sweat gland numbers are established by 2–3 years of age, so children have more sweat glands per surface area than young adults. However, these glands are smaller and less sensitive to thermal stress. In addition, the sweat of children tends to be more concentrated with electrolytes. This increases the potential for electrolyte imbalances for children during exercise.

Key Point

Children tend to have a lower sweat rate than adults, thereby increasing their potential for heat-related complications during physical activity in the heat.

CLOTHING AND SWEATING.   Uniforms and protective gear can have a significant impact on sweat rate. Darker colors such as black and dark blue absorb more radiant energy and increase the reliance on sweating to dissipate the heat. This makes the light blue and white colors of the Miami Dolphins better for playing on hot, sunny days than the black uniforms of the Oakland Raiders. Meanwhile, helmets and protective padding consisting of foam and other materials can restrict heat loss by convection and radiation. Once again, this increases the reliance on sweating to dissipate heat. This is particularly relevant to football and ice hockey players. Other sports such as lacrosse also require helmets and padding. However, they differ in that a greater percentage of the body is either directly exposed to the environment or clad in unpadded and "breathable" materials that are more favorable for heat release.

Adaptation in Sweat Rate

Is there a maximal sweat rate that a person can achieve during exercise? If so, can he or she experience a training-induced adaptation that allows even more sweating? In response to regular endurance exercise, sweat rate is increased during exercise relative to the level of exertion and climate. Because the number of sweat glands in a given area of skin is assumed to remain constant, the existing sweat glands must adapt to become "better sweaters." This occurs by increasing the production rate of primary sweat solution. In addition, sweat glands adapt to reabsorb more sodium and chloride from sweat during times of copious sweat production. This helps to protect an individual from losing too much sodium and potassium. Thus the net effect of the sweating adaptations is a greater production (volume) of a less concentrated or dilute sweat solution. This allows for greater heat dissipation and conservation of electrolytes but increases the potential for dehydration if sweat water loss is not compensated for by water intake.

SWEAT RATES OF LAND VERSUS WATER ATHLETES. In sports that take place in water, such as swimming and water polo, there is an increased opportunity for heat loss by convection. This is because heat is more easily dissipated into water than air. The rate of heat loss by convection when immersed in water depends on the temperature difference between the person and the water. The water temperature for indoor water sports tends to be at least 15–20°F below body temperature. Outdoor pools, lakes, rivers, and the ocean can vary more in temperature. Researchers have determined that sweat rates for water athletes are lower than for land athletes.[15,16] Therefore, water athletes may not experience the same degree of sweating adaptation as land athletes because they are less reliant on sweating for heat release during sport training at the same workload.

Key Point

Land-sport athletes, such as runners, may experience greater adaptations in sweating than water-sport athletes such as swimmers and water-polo players.

Estimation of Sweat Volume and Rate

How are sweat volume and rate estimated in order to plan for appropriate fluid consumption during exercise? Sweat volume and rate can be estimated by direct sampling using a container to capture sweat. The volume of sweat captured can then be extrapolated to total body surface area. However, sweat rate is not the same throughout the skin surface of the body. Therefore multiple sites of sweat collection are necessary and the volumes are pooled and averaged to increase the precision of the estimate.

A simpler method to estimate sweat losses involves comparing an athlete's body weight before and after exercise. Assuming that the weight loss during exercise is attributable to sweating, the weight loss is converted to

Key Point

The difference in body weight before and after exercise can be used as an estimate of sweat and total body water loss during exercise.

water volume (1 kg of water = 1 L of water). Sweat volume can then be divided by the exercise time to estimate sweat rate (ml/min or L/hour). Body weight change tends to overestimate sweat volume because some weight loss is attributable to water losses in breath. Also, reduced energy stores and electrolyte loss can make a minor contribution to weight loss during sustained exercise. However, sweating accounts for the majority of weight loss during exercise. The accompanying feature Nutrition in Practice: Estimating Sweat Loss provides some simple steps for

estimating sweat loss based on exercise-induced weight change.

■ EXERCISE-INDUCED SWEATING IN REVIEW

- Land athletes sweat more than water athletes, and children have lower relative sweat volumes and more concentrated sweat than adults.
- Exercise training can improve sweating efficiency by increasing sweat rate and decreasing electrolyte content.
- Sweat volume and rate can be estimated by comparing body weight before and after exercise, with certain considerations.

NUTRITION IN PRACTICE   **Estimating Sweat Loss**

An athlete can estimate sweat volume and rate by measuring the difference between body weight just prior to and after exercise. To do so, certain steps should be taken that increase the accuracy of the estimation.

1. *Preparation:* Note the time of day and season and the elapsed time and intensity of exercise. Also, note the temperature and humidity prior to and after the exercise bout. Fluids should be stored (sealed) in volumetric containers. (For bouts of less than an hour, pure water can be used if desired. For longer bouts a beverage that is 6–8% carbohydrate should be considered.) Use whatever will be used during competition. Empty volumetric containers need to be available to collect urine voided during and after the bout of exercise.

2. *Pre-exercise:* The athlete needs to empty the bladder and wipe the body with a towel to remove any pre-exercise sweat. Body weight should be determined on a calibrated scale. The athlete should be naked, except for a dry towel or shorts. Because the workout clothing will absorb sweat, which would change its weight postexercise, it should not be worn for either of the weighings.

3. *Postexercise:* Immediately after completing the exercise bout, the athlete needs to wipe the body of sweat, urinate into a volumetric container, and then weigh in wearing the towel or clothing used for the first weighing.

Your estimated sweat rate can be calculated as follows:

Total estimated sweat loss = $(A - B) + (C + D)$

where:

A = weight before exercise
B = weight after exercise
C = water consumed during exercise (accounting for weight contribution of carbohydrate and other ingredients)
D = water urinated during and after exercise

Sweat rate (L/hr or ml/min) = $\dfrac{\text{total estimated sweat loss}}{\text{elapsed time of exercise}}$

(This method does not account for breathing water losses or water generated during energy metabolism.)

Example

Jim is a soccer player. His body weight was 185 lb before and 182 lb after 2 hours of practice. Also, during practice he drank the equivalent of 1 L of water as sport drink and urinated only 100 ml after practice.

A = 185 lb (84.1 kg)
B = 182 lb (82.7 kg)
C = 1 L of H_2O = 1 kg (1000 g or ml)
D = 100 ml = 0.1 kg

Total estimated sweat loss:
$(84.1 \text{ kg} - 82.7 \text{ kg}) + (1 \text{ kg} + 0.1 \text{ kg}) = 2.5 \text{ kg} (2400 \text{ ml})$

Estimated sweat rate:
2.5 L/2 hr = 1.25 L/hr = 0.3 L/15 min

To balance sweat water, Jim can attempt to drink 300 ml every 15 minutes, which is about 10 oz (1¼ cups).

■ HYDRATION STATUS AND PERFORMANCE

Decreased body water status (dehydration) is one of the most significant nutritional factors that can reduce physical performance. How does dehydration have such an effect? Also, how can an athlete gauge hydration status during and between training sessions? Athletes need to understand the importance of achieving optimal hydration status prior to, as well as during physical activity. All too often endurance athletes and athletes participating in sports associated with heavy sweating, such as soccer, tennis, and football, experience at least some dehydration during training and competition. Other athletes such as wrestlers, body builders, and fitness competitors sometimes deliberately restrict their fluid consumption prior to competition to fit into a weight category or to enhance aesthetic presentation. However, voluntary reduction in body water status may cause health problems and is generally discouraged.[17]

Hydration Status Assessment

What are the most common methods used by athletes to estimate their hydration status? Thirst perception and evaluation of the volume and character of the urine are perhaps the most common. But neither is perfect and both raise concerns. For instance, thirst is a symptom of dehydration and may not be perceived until body weight has been reduced by 1% due to water loss. Thus, thirst reflects an existing level of dehydration, and as discussed below, even subtle reductions in hydration status can negatively influence athletic performance. In addition, individuals vary in their sensitivity to hydration status and thirst perception, and the thirst mechanism may be blunted in some athletes during strenuous exercise. So thirst should not be used as a gauge for fluid consumption during exercise.

During recovery periods, thirst can subside prior to complete rehydration, especially if water is consumed by itself following exercise that produces heavy sweat loss. Without recovery of electrolytes (sodium and chloride), the osmolality of the extracellular fluid decreases with water consumption. This can alleviate thirst prior to complete rehydration. Therefore, water consumption is encouraged for several hours following exercise, in conjunction with food or beverages that provide electrolytes.

Urine volume and character is a reasonable estimator of hydration status. Urine volume that matches fluid intake is a good sign. However, when exercise results in a significant loss of body water, it can take hours for water loss to be recovered and completely equilibrate throughout the body. During this time urine production may match water intake even though complete rehydration has not been achieved. Researchers have also noted that urine can be clear and odorless, suggestive of complete recovery of water loss, despite incomplete rehydration.[18]

Key Point

Thirst and urine volume and characteristics are often used to gauge hydration status, but they are not perfect gauges.

Exercise-Induced Dehydration

Body water loss is rapid during exercise in the heat and is often not matched by an athlete's fluid consumption. As body water is lost, athletic performance becomes compromised in a relative manner. What physiological operations are affected by dehydration, thereby influencing physical performance? In a dehydrated state, less water is available to absorb and transport heat from muscle and the body core. As a consequence, body core temperature increases and performance is hindered. Some research suggests that hyperthermia, rather than altered energy metabolism, is the main factor underlying the early fatigue associated with dehydration during prolonged exercise in the heat.[18,19] The recorded temperatures associated with fatigue were 104.3°F (esophagus), 105.4°F (muscle), and 98.8°F (skin).[19]

Excessive sweating can also result in electrolyte disturbances. Sodium levels in sweat vary between individuals, based on sweat rate and level of adaptation, but might approximate 50 mmol/L, or about 1 g of sodium per liter of sweat. **Hyponatremia** is a state in which the concentration of sodium in the blood is below 130 mmol/L. Disturbances in the proper functioning of the nervous system and muscle typically result at this lowered concentration of sodium in the extracellular fluid. Hyponatremia and electrolyte disturbances resulting from sweating are discussed in greater detail in Chapter 10.

CARDIOVASCULAR CONSEQUENCES OF DEHYDRATION. Fluid flux from the plasma during the early phase of endurance exercise is important to support efficient sweating. If sweating is mild or if fluid consumption is adequate, plasma volume can be stabilized during exercise. However, when sweating is heavy and is uncompensated by fluid consumption, plasma volume continues to decrease. As a consequence, less blood is returned to the heart, and in turn, stroke volume is reduced. In an attempt to maintain cardiac output, heart rate is accelerated. As shown in Figure 8-12, when plasma volume is further reduced, heart rate maximizes and cardiac output peaks. Cardiac output begins to decrease with further reductions in plasma volume.

Reduced cardiac output decreases athletic performance by delivering less blood and nutrients to working muscle. Also, when blood flow to working muscle tissue is decreased, there is a reduced ability to transport heat and lactic acid away from the muscle fibers. Extreme reductions in plasma volume threaten survival as cardiac output becomes too low to circulate minimal requirements of blood to vital organs. This condition, called **cardiogenic shock,** prevents organs from performing basic homeostatic operations. Rarely would an athlete still be engaged in training or competition at this point.

Key Point

Dehydration can impair athletic performance by decreasing heat tolerance and reducing cardiac output to working muscle.

METABOLIC CONSEQUENCES OF DEHYDRATION. Dehydration can reduce blood flow to working muscle and influence substrate utilization. This may lead to an increased use of carbohydrate and reduced use of fat by working muscle. For example, researchers evaluated cyclists riding at 60% VO_2max on two different occasions; once while they were provided adequate fluids to prevent dehydration and another time when they were not.[20] When the cyclists were not provided fluids, blood flow to their legs was lower and glycogen breakdown and lactate content were greater later in the ride in comparison to when they were provided sufficient fluids. It stands to reason that an increased production of lactate could result in a reduced work capacity. Also, increased glycogen breakdown exhausts glycogen stores more quickly and promotes earlier fatigue.

Key Point

Dehydration may result in increased carbohydrate utilization and lactate generation by working muscle increasing the potential for premature fatigue.

Hypohydration

Athletes participating in sports such as wrestling, lightweight crew, judo, and boxing often attempt to dehydrate in order to "make weight" and compete within a lower weight class. The term **hypohydration** is often used to describe voluntary efforts to reduce body water levels. Hypohydration efforts focus on restricting water intake as well as increasing water loss from the body with deliberate exposure to high heat (such as a sauna), sodium restriction, and possibly the use of diuretics. As mentioned, reduction of body water by 1 L is equivalent to 2.2 lb (1 kg). These athletes then attempt to rehydrate between the weigh-in qualification and actual competition. Other sports, such as bodybuilding and fitness competitions, are also associated with precompetition hypohydration. Here again fluid restriction is practiced, along with restrictions

Key Point

Some athletes trying to "make weight" may deliberately dehydrate themselves. However, this can reduce their level of performance and increase the risk of heat intolerance during competition.

on sodium and carbohydrate consumption. The athletes endure these regimens in an attempt to draw water out of muscle tissue to enhance their striated appearance (muscular definition).

ATHLETIC PERFORMANCE CONSIDERATIONS OF HYPOHYDRATION. Dehydration induced by hypohydration practices can have a different influence on physical performance than exercise-induced dehydration. First, the effects of dehydration may be experienced at the onset of physical activity. This can result in reduced performance even in shorter duration sports. For instance, when runners performed 1.5-, 5-, and 10-km runs in either a well-hydrated state or a partially hypohydrated state (2% reduction in body weight as water), their running speed was significantly lower at distances of 5 and 10 km, and a similar trend was observed during the 1.5-km run.[21] Moderate hypohydration resulted in reduced performance in 20-second sprinting bouts separated by 100 seconds of rest.[22] The results of other research studies also demonstrated that hypohydration (a reduction of 2–4% in body weight as water) reduces maximal aerobic power.[23–25]

In addition to dehydration, individuals must consider the potential consequence of other hypohydrating practices on sport performance. For instance, if carbohydrate is also restricted, then performance in higher intensity sports such as wrestling may be compromised. Restriction of sodium can eventually lead to electrolyte imbalances, as discussed in Chapter 10.

HEALTH CONSIDERATIONS OF HYPOHYDRATION. In addition to reducing physical performance, hypohydration can influence the general health of individuals. Hypohydration-induced dehydration can become critical when it compromises thermal regulation during activity in hot environments and causes cardiovascular complications, as discussed above. Much of the attention to hypohydration practices has focused on wrestlers, especially after a college wrestler died in 1997 from associated practices. He experienced kidney and heart failure while working out in a wet suit in a 92°F room.[26] Other deaths have also been associated with dangerous hypohydration practices by wrestlers. Coaches and wrestlers should be well educated on the dangers of severe dehydration and electrolyte imbalances. The American College of Sports Medicine

(ACSM) published a position stand in 1996 regarding weight loss in wrestlers.[17]

REHYDRATION AFTER HYPOHYDRATION. The time period between a weigh-in and competition determines whether complete rehydration is possible. For instance, youth football leagues are separated into weight classes and often the weigh-in takes place just an hour or two prior to a game. A high school wrestling weigh-in may occur the morning of a match, leaving only 7–8 hours for rehydration. Weigh-in for collegiate wrestling and professional boxing tends to occur 24 hours prior to competition. Complete recovery of body water can take hours depending on the extent of dehydration and other weight reduction efforts. In addition, partial and full rehydration does not guarantee a complete recovery of performance. This was demonstrated in a study involving rowers who failed to perform at their well-hydrated level after they were hypohydrated and then provided fluids.[27]

Key Point

Hypohydration to reduce body weight can impair performance if body water is not recovered prior to competition.

Hyperhydration

Hyperhydration refers to an attempt to begin an exercise bout with a slight surplus of body water. This would allow for a greater volume of sweat loss prior to a reduction in performance. Also, beginning exercise with a slightly expanded plasma volume might provide a slight buffer against detrimental reductions in plasma volume typically experienced during sustained activity. Hyperhydration can potentially benefit athletes during endurance exercise in a warmer environment. Athletes competing in intermittently high intensity sports for a couple of hours, with less opportunity to consume adequate fluids, might also consider hyperhydration. The sports usually associated with hyperhydration include distance running, soccer, and tennis.

Typically hyperhydration involves the consumption of 8–16 oz (250–500 ml) of fluid within 15 minutes of the onset of exercise. This is a gamble as fluid in the stomach, as well as urine production and the need to void the bladder early in competition, are potential problems. To enhance hyperhydration efforts, some athletes have experimented with nutrition supplements such as glycerol. Although glycerol supplementation might enhance hyperhydration efforts, certain issues need to be considered, as discussed in Chapter 11.

■ HYDRATION STATUS AND PERFORMANCE *IN REVIEW*

- Fluid consumption by individuals usually does not match water losses via sweating during endurance training and some sports in warmer environments.
- Thirst, urine volume, and urine characteristics can be used as indicators of hydration status but are not absolute.
- Hypohydration practices can impair performance even if an individual accomplishes complete rehydration prior to competition.
- Hyperhydration may enhance performance in the heat when the maintenance of optimal hydration status is unlikely due to a high sweat rate.

■ PRACTICAL GUIDELINES FOR WATER CONSUMPTION

For decades hydration status was not recognized as important during activity. Athletes were not encouraged to drink water during practices and some coaches even believed that drinking water during practice or games demonstrated mental weakness. Today hydration status is viewed as a crucial component of successful performance. Water bottles and coolers can be found on the sidelines at practices and games of sports in which sweating is characteristic. Recommendations for the amounts and type of fluid ingestion before, during, and after exercise depend on the nature of the sport and the environmental conditions at hand.[28,29] What follows are general guidelines for fluid ingestion and composition, which can serve as the basis for experimentation by athletes.

Water Consumption Before Exercise

All athletes are encouraged to drink generous amounts of water during training, and in particular, the day prior to competition. In addition, athletes are encouraged to drink 13–20 oz (400–600 ml), or about $1\frac{1}{2}$–$2\frac{1}{2}$ cups, of fluid 2–3 hours prior to exercise.[28] This time period allows for equilibration of ingested water throughout the various body fluid compartments and for urine formation and removal of excess water. Larger athletes may consume a greater quantity.

It is important for athletes to experiment with fluid volume as well as composition consumed prior to exercise. For instance, bigger athletes, such as football linebackers and linemen, may need to drink more fluid than the general recommendations above. Also, the fluid consumed prior to exercise may be a source not only of water but of carbohydrate and electrolytes. In fact, solutes such as carbohydrate and electrolytes enhance the retention of water prior to exercise. Sport drinks containing 4–8% carbohydrate are preferred and are discussed further in Chapter 11.

The time of day of exercise can also be an important factor in achieving optimal hydration status prior to exercise. For instance, athletes who train shortly after waking in the morning need to pay attention to what and how much they drink the night before as well as that morning. These athletes may want to experiment with drinking 16–32 oz (500–1000 ml), or about 2–4 cups, of fluid 1 hour prior to exercise. This recommendation is higher than above because of body water loss during sleep, without fluid consumption. Once again, athletes must experiment with various fluid levels during training to reduce the possibility of having to urinate early in competition.

Key Point

30 ml = 1 oz   240 ml = 8 oz (1 cup)
1000 ml (1 L) = 35 oz (4½ cups)

Water Consumption During Exercise

Athletes participating in a variety of sports often fail to maintain optimal hydration status during practice, training, or competition. Quite simply, fluid intake fails to match water loss during exercise. Endurance athletes such as runners or cross-country skiers especially are at higher risk of dehydration and reduced performance. For these athletes the motion of the sport can inhibit effective water consumption efforts. During competition, these athletes often bypass hydration opportunities in order to maintain velocity. The limited upper body activity of cycling makes fluid consumption easier, and cycling makes it easier to carry water bottles or wear a bladder-containing backpack (such as a camelback style).

For sports such as soccer, maintaining optimal hydration status can be a concern as well. Soccer is a fairly continuous game and players may not have the opportunity to consume adequate levels of fluid during game situations. During stoppages of play, these athletes should seek fluids. Also, because player substitutions have become more common at the scholastic and collegiate levels, players now have more opportunity to drink fluids. Tennis players also have limited opportunity to drink fluid during play. Players are allowed brief periods of rest at crossovers and in between sets and should be encouraged to consume fluids. Still, competitive play is often outside in warmer climates with a match lasting a couple of hours, and in such situations sweat water loss tends to exceed fluid recovery efforts. Consuming fluid during exercise can help prevent the physiological problems of dehydration and minimize the perceived effort of exercise exertion.[30]

A general recommendation is to drink a small amount of fluid: 6–12 oz (150–350 ml), or about ¾–1½ cups, of fluid every 15–20 minutes during exercise. This should add up to 20–40 oz (600–1200 ml) or 2½–5 cups of fluid per hour

and can match water losses when sweating is mild or moderate. However, additional fluid may be necessary when the exercise takes place in a very warm environment and sweating is heavier.

Fluid consumption may be enhanced when the fluid is cooler than ambient temperature and is flavored. In addition, the carbohydrate content should be appropriate for the physical activity and the electrolyte content should be related to the level of sweating. The appropriate composition of a sport drink is discussed in Chapter 11. However, a carbohydrate content of 6–8% would be sufficient. This would provide 14–19 g of carbohydrate in an 8-oz cup of fluid, or approximately 55–80 g/L. Also, glucose, sucrose, and maltodextrins are preferred over fructose, as discussed in Chapters 4 and 11. The level of sodium should be 0.5–0.7 g/L, as discussed in more detail in Chapter 11.[28]

Water Consumption After Exercise

It is not uncommon for endurance athletes as well as soccer, tennis, and football players to experience a reduction of 1–2% or more in body weight during practice and competition. In addition, it can take several hours to restore body water in all fluid compartments. Therefore, the goal for water recovery should be to provide adequate amounts of fluid at regular intervals. Athletes can begin by drinking 16½–33 oz (500–1000 ml), or about 2–4 cups, of fluid during the first 30 minutes after exercise and then consume at least 1 L every 1–2 hours thereafter until consumption has matched 150% of sweat weight loss to allow for complete rehydration. Fluids containing energy and electrolytes serve an athlete better than water alone, especially when food is not consumed after exercise. Carbohydrate and electrolytes promote the recovery of these nutrients in the body. Also, recovery of muscle glycogen stores helps draw water into skeletal muscle, and electrolytes enhance the recovery of extracellular water.

■ PRACTICAL GUIDELINES FOR WATER CONSUMPTION *IN REVIEW*

- Generous fluid consumption is encouraged during training and especially the day prior to exercise.
- Drinking 400–600 ml of water is recommended for athletes 2–3 hours prior to exercise to optimize hydration status and allow time for voiding excessive water as urine. Bigger athletes might want to experiment with greater levels.
- Drinking at least 150–350 ml of fluid is recommended during every 15–20 minutes of exercise.
- Chilling fluids and adding sodium and carbohydrate may promote greater consumption during exercise.

IN FOCUS   WATER BALANCE OF AN ATHLETE

Water balance is fundamental to normal function as well as to optimal athletic performance. Water loss, via urine, sweating, feces, and breath, must be balanced by water intake and metabolic generation to prevent dehydration. Athletes lose more body water than nonathletes due to increased volumes of sweat and breath. Their enhanced energy metabolism also allows for an increased generation of water in cells. However, this cannot compensate for enhanced water loss in athletes. Therefore increased water loss must be compensated for by increased consumption of beverages and to a lesser degree food. To demonstrate the special water balance for an athlete, Table A presents water losses of an endurance athlete (runner) compared to a nonathlete. To accommodate increased levels of water loss, the runner consumes more fluid between training sessions as well as during training.

Table A

Example of Water Balance for an Athlete and a Nonathlete

	Nonathlete	Athlete: Runner
Water Intake and Metabolic Generation		
Water in foods	1 L	1.5 L
Beverages (meals, snacks, fluids)	1.5 L	2.0 L
Water in sport drink consumed during exercise[a]		1 L
Metabolic generation of water	0.35 L	0.5 L
Total Water Provided	**2.85 L**	**5.0 L**
Water Loss		
Breath	150 ml	225 ml
Feces	100 ml	125 ml
Sweat	500 ml	2600 ml
Urine	2100 ml	2000 ml
Total Water Loss	**2.85 L**	**5.0 L**

[a]Exercise consists of an 8-mile training run.

Conclusions

Water is the ideal medium for the human body because it is an excellent solvent and has a high specific heat. Water is the basis of sweat and urine, which remove waste and excesses from the body. Water moves in and out of cells and between tissue and the blood as dictated by hydrostatic and osmotic pressures. Plasma volume can decrease during exercise as water fluxes into tissue and working muscle fibers. Critical reduction of blood volume during exercise can compromise performance by decreasing heat tolerance and blood delivery to tissue. Endurance training expands blood volume and results in adaptations that increase sweat volume and decrease its electrolyte composition. Athletes are encouraged to weigh themselves before and after exercise to estimate water losses during exercise. Optimal hydration status is vital to physical performance, and attention must be paid to water consumption before, during, and after exercise.

STUDY QUESTIONS

1. What are some of the unique properties of water that make it an ideal medium for the human body?
2. How is water distributed throughout the body, and what are some physiological roles of water?
3. What is the composition of sweat during light and heavy periods of sweating?
4. Does chronic exercise training influence sweat rate and composition?
5. How does endurance exercise influence plasma volume, and what factors are important in stabilizing plasma volume during exercise?
6. How can athletes estimate sweat volume?
7. How does dehydration influence athletic performance, and how might dehydration become detrimental to vitality?
8. What are the potential benefits and considerations of hyperhydration, and what types of athletes might benefit from it?
9. What are the common methods of hypohydration, and what are the potential performance detriments?
10. What are practical recommendations for water consumption before, during, and after exercise?

REFERENCES

1. Wildman REC, Medeiros DM. Water. In: *Advanced Human Nutrition*, Boca Raton, FL: CRC Press, 2000.
2. Bilz S, Ninnis R, Keller U. Effects of hypoosmolality on whole-body lipolysis in man. *Metabolism* 48(4):472–476, 1999.
3. Berneis K, Ninnis R, Haussinger D, Keller U. Effects of hyper- and hypoosmolality on whole body protein and glucose kinetics in humans. *American Journal of Physiology* 276:E188–E195, 1999.
4. Maw GJ, Mackenzie IL, Taylor NA. Human body-fluid distribution during exercise in hot, temperate and cool environments. *Acta Physiologica Scandinavica* 163(3):297–304, 1998.
5. Myhre LG, Hartung GH, Nunneley SA, Tucker DM. Plasma volume changes in middle-aged male and female subjects during marathon running. *Journal of Applied Physiology* 59(2):559–563, 1985.
6. O'Toole ML, Paolone AM, Ramsey RE, Irion G. The effects of heat acclimation on plasma volume and plasma protein of females. *International Journal of Sports Medicine* 4(1):40–44, 1983.
7. Maw GJ, Mackenzie IL, Taylor NA, Redistribution of body fluids during postural manipulations. *Acta Physiologica Scandinavica* 155(2):157–163, 1995.
8. Oscai LB, Williams BT, Hertig BA. Effect of exercise on blood volume. *Journal of Applied Physiology* 24:622–628, 1968.
9. Carroll JF, Convertino VA, Wood CC, Lowenthal DT, Pollack ML. Effect of training on blood volume and plasma hormone concentrations in the elderly. *Medicine and Science in Sports and Exercise* 27:79–85, 1995.
10. Convertino VA. Blood volume: its adaptation to endurance training. *Medicine and Science in Sports and Exercise* 23: 1338–1345, 1991.
11. Gillen CM, Lee R, Mack GW, Tomaselli CM, Nishezasa T, Nadel ER. Plasma volume expansion in humans after a single intense exercise protocol. *Journal of Applied Physiology* 71:1914–1920, 1991.
12. Shoemaker JK, Green HJ, Coates J, Ali M, Grant S. Failure of prolonged exercise training to increase red cell mass in humans. *American Journal of Physiology* 270(1 Pt 2): H121–H126, 1996.
13. Luetkemeier MJ. Dietary sodium intake and changes in plasma volume during short-term exercise training. *International Journal of Sports Medicine* 16(7):435–438, 1995.
14. Falk B. Effects of thermal stress during rest and exercise in the paediatric population. *Sports Medicine* 25(4):221–240, 1998.
15. Lemon PW, Deutsch DT, Payne WR. Urea production during prolonged swimming. *Journal of Sports Science* 7(3):241–246, 1989.
16. McMurray RG, Horvath SM. Thermoregulation in swimmers and runners. *Journal of Applied Physiology* 46(6): 1086–1092, 1979.
17. American College of Sports Medicine. Position stand: weight loss in wrestlers. *Medicine and Science in Sports and Exercise* 28(2):ix–xii, 1996.
18. Kovacs EM, Senden JM, Brouns F. Urine color, osmolality and specific electrical conductance are not accurate measures of hydration status during postexercise rehydration. *Journal of Sports Medicine and Physical Fitness* 39(1):47–53, 1999.
19. Gonzalez-Alonso J, Teller C, Andersen SL, Jensen FB, Hyldig T, Nielsen B. Influence of body temperature on the development of fatigue during prolonged exercise in the heat. *Journal of Applied Physiology* 86(3):1032–1039, 1999.
20. Gonzalez-Alonso J, Calbet JA, Nielsen B. Metabolic and thermodynamic responses to dehydration-induced reductions in muscle blood flow in exercising humans. *Journal of Physiology (London)* 15(520 Pt 2):577–589
21. Sawka MN, Pandolf KB. Effects of body water loss on physiological function and exercise performance. Fluid homeostasis during exercise. In: *Perspectives in Exercise Science and Sports Medicine*, vol. 3, Indianapolis: Benchmark Press, pp. 1–30, 1990.
22. Maxwell NS, Gardner F, Nimmo MA. Intermittent running: muscle metabolism in the heat and effect of hypohydration. *Medicine and Science in Sports and Exercise* 31(5):675–683, 1999.
23. Buskirk ER, Iampietro PF, Bass DE. Work performance after dehydration: effects of physical conditioning and heat acclimation. *Journal of Applied Physiology* 12:189–194, 1958.

24. Webster S, Rutt R, Weltman A. Physiological effects of a weight loss regimen practiced by college wrestlers. *Medicine and Science in Sports and Exercise* 22:229–234, 1990.

25. Craig EN, Cummings EG. Dehydration and muscular work. *Journal of Applied Physiology* 21:670–674, 1966.

26. Berardot D. Power sports (wrestling). In: *Nutrition for Serious Athletes*, Champaign, IL: Human Kinetics, 2000.

27. Burge CM, Carey MF, Payne WR. Rowing performance, fluid balance, and metabolic function following dehydration and rehydration. *Medicine and Science in Sports and Exercise* 25(12):1358–1364, 1993.

28. American College of Sports Medicine. Joint position statement: nutrition and athletic performance. American College of Sport Medicine, American Dietetic Association, and Dietitians of Canada. *Medicine and Science in Sports and Exercise* 32(12):2130–2145, 2000.

29. Latza WA, Montain SJ. Water and electrolyte requirements for exercise. *Clinical Sports Medicine* 18(3):513–524, 1995.

30. Burke LM. Nutritional needs for exercise in the heat. *Comparative Biochemistry and Physiology—Part A: Molecular & Integrative Physiology* 128(4):735–748, 2001.

VITAMINS AND EXERCISE

Chapter Objectives

- Discuss the food sources and functions of vitamins.

- Provide recommendations for dietary intake and describe how vitamins are involved in health and disease prevention.

- Describe the intake level of vitamins by active people.

- Discuss how vitamins influence physical activity and sports performance.

Personal Snapshot

© PhotoDisc/Getty Images

Melissa is a 32-year-old marathoner who wants to improve her personal record (PR) in the upcoming marathon season. She recently read an article about the relationship between endurance training and the production of free radicals. The harder one trains, the greater the production of these potentially damaging substances, according to the article. It also mentioned that although certain foods are good sources of antioxidants, such as vitamins C and E, which protect against free radical damage, they are not enough, so serious endurance athletes should bolster their antioxidant intake with supplements. Could this be the reason why Melissa has failed to improve her PR in the past three seasons? What is the current expert consensus on this matter? Should Melissa take antioxidant supplements or not?

Vitamins are organic substances essential to the diet. They are either water-soluble or fat-soluble, as shown in Figure 9-1. Water-soluble vitamins are subclassified as either B-complex vitamins (B-vitamins) or vitamin C.

Optimal vitamin status is vital to exercise performance, training, adaptation, and general health and fitness. For instance, muscle fibers are not able to process energy nutrients without thiamin, riboflavin, niacin, biotin, and pantothenic acid (see Figure 9-2). These vitamins are needed for the oxidation of carbohydrate and fat that supports exercise. Vitamin B_6 is needed for efficient amino acid metabolism as well as the breakdown of glycogen. In addition, as endurance exercise increases free radical production in muscle and other tissue, certain vitamins such as C and E and vitamin-related molecules such as carotenoids may serve a protective role. Other vitamins, such as folate and vitamins B_6 and B_{12}, are important in red blood cell (RBC) formation, a process fundamental to endurance exercise.

Athletes who consume an appropriate level of energy for weight maintenance and eat a variety of foods, including whole grains, fruits, vegetables, and animal products, would not have difficulty getting at least the current Recommended Dietary Allowance (RDA) or Adequate Intake (AI) levels for vitamins, as demonstrated in Nutrition in Practice: Getting Enough Vitamins. This does not mean

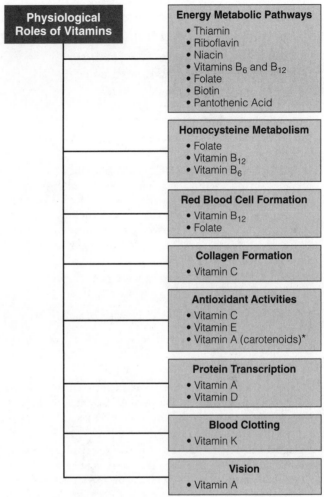

Figure 9-2
Classification of vitamins based upon function. *Some carotenoids can be converted to vitamin A.

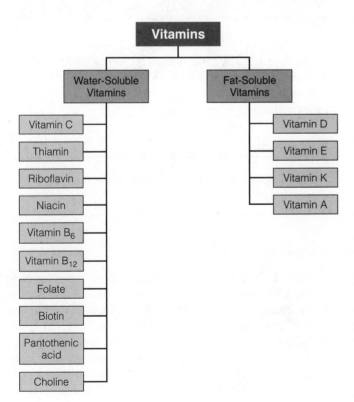

Figure 9-1
Classification of vitamins based on water solubility.

that the vitamin needs of athletes are the same as those of the general population. It is still unclear whether exercise training increases the requirements for vitamins, but it is likely that the need for certain vitamins involved in energy metabolism is greater for athletes (Table 9-1). Also, it is an unresolved issue whether higher levels of antioxidant vitamins (vitamin C and E) are required by endurance athletes, as discussed in the feature In Focus: Antioxidant Supplementation by Athletes at the end of this chapter. In light of the greater energy consumption of athletes, many sport nutritionists contend that a balanced diet can provide the additional amounts of vitamins that athletes may need, making supplements unnecessary. However, not all athletes consume a balanced diet, and as discussed in this chapter, many fall short of meeting the Dietary Reference Intake (DRI) levels. This chapter provides a basic overview of the food sources and physiological roles of each vitamin with special attention to active people.

Table 9-1

Factors That Could Influence Vitamin and Mineral Intake, Absorption, Metabolism, and Excretion Related to Physical Activity

Changes in eating patterns and/or volumes	• Athletes who chronically restrict total food intake to manipulate body weight
	• Athletes who alter the macronutrient ratio of their diet, which could influence the level of vitamins (such as hypohydration involving carbohydrate restriction by bodybuilders and wrestlers)
	• Athletes who focus on particular food groups (as in very high protein diets) might limit other foods, thereby reducing the intake of certain vitamins
	• Athletes who purposely restrict the intake of one or more food groups (such as vegetarians)
Greater demands due to augmented operations	• Increased energy metabolism • Increased oxidative stress • Increased recovery and adaptive effort
Changes in nutrient metabolism and excretion	• Increased or decreased urinary loss of vitamins and minerals • Increased loss of nutrients in sweat (such as sodium, chloride, potassium, calcium, iron)

Key Point

Vitamins include vitamins A, D, E, K, C, B_6, B_{12}, thiamin, riboflavin, niacin, folic acid, biotin, pantothenic acid, and choline.

Table 9-2

Approximate Vitamin C Content of Select Foods

Food	Vitamin C (mg)
Fruits	
Papaya (1 med)	188
Orange juice, fresh (1 c)	120
Kiwi (1)	105
Grapefruit juice, fresh (1 c)	95
Cranberry juice cocktail (1 c)	90
Orange (1)	85
Strawberries, fresh (1 c)	80
Cantaloupe (¼)	65
Grapefruit (1)	50
Raspberries, fresh (1 c)	30
Watermelon (1 c)	15
Vegetables and Other	
Green peppers (½ c)	95
Cauliflower, raw (½ c)	75
Broccoli (½ c)	70
Brussels sprouts (½ c)	65
Collard greens (½ c)	45
Cauliflower, cooked (½ c)	30
Potato (1)	25
Tomato (1)	25

better sources than cooked ones. Absorption of vitamin C is related to consumption level in the meal; as the level increases, less is absorbed. This is one way the body protects against toxicity.

Key Point

Better sources of vitamin C include fresh fruits and vegetables.

■ VITAMIN C

Vitamin C has long been popular and more people can name good food sources of this vitamin than any other. Furthermore, vitamin C has a long history as a supplement for athletes. This has resulted in more research involving vitamin C and exercise performance than most other vitamins. Recommendations (RDAs) for vitamin C are 75 and 90 mg for adult females and males, respectively.

Good sources of vitamin C include fresh fruits and vegetables, especially citrus fruits, papaya, strawberries, watermelon, broccoli, plantains, cauliflower, and green and red peppers (Table 9-2). Because vitamin C is not stable in heat, fresh fruits and vegetables are considered

Vitamin C Functions

Vitamin C is found in most tissues, with greater concentrations in more metabolically active tissue such as the heart, brain, pancreas, adrenal glands, thymus, and lungs. By and large the activity of vitamin C is related to its ability to donate electrons. Ascorbic acid (ascorbate) is the reduced state and dehydroascorbic acid is the oxidized state (Figure 9-3). By donating electrons, vitamin C participates in many metabolic processes. Once vitamin C is oxidized it can be recycled to its reduced state by compounds such as lipoic acid and glutathione.

Vitamin C works with key enzymes in the production of collagen, the most abundant protein in the body.

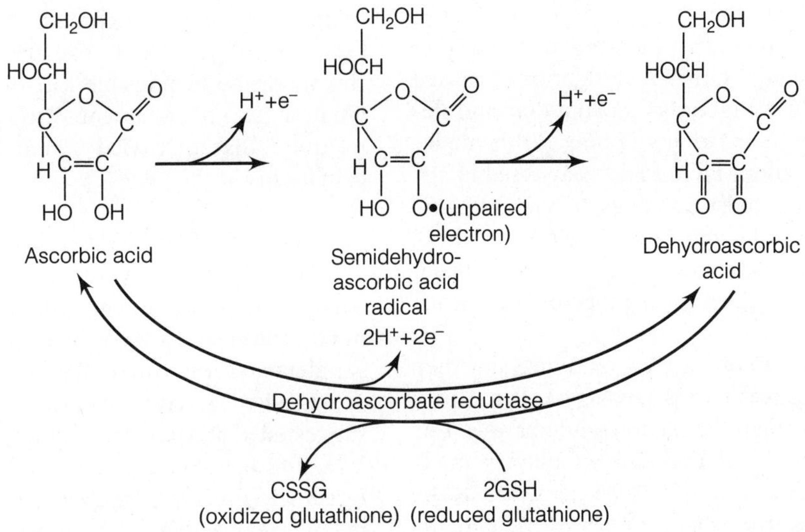

Figure 9-3
The interconversion of vitamin C forms.

Collagen provides much of the structure to connective tissue, so it is vital to bone, tendons, and ligaments. This makes vitamin C particularly important to athletes to minimize sprains, strains, and fractures. It is also an important antioxidant, and it participates in the formation of catecholamines (norepinephrine and epinephrine), carnitine, and maybe testosterone. In the digestive tract, it can enhance iron absorption. These properties make vitamin C supplements attractive for those who want to enhance their physical performance.

Key Point

Vitamin C is an antioxidant and is involved in making epinephrine, carnitine, and testosterone, leading to the notion that taking supplements could enhance athletic performance.

VITAMIN C STATUS ASSESSMENT. Classically, assessing a person's vitamin C status involved measuring serum levels. More recently researchers have begun to measure the level of vitamin C in leukocytes such as neutrophils, eosinophils and basophils, lymphocytes, and monocytes.[1] This is because serum vitamin C levels are more related to recent dietary intake, whereas leukocyte levels are better indicators of cellular stores and total body pool of vitamin C. After adjusting for dietary intake, women tend to have higher tissue and body fluid levels of vitamin C than men.[1] Also, although most athletes do not smoke, some do, and smokers tend to have lower serum and leukocyte vitamin C levels than nonsmokers whose dietary intake is similar.[2]

Vitamin C Imbalance

Reduced vitamin C status can affect many physiological systems. The negative influence on connective tissue results in the most dramatic signs and symptoms. Other functions of vitamin C become impaired as well. Although extreme vitamin C deficiency is unlikely in industrialized societies, people in general and athletes in particular should assess their diet for vitamin C adequacy to protect against marginal deficiency.

Vitamin C's popularity as a supplement raises concerns regarding toxicity, and a Tolerable Upper Intake Limit (UL) has been set at 2000 mg (2 g) for adults. Even though the efficiency of absorption decreases with higher levels in the digestive tract, more vitamin C is still absorbed. Higher levels also mean a proportionate increase in urinary loss of vitamin C and its metabolites. One concern about consuming gram-size doses of vitamin C is the production of **oxalate**, a principal metabolite of vitamin C. Oxalate is a component of the predominant type of kidney stones (calcium oxalates), so people taking high doses of vitamin C might be placing themselves at risk. The research in this area is again inconclusive. However, most health practitioners agree that people prone to forming kidney stones should avoid gram doses of vitamin C supplements.

Vitamin C Intake by Athletes

Several nutritional surveys of athletic groups have revealed that their vitamin C intake is generally above recommendations (RDA). For instance, it was reported that the average daily vitamin C intake of male and female marathon runners was 147 and 115 mg, respectively.[3] Also, researchers reported that the diet (food only) of high school football players provided an average of roughly 180 mg of vitamin C daily.[4] Other investigations assessed the vitamin C intake of competitive bodybuilders and determined that the average intake was 272 mg for the males and 196 mg for the females.[5] Field athletes assessed in another study also appeared to eat an adequate amount of vitamin C.[6] In a summary of about 20 reports on vitamin C intake among a variety of athletes and active people (such as ballet dancers and Navy Seals), the range of average intake was 95–529 mg daily.[7]

The results of such surveys are encouraging, but they demonstrate only statistical means (averages). Some participants consumed less than the recommended levels. For instance, more than 25% of the football players mentioned above consumed vitamin C at well below the RDA level, and the same was true of a group of wrestlers in another study.[4,8] Also, about 10–15% of the Navy Seals surveyed had an estimated vitamin C intake below the RDA,[9] and about half of female collegiate gymnasts and basketball players surveyed had an estimated vitamin C intake less than the RDA.[10] Thus, although a survey may show an average intake well above the RDA, as many as 10–25% of individuals within an athletic group may consume a level below the RDA.[7] However, vitamin C is a popular supplement or ingredient in supplement formulations, and athletes who use them might make up for a diet with otherwise inadequate vitamin C.

Key Point

The vitamin C intake and status of some athletes may be below recommendations and normal levels.

Vitamin C and Athletic Performance

The studies cited above compared the vitamin C intake of athletes and other active individuals to RDA levels, which were set for the general population. But do athletes need more vitamin C than the general population? This is uncertain. Researchers have reported that the plasma levels of vitamin C in athletes and active people are similar to those of inactive people—for instance, the average plasma vitamin C level of the elite ballet dancers in one study was 0.81 mg/100 ml, whereas a level above 0.6 mg/100 ml is deemed normal.[11] However, as with intake levels it is important to evaluate not only statistical averages but also ranges and individual levels. In one study 12% of the athletes assessed had plasma vitamin C levels below normal despite an average intake that was above normal.[12] A factor that might be important is when the last workout occurred and whether there is a residual influence of that activity.[7] In one study, the level of serum vitamin C was reported to increase for several hours following intense bouts of training and competition.[13] In addition, urinary excretion of vitamin C can decrease following exercise; this might be responsible for some of the elevated serum vitamin C. This is an interesting finding, as it is logical to think that increased serum levels lead to increased urinary loss.

VITAMIN C SUPPLEMENTATION AND ATHLETIC PERFORMANCE. More exercise-based research has been performed involving vitamin C than any other vitamin. However, it is far from clear whether vitamin C supplementation can enhance athletic performance. As discussed in a recent review of research, earlier studies suggested that vitamin C supplementation might reduce VO_2 and heart rate at a given intensity level, reduce fatigue, and prevent a decrease in muscular endurance later in an exercise bout.[7] Also, after 1 g of vitamin C was provided daily to males for 2 weeks they had a reduced heart rate at various levels of work output and had an increased work output at 170 beats per minute.[14] But other studies have failed to determine that vitamin C supplementation increases athletic performance during running trials.[7]

One of the most important factors in designing the research protocol is to achieve normal vitamin C status prior to providing the test dosage of vitamin C. For instance, if vitamin C is supplemented after a diet assessment shows that intake is below recommendations, it is hard to determine whether a positive effect is a nutritional effect or a supplement effect.

Key Point

Vitamins C and E are antioxidants and are often targets for nutrition supplementation.

VITAMIN C AND ANTIOXIDANT PROTECTION IN ATHLETES. Exercise bouts of prolonged endurance or intermittent effort are believed to increase the production of free radicals.[15,16] Oxygen uptake may be enhanced 10–15 times, which would allow for increased free radical production. Vitamin C supplementation has been suggested to enhance performance in the long term by bolstering antioxidant defenses and decreasing exercise-induced free radical damage.[17,18] For endurance athletes and intermittent sport athletes with heavy O_2 demands (such as football, soccer, and ice hockey players), increased tissue vitamin C might decrease activity related to free radicals, allowing for more productive training and competition over longer stretches of time. On the other hand, some researchers suggest caution before recommending antioxidant supplements. The controversy is discussed in more detail in the In Focus at the end of this chapter.

■ VITAMIN C *IN REVIEW*

- Better food sources of vitamin C include fresh fruits and some vegetables.
- Vitamin C is required for collagen formation and the production of several molecules, including carnitine and epinephrine.
- Supplemental vitamin C does not appear to enhance performance in well-nourished individuals, but it might help limit free radical activity associated with sports with high O_2 demand.

■ THIAMIN (VITAMIN B_1)

Thiamin is available in foods and is found in human tissue as either free thiamin or one of three phosphorylated forms: thiamin monophosphate (TMP), thiamin diphosphate (TDP) or thiamin pyrophosphate (TPP), and thiamin triphosphate (TTP or TPPP). All of the thiamin in plant foods is in the form of free thiamin, whereas animal foods contain mostly the phosphorylated forms, predominantly TPP.

Thiamin is common in foods, although most contain low concentrations. Good sources of thiamin include brewer's yeast, wheat germ, liver, pork, whole grains, soy milk, legumes, nuts, dark green vegetables, and enriched grain products (Table 9-3). The RDA for thiamin is 1.1 and 1.2 mg for adult females and males, respectively. However, many researchers feel that thiamin requirements are better expressed relative to energy expenditure. Here 0.5 mg of thiamin would be required for every 1000 kcal. Thus athletes expending 3000–6000 kcal daily would require 1.5–3.0 mg/day.

Key Point

Whole grains and enriched products, dark green vegetables, nuts, and enriched grain products are good sources of thiamin.

Thiamin Functions

About half of thiamin in the body is in skeletal muscle. Other tissues of higher thiamin concentration include those with higher metabolic expenditures such as the heart, liver, brain, and kidneys. Thiamin transported in the blood is primarily bound to proteins and within RBCs, and it is lost from the body primarily in urine. It serves as a coenzyme in many key reactions in the energy pathways. For instance, thiamin is needed for the *pyruvate dehydrogenase* and *α-ketoglutarate dehydrogenase* reactions. As shown in Figure 9-4, these reactions take place in mitochondria and are crucial for aerobic energy metabolism of carbohydrates,

Table 9-3

Approximate Thiamin Content of Select Foods

Food	Thiamin (mg)
Meats	
Pork roast (3 oz)	0.8
Beef (3 oz)	0.4
Ham (3 oz)	0.4
Liver (3 oz)	0.2
Nuts and Seeds	
Sunflower seeds (¼ c)	0.7
Peanuts (¼ c)	0.1
Almonds (¼ c)	0.1
Grains and Products	
Bran flakes (1 c)	0.6
Macaroni (½ c)	0.1
Rice (½ c)	0.1
Bread (1 slice)	0.1
Wheat germ (¼ c)	0.5
Vegetables	
Peas (½ c)	0.3
Lima beans (½ c)	0.2
Corn (½ c)	0.1
Broccoli (½ c)	0.1
Potato (1)	0.1
Fruits	
Orange juice (1 c)	0.2
Orange (1)	0.1
Avocado (½ c)	0.1
Other	
Brewer's yeast (1 oz)	4.3

fat, and protein-derived amino acids. In addition, thiamin is needed at two steps in the pentose phosphate pathway (PPP). One of these reactions involves a *transketolase*, and assessing the activity of this enzyme in RBCs is widely used to assess thiamin status (see below). The PPP generates ribose-5-phosphate for ATP production.

Thiamin is also needed to break down branched chain amino acids (BCAAs)—leucine, isoleucine, and valine (Figure 9-4). The complete breakdown of BCAAs occurs in muscle (skeletal and heart), liver, kidneys, and skeletal muscle. BCAA catabolism is particularly important as an energy source during endurance exercise. Thiamin is also vital for the normal functioning of neural tissue. This role seems to extend beyond simple energy metabolism and involves effective transmission of impulses. The extent to which thiamin influences the cognitive and neuromuscular control aspects of performance in certain sports (such as

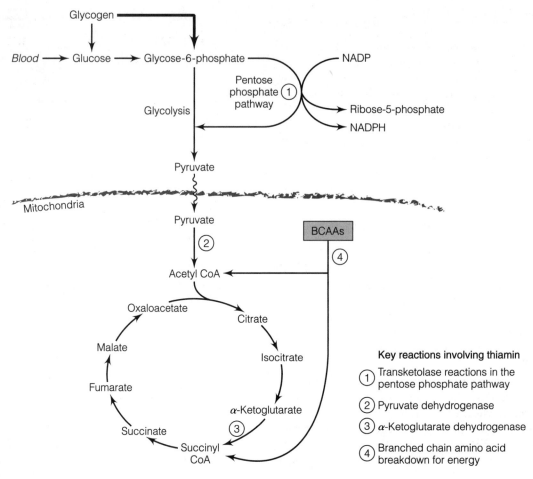

Figure 9-4
Thiamin is a coenzyme in energy pathways at the four points indicated.

rifle shooting, darts, horseshoes, bowling, and baseball) is still not completely understood.

Thiamin Imbalance

Thiamin deficiency is not a concern for most individuals. However, if thiamin were deficient from the diet for several weeks, signs and symptoms would appear. Among the tissues most sensitive to deficiency are neural tissue and muscle. More extreme thiamin deficiency can lead to neuromuscular abnormalities, weight loss, and possibly edema and enlargement of the heart. Excess thiamin (thiamin toxicity) is of little concern as it is promptly excreted in the urine. However, long-term thiamin intake of greater than 100 times recommendations has been associated with headaches, convulsions, weakness, allergic reactions, and irregular heart rhythms. A UL for thiamin has not been set.

Thiamin Intake by Athletes

Eating a variety of foods from a variety of food groups should provide ample thiamin to meet the needs of active people and athletes. However, researchers have reported the dietary intake of thiamin by female U.S. national figure skaters to be lower than the national average as reported by the National Health and Nutrition Examination Survey (NHANES III).[19] Also, some young gymnasts have dietary thiamin levels below recommendations.[20] Results of other studies suggest that the dietary intake of thiamin by athletic populations, such as Finnish male ski jumpers[21] and collegiate male cross-country runners,[22] meet recommendations. Supplements containing B-complex vitamins are popular with athletes and active populations, so supplemental thiamin might compensate for diet inadequacy. But some researchers determined that the diets of the athletes they assessed who used vitamin and mineral

Key Point

To determine if a vitamin supplement influences exercise-related parameters, participants must begin the study in good nutrition status.

supplements were already higher in thiamin than those of athletes who did not regularly use a supplement.[23]

Thiamin and Athletic Performance

Thiamin is fundamentally involved in exercise performance because it is a coenzyme for energy metabolism and ATP formation. However, exercise training has not been shown to alter indicators of thiamin status. For instance, a 24-week fitness training program did not change transketolase activity in RBCs,[24] and exercise training did not alter thiamin status in Finnish male athletes, including elite ski jumpers.[21] Even for pregnant women engaging in an aerobic exercise program (walking), thiamin status (erythrocyte transketolase activity) did not differ from that of pregnant women not participating in the exercise program.[25] Thiamin intake, rather than training itself, might better predict thiamin status in athletes (as suggested by a study involving Nordic skiers[26]).

Athletic performance can be significantly reduced by thiamin deficiency. For example, researchers depleted 24 healthy men of thiamin, riboflavin, and vitamin B_6 over an 11-week metabolic feeding period.[27] When comparing performance and metabolic measurements before and after vitamin depletion, maximal work capacity (VO_2max) was reduced by 12%, lactate threshold by 7%, VO_2 at lactate threshold by 12%, peak power by 9%, and mean power by 7%.

Key Point

Thiamin is often marketed as a performance-enhancing vitamin because of its involvement in energy metabolism.

THIAMIN SUPPLEMENTATION AND ATHLETIC PERFORMANCE. As with the other B-complex vitamins, thiamin supplementation has been promoted as a means of enhancing performance. However, supplementation of thiamin or potentially active metabolites has not been proved to enhance performance in well-nourished athletes. For instance, researchers provided well-trained athletes with 1 g of allithiamin (a thiamin derivative) in combination with pantothenic acid daily for 1 week prior to exercise testing.[28] Compared to a placebo trial, vitamin supplementation did not improve performance measures during a 50-km cycling bout (60% VO_2max) or during an ensuing 2-km time trial. In a related study, researchers determined that another thiamin derivative [thiamin tetrahydrofurfuryl disulfide (TTFD)] did not influence VO_2max, lactate threshold, or heart rate in the first part of a study or maximal 2-km cycling time in the second part.[29]

Researchers also reported that thiamin supplements, along with vitamin B_6 and riboflavin, failed to alter exercise-induced lactic acid production even though the participants presented what was probably marginal thiamin status prior to supplementation.[30] With regard to strength performance testing, researchers determined that oral allithiamin administration failed to positively influence isokinetic parameters of muscle performance and lactate accumulation prior to, during, and in recovery from isokinetic exercise.[31]

■ THIAMIN *IN REVIEW*

- Thiamin (vitamin B_1) is found in a variety of foods. The best sources are whole grains and enriched grain products, pork, liver, nuts, and green leafy vegetables.
- Thiamin pyrophosphate (TPP) is involved in several reactions of energy pathways, including the Krebs cycle, BCAA and pyruvate metabolism, and the pentose phosphate pathway.
- Thiamin supplementation does not appear to positively influence measures of performance and energy metabolism.

■ RIBOFLAVIN (VITAMIN B_2)

Riboflavin is an essential component of two coenzymes, *FAD (flavin adenine dinucleotide)* and *FMN (flavin mononucleotide)* (Figure 9-5), that are vital to energy metabolism. These are also the forms in which most riboflavin is found in foods. Good food sources of riboflavin include green leafy vegetables, beef liver, beef, and dairy products (Table 9-4); dairy being the most significant source in the American diet. The RDA for riboflavin is 1.1 and 1.3 mg for adult females and males, respectively. However, researchers have suggested that riboflavin requirements, like those for thiamin, might be more precise if they were expressed relative to energy expenditure. Here requirements would be 0.6 mg for every 1000 kcal. So athletes who expend 3000–6000 kcal/day might need 1.8–3.6 mg/day.

Key Point

Good sources of riboflavin include milk and dairy products, organ meats, whole grains and enriched grain products, eggs, and nuts.

Riboflavin Functions

Riboflavin is found in most tissues, more so in very active tissues such as those of the heart, liver, and kidneys. FAD serves as one of the electron carriers in oxidative (aerobic) energy metabolism. As energy molecules are broken down in metabolic pathways, electrons are removed (in oxidation)

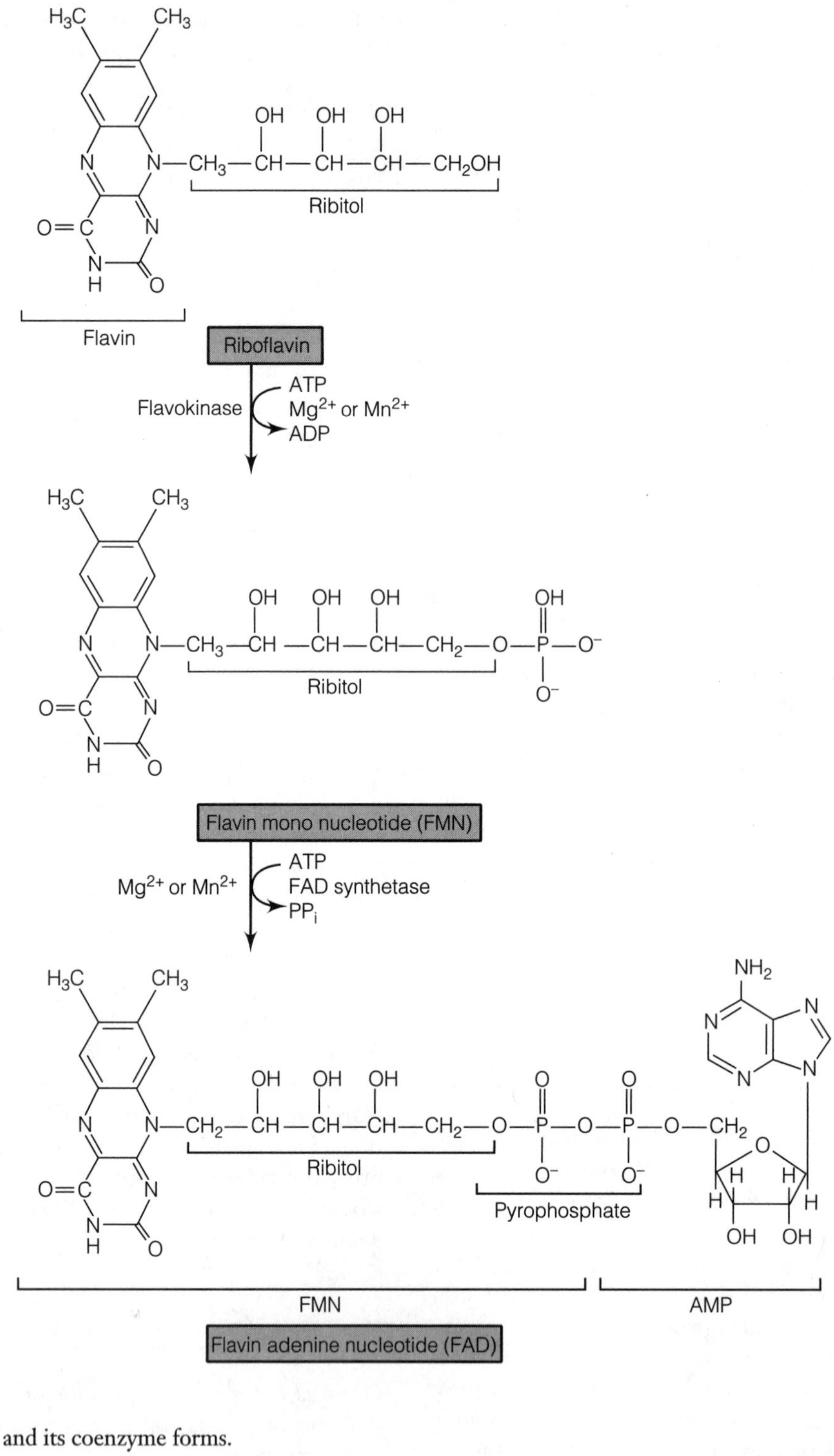

Figure 9-5
Structure of riboflavin and its coenzyme forms.

in key reactions. These electrons are transferred to the electron transport chain, which applies them to the process of ATP formation. FMN also functions in electron transfer as a key component of the electron transport chain. Beyond energy metabolism, FAD and FMN are used in many cell systems such as amino acid metabolism and steroid hormone metabolism.

Key Point

Riboflavin is vital for transferring electrons derived from energy pathways to the electron transport chain for aerobic ATP formation, making it vital to exercise.

Table 9-4

Approximate Riboflavin Content of Select Foods

Food	Riboflavin (mg)
Milk and milk products	
Milk, whole (1 c)	0.5
Milk, 2% (1 c)	0.5
Yogurt, low fat (1 c)	0.5
Milk, skim (1 c)	0.4
Yogurt (1 c)	0.1
Cheese, American (1 oz)	0.1
Cheese, cheddar (1 oz)	0.1
Meats	
Liver (3 oz)	3.6
Pork chop (3 oz)	0.3
Beef (3 oz)	0.2
Tuna (3 oz)	0.1
Vegetables	
Collard greens (½ c)	0.3
Broccoli (½ c)	0.2
Spinach, cooked (½ c)	0.1
Eggs	
Egg (1)	0.2
Grains	
Macaroni (½ c)	0.1
Bread (1 slice)	0.1

RIBOFLAVIN STATUS ASSESSMENT. Measuring the *erythrocyte glutathione reductase activity coefficient (EGRAC)* is often used to assess riboflavin status.[32] The key reaction involved is shown in Figure 9-6. The analysis is straightforward: blood is assessed with and without added FAD. As indicated by the equation in Figure 9-6, a higher EGRAC indicates reduced riboflavin status. It has been suggested that an EGRAC below 1.2 is acceptable, 1.20–1.39 marginal, and >1.40 deficient.[33] More conservative levels (EGRAC <1.25 for normal) have been recommended to account for experimental variability.[34] EGRAC is probably the most sensitive indicator of riboflavin status and is the method employed in most exercise-related studies involving riboflavin. Researchers should also determine the level of urinary riboflavin to support EGRAC measures when assessing riboflavin status.

Riboflavin Imbalance

Riboflavin deficiency rarely occurs by itself, but a person whose diet provides very little riboflavin would begin to show signs such as inflammation of the mouth and tongue after a few months. Other signs of severe riboflavin deficiency are dryness and cracking at the corners of the mouth, lesions on the lips, accumulation of fluid in tissue (edema), anemia, and neurological disorders including decreased coordination and mental confusion. Looking at the other side of riboflavin balance, excess riboflavin is rapidly voided in the urine. Thus the potential for riboflavin toxicity is low. But riboflavin intakes of more than 1000 mg/day can cause gastrointestinal discomfort. A UL for riboflavin has not been set.

Riboflavin Intake by Athletes

Some researchers have indicated that riboflavin intake among most athletes is adequate to meet general recommendations.[35] For instance, one German study involving athletes participating in a variety of sports met the current recommendations set by that country.[36] The results of this study also suggested that blood riboflavin levels correlated well with dietary levels. In another study, both male and female field athletes were assessed as consuming enough riboflavin.[6] However, some researchers have determined that riboflavin intake may be below recommended levels in some athletes engaged in sports associated with energy restriction and weight control.[37] Riboflavin is a popular supplement among athletes, usually in conjunction with other B-complex vitamins such as thiamin, niacin, pantothenic acid, and biotin.

Key Point

Athletes may need more riboflavin than the general population, according to some studies.

Riboflavin Status and Athletic Performance

Riboflavin is crucial to exercise energy metabolism. Research suggests that exercise can alter tissue levels of riboflavin. For instance, beginning an exercise program can increase EGRAC and decrease levels of riboflavin in the blood and urine.[34,38] Does this reflect an increased uptake of riboflavin into muscle tissue and increased riboflavin requirements due to exercise training? To address this question, researchers set out to estimate riboflavin requirements of young women participating in an exercise program.[34] The women participated in a 6-week program involving daily jogging and ingesting increasing levels of dietary riboflavin. The riboflavin intake required for an EGRAC of 1.25 was then determined in order to assess riboflavin requirement. The results suggested that healthy young women might need more riboflavin than the RDA standard, and that exercise further increases riboflavin requirements slightly. Somewhat similar results were produced in a study involving older women.[38] In this study,

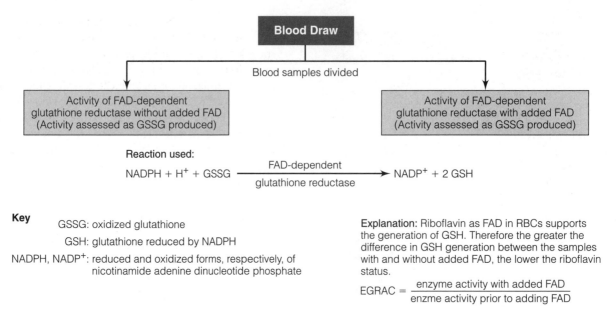

Figure 9-6
Basic protocol for assessing riboflavin status.

14 women 50–67 years of age participated in a 10-week study involving two riboflavin intake levels (0.6 and 0.9 μg/kcal). They trained on a cycle ergometer for 20–25 minutes six days a week at 75–85% of their maximal heart rate. Riboflavin status was assessed by measuring the EGRAC and urinary riboflavin excretion, and physical performance was assessed by VO$_2$max and anaerobic threshold. Exercise training did increase EGRAC and decreased urinary riboflavin status, which is suggestive of decreased body status, but increased riboflavin intake did not enhance improvements in endurance. That is, studies suggesting that exercise enhances riboflavin requirements failed to also show that increased dietary intake enhances performance. On the other hand, male athletes whose riboflavin intake met general recommendations (0.6 mg/1000 kcal) demonstrated normal EGRAC levels.[39] Researchers also reported that riboflavin status was not different from control values in female triathletes or tennis and track athletes.[40]

RIBOFLAVIN SUPPLEMENTATION AND ATHLETIC PERFORMANCE. Supplemental riboflavin does not seem to enhance performance in individuals who are in good status, but it may benefit those with poor riboflavin status. For instance, researchers provided 40 boys and girls (11–14.5 years old) who had poor vitamin status with a multivitamin and iron supplement (or placebo) for 5 weeks.[41] Those who received the supplement showed marked improvements in both riboflavin and vitamin C status. Furthermore, while the boys and girls not supplemented demonstrated a reduction in metabolic performance during treadmill running, supplementation prevented this from occurring. In another study involving

boys 12–14 years old, 19% had poor riboflavin status (EGRAC > 1.20).[42] When some of the boys were given 2 mg of riboflavin 6 days a week for 2 months, they tended to show improved VO$_2$max as riboflavin status improved. However, the results were not statistically significant.

RIBOFLAVIN *IN REVIEW*

- Riboflavin is found in whole grains and animal foods. Milk is a significant source.
- As a part of FAD and FMN, riboflavin participates in electron transfer in energy metabolism and the production of some amino acids and steroid hormones.
- Exercise may evoke shifts of riboflavin between tissues, but general riboflavin status is probably not affected and supplementation has not been shown to enhance exercise performance in well-nourished people.

NIACIN (VITAMIN B$_3$)

Niacin (vitamin B$_3$) occurs in two forms—*nicotinic acid* and *nicotinamide*—and is active in the body in two coenzyme forms, *nicotinamide adenine dinucleotide (NAD)* and *nicotinamide adenine dinucleotide phosphate (NADP)* (Figure 9-7). Niacin is found in a variety of foods; brewer's yeast, fish, pork, beef, poultry, mushrooms, legumes, and potatoes offer high content (Table 9-5). Plant foods contain nicotinic acid. Animal foods contain nicotinamide as well as NAD and NADP.

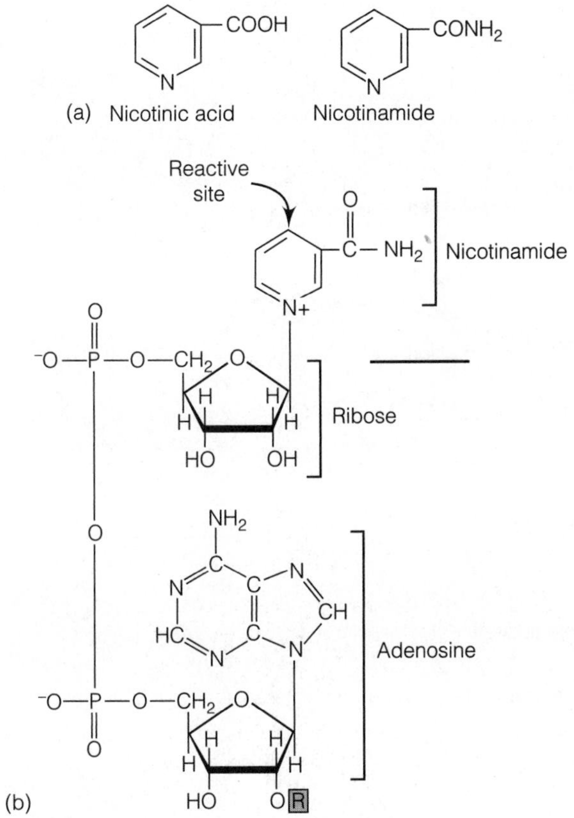

Figure 9-7

(a) Nicotinic acid and nicotinamide. (b) The structures of NAD and NADP.

The body can produce some niacin from the essential amino acid tryptophan, but it is an inefficient process: 60 mg of tryptophan are needed to synthesize 1 mg of niacin. Niacin recommendations are presented in milligrams, or as niacin equivalents (NE), which are units equivalent to a milligram of niacin to include tryptophan as a potential source. The RDA for niacin is 14 and 16 mg for adult females and males, respectively. As with thiamin and riboflavin, researchers speculate that estimated requirements for niacin would be more precise if based on energy expenditure (6.6 mg/1000 kcal). Thus the recommended dietary intake of niacin for athletes expending 3000–6000 kcal would be approximately 20–40 mg/day.

Key Point

Good food sources of niacin include meats and fish, eggs, legumes, nuts, mushrooms, and potatoes.

Niacin Functions

Most niacin functions involve electron transfer (redox). NAD and NADP are involved in more than 200 enzymatic

Table 9-5

Approximate Niacin Content of Select Foods

Food	Niacin (mg or NE)[a]
Meats and Seafood	
Liver (3 oz)	14.0
Tuna (3 oz)	10.3
Turkey (3 oz)	9.5
Chicken (3 oz)	7.9
Salmon (3 oz)	6.9
Veal (3 oz)	5.2
Beef, round steak (3 oz)	5.1
Pork (3 oz)	4.5
Haddock (3 oz)	2.7
Scallops (3 oz)	1.1
Nuts and Seeds	
Peanuts (1 oz)	4.9
Vegetables	
Asparagus (½ c)	1.5
Mushrooms, raw pieces (½ c)	1.4
Grains	
Wheat germ (1 oz)	1.5
Rice, brown (½ c)	1.2
Noodles, enriched (½ c)	1.0
Rice, white, enriched (½ c)	1.0
Bread, enriched (1 slice)	0.7
Milk and Milk Products	
Milk (1 c)	1.9
Cheese, cottage (½ c)	2.6
Other	
Brewer's yeast (1 oz)	10.7

[a] 1 NE = 1 mg niacin or 60 mg tryptophan.

reactions, including dehydrogenase reactions (such as pyruvate dehydrogenase). In these operations NAD is reduced to NADH, and NADP is reduced to NADPH. During carbohydrate oxidation NADH is produced in one reaction in glycolysis. Also, NADH is produced during the conversion of pyruvate to acetyl CoA in mitochondria as well as in three reactions of the Krebs cycle (Figure 9-8). One NADH is created for each "turn" of β-oxidation of fatty acids, as displayed in Figure 9-8. NADH is also produced during the metabolism of alcohol (ethanol).

Niacin is involved in cell processes that are not directly related to energy pathways. For instance, NAD is necessary to convert vitamin B_6 to pyridoxic acid, a primary excretory metabolite. Also researchers believe that

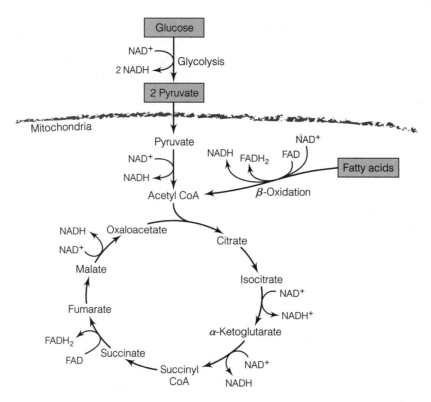

Figure 9-8
Involvement of NAD and FAD in oxidation reactions in glycolysis and Kreb's cycle as well as the conversion of pyruvate to acetyl CoA.

NAD may be a component of glucose tolerance factor (GTF). NADPH is needed for the synthesis of fatty acids and cholesterol, and it provides necessary electrons during the production of steroid hormones and deoxynucleotides for DNA.

Key Point

Like riboflavin, niacin is vital for transferring electrons derived from energy pathways to the electron transport chain for aerobic ATP formation. This makes niacin supplements attractive to athletes.

Niacin Imbalance

Symptoms of niacin deficiency might begin with decreased appetite, weight loss, and a general feeling of weakness. Aerobic ATP production suffers tremendously when niacin status is compromised. But when too much niacin (>100 mg) is ingested, it can cause vasodilation of the skin, leading to skin flushing and irritability. Higher doses of niacin are sometimes prescribed for people with elevated levels of blood lipids. The daily dose of >500 mg/day is split over the course of the day to reduce the occurrence and severity of side effects. A UL for niacin has been set at 35 mg/day for adults.

Niacin Intake by Athletes

Niacin intake levels of more active people and competitive athletes are probably at least adequate to meet RDA levels in light of their higher energy consumption. Male and female field athletes assessed in one study were found to consume adequate levels of niacin,[8] and similar findings were made in another study that assessed female athletes participating in a variety of sports.[43] One report involving young gymnasts found that the average dietary niacin levels were below recommendations, but in that study the potential conversion from tryptophan was not considered and there was no mention of enrichment and fortification of foods available to the participants.[20] For individuals who engage in energy restriction, special consideration must be made for niacin and other vitamin levels in the diet. Athletes need to be aware of foods rich in essential nutrients when restricting their energy intake and assess the composition of their diet to identify inadequacies.

Key Point

Niacin supplementation in megadoses might decrease athletic performance by inhibiting the mobilization of fatty acids from adipose tissue triglycerides.

Niacin and Athletic Performance

Niacin is crucial to optimal athletic performance because it is involved in both aerobic and anaerobic energy metabolism. However, excessive niacin in the form of nicotinic acid can work as a potent inhibitor of fatty acid release from adipose tissue. Thus supplementation of niacin at near or greater than gram doses may decrease FFA availability to working muscle and influence muscle fiber metabolism and performance. For instance, researchers reported that when nicotinic acid was ingested during cycling exercise, the rise in plasma FFA levels typically associated with endurance exercise was blocked.[44] Researchers also investigated the effects of nicotinic acid supplementation (3 g) on fuel utilization during exercise of trained runners.[45] They determined that nicotinic acid significantly impaired the typical rise in FFAs and glycerol associated with endurance bouts, and that the runners' respiratory exchange ratio (RER) became elevated, indicating a shift toward greater carbohydrate and less fatty acid utilization during the exercise. The results from these studies strongly suggest that, unless it is medically prescribed, athletes should not supplement niacin (nicotinic acid) at levels several times greater than the RDA. However, questions still remain regarding lower doses of niacin supplementation and performance. Interestingly, one study involving nicotinic acid supplementation indicated that the blunting of FFA release from adipose tissue was associated with an increase of three to six times in the level of growth hormone during exercise.[44]

■ NIACIN *IN REVIEW*

- Niacin is found in foods including meats, whole grains, mushrooms, and potatoes.
- Niacin is active as NAD, which is involved in energy metabolism, and NADP, which is more involved in synthetic operations.
- Niacin intake is believed to be adequate for most athletes, and low-dose supplementation probably does not enhance exercise performance (such as endurance) in a well-nourished individual.
- Higher dosages of niacin may decrease performance by reducing the mobilization of fatty acids.

■ BIOTIN

Biotin (Figure 9-9) is involved in several energy processes and is widely available at low levels in many foods. Among the best sources are liver, oatmeal, almonds, roasted peanuts, wheat bran, brewer's yeast, and molasses. Although biotin is not concentrated in them, milk and milk products are a significant source because of their prevalence in the diet. Some biotin is also derived from

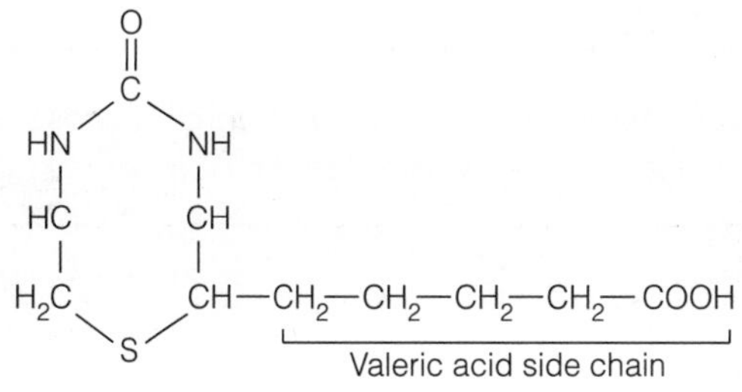

Figure 9-9
The structure of biotin.

bacterial synthesis in the colon. The AI for biotin is 30 μg for adults.

Biotin Functions

Although biotin is often overlooked, it is vital in the metabolism of carbohydrates, fatty acids, and the amino acid leucine in muscle and the liver. It serves as a coenzyme for a few enzymes, namely *acetyl CoA carboxylase, β-methylcrotonyl CoA carboxylase, pyruvate carboxylase,* and *propionyl CoA carboxylase,* whose functions are described in Table 9-6. Tissues with higher energy demands such as muscle, brain and liver have greater biotin content. Biotin and its metabolites (bisnorbiotin, biotin sulfone, and biotin sulfoxide) are primarily excreted in urine.

Table 9-6

Functions of Enzymes Related to Biotin

Enzyme Requiring Biotin as a Coenzyme	Activity of Enzyme
Acetyl CoA carboxylase	A rate-limiting step in fatty acid synthesis
Pyruvate carboxylase	Conversion of pyruvate to oxaloacetate (OAA) to maintain Krebs cycle intermediates and gluconeogenesis
Propionyl CoA carboxylase	Complete oxidation of odd-chain-length fatty acids; complete metabolism of the BCAAs isoleucine and methionine
β-Methylcrotonyl CoA carboxylase	Necessary for the complete breakdown of leucine (essential amino acid)

Key Point

Biotin is involved in the production of glucose and fatty acids, and in the breakdown of leucine.

Biotin Imbalance

Documented cases of independent biotin deficiency are rare because biotin is generally available in foods and is also produced by intestinal bacteria. However, a restrictive diet might decrease biotin status over time, especially when combined with certain antibiotics. Also, egg whites contain a protein that inhibits biotin absorption. Fortunately, heating egg whites inactivates this protein, so athletes eating lots of eggs or egg whites need not be concerned. Biotin toxicity is also rare, as loss through urine hinders excessive accumulation of biotin in tissue. A UL for biotin has not been set.

Biotin and Athletic Performance

Biotin has received little attention with regard to athletic performance. However, biotin status could be a major factor at several points in energy metabolism during exercise. Pyruvate carboxylase activity is important during gluconeogenesis in the liver during endurance exercise, so biotin helps prevent hypoglycemia and central fatigue during prolonged exercise. In skeletal muscle, biotin aids in the complete breakdown of isoleucine, a BCAA believed to be an important energy source during endurance activity.

Whether exercise influences biotin status remains to be conclusively determined. In one study of physically active men, baseline measures of blood biotin were considered normal.[46] However, when the participants were provided a multivitamin/mineral supplement that included 200 mg of biotin (396% RDA), blood biotin levels were increased. Urinary biotin levels remained elevated during the supplementation and remained elevated as blood biotin levels returned toward baseline levels. Therefore, biotin status can be enhanced via supplementation.

BIOTIN SUPPLEMENTATION AND PERFORMANCE. Although it seems that supplementation can transiently increase biotin availability, it is not known whether this has a positive influence on performance; studies investigating the possibility are lacking. Yet biotin is commonly a part of multivitamin supplements. In one study, a variety of athletes (basketball, gymnastics, rowing, swimming) were provided a multivitamin/mineral supplement containing biotin or a placebo.[39] Diets (foods alone) were monitored during the trial to make sure that recommended levels were met without supplementation. No enhancement

in performance was reported for the supplemented group. Based on the limited information available, it does not seem that biotin has ergogenic potential, but more research is necessary.

■ BIOTIN *IN REVIEW*

- Biotin is found in a variety of foods and is also produced by bacteria in the colon.
- Biotin is a coenzyme for a few carboxylase enzymes that are involved in gluconeogenesis, fatty acid synthesis, and the metabolism of odd-chain-length fatty acids and certain amino acids for energy.
- The relationship between biotin metabolism and exercise is poorly researched; however, the current consensus is that supplementation does not favorably influence performance and energy metabolism in a well-nourished individual.

■ PANTOTHENIC ACID

Along with biotin, pantothenic acid is one of the least recognized vitamins, but it is a component of two crucial molecules involved in energy metabolism. Pantothenic acid is widely available in foods. Good sources include egg yolk, organ meats and meats, tomato, whole grain products, legumes, broccoli, milk, sweet potatoes, mushrooms, avocados, molasses, and royal jelly from bees. Pantothenol and salts of pantothenate (such as calcium pantothenate) are typically used in supplements. The AI for pantothenic acid is 5 mg/day for adult men and women.

Pantothenic Acid Functions

Pantothenic acid is a component of two special molecules that affect the metabolism of carbohydrate, protein, and fatty acids. These molecules are *coenzyme A (CoA)* (Figure 9-10) and *acyl carrier protein (ACP)*. The attachment of CoA (in combined forms such as acetyl CoA, succinyl CoA, propionyl CoA, and malonyl CoA) is said to activate certain molecules, allowing them to be further metabolized. In addition, fatty acids must be activated prior to their use in β-oxidation, which is of particular importance during exercise. The second molecule containing

Key Point

Pantothenic acid is a vital component of coenzyme A (CoA), making it fundamentally important in exercise energy metabolism.

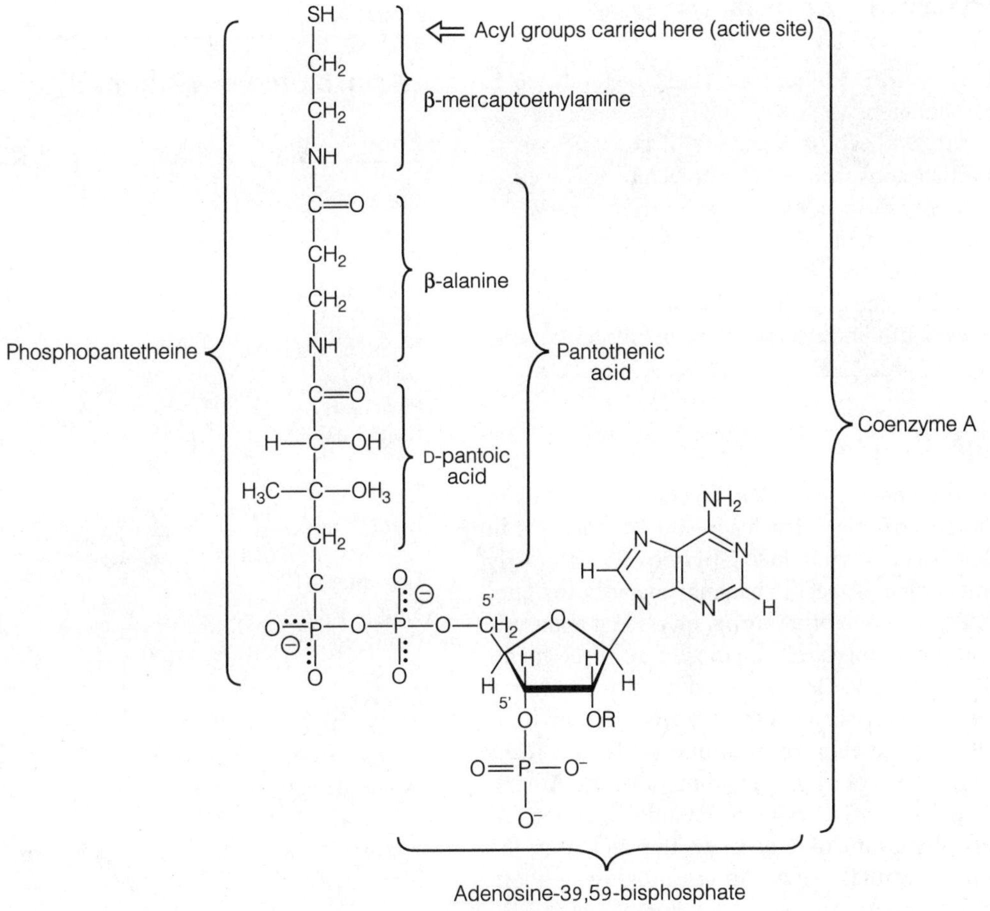

Figure 9-10
Structure of coenzyme (includes pantothenic acid).

pantothenic acid, acyl carrier protein (ACP), is needed in a preliminary step in fatty acid synthesis as a component of the large enzyme complex *fatty acid synthase (FAS)*.

Pantothenic acid is found in the blood largely as part of coenzyme A in RBCs, and some free pantothenic acid is dissolved in the plasma. For the most part, pantothenic acid is removed from the body in urine.

Pantothenic Acid Imbalance

Deficiency of pantothenic acid is rare because it is contained in so many foods. Reduced pantothenic acid status would logically influence the ability to derive energy from carbohydrate, fat, protein, and alcohol because of its involvement in coenzyme A. Thus symptoms might include feeling sluggish and fatigued, and there could be neuromuscular disturbances such as decreased coordination. Excess pantothenic acid is also unusual because it is efficiently eliminated in urine. A UL for pantothenic acid has not been set.

Pantothenic Acid and Athletic Performance

The importance of normal pantothenic acid status to athletic performance is undeniable, but whether pantothenic acid supplements enhance performance has not been thoroughly tested. As a component of coenzyme A it is pivotal in the oxidation of fatty acids, carbohydrates, and amino acids during exercise. However, when researchers gave 1 g of pantothenic acid to runners for 2 weeks, their performance was not enhanced and alterations in biochemical parameters were not observed.[47] In a related study involving competitive bicyclists, pantothenic acid, pantetheine (pantothenic acid derivative), and allithiamin supplements failed to cause positive changes in blood lactate levels or performance measures during cycling trials.[28] Based on the limited information available, it does not seem that pantothenic acid has ergogenic potential by itself at modest doses, but future studies involving pantothenic acid alone and in combination with other substances would be worthwhile.

■ PANTOTHENIC ACID *IN REVIEW*

- Pantothenic acid is found in a variety of foods, so the risk of deficiency is low if energy needs are met.
- Pantothenic acid is a component of coenzyme A (CoA), which activates certain molecules involved in energy operations, and acyl carrier protein (ACP), which is involved in fatty acid synthesis.
- Intake of pantothenic acid is currently believed to be sufficient in active individuals, but research on how exercise influences its status is incomplete.

■ VITAMIN B₆

Vitamin B_6 is the general name for several compounds in food with similar structure that have similar functions in the body. One form of vitamin B_6, pyridoxine, is mostly found in plant foods, especially bananas, navy beans, and walnuts (Table 9-7). Two other forms (pyridoxal and pyridoxamine) and phosphorylated versions of all three forms of vitamin B_6 (Figure 9-11) are found mostly in animal foods, such as meats, fish, and poultry. As vitamin B_6 is closely tied to the metabolism of amino acids, the RDA for it is based on the typical protein content of the American diet. Approximately 0.016 mg of vitamin B_6 is considered necessary per gram of protein in the diet. For the typical daily protein intake of an American adult of about 100–125 g, this translates to about 1.6–2 mg of vitamin B_6. The RDA for vitamin B_6 is 1.3 mg for adult men and women ages 19–50. Athletes consuming higher protein diets might need more vitamin B_6, but this notion has not been thoroughly researched.

Table 9-7

Approximate Vitamin B₆ of Select Foods

Food	Vitamin B₆ (mg)
Meat and Seafood	
Liver (3 oz)	0.8
Salmon (3 oz)	0.7
Chicken (3 oz)	0.4
Ham (3 oz)	0.4
Hamburger (3 oz)	0.4
Veal (3 oz)	0.4
Pork (3 oz)	0.3
Beef (3 oz)	0.2
Eggs	
Egg (1)	0.3
Legumes	
Split peas (½ c)	0.6
Beans, cooked (½ c)	0.4
Fruits	
Banana (1)	0.6
Avocado (½ c)	0.4
Watermelon (1 c)	0.3
Vegetables	
Brussels sprouts (½ c)	0.4
Potato (1)	0.2
Sweet potato (½ c)	0.2
Carrots (½ c)	0.1
Peas (½ c)	0.1

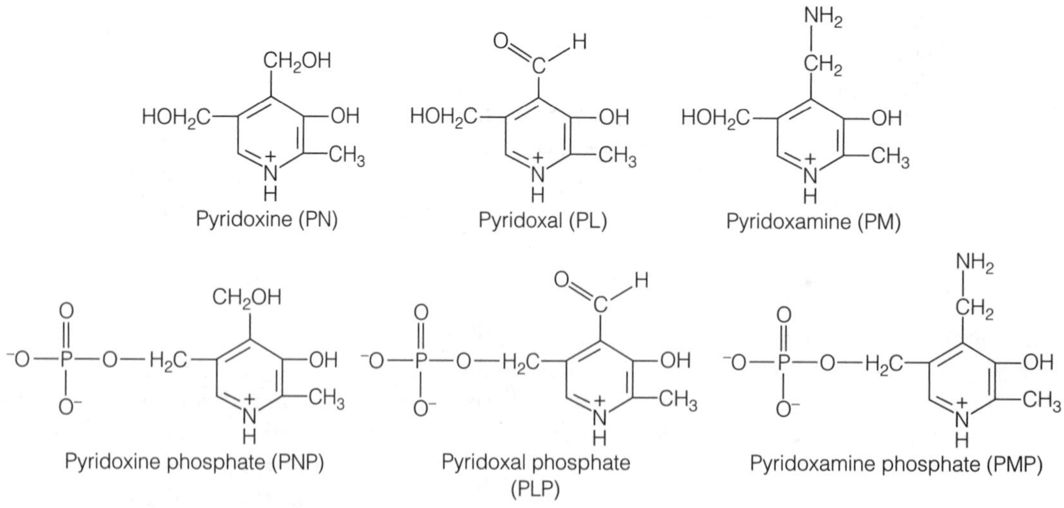

Figure 9-11
Vitamin B_6 forms.

Vitamin B$_6$ Functions and Metabolism

Vitamin B$_6$ is largely active in the phosphorylated form pyridoxal phosphate (PLP). PLP is involved in numerous cellular operations, including amino acid metabolism, where it has a role in transamination and deamination reactions (discussed in Chapter 4). These reactions allow the production of several nonessential amino acids as well as the breakdown of amino acids during fasting and endurance exercise. Vitamin B$_6$ is also involved with other tissue operations including niacin formation and glycogen breakdown. The breakdown of glycogen by the enzyme *glycogen phosphorylase* is influenced by PLP. Roughly 80% of the vitamin B$_6$ in an adult's body may be associated with skeletal muscle phosphorylase.

Vitamin B$_6$ (as PLP) is also involved in certain reactions that allow the formation of *γ-aminobutyric acid (GABA)* and catecholamines (such as epinephrine) from glutamic acid and tyrosine, respectively, as well as the production of serotonin from 5-hydroxytryptophan (Figure 9-12).[48] The involvement of vitamin B$_6$ in glycogen breakdown and the production of epinephrine and serotonin are of interest to exercise nutritionists. Efficient glycogen breakdown is fundamentally important to most competitive sports and higher intensity physical activity. The production of epinephrine and serotonin are of particular interest, as they would seem to have contrasting influences on exercise performance.

ASSESSMENT OF VITAMIN B$_6$ STATUS. Vitamin B$_6$ is primarily transported in the blood within cells such as RBCs or bound to the protein albumin. The liver and other tissues can convert vitamin B$_6$ to its metabolites 4-pyridoxic acid (4-PA) and pyridoxal lactone. These metabolites and some vitamin B$_6$ are excreted in the urine. Plasma PLP and total vitamin B$_6$ are commonly used to assess vitamin B$_6$ status. RBC vitamin B$_6$ content and transaminase activity or measurement of the urinary levels of metabolites such as 4-PA can also be used.

Several indicators should be used in concert to develop a more comprehensive understanding of status. Nutritional and physiological factors, such as high-protein diets, smoking, and age, can influence plasma PLP levels. Also, it is now known that oral contraceptive use by females can increase such transaminases but decrease vitamin B$_6$ pools, particularly in the blood. Therefore exercise studies involving females of reproductive age need to consider the use of oral contraceptives if RBC transaminase activity is measured as an indicator of vitamin B$_6$ status.

Key Point

Vitamin B$_6$ is fundamentally important in amino acid and glycogen metabolism, so it has a strong role in efficient energy metabolism during exercise.

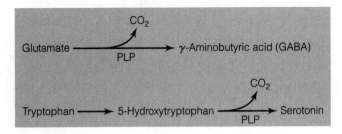

Figure 9-12
The formation of several amino acid derivatives such as the neurotransmitters GABA and serotonin require vitamin B$_6$ as pyridoxal phosphate, PLP).

Vitamin B$_6$ Imbalance

Deficiency of vitamin B$_6$ is unlikely when a diet provides adequate energy and includes a variety of foods including whole grain products. However, as with other vitamins, there is some concern during prolonged energy restriction to facilitate weight loss. In vitamin B$_6$ deficiency amino acid metabolism would be greatly restrained, leading to poor protein synthesis and turnover. The synthesis of hemoglobin, white blood cells, and many neurotransmitters would also be hindered. Excessive consumption of vitamin B$_6$ via supplementation can cause neurological disturbances. Long-term effects of taking 2 g/day or more include symptoms such as pins-and-needles feelings in the feet. A UL for vitamin B$_6$ has been set at 100 mg for adults.

Vitamin B$_6$ Intake of Active People

Many studies suggest that males are more likely than females to have adequate dietary intakes of vitamin B$_6$, especially females with inadequate energy intakes.[49,50] Assessment of the dietary intake of highly trained female cyclists revealed that more than a third of them consumed less than 67% of the RDA for vitamin B$_6$.[51] Similar findings were reported with female heavyweight rowers, who consumed an average of 2633 kcal/day and adequate protein but less than two-thirds of the RDA for vitamin B$_6$.[52] In a study involving male German speed skaters, many did not consume recommended levels.[50] However, in other studies male and female field athletes[6] were found to eat adequate vitamin B$_6$, as did collegiate male cross-country runners.[22]

Researchers have pointed out that vitamin B$_6$ is among the nutrients that are commonly consumed in inadequate quantities through food alone, which may particularly affect elderly people who exercise.[53] The other nutrients that are often lacking included vitamin B$_{12}$, calcium, and vitamin D. A supplement containing vitamin B$_6$ should be considered if diet alone is inadequate in meeting recommended levels. Virtually all vitamin supplements purported to increase energy metabolism contain the B-complex vitamin group, including vitamin B$_6$.

Vitamin B₆ and Athletic Performance

The involvement of vitamin B₆ in amino acid and glycogen metabolism makes it crucial to athletic performance. Researchers set out to determine the vitamin B₆ status of endurance athletes before, at the finish, and 2 hours after running a marathon.[54] Whole blood and serum vitamin B₆ levels, urine 4-PA levels, and alanine transaminase (ALT) activity in RBCs were used to assess status. Before the marathon, serum concentrations of vitamin B₆, ALT, and 4-PA were higher than reference values and only the total blood vitamin B₆ levels were below normal. The researchers estimated that the marathon race resulted in an average loss of approximately 1 mg of vitamin B₆ per participant. Elsewhere, in the study involving boys and riboflavin discussed earlier, researchers also assessed vitamin B₆.[42] Here it was determined that approximately one-fourth had poor vitamin B₆ status. When these individuals received 2 mg of vitamin B₆ regularly for 6 weeks, their physical work capacity (VO₂max) improved as their status improved.

In another study, researchers investigated the influence of vitamin B₆ depletion and supplementation and found that it may have applications to athletes. Their findings may also provide useful information regarding the assessment of vitamin B₆ status.[55] College-age males received a diet low in vitamin B₆ for 6 weeks, followed by a self-selected diet supplemented with vitamin B₆ (pyridoxine HCl). During depletion, excretion of 4-PA rapidly adjusted to approximate the intake. Plasma PLP averaged 81, 9, and 455 nmol/L at the end of the baseline, depletion, and supplementation periods, respectively. Muscle PLP and total vitamin B₆ values changed little throughout the dietary manipulation. The results suggest that vitamin B₆ pools in skeletal muscle are resistant to depletion, and

that depletion or supplementation does not initially affect the vitamin B₆ content of muscle. Because exercise training was not a component of this study, it is difficult to know how this information applies to athletes.

VITAMIN B₆ SUPPLEMENTATION AND ATHLETIC PERFORMANCE.

Vitamin B₆ plays a key role in several events associated with exercise energy metabolism.[47] First, vitamin B₆ is involved in the breakdown of glycogen stores. Second, as shown in Figure 9-13, the breakdown of amino acids in skeletal muscle during endurance bouts involves the removal of amino groups via PLP-dependent transaminases. Third, the formation of alanine involves PLP-dependent transaminases. Fourth, PLP is required to manufacture carnitine, which is needed to shuttle longer chain fatty acids across the mitochondrial inner membrane for oxidation. These functions have led to the notion that athletic performance might be enhanced by supplemental vitamin B₆.

In a study involving swimmers, researchers provided 51 mg of vitamin B₆ (pyridoxine HCL) daily for 6 months, but this did not improve the athletes' performance time in 100-yard sprints.[56] Nor did vitamin B₆ supplementation (20 mg/day) lengthen time till exhaustion in cyclists, although plasma FFAs were lower before, during, and after the exercise trial in comparison to a control (placebo) group.[57,58] In another study involving young and postmenopausal women (trained and untrained), vitamin B₆ supplementation was associated with decreased circulating FFAs during a 20-minute cycling bout at 80% VO₂max.[59]

Currently the research suggests that vitamin B₆ supplementation is not necessary for athletes to achieve good or "normal" levels if a balanced diet with adequate energy is consumed. However, several questions remain about the

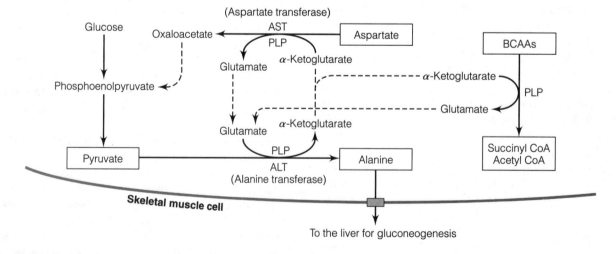

Figure 9-13

Transamination reactions involved in the breakdown of BCAAs and aspartate in skeletal muscle during fasting and endurance exercise. The aminotransferases involved are dependent on vitamin B₆ as pyridoxal phosphate, PLP.

influence of other dietary factors, such as protein and vitamin B_6 status indicators. It is also uncertain how supplementation of vitamin B_6 affects glycogen, fat, and BCAA metabolism as well as the production of serotonin during exercise. Perhaps, as with the other vitamins, interest in supplementation studies involving vitamins has waned in light of higher profile supplements such as creatine.

■ VITAMIN B_6 *IN REVIEW*

- Vitamin B_6 is provided in the diet mostly in animal foods and is mildly sensitive to cooking processes such as prolonged heat.
- Vitamin B_6 is principally involved in amino acid metabolism (such as transamination) and in glycogen breakdown.
- Optimal vitamin B_6 is important in support of efficient exercise metabolism by aiding alanine, glutamate, and glutamine formation during gluconeogenesis and amino acid oxidation in muscle fibers.
- It is unclear what influence vitamin B_6 supplementation may have on glycogen breakdown in prolonged exercise performance, but supplementation at high levels is not common among athletes.

■ FOLATE (FOLIC ACID)

The word *folate* is derived from the Latin term *folium*, as is the word *foliage*. Indeed, plants are among the better sources for this vitamin, but folate is also available from animal sources (Table 9-8). Good sources include vegetables such as asparagus, mushrooms, brussels sprouts, broccoli, turnip greens, lima beans, peas, nuts, some citrus fruits and juices, organ meats, and whole grain products. Refined grain products must be fortified with folate in the United States, and some manufacturers have fortified food products such as fruit juices. The RDA for folate is 400 µg daily for adults. The structure of folate is shown in Figure 9-14.

Folate Functions and Metabolism

The active form of folate is tetrahydrofolate (THF). THF accepts single-carbon groups from other molecules and, functioning as a coenzyme, helps incorporate them into molecules undergoing synthesis. Among the most significant of these molecules are building blocks of DNA. Folate is particularly important in cell types that reproduce rapidly, such as RBCs and cells in the skin and digestive tract. Folate is also involved in transferring single-carbon molecules in the processing of certain amino acids.

Because folate is fundamentally involved in DNA production and thus reproduction of cells, periods of the life span when rapid growth occurs demand higher folate intake. During pregnancy, a woman's diet must provide additional folate to assist in the rapid production of cells of the fetus and expanded tissue (blood, placenta). Optimal folate status prior to conception is important in minimizing the risk of birth defects (see Chapter 15). All prenatal vitamin supplements contain folate.

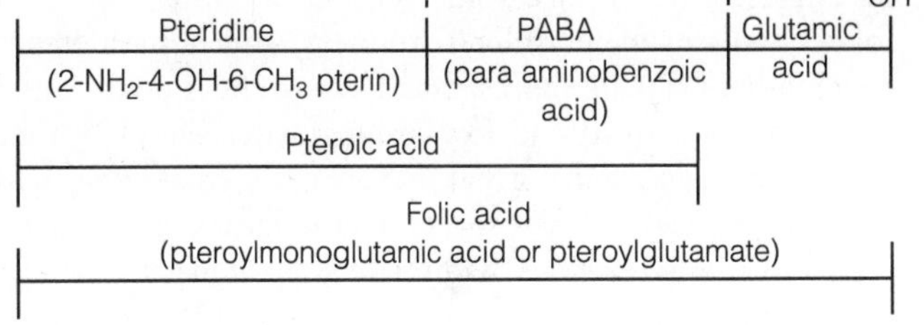

Table 9-8

Approximate Folate Content of Select Foods

Food	Folate (µg)
Vegetables	
Asparagus (½ c)	120
Brussels sprouts (½ c)	116
Black-eyed peas (½ c)	102
Spinach, cooked (½ c)	99
Lettuce, romaine (1 c)	86
Lima beans (½ c)	71
Peas (½ c)	70
Sweet potato (½ c)	70
Broccoli (½ c)	43
Fruits	
Cantaloupe (¼)	100
Orange juice (1 c)	85
Orange (1)	60
Grains	
Oatmeal (½ c)	95
Wild rice (½ c)	40
Wheat germ (1 tbs)	20

Figure 9-14
Folate structural components (pteridine, PABA, glutamic acid).

Key Point

Folate is crucial for effective RBC formation; thus adaptation to aerobic training might be hindered if folate status is compromised.

Folate Imbalance

The potential for folate deficiency has been reduced in industrialized countries due to enrichment and fortification efforts, supplementation, and public education. Poor folate status can decrease the efficiency of RBC formation and cell production. Excessive folate is difficult to consume in foods alone and is excreted in the urine. The level of folate in nutrition supplements is often limited based on the relationship between folate and vitamin B_{12} status. A UL for folate has been set at 1000 µg for adults.

Folate Intake by Athletes

A research review published a little over a decade ago stated that folate and vitamins E, B_6, and B_{12} are the vitamins most likely to be deficient from athletes' diet if they restrict their energy intake to maintain or reduce body weight.[60] Folate intake of more than one-third of highly trained female cyclists in one study was well below the RDA.[61] In another study it was determined that male and female field athletes consumed adequate amounts of folate.[6] Similarly, adequate folate consumption has been reported among elite Finnish ski jumpers[21] and male cross-country runners during the competitive season.[22] Folate in multivitamin supplements also reduces the number of athletes whose intake does not provide at least the RDA level.

Folate and Athletic Performance

There are several reasons why folate requirements might be higher during exercise training: the need to produce more RBCs, increased formation of nucleotides, and increased metabolism of certain amino acids during enhanced protein turnover. However, folic acid metabolism in relation to exercise training has not been thoroughly investigated. There are some reports of similar levels of folate status in male athletes (runners) and nonathletes.[62] In a study of female marathoners, researchers determined that roughly a third had low serum folate levels.[61] Other researchers reported normal indicators of folate status in athletes in Ireland.[63] But several of the female athletes assessed in the latter study had folate status measures at the lower end of the standard range.

FOLATE SUPPLEMENTATION AND ATHLETIC PERFORMANCE. Supplementation of folate in athletes with lower-than-normal folate status has not been shown to improve performance, despite improvements in measures of folate status. For instance, researchers found that folate status indicators improved following several months of supplementation,[12] but another study focused on performance found no ergogenic effect from a multivitamin and mineral supplement that included folate.[64,65] When female marathoners with low serum folate levels were supplemented with 5 mg/day (>10 times RDA) for 10 weeks, they did not demonstrate improvements in treadmill running performance or metabolic measures.[61] More research is needed to conclusively determine the relationship between this vitamin and exercise performance.

■ FOLATE *IN REVIEW*

- Folate is found in a variety of plant sources and organ meats.
- Folate is involved in the transfer of single-carbon groups, as in the making of nucleic acids.
- Some research shows that many athletes consume less folate than the RDA. However, little is known about the effect of exercise on folate metabolism, and researchers have not demonstrated a performance-enhancing effect with supplementation, even when folate is below normal.

■ VITAMIN B_{12}

Vitamin B_{12} is also called cobalamin because it contains an atom of cobalt (Figure 9-15). In the diet, vitamin B_{12} is found only in foods of animal origin (Table 9-9). The best dietary sources are meats, fish, poultry, shellfish, eggs, milk, and milk products. The vitamin B_{12} content in these foods is modest but compatible with needs. As discussed in Chapter 15, vegetarians need to seek out vitamin B_{12}–fortified foods that satisfy their dietary requirements, or they need to acquire this vitamin in a supplement. The RDA for vitamin B_{12} is 2.4 µg for adults; a typical diet contains three times that much. Recommendations for athletes have not yet been made.

Vitamin B_{12} Functions

Vitamin B_{12} is directly involved in the metabolism of folate; too little vitamin B_{12} can inhibit folate metabolism to the point of folate deficiency. As shown in Figure 9-16, vitamin B_{12} is required for the transfer of a methyl group to homocysteine to generate methionine. This reaction converts or "recycles" an inactive form of folate (N^5methyl THF) back to an active form. Without adequate vitamin B_{12}, folate remains trapped in the inactive form, which has led to the phrase *methyl-folate trap*. The recycling of folate, via vitamin B_{12}, ultimately decreases the dietary requirement for folate.

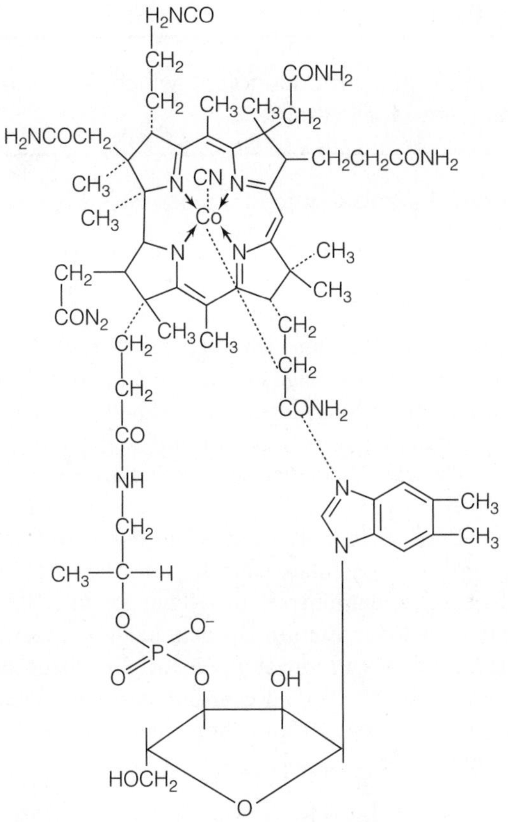

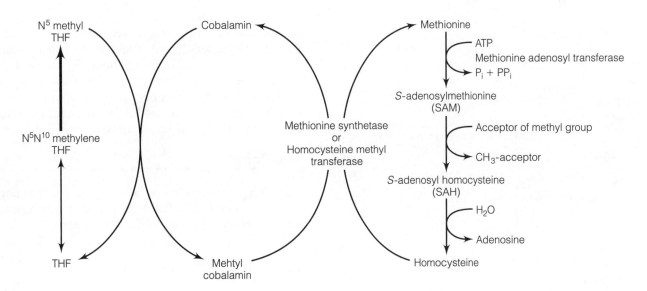

Figure 9-15
Vitamin B$_{12}$ (cobalamin) structure with atom of cobalt in its center.

Table 9-9

Approximate Vitamin B$_{12}$ of Select Foods

Food	Vitamin B$_{12}$ (μg)
Meats	
Liver (3 oz)	6.8
Trout (3 oz)	3.6
Beef (3 oz)	2.2
Clams (3 oz)	2.0
Crab (3 oz)	1.8
Lamb (3 oz)	1.8
Tuna (3 oz)	1.8
Veal (3 oz)	1.7
Hamburger (3 oz)	1.5
Eggs	
Egg (1)	0.6
Milk and Milk Products	
Milk, skim (1 c)	1.0
Milk, whole (1 c)	0.9
Yogurt (1 c)	0.8
Cottage cheese (1/2 c)	0.7
Cheese, American (1 oz)	0.2
Cheese, cheddar (1 oz)	0.2

Figure 9-16
The resynthesis of methionine from homocysteine, showing the roles of folate and vitamin B$_{12}$.

Vitamin B_{12} is also required for the breakdown of certain amino acids and fatty acids (odd-chain-length) for ATP production. In this complex reaction, propionyl CoA is converted to succinyl CoA, an intermediate of the Krebs cycle (Figure 9-17). In this capacity, vitamin B_{12} assists gluconeogenesis. Vitamin B_{12} also appears vital in maintaining the special insulating wrapping around nerve cells, called **myelin**. In contrast to other water-soluble vitamins, vitamin B_{12} losses from the body are small and occur primarily through the feces. Small quantities of vitamin B_{12} enter the digestive tract daily as part of bile released during meals. Most of this vitamin B_{12} is then reabsorbed from the digestive tract, as only about 0.1% of vitamin B_{12} stores are excreted daily.

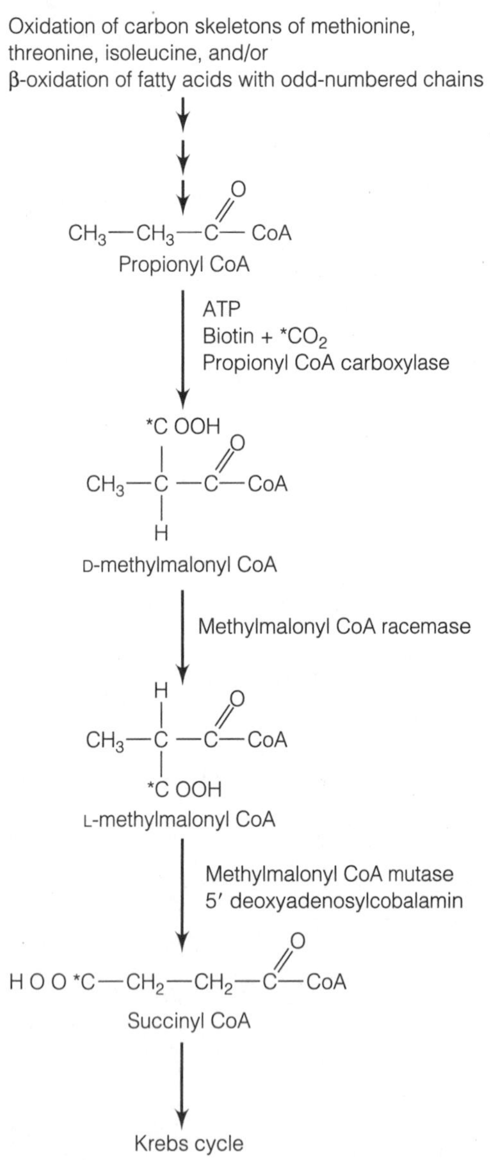

Oxidation of carbon skeletons of methionine, threonine, isoleucine, and/or β-oxidation of fatty acids with odd-numbered chains

$CH_3-CH_3-\overset{\displaystyle O}{\overset{\|}{C}}-CoA$
Propionyl CoA

ATP
Biotin + *CO_2
Propionyl CoA carboxylase

$\begin{array}{c} *C\,OOH \\ | \\ CH_3-C-\overset{\displaystyle O}{\overset{\|}{C}}-CoA \\ | \\ H \end{array}$
D-methylmalonyl CoA

Methylmalonyl CoA racemase

$\begin{array}{c} H \\ | \\ CH_3-C-\overset{\displaystyle O}{\overset{\|}{C}}-CoA \\ | \\ *C\,OOH \end{array}$
L-methylmalonyl CoA

Methylmalonyl CoA mutase
5′ deoxyadenosylcobalamin

$H\,O\,O\,*C-CH_2-CH_2-\overset{\displaystyle O}{\overset{\|}{C}}-CoA$
Succinyl CoA

Krebs cycle

Figure 9-17
Role of vitamin B_{12} in oxidation of odd-numbered-chain fatty acids and selected amino acids.

Key Point

Vitamin B_{12} helps recycle folate, which decreases the dietary need for folate.

Vitamin B_{12} Imbalance

An individual with good vitamin B_{12} stores could theoretically eat a diet lacking it for years before showing signs of deficiency. This means that a strict vegetarian eating only plant-derived foods, and who became a vegetarian later in life, might not show signs of vitamin B_{12} deficiency for a long time. How long depends on how much vitamin B_{12} storage existed prior to becoming vegetarian, and the rate of loss. If a child is raised vegetarian, the option of a vitamin B_{12} supplement should be discussed with a pediatrician. Poor vitamin B_{12} status can decrease folate recycling and reduce folate status. This can decrease cell production, as demonstrated by changes in RBCs. In severe folate and/or vitamin B_{12} deficiency, *macrocytic megaloblastic anemia* can develop, as might disturbances in nervous tissue function. Regarding excessive vitamin B_{12} intake and toxicity, not much information is available, so a UL cannot be set.

Vitamin B_{12} Intake by Athletes

Vitamin B_{12} consumption might be a concern for some active people and athletes. Researchers identified vitamin B_{12} as a particular concern for elderly people who are active,[53] and more than a third of highly trained female cyclists failed to consume at least the RDA.[51] Researchers also determined that a significant portion of female heavyweight rowers in one study failed to consume the RDA level for vitamin B_{12}.[52] It has been suggested that poor dietary intake and status of certain nonenergy nutrients, including vitamin B_{12}, might contribute to menstrual abnormalities in some female athletes who do not consume adequate energy. The level of vitamin B_{12} in the diet of athletes should be assessed, especially for those who follow restrictive vegetarian diets or simply eat less animal-based food. Vitamin B_{12} is often added to nutrition supplements marketed to athletes, which can enhance existing diet levels.

Key Point

Vitamin B_{12} intake might be below recommendations for some athletes who follow a restrictive vegetarian diet and do not regularly use a supplement.

Vitamin B_{12} and Athletic Performance

Vitamin B_{12} is important to athletic performance for several reasons: it is involved in recycling folate, in neural sheathing for effective neuromuscular communication, and in metabolism of amino acids. However, little is known about the vitamin B_{12} status of athletes. Based on urinary and blood values, researchers have reported that vitamin B_{12} status in athletes may be similar to that in nonathletes.[62,63] Supplementation with vitamin B_{12} might increase these indicators by about 50% in active men. But whether increased dietary intake of vitamin B_{12} can enhance performance is unknown. Research to date suggests that supplementation does not enhance performance in well-nourished athletes. For instance, vitamin B_{12} supplementation for 6 weeks did not influence VO_2max or measures of strength and coordination.[66] When vitamin B_{12} was provided in a B-complex supplement to college-aged males for 6 weeks it had no significant influence on measures of endurance.[67]

■ VITAMIN B_{12} *IN REVIEW*

- Vitamin B_{12} is found only in animal foods. Because the rate of excretion is normally low, the development of deficiency in a previously well-nourished adult can take years.
- Vitamin B_{12} is involved in folate recycling and has lesser-defined roles in neural tissue. Deficiency results in megaloblastic anemia and profound inhibition of appropriate central nervous system development and function.
- The relationship between vitamin B_{12} metabolism and exercise is not well researched.

■ CHOLINE

Choline is a recent addition to the list of vitamins, as it appears to be essential at least for those people who have a reduced capacity to synthesize it in cells. Good sources of choline include milk, liver, eggs, and peanuts. The AIs for adult males and females are 550 and 425 mg, respectively. More investigation is required on the relationship between choline and exercise before recommendations can be developed that are specific to athletes.

Choline Functions and Metabolism

Choline is a derivative of the amino acid methionine and is a component of phosphatidylcholine (lecithin), an important phospholipid in cell membranes and also a component of acetylcholine, the neurotransmitter at neuromuscular junctions. Choline can be converted to betaine, which is involved in single-carbon-unit transfers such as in homocysteine metabolism. The body can form choline from methionine to meet some of its needs.

Choline Imbalance

Choline deficiency has been reported in individuals who have a decreased capacity to synthesize it in their cells. Diminished plasma choline and phosphatidylcholine concentrations and signs of liver damage are among the effects of choline deficiency. Signs of choline toxicity include a fishy body odor, diarrhea, increased sweating and salivation, low blood pressure (hypotension), and liver dysfunction. A UL for adults has been set at 3.5 g/day.

Choline and Athletic Performance

Choline has several applications to exercise. First, it is a component of acetylcholine, and optimal levels of acetylcholine in motor neurons promote efficient stimulation of skeletal muscle activity. Second, it is a component of phosphatidylcholine, so choline supplementation may enhance the integrity of muscle fiber membranes as well as

NUTRITION IN PRACTICE   **Consuming Enough Vitamins**

A common theme throughout this chapter is that if an athlete's diet contains a variety of foods at an energy level appropriate for at least weight maintenance, the vitamin intake should easily exceed recommendations for the general population. However, some sports such as wrestling, gymnastics, and bodybuilding are associated with energy restriction, which could lead to vitamin inadequacies. Athletes also need to consider how their food preferences and other factors affect the vitamin content of their diet.

Table A presents a 24-hour diet recall for an amateur runner along with the vitamin analysis. This diet is also used in Chapter 10 to assess the adequacy of mineral intake.

Athletes who have access to nutrition assessment software (such as Food Processor or Nutritionist) can easily input their foods consumed and monitor their intake during the season as well as the off-season. Utilizing nutrition assessment software, it was determined that the vitamin content of the 24-hour diet recall in Table A provided several times the RDA for most vitamins, as shown in Figure A. Athletes can also gauge whether their diet provides adequate levels of vitamins by assessing the general qualities of their nutritional intake. Their diet should adhere to general recommendations for health discussed in Chapter 1. A checklist of basic assessment questions follows.

CHECKLIST FOR PROVISION OF VITAMINS
Does the athlete's diet and training regularly provide the following?

- Vitamins A and D fortified milk and/or dairy foods
- Several (five or more) servings of fresh fruits and vegetables

Consuming Enough Vitamins (CONTINUED)

- Whole grain products
- Legumes and nuts
- Fish and leaner meats
- Vitamin-fortified foods such as cereals and sport bars
- Adequate energy for weight maintenance

- Limited (but some) energy from fat and from a variety of sources
- Only a limited contribution to carbohydrate made by sweets (candies, soda)
- Frequent exposure to sunlight

Table A   **24-Hour Diet Recall of a 34-Year-Old Male Who Runs 30–45 Miles/Week**

Meal/Snack	Food	Notable Vitamin Content
7:00 A.M. **Breakfast**	1 large banana	
	1 c fresh orange juice	C, FA
	2 c Wheaties (with the banana)	Vitamin fortified
	1 c skim milk	A, D, B$_2$
	2 c coffee	
	1 tbs creamer	
10:30 A.M. **Midmorning Snack**	1 c vanilla yogurt	
	½ c granola cereal (in yogurt)	Vitamin fortified
	8 oz grapefruit juice	
1:00 P.M. **Lunch**	Turkey sandwich:	
	2 slices whole wheat bread	B$_1$, B$_2$, PA, Biot
	6 oz turkey breast	B$_3$, B$_6$, PA
	3 oz reduced-fat provolone cheese	A, D
	1 tbs mustard	
	2 slices tomato	C, carotenoids
	2 c low-fat chocolate milk	A, D, B$_2$, B$_{12}$, PA
	1 orange (med)	C, FA
	12 baby carrots	β-Carotene
5:00 P.M. **Dinner**	8 oz salmon steak	B$_2$, B$_6$, B$_{12}$, D, FA, PA
	1 med. potato (baked)	B$_1$, B$_3$, C, PA
	1 tbs sour cream (on potato)	
	1 c steamed broccoli	C
	2 c green salad:	
	1 c lettuce (romaine)	FA, PA
	½ tomato	C, FA
	¼ cucumber	
	1 tbs low-fat ranch dressing	
	1 bran dinner roll	B$_1$, B$_2$, B$_3$, PA
	1 c applesauce (with cinnamon)	
8 P.M. **Snack**	1 sport bar (high protein, fortified)	Vitamin fortified
	1 c vanilla frozen yogurt	B$_2$, B$_3$, B$_{12}$, PA

Abbreviations: A, D, E, K, C, B$_1$, B$_2$, B$_3$, B$_6$, B$_{12}$ = vitamins A, D, E, K, C, thiamin, riboflavin, niacin, B$_6$, B$_{12}$; FA = folic acid (folate), PA = pantothenic acid, Biot = biotin.

Notes:
- 1.5 L of H$_2$O was consumed during the day.
- Ran 6 miles at 4 P.M.
- Vitamins are indicated if the food provides ≥5% of RDA.
- Biotin and vitamin K are produced in the lower digestive tract by bacteria.
- Some vitamin D is made via exposure to UV light.
- Energy intake of 3700 kcal matches estimated expenditure.

Consuming Enough Vitamins (CONTINUED)

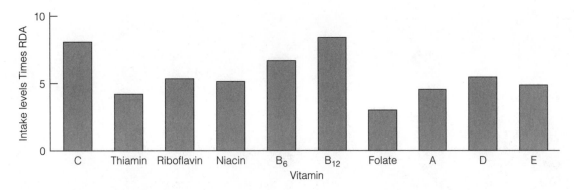

Figure A

Vitamin intake based on the information reported in the runner's 24-hour diet recall. The graph shows diet content relative to current RDAs. For instance, the runner's diet provides vitamin C at a level eight times greater than the RDA. Note: The inclusion of vitamin-fortified breakfast cereals and a sport bar made a significant contribution to the level of vitamins and should always be considered.

speed recovery and adaptive processes involving cell membranes. Research has shown that plasma choline levels can decrease following an endurance bout of exercise,[68] and supplementation seems to thwart this decline.[69] Despite inhibiting reductions in plasma choline during exercise, choline supplementation does not improve endurance performance.[70,71]

Choline supplementation does not appear to have an acute impact on endurance exercise performance. However, choline in combination with carnitine and caffeine improved VO_2max and increased fatty acid utilization in one test using rats.[72] But choline and carnitine were not tested independently in this study, so the independent contribution of choline remains unknown. Like several other purported ergogenic agents proven to be ineffective as independent supplements, choline's role as a synergistic factor remains to be seen. Furthermore, the potential repair benefit of choline supplementation by athletes experiencing frequent muscle damage during training (as in eccentric resistance training and marathons) and its effects on immunocompetence require further study.

■ CHOLINE *IN REVIEW*

- Choline has recently been added to the list of vitamins. Good sources include eggs, milk, liver, and peanuts.
- Choline is a component of acetylcholine and phosphatidylcholine.
- Choline levels in the plasma can decrease as a result of exercise, but short-term supplementation does not increase acute performance.

■ VITAMIN A

Vitamin A exists in three forms, collectively known as the **retinoids** (Figure 9-18a–c). Retinols are the alcohol form, retinals are aldehyde forms, and retinoic acids are the acid forms. These substances are found only in animal-based foods. The best sources are liver, fish and fish liver oils, and egg yolks. Vitamin A–fortified milk and derived dairy products, as well as fortified margarine, are among the major contributors of vitamin A to diets in the United States.

Plants do not make retinoids, but they do make molecules called **carotenoids** (including the three shown in Figure 9-18d–f). Carotenoids include molecules such as lycopene, α-carotene, β-carotene, γ-carotene, lutein, capsanthin, cryptoxanthin, zeaxanthin, and astaxanthin. Several of these are referred to as **provitamin A** because they can be split to yield retinol. The most abundant provitamin A carotenoid in common foods is β-carotene. Other forms such as α-carotene and γ-carotene are also nutritionally significant. Lycopene (another carotenoid) is found in tomatoes and tomato products (ketchup, salsa, and tomato sauce) but it is not converted to vitamin A.

Plant foods are the primary source of carotenoids, especially orange and dark green vegetables and some fruits such as squash, carrots, spinach, broccoli, papaya, sweet potatoes, pumpkin, cantaloupe, and apricots. Table 9-10 lists foods with vitamin A and carotenoids. Americans tend to consume 1–3 mg of carotenoids daily; other populations, whose fruit and vegetable intake is higher, consume more. The RDAs for vitamin A are stated in micrograms or as retinal activity equivalents (RAEs), where 1 RAE is equal to 1 μg of retinal or 12 μg of β-carotene or 24 μg of α-carotene or β-cryptoxanthin. The recommendation is stated as an RAE because of the

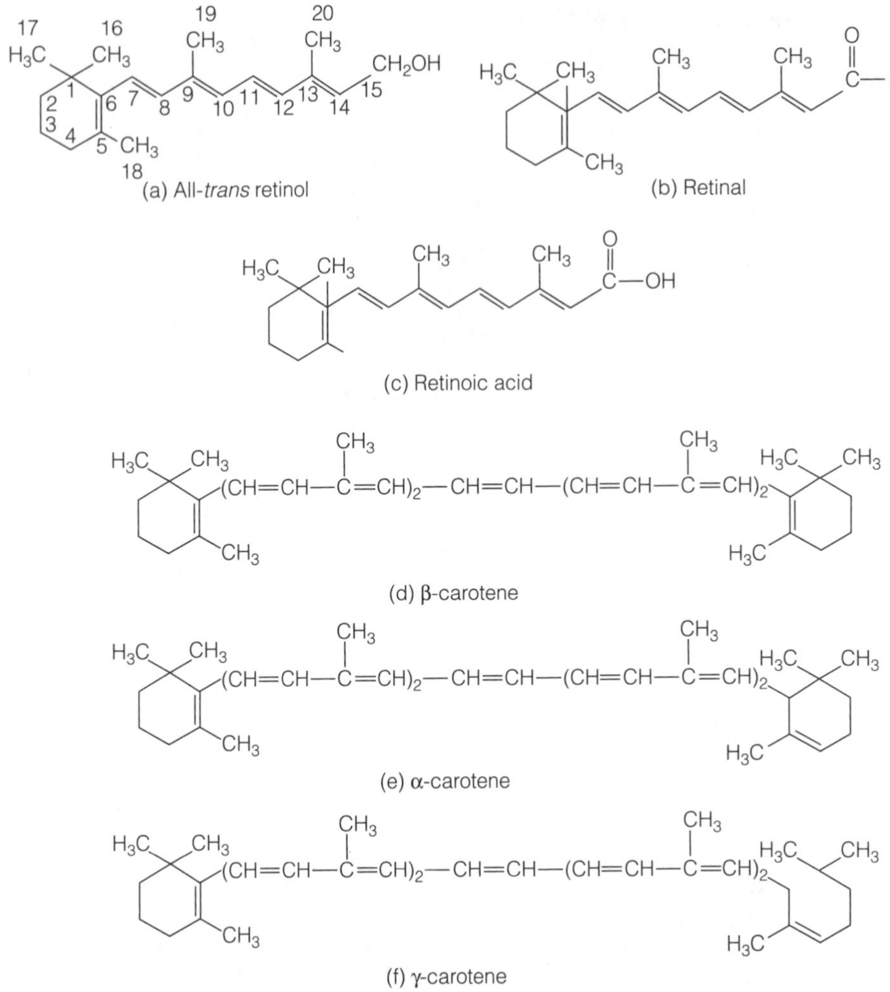

Figure 9-18
Vitamin A and carotenoid structures.

potential conversion of some carotenoids to vitamin A. Carotenoids are the predominant source of vitamin A for many individuals, especially vegetarians and those who refrain from consuming dairy products and other vitamin A–fortified foods. The RDA is 700 and 900 μg or RAEs for adult women and men, respectively. Very little is known about vitamin A requirements for athletes.

Vitamin A Functions and Metabolism

Vitamin A is involved in visual processes and DNA transcription; the carotenoids are antioxidants. Vitamin A in the form of 11-*cis* retinal helps generate nerve impulses from the eye to the optic center of the brain. Vitamin A is also involved in **cell differentiation**, the process by which generalized cells are changed into specialized cells with certain characteristics and properties. To have this effect, vitamin A must interact with a receptor. Researchers have

identified two types of retinoic acid receptors (RXR and RAR) in the nucleus of cells. The interaction of retinoic acid with these receptors influences the production of mRNA from genes. As many of these genes code for proteins that are involved in the cell cycle, vitamin A is a key factor in regulating cell turnover.

As discussed in the feature In Focus: Antioxidant Supplementation for Athletes at the end of this chapter, it is now clear that free radicals are involved in many disease processes and that production of free radicals increases during cardiorespiratory exercise. Carotenoids from plants serve as valuable antioxidants in human tissue. For example, β-carotene and lycopene interact with single-oxygen radicals, and β-carotene can squelch peroxyl radicals (O_2^{2-}). Carotenoids can do this because of their double bond system. But it is debatable whether people (even athletes) need carotenoid supplements, as long as their diet includes a variety of fruits and vegetables.

Table 9-10

Approximate Vitamin A Content of Select Foods

Food	Vitamin A (μg or RAE)[a]
Vegetables	
Pumpkin, canned (1/2 c)	1350
Sweet potato, canned (1/2 c)	1000
Carrots, raw (1/2 c)	1900
Spinach, cooked (1/2 c)	740
Broccoli, cooked (1/2 c)	110
Winter squash (1/2 c)	55
Green peppers (1/2 c)	40
Fruits	
Cantaloupe (1/4)	430
Apricots, canned (1/2 c)	210
Nectarine (1)	101
Watermelon (1 c)	60
Peaches, canned (1/2 c)	50
Papaya (1/2 c)	20
Meats	
Liver (1 oz)	3000
Salmon (3 oz)	55
Tuna (3 oz)	15
Eggs	
Egg (1)	85
Milk and milk products	
Milk, skim (1 c)	150
Milk, 2% (1 c)	140
Cheese, American (1 oz)	80
Cheese, Swiss (1 oz)	65
Fats	
Margarine, fortified (1 tsp)	45
Butter (1 tsp)	40

[a]RAE = retinol activity equivalents

Vitamin A and carotenoids are transported in the blood aboard lipoproteins and *retinol binding protein (RBP)*. Chylomicron remnants deliver the absorbed vitamin A and carotenoids to the liver. The liver cells can attach retinol to a plasma protein that carries vitamin A in circulation. Carotenoids circulate as components of lipoproteins; the main carotenoids in circulation are α-carotene, β-carotene, lycopene, cryptoxanthin, lutein, and zeaxanthin. About 90% of the vitamin A stored in the body is found in the liver, mostly in specialized lipid-storing cells called *stellate cells*. Vitamin A is removed from the body in bile (70%) and urine (30%). Carotenoid metabolites are also added to bile for excretion.

Key Point

Carotenoids are antioxidants that may counter free radicals, which are more heavily produced during exercise. But the relationships between carotenoids and exercise require more research.

Vitamin A Imbalance

Vitamin A deficiency is perhaps the leading cause of nonaccidental blindness in children worldwide. It is estimated that hundreds of thousands of school-age children go blind each year due to vitamin A deficiency. Common signs of poor vitamin A status include night blindness and Bitot's spots on the eyes (these look like foamy, whitish material in the conjunctiva of the eye).

Toxicity occurs when vitamin A is ingested at levels greater than 10 times the RDA (~10 mg/day). Signs of early toxicity include decreased appetite; dry, itchy, flaky skin; headache; hair loss; bone and muscle pain; ataxia; nausea; vomiting; and dry mouth.[1] Eye irritation and *conjunctivitis* may also develop. During pregnancy, reabsorption of the fetus, spontaneous abortion, or birth defects may occur, or the child may later develop learning disabilities. There do not appear to be similar concerns associated with high carotenoid intake levels A. UL for vitamin A has been set 3000 μg for adult men and women.

ACCUTANE.   The vitamin A form 11-*cis* retinoic acid (isotretinoin, Accutane) is used to treat severe cystic acne. This drug is typically prescribed at daily dosages of 0.5–1.0 mg/kg of body weight, with a maximum of 2 mg/kg of body weight. The side effects are similar to those mentioned above; clinical signs include proteinuria, hematuria, hyperuricemia, hypertriglyceridemia, and hyperglycemia. This treatment is an example of the use of toxic levels of a vitamin metabolite to treat a medical condition and is potentially dangerous.

Vitamin A Intake by Athletes

Some groups of athletes appear to get plenty of vitamin A in the diet; others fall below RDA standards. One study evaluated the diet of female heavyweight rowers and determined that they met recommendations.[52] That same year researchers assessed the nutrient intake of male cross-country runners during their competitive season and reported adequate intakes—in fact, the runners' vitamin A intake was approximately double the RDA.[22] But a study of young gymnasts revealed dietary vitamin A levels below recommendations.[20] However, this group of athletes is known for energy restriction, which leads to reduced food intake of many nutrients.

To make the data more relevant, sport nutritionists should consider both individual intake levels and group average intake in survey research related to vitamin status. For instance, one study reported that vitamin A intake among Tour de France cyclists averaged 1300 μg (or RAEs) daily during the 3-week competition.[73] But even one standard deviation below this statistical average is under recommended levels, meaning that about a third of cyclists may not have met RDA standards. However, storage in the liver could be ample enough that daily ingestion is not needed. Study design should also consider all dietary sources of vitamin A. Often the conversion from carotenoids is not considered. At this time very little is known about the influence of exercise on the conversion of carotenoids to vitamin A.

Vitamin A Metabolism and Exercise

It is unclear how exercise influences the circulation and storage of vitamin A. Studies in the 1940s and 1950s suggested that blood vitamin A levels can be increased at the end of an exercise bout and that some of the vitamin A is derived from liver stores.[74] Some confirmation comes from a more recent study revealing that marathoners have an 18% increase in blood levels of vitamin A following a marathon, which may be only partly accounted for by fluid shifts during the race.[75]

Among the lesser-known functions of vitamin A is in the maintenance of and assistance to the immune system. How these mechanisms work are uncertain, but they might have application to exercise. Particularly intriguing is the relationship between vitamin A status and postexercise changes in immune function. As discussed in Chapter 1, some researchers believe (but have not conclusively proved) that mild endurance training may reduce the incidence of sickness such as the common cold and upper respiratory tract infections. More extreme bouts of training and competition, such as marathons or longer races, as well as exhaustive endurance cycling and triathlons, may render athletes more prone to upper respiratory tract infections. Researchers have clearly shown that a key function of the immune system (involving natural killer cells) is greatly reduced by such activities. Whether and how vitamin A is involved remains uncertain.

Very little is known about the relationship between accutane use and athletic performance, a relevant issue given that most accutane users are in their teens and 20s, the usual ages of people who train and compete extensively. Currently accutane is not banned by sports-governing organizations.

Carotenoids, Antioxidation, and Exercise

Supplements of carotenoids, particularly β-carotene, are now used by many endurance athletes to decrease free radical activity during and after exercise. There is reasonable evidence that antioxidant supplements containing β-carotene can reduce markers of free radical activity, but

it is not yet conclusive. Arguments for and against the use of antioxidant supplements by athletes are discussed further in the In Focus feature at the end of this chapter. Because little is known at this time about how exercise influences the conversion of carotenoids to vitamin A, future research should address this issue.

■ VITAMIN A *IN REVIEW*

- Vitamin A is found in fortified dairy products, egg yolks, and fish livers and their oils. Some vitamin A can be formed by the conversion of carotenoids from plant foods.
- Vitamin A interacts with a receptor in the nucleus of cells and influences the differentiation of tissue. Vitamin A is also essential to proper vision.
- The relationship between exercise and vitamin A is not well researched, and most athletes do not supplement vitamin A in high doses because of toxicity concerns. More research is being conducted on carotenoids because of their antioxidant properties, but their actual protective potential has yet to be determined.

■ VITAMIN D

Vitamin D (cholecalciferol or D$_3$) is unique because the skin can produce adequate quantities with adequate exposure to sunlight, more specifically ultraviolet (UV) radiation. This means that there are two possible means for meeting vitamin D needs—diet and exposure to the sunlight. Vitamin D is also unique because it can be considered a hormone; it is produced and released by specific tissues, then circulates and interacts with a specific receptor type called vitamin D receptor (VDR). Production of vitamin D by exposure to UV radiation poses a risk for skin cancer if exposure is too great. Or UV exposure may be too limited in colder climates or where smog, heavy ozone layers, or heavy clouds prevent UV light from reaching the ground.

As shown in Table 9-11, most vitamin D in the diet is derived from animal foods such as eggs, liver, fatty fish (salmon, herring, tuna), and fortified milk and other dairy foods. Because margarine replaces butter in many diets, it is also fortified with vitamins A and D. The AI for vitamin D for adults is 5–10 μg daily. Because the relationship

Key Point

The skin produces vitamin D when exposed to ultraviolet light (such as from the sun or a tanning bed), but the risk of skin cancer increases from exposure to UV. Diet can provide adequate levels of vitamin D.

Table 9-11

Approximate Vitamin D Content of Select Foods

Food	Vitamin D (µg)
Milk	
Milk, all (1 c)	2.5
Fish and seafood	
Salmon (3 oz)	8.5
Tuna (3 oz)	3.5
Shrimp (3 oz)	3.0
Meats	
Beef liver (3 oz)	1.0
Eggs	
Egg (1)	0.7

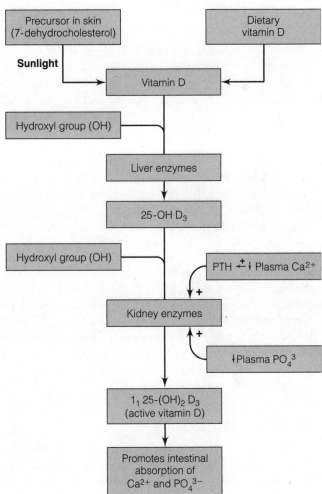

Figure 9-19
Interactions between PTH and vitamin D in controlling plasma calcium.

between vitamin D and exercise is unclear, no recommendations have yet been developed for athletes.

Vitamin D Metabolism and Functions

The vitamin D absorbed from the digestive tract and made in the skin is generally inactive. The inactive form, called cholecalciferol, must circulate to the liver and then the kidneys to be fully activated. Chylomicrons transport most of the vitamin D absorbed by the digestive tract, and *vitamin D binding protein (DBP)* transports most of the vitamin D produced in the skin and from the liver and kidneys after it is modified. The steps of vitamin D modification are presented in Figure 9-19. The system is tightly regulated to reduce potential toxicity. One of the most significant regulatory steps involves the enzyme 1-α-hydroxylase (1-α-OHase), which is responsible for the final step in producing the most potent form of vitamin D [1,25-dihydroxycholecalciferol, or 1,25-$(OH)_2,D_3)$]. The activity of the enzyme is enhanced by parathyroid hormone (PTH) and insulin-like growth factor-I (IGF-I), and inhibited as serum calcium and phosphorus levels become elevated. Another term for the most active form of vitamin D is *calcitriol*.

The most recognizable function of vitamin D is its involvement in calcium metabolism; VDR is found in bone tissue, intestines, and the kidneys. When the level of calcium decreases in the blood (hypocalcemia), more parathyroid hormone (PTH) is released. PTH then increases the activity of 1-α-OHase activity in the kidneys, resulting in the formation of more active vitamin D. Active vitamin D then returns blood calcium levels to the optimal level by increasing the absorption of calcium from the intestines. Vitamin D's role in this process is to

increase the production of the proteins responsible for the transport of calcium and phosphate into the cells lining the small intestine. These proteins include calbindin and an intracellular membrane calcium binding protein (IMCBP).

Vitamin D is believed to increase the activity of an enzyme called alkaline phosphatase, which increases the availability of phosphorus in the small intestine for absorption, and for phosphate transport proteins. In bone, vitamin D may promote the production of osteoclasts instrumental in mobilizing calcium and phosphate from bone mineral salts. The role of vitamin D in bone formation and maintenance is discussed in more detail in the Chapter 10 feature In Focus: Osteoporosis: More Than a Mineral Problem. VDR is also found in the stomach, heart, brain, and skeletal muscle, leading some researchers to investigate the relationship between sport training and vitamin D metabolism.

Key Point

Vitamin D produced by the skin or absorbed from the diet requires two steps of modification to become active.

Vitamin D Imbalance

Poor vitamin D status during childhood can result in reduced mineralization of growing bones, leading to malformed bones. Vitamin D deficiency during adulthood reduces calcium and phosphate absorption. As bone turnover occurs the matrix is preserved, but progressively loses its mineralization. Vitamin D deficiency, despite poor dietary intake, can be avoided by adequate exposure to sunlight. Too much vitamin D consumption can lead to excessive absorption of calcium and phosphate from the digestive tract and decreased urinary excretion. Elevated levels of these minerals in circulation can lead to the formation of calcium phosphate crystals in soft tissue such as muscle. A UL of 50 μg has been set for adult men and women.

Vitamin D Intake by Athletes

Vitamin D consumption below recommended levels has been reported for some athletic populations. For example, it was below the RDA for national figure skaters, according to one study.[19] In a study in Finland involving young female athletes it was determined by food frequency questionnaires (FFQs) and 4-day food records that their vitamin D intake was not unlike that of nonathletic girls.[76] However, the average vitamin D intake in both groups was still below recommendations. One Norwegian study involving female elite athletes suffering from eating disorders and meeting the criteria for *anorexia nervosa, anorexia athletica* (similar to female athlete triad, discussed in the Chapter 15 In Focus), or *bulimia nervosa* also found vitamin D intake levels below recommendations.[77] Athletes who limit their consumption of milk and dairy products are at particular risk of inadequate vitamin D levels.

One advantage athletes might have over less active people is that they may spend more time outside to train and compete, giving them greater exposure to sunlight for internal vitamin D formation. But not all sports are played outside, and people living in regions of the world with longer winters and less sunlight may still experience reduced vitamin D status.

Vitamin D and Athletic Performance

Vitamin D has not been viewed as one of the vitamins that can directly enhance sport performance as it is not fundamentally involved in energy metabolism, antioxidant activity, and the like. Therefore research investigation into the relationship between exercise and vitamin D has not been plentiful. As discussed in the Chapter 10 In Focus feature on osteoporosis, weight-bearing exercise mechanically stresses bone and may improve bone density in younger people and lessen bone loss as people age. Research indicates that good vitamin D status, along with that of calcium and other nutrients, is necessary to optimize the beneficial effect of resistance exercise on bone density,[78] especially in older people and postmenopausal women. Furthermore, weight training increases muscular strength and balance, which reduces the frequency of falls and thus the potential for fracture.[79] Researchers have determined that resistance-trained individuals tend to have higher levels of vitamin D [1,25-$(OH)_2D_3$ and 25-$(OH)D_3$] than untrained individuals.[80] This might allow for greater calcium availability to enhance bone density.

Vitamin D supplements are not usually sought specifically by athletes, but the vitamin is common in multiple vitamin/mineral pills. Vitamin D supplements have been discouraged unless dietary intake cannot be improved by milk and dairy consumption.

■ VITAMIN D *IN REVIEW*

- The most significant sources of vitamin D are fortified dairy products, margarine, and exposure to sunlight.
- Vitamin D interacts with a nuclear receptor and increases the production of proteins involved in calcium homeostasis.
- The relationship between vitamin D and exercise has not been investigated thoroughly, but preliminary research suggests that dietary vitamin D intake levels may be below recommendations for some athletes. This may not be a concern if their training and competition involves sunlight exposure.
- Because of toxicity issues, the use of vitamin D supplements at several times the RDA is not common among athletes and not recommended by sport nutritionists.

■ VITAMIN E

There are two families of vitamin E: the *tocopherols* (α, β, γ, δ) and the *tocotrienols* (α, β, γ, δ), illustrated in Figure 9-20. The best sources of vitamin E are plant oils such as cottonseed, corn, sunflower, safflower, soybean, and palm oils, and oil-derived products such as margarine, shortenings, and mayonnaise. Nuts and wheat germ and its oil are also rich sources, and some fruits and vegetables such as peaches and asparagus are fair sources as well. Fish contain appreciable amounts of vitamin E depending on their

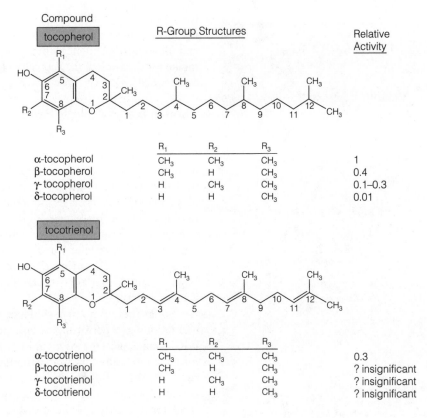

Figure 9-20

Structure and activity of natural tocopherols and tocotrienols. Activity levels are relative to α-tocopherol, the most active form of vitamin E. (*Source*: Diplock AT. Vitamin E. Fat-Soluble Vitamins (Diplock AT, ed.), Lancaster, PA: Technomic, p. 156, 1984.)

---

diet (Table 9-12). Recommendations for vitamin E are 15 mg/day of α-tocopherol (α-TE) or the equivalent activity in other forms for men and women. Figure 9-20 shows the relative activity of the various forms. It is not known whether athletes require more vitamin E.

Vitamin E Functions and Metabolism

Vitamin E is found in most tissues, including the adrenal glands, lungs, brain, muscle, adipose tissue, and the liver and blood (cells and lipoproteins). Vitamin E is well known as an antioxidant that helps protect cell membranes, particularly unsaturated fatty acid components of phospholipids, against oxidation by free radicals. This includes not only the plasma membrane but also organelle membranes. For instance, the membranes of the mitochondria and endoplasmic reticulum contain a higher concentration of unsaturated fatty acids and are therefore at greater risk of free radical peroxidation.[1]

Vitamin E status can be assessed in the plasma and in RBCs.[1] In addition, researchers have noted that to evaluate vitamin E status, tocopherol levels of platelets might be a sensitive indicator and supported by tocopherol levels of RBCs, lymphocytes, or plasma lipoproteins. Assessment of the vitamin E content of muscle is useful in exercise-based evaluations. Vitamin E is lost through the skin and in feces and urine. The vitamin E content of feces reflects unabsorbed vitamin E from the diet and bile.

Vitamin E Imbalance

As with other vitamins, a deficiency of vitamin E can result from poor dietary intake, decreased absorption, or increased excretion. Because vitamin E is a lipid it relies on digestive events associated with lipids for proper absorption. These include the presence of bile and the formation of chylomicrons in the cells lining the small intestine. If a person has reasonably good vitamin E stores but later has a deficient diet or experiences malabsorption of lipids, it could take many months or years before plasma vitamin E levels fall to critical levels. Vitamin E deficiency can result in the destruction of red blood cell membranes and hemolytic anemia. Degeneration of neuronal and muscular membranes may result in ataxia and muscular weakness.[1]

Unlike excess consumption of vitamins A and D, too much vitamin E is fairly nontoxic; it is thought to be one of the least toxic of all vitamins. Intakes as high as 500–800 mg of α-TE for several months to years have not resulted in significant effects. However, gram doses may result in fatigue, muscle weakness, and gastrointestinal

Table 9-12

Approximate Vitamin E Content of Select Foods

Food	Vitamin E Content (mg α-TE)
Oils	
Oils (1 tbs)	7
Margarine (1 tbs)	4
Nuts and Seeds	
Sunflower seeds (1/4 c)	25
Almonds (1/4 c)	12
Peanuts (1/4 c)	5
Cashews (1/4 c)	1
Vegetables	
Sweet potato (1/2 c)	7
Collard greens (1/2 c)	3
Asparagus (1/2 c)	2
Spinach, raw (1/2 c)	1.5
Grains	
Wheat germ (1 T)	2
Bread, whole wheat (1 slice)	2
Bread, white (1 sl)	1
Seafood	
Crab (3 oz)	4.5
Shrimp (3 oz)	3.5
Fish (3 oz)	2.5

distress. Such doses might impede vitamin K absorption as well as vitamin D's involvement in bone mineralization. A UL for vitamin E has been set at 15 mg for adult men and women.

Key Point

Vitamin E is important to athletes to help manage exercise-induced free radical production.

Vitamin E Intake by Athletes

Vitamin E intake by elite and power athletes is probably adequate in most cases because of high energy intake. However, some reports indicate that certain athletic populations consume lower-than-recommended levels of vitamin E. In one study, researchers reported that the average vitamin E intake of U.S. national figure skaters was not only below RDA levels but below general population levels (NHANES III).[19] Also, more than one-third of highly trained female cyclists failed to consume the RDA level

for vitamin E.[51] Vitamin E is a popular nutritional supplement, and more than 50% of athletes use a multivitamin and mineral supplement. Athletes who restrict their dietary intake for weight loss may consume less vitamin E than recommended, as might those who are on a very low-fat diet.

Vitamin E and Athletic Performance

Exercise might influence the vitamin E content of tissue. In one study, plasma vitamin E levels dropped by 33% in men in response to cardiorespiratory training and then rebounded within 24 hours.[81] Another study suggested that plasma vitamin E levels do not change but that RBC vitamin E levels decrease.[82] A higher dietary intake can increase the vitamin E content of tissue, including skeletal muscle.[83] These higher initial levels might be a factor in shifts in vitamin E during exercise, although many questions remain. Other factors could include age, gender, and level of training.

So far there is no firm evidence that vitamin E supplementation has a performance-enhancing effect on athletes. For instance, 400 mg of vitamin E failed to improve performance measures during swimming (400 m) and 1-mile runs.[84,85] Although vitamin E supplementation may not affect the performance of a well-nourished athlete in the short term, it may bolster antioxidant defenses and lessen exercise-induced free radical damage.[86] Many studies indicate that optimal vitamin E status is important to minimize exercise-induced free radical activity, and some researchers have recommended vitamin E supplements for endurance athletes. Other researchers question whether antioxidant supplementation is wise for otherwise well-nourished athletes. Some of the adaptive operations that occur in skeletal muscle and other tissues may be redux sensitive, so that vitamin E supplementation might hinder adaptive processes. However, an argument can also be made that limiting free radical activity may have health-promoting properties with regard to DNA damage and LDL oxidation. (See the In Focus at the end of the chapter.)

■ VITAMIN E *IN REVIEW*

- Vitamin E is a group of related molecules (tocopherols and tocotrienols) with similar properties of antioxidation. They associate with lipid portions of cells (such as membranes) and are transported in the blood aboard lipoproteins (such as LDLs).
- Vitamin E status and dietary levels are related to decreased free radical activity and the incidence of heart disease and some cancers.
- Vitamin E does not have an acute ergogenic effect, but athletes with normal to elevated vitamin status (supplementation) may have decreased free radical activity associated with endurance and intermittent exercise.

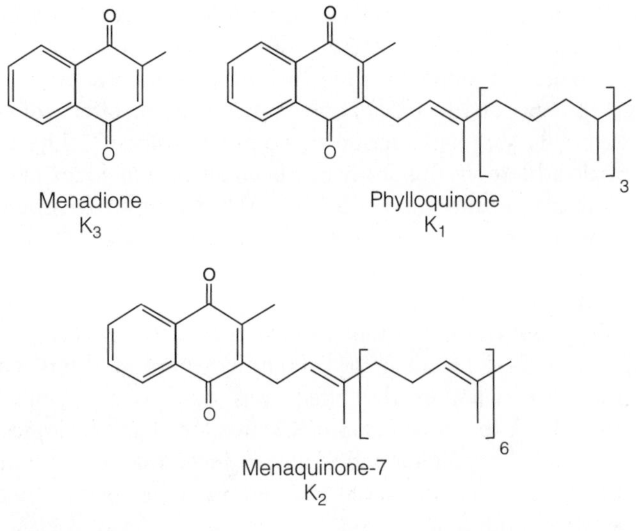

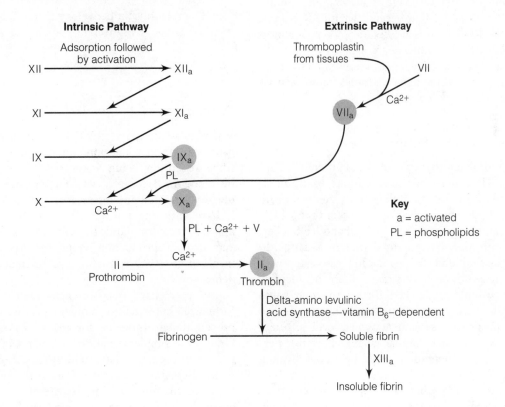

Figure 9-21
Biologically active forms of vitamin K.

Menadione
K_3

Phylloquinone
K_1

Menaquinone-7
K_2

Menadione is the synthetic version of vitamin K and is primarily used in nutrition supplements and fortified foods. Plant-based foods such as spinach, broccoli, brussels sprouts, cabbage, lettuce, and kale are rich in vitamin K. Cereals, meats, nuts, legumes, dairy products, and fruits also provide some vitamin K. The AI for adults is 90 and 120 µg for younger adult men and women, respectively. The relationship between vitamin K and exercise requires more investigation before requirements for athletes become clear.

Vitamin K Functions and Metabolism

Vitamin K is essential for proper clotting of the blood, but its involvement does not actually occur in the blood. A series of activation reactions with **clotting factors** is required for blood to properly coagulate and form a clot. As depicted in Figure 9-22, four clotting factors require

■ VITAMIN K

Vitamin K (Figure 9-21) is a group of molecules that have the same activity. Green plants contain *phylloquinone* and bacteria in the large intestine produce *menaquinones*.

Key Point

Vitamin K is important to blood clotting and perhaps bone integrity, but its relationship to exercise still needs research.

Figure 9-22
The activation of prothrombin and the roles of vitamin K–dependent clotting factors. Circled letters require vitamin K for their formation.

vitamin K. These and other clotting factors are produced in the liver. Before they are released, vitamin K is involved in a posttranslational modification of these proteins. In addition to these clotting factors, other vitamin K–dependent proteins have been identified in tissues such as bone and kidneys.[1] Vitamin K circulates mostly as a component of lipoproteins; chylomicrons carry vitamin K from the digestive tract and VLDLs circulate it from the liver. Vitamin K is primarily lost from the body in the urine and feces.

Vitamin K Imbalance

The potential for vitamin K imbalance is low for most people because it is abundant in food and is absorbed from intestinal bacterial synthesis. A restricted diet combined with the use of antibiotics can place an individual at greater risk of deficiency. High doses of vitamin K are relatively nontoxic in natural forms. However, toxicity from chronic use of excessive vitamin K in the form of synthetic menadione has been reported. A UL for vitamin K has not been set.

Vitamin K Intake by Athletes

Not much information is available about vitamin K intake by athletes. It is assumed that because vitamin K is in so many foods, the RDA level is at least met. There is almost no information on the influence, if any, of exercise on vitamin K metabolism, storage, or excretion. However, one study that assessed elite female athletes found their vitamin K status was below normal.[87]

Vitamin K and Athletic Performance

It does not seem likely that athletes need vitamin K supplementation.[88] But there may be exceptions, based on vitamin K metabolism and the influence of exercise on it. For instance, those who participate in sports that lead to frequent cuts and bruises, such as football and ice hockey, should make sure they have optimal vitamin K status. In addition, because athletic training is often associated with greater bone mass, the role of vitamin K in the adaptive process of bone-building deserves further research, especially if it can be maximized and perhaps mimicked in less active people prone to degenerative bone disorders. There is also the possibility that optimal vitamin K status, perhaps involving supplementation, could help bone healing after fracture.

As discussed in the Chapter 10 In Focus on osteoporosis, some female athletes experience reduced estrogen production and fail to menstruate (amenorrhea). This can lead to lowered peak bone mass (PBM) and bone material loss at a relatively young age. Because increased intake of vitamin K has been associated with an increase in serum markers for bone formation and a decrease in urinary markers for bone breakdown and calcium excretion in postmenopausal women, one study tested the effects of vitamin K supplementation in younger women.[85] The female athletes in this study had been amenorrheic for more than a year and were provided with a vitamin K supplement (10 mg/day) for a month. The results: a 15–20% increase in bone formation markers and a 20–25% decrease in markers of bone breakdown. These results are encouraging, but it is important to note that these females were vitamin K deficient prior to supplementation. Thus it is not clear whether the effect was nutritional, bringing them back to better vitamin K status, or pharmacological. Vitamin K supplementation also reduced markers of bone breakdown and increased markers of bone formation in another study.[89]

■ VITAMIN K *IN REVIEW*

- Vitamin K is found in a variety of foods. It is involved in activating clotting factors and in modifying a couple of other proteins after they are made in tissue. Poor vitamin K status is associated with decreased blood-clotting capability.
- The relationship between vitamin K and exercise is not properly researched, but vitamin K may maintain and increase bone density.

Conclusions

Vitamins are dietarily essential, non-energy-providing organic substances commonly found in foods. They are classified based on being water or fat soluble as well as by function. B-complex vitamins are involved in energy metabolism, functioning as coenzymes. Vitamins C and E have antioxidant properties. Vitamins A and D have nuclear receptors and directly influence transcription and protein production. Vitamins B_{12} and folate are involved in RBC formation, so anemia indicates their deficiency.

Vitamins have long been associated with athletic performance, but vitamin supplements have not generally proved to be ergogenic for the well-nourished athlete. Gram dose supplementation of niacin can inhibit performance by hindering fatty acid mobilization from adipose tissue. Researchers are focusing on the potential for antioxidant vitamins to minimize the effects of free radicals during and after exercise, which could have long-term benefits for performance and general health. Debate on the appropriate level of antioxidant vitamin consumption continues.

IN FOCUS   ANTIOXIDANT SUPPLEMENTATION BY ATHLETES

Certain types of sports and physical activity are associated with increased production of free radicals. These sports have high VO_2 demands, such as endurance sports (cycling, running) and maybe higher intensity intermittent sports (soccer, lacrosse). Because direct measurement of free radicals in tissue is difficult, most of the assumptions are based on the measurement of "markers" of free radical activity.

Because damage to cellular structures is perceived to possibly influence acute performance as well as recovery processes that could affect chronic performance, antioxidant supplementation has become fairly common in athletic populations. However, a consensus does not exist within the scientific community as to whether well-nourished athletes should use antioxidant supplements. Much of the controversy hinges on the possibility that muscle function and adaptation to training might actually be impaired if antioxidant status is too high because of aggressive supplementation.

Free Radicals

Understanding the nature of a free radical entity is made easier by a quick review of basic chemistry. Technically speaking, a free radical contains one or more electrons in an unpaired spin state. This renders the substance highly reactive. Free radicals act primarily as oxidants, meaning that they endeavor to steal electrons from other molecules to pair their unpaired electrons. The primary targets for free radicals in cells include unsaturated fatty acids, nucleic acids (DNA and RNA), and proteins. Thus free radicals attack the very structural and functional basis of tissue.

Many free radicals are oxygen based and are created within cells or brought into the body via inhalation. These free radicals include superoxide ($O_2 \bullet^-$), hydroxyl radical (HO•) and peroxyl radical (H_2O_2) and are referred to as reactive oxygen species (ROS). Although other types of free radicals exist, the focus here is on ROS in light of the increased VO_2 associated with exercise.

Measurement of Free Radical Activity

Once formed, free radicals can act very quickly. This makes direct measurement of free radical levels difficult in a laboratory. However, free radical activity can be estimated by measuring the products of free radical interactions, such as damaged molecules and by-products. For instance, attack of free radicals on cell membrane lipids increases the production of lipoperoxides (such as malondialdehyde and hydroperoxides). Diagnostic test systems are also available, such as TBARS (which stands for *thiobarbituric acid reactive substances*) to help estimate free radical activity on fatty acids. DNA metabolites such as 8-hydroxy-deoxyguanosine (8-OHdG) can also be measured in the plasma or urine to indicate free radical attack.

General Antioxidant Systems

The primary antioxidant systems employed by cells are both enzymatic and nonenzymatic. Enzyme systems such as superoxide dismutases (SODs), glutathione peroxidase (GSH-Px), and catalase are complemented by nonenzymatic antioxidants such as carotenoids, vitamins C and E, ubiquinone, glutathione, uric acid, and lipoic acid. The nature and function of these substances have been detailed in this chapter as well as Chapters 10 and 11 and are summarized in Table A, along with some of the most prominent cellular and tissue locations of these antioxidants.

Free Radical Activity and Cardiorespiratory Training

Free radical production is thought to increase during sustained exercise at higher VO_2 levels, but there are several complications with pinpointing such increases. It is not clear what the best indicators might be for free radical activity and damage in muscle. Also, several different tissues might be responsible for extra free radicals during exercise. The influence of the nature of training (workload) and training history need to be considered as well. For instance, the results of one study suggested that isolated bouts of either running or cycling might not result in increased levels of urinary 8-hydroxy-deoxyguanosine (8-OHdG) for 24 hours in healthy trained and untrained males.[90] This does not mean that increased levels of free radicals were not generated during the exercise bouts, only that appreciable DNA damage did not occur. The results of another study also suggested that an isolated bout of exercise may not induce significant DNA damage.[91] Findings of a third study suggested that it may take a couple of training sessions to result in increased DNA damage.[92]

It may be that a single isolated bout of exercise in an untrained, well-nourished individual may not result in an unmanageable level of free radical activity. The rationale for this thinking involves two factors. First, the untrained individual may not be able to

ANTIOXIDANT SUPPLEMENTATION BY ATHLETES (CONTINUED)

| Table A | Prominent Antioxidants and Their Activities |

Antioxidant	Activity
Vitamin C	Water-soluble antioxidant found in the cytoplasm of cells can inactivate a variety of free radicals and helps regenerate vitamin E after it is used.
Vitamin E	Being lipid-soluble, it protects lipid portions of membranes.
Glutathione peroxidase (GSH-Px)	Selenium-containing enzyme transfers electrons from glutathione to free radicals.
Superoxide dismutase (SOD)	Family of intracellular and extracellular enzymes convert superoxide to hydrogen peroxide, which can then be broken down by GSH-Px.
Ubiquinone (coenzyme Q)	Free ubiquinone in mitochondria functions as an antioxidant.
Carotenoids	Lipid-soluble antioxidants are transported in the blood via lipoproteins and may help protect LDL from oxidation. Most prevalent are β-carotene and lycopene.
Lipoic acid	Sulfur molecule associates with both membranes and intracellular fluid. It is important in recycling vitamin C.
Uric acid	Though not often considered an antioxidant, uric acid is a chief antioxidant factor in saliva.

sustain a workload requiring VO_2 use that would produce free radicals at a level that overwhelms basal antioxidant systems during their first workout. Second, an untrained person may have a VO_2max that is not supportive of a higher degree of free radical generation. This is to say that as an individual trains regularly, cardiovascular and metabolic adaptations allow for a greater VO_2max as well as a greater workload, VO_2, and free radical production.

Perhaps in support of the first notion, researchers have found that after a single bout of prolonged or exhaustive exercise, antioxidant capacity may be decreased in tissue such as the liver and heart. This may represent an exercise-induced expenditure of some of the basal antioxidant capabilities. For instance, several antioxidant-related factors (glutathione, ascorbic acid, vitamin E) might need to be reduced in order to have antioxidant activity again. As an individual's performance potential improves with regular training, the oxidative stress is enhanced in relation to the increasing work output of an exercise bout. In addition, some scientists have speculated that frequent training sessions might allow for a carryover effect, although this has not been clearly demonstrated.

Regular training probably induces changes in the redox state of muscle fibers and other tissue experiencing greater VO_2 during exercise (such as heart, lung, and liver). Such alterations probably put into play events that allow for an adaptation of antioxidant capabilities. For instance, the results of one study suggested that regular training might lead to a reduction in DNA damage markers.[93] A normalization of markers of oxidative stress and DNA damage was also suggested by another study.[94] Another piece of supporting evidence is that the urinary levels of 8-OHdG in long-distance runners were determined to not differ from control levels.[95] The results of these and other studies suggest that chronic training results in adaptive mechanisms that attempt to deal with the routine generation of free radicals.[96,97]

Adaptations to Antioxidant Systems

Muscle and other tissues adapt to free radical stress by bolstering these antioxidant protective mechanisms. Not all research studies have reached the same conclusions, which is probably the result of differences in exercise protocols, the nutritional status of the participants, and their level of training. Table B presents some of the anticipated adaptations in antioxidant capabilities associated with endurance training.

Adaptations in the antioxidant capabilities of different tissues in response to endurance training have been assessed in rats. In one study, researchers determined that liver glutathione peroxidase and glutathione reductase and total antioxidant capacity were increased, as was lipid peroxidation at rest.[98] In another study, total antioxidant capacity was increased in the heart after exercise, and plasma membrane integrity was increased and lipid peroxidation at rest was decreased after 10 weeks of training.[99]

Antioxidant Supplementation for Athletes

Many scientists advocate the use of antioxidant supplements as a means of reducing free radical activity and cellular damage. In fact, some scientists have made recommendations for athletes for some antioxidant

| Table B |

Anticipated Changes in Skeletal Muscle Antioxidant Capabilities Associated with Endurance Training

- Endurance training has been shown by some researchers to enhance SOD activity. In addition, a time and intensity relationship seems to exist that dictates the degree of enhancement.
- Endurance training enhances GSH-Px activity in trained muscle, and both cytosolic and mitochondrial GSH-Px activity increase (more in mitochondria).
- Endurance training does not seem to enhance catalase levels as it does for SOD and GSH-Px. This may be related to an increased dedication of iron toward formation of RBCs, myoglobin, and cytochromes.
- Glutathione levels may actually increase in more aerobic muscle fibers (Type I) in response to endurance training.
- Ubiquinone levels may be enhanced in more aerobic muscle fibers in response to endurance training (at least in rats).

Source: Adapted from reference 97.

substances.[98] However, other scientists question if this is in the best interest of the athlete. It may be that the development of free radicals and cellular redox stress are fundamentally important to the adaptive mechanisms mentioned above. In addition, some research findings suggest that supplementation of antioxidant nutrients may influence the redox events associated with muscle contraction and thus affect physical efforts. For instance, one research study suggests that force generation may be lessened after supplementation.[100]

When professional basketball players were given supplements containing vitamin E, vitamin C, and β-carotene for a month, they had lower levels of total lipoperoxides, a marker for free radical activity, as well as a lower ratio of lipoperoxide to total antioxidant status (TAS).[101] In another study, vitamin E supplementation inhibited the increase in malondialdehyde (MDA) levels following exercise.[102] MDA is used as a marker of lipid peroxidation. Vitamin E supplementation was also noted to decrease the level of muscle enzymes in the blood relative to measurements made without supplementation. The presence of muscle enzymes indicates cellular leakage due to the breakage of muscle cell plasma membranes (sarcolemma). Vitamin E supplementation has also been associated with a reduced level of DNA damage in white blood cells.[92]

Conclusions

At this time it cannot be broadly recommended that athletes in endurance and intermittent sports aggressively supplement antioxidant nutrients. However, athletes should strive to make sure their diet is rich in fruits, vegetables, and whole grains that are adequate in iron, copper, selenium, zinc and manganese. If an antioxidant supplement is to be utilized there are a couple of things to keep in mind. Antioxidant proteins such as glutathione peroxidase and SOD are digested to amino acids and minerals, so expensive antioxidant enzyme preparations may be a waste of money. However, factors such as lipoic acid, ubiquinone, and glutathione should be absorbed intact.

Many scientists believe that some of the health-protective effects of regular cardiorespiratory exercise are due to its enhancement of antioxidant capability throughout the day, while exercise is a small part of that time. Thus, despite increased free radical generation and activity during exercise, most of the day is spent in rest with enhanced antioxidant status. Thus the benefits of cardiorespiratory training on HDL-cholesterol and maybe blood pressure would be complemented by an enhanced basal antioxidant capability.

STUDY QUESTIONS

1. How are vitamins defined and classified?
2. Which vitamins are involved in antioxidant functions in the body?
3. Which vitamins are directly involved in energy metabolism?
4. Which vitamins are particularly important to adaptation in blood expansion during endurance training?
5. How is vitamin B_6 important to endurance exercise?
6. How is vitamin C important to recovery from a sport injury?
7. How might supplementation of niacin be detrimental to aerobic exercise metabolism?
8. Which vitamins might be important for normal growth of bone as well as adaptive processes in bone in response to exercise training?
9. Are the requirements for any of the vitamins increased due to exercise training?
10. What athletes are at risk for not consuming adequate levels of vitamins?

REFERENCES

1. Lee RD, Nieman DC. Laboratory assessment of nutrition status. In: *Nutritional Assessment*, Madison, WI: Brown & Benchmark, 1993.

2. Jacob RA. Vitamin C. In: *Modern Nutrition in Health and Disease*, 9th ed. (Shils ME, Olson JA, Shike M, Ross CA, eds.), Baltimore: Williams & Wilkins, 1999.

3. Nieman DC, Butler JV, Pollett LM, Dietrich SJ, Lutz RD. Nutrient intake of marathon runners. *Journal of the American Dietetics Association* 89:1273–1278, 1989.

4. Hickson JF, Duke MA, Risser WL, Johnson CW, Palmer R, Stockton JE. Nutritional intake from food sources of high school football athletes. *Journal of the American Dietetics Association* 87:1656–1659, 1987.

5. Bazzare TL, Kleiner SM, Ainsworth BE. Vitamin C intake and lipid profiles of competitive male and female bodybuilders. *International Journal of Sports Nutrition* 2:260–271, 1992.

6. Faber M, Benade AJ. Mineral and vitamin intake in field athletes (discus-, hammer-, javelin-throwers and shotputters). *International Journal of Sports Medicine* 12(3):324–327, 1991.

7. Keith RE. Ascorbic acid. In: *Sports Nutrition: Vitamins and Trace Elements* (Wolinsky I, Driskell JA, eds.), Boca Raton, FL: CRC Press, 1997.

8. Steen SN, McKinney W. Nutritional assessment of college wrestlers. *The Physician and Sportsmedicine*, 14:100–103, 1986.

9. DeBolt JE, Singh A, Day BA, Deuster PA. Nutrition survey of the US Navy SEAL trainees. *American Journal of Clinical Nutrition* 48:1316–1323.

10. Hickson JF, Schrader J, Trischler LC. Dietary intakes of female basketball and gymnastics athletes. *Journal of the American Dietetics Association* 86:251–253, 1986.

11. Cohen JL, Potosnak L, Frank O, Baker H. A nutritional and hematologic assessment of elite ballet dancers. *The Physician and Sportsmedicine* 5:43–48, 1985.

12. Telford RD, Catchpole EA, Deakin V, McLeay AC, Plank AW. The effect of 7 to 8 months of vitamin/mineral supplementation on the vitamin and mineral status of athletes. *International Journal of Sport Nutrition* 2(2):123–134, 1992.

13. Koz M, Erbas D, Bilgihan A, Aricioglu A. Effects of acute swimming exercise on muscle and erythrocyte malondialdehyde, serum myoglobin, and plasma ascorbic acid concentrations. *Canadian Journal of Physiology and Pharmacology* 70(10):1392–1395, 1992.

14. Howald H, Segesser B, Korner WF. Ascorbic acid and athletic performance. *Annals of the New York Academy of Sciences* 30;258:458–464, 1975.

15. Davies KJ, Quintanilha AT, Brooks GA, Packer L. Free radicals and tissue damage produced by exercise. *Biochemical and Biophysical Research Communications*, 31;107(4):1198–1205, 1982.

16. O'Neill CA, Stebbins CL, Bonigut S, Halliwell B, Longhurst JC. Production of hydroxyl radicals in contracting skeletal muscle of cats. *Journal of Applied Physiology* 81(3):1197–1206, 1996.

17. Packer L. Oxidants, antioxidant nutrients and the athlete. *Journal of Sports Science* 15(3):353–363, 1997.

18. Clarkson PM. Antioxidants and physical performance. *Critical Reviews in Food Science and Nutrition* 35(1–2):131–141, 1995.

19. Ziegler PJ, Nelson JA, Jonnalagadda SS. Nutritional and physiological status of U.S. national figure skaters. *International Journal of Sport Nutrition* 9(4):345–360, 1999.

20. Ersoy G. Dietary status and anthropometric assessment of child gymnasts. *Journal of Sports Medicine and Physical Fitness* 31(4):577–580, 1991.

21. Rankinen T, Lyytikainen S, Vanninen E, Penttila I, Rauramaa R, Uusitupa M. Nutritional status of the Finnish elite ski jumpers. *Medicine and Science in Sports and Exercise* 30(11):1592–1597, 1998.

22. Niekamp RA, Baer JT. In-season dietary adequacy of trained male cross-country runners. *International Journal of Sport Nutrition* 5(1):45–55, 1995.

23. Bazzarre TL, Scarpino A, Sigmon R, Marquart LF, Wu SM, Izurieta M. Vitamin-mineral supplement use and nutritional status of athletes. *Journal of the American College of Nutrition* 12(2):162–169, 1993.

24. Fogelholm M. Micronutrient status in females during a 24-week fitness-type exercise program. *Annals of Nutrition and Metabolism* 36(4):209–218, 1992.

25. Lewis RD, Yates CY, Driskell JA. Riboflavin and thiamin status and birth outcome as a function of maternal aerobic exercise. *American Journal of Clinical Nutrition* 48(1):110–116, 1988.

26. Fogelholm M, Rehunen S, Gref CG, Laakso JT, Lehto J, Ruokonen I, Himberg JJ. Dietary intake and thiamin, iron, and zinc status in elite Nordic skiers during different training periods. *International Journal of Sport Nutrition* 2(4):351–365, 1992.

27. van der Beek EJ, van Dokkum W, Wedel M, Schrijver J, van den Berg H. Thiamin, riboflavin and vitamin B_6: impact of restricted intake on physical performance in man. *Journal of the American College of Nutrition* 1994;13:629–640.

28. Webster MJ. Physiological and performance responses to supplementation with thiamin and pantothenic acid derivatives. *European Journal of Applied Physiology* 77(6):486–491, 1998.

29. Webster MJ, Scheett TP, Doyle MR, Branz M. The effect of a thiamin derivative on exercise performance. *European Journal of Applied Physiology* 75(6):520–524, 1997.

30. Fogelholm M, Ruokonen I, Laakso JT, Vuorimaa T, Himberg JJ. Lack of association between indices of vitamin B1, B2, and B6 status and exercise-induced blood lactate in young adults. *International Journal of Sport Nutrition* 3(2):165–176, 1993.

31. Doyle MR, Webster MJ, Erdmann LD. Allithiamine ingestion does not enhance isokinetic parameters of muscle performance. *International Journal of Sport Nutrition* 7(1):39–47, 1997.

32. Lewis RD. Riboflavin and niacin. In: *Sport Nutrition: Vitamins and Trace Elements* (Driskell JA, Wolinsky I, eds.), Boca Raton, FL: CRC Press, 1997.

33. Glatzle D, Korner WF, Christeller S, Wiss O. Method for the detection of a biochemical riboflavin deficiency.

Stimulation of NADPH2-dependent glutathione reductase from human erythrocytes by FAD. In vitro investigations on the vitamin B_2 status in healthy people and geriatric patients. *International Journal of Vitamin Research* 40:166–183, 1970.

34. Belko AZ, Obarzanek E, Kalkwarf HJ, Rotter MA, Bogusz S, Miller D, et al. Effects of exercise on riboflavin requirements of young women. *American Journal of Clinical Nutrition* 37(4):509–517, 1983.

35. Short SH. Surveys of dietary intake and nutritional knowledge of athletes and their coaches. In: *Nutrition in Exercise and Sport* (Wolinsky I, Hickson JF, eds.), Boca Raton, FL: CRC Press, 1994.

36. Rokitzki L, Sagredos A, Keck E, Sauer B, Keul J. Assessment of vitamin B_2 status in performance athletes of various types of sports. *Journal of Nutrition Science and Vitaminology* (Tokyo) 40(1):11–22, 1994.

37. Kirchner EM, Lewis RD, O'Connor PJ. Bone mineral density and dietary intake of female college gymnasts. *Medicine and Science in Sports and Exercise* 27:542–549, 1995.

38. Winters LR, Yoon JS, Kalkwarf HJ, Davies JC, Berkowitz MG, Haas J, Roe DA. Riboflavin requirements and exercise adaptation in older women. *American Journal of Clinical Nutrition* 56(3):526–532, 1992.

39. Guilland J, Panaranda T, Gallet C, Boggio V, Fuchs F, Klepping J. Vitamin status of young athletes including the effects of supplementation. *Medicine and Science in Sports and Exercise* 21:441–449, 1989.

40. Keith RE, Alt LA. Riboflavin status of female athletes consuming normal diets. *Nutrition Research* 11:727–735, 1991.

41. Suboticanec K, Stavljenic A, Schalch W, Buzina R. Effects of pyridoxine and riboflavin supplementation on physical fitness in young adolescents. *International Journal for Vitamin and Nutrition Research* 60:81–88, 1990.

42. Powers HJ, Bates CJ, Lamb WH, Singh J, Gelman W, Webb E. Effects of a multivitamin and iron supplement on running performance in Gambian children. *Human Nutrition and Clinical Nutrition* 39C:427–437, 1985.

43. Nutter J. Seasonal changes in female athletes' diets. *International Journal of Sport Nutrition*, 1(4):395–407, 1991.

44. Murray R, Bartoli WP, Eddy DE, Horn MK. Physiological and performance responses to nicotinic-acid ingestion during exercise. *Medicine and Science in Sports and Exercise*, 27(7):1057–1062, 1995.

45. Heath EM, Wilcox AR, Quinn CM. Effects of nicotinic acid on respiratory exchange ratio and substrate levels during exercise. *Medicine and Science in Sports and Exercise* 25(9):1018–1023, 1993.

46. Singh A, Moses FM, Deuster PA. Vitamin and mineral status in physically active men: effects of a high-potency supplement. *American Journal of Clinical Nutrition* 55(1):1–7, 1992.

47. Nice C, Reeves AG, Brinck-Johnsen T, Noll W. The effects of pantothenic acid on human exercise capacity. *Journal of Sports Medicine and Physical Fitness*, 24(1):26–29, 1984.

48. Hartvig P, Lindner KJ, Bjurling P, Laengstrom B, Tedroff J. Pyridoxine effect on synthesis rate of serotonin in the monkey brain measured with positron emission tomography. *Journal of Neural Transmission and Genetic Sectors* 102(2):91–97, 1995.

49. Manore MM. Vitamin B_6 and exercise. *International Journal of Sport Nutrition* 4(2):89–103, 1994.

50. Rokitzki L, Sagredos AN, Reuss F, Cufi D, Keul J. Assessment of vitamin B_6 status of strength and speedpower athletes. *Journal of the American College of Nutrition* 13(1):87–94, 1994.

51. Keith RE, O'Keeffe KA, Alt LA. Dietary status of trained female cyclists. *Journal of the American Dietetic Association*, 89:1620–1623, 1989.

52. Steen SN, Mayer K, Brownell KD, Wadden TA. Dietary intake of female collegiate heavyweight rowers. *International Journal of Sport Nutrition* 5(3):225–231, 1995.

53. Sacheck JM, Roubenoff R. Nutrition in the exercising elderly. *Clinical Sports Medicine* 18(3):565–584, 1999.

54. Rokitzki L, Sagredos AN, Reuss F, Buchner M, Keul J. Acute changes in vitamin B6 status in endurance athletes before and after a marathon. *International Journal of Sport Nutrition*, 4(2):154–165, 1994.

55. Coburn SP, Ziegler PJ, Costill DL, Mahuren JD, Fink WJ, Schaltenbrand WE, et al. Response of vitamin B-6 content of muscle to changes in vitamin B-6 intake in men. *American Journal of Clinical Nutrition* 53(6):1436–1442, 1991.

56. Lawrence JD, Smith JL, Bower RC, Riehl WP. The effect of alpha tocopherol (vitamin E) and pyridoxine HCl (vitamin B6) on the swimming endurance of trained swimmers. *Journal American College of Health Association* 23(3):219–222, 1975.

57. Virk RS, Dunton NJ, Young JC, Leklem JE. Effect of vitamin B_6 supplementation on fuels, catecholamines, and amino acids during exercise in men. *Medicine and Science in Sports Exercise* 31(3):400–408, 1999.

58. Virk R, Dunton NJ, Leklem JE. The effect of vitamin B_6 supplementation during exhaustive exercise, *FASEB* 6:A:1374, 1992.

59. Manore MM, Leklem JE. Effect of carbohydrate and vitamin B6 on fuel substrates during exercise in women. *Medicine and Science in Sports and Exercise* 20(3):233–241, 1988.

60. Haymes EM. Vitamin and mineral supplementation to athletes. *International Journal of Sport Nutrition* 1(2):146–169, 1991.

61. Matter M, Stittfall T, Graves J, Myburgh K, Adams B, Jacobs P, Noakes TD. The effect of iron and folate therapy on maximal exercise performance in female marathon runners with iron and folate deficiency. *Clinical Science* 72(4):415–422, 1987.

62. Brotherhood J, Brozovic B, Pugh LG. Haematological status of middle- and long-distance runners. *Clinical Science and Molecular Medicine* 48(2):139–145, 1975.

63. Barry A, Cantwell T, Doherty F, Folan JC, Ingoldsby M, Kevany JP, et al. A nutritional study of Irish athletes. *British Journal of Sports Medicine* 15(2):99–109, 1981.

64. Weight LM, Myburgh KH, Noakes TD. Vitamin and mineral supplementation: effect on the running performance of trained athletes. *American Journal of Clinical Nutrition* 47(2):192–195, 1988.

65. Weight LM, Noakes TD, Labadarios D, Graves J, Jacobs P, Berman PA. Vitamin and mineral status of trained athletes

including the effects of supplementation. *American Journal of Clinical Nutrition* 47(2):186–191, 1988.

66. Tin-May-Than, Ma-Win-May, Khin-Sann-Aung, Mya-Tu M. The effect of vitamin B12 on physical performance capacity. *British Journal of Nutrition* 40(2):269–273, 1978.

67. Read MH, McGuffin SL. The effect of B-complex supplementation on endurance performance. *Journal of Sports Medicine and Physical Fitness* 23(2):178–184, 1983.

68. Buchman AL, Jenden D, Roch M. Plasma free, phospholipid-bound and urinary free choline all decrease during a marathon run and may be associated with impaired performance, *Journal of the American College of Nutrition* 18:598–601, 1999.

69. Buchman AL, Awal M, Jenden D, Roch M, Kang SH. The effect of lecithin supplementation on plasma choline concentrations during a marathon. *Journal of the American College of Nutrition* 9:768–778, 2000.

70. Spector SA, Jackman MR, Sabounjian LA, Sakkas C, Landers DM, Willis WT. Effect of choline supplementation on fatigue in trained cyclists. *Medicine and Science in Sports Exercise* 27:668–678, 1995.

71. Warber JP, Patton JF, Tharion WJ, Zeisel SH, Mello RP, Kemnitz CP, Lieberman HR. The effects of choline supplementation on physical performance. *International Journal of Sport Nutrition and Exercise Metabolism* 10:170–178, 2000.

72. Sachan DS, Hongu N. Increases in VO(2)max and metabolic markers of fat oxidation by caffeine, carnitine, and choline supplementation in rats, *Journal of Nutritional Biochemistry* 11:521–529, 2000.

73. Saris WH, van Erp-Baart MA, Brouns F, Westerterp KR, ten Hoor F. Study on food intake and energy expenditure during extreme sustained exercise: the Tour de France. *International Journal of Sports Medicine* 100172-4622:S26–S31, 1989.

74. Stacewicz-Sapuntzakis M. Vitamin A and carotenoids. In: *Sports Nutrition: Vitamins and Trace Elements* (Wolinsky I, Driskell JA, eds.), Boca Raton, FL: CRC Press, 1997.

75. Duthie GG, Robertson JD, Maughan RJ, Morrice PC. Blood antioxidant status and erythrocyte lipid peroxidation following distance running. *Archives of Biochemistry and Biophysics* 282(1):78–83, 1990.

76. Lehtonen-Veromaa M, Mottonen T, Irjala K, Karkkainen M, Lamberg-Allardt C, Hakola P, Viikari J. Vitamin D intake is low and hypovitaminosis D common in healthy 9- to 15-year-old Finnish girls. *European Journal of Clinical Nutrition* 53(9):746–751, 1999.

77. Sundgot-Borgen J. Nutrient intake of female elite athletes suffering from eating disorders. *International Journal of Sport Nutrition* 3(4):431–442, 1993.

78. Lewis RD, Modlesky CM. Nutrition, physical activity, and bone health in women. *International Journal of Sport Nutrition* 8(3):250–284, 1998.

79. Reid IR. Therapy of osteoporosis: calcium, vitamin D, and exercise. *American Journal of Medical Science* 312(6):278–286, 1996.

80. Bell NH, Godsen RN, Henry DP, Shary J, Epstein S. The effects of muscle-building exercise on vitamin D and mineral metabolism. *Journal of Bone Mineral Research* 3(4):369–373, 1988.

81. Ashmaig ME, Starkey BJ, Ziada AM, Amro A, Sobki S, Ferns GA. Changes in serum concentration of antioxidants following treadmill exercise testing in patients with suspected ischaemic heart disease. *International Journal of Experimental Pathology* 82(4):243–248, 2001.

82. Kawai Y, Shimomitsu T, Takanami Y, Murase N, Katsumura T, Maruyama C. Vitamin E level changes in serum and red blood cells due to acute exhaustive exercise in collegiate women. *Journal of Nutrition Science and Vitaminology* (Tokyo) 46(3):119–124, 2000.

83. Meydani M, Evans WJ, Handelman G, Biddle L, Fielding RA, Meydani SN, et al. Protective effect of vitamin E on exercise-induced oxidative damage in young and older adults. *American Journal of Physiology* 264(5 Pt 2):R992–Rc998, 1993.

84. Sharman IM, Down MG, Norgan NG. The effects of vitamin E on physiological function and athletic performance of trained swimmers. *Journal of Sports Medicine and Physical Fitness* 16(3):215–225, 1976.

85. Sharman IM, Down MG, Sen RN. The effects of vitamin E and training on physiological function and athletic performance in adolescent swimmers. *British Journal of Nutrition*, 26(2):265–276, 1971.

86. Takanami Y, Iwane H, Kawai Y, Shimomitsu T. Vitamin E supplementation and endurance exercise: are there benefits? *Sports Medicine* 29(2):73–83, 2000.

87. Craciun AM, Wolf J, Knapen MH, Brouns F, Vermeer C. Improved bone metabolism in female elite athletes after vitamin K supplementation. *Archives of Biochemistry and Biophysics* 282(1):78–83, 1990.

88. Lewis NM, Frederick AM. Vitamins D and K. In: *Sports Nutrition: Vitamins and Trace Elements* (Wolinsky I, Driskell JA, eds.), Boca Raton, FL: CRC Press LLC, 1997.

89. Vermeer C, Gijsbers BL, Craciun AM, Groenen-van Dooren MM, Knapen MH. Effects of vitamin K on bone mass and bone metabolism. *Journal of Nutrition* 126 (4Suppl):1187S–1191S, 1996.

90. Sumida S, Okamura K, Doi T, Sakurai M, Yoshioka Y, Sugawa-Katayama Y. No influence of a single bout of exercise on urinary excretion of 8-hydroxy-deoxyguanosine in humans. *Biochemistry and Molecular Biology International* 42(3):601–609, 1997.

91. Sumida S, Doi T, Sakurai M, Yoshioka Y, Okamura K. Effect of a single bout of exercise and beta-carotene supplementation on the urinary excretion of 8-hydroxy-deoxyguanosine in humans. *Free Radical Research* 27(6):607–618, 1997.

92. Hartmann A, Niess AM, Grunert-Fuchs M, Poch B, Speit G. Vitamin E prevents exercise-induced DNA damage. *Mutation Research* 346(4):195–202, 1995.

93. Radak Z, Pucsuk J, Boros S, Josfai L, Taylor AW. Changes in urine 8-hydroxydeoxyguanosine levels of super-marathon runners during a four-day race period. *Life Sciences* 24;66(18):1763–1767, 2000.

94. Okamura K, Doi T, Hamada K, Sakurai M, Yoshioka Y, Mitsuzono R, et al. Effect of repeated exercise on urinary 8-hydroxy-deoxyguanosine excretion in humans. *Free Radical Research* 26(6):507–514, 1997.

95. Pilger A, Germadnik D, Formanek D, Zwick H, Winkler N, Rudiger HW. Habitual long-distance running does not

enhance urinary excretion of 8-hydroxydeoxyguanosine. *European Journal of Applied Physiology* 75(5):467–469, 1997.

96. Inoue T, Mu Z, Sumikawa K, Adachi K, Okochi T. Effect of physical exercise on the content of 8-hydroxydeoxyguanosine in nuclear DNA prepared from human lymphocytes. *Japan Journal of Cancer Research* 84(7):720–725, 1993.

97. Powers SK, Lennon SL. Analysis of cellular responses to free radicals: focus on exercise and skeletal muscle. *Proceedings of the Nutrition Society* 58:1025–1033, 1999.

98. Shigenaga MK, Gimeno CJ, Ames BN. Urinary 8-hydroxy-2'-deoxyguanosine as a biological marker of in vivo oxidative DNA damage. *Proceedings of the National Academy of Sciences USA* 86(24):9697–9701, 1989.

99. Takanami Y, Iwane H, Kawai Y, Shimomitsu T. Vitamin E supplementation and endurance exercise: are there benefits? *Sports Medicine* 29(2):73–83, 2000.

100. Coombes JS, Powers SK, Rowell B, Hamilton KL, Dodd SL, Shanely RA, et al. Effects of vitamin E and alpha-lipoic acid on skeletal muscle contractile properties. *Journal of Applied Physiology* 90(4):1424–1430, 2001.

101. Schroder H, Navarro E, Mora J, Galiano D, Tramullas A. Effects of alpha-tocopherol, beta-carotene and ascorbic acid on oxidative, hormonal and enzymatic exercise stress markers in habitual training activity of professional basketball players. *European Journal of Nutrition* 40(4):178–184, 2001.

102. Sumida S, Tanaka K, Kitao H, Nakadomo F. Exercise-induced lipid peroxidation and leakage of enzymes before and after vitamin E supplementation. *International Journal of Biochemistry* 21(8):835–838, 1989.

10

MINERALS AND EXERCISE

Chapter Objectives

- Discuss the food sources and functions of the essential minerals.

- Provide recommendations for dietary intake and insight into the involvement of minerals in disease prevention and good health.

- Describe the dietary level of essential minerals by active people and compare it to what is needed.

- Discuss how minerals affect physical activity and sport performance.

Personal Snapshot

© Digital Vision/Getty Images

Carla, a 24-year-old fitness competitor, recently had the opportunity to have DXA performed to assess her body composition. She was pleased with her body fat level but shocked to learn that her bone mineral density was below average for her age. She trains and competes year-round and often restricts her energy intake to prepare for competition. In talking with her coach and nutritionist, Carla stated that she feels "run down" during workouts and has not menstruated regularly for 3 years. A nutrition assessment reveals that she does not eat meats and other animal foods and that her diet provides only about half of her need for calcium, iron, copper, and zinc. Could the energy restriction and other dietary inadequacies be involved in her poor bone mineral density? Might her dietary shortcomings be involved in the fatigue she sometimes experiences? What other problems might she experience?

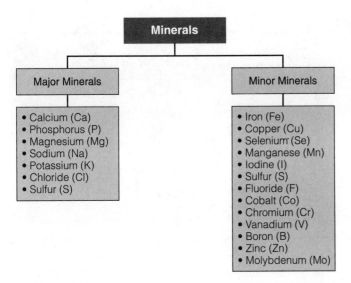

Figure 10-1
The major and minor minerals of the human body.

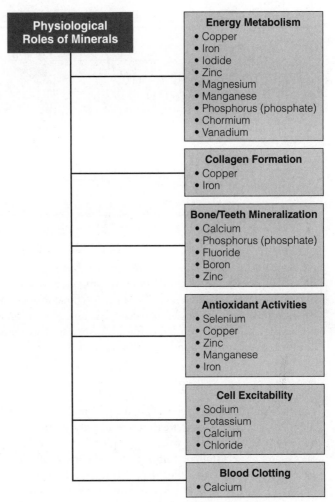

Figure 10-2
Classification of minerals based on function.

Minerals are widely known for their role in normal human function and health (Figures 10-1 and 10-2). For instance, most people would be able to tell you that calcium is important for bone hardness and that iron is a component of hemoglobin in the blood. But how important are minerals to athletic performance? Like vitamins, minerals are important in numerous ways for everyone, but athletes sometimes have additional needs for them. For example, iron in hemoglobin binds O_2 for transport to muscle tissue, and O_2 availability is critical to endurance performance. Iron and copper associated with the electron transport system assist in aerobic energy metabolism, and energy demands are significantly increased during exercise. Calcium evokes sarcomere contraction, and phosphate is a key component of high-energy phosphate molecules that power muscle contraction, namely ATP and creatine phosphate. In addition, several minerals, such as copper, iron, selenium, manganese, and zinc, are components of antioxidant enzyme systems that help control free radicals produced during exercise. (The increased production of free radicals during certain types of exercise is discussed in the Chapter 9 In Focus feature.)

This chapter provides an overview of the food sources and dietary recommendations for, and basic functions of, essential minerals (see Appendix A for the complete Dietary Reference Intake levels). Dietary levels of minerals consumed by active people are discussed, as well as how each mineral is involved in physical activity. The influence of mineral supplementation on athletic performance is also discussed, although phosphate, chromium, vanadium, and boron supplementation are described in greater detail in Chapter 11 because of their popularity and/or breadth of research base. Because of the association between minerals and bone and bone disorders, this chapter provides a closer look in the feature In Focus: Osteoporosis: Not Just a Calcium Problem.

■ CALCIUM (Ca)

Calcium (Ca^{2+}) is the most abundant mineral in the body; it accounts for about 40% of the total mineral mass of a typical adult, equating to about 1.5% of total body weight. Dairy products are the greatest overall contributors of calcium to the American diet—on average, more than 55% (Table 10-1). One cup of milk provides approximately 300 mg of calcium, as does ¾ cup of yogurt, 1½ oz of hard cheese, and 1¾ cups of ice cream. Other good sources of calcium are sardines, oysters, clams, tofu, molasses, almonds, calcium-fortified foods, and dark green leafy vegetables such as broccoli, kale, collards, and mustard and turnip greens. Stomach acidity and the presence of certain amino acids in the small intestine seem to enhance

Table 10-1 Calcium Content of Selected Foods

Food	Calcium (mg)	Food	Calcium (mg)
Milk and Milk Products		**Vegetables**	
Yogurt, low-fat (1 c)	448	Collard greens (½ c)	110
Milk, skim (1 c)	301	Spinach (½ c)	90
Cheese, Swiss (1 oz)	272	Broccoli (½ c)	70
Ice cream (1 c)	180		
Ice milk (1 c)	180	**Legumes and Products**	
Custard (½ c)	150	Tofu (½ c)	155
Cottage cheese (½ c)	70	Dried beans (½ c)	50
		Lima beans (½ c)	40

calcium absorption, which may mean absorption is more efficient during a protein-containing meal. But meals having a high phosphorus-to-calcium ratio might result in reduced absorption efficiency. The ratio of phosphorus to calcium in the diet should not exceed 2:1, and the optimal ratio is 1:1. The calcium RDA for adult men and women between the ages of 19 and 50 is 1000 mg.

Calcium Supplements

Calcium supplements are very popular. Usually people use them in an attempt to increase bone density or minimize bone calcium loss. Most calcium supplements are calcium salts (**salts** are crystalline compounds and consist of a negative and a positive ion other than hydrogen and hydroxide ions) and include amino acids or other substances such as carbonate and citrate. Calcium carbonate is more efficiently absorbed than calcium citrate, calcium acetate, calcium lactate, or calcium gluconate. Some food companies have begun fortifying fruit juices (such as orange juice) and other foods with calcium citrate malate (CCM). CCM may be even more efficiently absorbed than calcium carbonate.

Key Point

Calcium supplements are usually in the form of salts, such as calcium citrate, calcium carbonate, and calcium citrate malate.

Calcium Functions and Metabolism

Calcium is found throughout the body and serves both structural and functional roles. Roughly 99% is in bone and teeth (*hard tissue*) as mineral salts and crystals.[1] These complexes are *calcium phosphate* salt and *hydroxyapatite* crystals. In addition to making those tissues hard, they are a source of calcium and phosphorus when they are needed in circulation and in other tissues. Mobilization of mineral from bone tissue is referred to as **resorption.**

$CaHPO_4{\cdot}2H_2O$ and $Ca_3(PO_4)_2{\cdot}3H_2O$ — Calcium phosphates

$Ca_{10}(PO_4)_6(OH)_2$ — Hydroxyapatite

Approximately 1% of the calcium in the body is found in the blood (plasma) and "soft tissue," which includes skeletal and cardiac muscle, glands, and nerves. This calcium participates in second messenger systems, hormone and neurotransmitter release, and blood clotting. Calcium plays a pivotal role in the contraction of muscle tissue by initiating the process that allows for cross-bridging between actin and myosin, which leads to sarcomere contraction (see Chapter 2). It also helps initiate the action potential in the **sinoatrial node** of the heart that results in the electrical stimulus (action potential) that evokes a heartbeat.

Key Point

Calcium helps make bone and teeth hard and is involved in muscle contraction, blood clotting, and other functions.

BLOOD CALCIUM REGULATION AND EXCRETION. Two hormones [parathyroid hormone (PTH) and calcitonin] and vitamin D are directly involved in regulating the level of calcium in the blood. Circulating calcium is either dissolved in the plasma or associated with proteins such as albumin or anions (such as phosphates) and typically ranges between 8.8 and 10.8 mg/100 ml of blood. Because of the regulatory control, poor or excessive dietary levels of calcium rarely result in significant changes in blood calcium levels. However, with time these dietary imbalances can result in blood calcium alterations.

When blood calcium levels decrease, PTH is released into the circulation from the parathyroid gland. PTH increases the activation of vitamin D [1,25-$(OH)_2$,D_3 or calcitriol] in the kidneys, and along with vitamin D, PTH decreases the loss of calcium in the urine. In addition, vitamin D increases the absorption of calcium from the small intestine and resorption of calcium from bones. The net result is an increase in blood calcium content. On the other side of the balance, calcitonin promotes activities that are generally opposite to those of vitamin D and PTH; it is released from the thyroid gland when the level of blood calcium increases. Calcitonin decreases the release of calcium from bone and, in conjunction with calcium loss in urine, reduces blood calcium, returning it to the normal range.

Key Point

Parathyroid hormone, vitamin D, and calcitonin are involved in regulating blood calcium levels.

Calcium Imbalance

Inadequate intake of calcium during childhood results in poor calcium deposition into hard tissue during growth. Bone tissue that develops without adequate mineralization is subject to deformities, such as bowing of the femur and misshapen skull and rib cage. If poor calcium status continues throughout childhood it will result in a lower **peak bone mass (PBM)** at the culmination of bone development, usually in one's late 20s, rendering the bones more prone to osteoporosis later in life. During adulthood, inadequate calcium intake leads to excessive calcium resorption from bone and thus a too-low mineral content in bone, or **osteomalacia,** which renders the bone more susceptible to fracture. Vitamin D deficiency or kidney disorders can lead to a reduction in blood calcium by increasing calcium loss in the urine. Excessive sweating, such as by runners and tennis players, can also lead to appreciable losses of calcium. Sweat calcium levels may approximate 20 mg/L. Whatever the cause, a reduction in the level of circulating calcium can ultimately affect its concentration in tissue such as nerves, muscle, and glands. Thus, low blood calcium levels are associated with irritability of nervous tissue (the central nervous system) and skeletal muscle cramping.

Calcium supplementation of 1000 mg or more is sometimes advocated, but toxicity must be considered as dosage increases. The upper limit (UL) for calcium is 2500 mg. Although the efficiency of absorption decreases as intake increases, more calcium is still absorbed. Excessive calcium intake over time might lead to calcium salt deposition in softer tissue like skeletal muscle, heart, blood vessels, and lungs,[1] making them more rigid. In the urine, excess calcium can promote the formation of **calcium oxalate** kidney stones. Like other types of kidney stones, these stones form when the concentration of calcium and oxalate is great enough to allow for aggregation and the formation of crystals. In the digestive tract, excess calcium might decrease iron absorption.

Calcium Intake of Athletes and Other Active People

The involvement of calcium in muscle contraction is not an obscure physiological point. Many athletes know how it affects their performance from taking biology and physiology courses, reading magazine articles, or talking to teammates and friends. Still, some groups of athletes may not consume enough calcium. A nutritional assessment of elite athletes with cerebral palsy (CP) or who are otherwise physically disabled (Class IV) revealed that approximately half of the CP and two-thirds of the Class IV athletes did not consume the RDA level of calcium.[2] In other studies, researchers determined that many female athletes consume less than the RDA.[3] The results of one study involving gymnasts and speed skaters indicated that although the average calcium intake met recommendations, several of the participants were below recommended levels.[4] Athletes who restrict their diet to make a weight class may also risk substandard calcium intakes.[5] However, as more foods become fortified with calcium it is possible that this situation has improved. Reevaluation surveys that consider all calcium sources are warranted.

In one study in which male and female bodybuilders kept a 3-day food diary, their calcium intake was well below the RDA for both the males (433 ± 189 mg) and the females (478 ± 339 mg).[6] Certainly there are differences in the intake of many nutrients during noncompetition, precompetition, and competition periods for bodybuilders; perhaps no other athletic group manipulates their diet to such extremes. For instance, researchers reported that a group of male bodybuilders consumed an average of 2141 mg of calcium during early precompetition but reduced their intake to an average of 416 mg of calcium just prior to competition.[7] They also reported that female bodybuilders consumed an average of 705 mg of calcium during early precompetition but only 272 mg just before competition. All this was the result of the restrictive dietary practices of bodybuilders as they approach competition.

Key Point

Many athletes do not consume the recommended level of calcium; others rely on supplements to make up the shortcomings of their diet.

Calcium and Athletic Performance

Calcium is important to exercise in several ways. Optimal calcium is needed for both skeletal and heart muscle contraction, for efficient hormone and neurotransmitter activity during exercise, and for effective blood clotting at the site of minor hemorrhages sustained during practice and competition. Furthermore, optimal levels of calcium in the blood are important to maximize adaptive processes that affect bone density.[8] Because calcium is the predominant mineral in bone, changes in bone mineral mass reflect changes in bone calcium content.

Increased sweating during physical activity can lead to greater calcium losses. To prevent loss of calcium and other **bone mineral content** (**BMC,** the mass of all minerals in bone), this must be counterbalanced by increased absorption from the digestive tract, as demonstrated in a study of BMC in college basketball players over a 2-year period.[9] BMC was assessed using DXA (dual-energy X-ray absorptiometry, discussed in Chapter 7), and sweating was determined to be a significant route of calcium loss in these athletes. During the first year, total body BMC decreased by 6.1% and BMC of the legs decreased by 10.5%. When the players were provided calcium supplements in the second year, however, BMC increased along with lean body mass (LBM). The researchers concluded that bone loss is related to dietary calcium and that exercise can increase BMC only when calcium intake is sufficient to offset sweat losses.

Female athletes who restrict their energy intake and become very lean can become amenorrheic. Without the production of estrogen, they can experience significant loss of bone material, including calcium. Although exercise is able to have a positive influence on bone density, it cannot spare bone loss during amenorrhea. Amenorrhea is more common to gymnasts, distance runners, and fitness competitors who voluntarily restrict their food intake and/or demonstrate patterns of disordered eating patterns (see In Focus: Eating Disorders and the Female Athlete Triad in Chapter 15.)

It is not yet known whether dietary calcium needs are elevated for athletes. Calcium supplements are not necessary for athletes unless their diet provides inadequate calcium and dietary modification is unable to correct the inadequacy—unlikely now that many foods are calcium fortified. Manufacturers now fortify breads and fruit juices, and extra calcium is blended into yogurts. Candy-like supplements are available that contain the calcium equivalent of a glass of milk or more. Adequate calcium in the diet allows athletes to optimize peak bone mineral mass as positively influenced by physical training, so that they can perform at their highest level. See the In Focus at the end of this chapter for more discussion of the relationships between calcium, exercise, and bone.

■ CALCIUM AND EXERCISE *IN REVIEW*

- In food, calcium is found in dairy products, meats, and some vegetables.
- In the body, most calcium is found in hard tissue. The calcium in blood and soft tissues such as muscle and glands is important in functional and homeostatic operations (muscle contraction, blood clotting, and second messenger systems).
- Dietary calcium intake of some athletes is below recommendations, but dietary modification or numerous types of calcium supplements can correct this.
- Optimal calcium status might be crucial for bone remodeling and muscle hypertrophy induced by athletic training.

Table 10-2 **Phosphorus Content of Select Foods**

Food	Phosphorus (mg)	Food	Phosphorus (mg)
Milk and Milk Products		**Grains**	
Yogurt (1 c)	327	Bran flakes (1 c)	180
Milk (1 c)	250	Bread, whole wheat (1 slice)	52
Cheese, American (1 oz)	130	Noodles, cooked (½ c)	47
		Rice, cooked (½ c)	29
Meats and Alternatives		Bread, white (1 slice)	24
Pork (3 oz)	275		
Hamburger (3 oz)	165	**Vegetables**	
Tuna (3 oz)	162	Potato (1)	101
Lobster (3 oz)	125	Corn (½ c)	73
Chicken (3 oz)	120	Peas (½ c)	70
		Broccoli (½ c)	**54**
Nuts and Seeds		**Other**	
Sunflower seeds (¼ c)	319	Milk chocolate (1 oz)	66
Peanuts (¼ c)	**141**	Cola (12 oz)	51
Peanut butter (1 tbs)	61	Diet cola (12 oz)	45

■ PHOSPHORUS (P)

Phosphorus is the sixth most abundant element in the body by mass. After calcium it is the most abundant mineral. In food and body tissue it takes the form of **phosphate** ($H_2PO_4^{2-}$). The RDA for phosphate is 700 mg. The usual intake for men is about 1500 mg daily; for women it is about 1000 mg. Athletes tend to consume more, as they generally eat more food to balance their increased energy expenditure.

Many foods provide phosphate; some of the better sources are meat, poultry, eggs, fish, milk and milk products, cereals, legumes, grains, and chocolate (Table 10-2). Coffee and teas also provide some phosphate, and many soft drinks (colas) contain phosphoric acid. Much of the phosphate in cereal grains is in the form of **phytate** molecules. Phytate is *inositol hexaphosphate*, which, as shown in Figure 10-3, is inositol (carbohydrate) with up to six phosphate groups esterified to the carbon atoms in the ring. In the digestive tract phytate can interact with positively charged minerals such as calcium, magnesium, copper, zinc, and iron. This can decrease the efficiency of absorption of these minerals.

Phosphate Functions and Metabolism

Phosphate has numerous roles in the body. As much as 85% of the phosphate in the body is found in the skeleton and teeth as a component of calcium phosphate and hydroxyapatite, helping to make those tissues hard. The remaining 15% is found in other tissues as a component of

Phytate

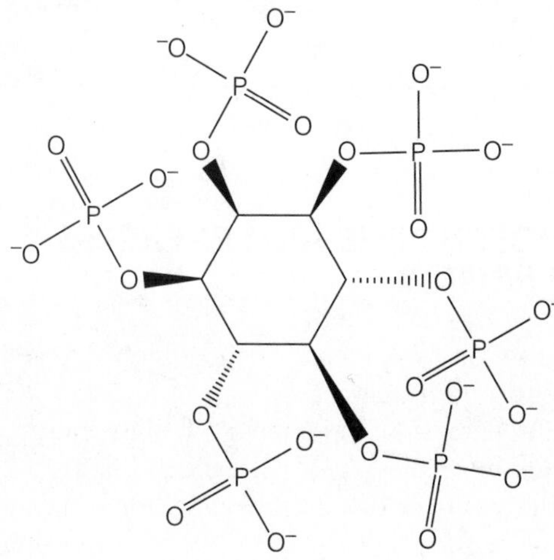

Figure 10-3
Phytate molecule (bearing numerous negative charges) found in plants. It can interact with positively charged minerals (such as Ca^{2+}, Mg^{2+}, Zn^{2+}) in the digestive tract and decrease their absorption.

phospholipids in cell membranes and in lipoprotein shells. Phosphate is also part of the high-energy phosphate molecules, namely adenosine triphosphate (ATP), guanosine triphosphate (GTP), and creatine phosphate (CP) (Figure 10-4). Approximately 30% of the phosphate in the circulation is freely dissolved (HPO_4^{2-} and $H_2PO_4^-$). Phosphate dissolved within the circulation and tissue can serve as a buffer against the accumulation of hydrogen ions (H^+) and a lowering of the pH. This function has been of particular interest to exercise researchers and certain types of athletes.

Key Point

Phosphate is a component of ATP and creatine phosphate, the most important anaerobic energy resources for muscle tissue.

Phosphate Intake by Athletes

Most experts today think that athletes whose diet contains a variety of foods consume enough phosphate to have optimal body levels. Some athletes also consume phosphate as part of sport supplements. Phosphate is important to athletic performance in many ways, so there has been considerable experimentation with phosphate supplements and loading procedures. Phosphate is often cosupplemented with related substances such as creatine. But if phosphate supplementation is used, its potential impact on calcium metabolism and other physiological processes must be considered. Long-term phosphate supplementation or routine periods of phosphate loading can decrease the calcium-to-phosphate ratio of the diet and could have a detrimental influence on calcium status. Also there is a potential for gastrointestinal distress (such as diarrhea, nausea, vomiting, and stomach pain), so the timing and dosage levels of phosphate loading are important. The UL for phosphate has been set at 4000 mg for adults.

Phosphate and Athletic Performance

Among the reasons athletes supplement phosphate, perhaps the foremost, is that supplementation might enhance the level of ATP and CP in muscle fibers, leading to greater power. However, research has yet to prove that phosphate supplementation by itself will work that way. There is also speculation that phosphate supplementation might increase the buffering capacity of skeletal muscle and the blood, delaying the onset of muscular fatigue associated with excessive lactic acid production. This could apply to very high intensity activities such as

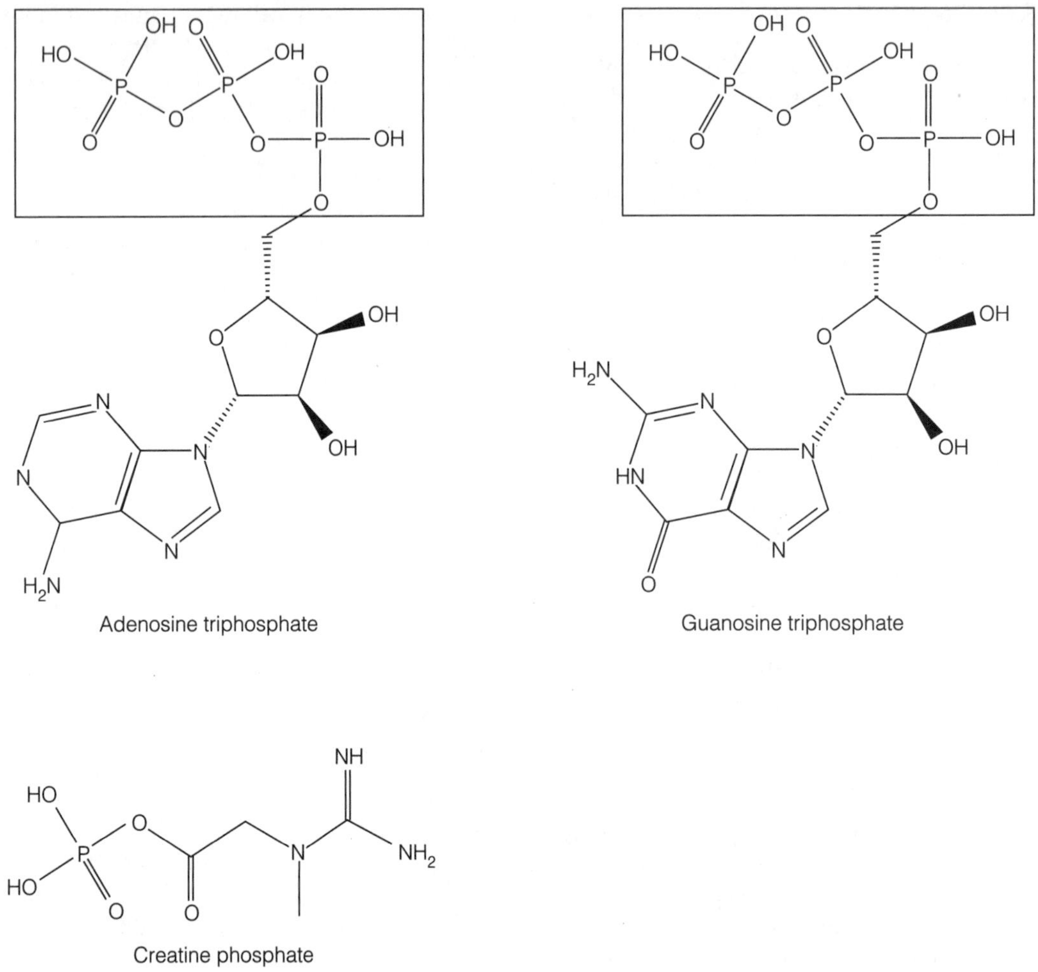

Adenosine triphosphate

Guanosine triphosphate

Creatine phosphate

Figure 10-4

The high-energy phosphates created within cells to power operations. (Note: creatine phosphate is created using ATP and serves as a mechanism for resynthesizing ATP when needed.)

weight training and high-power efforts (such as strong-man events) as well as all-out sprinting (running, cycling, swimming) or shorter distance runs (mile runs and the like). Furthermore, phosphate supplementation might enhance performance by increasing the level of 2,3-diphosphoglycerol (2,3-DPG) in red blood cells (RBCs).[10] For higher intensity, shorter duration endurance efforts such as mile runs, more 2,3-DPG might reduce lactate accumulation by enhancing O_2 release to tissues. It has been known for some time that 2,3-DPG shifts the O_2 dissociation curve in the direction of increased O_2 release to tissue such as muscle (see Figure 2-28). Precisely how phosphate is involved is not certain; see Chapter 11 for more details. But today it simply is not prudent to recommend phosphate supplementation or loading until more is known about its impact on calcium metabolism.

■ PHOSPHORUS AND EXERCISE *IN REVIEW*

- Phosphorus is widely distributed in foods and body tissue as phosphate.
- Phosphate is a component of high-energy phosphate molecules (ATP, CP, and GTP) and of phospholipids in cell membranes and lipoprotein shells.
- Athletes probably take in enough phosphate if their diet provides adequate energy and includes a variety of foods.
- Phosphate supplementation (loading) might enhance performance in sports that require athletes to cross the lactate threshold; however, it is not recommended because too little is known about its effect on calcium metabolism.

■ SODIUM, POTASSIUM, AND CHLORIDE

Sodium, potassium, and chloride are discussed together here, as they are interrelated in many ways. These minerals are the *electrolytes* often mentioned in television commercials and magazine articles (although other electrolytes are also dissolved in body fluids). Most foods are naturally low in sodium and chloride. However, because sodium chloride (NaCl)—table salt—is a popular flavor enhancer and food preservative, many processed foods are significant dietary providers of these minerals (Table 10-3 shows the sodium content of many foods). Adding salt during cooking and at the table also contributes sodium and chloride to the diet, as does drinking water, certain medicines, mouthwash, and toothpaste (as sodium fluoride). Among the foods that contribute the most sodium to the American diet are luncheon meats, snack food (such as chips and pretzels), french fries, hot dogs, cheeses, soups, and gravies. Food manufacturers in the United States are required to list the sodium content on the Nutrition Facts found on the food label. In addition, any statements made by the food manufacturer on the label regarding the sodium content of the product must follow specific criteria (Table 10-4).

Potassium is found in most natural foods, especially fresh fruits and vegetables and their juices (Table 10-5). Tomatoes, carrots, potatoes, beans, peaches, pears, squashes, oranges, and orange juice are all good sources of potassium. Milk, meats, whole grains, coffee and tea are lesser but still significant sources. Many athletes refer to bananas as "potassium sticks"; however, their potassium content is average compared to other fruit and vegetable sources. Unlike sodium, potassium is not a common component of food additives, although potassium chloride (KCl) is sometimes used as a salt substitute. Adult recommendations for sodium, potassium, and chloride are 500, 2000, and 750 mg, respectively.

Key Point

Most foods are naturally low in sodium and chloride. Potassium is found in most natural foods; fruits and vegetables are among the better sources.

Sodium, Potassium, and Chloride Functions and Metabolism

Sodium, potassium, and chloride are dissolved throughout the extracellular and intracellular compartments. Figure 10-5 presents the concentrations of the major electrolytes dissolved in the intracellular and extracellular fluids. Sodium is approximately 14 times more concentrated in extracellular fluid than in intracellular fluid. Chloride tends to follow the movement of sodium as a mechanism to balance electric charges; about 88% of the chloride in

Table 10-3 **Sodium Content of Selected Foods**

Food	Sodium (mg)	Food	Sodium (mg)
Meats and Alternatives		**Other**	
Corned beef (3 oz)	808	Salt (1 tsp)	2132
Ham (3 oz)	800	Pickle, dill (1)	1930
Fish, canned (3 oz)	735	Broth, chicken (1 c)	1571
Sausage (3 oz)	483	Ravioli, canned (1 c)	1065
Hot dog (1)	477	Broth, beef (1 c)	782
Bologna (1 oz)	370	Gravy (¼ c)	720
		Italian dressing (2 tbs)	720
Milk and Milk Products		Pretzels (salted), thin (5)	500
Cream soup (1 c)	1070	Olives, green (5)	465
Cottage cheese (½ c)	455	Pizza, cheese (1 slice)	455
Cheese, American (1 oz)	405	Soy sauce (1 tsp)	444
Cheese, Parmesan (1 oz)	247	Bacon (3 slices)	303
Milk, skim (1 c)	125	French dressing (2 tbs)	220
Milk, whole (1 c)	120	Potato chips (10)	200
		Ketchup (1 tbs)	155
Grains		Bagel (1)	260
Bran Flakes (1 c)	363		
Corn Flakes (1 c)	325		
English muffin (1)	203		
Bread, white (1 slice)	130		
Bread, whole wheat (1 slice)	130		
Crackers, saltines (4 squares)	125		

Table 10-4

Labeling Guidelines for Sodium Content

Label Claim	Sodium Content (Per Serving)
"Sodium free"	Must contain <5 mg per serving
"Very low sodium"	Must contain ≤35 mg sodium per serving
"Low sodium"	Must contain ≤145 mg sodium per serving
"Reduced sodium"	75% reduction in sodium content
"Unsalted"	No salt added to the recipe
"No added salt"	No salt added to the recipe

Key Point

Electrolytes allow for the excitation of skeletal muscle fibers that leads to the release of calcium into muscle cells, which in turn invokes contraction.

the body is in extracellular fluid. Meanwhile, 97–98% of the body's potassium is located within intracellular fluid, which has a 10-fold greater concentration than the extracellular fluid. The concentration differences of these **ions,** or charged molecules, makes them the most abundant **anions** and **cations** (negative and positive ions, respectively) in their respective fluids. Because the concentration of sodium and chloride is much higher than potassium in the plasma, sweat contains more sodium and chloride. They are also primary factors in maintaining the fluid volume of the blood.

Sodium, potassium, and chloride are vital to **excitable tissue,** muscular and neural tissue that responds electrically to stimuli, for the generation of *action potentials.* The action potential is then propagated along the plasma membrane and ultimately results in the events characteristic of neurons (nerve impulse conduction) and muscle cells (contraction). Chloride is also a component of HCl in

stomach juice and is involved in the **chloride shift** into RBCs that is associated with the activity of *carbonic anhydrase* (Figure 10-6). Carbonic anhydrase is a key mechanism involved in transporting CO_2 from tissue to the lungs for exhalation.

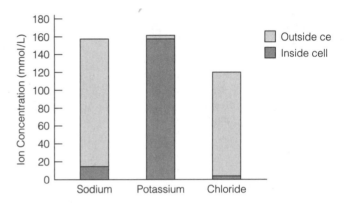

Figure 10-5
Relative difference in the concentration of sodium, potassium and chloride across the plasma membrane of cells. Cells must expend a lot of energy (ATP) to maintain these differences by pumping operations via proteins in the plasma membrane.

Table 10-5   ## Potassium Content of Selected Foods

Food	Potassium (mg)	Food	Potassium (mg)
Vegetables		**Meats**	
Potato (1)	780	Fish (3 oz)	500
Squash, winter (½ c)	327	Hamburger (3 oz)	480
Tomato (1 med.)	300	Lamb (3 oz)	382
Celery (1 stalk)	270	Pork (3 oz)	335
Carrots (1)	245	Chicken (3 oz)	208
Broccoli (½ c)	205		
		Grains	
Fruit		Bran buds (1 c)	1080
Avocado (½ c)	680	Bran flakes (1 c)	248
Orange juice (1 c)	469	Raisin bran (1 c)	242
Banana (1)	440	Wheat flakes (1 c)	96
Raisins (¼ c)	370		
Prunes (4)	300	**Milk and milk products**	
Watermelon (1 c)	158	Yogurt (1 c)	531
		Milk, skim (1 c)	400

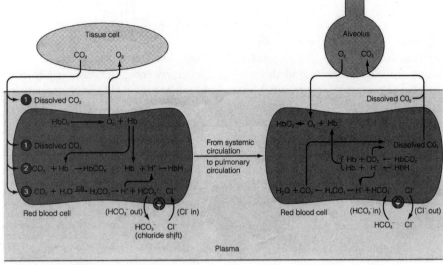

Figure 10-6

Transport and exchange of carbon dioxide and oxygen. Gas exchanges occurring (a) at the tissues and (b) in the lungs. Carbon dioxide is transported primarily as bicarbonate ion (HCO_3^-) in plasma; smaller amounts are transported bound to hemoglobin (indicated as $HbCO_2$) in the red blood cells or in physical solution in plasma. Nearly all the oxygen transported in blood is bound to hemoglobin as oxyhemoglobin (HbO_2) in red blood cells. Very small amounts are carried dissolved in plasma. (Reduced hemoglobin is indicated as HHb.) The relative sizes of the arrows indicate the proportionate amounts of O_2 and CO_2 moved by each method. (*Source*: Fig 23.20, p. 753 from Human Anatomy and Physiology, 2nd ed. by Elaine N. Marieb. Copyright © 1992 by The Benjamin/Cummings Publishing Company, Inc. Adapted by permission of Pearson Education, Inc.)

REGULATION OF BLOOD SODIUM, CHLORIDE AND, POTASSIUM LEVELS.

The efficiency of sodium, chloride, and potassium absorption from the digestive tract is very high (>90%). Therefore, the kidneys are mainly responsible for regulating the level of these minerals in the extracellular fluid. Approximately two-thirds of the reabsorption of sodium occurs in the proximal convoluted tubules, in association with the reabsorption of glucose and amino acids.[11] Reabsorption of the remaining one-third of sodium in the later segments of the renal tubule system is regulated by *aldosterone* from the adrenal glands. As discussed in Chapter 8, the release of aldosterone is itself regulated by the osmolality of the extracellular fluid. Chloride absorption parallels sodium absorption to maintain electrical neutrality. But *atrial natriuretic peptide (ANP)* opposes the activities of aldosterone. ANP levels in the blood increase in response to increased blood pressure, and ANP promotes several events that lower blood pressure. Among other things, it inhibits sodium reabsorption in the latter part of the nephron tubule system. Potassium is reabsorbed at a constant rate in the proximal convoluted tubule without hormonal regulation. However, it is pumped into the distal convoluted tubule and collecting duct. How much potassium is secreted depends on the status of potassium in the body.

Sodium, Potassium, and Chloride Imbalance

Unlike some other essential nutrients, signs of a sodium, potassium, or chloride deficiency can develop within a day or so. Most people with a well-balanced diet need not be concerned about their intake of these electrolytes. If there is an imbalance of these minerals, it is probably due to excessive losses rather than inadequate intake. For instance, copious sweating for a long time leads to disturbances in electrolyte status. As detailed in Chapter 8, the sodium and chloride concentration of primary sweat is about half of that found in the plasma. During very light (insensible) sweating at rest and light sweating during activity, many of the electrolytes in sweat are reabsorbed by the cells lining the tubules leading to the skin (see Figure 8-8). The final electrolyte content of the sweat reaching the skin surface is very low. However, during periods of copious sweating, such as exercising in a hot environment, primary sweat flows through the tubule too quickly for efficient electrolyte reabsorption. Those who ingest only water during exercise could be at risk of developing **hyponatremia** (low blood sodium level). A beverage containing electrolytes could prevent this.

An adult has a sodium level of about 1.8 g/kg of fat-free body weight that accounts for about 90% of the total cation content of the body. Thus a lean 154-lb (70-kg)

Key Point

Aldosterone regulates sodium reabsorption in the latter segments of the nephron.

adult male runner should have a little more than one-quarter pound of sodium in his body. Although sodium is rarely discussed in relation to bone, approximately 30–35% of body sodium is in the skeleton, and this storage is an important safeguard against hyponatremia.

On the other side of the imbalance coin, electrolyte toxicity is rare when the kidneys are operating efficiently. However, people who consume a salty diet should drink more water (water is attracted to sodium, so more water may be excreted along with excess sodium in the urine). An elevated blood potassium level **(hyperkalemia)** affects the proper functioning of excitable tissues, especially the heart and brain.

Sodium, Potassium, and Chloride Intake by Athletes

Athletes tend to have higher energy requirements and therefore consume more food, so their diets should provide more than average amounts of sodium, potassium, and chloride. Many athletes also eat sport bars, which typically contain 200 mg of sodium and 360 mg of potassium or more. Fluid- and electrolyte-replacement beverages also contain these minerals. For instance, 8 oz of many popular sport drinks (containing 6–8% carbohydrate) provide roughly 30 mg of potassium and 110 mg of sodium. Chloride content is not specified on the Nutrition Facts of food labels, but it is a component of "salt" in the ingredient list.

SALT TABLETS. Ingesting *salt tablets* before competing on a hot day used to be common among athletes. But it is usually not a good idea. Salt tablets cause cramping and diarrhea in some people, and they may add more sodium to the body than is lost in sweat. This can increase the concentration of sodium in the blood **(hypernatremia)**. Hypernatremia alters the function of excitable tissue (muscle and neurons), and when the kidneys attempt to get rid of the extra sodium, water is lost as well. This could be extremely detrimental if an individual is already dehydrated.

Sodium, Potassium, and Chloride and Athletic Performance

Because sodium, potassium, and chloride are involved in the function of muscle and nervous tissue, an imbalance could affect performance. The initial signals (action potential) for the contraction of muscle tissue begin in the motor cortex of the brain. The electrical impulse is then relayed along a series of neurons terminating at neuromuscular junctions. The stimulation of a muscle fiber also requires the development of an action potential, which then moves over the muscle fiber and promotes uniform contraction. Physical activity increases this electrical activity in the neuromuscular system and also in the heart. Imbalances of sodium, potassium, and chloride could decrease the efficiency of these electrical events.

Among other effects of electrolyte losses during exercise, a reduction in the sodium content of the extracellular fluid could decrease the total fluid volume of that compartment and potentially result in **hypovolemia,** reduced blood volume. Hypovolemia can lead to an even greater heart rate during exercise. Optimal sodium balance is also important for efficient glucose and amino acid reabsorption in the kidneys during exercise. In addition, exercise greatly increases the production of CO_2 from working muscle tissue, and optimal chloride balance aids in the removal of CO_2 from the body.

■ SODIUM, POTASSIUM, AND CHLORIDE AND EXERCISE *IN REVIEW*

- Sodium, potassium, and chloride are the major electrolytes in the extracellular and intracellular fluids.
- These electrolytes are vital for proper functioning of excitable tissue such as muscle and nervous tissues.
- Sodium is a chief regulator of plasma volume. Chloride is a component of HCl in stomach acid and is involved in carbonic anhydrase activity to transport CO_2 from the tissues to the lungs.
- Most athletes get enough sodium, potassium, and chloride through a balanced diet, but imbalances may occur in athletes who sweat excessively or have eating disorders.

■ MAGNESIUM (Mg)

Magnesium is found in nature in a divalent (Mg^{2+}) state. It is found in many foods; good sources are whole grain cereals, nuts, legumes, spices, seafood, coffee, tea, and cocoa (Table 10-6). Certain processing techniques, such as the refining of wheat, rye, oats, and barley and the polishing of rice, may result in significant losses of magnesium. Magnesium may also be depleted during cooking because it can dissolve into water during boiling. The RDA for magnesium for adult men and women ages 19–30 is 400 and 310 mg, respectively, and 420 and 320 mg for those over 31.

Magnesium Functions and Metabolism

The total magnesium in most adults is a little more than an ounce (~30 g). About 60% of it is in bone. Magnesium is vital for the proper functioning of more than 300 reaction systems, including those involved in protein synthesis in muscle. For example, magnesium appears to be necessary for transfer RNA to attach to its respective amino acid and for linking amino acids together to form proteins.

Because of its strong positive charge, magnesium can interact with the phosphate tail of ATP, as shown in

Table 10-6 **Magnesium Content of Select Foods**

Food	Magnesium (mg)	Food	Magnesium (mg)
Legumes		**Vegetables**	
Lentils, cooked (½ c)	134	Bean sprouts (½ c)	98
Split peas, cooked (½ c)	134	Black eyed peas (½ c)	58
Tofu (½ c)	130	Spinach, cooked (½ c)	48
		Lima beans (½ c)	32
Nuts			
Peanuts (3 c)	247	**Milk and Milk Products**	
Cashews (3 c)	93	Milk (1 c)	30
Almonds (3 c)	80	Cheddar cheese (1 oz)	8
		American cheese (1 oz)	6
Grains			
Bran buds (1 c)	240	**Meats**	
Rice, wild, cooked (½ c)	119	Chicken (3 oz)	25
Wheat germ (2 tbs)	45	Beef (3 oz)	20
		Pork (3 oz)	20

Figure 10-7. This stabilizes the phosphate tail and improves the efficiency of the enzymes that split ATP (ATPases) to release energy. A significant portion of the magnesium in muscle cells is associated with ATP and creatine phosphate.[12] In muscle fibers, magnesium associates with troponin, myosin, and actin in the cytosol. It is also found in the sarcoplasmic reticulum in association with calcium pumps (ATPases) and other proteins such as *calmodulin* and *parvalbumin*, both of which are calcium-sensor proteins.[13–15] This makes magnesium particularly important for the efficient contraction/relaxation cycling of muscle fibers. These various roles in muscle have made magnesium of particular interest to exercise researchers and supplement manufacturers, to the extent that magnesium supplementation has been experimented with (and marketed as) a possible way to enhance protein synthesis in muscle.

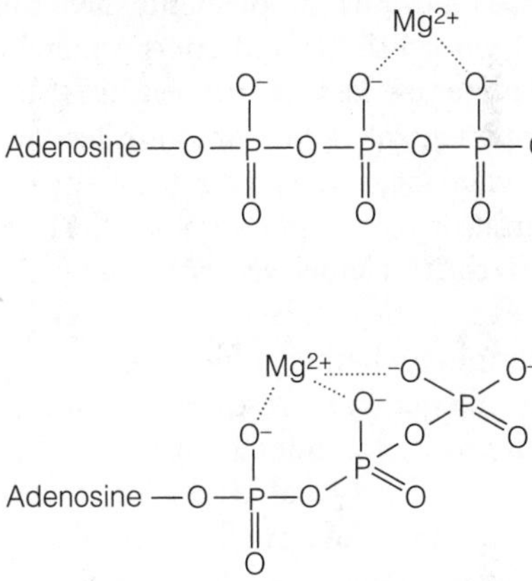

Figure 10-7
Magnesium association with ATP. These two forms stabilize ATP for energy-producing interactions with ATPases.

Key Point

Magnesium (Mg^{2+}) interacts with the phosphate groups of ATP and creatine phosphate, increasing the efficiency of their use in muscle and other tissues.

Magnesium Imbalance

Because magnesium is involved in ATP utilization, deficiency can alter energy-requiring processes such as the maintenance of electrolyte concentrations across cellular membranes. Thus, the proper function of muscle cells and neurons is jeopardized by magnesium deficiency, compromising athletic performance by causing muscle weakness. Cramps, spasms, or tremors may occur in cases of extreme

deficiency. Even subtle alterations in blood magnesium content can affect the release of parathyroid hormone (PTH) and its activity. Excess magnesium in the body is rare among people with a normal diet, as it is readily excreted in the urine. However, large doses of supplemental magnesium are not recommended. The UL for magnesium is set at 350 mg.

Magnesium Intake by Athletes

Whether or not athletes get enough magnesium in their diet has to do with the individual and, sometimes, the sports in which they participate (high sweating, for example, can deplete magnesium). A diet that includes a variety of foods and enough energy for weight maintenance usually provides adequate magnesium. However, energy-restrictive diets may not. One study of female gymnasts found that they had an average magnesium intake well below the RDA.[16] Other studies of female basketball players[17] and triathletes of both genders[18] reported less than adequate levels.

Magnesium and Athletic Performance

Magnesium is fundamental to exercise performance because of its involvement in ATP utilization and in numerous key enzymes in energy pathways and other systems. Magnesium status is probably normal for most athletes,[19] but it might not correspond to dietary levels. For example, many of the female athletes assessed in one study had normal magnesium status despite consuming a diet that provided less than recommended levels.[17]

MAGNESIUM SUPPLEMENTATION AND PERFORMANCE. Magnesium supplements are touted as enhancing protein synthesis and energy metabolism, or to replace the mineral's loss from sweating. (Magnesium is a normal component of sweat, although less abundant than sodium and chloride, but as the rate of sweating increases the concentration of magnesium in the sweat also increases.[20]) If these claims are true, supplements would benefit athletes.

To determine whether athletes could benefit from a magnesium supplement, researchers provided 500 mg of magnesium oxide to athletes with low to normal serum magnesium levels for 3 weeks.[21] The researchers reported that supplementation did not affect serum or cellular magnesium concentrations, nor did it positively influence exercise performance. Other researchers studied how supplementing magnesium oxide at a level of 212 mg daily for 4 weeks affected anaerobic and aerobic capabilities in active women.[22] The results suggested that magnesium supplementation increased the level of magnesium in the blood, but did not improve performance on anaerobic and incremental (aerobic) treadmill tests. Another study indicated that dosing athletes for 12 weeks with multivitamin and mineral supplements that included magnesium failed to improve aerobic performance.[23]

Little research has been conducted to show whether strength and power athletes benefit from magnesium supplementation. One study investigated the effects of magnesium supplementation on strength development in untrained males during a 7-week strength training program.[24] The results suggest that magnesium supplementation in combination with resistance training, as opposed to training alone, can lead to greater gains in absolute and relative torque adjusted for both body weight and lean body mass. More research on magnesium, especially as it affects muscle mass and strength development, is needed.

■ MAGNESIUM AND EXERCISE *IN REVIEW*

- Magnesium is a divalent cation (Mg^{2+}) found in many foods. Whole grains, legumes, and leafy vegetables are good sources.
- Magnesium can interact with the phosphate tails of ATP and improve the efficiency ATP splitting for energy.
- Magnesium supplementation for athletes requires further study.

■ IRON (Fe)

Iron is one of the most widely recognized minerals among athletes and people in general. For athletes, especially female runners, iron status is a great concern. Iron in foods is often classified as heme iron or nonheme iron. Heme iron is found only in animal foods; approximately half the iron in meat, fish, and poultry is part of heme. Nonheme iron is in both plant and animal foods. Heme iron is absorbed more efficiently than nonheme iron, but several food and digestion factors influence the absorption efficiency for nonheme iron. For instance, vitamin C appears to enhance absorption, as does the presence of meat, fish, or poultry and hydrochloric acid in stomach juice (see Table 10-7). Good sources of iron are listed in Table 10-8; they include meats, whole grains, and fortified foods. The RDA for adult men and women ages 19–50 is 8 and 18 mg, respectively.

Key Point

Iron status in the body is primarily regulated at the point of absorption. This makes it different from sodium, potassium, chloride, and magnesium, whose status is regulated by urinary excretion.

Table 10-7

Factors Influencing the Efficiency of Iron Absorption

↑ Absorption of Iron	↓ Absorption of Iron
• Vitamin C at the same meal	• Phytate, oxalates from plants
• Normal stomach acid production	• Tannins such as from tea
• Increased iron need (growth, pregnancy, poor status)	• Decreased stomach acid production or the use of antacid medication
• Meat, fish, poultry at same meal	

Table 10-8	**Iron Content of Select Foods**			

Food	Iron (mg)	Food	Iron (mg)
Meats and alternatives		**Grains**	
Liver (3 oz)	7.5	Breakfast cereal (1 c)[a]	4–18
Round steak (3 oz)	3	Oatmeal (2 c)[a]	8
Hamburger, lean (3 oz)	3	Bagel (1)	1.7
Baked beans (½ c)	3	English muffin (1)	1.6
Pork (3 oz)	2.7	Bread, rye (1 slice)	1
White beans (½ c)	2.7	Bread, whole wheat (1 slice)	0.8
Soybeans (½ c)	2.5	Bread, white (1 slice)	0.6
Fish (3 oz)	1		
Chicken (3 oz)	1	**Vegetables**	
		Spinach (½ c)	2.3
Fruits		Lima beans (½ c)	2.2
Prune juice (½ c)	4.5	Peas, black-eyed (½ c)	1.7
Apricots, dried (½ c)	2.5	Peas (½ c)	1.6
Prunes (5 med.)	2	Asparagus (½ c)	1.5
Raisins (¼ c)	1.3		
Plums (3 med.)	1.1		

[a]Iron-fortified

Iron Functions and Metabolism

An adult has about 2–5 g of iron in the body depending on diet, gender, size, muscularity, and menstrual status (Figure 10-8). Tissue iron is a component of heme-containing structures such as cytochromes, hemoglobin, and myoglobin, of nonheme enzymes such as catalase, of storage proteins such as **ferritin** and **hemosiderin,** and of the iron

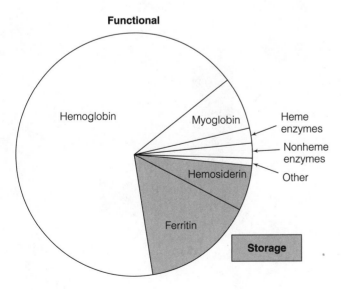

Figure 10-8

Approximate distribution of iron in functional and storage molecules in the body.

transport protein **transferrin.** About two-thirds of the iron in the body is in hemoglobin (Figure 10-8). The hemoglobin molecule contains four atoms of iron that bind O_2 for efficient transportation in the circulation. Muscle myoglobin accounts for nearly 10% of iron in the body, and heme and nonheme enzymes account for 2–4%. Myoglobin is an O_2 storage protein in muscle and is particularly concentrated in Type I and IIa muscle fibers as well as heart muscle cells. Myoglobin serves as a small O_2 reserve for muscle and to increase the rate of O_2 transfer from capillaries into muscle cells. As discussed in Chapter 5, the myoglobin content of muscle fibers increases with endurance training.[25,26]

Iron is also found within the heme portion of cytochromes, which are key components of the electron transport chain in mitochondria and other redux systems, such as the cytochrome P_{450} system in the endoplasmic reticulum. The cytochrome P_{450} system oxidizes compounds including alcohol, drugs, and carcinogens. Iron is also part of other complexes associated with the electron transport system, such as *NADH dehydrogenase, succinate dehydrogenase*, and *ubiquinone-cytochrome c reductase*.[1] However, these structures do not contain heme, so iron is utilized in a different manner. In addition, iron is a component of a couple of heme-containing enzymes (including catalase) involved in antioxidant systems.[1]

IRON STORAGE AND TRANSPORT. Iron is a highly reactive atom and is generally not found free and loose within cells. Ferritin is the primary iron storage protein; it is contained in many tissues, especially the liver, bone

Key Point

Iron is a component of hemoglobin, myoglobin, and cytochromes, all of which are involved in aerobic energy metabolism and are crucial during exercise.

marrow, intestine, and spleen. The molecular design of ferritin is like that of a container, giving it a capacity for more than 4000 iron atoms. The level of ferritin in the serum is an indicator of tissue iron status: 1 μg/L approximates 10 mg of tissue iron stores. Women typically have ≥12 μg/L and men ≥15 μg/L. Some cells also contain the iron storage protein hemosiderin, thought to be a spin-off molecule of ferritin that is produced when iron stores increase.

Iron is transported in the plasma in association with transferrin. Transferrin is produced in the liver and has a variable capacity for iron. How much potential "room" is available on transferrin indicates iron status. This is called **total iron binding capacity (TIBC).** So ferritin, transferrin, and TIBC can be used to gauge iron status.

Iron Imbalance

Iron deficiency occurs when iron stores become depleted and anemia results. However, this type of anemia is different from folate and/or vitamin B_{12} deficiency in that the RBCs are small and pale due to reduced hemoglobin content. This type of anemia is called **hypochromic microcytic anemia. Anemia,** which is the inability of hemoglobin content to meet body demand for oxygen, is often defined as hemoglobin levels of <7 g/100 ml of blood. Normal hemoglobin levels for men and women are ≥14 and 12 mg/100 ml of blood, respectively. Borderline anemia is defined as lowered hemoglobin levels but above 7 g/100 ml of blood.

Iron status becomes compromised prior to the development of anemia. As shown in Figure 10-9, the level of ferritin in serum (or plasma) is one of the first signs of decreasing tissue iron status. Therefore it is regarded as perhaps the most sensitive indicator of iron status, and nutritional assessment of athletes should include this measure.

Iron toxicity is also problematic. Because iron is an oxidant, too much iron can result in free radical activity.

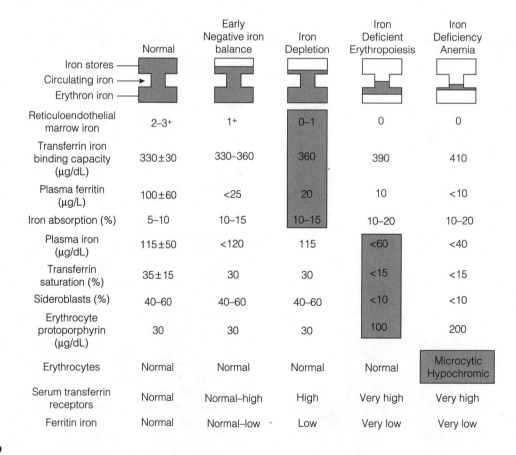

	Normal	Early Negative iron balance	Iron Depletion	Iron Deficient Erythropoiesis	Iron Deficiency Anemia
Iron stores / Circulating iron / Erythron iron					
Reticuloendothelial marrow iron	2–3+	1+	0–1	0	0
Transferrin iron binding capacity (μg/dL)	330±30	330–360	360	390	410
Plasma ferritin (μg/L)	100±60	<25	20	10	<10
Iron absorption (%)	5–10	10–15	10–15	10–20	10–20
Plasma iron (μg/dL)	115±50	<120	115	<60	<40
Transferrin saturation (%)	35±15	30	30	<15	<15
Sideroblasts (%)	40–60	40–60	40–60	<10	<10
Erythrocyte protoporphyrin (μg/dL)	30	30	30	100	200
Erythrocytes	Normal	Normal	Normal	Normal	Microcytic Hypochromic
Serum transferrin receptors	Normal	Normal–high	High	Very high	Very high
Ferritin iron	Normal	Normal–low	Low	Very low	Very low

Figure 10-9

Sequential changes in iron status associated with iron depletion. (*Source:* Adapted with permission from Victor Herbert, "Recommended Dietary Intakes (RDI) of Iron in Humans," *American Journal of Clinical Nutrition* 1987, 45:679–86. © American Society for Clinical Nutrition and Victor Herbert, *Journal of Nutrition*, 1995, 126:1213S–20S.)

Key Point

Serum ferritin is a sensitive indicator of iron status. Hemoglobin and hematocrit are late indicators of iron status, as they can remain within a normal range despite reductions in tissue iron levels.

Iron toxicity leads to problems in many organs, especially the liver. The UL for iron is set at 45 mg for adults.

Iron Intake by Athletes

Athletes with a high training volume may need to pay particular attention to iron intake,[27] based on reports of low iron intake among athletes, especially runners.[28–30] Diet records of National Figure Skating Championship competitors indicated that iron intake was below recommendations, although their biochemical indices of iron status were normal.[31] But another study indicated that despite a higher intake of iron by athletic boys, there was no difference in iron status indicators.[32] Researchers in Poland reported that serum ferritin levels were associated more with heme iron intake than total iron intake.[33] These studies demonstrate the complexity of iron status and how it is influenced by diet and exercise.

For strictly vegetarian athletes, iron levels and bioavailability may be a concern, but the evidence for or against the notion is mixed.[34] **Bioavailability** refers to how much of a food's nutrient (such as iron) is absorbed into the body and utilized. Nutrition considerations for vegetarians are discussed in Chapter 15.

Iron and Athletic Performance

Iron is important to exercise performance in many ways, including its involvement in O_2 transport and energy metabolism. Therefore, reduced iron status could lead to decreased athletic performance, particularly in sports that rely heavily on aerobic energy metabolism. When hemoglobin levels are reduced, the O_2-carrying ability of the blood is reduced. Ultimately, less O_2 delivery to cells can result in general lethargy and early fatigue during exercise. In addition, poor iron status could compromise the electron transport chain, which could reduce aerobic ATP generation during exercise.

Reduced iron status in athletes can come about by poor dietary intake or increased iron loss. Reduced iron status is a commonly cited nutritional concern for female athletes who are menstruating. Active females of reproductive age can experience significant losses of iron during menstruation, depending on menstrual blood volume and normal iron status. For instance, menstrual volume of 100–200 ml may lead to the loss of a couple milligrams of iron.

Because training and competition for many sports result in heavy sweating and because iron is found in sweat, researchers wonder if this has a negative impact on iron status. Sweat iron concentration has been reported between 0.03 and 0.5 mg/L.[30] However, like other minerals, the concentration of iron in sweat can decrease with chronic training as an athlete adapts to produce less concentrated sweat. Although sweat is a route of iron loss, which in theory could lead to reduced iron status, it is not considered a significant concern.[35] Iron loss may also result from intestinal bleeding, and from foot-strike injuries such as intravascular hemolysis in runners, in which some RBC destruction occurs during the repeated and forceful impact of the runner's feet striking the ground. Better cushioning in running shoes may prevent some of the destruction.

Endurance athletes often train at higher altitudes if their sport involves competition at similar altitudes in an attempt to increase RBC formation. For instance, many top cyclists in the United States train in the Rocky Mountains during the warmer seasons. It seems that endurance athletes experience a peak in the formation of new RBCs after about 8–10 days of training at a moderate altitude.[36] Furthermore, an athlete can expect a 1–4% increase in hemoglobin concentration after approximately 3 weeks of training at altitude. Increased levels of **erythropoietin,** a hormone produced by the kidneys, stimulate RBC formation. Several factors modulate erythropoietin production; the most important nutritional factor in altitude-induced erythropoiesis is iron availability.[36] During adaptation, iron stores might need to be mobilized to support erythropoiesis. For instance, when swimmers trained at an altitude of about 2.2 km (~7300 ft) the increase in RBC percentage (hematocrit) was associated with a decrease in serum ferritin levels, which then recovered over time.[37] These findings suggest that training at an altitude high enough to evoke erythropoiesis calls upon iron stores in tissue such as the liver to support RBC formation.

Key Point

Marginal to deficient iron status is one of the most prevalent nutrition concerns for athletes, especially menstruating women.

IRON DEFICIENCY ANEMIA AND PERFORMANCE. As mentioned above, anemia is a late indicator of poor iron status. This means that a person already has reduced iron stores prior to developing iron deficiency anemia (IDA). Therefore, individuals would be marginally iron deficient before becoming anemic, and the question is, would this be enough to reduce athletic performance? Many athletes have reduced ferritin levels without significant reductions in hematocrit and hemoglobin. Therefore, lowered ferritin as supported by increased TIBC can alert an athlete and a coach to the potential for reduced performance. For instance, in one study young swimmers (boys 10–12 years old) had reduced serum ferritin levels during their competitive season even though their diet iron level was similar to age-matched inactive boys.[38]

Individuals engaged in a training program often develop pseudoanemia, which can be misinterpreted.[39] **Pseudoanemia** is a reduction in the hematocrit and the concentration of blood hemoglobin that is caused by **hemodilution** during expansion of plasma volume associated with training, not an actual blood cell reduction. If training ceases, these parameters are normalized within a few days. Pseudoanemia is also corrected when the production of RBCs increases, which is a typical adaptation to endurance training.

IRON SUPPLEMENTATION AND PERFORMANCE. Mineral-containing supplements are common among athletic populations. For instance, a survey of competitive Norwegian skiers (alpine and cross-country) and power athletes (boxers, weightlifters, and track and field athletes) revealed that 84% used one or more micronutrient supplements regularly.[40] However, 94% of the power sport athletes (but only 7% of the female and 37% of the male alpine skiers) used an iron-containing supplement. An American survey of varsity collegiate athletes indicated that a higher percentage of the women used an iron or iron-containing supplement than the men.[41]

Logically, athletes presenting the clinical signs of anemia related to iron deficiency would demonstrate improved athletic performance with iron supplementation if their sport were endurance in nature. The training effects for power and strength sports should also be enhanced by eliminating iron deficiency. However, the evidence is not clear-cut for such effects from iron supplementation in marginally iron-deficient athletes.[42] When female distance runners with low iron status, but not IDA, were provided with an iron supplement for 8 weeks they had improved ferritin levels and a reduced TIBC, both indicative of improved iron status.[43] However, there was no difference in VO₂max and lactate levels in comparison to a placebo group. In a somewhat similar study involving nonanemic, iron-depleted competitive female runners, iron supplementation did not improve metabolic parameters associated with running performance.[28] In another study involving iron supplementation (intramuscular injection) of iron-depleted, nonanemic elite female netballers, serum ferritin levels improved but not athletic performance.[44] Athletic performance was evaluated by a vertical jump test, a sprint test on a cycle ergometer, and a 20-minute multistage shuttle run, and results were compared to a placebo trial.

The studies discussed so far have used iron-compromised athletes already engaged in training to see if

Key Point

Iron deficiency hinders exercise performance as well as adaptation to endurance training, either at sea level or at higher altitude.

iron supplementation leads to improvements in performance. In another line of investigation, researchers studied previously untrained, iron-depleted but nonanemic women to find the effects of iron supplementation on adaptation to aerobic training.[45] In this study, iron supplementation increased serum ferritin and decreased TIBC and led to greater improvements in performance in a 15-km run in comparison to a placebo trial. These results suggest that endurance adaptation occurs more quickly in individuals in better iron status at the onset of training.

■ IRON AND EXERCISE *IN REVIEW*

- Iron in foods occurs in heme and nonheme structures; heme iron is absorbed more efficiently.
- Iron is a component of hemoglobin and myoglobin and binds O_2. It is also a component of cytochromes and nonheme enzyme complexes.
- Ferritin and serum iron levels and TIBC are reduced prior to anemia.
- Athletic performance is hindered by iron deficiency; however, iron supplementation does not necessarily improve performance in individuals with marginal iron status.

■ ZINC (Zn)

Zinc (Zn^{2+}) is the most ubiquitous mineral in the body and is involved in nearly every human operation. In plants and animals zinc is greatly associated with amino acids and proteins, so foods with high protein content are presumably richer sources of zinc. In general, animal foods are better sources of zinc than plants, and the best sources of zinc include organ meats, other red meats, and seafood (especially oysters and mollusks). Poultry, pork, milk and milk products, whole grains (especially the germ and bran), and leafy and root vegetables are also respectable contributors of zinc to the diet (Table 10-9).

Zinc derived from animal tissue (meats) is better absorbed than zinc from plant tissue. The absorption of zinc from meats might be enhanced by the greater availability of certain amino acids during simultaneous protein digestion. Also, the absorption of zinc from plant tissue is probably hindered by phytates (see Figure 10-3). Other plant tissue substances that might reduce zinc absorption include oxalates found in many vegetables, fruits, and

Table 10-9   **Zinc Content of Various Foods**

Foods	Zinc (mg)	Food	Zinc (mg)
Meats and Alternatives		**Legumes**	
Liver (3 oz)	4–5	Dried beans, cooked (½ c)	1
Beef (3 oz)	4	Split peas, cooked (½ c)	1
Crab (½ c)	3–4	**Nuts and Seeds**	
Lamb (3 oz)	3–4	Pecans (½ c)	2
Pork (3 oz)	2–3	Cashews (¼ c)	1–2
Chicken (3 oz)	2	Sunflower seeds (¼ c)	1–2
		Peanut butter (2 tbs)	~1
Grains			
Wheat germ (2 tbs)	2–3	**Milk and Milk Products**	
Oatmeal, cooked (1 c)	1	Cheddar cheese (1 oz)	1
Bran flakes (1 c)	1	Milk, whole (1 c)	~1
Rice, brown, cooked (2 c)	~½	American cheese (1 oz)	~1
Rice, white (2 c)	~½		

chocolate, and tannins found in tea. The RDA for zinc is 8 and 11 mg for adult females and males, respectively.

Zinc Functions and Metabolism

Depending on size and body composition, a typical adult might contain 1.5–2.5 g of zinc. Every cell contains zinc, and tissues such as muscle, kidneys, bone, and the liver have higher contents. Skeletal muscle and bone contain roughly 60% and 30% of all the zinc, respectively. Zinc is a cofactor for as many as 200 enzymes and is involved in pH regulation, ethanol (alcohol) metabolism, bone matrix mineralization, protein digestion, heme production, antioxidant operations, immunity, and protein and nucleic acid metabolism. Table 10-10 lists several zinc-dependent enzymes and their function. These enzymes include *carbonic anhydrase, alcohol dehydrogenase, pyruvate dehydrogenase, lactate dehydrogenase, superoxide dismutase (SOD),* and *DNA* and *RNA polymerases.* Zinc is also a component of proteins (zinc finger proteins) that interact with DNA in the nucleus of cells and influence the production of certain proteins.

Zinc is excreted from the body in several ways. The feces are the primary route of excretion, containing

Table 10-10   **Zinc-Dependent Enzymes and Function**

Enzyme	General Function
Alcohol dehydrogenase	Oxidation of ethanol in the cytosol, primarily in hepatocytes
Lactate dehydrogenase	Reversible oxidation/reduction interconversion of pyruvate and lactic acid
Alkaline phosphatase	Primarily involved in mineralization of bone matrix
Angiotensin converting enzyme	Converts angiotensin I to angiotensin II
Carbonic anhydrase	Interconverts CO_2 and H_2O to carbonic acid to assist in the regulation of CO_2 content of blood and pH
Pyruvate dehydrogenase	Converts pyruvate to acetyl CoA within mitochondria
DNA and RNA polymerases	DNA replication and transcription
Superoxide dismutase (SOD)	Cytosolic antioxidant
Aspartate transcarbamylase	Pyrimidine synthesis
Carboxypeptidases A, B	Protein digestion
Phosphodiesterase	Cleave phosphodiester bonds (in nucleic acids)
Fructose-1,6-bisphosphatase	Gluconeogenesis
Leukotriene hydrolase	Eicosanoid metabolism
Elastase	Digestion of connective tissue elastin
Reverse transcriptase	DNA replication
Gustin	Taste acuity

unabsorbed dietary zinc as well as zinc secreted into the digestive tract via digestive juices. For instance, bile contains zinc, as do several digestive enzymes. The urine provides a secondary route for zinc excretion. Sweat, exfoliated skin, and hair loss also result in zinc removal from the body. Training and competing in sports that involve higher sweat rates result in greater zinc losses, as discussed below.

Key Point

Zinc is a component of about 200 enzymes and thus is instrumental in nearly every physiological function and is particularly involved in exercise performance, recovery, and adaptation.

Zinc Imbalance

Because zinc is involved in nearly all human functions, zinc deficiency results in abnormalities in many human operations. During adulthood, zinc deficiency can result in decreased immune capacity, skin rash, poor wound healing, reduced taste, hair loss, and decreased cognitive capabilities. On the other hand, zinc toxicity also produces several physiological abnormalities, some because excessive zinc leads to imbalances in other minerals. One of the best examples is copper. When excessive zinc, such as with supplemental doses of zinc, enters the cells lining the wall of the small intestine, those cells produce a protein called **metallothionein** (Figure 10-10). Although the function of metallothionein is to bind the additional zinc to prevent toxicity, it actually binds copper more aggressively. This means that chronic zinc supplementation can reduce copper absorption and potentially lead to copper deficiency. As little as 3–10 times the RDA for zinc reduces copper absorption. Thus if long-term zinc supplementation is necessary, copper status should be monitored by a physician and copper should be cosupplemented.

Over the past decade, zinc lozenges (zinc gluconate) have been marketed as a means of reducing the duration and severity of the common cold. Several clinical trials have been performed but it is still unclear whether zinc lozenges are helpful.[46–49] Here again, copper must be a foremost consideration. The UL for zinc has been set at 40 mg for adults.

Key Point

Excessive zinc consumption can reduce the absorption of copper.

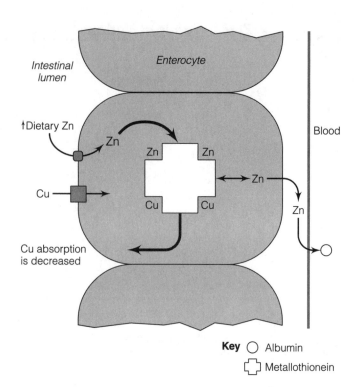

Figure 10-10
Excessive dietary zinc increases the level of metallothionein in absorptive cells.

Zinc Intake by Athletes

A diet containing a variety of foods and an energy level sufficient for weight maintenance for athletes should provide at least the RDA level for zinc. For instance, researchers reported that both male and female elite Nordic skiers consumed more zinc than the RDA, averaging 21.9 and 15.8 mg of zinc daily, respectively.[50] A study of puberty-age girls and boys participating in sports determined that the zinc intake of the boys was higher than that of age-matched boys not active in sports; the intake of the girls did not differ.[32] However, when reviewing studies involving nutritional assessment it is important to reflect on the distribution of dietary intake as well as the averages. Therefore, in some surveys showing that average zinc intake approximated the RDA level, many of the participants would get less than the RDA.[51–53] Also, some studies show low averages; male runners in one study averaged 7.4 mg of zinc daily, which is below their RDA level.[54] As part of a comprehensive nutrition assessment, athletes should understand the zinc level in their diet. This is especially true for those who restrict their energy level or follow more restrictive diet programs or philosophies. General guidelines for estimating the mineral adequacy of the diet are discussed in Nutrition in Practice: Eating Enough Minerals.

NUTRITION IN PRACTICE   **Eating Enough Minerals**

The Chapter 9 Nutrition in Practice provided a nutrition assessment of an actual diet recall of a 34-year-old male amateur runner. The ability of his nutritional intake to provide adequate quantities of the vitamins to support his athletic performance and general health was deemed more than satisfactory. Now it is time for a closer look at the mineral content of his diet.

As with vitamins, some minerals may be lacking in the diet of certain athletes. However, an athlete who eats a diet containing a variety of foods and adequate energy for

Table A   **24-Hour Diet Recall of a 34-Year-Old Male Who Runs 30–45 Miles/Week**

Meal/Snack	Food	Notable Mineral Content
7:00 A.M. **Breakfast**	1 large banana	K
	1 c fresh orange juice	K
	2 c Wheaties (with the banana)	Mineral-fortified
	1 c skim milk	Ca, P
	2 c coffee	
	1 tbs creamer	
10:30 A.M. **Midmorning Snack**	1 c vanilla yogurt	Ca, Mg, P, K, Se, Zn, Na
	½ c granola cereal (in yogurt)	Mineral-fortified
	8 oz grapefruit juice (calcium fortified)	Ca
1:00 P.M. **Lunch**	Turkey sandwich:	
	2 slices whole wheat bread	Mn, Se, Na
	6 oz turkey breast	Fe, Mg, K, Se, Zn
	3 oz reduced-fat provolone cheese	Ca, Na
	1 tbs mustard	Na
	2 slices tomato	
	2 c low-fat chocolate milk	Ca, Cu, Mg, Mn, P, Na
	1 orange (medium)	K
	12 baby carrots	K
5:00 P.M. **Dinner**	8 oz salmon steak	Cu, P, K, Se, Zn
	1 med. potato (baked)	Cu, Fe, Mg, Mn, P, K
	1 tbs sour cream (on potato)	
	1 c steamed broccoli	Ca
	2 c green salad:	
	1 c lettuce (romaine)	Mn
	½ tomato	
	¼ cucumber	
	1 tbs low-fat ranch dressing	Na
	1 bran dinner roll	Se
	1 c applesauce (with cinnamon)	
8 P.M. **Snack**	1 sport bar (high protein, fortified)	Mineral-fortified
	1 c vanilla frozen yogurt	Ca, Mg, P, Zn, Na

Abbreviations: Ca, P, K, Na, Mg, Fe, Se, Cu, Zn = calcium, phosphorus, potassium, sodium, magnesium, iron, selenium, copper and zinc.

Notes:
- 1.5 L of H_2O was consumed during the day.
- Ran 6 miles at 4 P.M.
- Minerals are indicated if the food provides ≥5% of RDA (or ESAADI).
- Energy intake of 3700 kcal matches estimated expenditure.

Eating Enough Minerals (CONTINUED)

at least weight maintenance should not be at risk for mineral deficiencies. Athletes and coaches can gauge the ability of the diet to provide adequate levels of minerals by assessing the general qualities of the nutritional intake. To perform the assessment they can use checklists and seek the technical assistance of a registered dietitian (R.D.) and/or credible nutrition assessment software and websites, as discussed in the last chapter. For instance, using nutrition assessment software to analyze the runner's 24-hour diet recall (Table A), the mineral content relative to the RDA or AI is presented in Figure A. In general, to meet mineral requirements the diet should adhere to sound recommendations for health as discussed in Chapter 1. A checklist of basic assessment questions follows.

CHECKLIST FOR MINERALS CONSUMPTION. Does the athlete's diet and training regularly provide the following?

- Milk and dairy foods
- Several (five or more) servings of fresh fruits and vegetables
- Whole grain products, legumes and nuts
- Lean meats and fish
- Mineral-fortified foods such as cereals and sport bars
- Iodinized salt and fluoridated water
- Adequate energy for weight maintenance with < 30% of energy from fat
- Only a limited contribution to carbohydrate made by sweets (candies, soda)

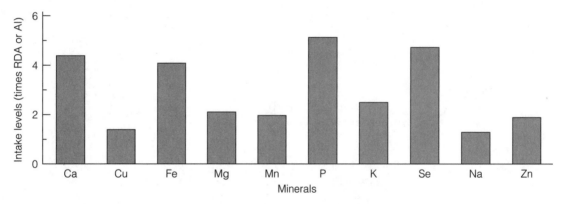

Figure A

Mineral intake based on the information reported in the runner's 24-hour diet recall relative to current RDAs. Note: The inclusion of vitamin-fortified breakfast cereals and a sport bar made a significant contribution to these levels. Also, the calcium-to-phosphate ratio is about 1:1, which is considered desirable.

It has been speculated that athletes consuming a vegetarian diet might be at risk of poor zinc status. This is because meats are a good source of zinc, and plants are endowed with phytates and other potentially zinc-chelating substances. However, there is not convincing evidence that vegetarian athletes are more prone to poor zinc status than athletes who consume animal foods.[34] Athletes who restrict energy intake (for example, many wrestlers) may be prone to poor intake of zinc and many other nutrients. Considerations for active people who are vegetarian are discussed further in Chapter 15.

Key Point

Like many minerals, zinc is likely to be adequate in a diet that is balanced and provides enough energy to at least maintain body mass.

Zinc and Athletic Performance

Zinc is important to exercise performance, recovery, and adaptation for several reasons including protein synthesis, regulation of pH, and antioxidant systems.[55] Since plasma zinc level is considered a sensitive indicator of zinc status, researchers set out to determine if exercise influences plasma zinc content. Several studies have indicated that the average plasma zinc levels are not reduced in a variety of sport groups.[50,56–58] However, again one must consider not only the study group's average but also the percentage of individuals below the normal range. For instance, nearly one-fourth of the male runners surveyed in one study had plasma zinc levels below 65 µg/100 ml, which was used as the standard for good zinc status.[59] Similar results were determined elsewhere in a study involving both male and female athletes.[60] Several mechanisms might explain how exercise training can influence plasma zinc levels in some athletes. These include hemodilution during training-induced plasma expansion, redistribution in tissue as a result of training, and increased zinc loss during and after exercise.[56,61] Sweat zinc losses might be an important factor in some athletes. It has been estimated that sweat zinc content may be as high as 1.15 mg/L,[62] and researchers have reported that serum zinc levels may be reduced after about one week of exposure to a hot environment.[63] The zinc content of sweat has also been determined to be higher early in exercise and similar between genders.[64] Thus sweat zinc losses may contribute to the seemingly reduced zinc status observed in some athletes. However, sweat zinc levels might decrease with exercise training based on heat acclimatization. Increased urine zinc levels have also been reported after strenuous exercise, which may contribute to reduced plasma zinc levels.[56,61,65,66]

Thus zinc status appears to be reduced in some although not most athletes.[56,61] The next question is, how does reduced status relate to performance? This area has not been thoroughly investigated and certainly requires more attention. For instance, research findings suggest that reduced zinc status may influence metabolic and performance aspects in soccer players.[67] During exercise testing, the players with low serum zinc levels demonstrated a lower power output and greater increase in blood lactate during exercise. They also seemed less able to effectively regulate plasma glucose levels and exhibited a tendency toward hypoglycemia. The study also suggested that zinc status might influence blood flow properties during exercise by either directly or indirectly modifying RBC flexibility. Another study found lower serum zinc levels in 20 male and female gymnasts (12–15 years old) than in age-matched inactive children.[68] The female gymnasts had lower zinc levels than the males, and zinc levels correlated to isometric adductor strength, which suggests improved muscular performance in the gymnasts with better zinc status.

POSTEXERCISE TIMING OF ZINC STATUS ASSESSMENT. Other potential mechanisms for lower plasma zinc levels must also be considered, including the timing of the measurement. An investigative protocol must control for the most recent training bout. For example, the level of zinc in a blood sample from a runner 1–2 hours after the last training session may differ from a blood sample taken 1–2 days after that run. Physical activity appears to cause significant redistribution of certain trace elements between body stores, the bloodstream, and tissues.[69] For instance, exercise has been reported to increase the production of metallothionein, the zinc binding protein, in the liver of rats,[70] and this also may occur in people. This mechanism has been suggested to be protective as it is associated with reduced markers of liver lipid peroxidation that may occur with intense exercise. Some reports suggest that plasma zinc levels may decrease for a couple of hours after exercise, but others do not.[69,71] The level of

training may also be a factor in changes of zinc status indicators after exercise.

Another potential factor that can influence plasma zinc levels is a training-associated increase in the zinc content of blood cells. For example, one study involving marathoners determined that the total zinc content of blood cells was roughly 30% higher in marathon runners than in sedentary people.[72] These results suggest that exercise training or an acute bout of endurance activity can influence zinc distribution. Training level may also influence zinc status indices. For instance, it was reported that in untrained people plasma zinc levels were reduced for a couple of days following 40 minutes of bench stepping exercise.[73] Neither concentric nor eccentric resistance exercises were noted to result in changes in plasma zinc.[74]

Key Point

Poor zinc status can result in poor athletic performance stemming from cardiovascular and muscular dysfunctions.

ZINC SUPPLEMENTATION AND PERFORMANCE. Zinc supplementation in conjunction with supplementation of other nutrients has been associated with positive effects on activity and performance. For instance, triathletes were able to improve their performance after utilizing an energy and micronutrient supplement containing zinc.[75] Prior to utilizing the supplement the triathletes were consuming less than recommended levels for energy, carbohydrate, zinc, and chromium. Therefore it is difficult to conclude that the zinc was directly responsible for the improved performance. However, children supplemented with vitamins A, D, and E and thiamin, riboflavin, and pyridoxine presented higher activity levels when the supplement also contained zinc.[76]

■ ZINC AND EXERCISE *IN REVIEW*

- Zinc is found in a variety of foods; however, zinc in animal sources like meats seems to be more concentrated and bioavailable.
- Zinc is a component of numerous enzymes including many involved in exercise metabolism and recovery operations.
- Zinc supplementation must consider copper status.

■ COPPER (Cu)

The richest sources of copper include organ meats, shellfish, nuts, seeds, legumes, dried fruits, and certain vegetables such as spinach, peas, and potato varieties

Table 10-11 **Copper Content of Select Foods**

Food	Copper (mg)	Food	Copper (mg)
Liver, beef (3 oz)	2–3	Cocoa powder (2 T)	0.4
Cashews, dry roasted (¼ c)	0.8	Prunes, dried (10)	0.4
Black-eyed peas (½ c)	0.7	Salmon, baked (3 oz)	0.3
Molasses, blackstrap (2 T)	0.6	Pizza, cheese (1 slice)	0.1
Sunflower seeds (¼ c)	0.6	Bread, whole wheat (1 slice)	0.1
V8 drink (1 c)	0.5	Milk chocolate (1 oz)	0.1
Tofu, firm (½ c)	0.5	Milk, 2% (1 c)	0.1
Beans, refried (½ c)	0.5		

(Table 10-11). Unlike most other nutrients, a little bit of copper absorption can take place across the wall of the stomach; however, the majority is left to the small intestine. Substances such as vitamin C, fiber, zinc, and bile in excessive amounts can decrease the efficiency of copper absorption. The RDA for copper is 900 μg.

Copper Functions and Metabolism

The total amount of copper in an adult is only about 0.1 g. Unlike zinc, copper serves as a cofactor for only a few enzymes. However, these enzymes are involved in extremely important operations such as collagen formation, electron transport, and ferroxidation. **Ferroxidation** is the oxidation of iron, converting it from ferrous to ferric form, which is necessary for iron to interact with its transport protein (transferrin) and circulate. The oxidation is accomplished by **ceruloplasmin**, a copper transporter that also has enzyme activity. Copper is also a component of *superoxide dismutase (SOD)*, an important antioxidant enzyme. The copper-endowed enzymes and their functions are listed in Table 10-12.

Copper Imbalance

Too little copper in the diet can reduce SOD levels in tissue, thus increasing the likelihood of free radical damage. In addition, poor copper status can result in inefficiencies in collagen formation, aerobic energy metabolism, and iron transport in the blood. On the other side of the balance, copper toxicity can result in liver disease. The UL for copper has been set at 10,000 μg for adults.

Copper Intake by Athletes

Copper intake has been estimated to be below recommended levels for some athletes but not others. For instance, the average copper intake was above recommended levels for female basketball players and runners but not for handball players in one study.[17] This research was conducted in Spain, as was another study involving athletic boys and girls. In this study, boys who played ice hockey were reported to have a higher copper intake than boys

Table 10-12

Copper-containing Enzymes and General Functions

Copper-Containing Proteins	General Function
Superoxide dismutase	Antioxidant (scavenges the superoxide radical)
Dopamine β-hydroxylase	Formation of epinephrine and norepinephrine from dopamine
Ceruloplasmin	Transports copper; oxidizes iron for transport on transferrin
Amine oxidases (not all)	Metabolism of physiologically active amines (such as tyramine and polyamines)
Lysyl oxidase	Collagen formation by fibroblastic cells
Cytochrome *c* oxidase	Electron transport in mitochondria
Tyrosine hydroxylase	Dopamine and melanin formation

who were not playing sports.[77] However, the athletic boys presented lower serum levels of copper as well as ferritin (the iron status indicator) while having higher zinc levels. The lower copper level was not low enough to be considered deficient. For athletic girls (gymnasts, runners, figure skaters) in this study, the copper intake and serum copper levels were not different from those of age-matched girls who did not participate in sports. When diet records of competitive bodybuilders were assessed, the females had a copper intake below recommendations while the males did not.[78] Some other findings: marathon runners in Italy were assessed to meet their copper recommendations[69] as were highly trained female runners.[52,79] More than half of recreational triathletes had dietary intakes of copper below recommendations, but many others in the group utilized a supplement containing copper, which improved their intake to levels exceeding recommendations.[18]

Copper and Athletic Performance

Four of the copper-containing enzymes stand out with regard to athletic performance (see Table 10-12).[55] The first is cytochrome c oxidase, which is involved in electron transfer to oxygen at the end of the electron transport chain. Second, without the actions of ceruloplasmin, iron is not effectively transported to bone marrow and anemia might develop, which would reduce the O_2-carrying capacity of the blood and decrease aerobic performance. The effective transportation of iron to bone marrow and muscle tissue also allows for efficient adaptation to exercise. Third, dopamine β-hydroxylase is needed to properly form epinephrine and norepinephrine. Fourth, lysyl oxidase is involved in the production of collagen, which is important to bone and muscle adaptations in response to exercise.[8]

Some earlier studies suggested that copper status was not altered by exercise.[79] However, other studies suggest that chronic strenuous exercise can negatively influence copper status. For instance, researchers reported that the runners they assessed had lower plasma copper and ceruloplasmin activity than nonrunners.[80] Ceruloplasmin level and its activity were also lower in male soccer players in comparison to controls.[81] Whether exercise alters blood copper levels is unclear and might depend on the indicator used to assess copper status and the level of copper in the diet.[82] For instance, in one study, both the level of diet copper intake and plasma copper were higher for female runners than for inactive women; however, their RBC copper levels were reduced.[83] Another study found that male swimmers did not experience a reduction in blood copper levels during their competitive season, and that their consumption of copper increased.[53] The researchers concluded that the level of dietary intake may be an important factor in the maintenance of copper status throughout the season. Also, the level of copper has been noted to increase after a few days of strenuous physical activity.[66] Elsewhere, researchers found that the level of urinary and intestinal excretion of copper was increased during strenuous exercise and that this was coupled with lower than normal copper status.[84]

Key Point

Though not typically associated with athletic performance, copper is a component of several enzymes (cytochrome c oxidase, ceruloplasmin, dopamine β-hydroxylase, and lysyl oxidase) crucial to exercise metabolism as well as recovery.

COPPER SUPPLEMENTATION BY ATHLETES. Many people use a multivitamin and mineral supplement to make up for inadequacies in their diet. However, one study found that dietary copper levels were similar between those who use a multivitamin and mineral supple-

ment and those receiving a placebo.[85] Also, after 7–8 months of receiving a vitamin and mineral supplement that included copper, athletes did not demonstrate an increase in plasma copper levels,[86] and similar results have been reported elsewhere.[87] For example, in one study involving highly trained endurance runners with poor copper status, daily supplementation of copper (copper carbonate at 0.09 mg/kg of body weight) failed to prevent negative copper balance and copper losses in feces and urine.[88] It appears that some athletes engaged in heavy endurance training and competitive schedules might experience reduced copper status and that supplementation might not be able to completely compensate for losses. Also, there is not enough evidence to suggest that supplementation would enhance performance if copper status is not initially below normal.[61]

■ COPPER AND EXERCISE *IN REVIEW*

- Copper is a component of several enzymes including cytochrome c oxidase, ceruloplasmin, and dopamine β-hydroxylase.
- Although poor copper status might negatively affect exercise, especially endurance efforts, very little research is available to demonstrate the relationship between copper status and metabolism and exercise performance.

Iodide (I)

Iodine occurs in nature as an anion (I^-) and thus is accurately referred to as iodide, just as chlorine is called chloride. Seafood is typically listed among the better sources of iodide, and freshwater fish and dairy foods are fair sources of iodide (Table 10-13). Iodide deficiency has for the most part been eradicated from many regions of the world including the United States, where iodide is added to salt typically at one part NaI for every 1000 parts NaCl. The RDA for iodide is 150 μg for adults.

Iodide Functions and Metabolism

Approximately 75–80% of the iodide in the body is concentrated in the thyroid gland. Iodide is used to make thyroid hormone, of which there are two. **Thyroxine (T_4)** contains four atoms of iodide and **triiodothyronine (T_3)** has three. Thyroid hormone affects most tissues in the body (Table 10-14); exceptions are the brain, testes, uterus, spleen, and the thyroid gland itself. Among its most recognizable effects is increasing cell metabolism, and thyroid hormone was once commonly prescribed to treat obesity. Iodide is removed from the body as part of stool (via bile) and sweat; for excessive iodide the urine is the primary means of excretion.

Table 10-13 **Iodide Content of Select Foods**

Food	Iodide (μg)	Food	Iodide (μg)
Salt, iodized (1 tsp)	400	Egg (1)	18–26
Haddock (3 oz)	104–145	Cheddar cheese (1 oz)	5–23
Cottage cheese (½ c)	26–71	Ground beef (3 oz)	8
Shrimp (3 oz)	21–37		

Table 10-14 **Some of the Effects of Thyroid Hormone on Specific Human Mechanisms**

Mechanism or Tissue	Effect of Thyroid Hormone
Carbohydrate metabolism	• Stimulates glucose absorption as well as uptake by cells • Enhances carbohydrate metabolism especially glycolysis and gluconeogenesis • Enhances insulin release
Fat metabolism	• Enhances fat mobilization from adipocytes • Increases FFA content of plasma and increases fatty acid oxidation in cells • Decreases plasma cholesterol probably by increasing bile cholesterol content and subsequent fecal loss
Protein synthesis	• Increases general protein synthesis; however, excessive amount result in protein catabolism
Basal metabolism	• Increases general metabolism of most cells • Lack of thyroid hormone results in a 50% reduction in basal metabolic rate
Cardiovascular system	• Increased heart rate and stroke strength • Increases blood volume slightly • Blood pressure is generally unchanged; systolic pressure can be elevated and diastolic relatively decreased
Respiration	• Increased respiration associated with increased cellular metabolism
Feeding/digestion	• Increased appetite and food intake • Increased rate of digestive juice secretion and motility of digestive tract • Lack of thyroid hormone results in constipation
Central nervous system	• Increases rapidity of elation • Excessive amounts result in nervousness and anxiety
Endocrine glands	• Increases rate of most endocrine secretions
Skeletal muscle	• Increases vigor of contraction

Iodide Imbalance

Decreased availability of iodide to the thyroid gland limits thyroid hormone production. For children, iodide deficiency would become obvious as measures of their growth and maturation would lag behind those of other children. For adults, iodide deficiency lowers thyroid hormone levels, which reduces the functional efficiency of most tissues in the body. On the other side of the balance, iodide toxicity results in elevated levels of thyroid-stimulating hormone, the hormone that stimulates the release of thyroid hormone into circulation. The UL has been set at 1100 μg for adults.

Iodide Intake by Athletes

Iodide intake is often overlooked in the athletic population. Part of this may be due to the high availability of iodized salt in the American population. However, some of the reports of dietary iodine levels in athletes indicate low levels. For instance, researchers assessed national competitive figure skaters, ages 11–18, and determined

Key Point

Iodide is a component of thyroid hormone, which is one of the most important hormones dictating energy metabolism.

that the females, but not the males, seemed on average to consume the recommended levels of iodide.[31] A study of young weightlifters (ages 14–17) found that their iodide intake levels were also below recommendations.[89] As with the other micronutrients, assessment of iodide intake must include supplements and sport foods. For instance, popular sport bars can contain approximately 50 μg of iodide.

Iodide and Athletic Performance

Iodide is important to exercise performance as a component of thyroid hormone, which is a factor in protein synthesis in skeletal muscle, energy expenditure, and weight control. Despite its fundamental involvement in exercise-related processes, relatively little research has focused on iodide and physical activity. As with many other minerals, chronic training and competition associated with excessive sweating increase the opportunity for iodide losses. Preliminary studies of athletes suggest that the iodide content of sweat remains fairly constant regardless of iodide status and intake.[90] The mean iodine level in sweat was noted to be about 37 μg/L, and the total iodide losses in sweat were related to total sweat volume. Urinary iodide concentration is considered to be more reflective of dietary and biological status.

■ IODIDE AND EXERCISE *IN REVIEW*

- Iodized salt is a primary contributor of iodide to the diet.
- Iodide is a key component of thyroid hormone that is one of the most influential factors related to energy metabolism in cells.
- The influence of exercise on iodide metabolism requires further research.

■ SELENIUM (Se)

The selenium content in natural foods is often a reflection of the Se content of the soil in which plants were grown and where animals grazed. Animal products, including seafood, tend to be better sources of dietary selenium than plants; Table 10-15 presents the selenium content of select foods. The RDA for selenium is 55 μg for adult males and females.

Selenium Functions and Metabolism

An adult's body may contain about 15 mg of selenium. Selenium is a component of the enzyme *glutathione peroxidase*, which is involved in antioxidant processes. Glutathione peroxidase helps inactivate free radical substances such as hydrogen peroxide (H_2O_2) and organic peroxides (Figure 10-11).[1] Like other free radicals, these substances either directly or indirectly lead to oxidative damage in cells. In addition, selenium is probably involved in the deiodination of T_4 in cells to form T_3. Thus, deficiency would be expected to decrease not only antioxidant processes but possibly metabolism and other thyroid hormone–dependent events.

Selenium Imbalance

Selenium deficiency is rare, although people living in regions with low soil and water content of selenium are at risk. The signs and symptoms of moderate selenium deficiency are vague; however, it is likely that physiological alterations would be related to reduced antioxidant capacity and efficiency of thyroid hormone metabolism. The effects of selenium toxicity have not been well studied in humans. However, selenium intakes greater than 750 g daily for several weeks may result in hair and nail loss, fatigue, nausea and vomiting, and hindrance in proper protein manufacturing. The UL for selenium has been set at 400 μg for adults.

Selenium and Athletic Performance

Selenium is rarely considered as a performance-enhancing supplement unless the focus is related to reducing free radical activity associated with endurance activity. For instance, young competitive road cyclists were recruited to test the antioxidant potential of a compound (tablet form) containing selenium, vitamin E, glutathione, and cysteine.[91] Supplementation led to a decrease in the level of malondialdehyde, which is a commonly used marker of lipid peroxidation. Elsewhere, male and female swimmers provided with a selenium supplement (100 μg/day) also had decreased malondialdehyde levels after a training session.[92] Similar results were also found in weightlifters and rowers.[93] In another study, young men were provided an antioxidant supplement mixture containing vitamin E, β-carotene, ascorbic acid, selenium, lipoic acid, and other

Table 10-15 **Selenium Content of Select Foods**

Food	Selenium (μg)	Food	Selenium (μg)
Snapper, baked (3 oz)	148	Sunflower seeds (¼ c)	25
Halibut, baked (3 oz)	113	Granola (1 c)	23
Salmon, baked (3 oz)	70	Ground beef (3 oz)	22
Scallops, steamed (3 oz)	70	Chicken, baked (3 oz)	17
Clams, steamed (20)	52	Bread, whole wheat (1 slice)	16
Oysters, raw (¼ c)	35	Egg (1)	12
Molasses, blackstrap (2 tbs)	25	Milk, 2% (1 c)	6

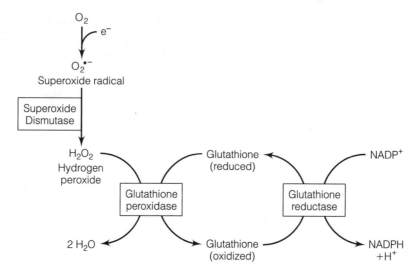

Figure 10-11
The role of glutathione peroxidase, which contains selenium, in an antioxidant process.

factors in an attempt to reduce oxidative stress during 24 days of cold-weather field training at a moderate altitude.[94] Overall, the supplement was not successful. However, the researchers noted that those men with reduced antioxidant capacity at the beginning of the study did respond to supplementation, which suggests that preliminary status is an important factor for determining supplement need. These results indicate that selenium supplementation may allow for a reduction in exercise-induced free radical activity and that preliminary status influences the magnitude of response. However, some researchers feel that more research is needed before antioxidant supplements can be recommended (see In Focus: Antioxidant Supplementation by Athletes in Chapter 9).

■ SELENIUM AND EXERCISE *IN REVIEW*

- Selenium is a key component of glutathione peroxidase, a sulfur-containing antioxidant enzyme.
- Selenium supplementation to enhance antioxidant activities in athletes remains somewhat controversial.

■ MANGANESE (Mn)

Whole-grain cereals, fruits and vegetables, legumes, nuts, tea, and leafy vegetables are good food sources of manganese. Animal foods are generally poor contributors of manganese. Additional substances in plants, such as fiber, phytate, oxalate along with excessive calcium, phosphorus, and iron can decrease manganese absorption. The RDA for manganese is 1.8 and 2.3 mg/day for adult females and males, respectively.

Manganese Functions and Metabolism

Manganese is involved in several general functions in the cells. First, manganese can interact with specific enzymes to increase their activity. These manganese-activated enzymes are involved in many processes including protein digestion and the making of glucose from certain amino acids and lactate (gluconeogenesis). Second, manganese is a structural component of many enzymes. These enzymes are involved in processes including urea formation, glucose formation, and antioxidant activities. Manganese is a component of one form of SOD. Manganese may also be involved in the activity of some hormones; research in this area is ongoing.

Manganese Imbalance

Manganese deficiency is rare and poorly understood; however, nausea, vomiting, dermatitis, decreased growth of hair and nails, changes in hair color may result. Manganese toxicity is also rare, although miners inhaling manganese-rich dust can experience Parkinson-like symptoms. The UL has been set at 11 mg of manganese for adults.

Manganese and Athletic Performance

The impact of manganese status on exercise performance is largely unknown. However, because of its role in the SOD antioxidant system, manganese might be important to managing free radical activity associated with some sports. In addition, optimal manganese status in the liver might be crucial to effective gluconeogenesis during endurance efforts. Exercise studies with rodents have focused more on antioxidant activities and are discussed in the In Focus feature in Chapter 9. Otherwise, studies addressing manganese status and athletic performance are lacking. This is also true for manganese supplementation, either as an independent supplement or in a combination supplement.

Table 10-16 Chromium Content of Select Foods

Food	Chromium (μg)	Food	Chromium (μg)
Meats		**Fruits and vegetables**	
Turkey ham (3 oz)	10.4	Broccoli (½ c)	11.0
Ham (3 oz)	3.6	Grape juice (½ c)	7.5
Beef cubes (3 oz)	2.0	Potatoes, mashed (1 c)	2.7
Chicken (3 oz)	0.5	Orange juice (1 c)	2.2
		Lettuce, shredded (1 c)	1.8
		Apple, unpeeled (1 c)	1.4
Grain Products			
Waffle (1)	6.7		
English muffin (1)	3.6		
Bagel, egg (1)	2.5		
Rice, white (1 c)	1.2		
Bread, whole wheat (1 slice)	1.0		

■ MANGANESE AND EXERCISE *IN REVIEW*

- Manganese is a key component of enzymes involved in antioxidant processes as well as gluconeogenesis.
- Research is limited related to the relationship between exercise and mangenese status and metabolism.

■ CHROMIUM (Cr)

Chromium is found in foods and in the body in the form of trivalent chromium (Cr^{3+}). Whole grains and meats are better sources (Table 10-16). Vegetation grown in chromium-rich soils may also make a significant contribution to the diet. Many multivitamin and mineral supplements include chromium, as do most "fat burner" or "metabolism booster" supplements. The AI for chromium has been set at 25 and 35 μg for males and females, respectively, between the ages of 19 and 50.

Chromium Function

Chromium is a key component of glucose tolerance factor (GTF). GTF is a molecular complex that appears to increase the efficiency of interaction between insulin and its receptor on the plasma membrane of cells. Therefore, chromium is vital for effective metabolism of glucose and insulin sensitivity.

Chromium Imbalance

Poor chromium status is likely to affect the utilization of glucose; however, this does not mean that above-normal chromium status enhances glucose utilization. Excessive chromium intake, which is more likely to occur via supplementation, can cause kidney disorders and possibly failure. The UL for chromium has not been determined.

Chromium and Athletic Performance

Chromium has been and continues to be a popular supplement. It has been touted as a means of increasing LBM and strength and decreasing body fat. In light of its popularity chromium is discussed in Chapter 11. As a preview, early research findings were suggestive that chromium picolinate could enhance LBM and strength; however, more recent studies have not supported the earlier research efforts.[95] Because the UL has not been determined at this time, some caution must be applied during supplementation.

■ CHROMIUM AND EXERCISE *IN REVIEW*

- Chromium appears to be important in the regulation of blood glucose as a component of glucose tolerance factor (GTF).
- Chromium has been one of the most popular sport supplements and has been touted to reduce body fat and increase LBM and strength; however, researchers have failed to demonstrate that supplemental chromium is effective in promoting these changes.

■ BORON (B)

Fruits, leafy vegetables, nuts, and legumes are rich sources of boron; meats are poor sources. Beer and wine are also respectable sources. Although not yet established, the human requirement for boron is believed to about 1 mg daily.

Boron Function

In the body, the largest concentration of boron is in bone, also the site of greater activity. Although the exact mechanism(s) remains a mystery, boron seems to affect certain factors in calcium metabolism such as vitamin D and magnesium. This area has been receiving more attention as researchers address bone diseases.

Boron Imbalance

Boron deficiency may result in an increased urinary loss of calcium and magnesium, presumably from their storage sites, which are primarily in bone. Although research results are not conclusive, boron deficiency may be linked to decreased bone mineral density. On the other side of the balance, supplementation involving large doses of boron (>100 mg) may induce nausea, vomiting, lethargy, and an increased loss of riboflavin. The UL for boron has been set at 20 mg for adults.

Boron and Athletic Performance

At one time boron received a lot of attention and its supplementation was marketed as a possible means for increasing testosterone levels.[95] Although this claim has not been supported by research with bodybuilders and other athletes, it is still being sold, and thus a more detailed discussion can be found in Chapter 11. However, the importance of boron in the formation and maintenance of bone and the influence of exercise requires research investigation.[96]

■ BORON AND EXERCISE *IN REVIEW*

- Boron appears to be involved in bone metabolism, as deficiency has been associated with reductions in bone density.
- Researchers have speculated that some of the bone-protective properties of boron may be related to steroid hormones.
- Boron has enjoyed sport supplement popularity as a potential mechanism of increasing testosterone levels in men; however, researchers have failed to demonstrate that this occurs.

■ VANADIUM (V)

The essentiality of vanadium has long been questioned. Although it is present in tissue throughout the body, researchers have not identified a necessary function that is truly dependent on vanadium. Breakfast cereals, canned fruit juices, fish sticks, shellfish, vegetables (especially mushrooms, parsley, and spinach), sweets, wine, and beer appear to be good sources. A human requirement for vanadium has yet to be established; however, any requirement is likely to approximate a few to a hundred micrograms.

Vanadium Function and Metabolism

Vanadium is present in trace concentrations in most organs and tissues of the body, and it appears to affect glucose metabolism in a manner similar to insulin. Promising research with diabetic animals has suggested that vanadium therapy may control high blood glucose levels. Yet, an essential role for vanadium in humans awaits discovery.

Vanadium Imbalance

Vanadium deficiency can result in reductions in growth rate, infancy survival, and hematocrit in animals, and it may alter the activity of the thyroid gland and its ability to properly utilize iodide. Conversely, signs of vanadium toxicity such as a green tongue, diarrhea, abdominal cramping, and alterations in mental functions have been reported in humans ingesting greater than 10 mg of vanadium daily over extended periods. No RDA or AI has been determined; the UL has been set at 1.8 mg for adults.

Vanadium and Athletic Performance

Vanadium has enjoyed popularity as a sport supplement as it has been touted to simulate the actions of insulin. As insulin is involved in promoting amino acid uptake into skeletal muscle cells and possibly decreasing the breakdown of protein, it is claimed to enhance muscle mass development.[95] In light of its commercial presence, the ergogenic potential of vanadium is discussed in more detail in Chapter 11. As a preview, research studies supportive of vanadium supplementation are generally lacking.[95]

■ VANADIUM AND EXERCISE *IN REVIEW*

- Vanadium may be fundamentally important in the regulation of blood glucose.
- Vanadium has enjoyed popularity as a sport supplement; however, supportive research is lacking.

Conclusions

Minerals are broadly classified as either major and minor minerals based on the contribution to human mass. The function of some minerals is as electrolytes, and others are components of enzymes. Minerals are thus involved in a broad range of normal function including muscle contraction and antioxidant activity. The intake of several minerals is below recommended levels for some athletic populations; however, each athlete must be assessed individually. Menstrual losses and foot-strike RBC destruction may increase iron losses, making iron a special consideration for female runners of reproductive age. Magnesium and zinc supplementation require further evaluation for performance enhancement; however, high zinc intake is a concern because of its relationship to copper status. Chromium, vanadium, and boron have been fairly popular nutrition supplements in athletic populations and are discussed in Chapter 11. However, independent supplementation of these minerals has not proven to enhance performance.

 **IN FOCUS**   OSTEOPOROSIS: NOT JUST A CALCIUM PROBLEM

Osteoporosis is a progressive disorder in which bone tissue density is reduced to the point that bone is quite susceptible to fracture and the quality of life is affected. Active people and athletes are not completely resistant to osteoporosis. Therefore, everyone should understand the risk factors associated with osteoporosis and how best to prevent or slow its development. As discussed in Chapter 1, people tend to achieve peak bone mass (PBM) in their late 20s and generally lose bone material in the decades that ensue. Nutrition and physical activity are the most controllable factors that influence the development of PBM and the rate of bone material loss, if any, in the ensuing years. This feature looks more closely at the involvement of these factors in the development and maintenance of healthy bone.

All too often, osteoporosis is presented to the public as a problem resulting from poor bone mineral content and more specifically, calcium. When calcium is supplemented, the impact usually is not as significant as expected unless the individual's diet was lacking in calcium. This suggests that although the level of dietary calcium is important, it is often not the sole root of the problem. Several nutrients play a role in the development and maintenance of healthy bone tissue. These are presented in Table A and include vitamin C, copper, iron, phosphorus, zinc, protein, and maybe boron and magnesium. Along with good nutritional status, one of the most important factors that regulate bone density is physical activity. Physical activity provides mechanical stress on bone and promotes bone development processes. Physical activity has different effects on bone depending on its intensity, frequency, duration, and the age at which it is started. The anabolic effect on bone is greater in adolescence and as a result of weight-bearing exercise. Adequate intakes of calcium appear necessary for exercise to have its bone-stimulating action.

Table A   Key Nutrients Involved in Bone Formation and Maintenance

Nutrient	Role in Bone Formation or Maintenance
Calcium and phosphorus	Provides hardness to bone in the form of hydroxyapatite and calcium phosphate crystals.
Zinc	Important in bone remodeling as a component of *alkaline phosphatase*.
Magnesium	Approximately 60% of body magnesium is found in bone. May provide support in bone rigidity.
Copper	Important to collagen formation as a component of lysyl hydroxylase, which is involved in posttranslational modification of collagen fibrils before they are released into the matrix.
Iron	Important to collagen formation as a component of prolyl hydroxylase, which is involved in posttranslational modification of collagen fibrils before they are released into the matrix.
Boron	Important in bone formation and maintenance, although the mechanism remains to be determined.
Vitamin C	Important to collagen formation as a component as a reducing factor for lysyl and prolyl hydroxylase. Vitamin C deficiency decreases the production of viable collagen.
Vitamin D	Responsible for making calcium available for bone formation and maintenance by increasing intestinal absorption as well as decreasing calcium loss in urine. Therefore the role of vitamin D in bone formation is indirect but fundamentally important.
Vitamin K	Important in bone formation and maintenance, although the mechanism remains to be determined. Increased intake of vitamin K is associated with increased markers of bone formation and decreased serum markers of breakdown.

Although calcium is often portrayed as the basis of bone tissue, the real foundation is a collagen network. A healthy collagen system provides bone's strength and allows mineral crystals, such as hydroxyapatite, to adhere and harden bone. If bone was a house, collagen would be the wooden frame and the mineral crystals would be the drywall. Therefore, building healthy bone begins with attention to collagen formation. As collagen fibers are formed, calcium phosphate salts can begin to be deposited on the surfaces of these protein structures. With time and the association of more and more calcium phosphate salts, hydroxyapatite crystals are formed. These platelike crystals orient themselves along lines of mechanical stress and serve to strengthen bone against torque and trauma.

Several nutrients are important in collagen formation including protein, vitamins C and D, zinc, copper, and iron. Protein provides amino acids for building collagen fibrils. Vitamin C, copper, and iron are involved in posttranslational modifications that allow collagen fibers

OSTEOPOROSIS: NOT JUST A CALCIUM PROBLEM (CONTINUED)

to form. Along with collagen, structural carbohydrate-based molecules are found in bone. One type is proteoglycans, composed of proteins plus glycosaminoglycan (GAG) molecules such as *chondroitin sulfate*. *Hyaluronic acid* is another GAG found in bone; however, it is not bound to proteins. The most abundant bone minerals are calcium and phosphate, which are found complexed as salts (such as calcium phosphate) and crystals (hydroxyapatite). Vitamin D is a vital stimulator of calcium and phosphate absorption from the

digestive tract. In addition, bone contains magnesium, potassium, and sodium ions, which are found in association with hydroxyapatite and elsewhere.

Thus recommendations for achieving maximal peak bone density and preventing osteoporosis include eating a balanced diet to provide all the needed nutrients. In addition, physical activity is important to stimulate maximal bone development and promote bone maintenance or slow the loss of bone material after peak bone mass is achieved.

STUDY QUESTIONS

1. How are calcium and phosphorus important to athletic performance?
2. What is the principal means of regulating the levels of sodium, potassium, and chloride? How does sweating influence the rate of loss of these minerals?
3. In what ways is iron important to aerobic energy metabolism during exercise?
4. Which minerals might be a concern for athletes who produce higher sweat volume?
5. How is chloride involved in releasing excessive CO_2 produced during exercise energy metabolism?
6. Explain a few ways in which zinc is important to exercise performance and adaptation to training.
7. What roles does copper play in exercise energy metabolism and adaptation to exercise training?
8. What do selenium, copper, manganese, and zinc have in common in relation to aerobic exercise?
9. How are sodium, potassium, and chloride important to exercise performance?
10. What are the most important considerations for an athlete related to consuming enough minerals?

REFERENCES

1. Wildman REC, Medeiros DM. Major Minerals. In: *Advanced Human Nutrition*, Boca Raton, FL: CRC Press, 2000.
2. Kandiah J. Calcium and iron intakes of disabled elite athletes. In: *Sport Nutrition: Minerals and Electrolytes* (Kies CV, Driskell JA, eds.), Boca Raton, FL: CRC Press, 1995.
3. Clarkson P, Haymes E. Exercise and mineral status of athletes: calcium, magnesium, phosphorus, and iron. *Medicine and Science in Sports and Exercise* 27:831–845, 1995.
4. Webster B, Barr S. Calcium intake of adolescent female gymnasts and speed skaters: lack of association with dieting behavior, *International Journal of Sport Nutrition* 5:2–12, 1995.
5. Williams M. Minerals: the inorganic regulators. In: *Nutrition for Health, Exercise and Sport*, 5th ed., San Francisco: WCB McGraw-Hill, 1999.
6. Sandoval WM, Heyward VH, Lyons TM. Comparison of body composition, exercise and nutritional profiles of female and male body builders at competition. *Journal of Sports Medicine and Physical Fitness* 29(1):63–70, 1989.
7. Heywood VH, Sandoval WM, Colville BC. Anthropometric, body composition and nutritional profiles of body builders during training. *Journal of Applied Sport Science Research* 3:22–27, 1989.
8. Branca F, Valtuena S, Vatuena S. Calcium, physical activity and bone health—building bones for a stronger future. *Public Health Nutrition* 4(1A):117–123, 2001.
9. Klesges R, Ward KD, Shelton ML, Applegate WB, Cantler ED, Palmieri GM, et al. Changes in bone mineral content in male athletes, *Journal of the American Medical Association* 276:226–230, 1996.
10. Bremner K, Bubb WA, Kemp GJ, Trenell MI, Thompson CH. The effect of phosphate loading on erythrocyte 2,3-bisphosphoglycerate levels. *Clinica Chimica Acta* 323(1–2):111–114, 2002.
11. Sherwood L. The urinary system. In: *Human Physiology: From Cells to Systems*, 3rd ed., Belmont, CA: Wadsworth 1997.
12. Maughan D. Diffusible magnesium in frog skeletal muscle cells. *Biophysics Journal* 43:75–80, 1983.
13. Ekelund MC, Erdman KAP. Shortening induced deactivation of skinned fibres of frog and mouse striated muscle, *Acta Physiologica Scandinavica* 116:189–195, 1982.
14. Kitazawa T, Shuman H, Somlyo AP. Calcium and magnesium binding to thick and thin filaments in skinned muscle fibres: electron probe analysis. *Journal of Muscle Research and Cell Motility* 3:437–445, 1982.

15. Lopez JR, Alamo L, Caputo C, Vergara J, DiPolo R. Direct measurement of intracellular free magnesium in frog muscle using magnesium-selective microelectrodes. *Biochemistry and Biophysics* 804:1–10, 1974.

16. Loosli AR, Benson J, Gillien DM, Bourdet K. Nutritional habits and knowledge in competitive adolescent female gymnasts. *Physician and Sportsmedicine* 8:118–125, 1986.

17. Nuviala RJ, Lapieza MG, Bernal E. Magnesium, zinc, and copper status in women involved in different sports. *International Journal of Sport Nutrition* 9(3):295–309, 1999.

18. Worme JD, Doubt TJ, Singh A, Ryan CJ, Moses FM, Deuster PA. Dietary patterns, gastrointestinal complaints, and nutrition knowledge of recreational triathletes. *American Journal of Clinical Nutrition* 51(4):690–697, 1990.

19. Clarkson PM. Minerals: exercise performance and supplementation in athletes. *Journal of Sports Science*, 9(Spec No):91–116, 1991.

20. Shirreffs SM, Maughan RJ. Whole body sweat collection in humans: an improved method with preliminary data on electrolyte content. *Journal of Applied Physiology* 82(1):336–341, 1997.

21. Weller E, Bachert P, Meinck HM, Friedmann B, Bartsch P, Mairbaurl H. Lack of effect of oral Mg-supplementation on Mg in serum, blood cells, and calf muscle. *Medicine and Science in Sports and Exercise* 30(11):1584–1591, 1998.

22. Finstad EW, Newhouse IJ, Lukaski HC, Mcauliffe JE, Stewart CR. The effects of magnesium supplementation on exercise performance. *Medicine and Science in Sports and Exercise* 33(3):493–498, 2001.

23. Singh A, Moses FM, Deuster PA. Chronic multivitamin-mineral supplementation does not enhance physical performance. *Medicine and Science in Sports and Exercise* 24(6):726–732, 1992.

24. Brilla LR, Haley TF. Effect of magnesium supplementation on strength training in humans. *Journal of American College of Nutrition*, 11(3):326–329, 1992.

25. Braun LT. Exercise physiology and cardiovascular fitness. *The Nursing Clinics of North America* 26(1):135–147, 1991.

26. Hickson RC. Skeletal muscle cytochrome *c* and myoglobin, endurance, and frequency of training. *Journal of Applied Physiology* 51(3):746–749, 1981.

27. Maughan RJ. Role of micronutrients in sport and physical activity. *British Medical Bulletin* 55(3):683–690, 1999.

28. Powell PD, Tucker A. Iron supplementation and running performance in female cross-country runners. *International Journal of Sports Medicine* 12(5):462–467, 1991.

29. Castillo MC, Lapieza MG, Leon F, Nuviala RJ. Iron intake and pharmacologic supplements in medium and long-distance runners. *Sangre (Barcelona)* 41(3):195–200, 1996.

30. Fogelholm M. Inadequate iron status in athletes: an exaggerated problem? In: *Sport Nutrition: Minerals and Electrolytes* (Kies CV, Driskell JA, eds.), Boca Raton, FL: CRC Press, 1995.

31. Ziegler PJ, Nelson JA, Jonnalagadda SS. Nutritional and physiological status of U.S. national figure skaters. *International Journal of Sport Nutrition* (4):345–360, 1999.

32. Fogelholm M, Rankinen T, Isokaanta M, Kujala U, Uusitupa M. Growth, dietary intake, and trace element status in pubescent athletes and schoolchildren. *Medicine and Science in Sports and Exercise* 32(4):738–746, 2000.

33. Malczewska J, Raczynski G, Stupnicki R. Iron status in female endurance athletes and in non-athletes. *International Journal of Sport Nutrition and Exercise Metabolism* 10(3):260–276, 2000.

34. Nieman DC. Physical fitness and vegetarian diets: is there a relation? *American Journal of Clinical Nutrition* 70(3 Suppl):570S–575S, 1999.

35. Beard J, Tobin B. Iron status and exercise. *American Journal of Clinical Nutrition* 72(2Suppl):594S–597S, 2000.

36. Berglund B. High-altitude training. Aspects of haematological adaptation. *Sports Medicine* 14(5):289–303, 1992.

37. Roberts D, Smith DJ. Training at moderate altitude: iron status of elite male swimmers. *Journal of Laboratory and Clinical Medicine* 120(3):387–391, 1992.

38. Spodaryk K. Iron metabolism in boys involved in intensive physical training. *Physiology and Behavior* 1–15;75(1–2):201–206, 2002.

39. Shaskey DJ, Green GA. Sports haematology. *Sports Medicine* 29(1):27–38, 2000.

40. Ronsen O, Sundgot-Borgen J, Maehlum S. Supplement use and nutritional habits in Norwegian elite athletes. *Scandinavian Journal of Medicine and Science in Sports* 9(1):28–35, 1999.

41. Krumbach CJ, Ellis DR, Driskell JA. A report of vitamin and mineral supplement use among university athletes in a division I institution. *International Journal of Sport Nutrition* 9(4):416–425, 1999.

42. Banister EW, Hamilton CL. Variations in iron status with fatigue modeled from training in female distance runners. *European Journal of Applied Physiology* 54(1):16–23, 1985.

43. Klingshirn LA, Pate RR, Bourque SP, Davis JM, Sargent RG. Effect of iron supplementation on endurance capacity in iron-depleted female runners. *Medicine and Science in Sports and Exercise* 24(7):819–824, 1992.

44. Blee T, Goodman C, Dawson B, Stapff A. The effect of intramuscular iron injections on serum ferritin levels and physical performance in elite netballers. *Journal of Science and Medicine in Sport* 2(4):311–321, 1999.

45. Hinton PS, Giordano C, Brownlie T, Haas JD. Iron supplementation improves endurance after training in iron-depleted, nonanemic women. *Journal of Applied Physiology* 88(3):1103–1111, 2000.

46. Jackson JL, Lesho E, Peterson C. Zinc and the common cold: a meta-analysis revisited. *Journal of Nutrition* 130(5 Suppl):1512S–1515S, 2000.

47. Macknin ML. Zinc lozenges for the common cold. *Cleveland Clinic Journal of Medicine* 66(1):27–32, 1999.

48. Macknin ML, Piedmonte M, Calendine C, Janosky J, Wald E. Zinc gluconate lozenges for treating the common cold in children: a randomized controlled trial. *Journal of the American Medical Association* 24;279(24):1962–1967, 1998.

49. Marshall S. Zinc gluconate and the common cold. Review of randomized controlled trials. *Canadian Family Physician* 44:1037–1042, 1998.

50. Fogelholm M, Rehuven S, Gref CG, Laakso JT, Lehto J, Ruokonen I, Himberg JJ. Dietary intake and thiamin, iron and zinc status in elite Nordic skiers during different training periods. *International Journal of Sports Nutrition* 2:351–365, 1992.

51. Duester PA, Day BA, Singh A, Douglass L, Moser-Veillon PB. Zinc status of highly trained women runners and

untrained women. *American Journal of Clinical Nutrition* 49:1295–1302, 1989.

52. Deuster PA, Dyle SB, Moser PB, Vigersky RA, Singh A, Schoomaker EB. Nutritional survey of highly trained women runners. *American Journal of Clinical Nutrition* 45:954–962, 1986.

53. Lukaski HC, Hoverson BS, Gallagher SK, Bolonchuk WW. Physical training and copper, iron and zinc status in swimmers. *American Journal of Clinical Nutrition* 51:1093–1099, 1990.

54. Hackman RM, Keen CL. Changes in serum zinc and copper levels after zinc supplementation in running and non-running men. In: *Sport, Nutrition and Health*, Chapter 8 (Katch FI, ed.), Champaign, IL: Human Kinetics, 1986.

55. Speich M, Pineau A, Ballereau F. Minerals, trace elements and related biological variables in athletes and during physical activity. *Clinica Chimica Acta* 312(1–2):1–11, 2001.

56. Haymes EM. Trace minerals and exercise. In: *Nutrition in Exercise and Sport*, 3rd ed. (Wolinsky I, ed.), Boca Raton, FL: CRC Press, 1998.

57. Lukaski HC, Bolonchuk WW, Klevay LM, Milne DB, Sandstead HH. Maximal oxygen consumption as related to magnesium, copper, and zinc nutriture. *American Journal of Clinical Nutrition* 37(3):407–415, 1983.

58. Fogelholm GM, Himberg JJ, Alopaeus K, Gref CG, Laakso JT, Lehto JJ, Mussalo-Rauhamaa H. Dietary and biochemical indices of nutritional status in male athletes and controls. *Journal of the American College of Nutrition* 11(2):181–191, 1992.

59. Dressendorfer RH, Sockolov R. Hypozinemia in runner. *Physicians Sportsmedicine* 8(4):97–102, 1980.

60. Haralambie G. Serum zinc in athletes in training. *International Journal of Sports Medicine* 2:135–138, 1981.

61. Clarkson PM. Minerals: exercise performance and supplementation in athletes. *Journal of Sports Science* 9(Spec No):91–116, 1991.

62. Prasad AS, Schulert AR, Sandstead HH, Miale A. Zinc, iron, and nitrogen content of sweat in normal and deficient subjects. *Journal of Laboratory and Clinical Medicine* 62:84–91, 1963.

63. Uhari M, Pakarinen A, Hietala J, Nurmi T, Kouvalainen K. Serum iron, copper, zinc, ferritin, and ceruloplasmin after intense heat exposure. *European Journal of Applied Physiology* 51(3):331–335, 1983.

64. Tipton K, Green NR, Haymes EM, Waller M. Zinc loss in sweat of athletes exercising in hot and neutral temperatures. *International Journal of Sport Nutrition* 3(3):261–271, 1993.

65. Anderson RA, Polansky MM, Bryden NA. Strenuous running: acute effects on chromium, copper, zinc and selected clinical variables in urine and serum of male runners. *Biological Trace Element Research* 6:327–333, 1984.

66. Kikukawa A, Kobayashi A. Changes in urinary zinc and copper with strenuous physical exercise. *Aviation Space and Environmental Medicine* 73(10):991–995, 2002.

67. Khaled S, Brun JF, Micallel JP, Bardet L, Cassanas G, Monnier JF, Orsetti A. Serum zinc and blood rheology in sportsmen (football players). *Clinical Hemorheology and Microcirculation* 17(1):47–58, 1997.

68. Brun JF, Dieu-Cambrezy C, Charpiat A, Fons C, Fedou C, Micallef JP, et al. Serum zinc in highly trained adolescent gymnasts. *Biological Trace Element Research* 47(1–3):273–278, 1995.

69. Bordin D, Sartorelli L, Bonanni G, Mastrogiacomo I, Scalco E. High intensity physical exercise induced effects on plasma levels of copper and zinc. *Biological Trace Element Research* 36(2):129–134, 1993.

70. Shinogi M, Sakaridani M, Yokoyama I. Metallothionein induction in rat liver by dietary restriction or exercise and reduction of exercise-induced hepatic lipid peroxidation. *Biological Pharmacy Bulletin* 22(2):132–136, 1999.

71. Anderson RA, Bryden NA, Polansky MM, Deuster PA. Acute exercise effects on urinary losses and serum concentrations of copper and zinc of moderately trained and untrained men consuming a controlled diet. *Analyst* 120(3): 867–870, 1995.

72. Marrella M, Guerrini F, Solero PL, Tregnaghi PL, Schena F, Velo GP. Blood copper and zinc changes in runners after a marathon. *Journal of Trace Elements and Electrolytes in Health and Disease* 7(4):248–250, 1993.

73. Gleeson M, Almey J, Brooks S, Cave R, Lewis A, Griffiths H. Haematological and acute-phase responses associated with delayed-onset muscle soreness in humans. *European Journal of Applied Physiology* 71(2–3):137–142, 1995.

74. Nosaka K, Clarkson PM. Changes in plasma zinc following high force eccentric exercise. *International Journal of Sport Nutrition* 2(2):175–184, 1992.

75. Frentsos JA, Baer JT. Increased energy and nutrient intake during training and competition improves elite triathletes' endurance performance. *International Journal of Sport Nutrition* 7(1):61–71, 1997.

76. Sazawal S, Bentley M, Black RE, Dhingra P, George S, Bhan MK. Effect of zinc supplementation on observed activity in low socioeconomic Indian preschool children. *Pediatrics* 98(6 Pt 1):1132–1137, 1996.

77. Rankinen T, Fogelholm M, Kujala U, Rauramaa R, Uusitupa M. Dietary intake and nutritional status of athletic and nonathletic children in early puberty. *International Journal of Sport Nutrition* 5(2):136–150, 1995.

78. Kleiner SM, Bazzarre TL, Ainsworth BE. Nutritional status of nationally ranked elite bodybuilders. *International Journal of Sport Nutrition* 4(1):54–69, 1994.

79. Singh A, Deuster PA, Day BA, Moser-Veillon PB. Dietary intakes and biochemical markers of selected minerals: comparison of highly trained runners and untrained women. *Journal of the American College of Nutrition* 9(1):65–75, 1990.

80. Resina A, Fedi S, Gatteschi L, Rubenni MG, Giamberardino MA, Trabassi E, Imreh F. Comparison of some serum copper parameters in trained runners and control subjects. *International Journal of Sports Medicine* 11(1):58–60, 1990.

81. Resina A, Gatteschi L, Rubenni MG, Giamberardino MA, Imreh F. Comparison of some serum copper parameters in trained professional soccer players and control subjects. *Journal of Sports Medicine and Physical Fitness* 31(3):413–416, 1991.

82. Clarkson PM, Haymes EM. Trace mineral requirements for athletes. *International Journal of Sport Nutrition* 4(2):104–119, 1994.

83. Singh A, Deuster PA, Moser PB. Zinc and copper status in women by physical activity and menstrual status. *Journal of Sports Medicine and Physical Fitness* 30(1):29–36, 1990.

84. Nasolodin VV, Gladkikh IP, Meshcheriakov SI. Providing athletes with trace elements during intensive exercise. *Gigiena i sanitariia* 1:54–57, 2001.

85. Bazzarre TL, Scarpino A, Sigmon R, Marquart LF, Wu SM, Izurieta M. Vitamin-mineral supplement use and nutritional status of athletes. *Journal of the American College of Nutrition* 12(2):162–169, 1993.

86. Telford RD, Catchpole EA, Deakin V, McLeay AC, Plank AW. The effect of 7 to 8 months of vitamin/mineral supplementation on the vitamin and mineral status of athletes. *International Journal of Sport Nutrition* 2(2):123–134, 1992.

87. Weight LM, Noakes TD, Labadarios D, Graves J, Jacobs P, Berman PA. Vitamin and mineral status of trained athletes including the effects of supplementation. *American Journal of Clinical Nutrition* 47(2):186–191, 1988.

88. Zorbas YG, Charapakin KP, Kakurin VJ, Kuznetsov NK, Federov MA, Popov VK. Daily copper supplement effects on copper balance in trained subjects during prolonged restriction of muscular activity. *Biological Trace Element Research* 69(2):81–98, 1999.

89. Bauer S, Jakob E, Berg A, Keul J. Energy and nutritional intake in young weight lifters before and after nutritional counseling. *Schweizerische Zeitschrift fur Medizin und Traumatologie*(3):35–42, 1994.

90. Mao IF, Ko YC, Chen ML. The stability of iodine in human sweat. *Japan Journal of Physiology* 40(5):693–700, 1990.

91. Dragan I, Dinu V, Cristea E, Mohora N, Ploesteanu E, Stroescu V. Studies regarding the effects of an antioxidant compound in top athletes. *Revue Roumaine de Physiologie* 28(3–4):105–108, 1991.

92. Dragan I, Dinu V, Mohora M, Cristea E, Ploesteanu E, Stroescu V. Studies regarding the antioxidant effects of selenium on top swimmers. *Revue Roumaine de Physiologie* 27(1):15–20, 1990.

93. Dragan I, Ploesteanu E, Cristea E, Mohora M, Dinu V, Troescu VS. Studies on selenium in top athletes. *Physiologie* 25(4):187–190, 1988.

94. Schmidt MC, Askew EW, Roberts DE, Prior RL, Ensign WY, Jr, Hesslink RE, Jr. Oxidative stress in humans training in a cold, moderate altitude environment and their response to a phytochemical antioxidant supplement. *Wilderness Environment Medicine* 13(2):94–105, 2002.

95. Wildman REC, Cilibert LJ. Ergogenic aids. In: *Nutrition Reference for Pharmacists* (Wolinsky I, Williams L, ed.), Washington, DC: American Pharmaceutical Association, 2002.

96. Rico H, Crispo E, Hernandez ER, Seco C, Crespo R. Influence of boron supplementation on vertebral and femoral bone mass in rats on strenuous treadmill exercise. A morphometric, densitometric, and histomorphometric study. *Journal of Clinical Densitometry* 5(2):187–192, 2002.

11

SPORT FOODS, SUPPLEMENTS, AND ERGOGENIC AIDS

Chapter Objectives

- Provide a definition of sport foods, nutrition supplements, and ergogenic aids.

- Discuss regulatory issues associated with the marketing and promotion of supplements.

- Describe the composition of sport foods and rationale for their specific formulations.

- Discuss the types and properties of different nutrition supplements marketed to active people and athletes and provide a review of the current supportive and unsupportive research.

- Describe other methods applied by athletes to enhance athletic performance.

- Discuss the basis of different doping agents and typical application and concerns for safety.

Personal Snapshot

© Najlah Feanny/SABA/Corbis

Paul is an avid bodybuilder and would like to increase his lean body mass by 10–15 lb while becoming leaner. His buddies in the gym use supplements like creatine, androstenedione, and HMB. However, Paul is skeptical and would like to know more about the different supplements available on the market before trying them. For instance, which supplements are supposed to build muscle mass? How much will he need to take and what is the cost? Also, are they safe and what regulatory issues apply? Paul is confused and feels that he does not know whom to trust for unbiased information.

Individuals continue to seek an edge to achieve their performance and aesthetic goals. Capitalizing on this interest, the sports foods and supplement industry has grown into a gigantic industry. Products manufactured by companies such as MET-Rx, EAS, PowerBar, and Gatorade are some of the biggest sellers in this market. Sport foods come in the form of bars, shakes, drinks, and gels, whereas supplements are typically formulated in pills, capsules, or powders and contain one or more substances touted to improve performance and/or fitness.

How do manufacturers formulate sport food and supplements, and how much is known about the efficacy and safety of these products? Also, looking beyond commercial products, how else might athletes attempt to enhance their performance and/or fitness? The term *ergogenic aid* is often used to describe specific substances that are ingested, injected, or inhaled into the body in an attempt to improve athletic performance. This chapter provides a basic overview of important concepts related to sport nutrition foods and supplements as well as other efforts by individuals to gain a greater edge in achieving goals.

■ SPORT FOODS AND SUPPLEMENTS: CONCEPTS

What is the difference between a sport food and a supplement? How are they formulated, and what rules do manufacturers have to follow? Sometimes the difference between a sport food and a supplement is clear. For instance, androstenedione and its cousin 19-norandrostenedione (19-NOR) are clearly supplements, and a PowerBar, Clif Bar, and MET-Rx Protein Plus Bar are clearly sport foods. However, what term applies if a substance such as creatine or hydroxymethylbutyrate (HMB) is added to a sport bar, or protein powder is mixed with water? Clearly there are areas of overlap. Sport foods include bars, drinks, and gels. As a rule, sport foods differ from supplements in that they tend to be more complex and foodlike, containing one or more energy sources. Supplements tend to be limited to single nutrients or combinations of related substances. This section provides an overview of basic concepts related to sport foods and supplements, including their regulation.

The Sport Food and Supplement Market

The sport nutrition market exploded in the United States and other countries during the 1990s. The growth of this market has continued into the new century, including significant growth of all its components (Figure 11-1). For instance, Figure 11-2 shows that the sport supplement market in the United States grew by 8.7% between 2000 and 2001 as sales increased from $1.6 billion to $1.74 billion.[1] Meanwhile the sport drink industry raked in $2.4 billion during 2001.[1] As expected, Gatorade paced the industry, accounting for 80% of sales in that market.[1] The sport bar market experienced 21% growth between 2000 and 2001 and during that year sales in United States reached almost $1.4 billion.[1] All totaled, the sport nutrition market exceeded $6 billion in sales in 2002.

PRODUCT NAMES AND TRADEMARKS. Sport food and supplement manufacturers choose product names that are associated with a desired outcome. For example, Actisyn (made by SportPharma) sounds like a hybrid between actin and myosin, the most recognizable contractile proteins in muscle. The "syn" may also infer synthesis. SportPharma markets another product called Testosterogen, a hybrid of *testosterone* and *genesis*. Testosterone is the male sex hormone, which promotes muscle growth, and *genesis* means "to create from what was not there before." Testosterogen contains androstenedione, DHEA, tribulus, and two flavonoids, daidzein and chrysin, which are stated to maximize free testosterone levels in the blood.

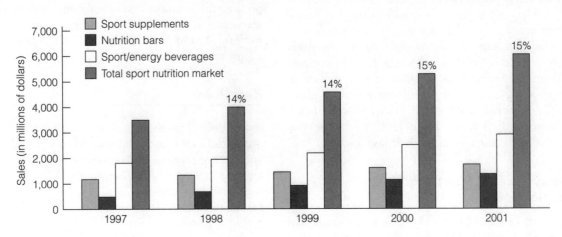

Figure 11-1
Growth of the sport nutrition market and its components in the United States with percentage growth from previous year.

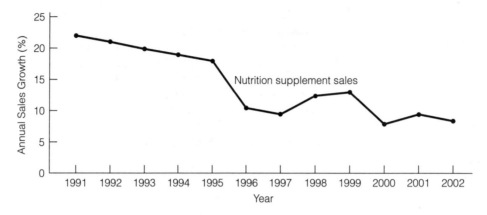

Figure 11-2
Percent change in sales of sport supplements in the United States during the 1990s and 2000–2002. (*Source*: reference 2.)

Key Point

The sport food and supplement industry generated approximately $6 billion in the United States during 2001.

Sport Food and Supplement Ingredients

Sport food and supplement manufacturers must follow the same food label guidelines as other food manufacturers. For instance, labels must list the ingredients in descending order of their contribution (mass) to the recipe. Therefore, energy-containing ingredients, which are typically provided in gram amounts, are listed before vitamins and mineral ingredients that are added in milligram and microgram quantities. Nonnutrient food additives are also included in the ingredient list.

The chemical properties of an ingredient may require that it be added to a recipe in a specific form. For example, chromium (Cr^{3+}) exists in a positive oxidation state and is **chelated** (electrically bound) to a negatively charged molecule or substance such as *picolinate* or *gluconate*. Another example is calcium (Ca^{2+}), which is attached to a negatively charged substance such as phosphate (PO_4^{3-}), citrate, carbonate, malate, or even negatively charged amino acids. Table 11-1 presents several popular sport bars and a listing of their ingredients.

Regulatory Issues of Sport Foods and Supplements

Manufacturers of sport foods and supplements are bound by regulatory guidelines as to what types of claims they can place on their food labels. General statements are commonly placed on food labels related to the "structure and function" of the product components. Manufacturers would be crossing the line if they put more aggressive outcome statements on the product label. For instance, the label for a high-protein sport bar cannot state that consuming this food "will increase muscle size and strength."

In the United States, the **Food and Drug Administration (FDA)** regulates nutrition supplements under a different set of regulations than those covering "conventional" foods and drug products (prescription and over-the-counter). Under the **Dietary Supplement Health and Education Act** of 1994 (DSHEA), the supplement manufacturer is responsible for ensuring that a product is safe before it is marketed. The FDA is responsible for taking action against any unsafe supplement product after it reaches the market. Generally, manufacturers do not need to register with the FDA or get FDA approval before producing or selling nutrition supplements. However, supplement manufacturers must make sure that product label information is truthful and not misleading. The postmarketing responsibilities of the FDA include monitoring product information, such as labeling, claims, package inserts, and accompanying literature. Another agency, the **Federal Trade Commission (FTC),** regulates nutrition supplement advertising.

Key Point

In the United States, nutrition supplements are not regulated in the same manner as foods and drugs and do not require FDA approval. The FDA monitors information accompanying the product, and the FTC regulates advertising.

Product Research and Potential Danger

Valid research is lacking for many supplements. For instance, one review of sport supplements revealed that only

Table 11-1   Ingredient Composition[a] and Carbohydrate-Protein-Fat (C:P:F) Ratio of Select Top-Selling Sport Bars

Sport Bar (Manufacturer)	C:P:F (energy ratio)[b]	Carbohydrate	Protein/ Amino acids	Fat/Lipid	Vitamins	Minerals	Additives
Promax (SportPharma) Flavor: double fudge brownie	47:30:16[a]	Sucrose, nonfat dry milk, beet sugar, oat fiber, beet fiber	Promax (whey protein concentrates, calcium caseinate, hydrolyzed whey protein concentrate, leucine, glutamine, valine, isoleucine, soy protein, nonfat dry milk, low-fat peanut butter)	Palm kernal oil, low-fat peanut butter	Ascorbic acid, pantothenate, pyridoxine, riboflavin, thiamin, vitamin D (D$_3$), folate, biotin	Tricalcium phosphate, magnesium oxide, calcium, zinc oxide, copper gluconate, chloride	Cocoa, vanilla, maltitol, guar gum, oat fiber, beet fiber, natural flavors
Ironman (Twin Lab) Flavor: cookie dough	40:30:30	High fructose corn syrup, sucrose, fructose, crisp rice	Ironman protein (soy protein isolate, calcium sodium caseinate, protein from peanut butter, amino acids from chromium amino acid chelates)	Lipid from peanut butter and soybeans, cocoa butter, palmitic acid	Ascorbic acid, vitamin A palmitate, manganese gluconate, tocopherol acetate, niacinamide, pyridoxine hydrochloride, vitamin D$_3$, riboflavin, thiamin hydrochloride, biotin	Tricalcium phosphate, calcium, sodium, sodium chloride, chloride, magnesium oxide, zinc oxide, copper gluconate, ferrous fumarate, potassium iodide	Lecithin (contains choline), natural flavors
Protein Plus (PowerBar) Flavor: chocolate fudge brownie	47:33:16[a]	Brown rice syrup, sugar, lactose from nonfat dry milk, high fructose corn syrup, fructose	PowerBar Trisource (whey protein isolates, calcium caseinate, soy protein isolate), whey powder Amino acids: glutamine, leucine, valine, methionine, isoleucine, aspartic acid	Fractionated palm kernal oil, cocoa powder	Vitamin C (ascorbic acid), vitamin E (α-tocopherol acetate), niacin, calcium pantothenate, vitamin B$_6$ (pyridoxine hydrochloride), riboflavin, thiamin hydrochloride, folic acid, biotin, vitamin B$_{12}$	Calcium phosphate, magnesium carbonate, zinc gluconate, iron (ferrous fumarate), copper gluconate, chromium aspartate, chloride	Cocoa powder, lecithin, vanilla flavor, glycerine, natural flavors, unsweetened chocolate, oat fiber Other: May contain traces of peanut or other tree nuts
Myoplex Plus Deluxe (EAS)	48:28:19	High-fructose corn syrup, sucrose, rice flour, lactose (nonfat dry milk and milk)	MyoPro (whey protein isolates, calcium caseinates, milk protein isolates), protein (rice flour), protein (nonfat dry milk, nonfat milk, and milk), glutamine, taurine	Low-fat cocoa, partially hydrogenated vegetable oil (cottonseed and soybean oils), sunflower oil, palmitic acid, unsweetened chocolate, milk fat (milk), conjugated linoleic acid (CLA)	Ascorbic acid, vitamin A palmitate, α-tocopherol acetate, calcium pantothenate, niacinamide, pyridoxine hydrochloride, vitamin D$_3$, riboflavin, thiamin hydrochloride, biotin, pyridoxine hydrochloride, phylloquinone, biotin	Calcium (calcium α-ketoglutarate and calcium pantothenate), dicalcium phosphate, calcium sodium (sodium RNA), potassium chloride and phosphate, sodium molybdate, vanadyl sulfate, ferric orthophosphate, copper gluconate, chloride,[4]	Low-fat cocoa, natural and artificial flavors, RNA, lecithin (contains choline)

Table 11-1   Ingredient Composition^a and Carbohydrate-Protein-Fat (C:P:F) Ratio of Select Top-Selling Sport Bars (cont.)

Sport Bar (Manufacturer)	C:P:F (energy ratio)^b	Carbohydrate	Protein/ Amino acids	Fat/Lipid	Vitamins	Minerals	Additives
Power Bar (Power Foods) Flavor: apple cinnamon	75:16:9	High-fructose corn syrup with grape and pear juice concentrate, oat bran, maltodextrin, rice crisps (milled rice, rice bran), brown rice, dried apples, cinnamon	Milk protein isolate, BCAAs (leucine, valine, isoleucine)	Glycerin, almond butter	Vitamin C (ascorbic acid), vitamin E acetate, vitamin B$_3$ (niacin), pantothenic acid (calcium pantothenate), vitamin B$_6$ (pyridoxine hydrochloride), vitamin B$_2$ (riboflavin), vitamin B$_1$ (thiamin hydrochloride), folic acid, biotin, vitamin B$_{12}$	manganese cyanate, zinc oxide, chromium citrate, potassium iodide Calcium phosphate, magnesium carbonate, zinc gluconate, iron (ferrous fumarate), copper gluconate, chromium aspartate	Natural flavors
Clif Bar (Clif Bar) Flavor: lemon poppyseed	64:16:20	Organic brown rice syrup, ClifPro (soy rice crisps, rice flour, malt extract, organic soy flour, organic roasted soybeans), organic rolled oats, Clif Crunch (apple fiber, oat fiber), soy white chocolate (organic evaporated cane juice, soy flour), organic evaporated cane juice, lemon puree from concentrate, glucose, pectin, fig paste	Soy protein isolate	Soy, seeds, may contain traces of peanuts and other nuts; flaxseed, cocoa butter, organic soy butter	Ascorbic acid (vitamin C), tocopherol acetate (vitamin E), beta carotene (vitamin A), phytonadione (vitamin K$_1$), biotin, niacinamide (vitamin B$_3$), calcium pantothenate, thiamin, cyanocobalamin, folic acid (vitamin B$_9$), riboflavin (vitamin B$_2$), pyridoxine hydrochloride (vitamin B$_6$)	Dicalcium phosphate, magnesium oxide, ferric orthophosphate (iron), zinc citrate, calcium pantothenate, potassium iodide, manganese gluconate, copper gluconate, sodium selenite, sodium molybdate	Natural flavors, lemon pieces, natural flavor, colored with annatto and turmeric

^{a}Based on labeling information.
^{b}Rounding may lead to minor inaccuracies.

21% of advertised products at that time had appropriate scientific information to support their claims.[2] This is not to say that these substances should be deemed worthless, only that they have not been properly tested to prove their efficacy. Researchers are attempting to test many of these substances; however, research funding is often lacking. Research into sport supplements is not considered high priority for funding by government agencies as they are not generally perceived as a benefit to human health or prevention of disease. Therefore many exercise nutrition laboratories work within limited budgets.

Because nutrition supplements do not require FDA approval, research identifying the risk of toxicity and the potential for adverse side effects is often lacking. For instance, episodes of sudden cardiac death have been attributed to ma huang (ephedra) use, and caution has been urged by researchers and health professionals.[3,4] Some may argue that these reports were individual case studies and not cause-and-effect trials, so it could be simply coincidence that the individuals who died were taking the supplement. Yet the growing number of medical reports associating ephedrine with fatalities seems to invalidate the possibility of mere coincidence.

Also, because the content of nutrition supplements is not routinely monitored, purity is a concern. Simply stated, a supplement might not contain exactly what it is supposed to contain. One reason could be that growing demand for supplements has, at times, exceeded the availability of ingredients. In response, some disreputable manufacturers might cheat the consumer by providing less of the active ingredient than stated on the label.

Key Point

Despite the large-scale presence of nutrition supplements, many of these products are marketed without significant research supportive of their benefit and/or safety.

Consumer Savvy in Purchasing Supplements

Caveat emptor, loosely translated as "let the buyer beware," certainly applies to sport supplements and sport foods. Like the marketing of so many other products, sport food and supplement manufacturers sell an image, goal, or dream. The sport supplement market also provides an excellent example of overly creative and at times fraudulent marketing, hinging on clever medical vocabulary and rhetoric, the impressive physiques of models and the testimonials of naturally gifted and highly trained athletes, and at times questionable or even fabricated research. This is not to say that all products falling into this broad area are bogus. However, the current consensus of sport researchers is that for a well-nourished athlete, most sport supplements have failed to make an appreciable difference

in physical status and performance in a consistent and predictable manner.

INFORMATION RESOURCES.　The consumer needs to be wary of misleading advertising and clever product names. For instance, in the previously mentioned review of sport products, 41% of the assessed products were determined to have no credible or applicable scientific backing of their claims.[2] Although 32% of the products did have some scientific backing of the purported effect, the claims were made in a misleading manner. The Internet has provided a huge opportunity for manufacturers to provide information about their products. A supplement company's website might offer testimonials and hyperlinks to information from research conducted in their "laboratory" or by their "scientists." It is important to keep in mind that testimonials are not research and that "research findings" cannot be taken seriously unless they are published in an appropriate peer-reviewed journal.

Key Point

Consumers must be savvy when considering the use of a sport nutrition supplement. Labeling and product information are often misleading.

■ SPORT FOODS AND SUPPLEMENTS: CONCEPTS *IN REVIEW*

- The sport food and supplement industry generates several billion dollars in sales annually.
- Sport foods tend to be more foodlike in that they provide energy and many other substances, whereas supplements contain one or just a couple of nutrients with a specific purpose (such as antioxidation or energy metabolism).
- Sport food and supplement manufacturers must follow the same guidelines as for other foods when developing product labels.
- The supplement industry must follow the labeling and marketing guidelines set in DSHEA.

■ SPORT FOODS: BARS, DRINKS, AND GELS

Sport foods include bars, drinks, and gels. The most salient feature of these products is that they provide energy nutrients for the purpose of providing energy. Some of these products are consumed before and during exercise, and others are meant to serve as partial or full meal replacements. This section provides an overview of some foundational concepts related to sport foods including composition, common use, and research into the capabilities of these products to provide nutrition and performance benefits.

Sport Bars and Shakes

Sport bars represent one of the fastest growing areas of the sport food industry. These products provide energy and other nutrients and their composition can vary tremendously based on the intended consumer and purpose of the sport food. For instance, some sport bars are marketed for their energy content, to support physical activity. Others are designed and marketed to be meal replacements or high protein sources. The energy nutrient formulation for some of the more popular sport bars is presented in Table 11-1.

Sport shakes usually contain the same ingredients as sport bars, so that sport bars could be viewed as dry shakes. Generally they are a formulation of carbohydrate, protein, fat, vitamins, minerals, and other nutrients and additives. Sport shakes come in powder or **ready-to-drink (RTD)** form and are purchased by athletes and active people as well as the general public for use as snacks, meal replacements, and exercise recovery drinks. Shakes can be made at home using common ingredients, and the feature Nutrition in Practice: How to Make a Sport Shake compares the nutritional composition of two home recipes to a commercial shake powder.

ENERGY COMPOSITION AND FIBER.   Carbohydrate is the energy foundation of many sport bars and shakes that are to be consumed before and after exercise. Corn syrup and high-fructose corn syrup (HFCS) are common carbohydrate ingredients and listed as the first ingredient for several sport bars (Table 11-1). Both corn syrup and HFCS are based on partially digested cornstarch; HFCS also contains fructose. Other carbohydrate ingredients include fruit juice concentrates (such as grape and pear juice concentrate), dried fruit (such as dried apple), oat bran, brown rice, and rice crisps. Many sport bars also contain fiber. Besides promoting health, the fiber content of sport bars could have athletic application, in that fiber could slow down the digestive process by slowing the rate of

NUTRITION IN PRACTICE   **How to Make a Sport Shake**

Sport shakes have become common components of the diet of athletes and people in general. They offer the opportunity to consume specific ratios and quantities of energy nutrients in a convenient manner. For instance, a packet of supplement powder is easily dumped in a glass or blender and mixed with water, milk, and other foods. Making a sport shake at home is simple and requires only a blender. The recipe ingredients could include foods that are known providers of specific energy nutrients or a combination of more than one energy nutrient. Egg substitutes such as Egg Beaters can serve as a potent protein resource for a shake, and nonfat dry milk provides carbohydrate plus protein. Flavorings can be added such as chocolate syrup, frozen yogurt, fruit chunks, jams and juices, and sugar substitutes. Table A shows the nutritional comparison of one supplement powder, MET-Rx Original Vanilla, and two homemade recipes. The protein in the homemade shakes is derived primarily from egg substitute. A vitamin/mineral supplement can be ingested along with the homemade sport shake to provide the vitamins and minerals in the commercial recipe.

Homemade Shake 1
3/4 c nonfat dry milk
8 oz egg substitute

1 tbs chocolate syrup
16 oz water (as ice or
   ice and water)*

Homemade Shake 2
10 oz egg substitute
1 med. scoop (4 oz or
   115 g) of nonfat

chocolate frozen yogurt
16 oz water (as ice or ice
   and water)*

---

*1 ice cube = 3/4 c fluid oz of water

Table A

Basic Nutritional Composition of a Commercial and Two Homemade Sport Shakes

Component	MET-Rx Original Vanilla	Homemade Power Shake 1	Homemade Power Shake 2
Energy	250 kcal	251 kcal	280 kcal
Protein	37 g	32 g	36 g
Carbohydrate	22 g	28 g	29 g
Fat	2 g	<1 g	2 g
Sodium	370 mg	780 mg	690 mg
Potassium	900 mg	650 mg	780 mg

emptying of the stomach. This would allow a slower and more even absorption of the carbohydrate and produce a lower and longer insulin response (glycemic response). On the other hand, fiber consumed just before and during exercise might be problematic for individuals prone to intestinal discomfort.

The protein component of sport bars is largely based on proteins isolated from milk and/or egg whites because of their higher biological value (see Chapter 5). Amino acids such as glutamine and BCAAs are often added to create a more desirable composition. Manufacturers of sport bars often trademark (™) their protein/amino acid source as a proprietary blend. Fat contributes energy, flavor and sensory aspects to sport bars although it is not the focus among the energy nutrient ingredients.

The energy nutrient ratios vary among sport bars depending on their purpose. Some sport bars have a higher ratio of carbohydrate to protein, providing a concentrated carbohydrate source in preparation for and recovery from exercise. Some sport bars (such as Power Bar) derive more than 60% of their energy content from carbohydrate, leading to a carbohydrate-to-protein ratio of 4:1 or more. In some other sport bars, protein accounts for more than half the energy and carbohydrate for about 20%. These bars are not designed to be an energy source during exercise but a protein resource for muscle protein synthesis.

Key Point

Manufacturers of sport bars and shakes often trademark the protein/amino acid source (proprietary blend).

VITAMINS, MINERALS, AND OTHER INGREDIENTS. Vitamins and minerals are typically added to sport bars, especially those vitamins directly involved in energy metabolism (such as B-complex vitamins, magnesium, iron, zinc, and copper). Adding nutrients such as vitamins C and E, copper, iron, zinc, manganese, lipoic acid, and glutathione often reflect an attempt to optimize antioxidant status (see In Focus: Antioxidant Supplementation by Athletes in Chapter 9). In addition, some sport bars contain substances that are common individual supplements such as creatine, carnitine, and HMB.

Key Point

Sport shakes can be made at home using ingredients such as nonfat dry milk, an egg substitute such as Egg Beaters, and skim milk.

Sport Drinks

Sport drinks are popular among a broad range of athletes. Researchers have demonstrated that carbohydrate-containing sport drinks can enhance performance during endurance and intermittent higher intensity sports (such as ice hockey) and may also benefit serious weightlifters. Sport drinks may be broken into two categories: (1) the classic **fluid and electrolyte replacement (FER)** sport drink in which the carbohydrate content is relatively low, which is more appropriate during exercise; and (2) a higher carbohydrate formulation that is more appropriate for consumption after training sessions and in preparation for upcoming competition. The second type is sometimes referred to as recovery or loading beverages. The composition of two popular sport drinks is presented in Table 11-2 and described below.

CARBOHYDRATE.   Carbohydrate is typically provided as glucose, sucrose, fructose, corn syrup, maltodextrins, and glucose polymers. These carbohydrates usually make up about 4–8% of an FER sport drink and >10% of a postexercise sport drink. One of the most important considerations with regard to carbohydrate percentage is how it influences the rate of gastric emptying. As solutions exceed 8% carbohydrate, the rate of gastric emptying begins to slow. FER sport drinks are more appropriate for endurance events such as distance running and cycling as well as intermittent sports such as ice hockey, tennis, lacrosse, field hockey, and soccer. The carbohydrate-concentrated loading and recovery sport drink would be more appropriate during recovery from strenuous resistance training and endurance events or perhaps in

Table 11-2

Comparison of Energy and Carbohydrate Content of Gatorade and Energy Drink (by Gatorade)

Nutrition Facts (240 ml or 8 oz)	Fluid and Electrolyte Replacement Drink (Gatorade)	High-Carbohydrate Drink (Energy Drink)
Total energy	50 kcal	210 kcal
Total carbohydrate	14 g	52 g
Sugars	14 g	28 g
Other carbohydrate	0 g	24 g
Protein	0 g	0 g
Fat	0 g	0 g
Sodium	110 mg	135 mg
Potassium	30 mg	70 mg

endurance event preparation (such as glycogen loading). As displayed in Table 11-2, these beverages might contain approximately 50 g of carbohydrate in an 8-oz (240-ml) serving, which is three to five times more concentrated than in an FER drink.

ELECTROLYTES AND OTHER NUTRIENTS.   The primary purpose of FER sport drinks is to replace water lost in sweat and to provide carbohydrate. FER drinks also provide sodium and chloride, which are the primary electrolytes in sweat. As discussed in Chapter 8, the greater the rate of sweating, the higher the concentration of sodium and chloride in sweat lost from the body. Although relatively little potassium is found in sweat, potassium is often added to sport drinks. As discussed in Chapters 8 and 10, potassium is in greater concentration in intra- than extracellular fluid, so its concentration in sweat tends to be lower than sodium and chloride. However, as for sodium and chloride, higher sweat rates do result in increased loss of potassium. Sport drinks may include other ingredients such as phosphorus, chromium, calcium, magnesium, iron, caffeine, and certain vitamins. The addition of most of these substances has not been proven to acutely improve performance.

Key Point

Sport drinks include the classic fluid and electrolyte replacement (FER) and recovery/loading beverages.

Sport Drink Recommendations

General recommendations for FER during exercise are 6–12 ounces (150–350 ml) every 15 minutes as tolerated.[5] More specific recommendations for FER consumption during training and competition depend on the intensity and duration of a sport and the athlete's size, level of training, and rate of sweating. The exercise intensity and level of training dictate the rate of carbohydrate utilization and thus when and how much carbohydrate is needed to maintain a high level of performance. Sweat loss is the primary determining factor in determining FER compensation. For instance, some sports and the environment in which they are performed are associated with heavy sweating (>2 L/hour), and others are associated with lower sweat rates. Every competitive athlete should have an understanding of his or her sweat rate during training sessions and competitions. The Nutrition in Practice: Estimating Sweat Loss in Chapter 8 describes a method of determining sweat volume and rate and can serve as one of the most accurate methods of estimating an athlete's fluid requirements during training and competition.

RECOMMENDATIONS FOR FER BASED ON SIZE OF AN ATHLETE.   Recommendations for sport drink consumption vary by athlete size. For instance, about 30–60 g of carbohydrate an hour is probably sufficient for an endurance athlete of average size (68 kg). This amount can be met by ingesting a 4–8% carbohydrate-containing solution at about 20–40 oz/hour (625–1250 ml/hour). Recommendations can be generated for larger and smaller endurance athletes by dividing their body weight in kilograms by 68 and then multiplying the result by the fluid recommendations for the 68-kg athlete. For instance, recommendations for a 110-lb (50-kg) female cyclist would be 460–920 ml of a 4–8% carbohydrate-containing beverage per hour of exercise.[6]

Examples:

$$50\text{-kg runner} \div 68 = 73.5 \times (625\text{–}1250 \text{ ml})$$
$$= 460\text{–}920 \text{ ml/hour}$$
$$75\text{-kg runner} \div 68 = 1.10 \times (625\text{–}1250 \text{ ml})$$
$$= 690\text{–}1375 \text{ ml/hour}$$
$$90\text{-kg runner} \div 68 = 1.32 \times (625\text{–}1250 \text{ ml})$$
$$= 825\text{–}1650 \text{ ml/hour}$$

Key Point

The intensity of exercise and the athlete's level of training, rate of sweat, and size are the primary factors determining the need for a sport drink.

Sport Drink Tolerance and Absorption

Several factors may influence the rate of absorption of sport drink ingredients and include the temperature and composition (osmolality and energy) of a sport drink. Sport-related factors such as bouncing of the torso in running might be a consideration as well. With regard to temperature, cooler solutions (5–15°C) may empty from the stomach more quickly than warmer or hot solutions.[6] In addition, cooler drinks are more enjoyable and perhaps promote greater consumption. Sport drinks are often kept in coolers and poured into cups for athletes to consume during endurance sporting events and on warmer days for sports such as soccer and football. However, it is important to keep in mind that melting ice in a cooler dilutes the sport drink.

Osmolality and energy content can influence stomach emptying. As the particle concentration within the stomach and small intestine exceed that in the extracellular fluid (~280–300 mOsm), water is drawn in by osmotic force. This increases the volume and slows the rate of emptying from the stomach and subsequent absorption. HFCS is considered a desirable carbohydrate ingredient because it has a lower osmolality than sucrose or glucose. In addition, a concentration of more than 8% carbohydrate in the stomach can slow down gastric emptying. This may be due to the release of the hormone cholecystokinin (CCK), which slows stomach emptying.

Sport researchers have long pondered whether exercise intensity might slow stomach emptying. For very high intensity exercise it probably does not matter because of the short duration. Athletes performing at higher intensities for a relatively short time (such as a 10-km run) might benefit more from consuming sport drink prior to competition or a training session. Mild to moderate submaximal exercise intensities ($\leq 70\%$ VO_2max) do not hinder gastric emptying even when the bout lasts several hours.

Sports involving greater torso movement (running) or a horizontal position (swimming) might make one more prone to **gastroesophageal reflux (GER),** the movement of stomach contents into the esophagus, during exercise. Researchers have looked into this matter by providing male triathletes different fluids (water and 7% carbohydrate solution) and assessing for GER symptoms during bouts of running and cycling.[7] GER was higher during bouts of running than cycling and occurred more frequently in both runners and cyclists who were provided the carbohydrate solution than in those given tap water. Thus the mode of exercise and type of beverage consumed seem to influence the occurrence of GER during exercise.

Key Point

Sport drinks are most rapidly absorbed if they are cool and have a low osmolality. Some gastroesophageal reflux can occur in sports involving more torso bouncing.

Sport Gels and Glucose Tablets

Sport gels and glucose tablets provide a readily available energy source in easy-to-carry packets and tubes. They can be carried in hand or tucked into running shorts, cycling shirt pockets, or a backpack (such as a camelback). For runners who have water stops (fountains) strategically planned throughout their distance training course, these products can be convenient. A packet of sport gel tends to provide around 100 kcal from carbohydrate. The primary carbohydrate ingredients are often maltodextrin and some fructose. Like many sport foods, gels and related products may venture beyond the simple provision of energy. For instance, some products include BCAAs, antioxidant vitamins, potential stimulants, and other substances such as caffeine, kola nut extract, and ginseng. Besides convenience, they offer an alternative fuel resource for people who might not like sport drinks. There is no physiological advantage to providing the same carbohydrate via a gel or tablet versus a sport drink. However, these carbohydrate-dense items need to be ingested in conjunction with an appropriate amount of water.

■ SPORT FOODS: BARS, DRINKS, AND GELS *IN REVIEW*

- Sport bars differ in energy content and ratio of energy nutrients depending on their purpose.
- The two types of sport drinks are fluid and electrolyte replacements (FER) and higher carbohydrate solutions to be consumed after intense training sessions or perhaps used in preparation for endurance events.
- The primary factors influencing the absorption of nutrients from sport drinks include the temperature and composition of the solution.
- Sport gels and glucose tablets can be a convenient source of carbohydrate to some athletes during exercise.

■ SPORT SUPPLEMENTS

Hundreds of different supplements are found in the sport nutrition market. Many of these substances are common nutrients or their derivatives, whereas others may seem more exotic. Sport supplements are designed and marketed as a means of enhancing acute performance; increasing size, lean body mass (LBM), and strength; decreasing body fat; and possibly enhancing the healing process related to sport injuries (see In Focus: Sport Injury and Nutrition Supplementation in this chapter). The amount of research on these substances varies tremendously. For instance, creatine is considered fairly well researched as a potential ergogenic aid, whereas ribose and glutamine are scantily researched despite their popularity.

Arginine, Ornithine, and Lysine

Lysine is an essential amino acid and arginine is considered semi-essential because it may become essential during periods of growth. Ornithine is a nonessential amino acid not found in proteins but important for efficient nitrogen removal in the urea cycle. Ornithine α-ketoglutarate (OKG) is ornithine bound with α-ketoglutarate (a Krebs cycle intermediate), which is supposed to enhance the efficiency of absorption. These amino acids can be purchased as individual supplements or in a combination sometimes marketed as "natural growth hormone."

As depicted in Figure 11-3, supplementation with arginine, lysine, and/or ornithine (or OKG) has been proposed to enhance growth hormone levels in circulation, leading to greater muscle mass development.[8] The original claims were based on clinical studies involving individuals who had sustained significant burns. For instance, when large doses of ornithine (as ornithine α-ketoglutarate) were ingested (20–30 g/day) or infused directly into the

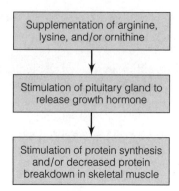

Figure 11-3
Proposed mechanism for supplementation of arginine, lysine, and ornithine on muscle mass development.

blood of burn patients, the level of circulating growth hormone was noted to increase.[9–11] In addition, burn patients established a positive nitrogen balance more quickly in the days that followed treatment,[9] and indicators of protein breakdown were also reduced during recovery.[10]

After the studies in which these amino acids were directly infused, researchers attempted to determine if oral supplementation of these amino acids could also raise growth hormone levels. In one study the combined daily supplementation of arginine and lysine (1.2 g each) increased the level of growth hormone, but the individual supplements did not.[12] Another study also suggested that a mixed supplement of arginine and lysine (1.5 g each) could raise growth hormone levels at rest.[13] However, in a study involving both younger and older men, arginine (5 g) failed to increase growth hormone levels for either age group.[14] Also, when competitive weightlifters were provided 2 g of arginine, ornithine, and lysine, growth hormone levels were not increased.[15] In a related study, daily supplementation with both arginine and lysine (2.4 g) or ornithine and tyrosine (1.85 g) failed to increase circulating growth hormone levels in male bodybuilders.[16] In yet another study, when arginine and glutamate were provided to highly trained cyclists, growth hormone levels did not differ from a placebo group at rest.[17]

As discussed in Chapter 12, strenuous exercise training can lead to a transient rise in growth hormone levels. So, one question that researchers have attempted to address is whether supplementation of these amino acids would enhance the increase in growth hormone in response to exercise training. At this time it does not seem so. For instance, even though growth hormone levels were raised in response to arginine and lysine supplementation (1.5 g each) in one of the studies mentioned above, when the supplement was instead taken prior to resistance training, growth hormone levels were not raised above the increase in growth hormone brought about by the exercise alone.[13] Furthermore, in another study, supplementation

was actually noted to hamper the typical elevation of growth hormone in response to a training ride.[14]

The research involving supplementation of arginine, lysine, and ornithine is somewhat contradictory. Although some studies report raised levels of growth hormone, more indicate that growth hormone levels are not increased. In addition, one study noting a positive effect of supplementation also noted that supplementation prior to exercise did not enhance the natural increase in growth hormone in response to the exercise bout.[13] The results of another research study suggested that supplementation may actually decrease the exercise-stimulated rise in growth hormone.[14] An important consideration regarding the efficacy of these supplements is that even in the studies that noted an increase in growth hormone, the research protocols did not extend to assessing changes in lean body mass, measures of strength, or insulin-like growth factor-1 (IGF-1). IGF-1 was noted to be unchanged in one study that failed to demonstrate a rise in growth hormone.[18] Finally, the research efforts involving burn patients provided these amino acids intravenously and at higher dosages. Oral ingestion of arginine, lysine, and ornithine requires that they must perfuse the liver before entering the general circulation. During this first pass the liver can aggressively remove all three of these amino acids. Therefore, the quantity of supplementation might be an extremely important factor. At this time supplementation of arginine, lysine, and ornithine to raise growth hormone levels is not supported.

Key Point

Arginine, ornithine, and lysine are marketed as a means of increasing growth hormone levels, but research so far has not supported this notion.

Aspartic Acid

Aspartic acid is a nonessential amino acid found in proteins and is also important in the formation of urea. As discussed in Chapter 5, the formation of urea serves as a mechanism for ridding the body of ammonia, a potentially toxic nitrogen-based molecule produced during the breakdown of amino acids. It is known that the concentration of ammonia in circulation tends to increase during prolonged strenuous exercise and has been associated with fatigue. Therefore it has been proposed that aspartic acid supplementation might enhance performance in certain activities by alleviating ammonia-related premature fatigue.

Aspartic acid supplementation has also been suggested as a possible means of slowing muscle glycogen depletion during endurance efforts. The idea is that aspartic acid could serve as a source of oxaloacetate (OAA), as shown in Figure 11-4. OAA is a key intermediate of the

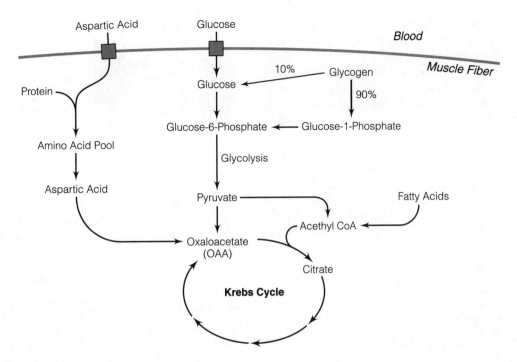

Figure 11-4
Proposed mechanism for aspartic acid to reduce glycogen breakdown by reducing the requirement for pyruvate-derived OAA. Pyruvate would be mostly derived from glycolysis.

Krebs cycle and combines with acetyl CoA derived from fatty acids and other fuel sources to form citrate. This would reduce the need for OAA derived from glycolysis, and thus the need for glycogen, during strenuous exercise. The result would be an enhancement in endurance performance by extending the time to fatigue associated with glycogen depletion.

In general, researchers have failed to demonstrate that aspartic acid supplementation is beneficial to performance. For instance, when 6 g of potassium-magnesium aspartate was provided to young males for a day prior to cycling at 75% VO_2max, there was not a positive change in the participants' time until exhaustion.[19] Aspartic acid supplementation also did not positively influence plasma ammonia concentrations or other factors related to energy metabolism, namely glucose, lactate, and FFA levels. Other research studies also suggest that aspartic acid supplementation does not improve submaximal endurance time.[20,21] In one of these studies arginine/aspartate supplements (15 g) were provided daily to runners for 2 weeks prior to a marathon. Although the level of these amino acids increased in the blood of the runners, they failed to provide a metabolic benefit during the run.[22] Shifting the focus from endurance to resistance exercise, researchers reported that 24-hour supplementation of aspartic acid prior to high-intensity resistance training failed to lower ammonia concentrations during and after a training bout.[21] Therefore, most of the research so far suggests

that aspartic acid supplementation does not positively influence athletic performance. More studies that apply different designs are needed before a concrete conclusion is drawn.

Key Point

Aspartic acid is often marketed as means of inhibiting or slowing the onset of fatigue during prolonged higher intensity exercise; however, researchers have not shown this to be so.

Boron

Boron is an essential mineral involved in bone mineral metabolism, and it has been speculated that at least some of its influence on bone is related to steroid hormone metabolism. Also, one study demonstrated that boron supplements were able to raise testosterone levels.[23] On this basis, boron supplementation was thought to be a means of increasing testosterone production, which would lead to gains in muscle mass and strength. However, what is often overlooked is that the participants of that study were postmenopausal women who received 3 mg of boron daily after following a low-boron diet.[23] Serum estrogen levels were also raised by the boron supplements.

In studies involving athletic populations, boron supplementation has not been proven to increase testosterone levels. For instance, male bodybuilders who took 7.5 mg of boron daily for 7 weeks had no changes in measures of testosterone, lean body mass, and strength versus training alone.[24,25] Other research efforts have failed to show that supplemental boron increases testosterone levels.[26,27] Also, a few studies have noted that estrogen levels might be raised as a result of boron supplementation.[28,29] One study that found a nonsignificant trend of increased plasma testosterone levels in response to boron supplementation also found significant increases in the level of plasma 17-β estradiol (estrogen) and in the ratio of estrogen to testosterone.[28] On the basis of these and other studies, boron supplementation does not appear to enhance athletic performance and cannot be advocated.

Key Point

Boron is often marketed as a means of increasing testosterone levels; however, researchers have not demonstrated this in target populations.

Branched Chain Amino Acids (BCAAs)

The branched chain amino acids are leucine, isoleucine, and valine, all of which are dietarily essential. They are found in proteins and are among the few amino acids that can be utilized for energy by skeletal muscle during endurance activity and extended fasting. They also appear to compete with tryptophan for transport across the blood-brain barrier. BCAA supplementation has been suggested as a mechanism for reducing central fatigue associated with endurance exercise.[6] This is based on the following concepts. First, as shown in Figure 11-5, BCAAs and tryptophan compete for the same transporter to cross the blood-brain barrier. Second, circulating BCAA levels decrease during endurance exercise. Third, tryptophan is used to make serotonin, which has been suggested to increase feelings of fatigue, and the rate-limiting step in the production of serotonin is tryptophan transport.[30,31] Therefore, supplementation of BCAAs before and/or during endurance exercise could minimize the amount of tryptophan that enters the brain during exercise, thereby inhibiting the onset of central fatigue.

So far most studies involving BCAA supplementation before or during endurance exercise suggest that supplementation does not improve or extend performance.[32–35] In one such study, cyclists were provided a 6% carbohydrate solution with either BCAAs or tryptophan during a cycling bout (70–75% maximal power output) to exhaustion.[32] Although consumption of the carbohydrate and BCAA solution increased plasma BCAA levels and reduced brain tryptophan uptake by 8–12%, time to exhaustion

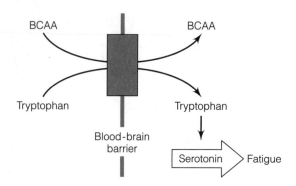

Figure 11-5

Tryptophan and the BCAAs compete for the same transport system to cross the blood-brain barrier. During prolonged exercise the level of circulating BCAAs decreases. Theoretically this could allow for a greater uptake of tryptophan and thus increased serotonin production in the brain, leading to fatigue.

during the exercise trial was not improved. In another study, BCAA supplementation failed to positively influence time until exhaustion of endurance-trained males cycling at 75% VO$_2$max.[33] In this study the cyclists began the bout with reduced muscle glycogen stores and were provided either flavored water or a 6% carbohydrate solution with or without BCAAs (7 g/L). Although the carbohydrate solution definitely extended the performance, the addition of the BCAAs did not further extend performance.

BCAA supplementation has been speculated to possibly limit glutamine reductions following an exhaustive exercise bout (such as a marathon or triathlon). Reduction in circulating glutamine has been associated with reduced immune system capabilities and increased incidence of infection following these strenuous exercise bouts. The research addressing the ability of BCAA supplementation to improve immune status and decrease the incidence of infection following exhaustive exercise bouts is preliminary, yet intriguing. For instance, BCAA supplementation has been noted to limit plasma glutamine reductions following an endurance bout (an Olympic distance triathlon) and to inhibit the reduction of interleukin-1 (IL-1) production that occurred in a nonsupplemented group.[36] The researchers also reported that those events were associated with decreased symptoms of infections.

Thus the notion that BCAA supplementation can improve endurance performance by extending time till fatigue does not seem to be strongly supported so far. The possibility that BCAA supplementation lessens the potential for illness after very strenuous exercise efforts requires further research. Additionally, newer roles for leucine (or BCAA) supplementation have been suggested, such as improving positive muscle protein turnover associated with exercise training, especially when coupled with insulin.[37,38] Researchers have also reported that BCAA supplementation reduced the level of plasma markers for muscle tissue

Key Point

The ability of BCAA supplementation to reduce central fatigue resulting from endurance exercise has not been proven.

damage in the days that followed a bout of endurance exercise.[39]

Caffeine

The term "caffeine" is used in a general sense to include caffeine and similar molecules such as theobromine and theophylline, as shown in Figure 11-6. Caffeine occurs naturally in plant sources such as coffee bean, tea leaf, kola nut, and cacao seed.[40] The average cup of coffee (6 oz) contains 50–150 mg of caffeine; tea has about 50 mg/cup, and caffeinated cola drinks have about 35 mg/12 oz. Caffeine promotes wakefulness and may also increase plasma epinephrine levels. Caffeine is believed to be potentially ergogenic for three reasons. First, caffeine and related substances stimulate the central nervous system (CNS) and possibly decrease perceived effort. This could increase mental alertness and focus on the exercise task at hand. Second, caffeine may have a direct effect on calcium transport and glycogen breakdown in muscle fiber, which may increase performance. Third, caffeine may stimulate the mobilization of fatty acids from adipose tissue, thereby increasing the availability of fuel to working muscle and sparing glycogen stores. As discussed below, caffeine is considered a doping agent by several sport governing bodies such as the International Olympic Committee (IOC). However, caffeine doping is defined as an unusually high level of intake, determined by finding a high urinary level of caffeine. This means that athletes can consume caffeinated beverages during training and before and during competition so long as the postcompetition urinary level is below a defined threshold.

A couple of early research studies strongly suggested that caffeine consumption could enhance performance in certain types of sports. For instance, when the athletes participating in one study ingested 330 mg of caffeine 1 hour before cycling at 80% VO_2max, their cycling time till exhaustion was extended by 14 minutes.[41] Another study conducted shortly thereafter found that 250 mg of caffeine ingested 1 hour before exercise and then every 15 minutes during the first $1\frac{1}{2}$ hours of an endurance bout could increase both work output and VO_2 by roughly 7%.[42] In addition, these studies suggested that fat utilization could be increased by approximately 30% as a result of caffeine ingestion. In the years that followed, some studies supported the earlier findings while others did not. It now appears that caffeine can have an ergogenic effect under some conditions and that individuals vary greatly in how they are affected.[43,44]

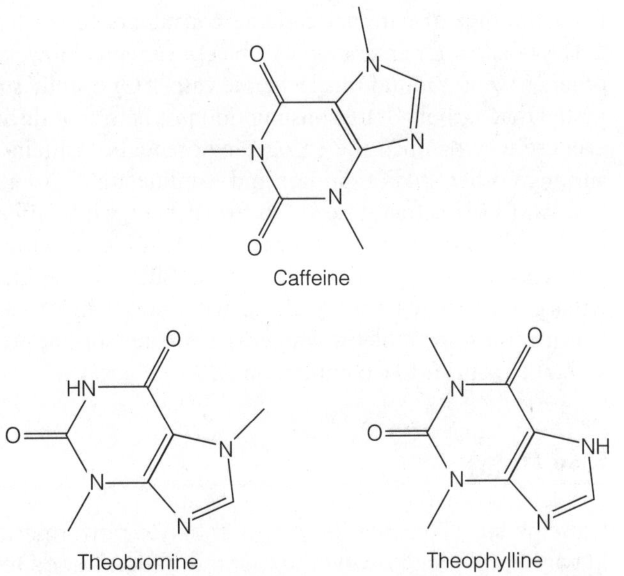

Figure 11-6
Caffeine and related molecules, theobromine and theophylline.

In several studies caffeine was provided at 3–13 mg/kg of body weight and yielded performance improvements of between 20 and 50% in elite and recreationally trained athletes running or cycling at 80–90% VO_2max.[45–48] Some athletes might also benefit from caffeine during even higher intensity bouts of exercise.[49–51] As an example, 150–200 mg of caffeine, as coffee, improved treadmill performance during a 1500-m trial by 4.2 seconds in well-trained runners.[51] Whether or not caffeine can improve sprinting efforts and power output is less clear and requires further assessment. For instance, two studies reported that caffeine consumption failed to improve maximal power output, muscular endurance, or total work in shorter tests.[50,52] On the other hand, other studies suggest that caffeine (250 mg) may increase maximum power output[53] and sprint swimming performance.[54]

In conclusion, caffeine may be ergogenic, but the results may be individualized and conditions seem to exist. Caffeine intake between 3 and 6 mg/kg of body weight did not result in urinary caffeine levels above the IOC acceptable limit and side effects were uncommon. With caffeine intake of ≥9 mg/kg of body weight, athletes can present urinary caffeine levels above the IOC acceptable levels for doping. Side effects such as dizziness, insomnia, headache, and gastrointestinal distress have also been noted at these higher doses.

Chronic caffeine consumption and other dietary factors may influence the ergogenic efficacy of supplementation. For instance, regular lower doses of caffeine intake (<9 mg/kg of body weight) may dampen the effect of caffeine on CNS stimulation and fat metabolism.[55] Some researchers have speculated that caffeine should be withdrawn from the diet of habitual caffeine consumers for at least 4 days to allow for an ergogenic effect of caffeine. However, others have not found this to be the case.[56] One study suggested that carbohydrate consumption just before or during exercise may dampen the ergogenic potential of caffeine,[57] although other studies did not find it influential.[58] An athlete who uses caffeine needs to experiment with caffeine levels and timing to judge how it influences his or her own performance. The potential side effects (such as sleepiness, lethargy) associated with caffeine withdrawal might be a consideration for athletes who cease caffeine consumption several days prior to a competition.

Key Point

Caffeine supplementation might enhance performance in some athletes; however, experimentation needs to occur during the off-season and athletes need to be aware of doping limits.

Carnitine

Carnitine is β-hydroxy-γ-trimethylammonium butyrate, which is found in a variety of tissues and is concentrated in muscle. Two amino acids, lysine and methionine, are used to make carnitine, and synthesis occurs in the liver and kidneys. As discussed in Chapter 6, carnitine transport mechanisms are necessary for movement of long chain fatty acids (LCFAs) into the mitochondrial matrix, the site of β-oxidation (see Figure 6-20). Shorter chain fatty acids are able to diffuse into the mitochondrial matrix without assistance.[40] Carnitine is often a component of "fat burner" supplements.

With regard to athletic performance, carnitine supplementation has been thought to potentially increase fat utilization, which could reduce the rate of glycogen loss and extend performance in endurance sports. Also, for sports with aesthetic considerations such as figure skating, gymnastics, and bodybuilding, enhanced fat utilization at rest and during exercise could enhance muscular appearance.

However, the results of nearly all of the experimental trials have not revealed a positive influence on either fatty acid utilization or glycogen sparing, nor does carnitine supplementation postpone the onset of fatigue. Furthermore, it probably does not enhance fat or weight loss or improve exercising VO$_2$max.[59–61] However, a couple of studies suggest some application of carnitine supplementation in exercise and sport. In one study, when moderately trained young men ingested 2 g of carnitine just prior to a maximal cycling bout there was an increase in both VO$_2$max and power output measurements.[62] In another study, resistance-trained men recovered more quickly from strenuous squatting exercise following carnitine supplementation (2 g daily for 3 weeks).[63] However,

since the bulk of the research so far does not support an ergogenic effect of carnitine it cannot be considered an effective independent supplement. Several questions remain, including whether changes in dosage level and timing might produce an effect and whether carnitine might support the actions of other supplements or pharmaceuticals, such as caffeine.[64]

Key Point

Nearly all of the research involving sport populations has failed to demonstrate that carnitine can enhance fatty acid utilization and improve performance or body composition. Future research should focus more on synergistic roles.

Choline

Choline, a derivative of the amino acid methionine, is a component of phosphatidylcholine (lecithin) and included in the list of vitamins (see Chapter 9). Choline has been suggested to be ergogenic for a few reasons. First, and probably most obvious, choline is a component of acetylcholine, the neurotransmitter in neuromuscular signaling. Perhaps increased acetylcholine levels in motor neurons would lead to more efficient stimulation of skeletal muscle activity. Second, as a component of phosphatidylcholine, choline supplementation may enhance the integrity of muscle fiber membranes and speed recovery and adaptive processes involving cell membranes.

Researchers have shown that plasma choline levels are decreased following an endurance bout of exercise[65] and supplementation seems to thwart this occurrence,[66] yet it does not improve endurance performance.[67,68] Therefore, it does not appear that choline supplementation has an acute impact on endurance exercise performance. However, choline in combination with carnitine and caffeine improved VO$_2$max and increased fatty acid utilization.[64] The test subjects were rats, and choline and carnitine were not tested independently in this study, so it is difficult to know what contribution was made by choline. Thus choline's role as a synergistic factor remains to be addressed by researchers.

Key Point

Choline is a component of acetylcholine and lecithin in cell membranes, and supplementation has not been proven to enhance acute performance.

Chromium

Chromium is an essential mineral and as a component of glucose tolerance factor (GTF), it is believed to be vital in the appropriate metabolism of energy nutrients. Chromium supplements are marketed largely as chromium picolinate as well as chromium nicotinate and chromium chloride. Picolinate is derived from tryptophan (an amino acid) and forms an ionic complex with chromium (Cr^{3+}). Picolinate and related substances are thought to chaperone chromium in the digestive tract for more efficient absorption. In doing so, these substances would decrease the potential interaction between chromium and other negatively charged substances in foods such as phytates.

Chromium supplementation is touted to increase metabolic rate and aid in weight reduction and increases in lean body mass. Chromium would accomplish these feats by enhancing the effects of insulin on muscle tissue. Insulin is considered to be generally anabolic by inhibiting protein breakdown and supporting protein synthesis. Such enhancement of muscle tissue could then lead to increased strength and power and increased energy expenditure and fat utilization at rest. Several well-designed research studies have been unable to demonstrate that chromium supplementation influences weight loss, LBM, or muscular strength. For instance, chromium supplements (3.3–3.5 μmol/day) provided to young men in combination with a weight training program did not result in greater changes in body composition and gains in strength versus the weight training alone.[69] In another study, football players received chromium picolinate (200 μg/day) for 9 weeks during a strength training program. Here again supplementation failed to independently influence skinfold measurements, percent body fat or lean body mass, and strength measurements.[70] Other well-designed research efforts have also reported that chromium supplementation failed to improve measures of strength or body composition when compared to a placebo trial.[71–74]

Chromium is probably not effective in increasing lean body mass and favorably altering body composition, or improving strength in either trained or untrained individuals, regardless of age. However, the potential impact of chromium supplementation in combination with other nutritional factors awaits thorough investigation. In addition, issues of toxicity still need to be addressed.[75–77] For athletes with insufficient chromium intake, supplementation may promote optimal insulin effectiveness and optimize carbohydrate and lipid metabolism by helping the individual to achieve "normal" chromium status. Therefore before a supplement study involving chromium or any other essential nutrient begins, participants need to demonstrate good status.

Key Point

Contrary to earlier hypotheses, chromium supplementation does not produce enhancements in LBM, strength, or performance.

Coenzyme Q₁₀ (Ubiquinone)

Coenzyme Q_{10} is often referred to as CoQ_{10} or ubiquinone. Ubiquinone may be found in all cells and is more concentrated within mitochondria. It is believed to exist attached to proteins of the electron transport chain as well as unattached and mobile. The free ubiquinone appears to function more as an antioxidant whereas the bound ubiquinone functions more as a component of the electron transport chain. As expected, more ubiquinone is found in Type I muscle fibers, and endurance training appears to increase its concentration in muscle.[78] Considering its functions, ubiquinone as a supplement might have two ergogenic effects. First, functioning as a component of the electron transport chain, ubiquinone might enhance aerobic energy metabolism in muscle. Second, functioning as an antioxidant, ubiquinone could help limit free radical stress during and perhaps after endurance exercise.[79]

Research supporting ubiquinone supplementation as a way to improve athletic performance is limited. In one study, Finnish top-level cross-country skiers who received 90 mg/day of supplemental ubiquinone had improved indices of physical performance including VO₂max.[80] However, another study that provided 100 mg/day of ubiquinone for 8 weeks to male cyclists failed to demonstrate improvements in cycling performance, VO₂max, or lipid peroxidation.[81] Likewise, when researchers provided 100 mg of ubiquinone in a daily supplement that also contained vitamin E (200 IU), inosine (100 mg), and cyto-chrome c (500 mg) for 4 weeks to triathletes, measures of endurance performance and metabolic parameters at exhaustion were unchanged.[82] Other research efforts have also failed to show that ubiquinone supplementation had a positive impact on aerobic performance or a desirable influence on indicators of energy metabolism in the plasma (such as lactate).[83–87]

Studies involving ubiquinone supplementation and antioxidant capacity have shown some positive results. For instance, when endurance athletes were provided 60 mg/day of ubiquinone in combination with vitamin E and vitamin C, the researchers reported improvements in their serum and LDL antioxidant potential.[88]

Key Point

Ubiquinone supplementation is touted to enhance aerobic energy metabolism as well as provide antioxidant protection.

So far the balance of research results does not support the notion that ubiquinone supplementation can improve endurance performance and favorably alter energy metabolism. The potential benefit of ubiquinone supplementation with regard to antioxidant activities requires more research. However, some scientists are not convinced that aggressive antioxidant supplementation is the best way to

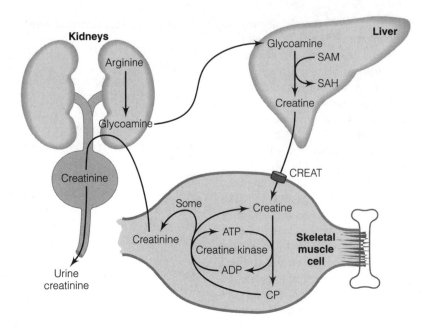

Figure 11-7
Production and transport of creatine into muscle cells.

enhance performance and adaptation processes, as some cellular activities and adaptation mechanisms are probably redux sensitive (see Chapter 9 In Focus: Antioxidant Supplementation by Athletes).

Creatine

Creatine is a molecule derived from arginine, glycine, and methionine and requires two organs (liver and kidneys) for production (Figure 11-7). As creatine circulates it is taken up by skeletal and cardiac muscle and other tissue including the brain by means of a creatine transport protein (CREAT).[89] Creatine is phosphorylated by creatine kinase in mitochondria to form creatine phosphate (CP), which then serves as a small high-energy phosphate reserve and a mechanism to rapidly regenerate ATP anaerobically (see Figure 2-15). Skeletal muscle may have a creatine content of about 3–5 g/kg. Therefore a 70-kg male might have a creatine pool of roughly 120 g or a little more than $\frac{1}{4}$ lb. Several exhaustive reviews of creatine metabolism are available.[90–94] A typical American diet might provide about 1–2 g of creatine daily, with meat eaters consuming more than vegetarians.

Creatine has been one of the most popular ergogenic supplements and is usually marketed in the form of creatine monohydrate. Creatine is touted to enhance muscle CP levels, thus increasing the anaerobic ATP regenerating potential of muscle fibers. This in turn would enhance quick-burst or high-power activity, such as sprinting and weight training. It is often recommended that creatine be ingested with carbohydrate for maximal tissue retention.

Some of the carbohydrate effect may be related to increased creatine transport into tissue in an insulin-related manner, although this is speculative. Initially, creatine "loading" followed by a "maintenance" intake were recommended; recommendations today vary. Loading doses approximate 20–25 g/day for 5–7 days, which is then followed by maintenance doses of roughly 3–10 g/day or more. Supplements are often broken up over the course of the day if a loading dose is used (for example, 5 g taken four times a day). Using lower doses of creatine (3–5 g/day) is more economical and has become an extremely popular protocol as an alternative to loading.

Initial recommendations for creatine loading and maintenance doses were based on earlier research efforts. In one study, the researchers provided 20 g of creatine daily for 6 days and observed a 20% rise in muscle creatine levels.[95] The rise in creatine levels could be maintained for 30 days by ingesting only 2 g per day. In addition, the researchers found that ingesting 3 g/day could also increase muscle creatine levels; however, this required a few weeks. Several other research efforts have also demonstrated increased levels of muscle creatine levels with supplementation.[96–98]

Individuals vary in their degree of change in muscle creatine level with supplementation. For instance, in one study five of the eight participants demonstrated a significant increase in muscle creatine (15–32%), yet the remaining three participants experienced a more modest increase of 5–7%.[98] The degree of change is believed to relate to the degree of performance improvement. Dietary creatine levels and internal production of creatine and

CREAT may account for some of the differences in creatine accumulation during supplementation. In addition, the effect of training on muscle fiber creatine content may partly explain why creatine supplementation has failed to produce an ergogenic response in some highly trained and elite athletes but has done so in others.[99–103]

Much of the research involving creatine supplementation and measures of strength and power has been positive, especially when utilizing the loading dosages.[91–94] For instance, one study revealed that creatine supplementation led to an extension of all-out treadmill running by 13% in sprinters.[100] In other studies, creatine supplementation increased measures of power output during repeated bouts of maximal isokinetic cycling[104] and greater knee extension torque production.[105] In addition, it led to a 6% increase in 1-RM free weight measurements[106] (1 repetition of maximum, or 1 RM, is the maximum amount of weight a person can lift) and static vertical jump (SVJ) power output and peak rate of force development for SVJ.[107]

Several studies have reported an increase in body mass and LBM and a decrease in fat mass in response to creatine supplementation.[108,109] The nature of these gains is not entirely clear. Certainly, elevated muscle creatine levels increase both LBM and total mass. Furthermore, increased intracellular (and extracellular) creatine exerts an osmotic effect. In muscle this would swell the muscle fibers and muscle tissue. Many researchers believe that creatine-induced swelling of skeletal muscle fibers leads to increased protein synthesis. Thus it is possible that osmotic swelling and increased overload have a synergistic influence on protein synthesis.

The research evidence gives more than adequate support for creatine supplementation as a means of potentially improving LBM, strength, and power. Key determinants of an individual's response (change in muscle creatine) to supplementation include diet and level of training. Well-trained athletes might not respond to supplementation or might experience less change than untrained individuals. Creatine supplementation is most applicable to sports involving sprinting and power aspects. For endurance athletes, supplementation may not improve endurance performance;[110,111] however, if a competition involves sprinting (such as cycling breakaways or final running sprints) creatine may be of benefit. However, athletes must also consider and monitor the potential weight gain that may accompany creatine supplementation.

Despite fairly clear professional recommendations, creatine users often listen more to anecdotal information in the gym, in magazines, or posted on websites. For instance, researchers noted a varied consumption among collegiate Division I athletes.[112] Personal admissions by some young men to the authors of this book revealed creatine supplementation of serious weightlifters at levels of ≥40 g/day. As creatine has become widely popular among athletes and noncompetitive weight trainers, issues of toxicity (either from creatine or contaminants) and effect on disease risk factors have been raised.[113,114] Although more research is needed in this area, studies of short duration (weeks) have not indicated potential harmful effects. For instance, creatine supplementation does not increase blood pressure or creatine kinase levels in the plasma (a marker for tissue change) nor does it affect performance of the kidneys.[109,115] Creatine supplementation (loading) did not influence heart size or performance in young men in one study.[116] In a retrospective assessment of long-term creatine supplementation, blood pressure and lipids did not differ between users and nonusers.[117] Another concern raised by some health professionals is that creatine loading may have been involved in the death of some athletes (such as wrestlers) during hypohydration efforts. Although enhanced creatine stores in muscle might have delayed fluid shifts from tissue to the blood, the primary problem is still dehydration.

Key Point

Creatine supplementation might enhance muscle creatine phosphate levels in some individuals and allow for enhanced power performance as well as increased LBM.

DHEA and Androstenedione

Dehydroepiandrosterone (DHEA), also known as prasterone, is a steroid hormone produced and released from the adrenal glands. DHEA can be sulfated (DHEAS), and both of these hormone forms circulate to tissues where they are converted into androgens (including testosterone) and/or estrogens (Figures 11-8 and 11-9). DHEA is derived from cholesterol and can be converted to androstenedione, which in turn can be converted to testosterone. DHEA circulates both independently and loosely bound to albumin and sex hormone–binding globulin (SHBG), whereas DHEAS circulates strongly bound to albumin. The ability of tissue to convert DHEA and androstenedione to testosterone and/or estrogens relies on the presence of steroidogenic and metabolizing enzyme systems. Both androstenedione and testosterone can be converted to estrogens via *aromatase* (Figure 11-9). Many tissues produce these converting enzymes including the gonads (testes and ovaries), liver, kidneys, and adipose and endometrial tissue. This allows the conversion of DHEA and androstenedione to the androgens and estrogens to be regulated at the tissue level. Skeletal muscle lacks the ability to convert androstenedione to testosterone.

Supplementation of DHEA and/or androstenedione has been marketed as a means of increasing testosterone

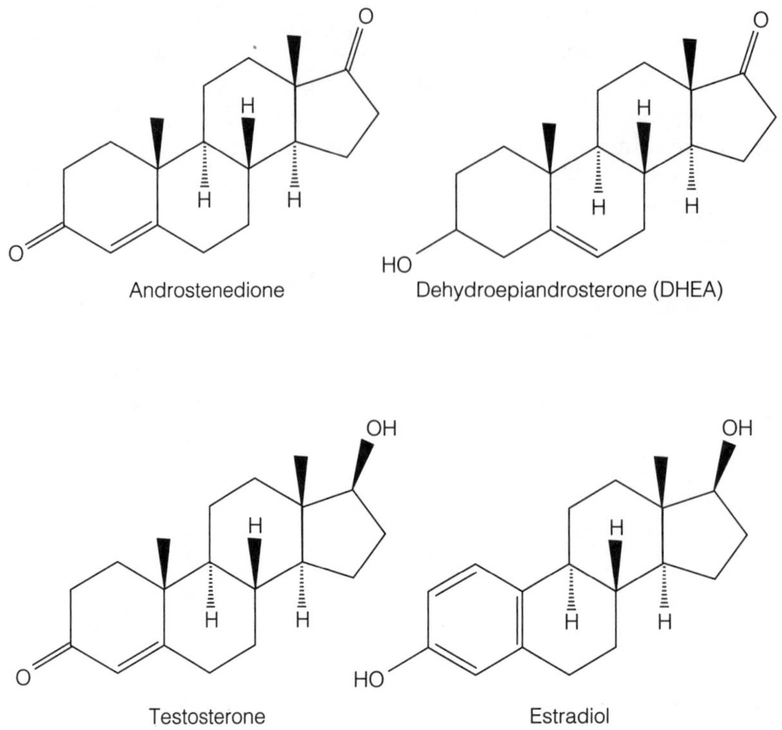

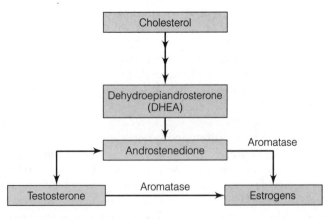

Figure 11-8
Molecular structures of androstenedione, DHEA, testosterone, and 17β-estradiol (an estrogen).

levels, which in turn could promote greater LBM, strength, and power. Researchers have demonstrated that androstenedione and DHEA supplementation can increase the level of androstenedione in the blood.[118,119] With regard to testosterone formation, the results of

Cholesterol

↓

Dehydroepiandrosterone (DHEA)

↓

Androstenedione — Aromatase →

Testosterone — Aromatase → Estrogens

Figure 11-9
Simplified reaction pathway for the production of popular steroid substances (DHEA and androstenedione) and derived hormones (testosterone and estrogens).

one study suggest that although supplementation of androstenedione at 100 mg did not alter serum testosterone levels, at 300 mg it elevated testosterone levels by 24%.[118] Therefore, if it occurs, the conversion of androstenedione to testosterone may be dose dependent. Another research team also reported that 100 mg of androstenedione (or DHEA) failed to influence testosterone levels or measures of LBM or strength.[120] Similar results were found when 150 mg of DHEA was provided to young men for 8 weeks.[121] Likewise, 5 days of androstenedione supplementation (100 mg) did not influence testosterone levels or measures of protein turnover; however, the level of estradiol was increased.[124]

One of the greatest concerns for androstenedione (and DHEA) supplementation for males is the conversion to estrogen. Even when testosterone levels were not altered in research involving lower levels of androstenedione supplementation (such as 100 mg/day), serum estrogen levels were noted to be increased.[118] Other studies have also suggested that androstenedione supplementation of greater than 200 mg can increase estrogen levels, either with or without concurrent positive influence on testosterone levels.[122] Conversion of androstenedione to estrogens may be related to body composition, as estrogens can be formed in adipose tissue. Therefore, special consideration should be applied to DHEA and androstenedione supplementation in obese men attempting to increase muscularity.[123] Furthermore, some supplement manufacturers advise that androstenedione should be complemented with the flavonoids chrysin and daidzein, purported to inhibit aromatization. However, a recent investigation revealed that this may not work.[124]

The use of DHEA and androstenedione supplementation to increase testosterone levels is cause for concern. First, DHEA and androstenedione do not appear to predictably elevate serum testosterone levels. This appears to be true in younger men as well as middle-aged men. Second, supplementation seems to elevate estrogen levels in males. The International Olympic Committee (IOC) and the National Collegiate Athletic Association (NCAA) currently ban DHEA and androstenedione from use in sports. Also, supplementation of these substances may influence the ratio of testosterone to epitestosterone (T/E), which is used to screen for steroid doping by organizations such as the IOC and NCAA.[125,126] Finally, a lowering of HDL-cholesterol levels has been noted in a few studies.[127,128] A low level of HDL-cholesterol is considered a risk factor for heart disease.

Key Point

DHEA and androstenedione do not effectively and predictably increase testosterone levels.

Glutamine

Glutamine is a nonessential amino acid and the most abundant amino acid in the body, including skeletal muscle and serum amino acid pools. It serves as an important fuel source for the intestines and carries nitrogen from skeletal muscle for subsequent removal from the body. Recently, glutamine has become extremely popular among weightlifters, as it is believed to increase protein synthesis and/or limit protein breakdown in skeletal muscle. Therefore, glutamine supplementation would (in theory) promote a positive protein balance in skeletal muscle in response to resistance training.[129] Also, as described above, the level of circulating glutamine in the plasma is known to decrease after a strenuous bout of exercise or as a cumulative effect during periods of frequent strenuous efforts.[36,130,131] This is believed to render an athlete more susceptible to microbial infection after isolated competitions (such as a marathon or triathlon) or a series of daily bouts (heavy daily training, cycling tours). Glutamine supplementation is believed by some to stabilize plasma glutamine levels during chronic training and to reduce sickness.

Research evidence that glutamine supplementation can enhance muscle protein gains in response to resistance training is lacking. For instance, glutamine supplementation (0.9 g/kg of LBM daily) failed to improve muscle protein turnover in young men training with weights.[132] Other researchers infused an amino acid mixture (with or without glutamine) directly into the blood of male and female participants for several hours while estimating the rate of muscle protein synthesis.[133] Although protein synthesis was enhanced by the amino acid infusion, there was not an additional enhancement with the mixture that included glutamine.

With regard to improving immune status or decreasing overtraining-related changes in the immune system, a few recent reviews have stated that the evidence supportive of glutamine supplementation is weak or at best equivocal.[134,135] Glutamine supplementation does seem to be effective in thwarting the reduction in plasma glutamine that typically develops during strenuous endurance activity.[130] However, this can occur in the absence of a positive influence on exercise-induced changes in the immune system.[130]

Key Point

Despite claims, glutamine supplementation has not been proven to effectively improve muscle protein production.

Glycerol

Glycerol is a small, simple molecule consisting of three carbons and three hydroxyl (OH) groups. Glycerol serves as the backbone for fat molecules and can be an energy resource during exercise and fasting. The use of glycerol for energy occurs largely through its conversion to glucose (via gluconeogenesis) by the liver (see Chapter 6). Glycerol has been touted as a possible ergogenic aid for two reasons. First, it is a substrate for gluconeogenesis and thus may become an important resource for glucose production during endurance bouts. Second, supplemented glycerol distributes evenly throughout body fluid and may provide an osmotic influence that could help an athlete hyperhydrate prior to exercise in a warmer environment. This would have special application to sports in which water intake typically fails to match water losses in sweat.

As discussed in Chapter 8, dehydration during exercise is known to reduce performance. Accordingly, protocols for pre-exercise water ingestion have been suggested in an attempt to induce hyperhydration and thwart the development of dehydration. However, the potentially beneficial hyperhydration effects of water consumption alone are often negated by the rapid diuretic effect. With this in mind, researchers began to seek an osmotically active substance to consume prior to exercise to enhance water retention. Glycerol has been suggested to have such an effect. Although not as mainstream as other supplements, glycerol is available commercially. Glycerol supplement manufacturers typically recommend around 1 g/kg of body weight with an additional 1.5 L of water taken 60–120 minutes prior to exercise.

While the first possibility mentioned above, a role in gluconeogenesis, has been generally dismissed because of the slow rate of glycerol conversion to glucose, the second ergogenic possibility has received some support via research findings. Researchers reported that glycerol-induced hyperhydration just before moderate exercise was more effective than water-induced hyperhydration in reducing the thermal stress associated with exercise in the heat.[136] The study provided 1 g of glycerol + 21.4 ml of H_2O per kilogram of body weight. In another study, researchers provided 1.2 g of glycerol + 26 ml of H_2O per kilogram of body weight to cyclists.[137] They noted that the glycerol supplementation extended cycling time till exhaustion (65% VO_2max) in a neutral laboratory environment, and that the athletes had a lower heart rate.

Based on the available research, it seems that glycerol-aided hyperhydration might have ergogenic potential, but more research is needed. Some researchers argue that if an athlete is able to maintain proper hydration status during exercise, hyperhydration does not provide a meaningful advantage.[138,139] Also, most of the research has involved an isolated exercise bout, which does not address questions about the effect of glycerol when several exercise bouts are performed over a day or two and sweating is copious. It seems that glycerol can indeed promote sustained hyperhydration states, which might be beneficial during training and competition of this nature.[140,141] However, because of the osmotic potential and the dosages involved, athletes wishing to try glycerol need to experiment with it during the off-season and during practices or training; some cramping, bloating, nausea, and headaches have been noted.[142]

Key Point

Glycerol may improve hyperhydration efforts before sports in which water loss greatly exceeds water intake; however, there is a potential for gastrointestinal discomfort.

HMB (β-Hydroxy-β-Methylbutyrate)

β-Hydroxy-β-methylbutyrate, commonly called HMB, is a metabolite of the essential amino acid leucine; its physiological significance remains to be clearly identified. HMB has become popular in the weight training community, as it is purported to increase net protein balance by decreasing muscle catabolism associated with heavy resistance training and to promote LBM augmentation. HMB is available to consumers as an independent supplement or as an ingredient of a combination supplement or sport food.

The number of studies to test the efficacy of HMB as an ergogenic substance is rapidly growing. The results are split between supporting and not supporting supplementation. In one supportive study, researchers provided 1.5 or 3 g of HMB daily to untrained males for 3 weeks during supervised strength training sessions lasting 1.5 hours and performed 3 days/week.[143] Compared to a placebo group, the HMB-supplemented group had lower levels of urinary 3-methyl histidine (3-MH) during the first 2 weeks of exercise and lower serum creatine kinase during the 3rd week. As discussed in Chapter 5, 3-MH is a marker of contractile protein breakdown, and creatine kinase is an indicator of muscle tissue damage. In addition, the amount of weight lifted was greater each week when the participants received HMB, suggesting a performance enhancement. In the second part of the study, participants received 3 g of HMB (or placebo) for 7 weeks in conjunction with weight training. Here, the HMB group was noted to have an increased fat-free mass at weeks 2, 4, 5, and 6 of the trial. In another study, 3 g of HMB was provide with or without creatine for 3 weeks and positive changes in strength and LBM were noted in either scenario.[144]

In another study, HMB supplementation at 3 g daily for 4 weeks was reported to increase upper body strength and minimize muscle damage when combined with an exercise program.[145] The subjects included both males and females of varied training status. In another study, HMB was provided to untrained college-age men at 38 or 76 mg/kg of body weight (roughly 3–6 g) daily while they undertook a resistance training program for 8 weeks.[146] Although changes in 1-RM were not reported between groups, the HMB-supplemented groups demonstrated greater increases in LBM and scored higher for peak isometric and various isokinetic torque measures. In addition, plasma creatine kinase levels were lower in the HMB-supplemented groups. Another study found a reduction in markers of muscle protein breakdown during resistance training.[147] Also, HMB supplementation was noted to possibly increase LBM and decrease fat mass in people over the age of 70 participating in a resistance training program.[148]

However, several other studies failed to find an ergogenic effect from HMB supplements. One study reported that both 3 and 6 g of HMB provided daily for 4 weeks to experienced weightlifters failed to reduce indicators of protein breakdown.[149] In addition, a positive influence on body composition and strength relative to a placebo trial were not found. A study involving collegiate football players taking 3 g of HMB daily whose strength training program was monitored by their strength and conditioning coach also failed to demonstrate a positive effect on strength (bench press, squats, and power cleans) and body composition.[150] A third study found that 6 weeks of HMB supplementation (3 g/day) did not influence changes in strength and body composition in response to resistance training in strength-trained athletes.[151] This study also suggested that the form (standard supplement capsule or time-release capsule) of HMB may not be a factor either. A fourth study also found no significant changes in body composition, including body fat and body weight. Also, short-term daily HMB supplementation of 40 mg/kg of body weight had no beneficial effect on a range of symptoms associated with eccentric muscle damage.[152]

The amount of published research involving HMB as an ergogenic aid is still limited but growing rapidly. So far it seems that any benefit from HMB supplementation would be related to strength and muscle mass development in untrained people, not already trained people and conditioned athletes. It also appears that supplementation of HMB at levels approximating 1.5–3 g/day are generally safe in the short term.[153,154]

Key Point

HMB, a derivative of leucine, may increase measures of strength and LBM in untrained people starting strength training, but not in already trained people.

Inosine

Inosine in the form of nucleotide inosine monophosphate (IMP) is used to make AMP and GMP, which in turn can be phosphorylated to the high-energy phosphate molecules ATP and GTP. IMP is made in many cells using ribose-5-phosphate via the pentose phosphate pathway and the amino acids aspartic acid, glutamine, and glycine with the assistance of folate.

Because ATP is derived from IMP, it seems logical that inosine supplementation might be able to enhance ATP levels in muscle cells. This would enhance maximal contractile activities, such as during weight training and sprinting, and increase performance and adaptation to exercise. Inosine is typically marketed for supplementation at 5–10 g/day.

In addition, inosine supplementation has been speculated to benefit performance by increasing the level of 2,3-diphosphoglycerate (2,3-DPG) in red blood cells (RBCs). Increased 2,3-DPG would shift the O_2-hemoglobin curve to the left and allow for more unloading of O_2 in muscle tissue, permitting greater aerobic ATP production and improved performance. This would especially be true for endurance sports performed around the lactate threshold.

In general, researchers have reported that inosine supplementation does not improve athletic performance.[155] For instance, researchers provided nine endurance athletes with 6 g of inosine (or placebo) for 2 days and measured their 3-mile run time, VO₂max, respiratory exchange ratio (RER), and rating of perceived exertion (RPE).[156] The inosine supplementation did not favorably alter any of the measures of metabolism or performance.

In a different study, male cyclists were provided with 5 g of inosine (or placebo) for 5 days before performing a 30-minute self-paced cycling bout at a constant load followed by a supramaximal sprint to fatigue.[157] The inosine supplementation failed to alter metabolic and performance parameters during the self-paced cycling bout, and the supramaximal cycling sprint time to fatigue was actually shorter than the placebo trial, suggestive of an ergolytic effect. Other research has also failed to determine a metabolic or performance benefit of inosine supplementation.[158]

Studies have also failed to demonstrate that inosine supplementation increases unloading of O_2 in muscle tissue by increasing 2,3-DPG in RBCs.[156–158] Furthermore, blood analysis revealed that uric acid levels may be increased due to inosine supplementation,[157,158] which raises concerns about the effect of chronic supplementation on health.

Thus it seems that inosine supplementation may not improve measures of aerobic performance and short-term power production and may actually have an ergolytic effect under some conditions.

Key Point

Inosine is often supplemented to increase ATP levels in muscle tissue; however, research to date does not indicate such an effect.

Medium Chain Triglyerides (MCT)

Medium chain triglycerides (MCTs) contain medium chain (length) fatty acids (MCFAs), which have 6 to 12 carbon atoms. Fat digestion releases these fatty acids, which can be absorbed directly into the portal vein, as they are water soluble. Thus MCFAs are available sooner after ingestion than long chain fatty acids because MCFAs bypass chylomicron formation (see Chapter 6). In addition, medium chain fatty acids may not require the carnitine transport system to cross the mitochondrial inner membrane like long chain fatty acids.

Because MCFAs can be absorbed and circulate dissolved in the plasma, they can be rapidly available to muscle. Furthermore, if transport into the mitochondria is a rate-limiting step in longer chain fatty acid oxidation, then MCFAs may enhance fat utilization by skipping this step. Increased fatty acid oxidation during endurance activity could in theory decrease the need for carbohydrate and thus slow glycogen depletion.

The results of some studies do indeed suggest that MCFAs derived from MCT supplementation prior to an endurance bout may serve as an energy source during endurance exercise, especially when consumed with carbohydrate.[159,160] Yet the contribution of pre-exercise MCT supplements to fuel used during exercise is still small.[161,162]

Even when fat oxidation is increased during exercise in glycogen-compromised athletes, MCTs still fail to make a large contribution to total energy expenditure (maximum 3–7%).[161] In addition, MCT supplementation (alone or with carbohydrate) has not consistently been shown to alter carbohydrate oxidation or the breakdown and oxidation of muscle glycogen.[163–165]

A few research efforts have attempted to determine whether MCT supplementation can influence performance. For instance, researchers provided a placebo, carbohydrate, or carbohydrate and MCTs to cyclists during a 100-km time trial ride.[163] The researchers reported that the carbohydrate enhanced performance in comparison to the placebo, but that the addition of MCT did not make a difference. In a related study, athletes cycled for 2 hours at 63% VO$_2$max and then performed a simulated 40-km time trial while consuming carbohydrate or carbohydrate plus MCT.[165] Here again the carbohydrate supplementation improved measures of performance, but the MCT did not provide an additive effect. Other research studies have also failed to determine an ergogenic effect of MCTs.[160] One review of the literature also concluded that there was very little convincing scientific evidence to recommend MCT use to improve exercise performance.[166]

At this time the use of MCT as an ergogenic aid is very questionable. Clearly, ingested MCT can be used for fuel during exercise, but whether that affects the oxidation of carbohydrate, more specifically muscle glycogen, is in doubt. One of the largest concerns for MCT supplementation is the potential for gastrointestinal discomfort, including diarrhea.[167] Also, its theoretical use would seem to be limited to ultra-endurance athlete populations.

Key Point

MCTs do not require the carnitine transport system for use in muscle and have been touted to enhance fat utilization during exercise; however, research has not supported this notion.

Ribose

Ribose is a five-carbon carbohydrate. Ribose as ribose-5-phosphate is made in many cells via the pentose phosphate pathway. Ribose-5-phosphate can then serve as a building block for adenosine, which itself is a building block for ATP. It has been speculated that ribose supplementation could lead to increased ATP levels in skeletal muscle and thus greater power production during very high intensity activities. This enhanced overload could in turn lead to greater muscular adaptations (hypertrophy).

Although research addressing the potential application of ribose supplementation to physical performance is limited, a couple of human studies and a few studies performed on rats provide some insight. Ribose was provided as a supplement four times a day at 4 g/dose for 6 days to young males and females who were engaged in a strength training program.[168] Measures of strength were determined before and after the supplementation and training period along with ATP recovery activity assessed using muscle biopsies. Ribose supplementation did not result in greater measures of strength or postexercise tissue ATP recovery. In another study, 36 hours of ribose supplementation resulted in an inconsistent improvement of measures of power during repeated sprint trials.[169] More research would be necessary to determine the influence of ribose supplementation on muscle ATP content and measures of force and power production. Ribose cosupplemented with inosine and/or adenine could be more beneficial but requires more research.

Key Point

Ribose supplementation is marketed as a means of enhancing ATP levels in muscle fiber or increasing postexercise ATP recovery; however, research supportive of these notions is lacking.

Vanadium

Vanadium is a minor mineral and is believed to be nutritionally essential. It is believed to be involved in insulin metabolism and/or to function in some capacity that mimics the activity of insulin. Vanadium is typically marketed in the form of vanadyl sulfate. Like chromium picolinate, the positively charged vanadium is bound to the negatively charged sulfate molecule. Vanadium as a sport supplement was popularized in the 1990s as diabetic rats became more glucose tolerant when they were provided vanadium. Because vanadium seemed to somehow mimic the actions of insulin, it was speculated that vanadium supplements might prove anabolic in weight trainers and bodybuilders. Like chromium, vanadium was touted to enhance lean body mass by increasing amino acid uptake by muscle in conjunction with enhanced protein synthesis and inhibition of protein breakdown.

The little research done so far does not support the notion that vanadium supplements enhance LBM and strength in well-nourished people and athletes. Researchers provided individuals in a 12-week weight training study vanadyl sulfate (or placebo) at a level of 0.5 mg/kg of body weight daily.[170] They measured changes in 1 RM and 10 RM for the leg extension and bench press and assessed body composition using DXA scans. Vanadyl sulfate supplementation failed to improve strength gains or change body composition relative to the placebo group. In addition, two of the participants receiving the vanadyl sulfate supplements dropped out of the study due to side effects.

At this time there is very little research addressing the ergogenic impact of vanadium (vanadyl sulfate) supplementation. Furthermore, the results of the only study reviewed are not suggestive of possible ergogenic benefit. This notion is echoed in another review of popular nutrition supplements associated with muscle growth and resistance exercise.[108] With regard to the potential for vanadium toxicity, information is limited. Vanadium supplementation at 13.5 mg/day or 9 mg/day for 6 weeks and 16 months, respectively, were not reported to result in signs of toxicity.[171] However, larger doses have been reported to produce diarrhea, green tongue, gastrointestinal disturbances, and cramps.

■ SPORT SUPPLEMENTS *IN REVIEW*

- Sport supplements include substances purported to enhance performance and/or favorably alter body composition.
- Researchers have substantiated the efficacy of some supplements, including creatine and glycerol; however, practical issues and individuality are important considerations.
- Androstenedione and DHEA at higher doses may enhance testosterone levels but may enhance estrogen levels even more, making their use problematic.
- The research findings are equivocal about an independent ergogenic effect for many supplements, including BCAAs, arginine, lysine, ornithine, and negative for others, including carnitine.
- Despite the popularity of some supplements including ribose and glutamine, very little sport research has been conducted on them.
- Caffeine may enhance performance for some athletes; athletes who want to use it in competition need to experiment with its use and understand doping restrictions.

■ PHYSIOLOGICAL ERGOGENIC EFFORTS

Beyond supplements, athletes often experiment with other substances and methods that may have a more direct influence on physiological systems associated with exercise. These include bicarbonate and lactate loading and O_2 supplementation.

Bicarbonate Loading

Bicarbonate (HCO_3^-) is a naturally occurring buffer in the extracellular fluid. As shown in Figure 11-10, bicarbonate (as part of the carbonic anhydrase system) serves as a means of neutralizing metabolic acids (such as lactic acid) to maintain the optimal pH of the extracellular fluid. Ingestion of sodium bicarbonate ($NaHCO_3$), or *bicarbonate*

$$HCO_3^- + H^+ \longleftrightarrow H_2CO_3 \longleftrightarrow H_2O + CO_2$$

Figure 11-10
Bicarbonate (HCO_3^-) serves as a buffer against excessive acid production.

loading, is believed to enhance performance by delaying the onset of fatigue related to an increased lactic acid generation. The level used might be about 0.3 g/kg of body weight and a benefit might be observed in higher-intensity exercise bouts lasting 1–10 minutes (such as 400- to 3000-meter runs) or repeated bouts of high-intensity efforts.[172]

Researchers who assessed the findings of several studies reported that $NaHCO_3$ ingestion resulted in a more alkaline extracellular fluid.[173] However, the dosage level used in these studies was only moderately related to the increase in extracellular fluid pH and HCO_3^- level. Athletic performance was generally enhanced, but the range of effects was large. In the studies that measured time to exhaustion, there was an average increase in duration of 27%. However, the relationship between the change in extracellular pH and performance enhancement was not as strong as expected.

Not all studies have demonstrated a positive influence from bicarbonate loading.[173] The studies that did demonstrate a positive influence showed significant individuality in the results. At this time bicarbonate loading appears relatively safe, and athletic governing bodies (including the NCAA and IOC) have not banned bicarbonate loading. Perhaps the most significant consideration is the potential for gastrointestinal discomfort such as abdominal pain and diarrhea. Experimentation during the off-season and practices will help predict its potential for enhancing an individual's performance.

Key Point

Bicarbonate loading attempts to enhance the buffering capacity of the extracellular fluid during high-intensity exercise.

Phosphate Loading

Although it is uncommon, some athletes have tried to enhance performance by ingesting gram amounts of phosphate shortly before strenuous training or competition. This practice is called phosphate loading and is believed to increase the buffering capacity of skeletal muscle and the blood, thus prolonging the onset of muscular fatigue associated with excessive lactic acid production. This could have application to very high intensity activities such as weight training and high-power efforts (for example, strongman events) as well as all-out sprinting (running, cycling, swimming) or shorter distance runs (mile runs). Also, it has been speculated that phosphate supplementation might enhance performance by increasing the level of 2,3-diphosphoglycerol (DPG) in RBCs.[174] For higher intensity, shorter duration endurance efforts, such as mile runs, the increased presence of 2,3-DPG might reduce lactate accumulation by enhancing O_2 release to tissues.

Some research findings suggest that phosphate loading might enhance performance. For instance, in one study cyclists and triathletes were provided 1 g of sodium phosphate four times daily for 3 days prior to a 40-km time trial simulation or a ride to exhaustion.[175] Among other measures, these athletes demonstrated an improvement in anaerobic threshold, VO_2max, and enhanced endurance performance. In another study, researchers provided endurance athletes 1 g of sodium phosphate four times each day leading up a maximal running stress test and a 5-mile performance run.[176] The researchers reported that phosphate loading significantly increased resting and postexercise serum phosphate levels and increased VO_2max and anaerobic thresholds during a maximal running stress test. Some other trials have also reported that phosphate loading can positively influence measures of performance,[177,178] but others have not.[179,180] Differences in the methods of these studies may be important. For instance, in one of the studies failing to show performance enhancement, a single, large amount of phosphate (22.2 g calcium phosphate) was provided 90 minutes prior to exercise.[179]

The ergogenic potential of phosphate needs further research before recommendations can be made for its use. If phosphate supplementation is used, it must be very short term. Chronic phosphate loading decreases the calcium-to-phosphate ratio of the diet, which could be detrimental to calcium status. Also, the potential for gastrointestinal distress (diarrhea, nausea, vomiting, stomach pain) exists, so the timing of phosphate loading is an important consideration along with dosage level.

Key Point

Phosphate supplementation (loading) prior to competition has a long history although it is not clear whether it is beneficial, or worthwhile in light of the likelihood of gastrointestinal distress.

Lactate Loading

Lactic acid is formed during all types of exercise, and the rate of production is related to the intensity of the sport. Whether lactate accumulates in tissue reflects the balance between its rate of formation and its metabolism. In a

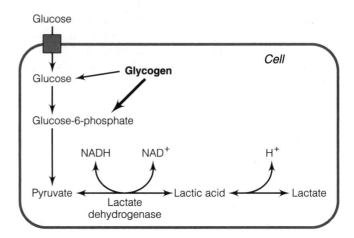

Figure 11-11
Lactate dehydrogenase is the mechanism by which lactate is created and metabolized.

bidirectional reaction, lactate dehydrogenase generates lactic acid from pyruvate and vice versa (Figure 11-11). As the formation of lactate is unavoidable, researchers have pondered ways to increase the metabolism of lactate back to pyruvate once it is formed. It has been speculated that providing lactate as a supplement prior to competition might lead to adaptations in lactate metabolism in muscle or other tissue. In turn, this could improve lactate metabolism during exercise and reduce the potential for muscular fatigue associated with lactic acid.

Despite fairly widespread knowledge of lactate loading, its efficacy has received little testing. In one study, participants were provided a lactate-containing drink twice a day for 3 weeks.[181] This protocol did not lead to improved lactate clearance during exercise. Other researchers have reported that lactate supplementation was unable to positively influence performance.[182] However, these results were contradicted by another study that demonstrated that lactate may indeed be beneficial by positively influencing pH.[183] Based on the limited research information available, lactate loading cannot be supported. Furthermore, the protocol would probably involve fairly high levels of lactate that could lead to intestinal discomfort.

Key Point

Lactate loading has been discussed for decades although very little research is available to support its use.

Oxygen Supplementation

Oxygen supplementation is one of the most recognizable ergogenic attempts during competition. O_2 tanks and canisters are prevalent on the sidelines of American football games and on ice hockey benches. Several sports magazines advertise small O_2 canisters weighing about a pound with a small face mask. O_2 supplementation is an attempt to maximize O_2 availability just prior to exercise and/or after an exercise bout to promote a quicker recovery. O_2 supplementation is attractive to endurance athletes when performing at or enduring transient periods above their lactate threshold (such as cycling climbs and sprints). It is also attractive to runners, cyclists, and climbers performing at higher altitudes.

Researchers provided professional soccer players with 100% O_2 during a recovery period following an exhaustive bout of treadmill running and prior to running a second bout.[184] They reported that the O_2 failed to improve measures during the recovery period as well as performance in the second exhaustive bout. Therefore, whether O_2 supplementation can benefit an athlete requires further research. It seems likely that for any ergogenic effect to occur the O_2 would need to be utilized within seconds of an exercise bout. Football in the United States is perhaps the sport most associated with O_2 supplementation, yet very little research has been performed on O_2 supplementation in this sport. One observation is that the time lapse between O_2 use on the bench and the return to play might be too great for an acute benefit. Many researchers think that bench time alone would probably allow for adequate recovery when breathing normally.

The use of O_2 supplementation during exercise would require an athlete to carry a canister, which would be difficult with the bounce of running and the "waggle" and hands-on attention of cycle sprinting. Supplementation just seconds prior to competition is also problematic. For instance, in the seconds just prior to many track events, the athletes are involved in warming up and getting in position.

O_2 canisters are marketed based on their size and practicality. However, as the canisters get smaller so does their O_2 content. Smaller canisters may last only 3–20 minutes. Some canisters are designed to be strapped on an individual's back (for a runner, cyclist, or cross-country skier, for example). Still, issues of practicality remain the primary obstacle. Breaking a true running stride to hold a canister mask to an athlete's face, carrying the canister in the hand or on the back, and the additional weight involved are among the primary factors hindering its acceptance.

Key Point

O_2 supplementation is touted to enhance performance or speed recovery; however, many practical issues limit its use in sports.

External Nasal Dilator Strips (ENDS)

External nasal dilator strips (ENDS) have been developed as a means of maximizing the nasal canal diameter during exercise and recovery periods. ENDS are marketed to athletes and also to the public as an aid for nasal breathing anomalies and to reduce snoring. During exercise, muscles (the alae nasi) dilate the nasal passageway, and the activity of these muscles appears to increase relative to airflow and drag in the passageway.[185] The question is whether spring-loaded ENDS allow for an even greater airflow through the nose during exercise, and whether this matters after the point in exercise when an athlete begins to depend more on mouth breathing. For sports such as American football and rugby athletes wear a mouthpiece to protect teeth, thus making ENDS more attractive.

Researchers studied the effects of ENDS in Caucasian and African American men during rest and after 15 minutes of exercise.[186] They measured changes in airway resistance and minimal cross-sectional area and reported that the ENDS improved the nasal valve opening by 21% for the participants as a whole and provided a 27% reduction in airway resistance in the Caucasian men. In the African American group, a wider range of resistance changes was observed. In fact, nasal resistance improved significantly in some subjects but worsened in others. In light of this great variability in the African American men, no significant change in nasal resistance occurred as a group.

Nasal anatomy varies from person to person and thus ENDS may not benefit everyone, so experimentation is necessary prior to competition. Also, many scientists question the true impact of ENDS during exercise because of the heavy reliance on mouth breathing during periods of greater O_2 demand. However, athletes of sports involving a mouthpiece may try keep their mouth shut as a matter of protection, which might make ENDS more useful.

■ PHYSIOLOGICAL ERGOGENIC EFFORTS *IN REVIEW*

- Bicarbonate loading attempts to enhance the acid buffering capacity of muscle and perhaps other tissues during higher intensity exercise. Results of studies indicate that the dosage level is important and the effects vary from athlete to athlete.
- Lactate loading attempts to improve the metabolism of lactate acid during higher intensity exercise; its application has not been thoroughly researched.
- O_2 supplementation may be considered for athletes performing at or above their lactate threshold or at higher altitudes; however, practical issues can be a major drawback.
- External nasal dilator strips (ENDS) may increase airflow through the nasal passageway for some athletes but not others.

■ SPORT DOPING

The IOC clearly states in Rule 29A of the Olympic Charter that "doping is forbidden" and that the "IOC Medical Commission shall prepare a list of prohibited classes of drugs and of banned procedures." In 1994 the IOC Medical Commission banned (1) the administration of substances belonging to selected classes of pharmacological agents and (2) the use of various doping methods. Thus, the IOC bans doping by way of pharmaceutical agents (including amphetamines and anabolic agents) as well as doping procedures such as blood doping. The drugs may not be used for training or competitive enhancement and/or recreationally/socially. Table 11-3 presents the 1994 IOC Medical Commission list of doping categories. The position of the International Amateur Athletic Federation (IAAF) on doping is that it is strictly forbidden. Rule 55 stated in the 1992–1993 IAAF Handbook states that any of the following constitute a violation:

1. A prohibited substance is found to be present in the athlete's body tissue or fluids.
2. An athlete takes advantage of a prohibited technique.
3. An athlete admits having used or taken advantage of a prohibited substance or prohibited technique.

As the most prestigious governing body, IOC provides the professional paradigm for drug/substance banning and testing. Although other governing bodies such as the NCAA, IAAF, National Football League, Major League Baseball, and National Basketball Association may have their own specific list of banned substances and methods, there is usually great similarity to the IOC list.

Table 11-3

IOC Medical Commission Doping Categories

I. Doping Classes
a. Stimulants
b. Narcotics
c. Anabolic agents
d. Diuretics
e. Peptide and glycoprotein hormones and analogs

II. Doping Methods
a. Blood doping
b. Pharmacological, chemical, and physical manipulation

III. Classes of Drugs Subject to Certain Restrictions
a. Alcohol
b. Marijuana
c. Local anesthetics
d. Corticosteroids
e. β-blockers
f. Specified β_2-agonists

Natural Versus Synthetic Compounds

Many doping agents are synthetic; others are naturally found in the body, which makes it difficult to determine doping. For instance, testosterone doping is currently determined by comparison to other hormones originating from the same and another tissue including luteinizing hormone (LH), which is produced by the pituitary gland, and epitestosterone (E), which is produced by the testes. For instance, a T/E ratio greater than 6:1 will bring about further investigation. Other substances on the doping list that are normally found in the body include growth hormone, dehydroepiandosterone (DHEA), and erythropoietin.

One of the major problems facing athletic regulatory committees is that laboratory synthesizing techniques often exceed the techniques of assessment. Many new substances are minor molecular modifications of banned substances to avoid detection. This puts the regulatory committees in what seems to be a perpetual state of catch-up. In addition, "masking" substances are becoming more common.

Timing of Testing for Doping

Another problem facing regulatory committees is when to test athletes. Decades ago, it was more common for doping substances (such as stimulants) to be used the day of competition. Thus, testing typically occurred after competition. Today, however, many doping agents are used during training or the off-season and discontinued prior to competition and testing. Protocols for doping agent use, including dosage level and schedules and when to discontinue use prior to competition to avoid detection, are available. The period between the cessation of a drug and competition is often called a "washout" period. To address this issue, testing for doping agents also occurs during preseason preparation (such as camps) and the off-season.

Key Point

Doping agents are used during training as well as competition, so testing should occur during the off-season as well as during the season.

Potential Danger of Doping

The experimentation with substances to enhance performance has brought on some tragedies. In one drug-related sport death in 1968, cyclist Arthur Linton collapsed and died as a result of strychnine toxicity.[187] During the 1904 Olympic Marathon, a British runner named Thomas Hicks may have been close to death as a result of strychnine toxicity. He won the marathon; however, he collapsed just after the race and required immediate medical attention. During the opening day of the 1960 Olympics in Rome, cyclist Knud Jenson collapsed and died during competition in the 100-km team time trial. His blood contained evidence of amphetamines and nicotinyl nitrate. In 1967 British cyclist Tommy Simpson died during the Tour de France, and amphetamines and alcohol were suspected.[187]

■ SPORT DOPING *IN REVIEW*

- A list of doping agents has been developed by sport governing organizations such as the IOC and NCAA.
- Synthetic analogs of naturally occurring chemicals, masking agents, and knowledge of washout periods have made detection difficult.
- Sport doping has a long history of tragedy including death.

■ DOPING DRUG CLASSIFICATION

Most of the attention to sport doping agents has focused on anabolic agents and amphetamines; however, these substances account for only a portion of the substances that have been banned by the IOC and NCAA (Tables 11-3 and 11-4). For instance, narcotics, diuretics, and many peptide and glycoprotein hormones and analogs have also been banned. Some of these substances (mainly stimulants) are components of common prescription and over-the-counter (OTC) products, and athletes may be unaware of them. The U.S. Olympic Committee (USOC) has developed the National Anti-Doping Program (NADP) to assist Olympic athletes in avoiding doping situations and to present the penalties associated with their detection. The USOC Drug Education Hotline (800-233-0393) can be contacted for further assistance.

Stimulants

Stimulants are drugs used to stimulate the CNS to increase alertness and prevent tiredness and fatigue. They are often applied by both endurance and strength/power athletes. They were originally used during competition in an attempt to enhance performance in strenuous events and/or reduce the sensitivity to pain. Today, however, many individuals use these substances during training as well. Among the list of stimulants are cocaine, caffeine, and methylphenidate HCl, more commonly known as Ritalin and often prescribed to treat attention deficit hyperactivity disorder (ADHD).

CAFFEINE.   Although caffeine is commonly used as an ergogenic aid, it is considered illegal by the IOC when the level ingested produces urinary levels greater than 12 $\mu g/ml$ following competition; the NCAA criterion is 15 $\mu g/ml$. This makes caffeine unusual in that some ingestion during training and prior to an event is not illegal. In fact, most individuals consuming caffeinated beverages prior to an event do not approach the legal limit following competition. Table 11-5 provides the caffeine content of common foods and beverages.

EPHEDRA (MA HUANG).   Ma huang is a Chinese herb that has been used for thousands of years to treat conditions such as asthma. The effects of ephedra are generally attributed to an alkaloid substance, ephedrine. Ephedrine is a **sympathomimetic** substance, meaning it mimics the effects of the sympathetic nervous system. Therefore, ephedrine may produce central nervous system (CNS) stimulation, **vasoconstriction** (constriction of blood vessels), elevated blood pressure, **bronchodilation** (dilation of airways in the lungs), and cardiac stimulation. Because of this, ephedrine and other sympathomimetics are banned by the IOC and other organizations such as the National Football League (NFL).

More than 20 years ago it was reported that although ephedrine (24 mg) increased heart rate and blood pressure during exercise, it did not lead to enhanced hand-eye coordination, muscle strength, endurance, power, or VO_2max.[188] In another study, ephedrine supplementation of 1 mg/kg of body weight failed to improve endurance performance of athletes.[189] More recently, however, researchers have reported that combining ephedrine with caffeine can enhance performance.[189,190] A synergistic relationship might be involved, as levels of these supplements leading to enhanced performance was lower than expected.[191] This is important because such side effects as vomiting have been noted during exercise trials involving higher levels of ephedrine and caffeine.

Ephedrine is mostly used in an attempt to enhance energy expenditure, fat utilization, and general leanness. Because it has the potential to affect the CNS, chronic supplementation and "self-prescription" of ephedrine should be evaluated for potential psychological or physiological addiction. One recent study of female athletes noted that most of those who reported ephedrine use experienced at least some adverse effects, and 19% displayed ephedrine dependence.[192]

Perhaps the greatest concern associated with ephedrine is that it is fairly common in over-the-counter remedies and prescription drugs. For instance, in the 1972 Olympic Games, American gold medallist Rick Demont had to return his medal and was disqualified from further competition after testing positive for ephedrine.[193] Ephedrine was a component of a prescribed antiasthma drug he was using at the time of testing. Athletes must be very cautious about ephedrine as they may ingest it unknowingly as part of various antiasthmatic, cold, or cough medications (Table 11-6). For instance, the urine of a Dutch professional cyclist was found to be positive for

Table 11-4 **Drugs Banned by the IOC and by NCAA Bylaw 31.2.3.1**

Stimulants

amiphenazole	cropropamide	fencamfamine	phenmetrazine
amphetamine	crothetamide	meclofenoxate	phentermine
bemigride	diethylpropion	methamphetamine	picrotoxine
benzphetamine	dimethylamphetamine	methylphenidate	pipradol
bromantan	doxapram	nikethamide	prolintane
caffeine[1]	ephedrine	pemoline	strychnine
chlorphentermine	ethamivan	pentetrazol	
cocaine	ethylamphetamine	phendimetrazine	

Anabolic Agents

anabolic steroids	dehydroepiandrosterone	methandienone	oxymesterone
androstenediol	(DHEA)	methenolone	oxymetholone
androstenedione	dihydrotestosterone	methyltestosterone	stanozolol
boldenone	(DHT)	nandrolone	testosterone[2]
clostebol	dromostanolone	norandrostenediol	
dehydrochlormethyl	fluoxymesterone	norandrostenedione	
testosterone	mesterolone	norethandrolone	

Other Anabolic Agents

clenbuterol

Substances Banned for Rifle Sports

alcohol	metoprolol	pindolol	timolol
atenolol	nadolol	propranolol	

Diuretics

acetazolamide	chlorthalidone	hydroflumethiazide	spironolactone
bendroflumethiazide	ethacrynic acid	methyclothiazide	triamterene
benzthiazide	flumethiazide	metolazone	trichlormethiazide
bumetanide	furosemide	polythiazide	
chlorothiazide	hydrochlorothiazide	quinethazone	

Street Drugs

heroin	marijuana[3]	THC (tetrahydrocannabinol)[3]

Peptide Hormones and Analogs

chorionic gonadotrophin (hCG, human chorionic gonadotrophin)
corticotrophin (adrenocorticotrophin, ACTH)
growth hormone (hGH, somatotrophin)
All the respective releasing factors of the above-mentioned substances also are banned.
erythropoietin (EPO)
sermorelin

Definitions of positive depends on the following:

1. For caffeine: if the concentration in urine exceeds 15 μg/ml (NCAA) or 12 μg/ml (IOC rules)
2. For testosterone: if the administration of testosterone or the use of any other manipulation has the result of increasing the ratio of the total concentration of testosterone to that of epitestosterone in the urine to greater than 6:1, unless there is evidence that this ratio is due to a physiological or pathological condition
3. For marijuana and THC: if the concentration in the urine of THC metabolite exceeds 15 ng/ml

Table 11-5 **Caffeine Content of Beverages, Foods, and Over-the-Counter Drugs**

Beverages and Foods	Average (mg)	Range (mg)	Drugs[a]	Average (mg)
Coffee (5-oz cup)			Cold remedies (standard dose)	
Brewed, drip method	130	110–150	Dristan	0
Brewed, percolator	94	64–124	Coryban-D, Triaminicin	30
Instant	74	40–108	Diuretics (standard dose)	
Decaffeinated, brewed or instant	3	1–5	Aqua-ban, Permathene H$_2$ Off	200
Tea (5-oz cup)			Pre-Mens Forte	100
Brewed, major U.S. brand	40	20–90	Pain relievers (standard dose)	
Brewed, imported brands	60	25–110	Excedrin	130
Instant	30	25–50	Midol, Anacin	65
Iced (12-oz can)	70	67–76	Aspirin, plain (any brand)	0
Soft drinks (12-oz can)			Stimulants	
Dr. Pepper	40		Caffedrin, NoDoz, Vivarin	200
Colas and cherry cola			Weight-control aids (daily dose)	
Regular		30–46	Prolamine	280
Diet		2–58	Dexatrim, Dietac	200
Caffeine-free		0–trace		
Jolt	72			
Mountain Dew, Mello Yello	52			
Fresca, Hires Root Beer, 7-Up,				
Sprite, Squirt, Sunkist Orange	0			
Cocoa beverage (5-oz cup)	4	2–20		
Chocolate milk beverage (8 oz)	5	2–7		
Milk chocolate candy (1 oz)	6	1–15		
Dark chocolate, semisweet (1 oz)	20	5–35		
Baker's chocolate (1 oz)	26			
Chocolate flavored syrup (1 oz)	4			

Note: A pharmacologically active dose of caffeine is defined as 200 milligrams.
[a]Because products change, contact the manufacturer for an update on products you use regularly.

norpseudoephedrine during a test for doping.[194] The cyclist had consumed a liquid herbal food supplement in which ephedra was listed as an ingredient. Interestingly, chemical analysis of several batches of the herb supplement revealed that the level of norpseudoephedrine was hundreds of times higher than ephedrine. The researchers concluded that an increased awareness and regulation of "banned" ingredients in herbal supplements is warranted.

Key Point

Ephedrine is a potentially dangerous and controversial substance that has not been shown to have an independent ergogenic influence.

AMPHETAMINES. Amphetamines include a variety of molecules that bear a similar structure, such as dextroamphetamine, methamphetamine, phenmetrazine, and methylphenidate. The manufacturing of amphetamines

Table 11-6

Foods and Drugs Containing Ephedrine and Related Compounds

Over-the-Counter Drugs	Beverages	Popular Supplements
Primatene CoTylenol	Herbal teas with ma huang	Metabolife Ripped Fuel[a]
Contac—Severe Cold and Flu Vicks NyQuil		Metabolic Enhancer[a] Metabolift[a]
Vicks Inhaler		
Alka-Seltzer Plus		
Bronkotabs		

[a]Twin Labs.

can be traced back to the 1920s. Dextroamphetamine sprays were prescribed as a nasal decongestants. Some amphetamines were supposedly used during World War II by naval and flight personnel to reduce fatigue and increase alertness during patrols. Amphetamines can affect athletic performance by directly interacting at sympathetic nerve terminals and indirectly by evoking a release of catecholamines. These drugs are often referred to as sympathomimetic amines as they mimic the effect of the catecholamines. As mentioned above, they can increase mental alertness and possibly contribute to competitiveness and aggression. However, side effects associated with amphetamine use include sweating, **insomnia** (inability to sleep well), **vertigo** (dizziness), and cardiac **arrhythmia** and **tachycardia** (irregular or too-fast heart rates). Also, psychological and physiological dependence may develop.

β_2-AGONISTS. Amphetamine substances that many people are unaware of in prescription drugs include the β_2-agonists widely used in the treatment of asthma as a component of inhaler mists. When ingested or injected, these drugs are purported to have stimulatory and anabolic properties and are often classified as both stimulants and anabolic agents. However, nasal inhalation is not associated with anabolic effects. Clenbuterol is a popular β_2-agonist used to treat asthma in Europe and in other countries. The FDA has not approved its use in the United States, leaving the black market as the means of obtaining this drug. Brand names for clenbuterol include Clenasma and Prontovent. A very similar compound is albuterol (salbutamol), which has been approved for prescription by the FDA and is marketed under brand names including Proventil and Ventolin.

Narcotic Agents

This group of drugs includes several substances derived from the Asian poppy. These *opiates* and related compounds include morphine. Their primary impact on an athlete is to alleviate moderate to severe pain during and after training and competition, that is, to function as **analgesics**. However, they can be addictive as well as fatal. A list of narcotic agents banned by the IOC and NCAA is presented in Table 11-4.

Anabolic Agents

The anabolic agents are a vast group of substances that typically fall into anabolic steroids and peptide hormones. Steroid hormones are derived from cholesterol; the most widely known is testosterone. The peptide hormones include anabolic agents such as human growth hormone (hGH), adrenocorticotrophic hormone (ACTH), human chorionic gonadotrophin (hCG), and luteinizing hormone (LH). In addition, many of the β_2-agonists (including clenbuterol) are also listed as anabolic agents, as long-term use can promote muscle development.

ANABOLIC STEROIDS. In 1935 testosterone was identified in a laboratory. Testosterone is part of a large family of molecules derived from cholesterol and has both anabolic and androgenic actions. *Anabolic* means "to stimulate growth," and in the case of supplements usually means increased protein synthesis, particularly in skeletal muscle. The **androgenic** effects of testosterone are fundamentally important in the development and maintenance of primary and secondary sexual characteristics.

Testosterone is produced in the gonads (testes and ovaries) and other tissues and is released into circulation shortly after it is made. Testosterone circulates in the blood, primarily bound to *sex hormone binding protein* (SHBP). On reaching various tissues, some of testosterone is converted to dehydrotestosterone (DHT), which competes with testosterone for the same receptors to have activity. Because the receptors bind DHT more aggressively than testosterone, DHT is considered more potent than testosterone. This has led to the notion that testosterone is a prohormone to DHT. However, the degree to which various tissues convert testosterone to DHT can differ tremendously. For instance, the testosterone activities in tissue such as the prostate, seminal vesicles, external genitalia, and genital skin is more influenced by DHT, as the conversion to DHT is relatively high in these tissues. In muscle, bone, kidneys, and the brain testosterone is more active, since conversion to DHT is much lower in these tissues. Because of its greater association with muscle tissue, testosterone is considered a greater doping problem than DHT, although both are banned by organizations such as the IOC and NCAA.

Testosterone stimulates protein production in muscle, as evidenced by increases in mRNA levels in muscle. Researchers demonstrated that testosterone increases net protein synthesis without substantial increase in amino acid transport into muscle tissue.[195] Thus, testosterone may increase protein production in muscle more by efficient reutilization of amino acid. The testosterone analog oxandrolone (Anavar) has also been reported to stimulate net protein synthesis in young men.[196]

In efforts to reduce the androgen properties of testosterone, numerous analogs were developed. Figure 11-12 demonstrates the structural similarity between testosterone and its analogs. When the structure of testosterone is changed so too are its properties. For instance, testosterone derivatives such as stanozolol (Winstrol) and oxandrolone are more resistant to degradation and perhaps more toxic to the liver. Some testosterone derivatives, such as testosterone enanthate and nandrolone (Durabolin and Deca-Durabolin), are lipid soluble and are more effective when introduced via injection, whereas others, such as mesterolone (Androviron and Proviron), are usually taken orally. Table 11-4 provides a listing of most of the anabolic steroids banned by the IOC and NCAA.

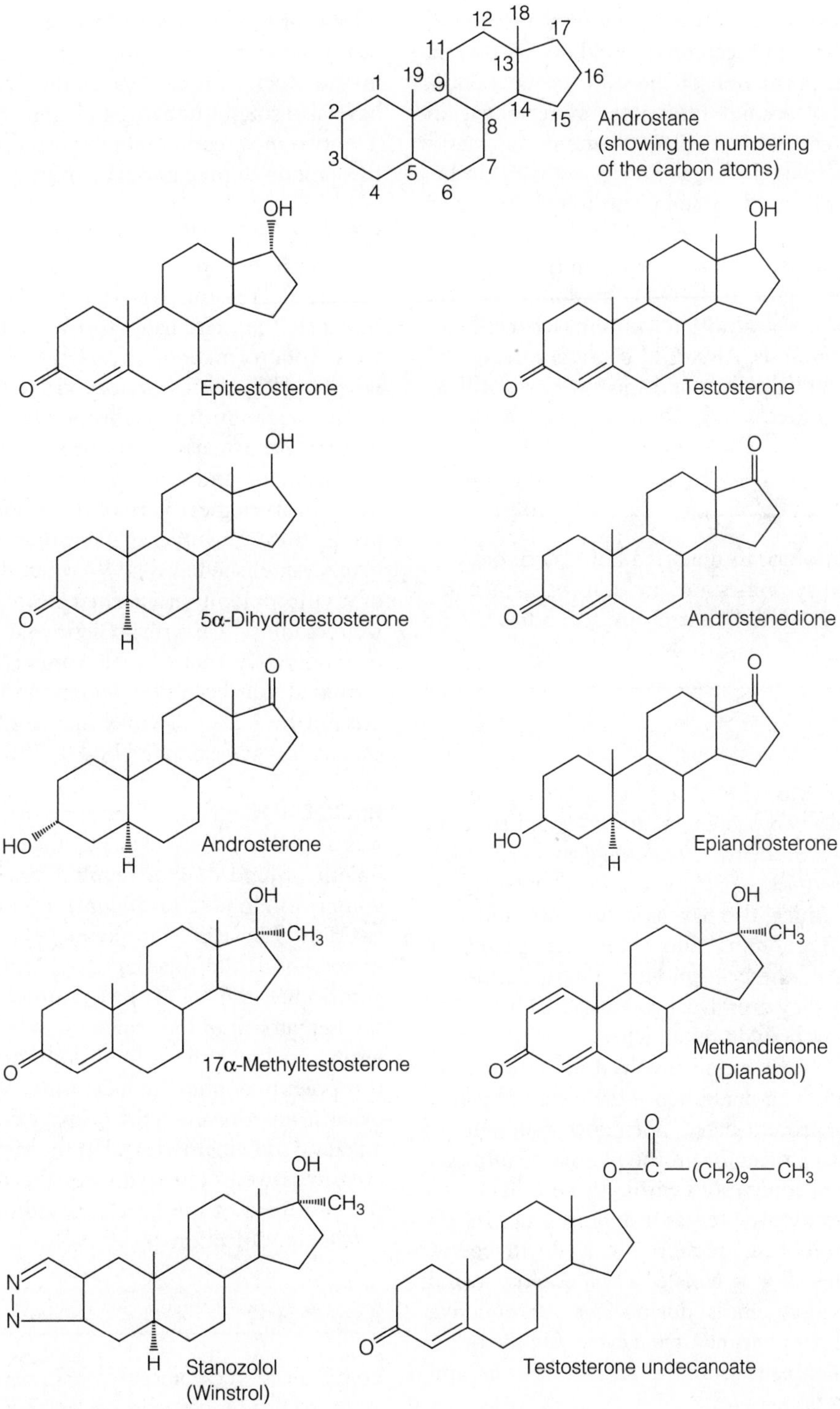

Figure 11-12
Testosterone and its analogs.

GROWTH HORMONE. If ever a hormone's name was able to capture the desires of many athletes seeking increased muscle mass, growth hormone (GH) is it! Growth hormone is made by the anterior pituitary gland of mammals; however, it is important to know that only GH from primates is active in humans. Therefore therapeutic GH must be obtained from humans (hGH), other primates, or synthesized via recombinant DNA technique.

Like all hormones, GH must act by first interacting with a receptor. The GH receptor is found on the plasma membrane, and the formation of the GH/receptor complex initiates several second messenger systems within those cells. Although GH is involved in numerous metabolic functions, its roles in stimulating growth and influencing the metabolism of protein, carbohydrates, and lipids are the most intriguing to athletes. Its growth-related effects are mediated by an increase in the release of insulin-like growth factor 1 (IGF-1). In doing so, GH stimulates the uptake of amino acids into muscle cells and increases protein synthesis. Also, GH decreases the use of glucose in muscle and the liver and increases the utilization of fat as a fuel source.

Key Point

Anabolic agents attempt to enhance LBM by directly stimulating protein synthesis and include testosterone and related steroids, growth hormone, and a few β_2-agonists.

Glycoproteins

Luteinizing hormone is produced by the anterior pituitary gland and comprises two protein subunits (α and β). The α subunit is also part of *thyroid stimulating hormone (TSH)*, *follicle stimulating hormone (FSH)*, and *human chorionic gonadotrophin (hCG)*, which also have two subunits. The differences in the β subunits allow for the particular properties of each of these hormones. These hormones are unusual in that they are glycoproteins; they have one or two short chain carbohydrates attached.

LH, FSH, and hCG are all involved in steroid hormone production in the gonads; hence the interest by athletes. All of these protein-based hormones function by first interacting with a receptor on the plasma membrane, which leads to specific operations within those cells.

FSH is not commonly used as a doping drug. Its effects on steroid metabolism are more involved with sperm formation for males. For a female, FSH doping would have significant consequences during her reproductive years as it would disrupt her menstrual cycle. On the other hand, LH for males can stimulate steroidogenesis and testosterone production.

Erythropoietin and Blood Doping

Oxygen delivery to tissues depends on red blood cells (RBCs). Therefore any way to increase the RBC content of the blood (hematocrit) could increase VO$_2$max and elevate the lactate threshold. This would allow endurance athletes to work at a higher intensity without fatiguing and possibly use less carbohydrate. For decades now, endurance athletes have known that training at altitude can naturally enhance hematocrit levels by increasing the

release of erythropoietin from the kidneys. In application of this concept, many professional cyclists choose to train in the Rocky Mountains of the United States. Athletes have also sought unnatural means of elevating hematocrit. The two most commonly discussed and practiced are erythropoietin doping and RBC reinfusion (blood doping).

ERYTHROPOIETIN. Erythropoietin is a hormone (also a glycoprotein) produced by the kidneys in response to varying degrees of **hypoxia,** a reduced level of O$_2$ in the blood. It then circulates to the bone marrow and stimulates RBC formation, or *erythropoiesis*. Erythropoietin can be created in the laboratory via recombinant DNA technique [recombinant erythropoietin (rEPO)]. Erythropoietin is particularly attractive to athletes like cyclists, marathoners, and cross-country skiers, as it can increase the O$_2$-carrying capacity of the blood by stimulating RBC production. The efficacy of erythropoietin doping was not firmly established until 1991 when the effects of low doses of erythropoietin on maximal treadmill time and VO$_2$max were studied.[197] Erythropoietin was provided for 6 weeks to moderately trained male runners. The researchers determined that both hemoglobin and hematocrit levels increased by 10%, VO$_2$max increased 6–8%, and time till exhaustion was extended by 11–17%.

BLOOD DOPING. Blood doping or RBC reinfusion was more common prior to the availability of erythropoietin. Blood doping requires that an athlete provide a volume of blood (250–500 ml) several weeks prior to competition. The RBCs are extracted and stored in a refrigerator, and the plasma is reinfused. As competition approaches the RBCs are reinfused, thereby increasing the hematocrit of the athlete. It is believed (with some scientific backing) that blood doping is as effective as erythropoietin doping in increasing VO$_2$max and time till exhaustion. However, the primary concern associated with an elevated hematocrit is that the viscosity of the blood increases as well. This increases the drag of the blood and the workload of the heart. In addition, infusion of RBCs can lead to infection.

Key Point

Erythropoietin doping and RBC reinfusion attempt to increase the O$_2$ carrying capacity of the blood. They are associated with sports such as distance running, cycling, and high-altitude sports such as cross-country skiing.

Alcohol

Alcohol is produced by the fermentation of carbohydrates in plants and provides 7 kcal/g. Wine contains about 12% alcohol and beer is about 4% alcohol. The alcohol content of distilled spirits such as whiskey, vodka, and rum is half of its *proof*. Therefore, a liquor that is 100 proof is

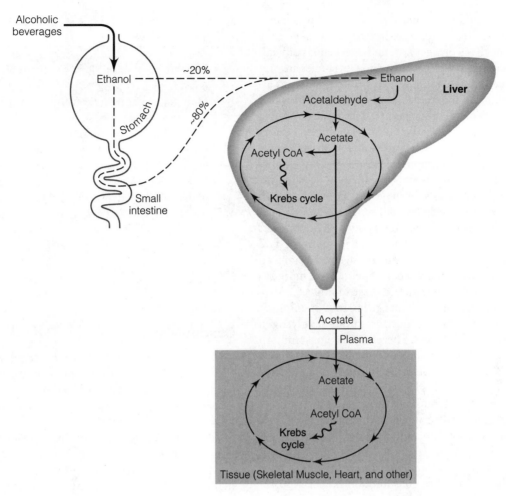

Figure 11-13
Metabolism of ethanol (alcohol).

50% alcohol. Because of its energy content and pharmacological effect on the CNS, alcohol is considered both a drug and a food. Once consumed, alcohol is rapidly absorbed in the stomach and small intestine and easily dissolves into the plasma. The rate of absorption is enhanced by gas molecules in the beverage (as in champagne, beer, and spirits mixed with a carbonated beverage) and by a higher concentration of alcohol, and can be slowed by the presence of food in the gut (for example, having wine with dinner). The liver is the primary site of metabolism via *alcohol dehydrogenase*, *microsomal ethanol oxidizing system (MEOS)*, and peroxisomes; alcohol dehydrogenase is the most significant. Alcohol is metabolized to acetaldehyde and then acetate. Much of the acetate actually diffuses from the liver and is metabolized by other tissue. In mitochondria, acetate is converted to acetyl CoA, which then enters the Krebs cycle (Figure 11-13).

Some athletes experiment with alcohol (ethanol or ethyl alcohol) prior to or during competition seeking greater confidence, relaxation, and the reduction of anxiety. Thus, alcohol can be viewed as more of a psychological

than a physiological ergogenic aid. However, alcohol might hinder exercise performance in several ways including the impairment of muscular and cardiovascular endurance as well as hand-eye coordination, accuracy, balance, and coordination. Anxiety or emotional tension is common prior to meaningful competition and it can adversely affect performance by influencing mental focus, concentration, and steadiness in limbs. Behavioral researchers are able to identify and measure anxiety by assessing muscular tremors via electromyography, testing for stress hormone levels in the blood and urine, observing restlessness, and having athletes perform paper and pencil tests. Anxiety correlates well with plasma epinephrine levels that in turn correlate well with pre-activity heart rate. High speed and dangerous activities can be associated with a higher pre-event heart rate. For instance, motor racing, downhill skiing, and leisure park rides elicited the highest pre-event heart rates; American football and soccer were at the lower end.[198]

Research studies involving alcohol and performance are difficult to carry out. Ideally, a treatment or variable within a study should be administered without the subjects knowing whether they are receiving the treatment or a placebo. Alcohol has a distinct taste and effect and is difficult to conceal or mimic in a placebo. Long-term studies are also complicated by variations in physiological adaptation to alcohol metabolism. In the studies, alcohol is usually provided according to body mass, although metabolism varies.

Studies show that alcohol consumption affects performance and has the most deleterious effects in sports requiring fast reactions, complex decision making, highly skilled actions, and hand-eye coordination.[198] For instance, motor racing skills become impaired as smooth control movements, such as skillfully steering a car through a pack, become jerky. One older study did report that in some cases moderate alcohol consumption could improve isometric strength, similar to the effect produced by cheering and vocal encouragement. However, more recent studies have failed to replicate these findings.[199] Also, alcohol ingestion within 24 hours prior to a training or competition can result in a reduction in muscle glycogen levels at the onset of exercise. Therefore the consumption of alcohol within a day or two of endurance competition is certainly not advised. As described several times in this text, depletion of muscle glycogen stores is paramount in the development of muscular fatigue. However, researchers demonstrated that alcohol consumption does not impair fat breakdown in adipose tissue.[200]

Alcohol consumption is common among athletes of varying cultural backgrounds. One important consideration is the effect of alcohol consumption on the restoration of fluid and electrolyte balance after exercise-induced dehydration. Studies suggest that alcohol has a negligible diuretic effect when consumed at levels of 2% in dilute solution after moderate dehydration.[201] However, drinks containing 4% alcohol or more tend to delay the recovery process. Thus, drinking extra water when also consuming alcoholic beverages after moderate dehydration is recommended, which means fluid volume will need to exceed the mass of weight loss during exercise-induced dehydration.

■ DOPING DRUG CLASSIFICATION IN REVIEW

- Regulatory bodies such as the IOC and NCAA provide lists of doping agents.
- Doping agents include stimulants, anabolic agents, narcotics, alcohol, and some physiological methods (including blood doping).
- Stimulants attempt to increase CNS activity whereas anabolic agents are used to enhance LBM.
- Erythropoietin and blood doping increase hematocrit (percentage of blood that is RBCs), which increases the O_2 carrying capacity of the blood.
- Alcohol can reduce anxiety for some athletes but will also impair performance.

Conclusions

Athletes continuously look for an edge, whether it is in foods that enhance training adaptations, or in substances such as creatine or androstenedione to enhance normal physiological processes. Sport foods include drinks, bars, semisolids (such as glucose tablets), and powders. Sport supplements are extremely popular and research suggests that several might enhance aspects of sport performance; others have proved to be ineffective by themselves. For instance, findings on creatine have often been favorable, but androstenedione and DHEA appear to be problematic and are discouraged. In addition, some athletes use substances such as anabolic agents or methods such as blood doping that unnaturally enhance LBM or performance. These substances and practices are considered doping by competitive sport governing bodies (including the IOC and NCAA), and athletes are subject to testing for their use and disqualification if it is found.

 IN FOCUS   SPORT INJURY AND NUTRITION SUPPLEMENTATION

Sport injury is a broad term. A simple definition is any injury that results from the physical activity associated with sport training or competition. A sport injury may be cellular in nature; that is, the physical and perhaps chemical events associated with muscle contraction may result in disruption and/or destruction of cellular structures. The more recognizable forms of sport injury involving strains, ruptures, tears, and fractures of connective tissue.

Sources of Sport Injuries

MUSCLE CONTRACTION DAMAGE IN WEIGHT TRAINING. Resistance training, especially eccentric contraction, is associated with structural damage to muscle fibers. As described in Chapter 2, eccentric muscle contraction involve a lengthening of a muscle while it is contracting.

SPORT INJURY AND NUTRITION SUPPLEMENTATION (CONTINUED)

Damage to cellular structures can be observed using microscopy. Disruption of the fine structure of Z-lines of sarcomeres ("Z-line streaming") is commonly observed. Researchers at Ohio University used electron microscopy to demonstrate some of the alterations (Figures A and B) that occurred during a strength training program.

MUSCLE CONTRACTION DAMAGE IN ENDURANCE EFFORTS.   The stress placed on muscle tissue by endurance efforts such as marathons can also result in cellular disruption. Again, researchers at Ohio University demonstrated this by extracting muscle tissue before and after a competitive marathon effort. Figures C, D, and E chronicle the changes observed in leg muscle tissue of a marathon runner.

FATIGUE AND INJURY.   Fatigue is a multifactorial process that takes place at a variety of sites in a contracting muscle. There are many different theories of what mechanisms are affected by prolonged muscular activity and how these mechanisms affect a muscle's ability to produce force. Fatigue may result from changes in the muscle itself (peripheral fatigue) or from changes in the neural input to the muscle (central and peripheral fatigue).

Muscle fatigue is commonly defined as a loss in force generation due to repetitive contraction, which leads to a decreased performance of a given task.[202] The nature of muscle fatigue is dependent on the type, duration, intensity, and mode of exercise, environmental factors, muscle fiber type composition, and level of fitness. Fatigue in high-intensity, short-duration exercise is surely dependent on factors that differ from those precipitating fatigue in endurance activity. Similarly, fatigue during tasks involving heavily loaded contractions (such as weightlifting) are likely to differ from that produced during relatively unloaded movement (running or swimming).[203]

The mechanisms that are responsible for muscle fatigue after short-duration, high-intensity exercise are complex and involve multiple factors. Recovery of force production usually has two components that are probably caused by separate mechanisms: (1) a rapidly reversible perturbation mediated by non-H^+ ions, which is likely related to changes in the processes coupling the excitation and contraction (E-C coupling) of muscle tissue, and (2) a slower change that is likely mediated by the H^+ and phosphate (P_i) ions.[203]

In prolonged endurance exercise, the depletion of body carbohydrate stores in the form of muscle glycogen frequently occurs and may be an important factor in fatigue. However, other factors must also be involved, because muscle glycogen depletion can exist without fatigue. It has also been speculated that the sarcoplasmic reticulum plays a role in this fatigue process.[203]

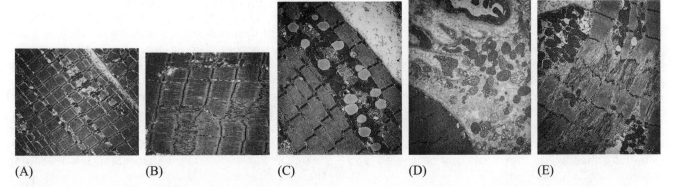

(A)   (B)   (C)   (D)   (E)

(A) Part of two muscle fibers from the vastus lateralis of a young male before a strength training program. Magnification 11,200×.
(B) A muscle fiber from the vastus lateralis of a young male after a 2-month strength training program. The myofibrils are disrupted and show Z-line streaming. Magnification 26,000×. **(C)** Muscle fiber from the gastrocnemius of a runner just prior to running a marathon. The fiber is filled with mitochondria, lipids (white spheres), and glycogen particles (the black dots in between the myofibrils). Magnification 28,000×. **(D)** Gastrocnemius muscle from a runner immediately after completing a marathon. A relatively normal muscle fiber is in the lower left. Most of the center of the photo contains part of a damaged muscle fiber. The sarcolemma is disrupted (it forms an irregular line running across the upper third of the photo). Most of the myofibrils are missing and the cytoplasm is filled with abnormal mitochondria. Magnification 28,000×. **(E)** Gastrocnemius muscle fiber taken the day after the marathon. The mitochondria are still present, but the glycogen and lipids are significantly reduced. The myofibrils are disrupted and show streaming of the Z-lines similar to the weight trainer (Figure B). Magnification 28,000×.

SPORT INJURY AND NUTRITION SUPPLEMENTATION (CONTINUED)

Sports medicine professionals agree that muscle fatigue may be a factor in muscle injuries. One researcher hypothesized over three decades ago that fatigue and inadequate warm-up led to muscle strains.[204] He observed that most muscle strains occur during the very early or late stages of a practice or competition.

According to injury surveillance by the National Collegiate Athletic Association (NCAA) for the 1998–99 season, injury rates for football increased in the second half of games and practices by approximately 20%.[205] Men's soccer had a 40% increase in the number of injuries in the second half and overtime of games, and men's hockey had an increase of injuries in the second and third periods of competition. In men's lacrosse the injury rate was observed to be 50% greater in the second half of practice than the first half. During the competitive season injury rates were 20% higher in the second halves of games and 50% greater during the last half of practice.

SPORT INJURY TO JOINT CONNECTIVE TISSUE.
Sport injury to joint connective tissue can involve ligaments, which hold bones together, and cartilage, which cushions and protects bone endings. Tears in ligaments such as the anterior cruciate ligament (ACL) and inflammation associated with joint cartilage are fairly common in sports. The ACL is responsible for holding bones together in the knee joint as well as guiding the motion of the joint. The cartilage (hyaline) that covers the endings of bone in the knee joint serves to facilitate gliding and sliding of the bone ends on each other. When cartilage is "torn" in the knee joint, it is usually the meniscus cartilage, which is not firmly attached to bone and serves as a shock absorber.

SPORT INJURY TO BONES AND TENDONS.
Fracturing of bones and tearing and rupturing of tendons can occur during sport training and competition. If a tendon ruptures and becomes detached from bone it will need to be reattached by a specialist. The fracturing of a bone will require proper setting and immobilization, after which the break can be healed and sealed.

Nutrition Supplements Used for Sport Injury Recovery

Optimal recovery from sport injury is based on appropriate therapeutic care. Optimal nutritional status of all factors associated with tissue repair and recovery supports appropriate recovery. These factors include sufficient energy, protein, vitamin C, copper, and iron for connective tissue formation, and calcium and vitamin D for bone formation.

Many athletes have experimented with several nutrition supplements in an attempt to minimize chronic inflammation and pain and/or hasten the recovery process. At this time, most of these supplements do not have adequate scientific backing derived from clinical trials to be deemed effective. Substances and theories that lack such investigation are discussed without citation of research.

CHONDROITIN SULFATE AND GLUCOSAMINE.
Chondroitin sulfate belongs to a special class of carbohydrate polysaccharides referred to as glycosaminoglycans or GAGs (see Chapter 4). GAGs are structural polysaccharides and are mostly found attached to proteins, forming proteoglycans. Glucosamine is a component of several GAGs including hyaluronic acid. Chondroitin sulfate and hyaluronic acid are found in connective tissue; in fact, chondroitin sulfate is the major GAG found in cartilage. Hyaluronic acid is found in cartilage and also in joint synovial fluid.

Supplementation of glucosamine and chondroitin sulfate is touted to reduce the degree of inflammation and soreness associated with joint inflammatory disorders. Some research suggests that this might indeed be the case for sufferers of osteoarthritis. For instance, supplemental doses have been purported to benefit individuals with osteoarthritis (OA) by enhancing the activity of chrondrocytes.[206–208]

Several recent reviews[209–211] have been published including a meta-analysis assessment[212] regarding the efficacy of glucosamine in treating OA. These researchers are also optimistic about the efficacy of glucosamine. However, numerous flaws in experimental designs of previous studies have been identified and need to be controlled in future investigations. One concern is that in some studies the participants were not "blinded," or kept from knowing what substance they received, which raises concerns about a potential placebo effect. One study challenged the effectiveness of glucosamine in reducing pain from OA of the knee.[213] Here neither the participants nor the researchers knew whether the participants were receiving the glucosamine (500 mg three times daily) or a placebo for 2 months. Thus it was a "double-blind" experiment, which reduces the possibility of a false positive effect. It was concluded that glucosamine was no better than placebo in reducing pain from OA of the knee in this group of patients. The future awaits better designed studies of glucosamine and chondroitin sulfate supplementation to determine the efficacy of these and related substances in rebuilding joint tissue after athletic injury, reducing athletic "wear and tear," and reducing OA-associated symptoms.

SPORT INJURY AND NUTRITION SUPPLEMENTATION (CONTINUED)

S-ADENOSYLMETHIONINE AND METHYLSULFONYL-METHIONINE. S-Adenosylmethionine (SAMe) and methylsulfonylmethionine (MSM) are derivatives of the amino acid methionine. SAMe has been purported to alleviate pain and inflammation while improving flexibility and mobility for osteoarthritis patients. It is also touted to cause fewer gastrointestinal side effects than other anti-inflammatory agents and may have antidepressant effects. SAMe is typically administered in doses of 200–1200 mg/day. MSM is believed to aid in the treatment of arthritis and other degenerative joint conditions. It is a methylated and sulfated version of the amino acid methionine, and as a sulfur compound, it may have some connection to the formation of sulfated GAGs in connective tissue. Whether or not SAMe and MSM have application to sport-related joint inflammation awaits thorough clinical investigation.

HYDROLYZED COLLAGEN PROTEIN. Hydrolyzed collagen protein (HCP) is also known as gelatin. Collagen is the chief structural protein that makes up human connective tissue including joint cartilage and bone. The hydrolyzed form is simply a modified version of the protein that has been partially broken down into smaller pieces by enzymes. HCP is purported to provide key amino acids that could aid in rebuilding damaged cartilage. European studies of the effect of HCP on joint pain and stiffness utilized 7–10 g/day. Research studies involving sport injuries and HCP supplementation are lacking. However, many researchers are not optimistic regarding the potential efficacy of HCP because the special hydroxylated amino acids (namely hydroxylysine and hydroxyproline) are formed post-translationally (see Chapter 5) and cannot be used to make new collagen.

OMEGA-3 POLYUNSATURATED FATTY ACIDS. Omega-3 polyunsaturated fatty acids (ω-3 PUFAs) may aid in reducing inflammation by increasing the level of anti-inflammatory eicosanoids and inhibiting the formation of the inflammatory ones. The ω-3 fatty acids EPA and DHA are found in fish oils that are often experimented with in joint inflammatory disorders. Dosage recommendations are usually 2–4 g/day for EPA and DHA and 8–12 g/day of fish oil. More research is required in this area and it is becoming clearer that a proper balance is required between the ω-3 and ω-6 PUFAs to keep inflammatory and anti-inflammatory efforts in greater harmony.

DEVIL'S CLAW AND WHITE WILLOW BARK. Devil's claw is an herb touted to alleviate pain in arthritic conditions. Reviews suggest it is more effective for joint pain in the lower back than in appendage joints. Devil's claw is usually administered as a 1000–3000 mg/day dose. White willow bark, sometimes called "nature's aspirin," contains salicin and is frequently used as a mild anti-inflammatory. Raw white willow powders are not believed to provide high enough concentrations of salicylates to provide much benefit and extracts are more commonly applied. Although information can be found on the Internet regarding these products, objective clinical research is lacking and the reader must exercise caution.

D-L-PHENYLALANINE. D-L-Phenylalanine (DLPA) is a synthetic form of phenylalanine. It has been used from time to time in relation to sport injuries to elevate mood and reduce pain. Phenylalanine is converted into tyrosine and then into the neurotransmitters dopamine and norepinephrine. These neurotransmitters are believed by some to have roles in pain signal transmission, and DLPA is also touted to increase brain endorphin levels. Objective clinical research is lacking with these products. Information can be found on the Internet regarding D-L-phenylalanine; however, objective clinical research is lacking and the reader must exercise caution.

BOSWELLIA SERRATA. Extracted from the sap of the boswellia tree, boswellia serrata is a traditional herbal treatment in India. It has been compared with conventional nonsteroidal anti-inflammatory drugs but touted not to cause gastrointestinal discomfort. Recommendations suggest 600–1200 mg/day of a 65% boswellic acid extract or 1200–2400 mg/day of a 35% extract divided into doses given two to four times per day. Caution must be applied when reading information regarding this product as it has not been thoroughly investigated by clinical researchers.

Acknowledgments

Contributing authors for this feature are Mike Higgins, PhD, Department of Health and Exercise Science, University of Delaware, Newark, and Lou Ciliberti, MS, Nutrition Program, University of Louisiana, Lafayette. The authors of *Sport and Fitness Nutrition* wish to also thank Dr. Robert S. Hikida of the Department of Biomedical Sciences at the Ohio University for providing the electron micrographs (Figures A–E).

STUDY QUESTIONS

1. What is the definition of an ergogenic aid?
2. What are general and specific recommendations for sport drink consumption by athletes before and/or during exercise?
3. What are the basic recipe ingredients of sport bars and ready-to-drink shakes?
4. What are the most important factors that can influence the rate of absorption of the ingredients of sport drinks?
5. Which sport supplements are believed to increase the level of growth hormone?
6. How might some individuals benefit from creatine supplementation?
7. What are the most important considerations for supplementation of DHEA and androstenedione?
8. How is the use of ergogenic agents regulated in organized sports, and what are the classes and examples of banned substances?
9. How might alcohol be used by athletes, and what are the most important considerations related to sport performance?
10. Which ergogenic substances and foods are associated with long-term supplementation and which are associated with more acute effects?

REFERENCES

1. Nutrition Business Journal. *NBJ's Supplement Business Report 2002*, October 2002.
2. Barron RL, Vanscoy GJ. Natural products and the athlete: facts and folklore. *Annals of Pharmacotherapy* 27(5):607–615, 1993.
3. Theoharides TC. Sudden death of a healthy college student related to ephedrine toxicity from a ma huang-containing drink. *Journal of Clinical Psychopharmacology* 17(5):437–439, 1997.
4. Mack RB. "All but death, can be adjusted". Ma Huang (ephedrine) adversities. *North Carolina Medical Journal* 58(1):68–70, 1997.
5. American College of Sports Medicine and American Dietetics Association and Dieitians of Canada. Nutrition and Athletic Performance. *Medicine and Science in Sports and Exercise* 32(12):176–192, 2000.
6. Wildman REC, Medeiros DM. Exercise and nutrition. In: *Advanced Human Nutrition*, Boca Raton, FL: CRC Press, 2000.
7. Peters HP, Wiersma JW, Koerselman J, Akkermans LM, Bol E, Mosterd WL, de Vries WR. The effect of a sports drink on gastroesophageal reflux during a run-bike-run test. *International Journal of Sports Medicine* 21(1):65–70, 2000.
8. Chromiak JA, Antonio J. Use of amino acids as growth hormone-releasing agents by athletes. *Nutrition* 18(7–8): 657–661, 2002.
9. Donati L, Ziegler F, Pongelli G, Signorini MS. Nutritional and clinical efficacy of ornithine alpha-ketoglutarate in severe burn patients. *Clinical Nutrition* 18(5):307–311, 1999.
10. De Bandt JP, Coudray-Lucas C, Lioret N, Lim SK, Saizy R, Giboudeau J, Cynober L. A randomized controlled trial of the influence of the mode of enteral ornithine alpha-ketoglutarate administration in burn patients. *Journal of Nutrition* 128(3):563–539, 1998.
11. Jeevanandam M, Petersen SR. Substrate fuel kinetics in enterally fed trauma patients supplemented with ornithine alpha ketoglutarate. *Clinical Nutrition* 18(4):209–217, 1999.
12. Isidori A, LoMonaco A, Cappa M. A study of growth hormone release in man after oral adminstration of amino acids. *Current Medical Research Opinion* 7(7):475–481, 1981.
13. Suminski RR, Robertson RJ, Goss FL, Arslanian S, Kang J, DaSilva S, Utter AC, Metz KF. Acute effect of amino acid ingestion and resistance exercise on plasma growth hormone concentration in young men. *International Journal of Sport Medicine* 7(1):48–60, 1997.
14. Marcell TJ, Taaffe DR, Hawkins SA, Tarpenning KM, Pyka G, Kohlmeier L, Wiswell RA, Marcus R. Oral arginine does not stimulate basal or augment exercise-induced GH secretion in either young or old adults. *Journal of Gerentology. Series A, Biological Sciences and Medical Sciences* 54(8):M395–399, 1999.
15. Fogelholm GM, Naveri HK, Kiilavouri KT, Harkonen MH. Low-dose amino acid supplementation: no effects on serum human growth hormone and insulin in male weightlifters. *International Journal of Sports Nutrition* 3(3): 290–297, 1993.
16. Lambert MI. Failure of commercial oral amino acid supplements to increase serum growth hormone concentrations in male body-builders. *International Journal of Sports Nutrition* 3(3):298–305, 1993.
17. Eto B, Le Moel G, Porquet D, Peres G. Glutamate-arginine salts and hormonal responses to exercise. *Archives of Physiology and Biochemistry* 103(2):160–164, 1995.
18. Corpas E, Blackman MR, Roberson R, Scholfield D, Harman SM. Oral arginine-lysine does not increase growth hormone or insulin-like growth factor-I in old men. *Journal of Gerontology* 48(4):M128–M133, 1993.
19. Maughan RJ. The effects of oral supplementation of salts of aspartic acid on the metabolic response to prolonged exhausting exercise in man. *International Journal of Sports Medicine* 4:119–123, 1983.
20. deHaan A, van Doorn JE, Westra HG. Effects of potassium + magnesium asparate on muscle metabolism and force development during short intensive static exercise. *International Journal of Sports Medicine* 6(1):44–49, 1985.
21. Tuttle JL, Potteiger JA, Evans BW, Ozmun JC. Effect of acute potassium-magnesium aspartate supplementation on ammonia concentrations during and after resistance

training. *International Journal of Sports Nutrition* 5(2): 102–109, 1995.

22. Colombani PC, Bitzi R, Frey-Rindova P, Frey W, Arnold M, Langhans W, Wenk C. Chronic arginine aspartate supplementation in runners reduces total plasma amino acid level at rest and during a marathon run. *European Journal of Nutrition* 38:263–273, 1999.

23. Nielsen FH, Hunt CD, Mullen LM, Hunt JR. Effect of dietary boron on mineral, estrogen, and testosterone metabolism in postmenopausal women. *FASEB Journal* 1:394–397, 1987.

24. Green NR, Ferrando AA. Plasma boron and the effects of boron supplementation in males. *Environmental Health Perspectives.* 102:73–77, 1994.

25. Ferrando AA, Green NR. The effect of boron supplementation on lean body mass, plasma testosterone levels, and strength in male bodybuilders. *International Journal of Sport Nutrition* 3:140–150, 1993.

26. Meacham SL, Taper LJ, Volpe SL. Effects of boron supplementation on bone mineral density and dietary, blood, and urinary calcium, phosphorus, magnesium, and boron in female athletes. *Environmental Health Perspectives* 102:79–82, 1994.

27. Meacham SL, Taper LJ, Volpe SL. Effects of boron supplementation on blood and urinary calcium, magnesium, and phosphorus and urinary boron in athletic and sedentary women. *American Journal of Clinical Nutrition* 61:341–345, 1995.

28. Naghii MR. The significance of dietary boron, with particular reference to athletes. *Nutrition Health* 13(1):31–37, 1999.

29. Samman S, Naghii MR, Lyons Wall PM, Verus AP. The nutritional and metabolic effects of boron in humans and animals. *Biological Trace Element Research* 66(1–3):227–235, 1998.

30. Wagenmakers AJ, Brookes JH, Coakley JH, Reilly T, Edwards RH, Exercise-induced activation of the branched-chain 2-oxo acid dehydrogenase in human muscle. *European Journal of Applied Physiology and Occupational Physiology* 59:159–167, 1989.

31. Paul GL, Gautsch TA, Layman DK. Amino acid and protein metabolism during exercise and recovery. In: *Nutrition in Exercise and Sport*, Boca Raton, FL: CRC Press, 1998.

32. van Hall G, Raaymakers JS, Saris WH, Wagenmakers AJ. Ingestion of branched-chain amino acids and tryptophan during sustained exercise in man: failure to affect performance. *Journal of Physiology* 486:789–794, 1995

33. Blomstrand E, Andersson S, Hassmen P, Ekblom B, Newsholme EA. Effect of branched-chain amino acid and carbohydrate supplementation on the exercise-induced change in plasma and muscle concentration of amino acids in human subjects. *Acta Physiologica Scandinavica* 153:87–96, 1995.

34. Varnier M, Sarto P, Martines D, Lora L, Carmignoto F, Leese GP, Naccarato R. Effect of infusing branched-chain amino acid during incremental exercise with reduced muscle glycogen content. *European Journal of Applied Physiology and Occupational Physiology* 69:26–31, 1994.

35. Madsen K, MacLean DA, Kiens B, Christensen D. Effects of glucose, glucose plus branched-chain amino acids, or placebo on bike performance over 100 km. *Journal of Applied Physiology* 81:2644–2650, 1996.

36. Bassit RA, Sawada LA, Bacurau RF, Navarro F, Costa Rosa LF. The effect of BCAA supplementation upon the immune response of triathletes. *Medicine and Science in Sports and Exercise* 32:1214–1219, 2000.

37. Anthony JC, Anthony TG, Kimball SR, Jefferson LS. Signaling pathways involved in translational control of protein synthesis in skeletal muscle by leucine. *Journal of Nutrition* 131:856S–865S, 2001.

38. De Palo EF, Gatti R, Cappellin E, Schiraldi C, De Palo CB, Spinella P. Plasma lactate, GH and GH-binding protein levels in exercise following BCAA supplementation in athletes. *Amino Acids* 20(1):1–11, 2001.

39. Coombes JS, McNaughton LR. Effects of branched-chain amino acid supplementation on serum creatine kinase and lactate dehydrogenase after prolonged exercise. *Journal of Sports Medicine and Physical Fitness*, 40:240–246, 2000.

40. Wildman REC, Medeiros DM. Nutraceuticals and nutrition supplements. In: *Advanced Human Nutrition*, Boca Raton, FL: CRC Press, 2000.

41. Costill DL, Dalsky GP, Fink WJ. Effects of caffeine ingestion on metabolism and exercise performance. *Medicine and Science in Sport and Exercise* 10:155–158, 1978.

42. Ivy JL, Costill DL, Fink WJ, Lower RW. Influence of caffeine and carbohydrate feedings on endurance performance. *Medicine and Science in Sport and Exercise* 11:6–11, 1979.

43. Graham, T.E., Spriet, L.L., Caffeine and exercise performance, *Sport Science Exchange* (Gatorade), 60(9):1–6, 1996.

44. Conlee RK. Amphetamine, caffeine and cocaine. In: *Ergogenics: Enhancement of Performance in Exercise and Sport* (Lamb DR, Williams MH, eds.), Indianopolis: Brown and Benchmark, 1991.

45. Graham TE, Spriet LL. Performance and metabolic responses to high caffeine dose during prolonged exercise. *Journal of Applied Physiology* 71:2292–2298, 1991.

46. Graham TE, Spriet LL. Metabolic, catecholamine and exercise performance responses to varying doses of caffeine. *Journal of Applied Physiology* 78:867–874, 1995.

47. Spriet LL, MacLean DA, Dyck DJ, Hultman E, Cederblad G, Graham TE. Caffeine ingestion and muscle metabolism during prolonged exercise in humans. *American Journal of Physiology* 262:E891–E898, 1992.

48. Pasman WJ, VanBaak MA, Jeukendrup AE, DeHaan A. The effect of different dosages of caffeine on endurance performance time. *International Journal of Sport Medicine* 16:225–230, 1995.

49. Trice I, Haymes EM. Effects of caffeine ingestion on exercise-induced changes during high-intensity, intermittent exercise. *International Journal of Sports Nutrition* 5:37–44, 1995.

50. Collomp K, Ahmaidi S, Audran M, Chanal J-L, Prefaut C. Effects of caffeine ingestion on performance and anaerobic metabolism during the Wingate test. *International Journal of Sports Medicine* 12:439–443, 1991.

51. Wiles JD, Bird SR, Hopkins J, Riley M. Effect of caffeinated coffee on running speed, respiratory factors, blood lactate and perceived exertion during 1500-m treadmill running. *British Journal of Sports Medicine* 26:166–170, 1992.

52. Williams JH, Signoille JF, Barnes, WS, Henrich TW. Caffeine, maximal power output and fatigue. *British Journal of Sports Medicine* 229:132–134, 1988.

53. Anselme F, Collomp K, Mercier B, Ahmaidi S, Prefaut C. Caffeine increases maximal anaerobic power and blood lactate concentration. *European Journal of Applied Physiology* 65:188–191, 1992.

54. Collomp K, Ahmaidi S, Chatard JC, Audran M, Prefaut C. Benefits of caffeine ingestion on sprint performance in trained and untrained swimmers. *European Journal of Applied Physiology* 64:377–380, 1992.

55. Graham TE, Rush JWE, VanSoeren MH. Caffeine and exercise: metabolism and performance. *Canadian Journal of Applied Physiology* 2:111–138, 1994.

56. VanSoeren MH, Sathasivam P, Spriet LL, Graham TE. Short term withdrawal does not alter caffeine-induced metabolic changes during intensive exercise. *FASEB Journal* 7:A518, 1993.

57. Weir J, Noakes TD, Myburgh K, Adams B. A high carbohydrate diet negates the metabolic effect of caffeine during exercise. *Medicine and Science in Sport and Exercise* 19:100–105, 1987.

58. Spriet LL. Caffeine and performance. *International Journal of Sports Nutrition* 5:S84–S99, 1995.

59. Heinenen OJ. Carnitine and physical exercise. *Sports Medicine* 22(2):109–132, 1996.

60. Colombani P, Wenk C, Kunz I, Krahenbuhl S, Kuhnt M, Arnold M, et al. Effects of L-carnitine supplementation on physical performance and energy metabolism of endurance-trained athletes: a double-blind crossover field study. *European Journal of Applied Physiology* 73(5):434–439, 1996.

61. Brass EP, Hiatt WR. The role of carnitine and carnitine supplementation during exercise in man and in individuals with special needs. *Journal of American College of Nutrition* 17(3):207–215, 1998.

62. Vecchiet L, Di Lisa F, Pieralisi G, Ripari P, Menabo R, Giamberardino MA, Siliprandi N. Influence of L-carnitine administration on maximal physical exercise. *European Journal of Applied Physiology* 61(5–6):486–490, 1990.

63. Volek JS, Kraemer WJ, Rubin MR, Gomez AL, Ratamess NA, Gaynor P. L-Carnitine L-tartrate supplementation favorably affects markers of recovery from exercise stress. *American Journal of Physiology Endocrinology and Metabolism* 282(2):E474–E482, 2002.

64. Sachan DS, Hongu N. Increases in VO(2)max and metabolic markers of fat oxidation by caffeine, carnitine, and choline supplementation in rats. *Journal of Nutritional Biochemistry* 11:521–529, 2000.

65. Buchman AL, Jenden D, Roch M. Plasma free, phospholipid-bound and urinary free choline all decrease during a marathon run and may be associated with impaired performance. *Journal of the American College of Nutrition* 18:598–596, 1999.

66. Buchman AL, Awal M, Jenden D, Roch M, Kang SH. The effect of lecithin supplementation on plasma choline concentrations during a marathon. *Journal of the American College of Nutrition* 9:768–778, 2000.

67. Spector SA, Jackman MR, Sabounjian LA, Sakkas C, Landers DM, Willis WT. Effect of choline supplementation on fatigue in trained cyclists. *Medicine and Science in Sports and Exercise* 27:668–678, 1995.

68. Warber JP, Patton JF, Tharion WJ, Zeisel SH, Mello RP, Kemnitz CP, Lieberman HR. The effects of choline supplementation on physical performance. *International Journal of Sport Nutrition and Exercise Metabolism* 10:170–178, 2000.

69. Lukaski HC, Bolonchik WW, Siders WA, Milne DB. Chromium supplementation and resistance training: effects on body composition, strength, and trace element status of men. *American Journal of Clinical Nutrition* 63(6):954–965, 1996.

70. Clancy SP, Clarkson PM, DeCheke ME, Nosaka K, Freedson PS, Cunningham JJ, Valentine B. Effects of chromium picolinate supplementation on body composition, strength, and urinary chromium loss in football players. *International Journal of Sport Nutrition* 4(2):142–153, 1992.

71. Naval Health Research Center, San Diego, CA. Effects of chromium picolinate on body composition. *Journal of Sports Medicine and Physical Fitness* 35(4):273–280, 1996.

72. Hallmark MA, Reynolds TH, DeSouza CA, Dotson CO, Anderson RA, Rogers MA. Effects of chromium and resistive training on muscle strength and body composition. *Medicine and Science in Sports and Exercise* 28 (1):139–144, 1996.

73. Hasten DL, Rome EP, Franks BD, Hegsted M. Effects of chromium picolinate on beginning weight training students. *International Journal of Sport Nutrition* 2(4):343–350, 1994.

74. Campbell WW, Joseph LJ, Davey SL, Cyr-Campbell D, Anderson RA, Evans WJ. Effects of resistance training and chromium picolinate on body composition and skeletal muscle in older men. *Journal of Applied Physiology* 86(1): 29–39, 1999.

75. Cerulli J, Grabe DW, Gauthier I, Malone M, McGoldrick MD. Chromium picolinate toxicity. *Annals of Pharmacotherapy* 32(4):428–431, 1998.

76. Stearns DM, Wise JP Sr, Patierno S, Wetterhan KE. Chromium (III) picolinate produces chromosome damage in Chinese hamster ovary cells. *FASEB Journal* 9:11643–1649, 1995.

77. Stearns DM, Belbruno JJ, Wetterhahn KE. A prediction of chromium (III) accumulation in humans from chromium dietary supplements. *FASEB Journal* 9:1650–1657, 1995.

78. Powers, SK, Lennon SL. Analysis of cellular responses to free radicals: focus on exercise and skeletal muscle. *Proceedings of the Nutrition Society* 58:1025–1033, 1999.

79. Kaikkonen J, Tuomainen TP, Nyyssonen K, Salonen JT. Coenzyme Q10: absorption, antioxidative properties, determinants, and plasma levels. *Free Radical Research* 36(4):389–397, 2002.

80. Ylikoski T, Piirainen J, Hanninen O, Penttinen J. The effect of coenzyme Q10 on the exercise performance of cross-country skiers. *Molecular Aspects of Medicine* 18(suppl):S283–S290, 1997.

81. Braun B, Clarkson PM, Freedson PS, Kohl RL. Effects of coenzyme Q10 supplementation on exercise performance, VO2max, and lipid peroxidation in trained cyclists. *International Journal of Sport Nutrition* 1:353–365, 1991.

82. Snider IP, Bazzarre TL, Murdoch SD, Goldfarb A. Effects of coenzyme athletic performance system as an ergogenic aid on endurance performance to exhaustion. *International Journal of Sport Nutrition* 2:272–286, 1992.

83. Laaksonen R, Fogelholm M, Himberg JJ, Laakso J, Salorinne Y. Ubiquinone supplementation and exercise capacity in trained young and older men. *European Journal of Applied Physiology and Occupational Physiology* 72:95–100, 1995.

84. Porter DA, Costill DL, Zachwieja JJ, Fink WJ, Wagner E, Folkers K. The effect of oral coenzyme Q10 on the exercise tolerance of middle-aged, untrained men. *International Journal of Sports Medicine* 16:421–427, 1995.

85. Weston SB, Zhou S, Weatherby RP, Robson SJ. Does exogenous coenzyme Q10 affect aerobic capacity in endurance athletes? *International Journal of Sport Nutrition* 7:197–206, 1997.

86. Nielsen AN, Mizuno M, Ratkevicius A, Mohr T, Rohde M, Mortensen SA, Quistorff B. No effect of antioxidant supplementation in triathletes on maximal oxygen uptake, 31P-NMRS detected muscle energy metabolism and muscle fatigue. *International Journal of Sports Medicine* 20:154–158, 1999.

87. Bonetti A, Solito F, Carmosino G, Bargossi AM, Fiorella PL. Effect of ubidecarenone oral treatment on aerobic power in middle-aged trained subjects. *Journal of Sports Medicine and Physical Fitness* 40:51–59, 2000.

88. Vasankari TJ, Kujala UM, Vasankari TM, Vuorimaa T, Ahotupa, M. Increased serum and low-density-lipoprotein antioxidant potential after antioxidant supplementation in endurance athletes. *American Journal of Clinical Nutrition* 65:1052–1056, 1997.

89. Dodd JR, Zheng T, Christie DL. Creatine accumulation and exchange by HEK293 cells stably expressing high levels of a creatine transporter. *Biochimica Biophysica Acta* 18:1472(1–2):128–136, 1999.

90. Benzi G. Is there a rationale for the use of creatine either as nutritional supplementation or drug administration in humans participating in a sport? *Pharmacological Research* 41:255–264, 2000.

91. Williams MH, Branch JD. Creatine supplementation and exercise performance: An update. *Journal of the American College of Nutrition* 17:216–234, 1998.

92. Kraemer WJ, Volek JS. Creatine supplementation. Its role in human performance. *Clinical Sports Medicine* 18:651–666, 1999.

93. Bolotte CP. Creatine supplementation in athletes: benefits and potential risks. *Journal of Louisiana State Medical Society* 150:325–337, 1998.

94. Jacobs I. Dietary creatine monohydrate supplementation. *Canadian Journal of Applied Physiology* 24:503–514, 1999.

95. Hultman E. Muscle creatine loading in men. *Journal of Applied Physiology* 81(1):232–237, 1996.

96. Green AL, Simpson EJ, Littlewood JJ, Macdonald IA, Greenhaff PL. Carbohydrate ingestion augments creatine retention during creatine feeding in humans. *Acta Physiologica Scandinavica* 158:195–202, 1996.

97. Green AL, Hultman E, Macdonald IA, Sewell DA, Greenhaff PL. Carbohydrate ingestion augments skeletal muscle creatine accumulation during creatine supplementation in humans. *American Journal of Physiology* 271:E821– E826, 1996.

98. Greenhaff PL. The effect of oral creatine supplementation on skeletal muscle phosphocreatine resynthesis. *American Journal of Physiology* 266:(Endocrinol. Metab. 29):E725–E730, 1994.

99. Burke LM, Pyne DB, Telford RD. Effect of oral creatine supplementation on single-effort sprint performance in elite swimmers. *International Journal of Sport Nutrition* 6:222–233, 1996.

100. Bosco C, Tihanyi J, Pucspk J, Kovacs I, Gabossy A, Colli R, et al. Effect of oral creatine supplementation on jumping and running performance. *International Journal of Sports Medicine* 18:369–372, 1997.

101. Mujika I, Padilla S. Creatine supplementation as an ergogenic acid for sports performance in highly trained athletes: a critical review. *International Journal of Sports Medicine* 18:491–496, 1997.

102. Mujika I, Chatard JC, Lacoste L, Barale F, Geyssant A. Creatine supplementation does not improve sprint performance in competitive swimmers. *Medicine and Science in Sports and Exercise* 28:1435–1441, 1996.

103. Greenhaff PL. The nutritional biochemistry of creatine. *Nutrition Biochemistry* 11:610–618, 1997.

104. Birch R. The influence of dietary creatine supplementation on performance during repeated bouts of maximal isokinetic cycling in man. *European Journal of Applied Physiology and Occupational Physiology* 69:268–270, 1994.

105. Greenhaff PL. Influence of oral creatine supplementation on muscle torque during repeated bouts of maximal voluntary exercise in man. *Clinical Science* 84:565–571, 1993.

106. Earnest CP. The effect of creatine monohydrate ingestion on anaerobic power indices, muscular strength, and body composition. *Acta Physiologica Scandinavica* 153:207–209, 1995.

107. Stone MH, Sanborn K, Smith LL, O'Bryant HS, Hoke T, Utter AC, et al. Effects of in-season (5 weeks) creatine and pyruvate supplementation on anaerobic performance and body composition in American football players. *International Journal of Sport Nutrition* 9(2):146–165, 1999.

108. Kreider RB. Dietary supplements and the promotion of muscle growth with resistance exercise. *Sports Medicine* 27:97–110, 1999.

109. Mihic S, MacDonald JR, McKenzie S, Tarnopolsky MA. Acute creatine loading increases fat-free mass, but does not affect blood pressure, plasma creatinine, or CK activity in men and women. *Medicine and Science in Sports and Exercise* 32:291–296, 2000.

110. Engelhardt M, Neumann G, Berbalk A, Reuter I. Creatine supplementation in endurance sports. *Medicine and Science in Sports and Exercise* 30;1123–1129, 1998.

111. Vandebuerie F, Vanden Eynde B, Vandenberghe K, Hespel P. Effect of creatine loading on endurance capacity and sprint power in cyclists. *International Journal of Sports Medicine* 19:490–495, 1998.

112. Greenwood M, Farris J, Kreider R, Greenwood L, Byars A. Creatine supplementation patterns and perceived effects in select division I collegiate athletes. *Clinical Journal of Sport Medicine* 10:191–194, 2000.

113. Balsom PD, Harridge SD, Soderlund K, Sjodin B, Ekblom B. Creatine supplementation per se does not enhance endurance exercise performance. *Acta Physiologica Scandinavica* 149:521–523, 1993.

114. Poortmans JR, Francaux M. Adverse effects of creatine supplementation: fact or fiction? *Sports Medicine* 30:155–170, 2000.

115. Poortmans JR, Francaux M. Long-term oral creatine supplementation does not impair renal function in healthy athletes. *Medicine and Science in Sports and Exercise* 31:1108–1110, 1999.

116. Wildman REC, Ciliberti L, Sanders C. Preliminary studies in echocardiography in men during creatine loading. *Journal of Nutraceuticals, Functional & Medical Foods* 4(2), 2003.

117. Schilling BK, Stone MH, Utter A, Kearney JT, Johnson M, Coglianese R, et al. Creatine supplementation and health variables: a retrospective study. *Medicine and Science in Sports and Exercise* 33:183–188, 2001.

118. Leder BZ, Longcope C, Catlin DH, Ahrens B, Schoenfeld DA, Finkelstein JS. Oral androstenedione administration and serum testosterone concentrations in young men. *Journal of the American Medical Association* 283(6):779–782, 2000.

119. Brown GA, Vukovich MD, Sharp RL, Reifenrath TA, Parsons KA, King DS. Effect of oral DHEA on serum testosterone and adaptations to resistance training in young men. *Journal of Applied Physiology* 87(6):2274–2283, 1999.

120. Wallace MB, Lim J, Cutler A, Bucci L. Effects of dehydroepiandrosterone vs androstenedione supplementation in men. *Medicine and Science in Sports and Exercise* 31(12):1788–1792, 1999.

121. Rasmussen BB, Volpi E, Gore DC, Wolfe RR. Androstenedione does not stimulate muscle protein anabolism in young healthy men. *Journal of Clinical Endocrinology and Metabolism* 85(1):55–59, 2000.

122. King DS, Sharp RL, Vukovich MD, Brown GA, Reifenrath TA, Uhl NL, Parsons KA. Effect of oral androstenedione on serum testosterone and adaptations to resistance training in young men: a randomized controlled trial. *Journal of the American Medical Association* 281(21):2020–2028, 1999.

123. Kley HK, Deselaers T, Peerenboom H, Kruskemper HL. Enhanced conversion of androstenedione to estrogens in obese males. *Journal of Clinical Endocrinology Metabolism* 51(5):1128–1132, 1980.

124. Brown GA, Vukovich MD, Reifenrath TA, Uhl NL, Parsons KA, Sharp RL, King DS. Effects of anabolic precursors on serum testosterone concentrations and adaptations to resistance training in young men. *International Journal of Sport Nutrition and Exercise Metabolism* 10(3):340–359, 2000.

125. Bosy TZ, Moore KA, Poklis A. The effect of oral dehydroepiandrosterone (DHEA) on the urine testosterone/epitestosterone (T/E) ratio in human male volunteers. *Journal of Analysis of Toxicology* 22(6):455–459, 1998.

126. Johnson R. Abnormal testosterone:epitestosterone ratios after dehydroepiandrosterone supplementation. *Clinical Chemistry* 45(2):163–164, 1999.

127. Broeder CE, Quindry J, Brittingham K, Panton L, Thomson J, Appakondu S, et al. The Andro Project: physiological and hormonal influences of androstenedione supplementation in men 35 to 65 years old participating in a high-intensity resistance training program. *Archives of Internal Medicine* 160:3093– , 2000.

128. Brown GA, Vukovich MD, Martini ER, Kohut ML, Franke WD, Jackson DA, King DS. Endocrine responses to chronic androstenedione intake in 30- to 56-year-old men. *Journal of Clinical Endocrinology and Metabolism*, 85:4074– , 2000.

129. Antonio J, Street C. Glutamine: a potentially useful supplement for athletes. *Canadian Journal of Applied Physiology* 24(1):1–14, 1999.

130. Krzywkowski K, Petersen EW, Ostrowski K, Kristensen JH, Boza J, Pedersen BK. Effect of glutamine supplementation on exercise-induced changes in lymphocyte function. *American Journal of Physiology and Cell Physiology* 281(4):C1259–1265, 2001.

131. Krzywkowski K, Petersen EW, Ostrowski K, Link-Amster H, Boza J, Halkjaer-Kristensen J, Pedersen BK. Effect of glutamine and protein supplementation on exercise-induced decreases in salivary IgA. *Journal of Applied Physiology* 91(2):832–838, 2001.

132. Candow DG, Chilibeck PD, Burke DG, Davison KS, Smith-Palmer T. Effect of glutamine supplementation combined with resistance training in young adults. *European Journal of Applied Physiology* 86(2):142–149, 2001.

133. Zachwieja JJ, Witt TL, Yarasheski KE. Intravenous glutamine does not stimulate mixed muscle protein synthesis in healthy young men and women. *Metabolism* 49(12):1555–1560, 2000.

134. Wagenmakers AJ. Amino acid supplements to improve athletic performance. *Current Opinion in Clinical Nutrition and Metabolism and Care* 2(6):539–544, 1999.

135. Williams MH. Facts and fallacies of purported ergogenic amino acid supplements. *Clinical Sports Medicine* 18(3):633–649, 1999.

136. Lyons T. Effects of glycerol-induced hyperhydration prior to exercise in the heat on sweating and core temperature. *Medicine and Science in Sports and Exercise* 22:477–483, 1990.

137. Monter P. Glycerol, hyperhydration and endurance exercise. *Medicine and Science in Sports and Exercise* 24:S157–S164, 1992.

138. Latzka WA, Sawka MN. Hyperhydration and glycerol: thermoregulatory effects during exercise in hot climates. *Canadian Journal of Applied Physiology* 25:536–545, 2000.

139. Hargreaves M. Pre-exercise nutritional strategies: effects on metabolism and performance. *Canadian Journal of Applied Physiology* 26(Suppl):S64–S70, 2001.

140. Koenigsberg PS, Martin KK, Hlava HR, Riedesel ML. Sustained hyperhydration with glycerol ingestion. *Life Science* 57:645–653, 1995.

141. Riedesel ML, Allen DY, Peake GT, Al-Qattan K. Hyperhydration with glycerol solutions. *Journal of Applied Physiology* 63:2262–2268, 1987.

142. Wagner DR. Hyperhydrating with glycerol: implications for athletic performance. *Journal of the American Dietetics Association* 99:207–212, 1999.

143. Nissen S, Sharp R, Ray M, Rathmacher JA, Rice D, Fuller JC, et al. Effect of leucine metabolite β-hydroxy-β-methylbutyrate on muscle metabolism during resistance training. *Journal of Applied Physiology* 81:2095–2104, 1996.

144. Jowko E, Ostaszewski P, Jank M, Sacharuk J, Zieniewicz A, Wilczak J, Nissen S. Creatine and beta-hydroxy-beta-methylbutyrate (HMB) additively increase lean body mass and muscle strength during a weight-training program. *Nutrition* 17(7–8):558–566, 2001.

145. Panton LB, Rathmacher JA, Baier S, Nissen S. Nutritional supplementation of the leucine metabolite beta-hydroxy-beta-methylbutyrate (HMB) during resistance training. *Nutrition* 16: 734–739, 2000.

146. Gallagher PM, Carrithers JA, Godard MP, Schulze KE, Trappe SW. Beta-hydroxy-beta-methylbutyrate ingestion, Part I: effects on strength and fat free mass. *Medicine and Science in Sports and Exercise* 32(12):2109–2115, 2000.

147. Knitter AE, Panton L, Rathmacher JA, Petersen A, Sharp R. Effects of beta-hydroxy-beta-methylbutyrate on muscle damage after a prolonged run. *Journal of Applied Physiology* 89(4):1340–1344, 2000.

148. Vukovich MD, Stubbs NB, Bohlken RM. Body composition in 70-year-old adults responds to dietary beta-hydroxy-beta-methylbutyrate similarly to that of young adults. *Journal of Nutrition* 131(7):2049–2052, 2001.

149. Kreider RB, Ferreira M, Wilson M, Almada AL. Effects of calcium beta-hydroxy-beta-methylbutyrate (HMB) supplementation during resistance-training on markers of catabolism, body composition and strength. *International Journal of Sports Medicine* 20(8):503–509, 1999.

150. Ransone J, Neighbors K, Lefavi R, Chromiak J. The effect of beta-hydroxy beta-methylbutyrate on muscular strength and body composition in collegiate football players. *Journal of Strength and Conditioning Research* 17(1):34–39, 2003.

151. Slater G, Jenkins D, Logan P, Lee H, Vukovich M, Rathmacher JA, Hahn AG. Beta-hydroxy-beta-methylbutyrate (HMB) supplementation does not affect changes in strength or body composition during resistance training in trained men. *International Journal of Sport Nutrition and Exercise Metabolism* 11(3):384–396, 2001.

152. Paddon-Jones D, Keech A, Jenkins D. Short-term beta-hydroxy-beta-methylbutyrate supplementation does not reduce symptoms of eccentric muscle damage. *International Journal of Sport Nutrition Exercise Metabolism* 11(4):442–450, 2001.

153. Gallagher PM, Carrithers JA, Godard MP, Schulze KE, Trappe SW. Beta-hydroxy-beta-methylbutyrate ingestion, part II: effects on hematology, hepatic and renal function. *Medicine and Science in Sports and Exercise* 32(12):2116–2119, 2000.

154. Nissen S, Sharp RL, Panton L, Vukovich M, Trappe S, Fuller JC Jr. Beta-hydroxy-beta-methylbutyrate (HMB) supplementation in humans is safe and may decrease cardiovascular risk factors. *Journal of Nutrition* 130(8):1937–1945, 2000.

155. Williams MH. Ergogenic and ergolytic substances. *Medicine and Science in Sports and Exercise* 24(9 Suppl): S344–S348, 1992.

156. Williams MH, Kreider RB, Hunter DW, Somma CT, Shall LM, Woodhouse ML, Rokitski L. Effect of inosine supplementation on 3-mile treadmill run performance and VO2 peak. *Medicine and Science in Sports and Exercise* 22(4):517–522, 1990.

157. Starling RD, Trappe TA, Short KR, Sheffield-Moore M, Jozsi AC, Fink WJ, Costill DL. Effect of inosine supplementation on aerobic and anaerobic cycling performance. *Medicine and Science in Sports and Exercise* 28(9):1193–1198, 1996.

158. McNaughton L, Dalton B, Tarr J. Inosine supplementation has no effect on aerobic or anaerobic cycling performance. *International Journal of Sport Nutrition* 9(4):333–344, 1999.

159. Jeukendrup AE, Saris WH, Schrauwen P, Brouns F, Wagenmakers AJ. Metabolic availability of medium-chain triglycerides coingested with carbohydrates during prolonged exercise. *Journal of Applied Physiology* 79(3):756–762, 1995.

160. Van Zyl CG, Lambert EV, Hawley JA, Noakes TD, Dennis SC. Effects of medium-chain triglyceride ingestion on fuel metabolism and cycling performance. *Journal of Applied Physiology* 80:2217–2225, 1996.

161. Jeukendrup AE, Saris WH, Van Diesen R, Brouns F, Wagenmakers AJ. Effect of endogenous carbohydrate availability on oral medium-chain triglyceride oxidation during prolonged exercise. *Journal of Applied Physiology* 80:949–954, 1996.

162. Massicotte D, Peronnet F, Brisson GR, Hillaire-Marcel C. Oxidation of exogenous medium-chain free fatty acids during prolonged exercise: comparison with glucose. *Journal of Applied Physiology* 73:1334–1339, 1992.

163. Angus DJ, Hargreaves M, Dancey J, Febbraio MA. Effect of carbohydrate or carbohydrate plus medium-chain triglyceride ingestion on cycling time trial performance. *Journal of Applied Physiology* 88:113–119, 2000.

164. Horowitz JF, Mora-Rodriguez R, Byerley LO, Coyle EF. Preexercise medium-chain triglyceride ingestion does not alter muscle glycogen use during exercise. *Journal of Applied Physiology* 88:219–225, 2000.

165. Goedecke JH, Elmer-English R, Dennis SC, Schloss I, Noakes TD, Lambert EV. Effects of medium-chain triacylglycerol ingested with carbohydrate on metabolism and exercise performance. *International Journal of Sport Nutrition* 9:35–47, 1999.

166. Hawley JA, Brouns F, Jeukendrup A. Strategies to enhance fat utilization during exercise. *Sports Medicine* 25:241–257, 1998.

167. Jeukendrup AE, Thielen JJ, Wagenmakers AJ, Brouns F, Saris WH. Effect of medium-chain triacylglycerol and carbohydrate ingestion during exercise on substrate utilization and subsequent cycling performance. *American Journal of Clinical Nutrition* 67:397–404, 1998.

168. Op't Eijnde B, Van Leemputte M, Brouns F, Van Der Vusse GJ, Labarque V, Ramaekers M, et al. No effects of oral ribose supplementation on repeated maximal exercise and de novo ATP. *Journal of Applied Physiology* 91(5): 2275–2281, 2001.

169. Berardi JM, Ziegenfuss TN. Effects of ribose supplementation on repeated sprint performance in men. *Journal of Strength and Conditioning Research* 17(1):47–52, 2003.

170. Fawcett JP, Farquhar SJ, Walker RJ, Thou T, Lowe G, Goulding A. The effect of oral vandyl sulfate on body composition and performance in weight-training athletes. *International Journal of Sport Nutrition* 6 (4):382–390, 1996.

171. Nielsen FH. Other trace elements. In: *Present Knowledge in Nutrition*, 7th ed. (Ziegler EE, Filer LJ Jr, eds.), Washington, DC: International Life Sciences Institute, 1996.

172. Horswill CA. Effects of bicarbonate, citrate, and phosphate loading on performance. *International Journal of Sport Nutrition* 5(Suppl):S111–S119, 1995.

173. Matson LG, Tran ZV. Effects of sodium bicarbonate ingestion on anaerobic performance: a meta-analytic review. *International Journal of Sport Nutrition* 3(1):2–28, 1993.

174. Bremner K, Bubb WA, Kemp GJ, Trenell MI, Thompson CH. The effect of phosphate loading on erythrocyte 2,3-bisphosphoglycerate levels. *Clinica Chimica Acta* 323(1–2): 111–4, 2002.

175. Kreider RB, Miller GW, Schenck D, Cortes CW, Miriel V, Somma CT, et al. Effects of phosphate loading on metabolic and myocardial responses to maximal and endurance exercise. *International Journal of Sports Nutrition* 2(1): 20–47, 1992.

176. Kreider RB, Miller GW, Williams MH, Somma CT, Nasser TA. Effects of phosphate loading on oxygen uptake, ventilatory anaerobic threshold, and run performance. *Medicine and Science in Sports and Exercise* 22(2): 250–256, 1990.

177. Stewart I, McNaughton L, Davies P, Tristram S. Phosphate loading and the effects on VO_2max in trained cyclists. *Research Quarterly in Exercise and Sport* 61(1):80–84, 1990.

178. Cade R, Conte M, Zauner C, Mars D, Peterson J, Lunne D, et al. Effects of phosphate loading on 2,3-diphosphoglycerate and maximal oxygen uptake. *Medicine and Science in Sports and Exercise* 16(3):263–268, 1984.

179. Galloway SD, Tremblay MS, Sexsmith JR, Roberts CJ. The effects of acute phosphate supplementation in subjects of different aerobic fitness levels. *European Journal of Applied Physiology* 72(3):224–230, 1996.

180. Duffy DJ, Conlee RK. Effects of phosphate loading on leg power and high intensity treadmill exercise. *Medicine and Science in Sports and Exercise*, 18:674–680, 1986.

181. Brouns F, Fogelholm M, van Hall G, Wagenmakers A, Saris WH. Chronic oral lactate supplementation does not affect lactate disappearance from blood after exercise. *International Journal of Sport Nutrition* 5(2):117–124, 1995.

182. Swensen T, Crater G, Bassett DR Jr, Howley ET. Adding polylactate to a glucose polymer solution does not improve endurance. *International Journal of Sport Nutrition* 15(7):430–434, 1994.

183. Fahey TD, Larsen JD, Brooks GA, Colvin W, Henderson S, Lary D. The effects of ingesting polylactate or glucose polymer drinks during prolonged exercise. *International Journal of Sport Nutrition* 1(3):249–256, 1991.

184. Winter FD, Snell PG, Stray-Gundersen J. Effects of 100% pure oxygen on performance of professional soccer players. *Journal of the American Medical Association* 262: 227–229, 1989.

185. Connel DC, Fregosi RF. Influence of nasal airflow and resistance on nasal dilator muscle activities during exercise. *Journal of Applied Physiology* 74(5):2529–2536, 1993.

186. Portugal LG, Mehta RH, Smith BE, Sabnani JB, Matava MJ. Objective assessment of the breathe-right device during exercise in adult males. *American Journal of Rhinology* 11(5): 393–397, 1997.

187. Verrokenm M. Drug use and abuse in sport. In: *Drugs in Sport*, 2nd ed. (Mottram DR, ed.), London: E & FN Spon, 1996.

188. Sidney KH, Lefcoe NM. The effects of ephedrine on the physiological and psychological responses to submaximal and maximal exercise in man. *Medicine and Science in Sports* 9: 95–103, 1977.

189. Bell DG, Jacobs I, Zamecnik J. Effects of caffeine, ephedrine and their combination on time to exhaustion during high-intensity exercise. *European Journal of Applied Physiology and Occupational Physiology* 77:427–438, 1998.

190. Bell DG, Jacobs I. Combined caffeine and ephedrine ingestion improves run times of Canadian Forces Warrior Test. *Aviation and Space Environmental Medicine* 70: 325–342, 1999.

191. Bell DG, Jacobs I, McLellan TM, Zamecnik J. Reducing the dose of combined caffeine and ephedrine preserves the ergogenic effect. *Aviation and Space Environmental Medicine* 71:415–423, 2000.

192. Gruber AJ, Pope HG. Ephedrine abuse among 36 female weightlifters. *American Journal of Addiction* 7:256–262, 1998.

193. DeMeersman R, Getty D, Schaefer DC. Sympathomimetics and exercise enhancement: all in the mind? *Pharmacolology, Biochemistry and Behavior* 28(3):361–365, 1987.

194. Ros JJ, Pelders MG, De Smet PA. A case of positive doping associated with a botanical food supplement. *Pharmaceutical World Science* 21(1):44–46, 1999.

195. Ferrando AA, Tipton KD, Doyle D, Phillips SM, Cortiella J, Wolfe RR. Testosterone injection stimulates net protein synthesis but not tissue amino acid transport. *American Journal of Physiology* 275(5 Pt 1):E864–E871, 1998.

196. Sheffield-Moore M, Urban RJ, Wolf SE, Jiang J, Catlin DH, Herndon DN, et al. Short-term oxandrolone administration stimulates net muscle protein synthesis in young men. *Journal of Clinical Endocrinology and Metabolism* 84(8):2705–2711, 1999.

197. Ekblom B & Berglund B. Effect of erythropoietin administration on maximal aerobic power. *Scandinavian Journal of Medicine and Science in Sports* 1:88–93, 1991.

198. Reilly T. Alcohol, anti-anxiety drugs in sport. In: *Drugs in Sport*, 2nd ed. (Mottram DR. ed.), London: E & FN Spon, 1996.

199. Ikai M, Steinhaus AH. Some factors modifying the expression of human strength. *Journal of Applied Physiology* 16, 157–161, 1961.

200. Juhlin-Dannfelt A, Jorfeldt L, Hagenfeldt L, Hulten B. Influence of ethanol on non-esterified fatty acid and carbohydrate metabolism during exercise in man. *Clinical Science and Molecular Medicine* 53(3):205–214, 1977.

201. Shirreffs SM, Maughan RJ. Restoration of fluid balance after exercise-induced dehydration: effects of alcohol consumption. *Journal of Applied Physiology* 83(4):1152–1158, 1997.

202. Bigland-Richie BR, Dawson NJ, Johansson RS, et al. Reflex origin for the slowing of motoneuron firing rates in fatigue of human voluntary contractions. *Journal of Physiology* 379:451–459, 1986.

203. Fitts RH. Mechanisms of Muscular Fatigue. In: *American College of Sports Medicine, Resource Manual for Guidelines for Exercise Testing and Prescription*, 2nd ed. Baltimore: Lea and Febiger, 1993.

204. Dorman P. A report on 140 hamstring injuries. *Australian Journal of Sports Medicine* 4:30–36, 1971.

205. NCAA Injury Surveillance Report. *NCAA*, 1998.

206. Pipitone VR. Chondroprotection with chondroitin sulfate. *Drug Experimental Clinical Research* 17(1):3–7, 1991.

207. Fillmore CM, Bartoli L, Bach R, Park Y. Nutrition and dietary supplements. *Physical Medicine and Rehabilitation Clinics of North America* 10(3):673–703, 1999.

208. Kelly GS. The role of glucosamine sulfate and chondroitin sulfates in the treatment of degenerative joint disease. *Alternative Medicine Review* 3(1):27–39, 1998.

209. Towheed TE, Anastassiades TP. Glucosamine and chondroitin for treating symptoms of osteoarthritis: evidence is widely touted but incomplete. *Journal of the American Medical Association* 15;283(11):1483–1484, 2000.

210. Gaby AR. Natural treatments for osteoarthritis. *Alternative Medical Review* 4(5):330–341, 1999.

211. Delafuente JC. Glucosamine in the treatment of osteoarthritis. *Rheumatic Diseases Clinics of North America* 26(1): 1–11, 2000.

212. McAlindon TE, LaValley MP, Gulin JP, Felson DT. Glucosamine and chondroitin for treatment of osteoarthritis: a systematic quality assessment and meta-analysis. *Journal of the American Medical Association* 15;283(11):1469–1475, 2000.

213. Rindone JP, Hiller D, Collacott E, Nordhaugen N, Arriola G. Randomized, controlled trial of glucosamine for treating osteoarthritis of the knee. *Western Journal of Medicine* 172(2):91–94, 2000.

CHAPTER 12

STRENGTH AND RESISTANCE EXERCISE AND TRAINING

Chapter Objectives

- Define and discuss the various forms of strength training equipment.

- Discuss energy systems that directly apply to strength training in skeletal muscle.

- Apply the training variables to strength training.

- Integrate the concepts of training and apply them to sport.

- Discuss age- and gender-related factors in strength training.

Personal Snapshot

Tom & Dee Ann McCarthy/Corbis

Steve is a 20-year-old college junior majoring in nutrition. He has been thinking about starting a personal-training business for part-time employment while he is in school. He has taken courses in anatomy, physiology, and exercise physiology but none that specifically addressed strength training. Steve feels that if he were to develop a greater understanding of strength training theory, it would give him a competitive edge as a personal trainer. He wants to know where he can find science-based information specific to strength training approaches, equipment, and program design, so he has decided to take an additional class in strength training theory taught by the head strength and conditioning coach on campus.

Figure 12-1
Strength training equipment.

Courtesy of Scot Larrimore and Joe Hoddinott

Health clubs and gyms typically provide free-weight equipment such as barbells, dumbbells, squat racks, and bench presses, and sometimes machines designed to train specific muscle groups (Figure 12-1). Other means are also available for strength training. But what training modality is best to achieve specific athletic goals? This chapter discusses many types of equipment and modalities used for strength training and also provides practical guidelines for developing a strength training program.

■ TYPES OF STRENGTH AND RESISTANCE TRAINING

To determine which modality is best to achieve specific goals, several questions must be asked. First, will this type of training or exercise produce the desired outcomes (such as. hypertrophy, strength, power)? Second, is strength increased throughout the range of motion (ROM), and does the exercise simulate desired athletic movements? Third, is the speed of the training similar to the speed of the desired results?

Free Weights (DCER)

Free weight refers to equipment such as barbells, dumbbells, bench presses, squat racks, and other pieces in which some type of barbell or dumbbell is used. The barbells are loaded with plates of various weights and then used for exercises such as the bench press, squat, shoulder press, and the various Olympic lifts and movements.

Isotonic is an older term used to indicate the use of constant weight during an exercise. It has been replaced

with a new term, **dynamic constant external resistance (DCER).** Because the weight remains constant while the forces exerted by the muscles (internal forces) change, DCER is a better descriptor of the activity. As discussed in Chapter 2, the internal forces change as the segments move through a ROM because of altering force arm lengths. For example, usually in weightlifting some parts of the lift (the beginning or end point) are easier than others. This is due to changing mechanical advantage as the bones (segments) are moved through a range of motion.

ADVANTAGES AND DISADVANTAGES OF FREE WEIGHTS. Free weights offer several advantages over machines. Because they involve no constraints from a machine or cables, free weights allow a full ROM for individuals of all sizes and shapes. Free weights also utilize total body training, as many lifts require multiple muscle groups to contract in a coordinated effort. Furthermore, the smaller stabilizing muscles help support the joint while allowing for proper movement mechanics. Many free-weight exercises mimic athletic and functional movements (such as picking up a box), and performing them may enhance athletic performance. Free weights are also inexpensive compared to machines.

There are some disadvantages, however. It takes longer to learn proper form and technique using free weights. With machines the movement patterns are predetermined, making them easier to learn. Also, the time spent exercising is generally greater using free weights because of the loading and unloading of the barbells. Finally, many isolation-type exercises (such as knee flexion and hip adduction and abduction) are difficult to mimic using free weights.

Key Point

The terms "free weights" and "isotonic resistance" have been replaced with the phrase *dynamic constant external resistance (DCER)*. DCER is a better descriptor because internal forces change throughout the ROM while external resistance remains constant.

Olympic Movements

The sport of Olympic lifting includes just two exercises, the clean and jerk, and the snatch (Figures 12-2, 12-3). Variations and partial movements of these two lifts have been devised for use in training. These exercises are called *Olympic movements* and include the hang clean, high pull, hang and high pull, hang and power snatch, snatch squat, front squat, push press, and a few others. Which of these movements work best for a given athlete depends on the desired outcome.

Figure 12-2
The clean and jerk exercise.

Figure 12-3
The snatch exercise.

Key Point

Olympic movements include the Olympic lifts (clean and jerk, snatch) and partial movements of these two lifts.

ADVANTAGES AND DISADVANTAGES OF OLYMPIC MOVEMENTS. The clean and jerk and the snatch involve many of the body's muscles. The ROM and speed of movement are controlled solely by the athlete, not a machine. By utilizing more musculature at once, the training economy (total time spent training) and musculature per repetition improves and potentially decreases the number of exercises, repetitions, and time needed for resistance training. In addition, Olympic movements may be of benefit by increasing flexibility and strength throughout the ROM. Furthermore, Olympic movements utilize higher power outputs due to their explosive nature that often simulates athletic and functional movements.

Olympic lifts and some of the movements are complex and it often takes time to learn proper form, so it may be awhile before benefits are realized. Because of the ballistic nature and overhead execution of these exercises, some coaches think they may endanger the athlete.

Key Point

By using more musculature at once, Olympic lifts and movements can reduce total training time, allowing for extended practice times, rest, and other commitments.

Powerlifting and Compound Exercises

Powerlifting exercises are different than Olympic lifting. The competition lifts include the front squat, back squat, bench press, and deadlift (Figures 12-4, 12-5, 12-6). Speed is not important, but the lifts require strength from a large portion of the body's musculature and thus generally assist in building overall body mass and symmetry.

Courtesy of Scot Larrimore and Joe Hoddinott

Figure 12-4a
The front squat exercise.

Courtesy of Scot Larrimore and Joe Hoddinott

Figure 12-4b
The back squat exercise.

Courtesy of Scot Larrimore and Joe Hoddinott

Figure 12-5a
The flat bench press exercise.

Figure 12-5b
The incline bench press exercise.

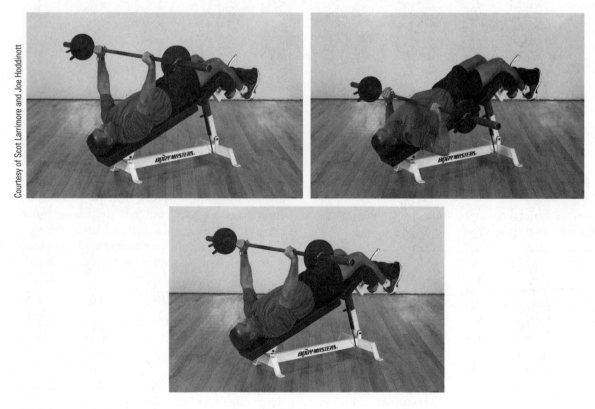

Figure 12-5c
The decline bench press exercise.

Courtesy of Scot Larrimore and Joe Hoddinott

Figure 12-6
The dead lift exercise.

Key Point

Competitive powerlifting exercises include the squat, bench press, and deadlift.

The bench press (flat, incline, decline) and squat exercises (front, back) are usually included with strength training programs. Compound or multijoint exercises (those that involve more than one joint and generally train larger muscle groups of the body in a coordinated effort) are similar to powerlifting exercises but are not necessarily used in competition. These are sometimes called core or structural exercises because of their basic functional movements Other examples of compound exercises include the shoulder press, bent-over rows, upright rows, chest press, leg press, and pull-ups. Many compound exercises can be performed with either free weights or machines, and their advantages and disadvantages are generally associated with those of free weights or machines.

Selectorized Machines

Selectorized machines are another type of DCER. They have a weight stack connected to the resistance arm (moving part) of the machine by a cable, belt, or other mechanical means (Figure 12-7). To select the amount of weight, a pin is inserted within the stack that determines the number of weight plates used. The weight stack moves vertically, so the net force of gravity remains constant.

Figure 12-7
Selectorized machines.

Figure 12-8
Oblique camshape.

The orientation of the cable and pulley system must be analyzed to be sure that a machine is a true DCER. For the external resistance to be constant, the pulley system must not change the length of the resistance arm. Therefore, the shape of the pulley or cam over which the cable or belt rotates must be circular rather than obliquely shaped (Figure 12-8). Circular pulleys do not affect the length of the resistance arm and thus keep the external resistance constant.

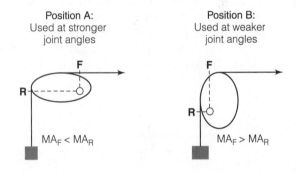

Figure 12-9
Mechanical advantage alterations by variable resistance equipment.

ADVANTAGES AND DISADVANTAGES OF SELECTORIZED MACHINES.

Some of the benefits of using selectorized equipment include the relative ease of isolating muscle groups, the short time needed to learn proper form and technique, and the ease of adjusting the weight. Disadvantages include cost, overly controlled movement patterns that lack similarity to actual athletic movements, and sometimes hindered speed of movement. Depending on the goal of a particular exercise or the program in general, the use of selectorized machines may or may not be optimal. For example, an athlete may prefer free weights because of their demand and use of stabilizing muscles. But an injured athlete may prefer the control of a machine in order to isolate a certain muscle. However, most strength training programs include both free-weight and selectorized exercises to address their many goals.

Variable Resistance Selectorized Machines

If the cam or pulley shapes of selectorized machines are not circular (are oblique, for example, as in Figure 12-9),

the external resistance is not constant due to the changing resistance arm length. This is called **variable resistance**. Several machine configurations can change the external load; the concept is to alter the weight load so it coincides with the strength curve for a particular movement. That is, at the point in the ROM where the body joint has its highest mechanical advantage, the machine should be configured to apply the most resistive force (the machine's weakest mechanical advantage). In theory, this optimizes strength gains throughout the full ROM (Figure 12-10). The trouble in designing these machines is correctly matching the strength curves for that movement.

There are three basic shapes of strength curves: ascending, descending, and bell shaped. Configuring a

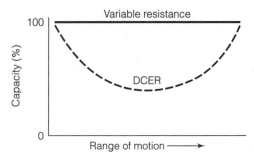

Figure 12-10
Variable resistance versus DCER. During DCER exercises the amount of weight used must be adjusted to the weakest point of the lift. With variable resistance exercises, the muscle(s) can work at 100% capacity throughout the range of motion.

Courtesy of Scot Larrimore and Joe Hoddinott

Figure 12-11
Plate-loaded equipment must have weight added to the machine to provide resistance.

machine to coincide with joint movement patterns while matching the load with the strength curve is often difficult. Furthermore, individuals differ in anatomical structure, so the strength curves for each person differ. No single machine is optimal for all users. Nearly all major fitness equipment manufacturers (such as Universal, Nautilus, Cybex, LifeFitness, Body Master, and Stairmaster) have variable resistance equipment in their product lines.

Key Point

Selectorized equipment machines have weight stacks. The amount of resistance can be selected by inserting a pin to include the desired number of weight plates. Some of these pieces have obliquely shaped cams and/or pulleys so that they offer variable resistance.

Plate-Loaded Equipment

Plate-loaded equipment may be variable resistance or DCER. **Plate loaded** simply means that weight must be manually added to the machine to provide resistance other than the weight of the machine framework itself, as shown in Figure 12-11. Common types include hip sleds, hack squats, ram racks, t-bar rows, and seated calf machines.

Some plate-loaded equipment can maintain a constant external resistance by following a linear path of motion. If the path of motion is rotational, the external resistance is not constant because of changes in the length of the resistance arm. Remember that the moment (resistance) arm is the perpendicular distance to the line of action. Gravity pulls in the vertical direction, so as rotation

occurs, the perpendicular distance from the axis to this vertical line of gravity changes. As demonstrated in Figure 12-12, as the resistance arm moves closer to the horizontal plane, length increases and thus the resistance increases. Machine configurations vary greatly, but these concepts should help in analyzing the appropriateness of a machine and using it more efficiently.

This rotational concept holds true for other exercises (such as the biceps curl) in which nonvertical (rotational) patterns of motion are executed. For simplicity this is often not considered in the analysis of free-weight movements, as most are performed with the weight moving vertically.

Among recent additions to the lines of plate-loaded equipment are several with *converging/diverging axes*. This simply means the pattern of motion brings each of the two isolateral (each arm or leg moves independently of the other arm or leg) movement arms toward or away from the midline of the body. These pieces of equipment generally follow a set arc path of movement so the external resistance changes throughout the rotation as mentioned above. The general concept for converging/diverging equipment is to coordinate the natural biomechanical movement patterns for a given joint with the strength curves of that joint. The advantage of the isolateral component is that one must control each appendage independently and thus optimize the strength for each.

Key Point

Plate-loaded equipment requires weight plates to be manually loaded. These machines can be DCER or variable resistance.

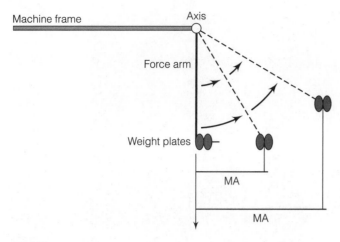

Figure 12-12
Changing mechanical advantage due to rotational movements. As the force arm segment is rotated toward the horizontal plane, the mechanical advantage is altered. The closer the segment moves toward the horizontal, the greater the resistive torque.

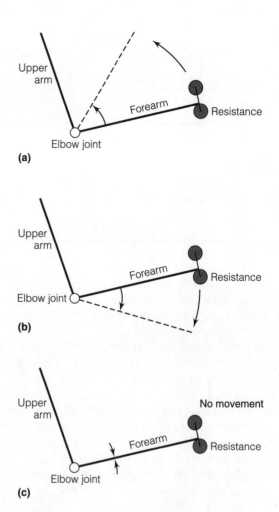

Figure 12-13
The three types of muscle actions diagrammed for the biceps brachii: (a) concentric muscle action, (b) eccentric muscle action, (c) isometric muscle action.

Isometrics

An isometric muscle action is a generation of muscular force without a change in joint angle; there is no movement. Isometric activities have been shown to increase strength, although the benefits are mainly for the particular angle of the muscular action. In other words, strength improves for that particular joint angle but not throughout the full ROM.

Types of isometric resistance can come in many forms: applying force to an immovable object such as a wall or against another body part (as in pressing hands together). Old magazine advertisements by Charles Atlas suggested that simply contracting muscles isometrically would build muscle mass and increase muscular strength. Since then, researchers have clearly demonstrated superior strength and mass gains with combinations of both concentric and eccentric muscle actions. However, isometric muscle actions play unique roles in the stabilization of joints during both dynamic and static activities. Figure 12-13 illustrates the different muscle actions.

Isokinetics

Isokinetic exercises are those in which the velocity or speed of the movement is controlled. The resistive force is directly proportional to the force applied because any force that would accelerate the lever arm beyond the set velocity would result in an increase in resistance. These types of exercises result in strength increases because maximal forces can be applied throughout the ROM. However, this maximal force is specific to the velocity of movement. Therefore, for the best results one should train at the specific velocity desired. For example, a baseball pitcher should use a velocity similar to actual pitching to achieve the best results. Isokinetic machines are common in rehabilitation facilities but have become scarce in athletic and recreational settings due to their cost and the difficulty of mechanically replicating the velocities and vast numbers of movement patterns of a pitcher's arm, much less other athletic movements.

Fluids

Although somewhat outdated, fluids have been utilized in the design of resistance equipment. Several machine manufacturers have used hydraulic (liquid) and pneumatic (gas) resistance. These systems operate by moving an object through this medium (Figure 12-14). A pin or piston attached to the lever arm is passed through a cylinder of this material. When the lever arm moves the pin, it is

Fluid resistance

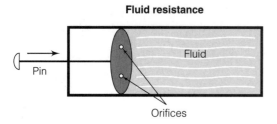

Figure 12-14
Fluid mechanics: A pin with two orifices being displaced through a fluid. Both the orifice size and viscosity of the fluid affect drag forces.

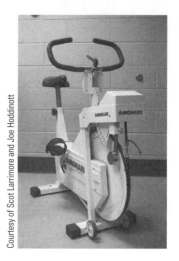

Courtesy of Scot Larrimore and Joe Hoddinott

Figure 12-15
Stationary bike with a friction belt.

pushed or pulled through the resistive substance and creates what is termed "drag." The shape of the piston, its surface area, and the viscosity of the fluid all have an effect on this resistive drag force.

Because of the fluid dynamics, the velocity for this type of exercise is generally moderate. These machines often have several settings that manipulate the piston orifice to alter the resistance. A major drawback is that this type of resistance training does not allow for eccentric muscle action. Research has shown that a combination of both concentric and eccentric muscle actions generally elicits the best results and simulates real-world activities. Table 12-1 provides the equation for determining fluid resistance.

Table 12-1 **Fluid Resistance Equation**
$F_r = kv$ F_r = fluid resistive force k = fluid characteristics, piston size, cylinder, orifice v = piston velocity

Friction

Another type of resistance works on the principle of friction. When two objects are pressed together, the friction between them opposes their movement. Equipment utilizing friction for the resistive forces often has belt or brake pad configurations made of leather or vinyl. Typical training devices using friction are wrist curl equipment and cycle ergometers. The wrist curl equipment is generally considered outdated, but many of the cycle ergometers (such as the bike in Figure 12-15) still use this type of resistance. The force needed for a given velocity is generally constant after static friction has been overcome. Table 12-2 gives the equation and components for determining frictional resistance.

Table 12-2 **Frictional Resistance Equation**
$F_r = kF_c$ F_r = resistive force k = coefficient of friction F_c = force compressing objects

Elasticity

Elasticity can provide yet another source of resistance. As the force of a rubber band holds items together, there is an equal and opposite force opposing this action. Exercise equipment has capitalized on this phenomenon and utilized surgical tubing, plastic rods (as in Bowflex and Soloflex equipment), and other material to provide resistance. As a component is stretched the resistive force is increased. Therefore, the amount of force needed is smallest at the outset and gradually increases as the material is stretched, bent, or deformed. Unfortunately, joint strength curves do not always have an ascending orientation, and using this form of resistance may not be optimal. Table 12-3 gives the equation for determining elastic resistance.

Table 12-3 **Elastic Resistance Equation**
$F_r = kx$ F_r = resistive force k = constant relative to elasticity of material x = distance the component is deformed

Key Point

Using elasticity for resistance creates an ascending force orientation, which may not be optimal for many joints.

Plyometrics (Stretch-Shortening)

Plyometrics or stretch-shortening exercises have been utilized to enhance muscular strength, power, and speed strength. The underlying premise of this type of training is to utilize the stretch of the series-elastic components and the *myototic stretch reflex* to enhance the force of the concentric muscle action. The myototic stretch reflex involves the **muscle spindle**, the spindle-shaped sensory receptor located within skeletal muscle that senses muscle "stretch." When the spindle is forcefully stretched, additional motor units are recruited to "fight back" against the stretch. Therefore, the nervous system is very active during the quick eccentric loading and if the shift to concentric muscular action is fast enough, this additional motor unit recruitment can be utilized.

Typical modes of plyometric training include jumping, depth jumping, bounding, medicine ball throwing, and many other ballistic movements (Figure 12-16). Due to the high impact and forces generated during this type of training it is recommended that the participant possess a solid base of strength before using this mode of training. Also, due to the nature of the movements, a thorough warm-up is needed. It is also recommended that plyometrics be used in conjunction with an overall strength training program.

Key Point

Many team sports (such as football and volleyball) are ballistic in nature, and their strength and conditioning programs include plyometric exercises.

Complex Training

Complex training is relatively new to the strength training realm, at least in the scientific literature, and combines

Courtesy of Scot Larrimore and Joe Hoddinott

Figure 12-16
Plyometric jumping exercise.

strength and power movements during the same training session.[1] The idea is to overload the fast-twitch muscle fibers with both high-force and high-speed movements in order to capitalize on these two specific muscle fiber characteristics. The prescribed order is to use a high-force movement such as a heavy squat and then follow it directly with a power movement such as a plyometric or sprint exercise. Following this order allows the speed movements to capitalize on the enhanced neural excitation from the previous heavy load exercise. For an upper body training session, a heavy bench press set could be followed directly by medicine ball throws or ballistic (clapping) push-ups.

Functional Training

Functional training refers to exercises and movements that simulate real-world activities. In other words, the body controls the ROM and force production without the assistance of a machine. Free-weight exercises are often considered functional exercises, along with yoga and mat-based exercises. Although there is some discrepancy in the terms used, the utilization of the musculature that controls and supports the spine (the "core") while executing movements seems to be the primary criterion for exercises to be considered "functional."

Functional training may have been one of the first concepts used for resistive training. In the "old days," large tires, sandbags, boulders, bricks, and logs were used for resistance training. They have gained popularity recently because they are often difficult to control and better simulate functional, real-world movements. Barbells and dumbbells have uniform shapes and are much easier to control than a sandbag. Real-world objects can also better simulate athletic movements where opponents and external forces are more dynamic than the uniform qualities of barbells.

The strongman competitions use large tires, logs, kegs, boulders, and so on in various events. These competitions are televised, and their popularity has heightened awareness of the benefits of using such devices and perhaps been the catalyst for the current functional strength training movement. These events exhibit the great strength and power these athletes possess. Several of the competitors are strength and conditioning coaches at major universities and clubs around the world, including the strongman profiled in the In Focus feature near the end of this chapter.

Calisthenics, Sprints, and Other Body Weight Exercises

Other types of exercises also contain components of strength. Traditional calisthenics using push-ups and abdominal crunches (Figures 12-17 and 12-18) utilize a

Figure 12-17
Push-up.

Figure 12-18
Abdominal crunch.

strength component. Depending on the participant's fitness level, these exercises may be more oriented to either muscular strength or muscular endurance. Also, the speed at which these exercises are performed is important to the type of forces created and thus the overloads and adaptations involved.

Utilizing sprinting and other various drills (such as lunges, sliding, squatting) can provide resistance through one's own body weight and improve movement components. The type, ROM, intensity, and speed at which these exercises are executed collectively determines the specific overload. A comprehensive training program usually includes some of these calisthenics and body weight exercises.

■ TYPES OF STRENGTH AND RESISTANCE TRAINING *IN REVIEW*

- Free-weight exercises provide a constant external resistance. However, the internal forces change throughout the ROM due to the changes in mechanical advantage.
- Olympic movements use free weights and include the two Olympic lifts (clean and jerk, snatch) and many partial movements of these lifts (such as hang clean, front squat).
- Powerlifts include the squat, bench press, and deadlift, which are included in most strength training programs.
- Selectorized equipment has a weight stack in which the amount of resistance is chosen (selected) by use of a push or weight pin.
- Variable resistance equipment has obliquely shaped pulleys and cams. The external resistance changes to coincide with the strength curves for a given exercise.
- Plate-loaded equipment can be either DCER or variable resistance based on configuration.
- Plyometric or stretch-shortening exercises use the stretch and increased neural activity of an eccentric loading exercise and apply it to a subsequent concentric contraction.

■ ENERGY SYSTEM ASPECTS OF STRENGTH TRAINING

What fuel sources are the major contributors during strength training? How does the intensity level of exercise make a difference? The energy for resistive and strength training activities is derived primarily from anaerobic systems. This is due to the extremely high ATP requirements necessary to generate large levels of force during activities of higher intensity and short duration. In accordance, skeletal muscle fibers adapt to complement hypertrophy with anaerobic energy metabolism capabilities.

ATP and Creatine Phosphate

ATP pools and creatine phosphate in muscle cells are the predominant sources of ATP during the first couple of seconds of high-intensity activities (see Figures 2-18 and 2-14). ATP levels are maintained during the first couple of seconds of very high intensity exercise (Figure 12-19), after which creatine phosphate levels are depleted and ATP levels fall in a corresponding manner. However, ATP levels are not depleted to the point of fatigue, indicating that other mechanisms are more responsible for muscle fatigue during very high intensity efforts. In re-

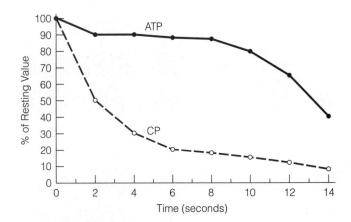

Figure 12-19
Creatine phosphate is the predominant ATP regenerating mechanism during the first couple of seconds of very high intensity exercise. Even at muscular fatigue, ATP is still available.

sponse to higher intensity training, the level of creatine phosphate increases; this adaptation is greater in Type II than Type I muscle fibers.

Anaerobic Glycolysis

Anaerobic glycolysis (Figure 12-20) is increased as very high intensity exercise is initiated. The contribution of anaerobic glycolysis to ATP regeneration during the first seconds is much less than that of creatine phosphate; however, it is an important system. As exercise continues, creatine phosphate levels are depleted and anaerobic glycolysis becomes the predominant ATP regenerating system. Consequently, the production of lactic acid becomes one of the performance-limiting factors. Anaerobic glycolysis occurs primarily in Type II fibers that have been recruited due to the high force requirement (Table 12-4). Some aerobic ATP is also generated because Type I fibers are always recruited first and some O_2 is available, but the contribution of the aerobic pathway is minimal. Total ATP production is partly dependent on the fitness level, frequency, intensity, time, and type of the activity as well as the hydration and energy status of the athlete.

LACTIC ACID PRODUCTION.   The generation of ATP via glycolysis and the subsequent lactic acid (lactate) production from pyruvate allows for rapid ATP production during high-intensity exercise (see Figure 12-20). Notice that the NADH needed to convert pyruvate to lactic acid was produced earlier in the glycolysis process. The *lactate dehydrogenase (LDH)* reaction thus regenerates NAD^+ for that early step in glycolysis. Without the rapid conversion (reduction) of pyruvate to lactic acid, glycolysis would be slowed, as NAD^+ would be limiting as a reactant.

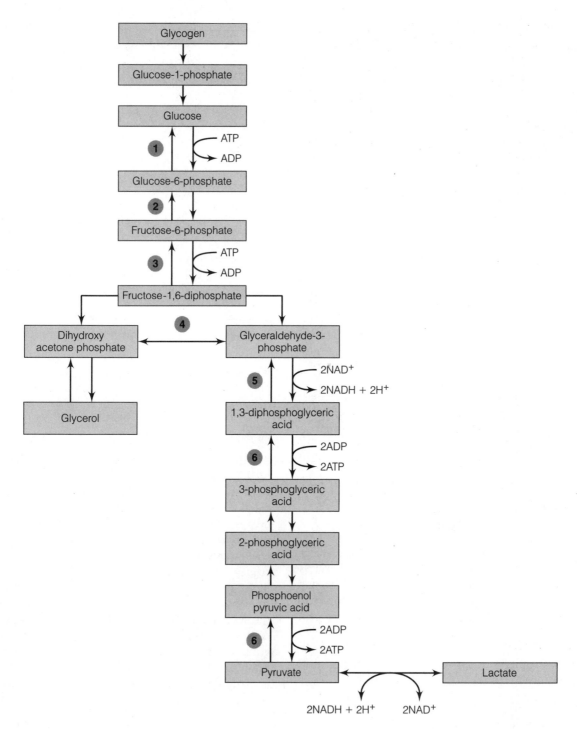

Figure 12-20
Anerobic glycolysis.

The greater reliance on anaerobic glycolysis and the generation of lactic acid for a rapid regeneration of ATP is paradoxical. On the one hand it is an extremely rapid anaerobic means of generating ATP, which is supportive of more powerful exercise efforts. On the other hand, the generation of lactic acid within the muscle tissue decreases its functional capacity by inhibiting cross-bridging, calcium binding, and enzymatic activity. As lactic acid production is directly related to exercise intensity, this energy system can sustain exercise efforts only briefly during a very powerful exercise.

Table 12-4

Energy Systems and Characteristics

Characteristic	Stored ATP and CP	Anaerobic Glycolysis	Aerobic
Duration	0–10 seconds	1–3 minutes	>16 minutes
Exercise-to-Rest Ratio	1:12 to 1:20	1:3 to 1:5	1:1 to 1:3
Intensity	High	Moderately high	Low to moderate
Muscle Fiber Type	Type IIb	Type IIA	Type I

■ ENERGY SYSTEM ASPECTS OF STRENGTH TRAINING *IN REVIEW*

- Anaerobic sources are the primary contributors of energy for high-intensity activity.
- ATP pools and creatine phosphate are the predominant sources of ATP during the first few seconds of high-intensity activity.
- Anaerobic glycolysis production increases as very high intensity exercise is initiated, and it becomes the predominant ATP regenerating system as creatine phosphate levels are depleted.
- Total ATP production is partly dependent on the fitness level, frequency, intensity, time, and type of the activity as well as the hydration and energy status of the athlete.

■ TRAINING CONSIDERATIONS: APPLICATION OF FITT

Now that the energy sources and types and forms of strength training have been discussed, the next step is to understand and apply the concepts of training. The discussion begins with the FITT (frequency, intensity, time, and type) principle, introduced in Chapter 3, since it provides the necessary framework from which physiological adaptations are derived.

Frequency of Strength Training

The frequency of resistance training may range from 1 to 7 days per week. Typical programs range from 2 to 4 days per week and address each muscle group at least twice a week. Increasing or decreasing the frequency of training can easily manipulate training volume (total workload). The frequency of training is partly determined by the intensity and duration of the activities performed. Those activities high on either the intensity or duration scale may require longer recovery times (2–3 days). Some bodybuilding programs may have periods when a specific muscle group is trained only once per week. However, that single day is usually very high on the intensity and/or volume scale. Most strength training programs use 1 or 2 days of rest between resistance workouts for any specific muscle group.

Intensity of Strength Training

The intensity required for strength training exercises is based on the number of repetitions of maximum (RM) that are to be performed. A *repetition of maximum* refers to the maximal number of repetitions that can be performed for a given weight. One repetition of maximum (1 RM) is the maximal amount of weight that can be lifted (see Sport in Practice: How to Estimate 1 RM). Five repetitions of maximum (5 RM) is the maximal amount of weight that can be used to execute five repetitions. A repetition is defined as one complete movement through the full ROM. The number of repetitions (reps) are generally expressed as the second set of numbers for an exercise prescription: For example: 2–3 (sets) × 10–12 (reps). Using a lower number of reps (1–5 RM) addresses muscular strength; a higher number (15–20 RM) addresses muscular endurance. Repetitions in the range of 8–12 RM are generally prescribed for muscle hypertrophy because total training volume is greater.

 SPORT IN PRACTICE   **How to Estimate 1 RM**

Trying different weights to determine what constitutes 1 RM for an athlete can be dangerous. Instead it can be predicted in order to prescribe appropriate intensity levels. The formula below can be used to estimate 1 RM after the person performs as many repetitions as possible at a lower than maximal weight. This formula is most accurate for a repetition range of 3–15.

$$1\ RM = [(0.03) \times (\text{no. of repetitions}) \times (\text{weight lifted})] + \text{weight lifted}$$

For example: 5 reps of 200 lb = $(0.03 \times 5 \times 200) + 200$ = estimated **1 RM of 230 lb.**

The use of multiple sets increases the training volume for a given intensity level and is generally used for increasing muscular strength, hypertrophy, and power. For example, 4 sets of 5 RM has a total training volume of 20 reps. However, because of the high intensity, rest must be taken after each set of five reps. By training in this manner, increases in the training volume for that specific intensity can be accomplished and thus create specific overload. Multiple sets have proven superior for gains of strength, power, and hypertrophy because of the increased volume relative to specific intensity levels.

Lower repetitions of maximum are used for high-intensity exercises and utilize high force or power outputs. Intensity level determines motor unit recruitment and thus the type of muscle fibers activated. High-intensity exercises utilize and train both the Type II and Type I muscle fibers, whereas lower intensity exercises address Type I more than Type II. The intensity level is also associated with specific energy systems and thus plays a role in determining the subsequent metabolic adaptations. Higher intensity exercises rely heavily on anaerobic energy systems, whereas those of lower intensity make more use of oxidative systems. Thus the metabolic adaptations associated with lower intensity training occur in Type I muscle fibers and are more aerobic in nature. Higher intensity strength training results in more adaptations in Type II muscle fibers and are more anaerobic.

Key Point

Lower numbers of repetition of maximum are used to increase strength and improve power.

Time of Strength Training

For weight training or resistance exercises, discussions of time usually refer to the number of weeks or months rather than the duration of a training session. Time for a single training session is determined by the type of exercises used, the number of sets and repetitions, and the lengths of the rest intervals. The amount of time for a given phase in a training program is based on the outcomes desired for that specific period. For example, a program may consist of 4–6 weeks of moderate-intensity exercises for increasing muscle size followed by 3–4 weeks of high-intensity exercises to improve strength.

In some circuit weight training programs, a specified time is spent at each station. For example, a circuit routine may consist of 15 different exercise stations with a participant performing each exercise for 30 seconds with 15 seconds of rest between stations. Choosing a weight load that can be performed continuously for the 30 seconds would be ideal, but this is often difficult to predict and accomplish. Therefore, the length of time spent performing an exercise determines the intensity level. After all, a person cannot perform 1 RM repetitions (reps) for 30 seconds. So in essence this is another way in which to manipulate intensity.

Type of Strength Training

Type refers to the specific modality or exercise performed. A variety of exercises, equipment types, and movements are used for strength training, and the differences between modes range from subtle to vast. The differences include such factors as the magnitude of forces, timing of force production, rate of force production, speed of contraction,

SPORT IN PRACTICE   **How to Periodize a Strength Training Program**

In order to continually accomplish overload relative to specific characteristics a periodization schedule is often organized. This not only enhances the specific component being addressed but also helps reduce overtraining and boredom.

Periodized programs for strength training are often organized into the following phases:

Hypertrophy → strength → power
→ peaking → active rest

Hypertrophy Phase
- 4–6 weeks of 3–6 sets × 8–12 reps
- Low to moderate intensity, high volume

Strength Phase
- 4–6 weeks of 3–6 sets × 3–6 reps

- Focus on using heavier weights (moderate to high intensity), low to moderate volume

Power Phase
- 4 weeks of 4–6 sets × 1–6 reps
- Focus on speed of contraction (lighter weights, high intensity), low volume

Peaking Phase
- 4 weeks of 2–4 sets × 1–6 reps
- Maintenance of both strength and power (high intensity), low volume

Active Rest
- 4 weeks
- Light exercising—unstructured
  Begin cycle all over again!

eccentric muscle action, stabilizing muscles used, ROM, and more.

Consequently, each modality of exercise elicits specific overload and subsequent adaptations in the muscle groups used, the number of motor units recruited, and the neural activation patterns. These subtle differences are often recognized when substitute exercises are chosen (such as a free-weight squat instead of a leg press machine). Muscle soreness often occurs when new exercises are incorporated into a training regimen, despite the previous use of similar exercises. Therefore, the specific type or mode of exercise must be considered in order to yield the desired adaptations.

Key Point

Adaptations are specific to the type of strength training exercises used. Differences include the magnitude of force, timing of force production, rate of force production, speed of contraction, eccentric muscle action, stabilizing muscles used, and ROM.

Periodization of Strength Training

The manipulations of FITT should evoke a training stimulus (overload). Overloads result in physiological adaptations and thus provide the necessary framework for designing training programs. A formalized adjustment or manipulations schedule that alters the overload is called **periodization,** or simply stated, systematic variation. See Sport in Practice: How to Periodize a Strength Training Program, for an outline of the method.

The hypertrophy phase is generally the first phase of a periodized program because it establishes basic strength levels and increases the cross-sectional area of the muscle tissue. This provides the needed groundwork for the following phases of strength and power improvement. The strength phase focuses on just that, strength. For this phase the intensity level increases and the overall training volume decreases. In the power phase, the athlete focuses on developing power. This is the phase where the speed of contraction is addressed. The power phase typically consists of lower training volumes with higher intensities. The loads used are generally decreased to foster and emphasize the speed of movement.

Following the power phase the athlete should be approaching peak performance levels; thus a peaking phase is utilized. This phase generally is low in volume but very high in intensity due to the emphasis on sport performance and the maintenance of strength and power. Depending on the exercise, either strength or speed is emphasized. If it is a speed movement (such as hang clean or hang snatch), the loads used are generally 65–75% of 1 RM. If strength is the component being addressed, then near maximal loads (90–100%) are used.

The peaking phase is the competition phase and the concept is to maintain the benefits gained throughout the strength training season. Following the peaking phase, a short period of active rest (recreational activities, light running) is implemented, allowing the athlete to recover and prepare to do it all over again.

Within each phase there is also planned variation, depending on the goals of the program. For example, within a hypertrophy phase lasting 6 weeks, adjustments in the volume performed each day and each week can be manipulated. Higher volumes are often performed early in the week to take advantage of weekend rest days. Different lifts and exercises can be utilized each day within the week. The combinations of exercises, sets, and reps are endless, but the goals of the program must continue to be met. Table 12-5 shows a set and repetition format for a periodized program.

Rest Periods and Strength Training

The amount of rest prescribed between sets of an exercise and between workouts is relevant to the adaptations desired. The rest periods vary depending on the goals of the strength training program and the conditioning level of the athlete. For a program aimed at strength improvements, long rest periods (2–3 minutes) between sets should be used to provide the necessary time to replenish the ATP-CP system, which in turn allows for maximal efforts on subsequent repetitions. If improving high-intensity endurance is the goal of the strength program, shorter rest periods such as 30 seconds or less are used. Logically, the rest periods between sets should coincide with the energy system being addressed.

For rest periods between workouts, the general guideline is 1 day or a total of 48 hours between workouts. This issue is still controversial and may depend on the intensity and/or duration of the activity. For example, for a conditioned participant, performing one set of 25 reps for six different exercises does not stress the musculature and require 48 hours of rest, especially if these exercises are regularly performed. Many construction workers and manual laborers perform low- to moderate-intensity exercises on a daily basis. Therefore, depending on the goals of the program and the subsequent intensity level, training every day may or may not be appropriate. Also, strength training

Table 12-5

Periodization for a Strength/Power Athlete

	Phase			
Factor	*Hypertrophy*	*Strength*	*Power*	*Peaking*
No. of weeks	6–10	4–8	3–6	1–6
Sets	3–6	3–6	3–5	1–3
Reps	8–15	2–6	2–3	1–3
Intensity	Low to moderate	High	High	Very high

routines often break the body into specific regions (such as biceps and back, chest and triceps, legs) so that training may take place on consecutive days because the muscle groups utilized and trained each day differ.

■ TRAINING CONSIDERATIONS AND FITT *IN REVIEW*

- The frequency of strength training is generally two to four times per week per muscle group.
- The intensity of strength training is determined by repetitions of maximum.
- Higher intensities (1–6 RM) generally address muscular strength adaptations.
- Improving power is generally addressed with low repetitions and a concentration on speed of movement.
- The duration or time component of strength training is usually associated with a specific training period (weeks, months).
- Different types of strength training exercises vary in the magnitude of forces produced, timing of force production, and ROM. Therefore it is important to compare the movement characteristics of the strength training exercises with those desired.
- Periodized programs should continuously overload the body, guard against overtraining, and improve motivation levels.
- Rest is important to the metabolic adaptations desired and the recovery from previous training bouts.

Key Point

Recovery from strength training bouts generally requires 48 hours of rest.

■ DESIGNING A STRENGTH TRAINING PROGRAM

Now that the types of equipment used and concepts of strength training are understood, it is time to put these together to form a strength training program. There are three major issues to consider regarding the extent of an individual's adaptations resulting from a program.

1. Each individual has a different genetic makeup and often a different potential for improvement. Therefore, each athlete will respond a little differently in quality and/or quantity to a training program.
2. The amount or magnitude of improvement is related to the amount of change possible for a given participant. In other words, one can expect larger improvements from beginners and smaller improvements from more advanced participants. This is the law of diminishing return.

3. The level and amount of adaptation depend on the degree to which the exercise prescription addresses the needs of the athlete. In other words, are the exercises and conditioning appropriate to achieve the desired outcomes?

Key Point

Individuals may differ in response to a given strength training program.

The design for an effective strength training program also needs to address the characteristics of the person's sport, the physical components of the goal, and certain training considerations. These factors in program design are summarized and discussed below.

Performance Characteristics and Nature of the Sport or Activity

Before determining what the specific goals are for an athlete and designing the training program, it is important to analyze the sport or activity. The questions listed below are designed to help determine and understand the nature of the activity. These questions may not be all-encompassing for any particular sport, but they provide a starting point for the information collecting process.

- What types of movement patterns are used in this activity and how complex are they?
- How long does the activity last and is it continuous or intermittent?
- How long are the season and the practice periods, and how frequent are competitions?
- What are the characteristics of the travel schedule?
- In what type of climates does this activity take place?

Key Point

Understanding the nature of the sport, length of season, and frequency of competitions is important in designing an effective strength training program.

Physical Characteristics

The next step in the program design process is to identify the specific physical components necessary for a given performance level. The questions below may not satisfy all sports but they initiate the thought process for assessing the physical attributes (levels of speed, strength, power, flexibility, jumping ability, agility, endurance, aerobic power, anaerobic power) necessary for participation in the activity.

- What are the primary and secondary muscle groups used for the sport/activity?

- What are the concerns for injury? What joints and tissues get injured and how can they be strengthened? Does the participant have any special injury considerations?

Training Considerations

After the physical components have been determined, the next step is to figure out how to improve them. Following are some questions to get started.

- What type of training protocol will elicit the desired outcomes?
- What frequency, intensity, time, and type of exercises should be used?
- What training volumes are appropriate and how will yearly training be structured?

For each desired physiological adaptation (such as improved power), the frequency, intensity, time, and type of activity must be considered. The type of exercise is the first step, as this may dictate intensity, frequency, and time. Choosing exercises that mimic or are similar in nature to the activity allows the greatest transfer of training effect. Other primary multijoint exercises and specific body part (isolation) exercises should be considered as well. Table 12-6 depicts the exercise categories and some common lifts associated with them.

Table 12-6

Exercise Categories and Associated Exercises

Olympic Lifts and Movements

Power clean	Clean and jerk
Snatch	Push press
Hang clean	Hang snatch
High pull	Snatch squats

Power Lifting Movements

Deadlift
Bench presses (flat, incline, decline)
Squats (back)

Multijoint and Compound Exercises

Squats (front, back)	Bench presses (flat, incline, decline)
Lunges	
Leg presses	Shoulder presses
Hip sleds	Pull-ups/downs
Rowing exercises	Bar dips
Bent-knee deadlifts (RDL)	Pull-overs

Isolation and Single-Joint Exercises

Biceps curls	Triceps extensions
Leg curls	Shoulder lateral/front raises
Leg extensions	Chest Fly
Back extension	Abdominal crunch
Heel raises	Neck exercises

Table 12-7

Exercise Order Based on Body Parts

1. Upper legs and hips	5. Arms
2. Back	6. Lower legs
3. Chest	7. Abdominals and low back
4. Shoulders	

Key Point

Structure the training program around the physical attributes desired.

EXERCISE ORDER.   The order of the chosen exercises also plays a role in a program's effectiveness. The exercises most important to the program should generally be executed first. The more complex structural and multijoint exercises should be completed prior to the isolation exercises. The higher power and higher speed movements (such as Olympic movements) should be performed early in the training session due to their use of the total body. Table 12-7 provides a typical and suggested order of training by body part. Notice that abdominal and low back exercises are the last muscle groups trained. This is due to their function in postural control of the spine; premature fatigue could increase the chance of injury.

TRAINING SCHEDULES.   In order to meet the goals of the training program, many different exercise schedules and protocols have been developed and used. Table 12-8 shows samples of different weight training protocols for specific body parts in terms of the frequency per week. Training programs that break the body into specific muscle groups for training on specific days are called **split routines.**

Key Point

The order of exercises is important and generally begins with large muscle groups and then moves to smaller groups.

TRAINING VOLUME.   Following the exercise selection and appropriate order, the overall volume and specific muscle group volume should be determined. This translates into determining the number of sets and repetitions for each exercise based on the specific adaptations desired (strength, power, speed, size), individual characteristics, and the phase of periodization.

REST PERIODS.   Along with the set and repetition prescription, the amount of rest between sets and between workouts must be decided. The amount of rest between

Table 12-8   Sample Training Schedules

3 Days per Week, Total Body   M, W, F or T, TH, S

Week	M	T	W	Th	F	Sa, Su
4 Days per Week, Split Routine						
All	L,B,Bi,Cf	C,S,T,A	Rest	L,B,Bi,Cf	C,S,T,AT	Rest, Rest
5 Days per Week, Split Routine						
1	L,Cf	C,T	Rest	S,A	B,Bi	L,Cf	Rest
2	Rest	C,T	S,A	Rest	B,Bi	L,Cf	C,T
3	Rest	Rest	S,A	B,Bi	Rest	L,Cf	C,T
4	S,A	Rest	Rest	B,Bi	L,Cf	Rest	C,T
5	S,A	B,Bi	Rest	Rest	L,Cf	C,T	Rest
6	S,A	B,Bi	L,Cf	Rest	Rest	C,T	S,A

Abbreviations:

A: abdominals	Cf: calves
B: back	L: legs
Bi: biceps	S: shoulders
C: chest	T: triceps

sets is primarily determined by the energy system being addressed and the specific adaptation desired. For example, longer rest periods (>2 minutes) are generally used with programs designed for improving maximal strength and power, whereas shorter rest periods (<2 minutes) are used for high-intensity endurance. For moderate-intensity muscular endurance adaptations, rest periods generally range from 30 to 60 seconds. The rest period between workouts is primarily determined by the recovery time needed. Higher intensity bouts of exercise and those with greater volumes may require longer recovery periods.

PRACTICAL CONSIDERATIONS.   Finally, the program design should implement all of the above items in a reasonable manner based on time constraints, facility space, equipment availability, travel schedule, and more. Table 12-9 summarizes the categories for designing a strength training program.

■ DESIGNING A STRENGTH TRAINING PROGRAM *IN REVIEW*

- Understanding the nature of the sport and its physical attributes is the first step in structuring the strength training program.
- Use the training variables (FITT) to address the physical characteristics desired.
- Consider the metabolic adaptations desired and structure the rest periods to coincide with them.
- Formulate a nutrition strategy based on the goals of the program.

Table 12-9

Determinants for Creating an Exercise Program

Step 1	Performance characteristics
Step 2	Physical characteristics
Step 3	Training program
Step 4	Nutritional considerations
Step 5	Practical considerations

■ AGE- AND GENDER-RELATED FACTORS IN STRENGTH TRAINING

When is it safe to begin strength training? Most young athletes are beginning to train prior to complete maturation. This may or may not be beneficial based on the type of training, the athlete's age, the particular sport, and more. On the other end of the age spectrum, the senior population is also rapidly adopting strength training as a mode of exercise. Females also have more opportunities in organized sports with the advent of Title IX, and their participation in other recreational activities has risen as well. Does strength training affect the genders differently? This section explores the ramifications of strength training for these populations.

Resistance Training and Children

All too often, children are considered as simply small adults when it comes to exercise training. However, children have growing bones, connective tissues, and muscles that require special consideration for participation in safe and effective strength training programs. The use of strength training programs for improving various physical attributes in children is supported by several major organizations, including the National Strength and Conditioning Association (NSCA), the American Orthopedic Society for Sports Medicine, the American College of Sports Medicine (ACSM), and the American Academy of Pediatrics.

BENEFITS OF STRENGTH TRAINING FOR CHILDREN. The benefits for children are similar to those for other populations: increases in muscular strength and muscular endurance, enhanced functional capacity, improved performance capacity, and decreases in the number or severity of injuries during physical activities. Several studies have shown that training results in strength gains in children, including prepubescent children, when compared to children of the same age who do not train.[2–7]

The increases in strength found in prepubescent children have generally been attributed to neural adaptations rather than muscle hypertrophy. The lack of muscle hypertrophy observed in prepubescent children is most likely due to preferential energy utilization and hormonal factors not present until or after adolescence. More specifically, the demands of growth and maturation often consume a good deal of the available energy and thus limit muscle hypertrophy. Also, due to a lack of the hormone testosterone, most prepubescent children do not have the capacity to significantly increase muscle mass. Therefore, gains in muscle mass should not be a primary goal in resistance training programs until the child has entered or passed through adolescence.

Muscular development has been correlated with skeletal age and therefore the potential for hypertrophy increases as one approaches adulthood. From birth to adulthood, muscle mass increases from approximately 25% to 40% of total body weight. This growth in muscle mass is primarily accomplished through muscle fiber hypertrophy caused by increases in contractile properties. Figure 12-21 outlines the associated factors for strength development in males from birth through adulthood.

A recent report on different types of weight training programs for children indicated that a higher volume of work (such as 1 set of 12–15 reps) was more beneficial in developing both muscular strength and endurance than higher intensity training (such as 1 × 6–8).[3] The study found this to be true for both leg extension and chest press tests. However, there was a difference in the level of improvement between the upper and lower body tests, which may indicate that optimal training protocols for strength and endurance vary for specific muscles.

The greater improvement found with the higher volume program may be due to the increased overload stimulus of performing more repetitions and thus better neurological adaptations (motor unit recruitment). This differs from an adult model, in which a lower number of repetitions is generally prescribed for muscular strength. Further research in this area is warranted to better understand the responses of specific muscle groups to various sets and repetition programs in children.

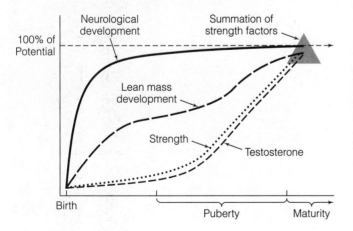

Figure 12-21
Summation of strength development factors in males.

Key Point

Gains in muscular strength in children are generally attributed to neural adaptations rather than muscular hypertrophy.

STRENGTH TRAINING AND THE NERVOUS SYSTEM IN CHILDREN. As children age, they develop more efficient movement patterns (coordination), become more agile, and generally improve their balance. This is a result of the developing nervous system. Nerve fibers need to become myelinated in order for fast transmission and communication of stimuli to occur. This myelination process occurs rapidly during childhood and continues through puberty. Although practicing certain movements may be helpful, a child's physical ability is limited by incomplete myelination.

Key Point

Improvements in performance and strength are often associated with the maturation process of the nervous system.

STRENGTH TRAINING AND ANAEROBIC CAPACITY IN CHILDREN. After exercise, levels of lactate in muscle tissue and blood are lower in children than in adults. This makes it seem that children have high anaerobic capacities, when in fact they have lower capacities than adults do. The relatively lower levels of tissue lactate suggest that glycolytic capacity is compromised in children. As discussed in Chapter 4, the enzyme *phosphofructokinase (PFK)* regulates a key reaction and thus is a rate-limiting step in glycolysis. Children have been reported to have lower concentrations of PFK than adults, which in turn affects glycolytic capacity. Additionally, anaerobic power has been shown to be lower in children. Despite this lower anaerobic capacity, children who undertake anaerobic strength training have shown increases in resting levels of phosphocreatine, ATP, glycogen, PFK activity, and maximal blood lactate levels.

Key Point

Children have lower anaerobic capacities than adults; however, strength training can still improve their anaerobic capacity.

STRENGTH TRAINING AND BODY COMPOSITION IN CHILDREN. At birth the fat percentage of total body weight is approximately 10–12%. At maturity body fat levels are typically increased to approximately 15% and

25% for males and females respectively. One major reason for the difference between the genders is the increased levels of estrogen in females starting with the onset of puberty. Subsequently, this increase in estrogen promotes fat deposition.

As discussed in greater detail in Chapter 7, fat tissue can expand by increasing individual cell size (hypertrophy) and cell number (*hyperplasia*). Hypertrophy accounts for the majority of adipose tissue expansion. Most adipocytes in fat tissue do not seem to have the ability to expand to the point of splitting. Periods of hyperplasia probably occur when small pockets of adipocyte stem cells divide due to signals from hormones, including some released by swollen adipocytes.[8]

The level of adipose tissue found in children is determined by a combination of factors that include exercise habits, health, heredity, and nutritional intake. Unfortunately, obesity among children is becoming more common and is now a major health concern, as discussed in Chapter 1. The increasing incidence of obesity in children stems from overconsumption of energy, especially energy derived from fast foods and snacks, combined with a decrease in overall physical activity.

STRENGTH TRAINING AND BONE DENSITY IN CHILDREN. The bone density of children may be enhanced through the muscular tension created by resistance training. This muscular tension combined with the load bearing of barbell and machinery can provide the needed stimuli for bone growth. Increasing bone density can be helpful in decreasing the chances of developing osteoporosis later in life. Also, any increase in bone density increases the body's structural integrity and total weight, which is particularly advantageous in the contact sports, such as football and hockey.

STRENGTH TRAINING AND INJURY IN CHILDREN. Many of a child's tissues are in the developmental stage and susceptible to injury. Of particular concern are the growth cartilage tissues found at the epiphyseal plates (growth plates), the epiphyses (the articular cartilage at joint surfaces), and the apophyseal (tendon) insertions (Table 12-10). To avoid injuring these tissues, it is extremely important to avoid maximal and/or near maximal lifts and those that are very ballistic in nature. The forces generated by these high-intensity movements may not be adequately attenuated by the developing musculature and thus may cause injury at these growth sites. Using correct lifting techniques, especially with free-weight exercises, greatly reduces the risk of injury to these tissues. To ensure a safe and effective weight training program, proper education, training, and supervision must be provided.

Table 12-10 **Some Injury Sites of Children**

Epiphyses: articular cartilage of joint surfaces
Epiphyseal plates: growth plates of the long bones
Apophyseal insertions: tendon insertion points

Key Point

Children have growing tissues that are susceptible to injury from high-intensity exercises.

Strength Training Program Design for Children

Developing a strength training program for children should follow the same steps as those for adults. However, the level of maturity both mentally and physically must be considered. The techniques of the exercises and proper use of the equipment must be clearly understood. Despite recent developments in the fitness industry, most resistance machines are still designed for average-sized adult users and do not properly accommodate children. Therefore, many of the exercises using body weight as the resistance are recommended, as are free-weight exercises performed at low to moderate intensities. Table 12-11 outlines some guidelines for strength training for children.

Although the same FITT concepts apply to children and adults, each exercise program design should be individualized. Variability within the same age, both mentally and physically, often occurs and therefore using a standardized protocol may not be appropriate.

Table 12-11

Basic Guidelines for Resistance Exercise Progression for Children

Age	Considerations
5–7	Introduce basic exercises with little or no weight such as calisthenics and partner exercises; develop concept of a training session; emphasize exercise techniques; keep volume low
8–10	Gradually increase number of exercises; practice exercise techniques for all movements; start gradual progressive overload techniques; keep exercises simple; slowly increase volume; monitor exercise tolerance carefully
11–13	Teach all basic exercise techniques; continue progressive overloading; emphasize exercise technique; introduce more advanced lifts with light or no weight
14–15	Progress to more advanced resistance exercise programs; add sport-specific components; emphasize technique; increase volume
16+	Entry level into adult programs after background experience has been gained

*Source: Adapted from Fleck SJ, Kraemer WJ. *Designing Resistance Training Programs*, 2nd ed., Table 10.2, p. 206, Champaign, IL: Human Kinetics, 1997.

Strength and Resistance Training for Women

An increasing number of women use resistance training as part of comprehensive conditioning programs. Also, female collegiate athletic programs are growing and the need to properly condition these athletes must be met. How do females differ from their male counterparts when it comes to strength training?

STRENGTH AND WOMEN.   The maximal mean total body strength of an average woman is 63.5% of an average man's.[9] This average was calculated using a variety of lifts on an absolute scale. Lower extremity strength tests in women showed a closer relationship (greater percentages) to the men than the upper extremity tests. However, when adjustments for total body weight and lean body mass are made, women are much closer to men in strength. Women's bench press strength has been reported as 46% of men's when adjusted for total body weight and 55% when using lean body weight.[10] For lower extremity strength, these numbers were much higher: 92% and 106% for total body weight and lean body mass, respectively. These data indicate that upper body strength in females is lower both absolutely and relatively than in males. However, lower body strength is similar and possibly greater in females when adjustments for body weight and lean body mass are made.

Key Point

Women are closer to men in absolute and relative strength when calculations account for lean body mass.

POWER OUTPUT FOR WOMEN.   In studies comparing power output between the genders, men had higher values for many tests even when lean body mass and other anthropometric characteristics were accounted for. The reasons are still being debated; one possible explanation is that the rate of force development and thus power is slower for the average woman.[11,12]

MUSCLE ADAPTATIONS IN RESPONSE TO STRENGTH TRAINING IN WOMEN.   Training effects appear to be similar for men and women. The manner in which muscle fibers (Types I, IIa, IIb) respond to resistance training are similar in men and women.[13–15] The rate of strength gains (% improvement) from weight training programs has also been found to be similar, although differences in absolute measures still exist.[10,13,16]

Muscle hypertrophy occurs within the various fiber types in women, but there are differences in the level of hypertrophy compared to men. These differences are generally attributed to hormone concentrations of testosterone, cortisol, and human growth hormone. Women generally have lower levels of these hormones and do not develop the mass observed in their male counterparts. However, there is a large variability within each gender, so some women naturally develop larger muscle masses than some men.

Despite the discrepancy in the level of muscle hypertrophy, both genders respond in a similar fashion to resistance training. Therefore, it seems logical to prescribe exercise in a similar fashion regardless of gender. Performing a needs analysis provides insight to the desired outcomes (such as increased speed, power, strength, size, or endurance) and the concepts of training are applied in the same way.

Key Point

In general, women respond to strength training similarly to men. Due to the level of certain hormones, the degree of muscle hypertrophy that can occur in most women is less than that in men.

CHANGES IN MENSTRUATION WITH STRENGTH TRAINING.   In many types of physical training, some women experience changes in their menstrual cycles. Although there is variability within each woman's menstrual cycle, physical exercise, including strength training, has been considered a contributing factor. The volume of the activity seems to be a primary cause of menstrual irregularities, particularly when the exercise volume creates a significant chronic energy deficit.[17] Once the activity volume levels are decreased or activity is stopped, menstrual cycles usually return to normal without any permanent damage to the reproductive system. However, amenorrheic women have been found to have bone density losses, which may increase the risk of developing osteoporosis.

On a more positive note, physically active women have reported decreases in premenstrual symptoms such as breast enlargement, appetite cravings, and mood swings.[18] Pain during menstruation (dysmenorrhea) has been reported to decrease among active women, and physically active women have shown increases in bone density, barring amenorrhea.

Resistance Training for Older Individuals

It has been established that the senior population, just as any other, can benefit from strength training programs. Improvements in muscular strength, muscular endurance, muscular hypertrophy, and functional capacities for daily life activities have all been reported from participation in resistance training. Unfortunately, the normal pattern for physical activity in adults declines with age. In industrialized societies, the level of regular physical activity declines soon after adulthood is reached. Improvements in technology have made it easier for people to avoid physical work.

Technology has also changed the fitness environment with specially designed machines and facilities. In fact, some people have become so accustomed to this environment that they do not exercise unless they have such things as air conditioning, music, TV, and automated machinery.

MUSCULAR STRENGTH IN OLDER INDIVIDUALS. As individuals age, levels of strength and endurance begin to decline. The rate of this decline has been reported to be 1–2% per year starting between the ages of 20 and 35.[10] This age-related decline in muscular strength is primarily due to decreases in muscle mass. It appears that the percentage of Type II muscle fibers decreases and the relative presence of Type I muscle fibers increases, although these changes are still debated. The suggested mechanism for this is a lack of neural activation of the Type II motor neurons. Without proper innervation, these muscle fibers tend to atrophy and may eventually die. Some other postulated reasons for loss in muscle strength in the elderly include the accumulated effects of chronic diseases, hormonal changes (decreased testosterone and growth hormone levels), lack of proper diet, and use of medications.

Despite the age-related loss of muscle mass, regular training can deter this process, at least in part, and help maintain muscle tissue and possibly strength. Neurologically, the aging process tends to slow down the capacity to process information and activate motor units. Physical activity can again help deter this process and maintain function.

Key Point

Regular strength training can help deter strength losses due to the aging process.

POWER PRODUCTION IN OLDER INDIVIDUALS. Similar to losses in strength, decreases in the production of power by muscle occur as one ages. Power may be more relevant in functional capacities than strength.[17] For example, movements needed to avoid falls, climb stairs, and lift objects all require a degree of power.[19] Therefore, training for power may enhance the performance of daily life skills for the elderly population.

MUSCLE ADAPTATIONS IN RESPONSE TO STRENGTH TRAINING IN OLDER INDIVIDUALS. Strength training for older people has been proven to increase muscular strength, muscular endurance, and muscular hypertrophy as long as the training overload stimulus is adequate. Additionally, decreases in percentage of body fat, intra-abdominal fat, and pain levels are expected. However, the level of adaptation may be somewhat lower than in a younger person.

STRENGTH TRAINING PROGRAM DESIGN FOR OLDER INDIVIDUALS.   On the basis of the physiological adaptations observed in the elderly, the training protocol

---

Table 12-12

Safety Considerations for Training the Senior Population

- Increase warm-up time to ensure the body is prepared for work.
- Increase rest periods (to 2–3 min) to ensure recovery from previous workload.
- Progress to multijoint, functional exercises when possible.
- Account for ROM differences to exercise through the pain-free range.
- Progress slowly in regard to intensity, duration, and volume (for example, 1 set × 8–12 reps).
- Be sure to use equipment with appropriate weight increments.
- Use extreme caution when using power oriented and ballistic movements.

---

and concepts remain consistent with any other population. Performing a needs analysis and then applying the FITT principle to meet the needs provides a sound framework. However, the considerations outlined in Table 12-12 should be taken into account when prescribing programs for seniors. Using these safety considerations and applying basic training concepts and principles can provide good results for older people. It is important to consider all health components of the participant prior to beginning an exercise program. If there are questions or concerns regarding the participant, a physician should be consulted.

■ AGE- AND GENDER-RELATED FACTORS IN STRENGTH TRAINING *IN REVIEW*

- Children are not just small adults. They have growing connective tissues that need to be considered when prescribing strength training programs.
- Strength improvements in youth are generally attributed to neurological improvements rather than muscle hypertrophy.
- Increasing lean body mass is difficult with children due to the large metabolic demand of the maturation process and low levels of testosterone.
- Muscular strength and muscular endurance adaptations from strength training programs are similar between the genders. However, muscle hypertrophy is generally lower in women as a result of lower concentrations of the hormones testosterone, cortisol, and human growth hormone.
- Strength training programs for the elderly can significantly improve muscular strength, muscular endurance, power, and body composition.

Conclusions

Strength training equipment comes in many forms, and understanding the benefits and limitations of each category of equipment should assist in choosing appropriate modalities. Using these modalities and incorporating the training variables (FITT) to address the exercise goals, a strength training program can be structured. The rest periods and intensity levels of the program can be designed to improve energy metabolism. Strength programs can improve many physical characteristics for children, the elderly, and both genders. Considering both the potential for improvement and the limitations for these groups should assist in developing safe and effective strength training programs.

IN FOCUS   MARK PHILLIPI, WORLD STRONGMAN COMPETITOR AND NATIONAL POWERLIFTING CHAMPION

Mark Phillipi has been a collegiate strength and conditioning coach for the past decade at the University of Nevada at Las Vegas (UNLV). Mark has had the good fortune, over the past decade and a half, to compete at an elite level in two of the more popular strength sports in the world today, powerlifting and the strongman competition. After completing a college football career in 1986, Mark began powerlifting and after 10 years of competing he won the national and world drug-free lifting titles in the 319-lb lifting class in 1996. Mark posted his best score at the national championships with a combined total of 2165 lb (984 kg). He wrapped up the title with an amazing final deadlift pull of 821 lb.

Mark enjoyed the sport and wanted to continue lifting so he sought out new challenges and found what he was looking for in the strongman competition. Mark competed in his first strongman event in 1997 and took second at a major Grand Prix event, the World Musclepower championship. His second-place showing put him ahead of several competitors who had finished in the finals of the World Strongest Man contest and also qualified him to compete in the American Strongest Man contest. Mark won the American title and went on to compete in the World's Strongest Man contest. Mark has made it to the finals twice and has been ranked as one of the top 10 strength athletes worldwide.

Although the training requirements have similarities, preparing for a powerlifting contest is quite different than preparing for a strongman event. Powerlifting is a sheer test of absolute strength in three lifts (squat, deadlift, bench press), whereas the strongman competition tests every facet of strength through the events used. Mark says that training is easier for powerlifting than for strongman events because the type of lifts are known in advance.

Mark uses many different types of powerlifting routines, incorporating several different schools of thought including periodization, speed and volume training, and high intensity. He uses unique programs for the squat, bench, and his specialty, the deadlift. He usually starts with a solid basic periodized program and then adds items to improve weak areas. He thinks this works better for the powerlifting competitions because they are more predictable than strongman contests, and progress can be easily monitored.

Strongman contests are becoming popular throughout the world. These events are a decathlon for strength athletes, testing every facet of strength: absolute strength, explosive strength, and strength endurance. Many of the events test several types of strength over a period of time; for example, the

MARK PHILLIPI, WORLD STRONGMAN COMPETITOR AND NATIONAL POWERLIFTING CHAMPION (CONTINUED)

farmer's walk (carrying heavy objects by handles over a distance) tests not only grip strength but leg endurance. Some events are extremely intense and can cause injury if used in training too frequently. Mark uses Saturdays for a special event day, practicing events that take a long time to set up.

A sound nutritional program is important to maximizing the benefits of intense workouts. Total recovery from intense workouts cannot take place unless proper nutrition is a priority. Mark has tried many types of diets and nutritional supplements in order to improve performance and has concluded that a solid diet is the first key to success. He eats a diet higher in protein than the normal diet and considers it vital to recovery. Extra protein is required to repair muscle tissue after intense weight training. Mark consumes approximately 0.75–1 g of protein per pound of body weight when training for a contest. Also, he does not monitor fat intake closely because he generally stays away from foods that are high in fat and those providing little nutritional benefit.

Mark's carbohydrate intake depends on his workout schedule. He tries to ingest carbohydrates immediately before and after workouts, especially on days that include strength endurance exercise. He also adds carbohydrates to his diet if he is trying to add weight for a contest. He might also reduce the level of carbohydrates if more speed events are in a particular contest and it is beneficial to be lighter and quicker.

Mark advocates only those supplements that are well researched. He uses supplements that are food-based or are naturally found in the diet but in low amounts so that consuming enough food to accommodate his nutrient requirements would be difficult. He emphasizes that he "sticks to the basics." A protein and carbohydrate supplement, creatine, and a vitamin and mineral packet make up his normal routine of supplements. He does not experiment with anything that sounds too good to be true because these are usually a waste of money or can be dangerous. There is no substitute for a solid diet, Mark says, so his nutritional focus starts with the food he eats.

Mark also feels that mental preparation is as important as physical preparation. He suggests that during a successful training cycle or a feeling of "being in the zone," an athlete try to analyze why things are going well in order to duplicate the actions that created the feeling in the next training period or competition.

Mark's final recommendation is to train hard, train smart, follow a solid diet, and keep an open mind to areas that can improve your performance. Leave nothing to chance and always wonder, "What is my opponent doing today? Did I outwork them?"

STUDY QUESTIONS

1. What are the components necessary to classify equipment as providing dynamic constant external resistance (DCER)?
2. Define and discuss the use of variable resistance equipment. How does it differ from DCER?
3. What are the pros and cons of using elasticity as a form of resistance?
4. What are the advantages of using free weights versus other forms of resistance?
5. What characteristics do "functional" exercises have, and what are the advantages of this form of resistance training?
6. What are plyometric exercises and how do they work?
7. What energy sources are used during strength training?
8. How is periodization accomplished and what are the benefits of this type of training regimen?
9. What are the implications of rest interval lengths between sets and workouts?
10. What are the benefits and limitations of strength training for youth?
11. How does gender affect strength training results?

REFERENCES

1. Ebben WP, Watts PB. A review of combined weight training and plyometric training modes: complex training. *Strength and Conditioning* 20(5):18–27, 1998.
2. Blimkie CJR. Age and sex-associated variation in strength during childhood: anthropometric, morphologic, neurologic, biomechanical, endocrinologic, genetic and physical activity correlates. In *Perspectives in Exercise Science and Sports Medicine*, vol. 2, *Youth Exercise and Sport* (Gisolfi C, Lamb D, eds.), pp. 99–163, Indianapolis: Benchmark Press, 1989.
3. Faigenbaum AD, Westcott WL, Loud RL, Long C. The effects of different resistance training protocols on muscular strength and endurance development in children. *Pediatrics* 104(1):1–14, 1999.

4. Faigenbaum AD, Zaichowsky L, Westcott W, Micheli L, Fehlandt A. The effects of twice a week strength training program on children. *Pediatric Exercise Science* 5:339–346, 1993.

5. Freedson PS, Ward A, Rippe JM. Resistance training for youth. *Advances in Sports Medicine and Fitness* 3:57–65, 1990.

6. Kraemer WJ, Fry AC, Warren BJ, Stone MH, Fleck SJ, et al. Acute hormonal responses in elite junior weightlifters. *International Journal of Sports Medicine* 13(2):103–109, 1992.

7. Sale DG. Strength training in children. In *Perspectives in Exercise Science and Sports Medicine*, vol. 2, *Youth Exercise and Sport* (Gisolfi C, Lamb D, eds.), pp. 165–216. Indianapolis: Benchmark Press, 1989.

8. Wildman REC, Medeiros DM. Body composition and obesity. In: *Advanced Human Nutrition*. Boca Raton, FL: CRC Press, 1999.

9. Laubach LL. Comparative muscular strength of men and women: a review of the literature. *Aviation, Space and Environmental Medicine* 47:53442, 1976.

10. Wilmore JH, Costill DL. *Physiology of Sport and Exercise*. Champaign, IL: Human Kinetics, 1994.

11. Komi PV, Karlsson J. Skeletal muscle fiber types, enzymes activities and physical performance in young males and females. *Acta Physiologica Scandinavica* 103:210–218, 1978.

12. Ryushi T, Hakkinen K, Kauhanen H, Komi PV. Muscle fiber characteristics, muscle cross-sectional area and force production in strength athletes, physically active males and females. *Scandinavian Journal of Sports Science* 10:7–15, 1988.

13. Cureton KJ, Collins MA, Hill DW, McElhannon FM. Muscle hypertrophy in men and women. *Medicine and Science in Sports and Medicine* 20:338–344, 1988.

14. Staron RS, Leonardi MJ, Karapondo DL, Malicky ES, Falkel JE, et al. Strength and skeletal muscle adaptations in heavy-resistance trained women after detraining and retraining. *Journal of Applied Physiology* 70:631–640, 1991.

15. Staron RS, Mallicky ES, Leonardi MJ, Falkel JE, Haferman F, Dudley GA. Muscle hypertrophy and fast fiber type conversions in heavy resistance trained women. *European Journal of Applied Physiology* 60:71–79, 1989.

16. Wilmore JH, Parr RB, Girandola RN, Ward P, Vodak PA, et al. Physiological alterations consequent to circuit weight training. *Medicine and Science in Sports* 10:79–84, 1978.

17. Fleck SJ, Kraemer WJ. *Designing Resistance Training Programs*, 2nd ed., Champaign, IL: Human Kinetics, 1997.

18. Prior JC, Vigna YM, Mckay DW. Reproduction for the athletic female: new understandings of physiology and management. *Sports Medicine* 14:190–199, 1992.

19. Bassey EJ, Fiatarone MA, O'Neill EF, Kelly M, Evans WJ, Lipsitz LA. Leg extensor power and functional performance in very old men and women. *Clinical Science* 82:321–327, 1992.

CHAPTER

13

ENDURANCE EXERCISE AND TRAINING

Chapter Objectives

- Define and describe aerobic and anaerobic exercise.

- Discuss the benefits of endurance training.

- Discuss the measures of cardiorespiratory function and intensity.

- Outline the concepts related to endurance training.

- Describe the acute responses and long-term adaptations to endurance training.

- Discuss fuel utilization and sources for aerobic activity under varying intensity levels.

- Discuss age- and gender-related differences regarding endurance training.

Personal Snapshot

Jose Galvez/PhotoEdit

Janice, a 30-year-old physical education instructor, has just been given the duties of coaching the track team at the local high school. Janice has experience in many sports and physical activities but wants to be sure that the training she is about to undertake with these teenage athletes makes sense. She remembers a small portion of the information from her exercise physiology class and has gained some insight from other coaches, but wants to brush up on her training concepts. She also wonders what nutritional factors might help the team members in their endeavors, so she decides to attend a Sport, Fitness, and Nutrition symposium to see what she can learn.

There is often some confusion in defining what constitutes cardiovascular, cardiorespiratory, endurance, or aerobic activity. These terms are often used interchangeably, and for the most part they refer to the same type of exercise from differing perspectives. For instance, *aerobic exercise* infers an energy metabolism perspective and may be more popular among fitness enthusiasts. *Endurance exercise* implies a duration or time perspective, and may be more applicable to competitive athletes. However, *cardiorespiratory endurance exercise* is probably the best descriptor because both the cardiovascular and respiratory systems are essential for sustained work. All three terms are used throughout this chapter and the book.

For the purposes of this text we define *cardiorespiratory exercise* as any activity that incorporates large muscle groups sustained in rhythmical activity for an extended time (>5 minutes). This includes activities such as walking, jogging, cycling, and even some forms of manual labor like gardening and cleaning. This chapter provides an overview of some basic concepts in cardiorespiratory exercise and training program development.

■ AEROBIC EXERCISE BASICS

What is *aerobic* activity? At what point, or more specifically, at what intensity level, does an activity become *anaerobic*? Exercise intensity can be placed on a scale, as shown in Figure 13-1. At lower intensities, oxygen availability is adequate to allow for complete oxidation of fuel sources by working muscle and other tissues. As intensity increases, O_2 consumption and availability increase in a related manner, but only to a point. When exercise intensity climbs even higher, muscle fibers have to rely on anaerobic systems to support the aerobic means of energy production, which have reached a plateau. When the intensity of exercise demands that the majority of ATP be generated by anaerobic mechanisms, the exercise is deemed anaerobic. The greater reliance on anaerobic mechanisms is reflected in increased production and accumulation of lactic acid (lactate). Lactic acid production

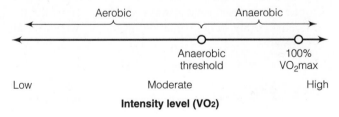

Figure 13-1
Intensity and energy continuum. The anaerobic threshold or lactate threshold represents a level of intensity at which the level of lactate in the blood rises significantly.

along with many other factors (nutritional state, fuel delivery, fitness level) is associated with muscular fatigue.

Anaerobic Threshold

Exercise physiologists refer to the intensity level associated with the accumulation of lactate in the blood as the **lactate** or **anaerobic threshold**. Below this intensity level, the lactate that is produced by working muscle and other cells (such as red blood cells) is efficiently oxidized by tissue such as the heart, liver, and inactive muscle cells. Exercise performed at intensities above the lactate threshold derives the additional ATP needed by anaerobic mechanisms. Thus, exercise becomes increasingly anaerobic as exercise intensity rises above the lactate threshold. The energy source utilized during sustained exercise above the lactate threshold is primarily carbohydrate, as discussed in Chapter 4.

Activities such as walking and gardening are considered aerobic since the intensity level is usually well below the lactate threshold. Running and cycling are also considered aerobic because they are generally performed at an intensity level that is below or close to the lactate threshold. However, when the intensity level for these exercises exceeds the anaerobic threshold, the increased lactate production will be a limiting factor to performance.

Endurance athletes must often perform at an intensity level at or just above the lactate threshold in an effort to be competitive. However, exceeding lactate threshold levels early in competition can be detrimental to performance. Therefore, competitive strategy needs to include the timing of high-intensity efforts (such as "sprints" or "kicks" in a 10-km race) to avoid excessive muscular fatigue. Endurance athletes train at or above anaerobic threshold in an effort to elevate the threshold level and adapt to better manage higher lactic acid levels.

Key Point

Competitive endurance athletes train at or just above the lactate threshold.

Endurance Training and Health-Promoting Adaptations

Aside from elevating the lactate threshold, endurance training has many health benefits. The most significant are improvements of the heart, lungs, and arterial system because any improvements to these can help reduce the chances of developing cardiovascular disease. As described in Chapter 1, heart disease is the number one cause of death in the United States and many other countries. Cardiorespiratory benefits of endurance training include decreased resting heart rate (HR) and blood pressure (BP) and an increase in stroke volume (SV). These cardiovascular aspects are defined in Chapter 2. Psychological benefits of cardiorespiratory training include decreased stress levels and an improved sense of well-being.

Cardiorespiratory training has also been correlated with decreases in the development of several cancers, obesity, diabetes, osteoporosis, and more. As the amount of time spent training increases, so do the benefits in health. However, the magnitude of improvement decreases as the potential for improvement decreases. In other words, for a similar percentage of improvement, an increased training frequency, intensity, or time is needed.

Key Point

Endurance training can help decrease the chances of developing cardiovascular disease, the number one killer in America.

LAW OF DIMINISHING RETURN AND ENDURANCE TRAINING. The concept of the law of diminishing return, introduced in Chapter 3, is again illustrated in Figure 13-2. Once initial thresholds have been met relative to frequency, intensity, and time, the return or response to increasing levels of these decreases. For example, if running time increases from 20 minutes to 30 minutes, the proportional improvement for the added 10 minutes is diminished compared to the return previously. This does not mean that improvements cease with increased efforts; it simply means that the proportional return for additional

training is less. This is also true for the intensity and frequency training variables.

The law of diminishing return also applies to a person who moves from a sedentary lifestyle to one of athletic performance. The potential for improvement of the sedentary person is initially high and then diminishes as improvements are made. Consequently, the level of benefit from a given amount of training begins to diminish as improvement increases. In other words, it gradually takes more effort to keep improving. For example, an elite marathon runner may have to train for a year in order to improve running times significantly, whereas a beginning runner can improve running times easily in a just a few weeks or months of training.

■ AEROBIC EXERCISE BASICS IN REVIEW

- Intensity level and subsequent energy sources determine when an activity is aerobic or anaerobic.
- The intensity level at which an aerobic activity becomes anaerobic is the lactate threshold.
- Aerobic training has numerous benefits that include decreases in risk for heart disease, resting heart rate, and blood pressure.
- As training progresses, the magnitude of improvements from a certain level of exercise decrease because of the diminished potential for improvement.

■ MEASURES OF CARDIORESPIRATORY EXERCISE INTENSITY

As noted above, the intensity of an exercise determines whether it is aerobic or anaerobic. **Intensity** is defined as *metabolic demand* or "how hard" the activity is being performed. Intensity of cardiorespiratory exercise can be assessed by measuring VO_2, HR, respiratory exchange ratio (RER), or respiratory quotient (RQ) and subjectively by a rating of perceived exertion (RPE) or the talk test. It should be noted that the maximum volume of oxygen consumed is expressed several ways which are generally used interchangeably. These include VO_2max, VO_2peak, and $VO_2reserve$.

Key Point

Intensity is the metabolic demand of exercise and is often measured by the volume of oxygen consumed (VO_2), heart rate (HR), respiratory exchange ratio (RER) or respiratory quotient (RQ), and rating of perceived exertion (RPE).

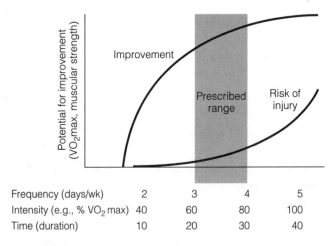

Frequency (days/wk)	2	3	4	5
Intensity (e.g., % VO_2 max)	40	60	80	100
Time (duration)	10	20	30	40

Figure 13-2
Concept of diminishing return. Increasing the frequency, intensity, and time of exercise has greater benefits in the low and middle ranges than in the upper range. The risk of injury also increases faster in the upper range.

Volume of Oxygen (VO_2)

The metabolic demand for a given endurance activity can be determined by the volume of oxygen consumed (VO_2), which is reported both in absolute terms (liters of oxygen/minute) and relative terms (ml/kg of body weight per minute, abbreviated ml/kg/min). The relative term adjusts for differences in body mass, allowing for comparisons of individuals of varying sizes. VO_2 has a positive linear relationship with work rate. Therefore an individual's VO_2 for walking should be lower than his or her VO_2 for running since the work rate is lower for walking. The maximal amount of oxygen consumed is termed *VO_2max* and is expressed in L/min or ml/kg/min. Intensity of cardiorespiratory exercise can also be assessed in metabolic equivalents (METs), calculated as a multiple of the resting VO_2 (1 MET is O_2 consumption of 3.5 ml/kg/min).

Key Point

VO$_2$ has a positive linear relationship with work rate.

The concept of VO_2 revolves around the ability of the body to take in and utilize oxygen in energy metabolism. More specifically, VO_2 measures how well oxygen can be inspired into the lungs, diffused into the pulmonary bloodstream, pumped by the heart to working muscle, and utilized to produce ATP. Thus many organs and tissues are associated with endurance activities, including the heart, lungs, blood, blood vessels, and skeletal muscles.

GRADED EXERCISE TEST. To determine VO_2 a **graded exercise test (GXT)** is typically conducted. These tests can be maximal or submaximal in nature, and they determine the VO_2 consumed at gradually increasing work rates. Maximal tests gradually increase work rates until VO_2 peaks and levels off; at this point the test is terminated. Submaximal tests increase work rates to a heart rate of approximately 85% of theoretical maximum. At this point the GXT is terminated and a maximum work rate is extrapolated from the data. From this estimate, VO_2 values can be calculated. It should be noted that each particular activity might provide differing, although similar, VO_2max values. GXTs determine physiological responses to work rates, assist in program design, and screen for cardiovascular disease.

MAXIMAL TESTS.   In a maximal GXT, the participant ventilates through a hose attached to a metabolic cart that determines the volume of oxygen used (O_2 inspired − O_2 expired = arterial-venous difference: A-V O_2 diff). A GXT also records HR and BP since these are basic indicators of cardiovascular health. The HR and SV of the heart determine cardiac output (Q), which indicates how much oxygen-endowed blood is being delivered per minute to working muscle tissue. Therefore $VO_2 = HR \times SV \times$ A-V O_2 diff.

The VO_2 and HR gradually increase with elevated work rates and finally plateau. At this plateau the delivery of oxygen can no longer meet the demands of the working muscle tissue and the GXT is ended. This point determines the maximal volume of oxygen that can be delivered and utilized (VO_2max) for that particular activity. It also determines the maximal attainable number of heartbeats/minute (HR_{max}). To determine the percentage of VO_2max use the following equation:

$$\%VO_2max = (VO_2 \text{ during exercise})/VO_2max \times 100$$

SUBMAXIMAL TESTS.   Although not as accurate as maximal tests, estimates or submaximal tests are often used to predict VO_2max and level of fitness. Submaximal tests are lower in cost, quicker, easier to administer, and safer. They can be conducted on a large scale and provide general information on cardiorespiratory function. For example, health clubs often use submaximal tests to provide guidelines for the exercise prescription and to screen for cardiovascular disease. The results from these tests do not need to be exact; general information will suffice. In contrast, an athletic team might use maximal tests in order to more precisely prescribe exercise and evaluate the present condition of an athlete. Because these individuals are training regularly, the risk of a cardiac incident from using a maximal GXT is reduced. Several methods used to estimate VO_2max or intensity levels are shown in Table 13-1.

Key Point

To determine VO$_2$max and HR$_{max}$ a graded exercise test (GXT) is often used. This not only determines values for these variables but also screens for cardiovascular disease.

Heart Rate (HR)

Because HR has a linear relationship with work rate and VO_2, it can be used as a predictor of cardiorespiratory exercise intensity. Several different methods have been used to predict HR_{max}. The straight-line method uses an equation ($HR_{max} = 220 -$ age) that accounts for the age of the participant. This serves as a general guideline because HR_{max} decreases with age. Another, more accurate estimate is the heart rate reserve (HRR) method. This equation ($HR_{max} - HR_{rest}$) is the difference between maximal and resting heart rate and determines the number of heartbeats per minute available for work. This method became popular because the percentage of HRR is

Table 13-1	**Definitions and Equations for Intensity Levels**

VO_2max (also termed $VO_{2reserve}$ or VO_2R): the maximum volume of oxygen that can be delivered to working muscle tissue

Volume of oxygen consumption: VO_2 = HR × SV × A-V O_2 diff.

Expressed in L/min, ml/kg/min, or METs (**1 MET** = 3.5 ml of O_2 per kg of body weight per minute) (A-V O_2 diff = O_2 inspired − O_2 expired)

Predicted heart rate maximum: HR_{max} in beats per minute (bpm) = 220 − age

Straight line target heart rate (THR) method: % of HR_{max}

Example: Age 40, THR of 60–80%

220 − 40 = 180 bpm HR_{max}

180 = .60 = 108 bpm

180 × .80 = 144 bpm   **THR Range = 108–144 bpm**

Heart rate reserve: HRR = HR_{max} − HR_{rest}

% of HRR = % VO_2max

Target heart rate (THR) = %HRR + HR_{rest}

Example: HR_{max} of 200 bpm, HR_{rest} is 70 bpm, THR 60–80%

200 − 70 = 130 × .60 = 78 bpm + 70 bpm = 148 bpm

200 − 70 = 130 × .80 = 104 bpm + 70 bpm = 174 bpm   **THR Range = 148–174 bpm**

Respiratory exchange ratio: RER = VCO_2/VO_2

1 L of O_2 = ~5 kcal of energy expenditure

Example: Caloric expenditure for a bout of exercise

VO_2 = 3.0 L/min; intensity level 60% of VO_{2max}; exercise times 20 min

3.0 L/min × .60 × 5 kcal/L × 20 min = 180 kcal

Stroke volume (SV): volume of blood ejected from the left ventricle

Cardiac output: Q = heart rate (HR) × stroke volume (SV)

Lactate threshold: = work rate at which blood lactate levels rise beyond resting levels.

approximately the same as the percentage of VO_2max.[1] Figure 13-3 depicts various areas of the body to measure HR.

Intensity levels can be determined by the percentages of these variables: VO_2max, HRR, or HR_{max}. The straight-line method for HR_{max} is the least accurate of these methods because it does not use any individual values, just standardized (average) numbers for HR. Some research findings suggest that straight-line HR_{max} underestimates VO_2 by 15%.[2] Therefore, one can expect the numerical values for %HR_{max} to be slightly different than those of %VO_2max or %HRR. Table 13-2 illustrates the intensity relationships of these three variables.

Key Point

Use of the straight-line method for determining HR is not as accurate as the HRR method because it does not account for individual differences in resting heart rate.

Respiratory Exchange Ratio (RER)

Along with VO_2 measures, the RER can be calculated. This is the ratio of the volume of CO_2 produced divided

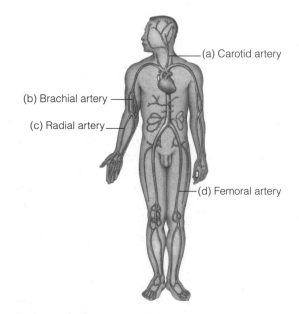

(a) Carotid artery

(b) Brachial artery

(c) Radial artery

(d) Femoral artery

Figure 13-3

Common sites for measuring heart rate. Carotid artery, brachial artery, radial artery, femoral artery.

Table 13-2

Relationship of %VO$_2$max, %HRR, %HR$_{max}$, MET

%VO$_2$max	%HRR	%HR$_{max}$	MET
50	50	66	5
55	55	70	5.6
60	60	74	6.2
65	65	77	6.8
70	70	81	7.5
75	75	85	8.1
80	80	88	8.7
85	85	92	9.3
90	90	96	10

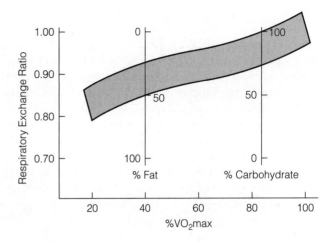

Figure 13-4
Relationship of exercise to energy yields from carbohydrate and fat. Estimated energy yields (%) of fat and carbohydrate in relation to nonprotein respiratory exchange ratios at various intensities (%VO$_2$max). Extended duration, level of training, and nutritional state can have significant effects on these energy yields.

by the volume of O$_2$ consumed (VCO$_2$/VO$_2$). The amount of oxygen needed to oxidize an energy molecule is proportional to the amount of carbon dioxide produced. As presented in the feature In Focus: Understanding Respiratory Exchange Ratio (RER) and Respiratory Quotient (RQ) in this chapter, the complete oxidation of a mole of glucose (C$_6$H$_{12}$O$_6$) liberates six moles of CO$_2$ and requires six moles of O$_2$ to serve as acceptors of the hydrogen atoms. Therefore, for glucose the exchange ratio is 1.0. For the utilization of fatty acid the RER drops considerably (to 0.71) because of the higher amount of oxygen necessary to oxidize a fatty acid. For a 50/50 mixture of fats and carbohydrate the RER is approximately 0.85. Figure 13-4 depicts how fuel sources and thus RER values generally change with exercise intensity. RER is also discussed relative to energy balance and body composition in Chapter 7.

Key Point

Intensity can be estimated by RER because the amount of CO$_2$ released and O$_2$ used indicates the type of fuel utilized (carbohydrate, fatty acid).

Rating of Perceived Exertion (RPE)

Another method assessing the intensity level of cardiorespiratory exercise is the use of a rating of perceived exertion (RPE) chart. This numbered chart correlates how a person feels while exercising with an estimate of HR. The chart uses the number range of 6–20 to represent perceived exertion from resting to a maximal level.[3] These numbers can simply be multiplied by 10 to approximate the HR for a given work rate. Therefore a rating of 7 correlates to a HR of 70 beats/minute. Table 13-3 provides a

Table 13-3

Borg Scale of Rating of Perceived Exertion (RPE)

Scale	Description of Perceived Exertion	Associated Heart Rate
6		~60 beats per minute (bpm)
7	Very, very light	~70 bpm
8		~80 bpm
9	Very light	~90 bpm
10		~100bpm
11	Fairly Light	~110bpm
12		~120bpm
13	Somewhat hard	~130bpm
14		~140bpm
15	Hard	~150bpm
16		~160bpm
17	Very hard	~170bpm
18		~180bpm
19	Very, Very hard	~190bpm
20		~200bpm

sample RPE scale with associated heart rates. A true scale would not disclose the numerical HR values. RPE charts are often used for GXTs, fitness classes, and other clinical settings.

Talk Test

Another practical, although not very precise, method used to gauge cardiorespiratory exercise intensity level is the "talk test." This method also helps determine an appropriate training level for general fitness. It can be used alone or in combination with other methods (such as HR), allowing individuals to assess their approximate intensity levels based on breathing and talking. If talking is labored or not possible during the activity, the intensity level is considered too high for general fitness purposes. Research studies have indicated better exercise program adherence when intensity levels are on the lower end of the appropriate training zone (50–65% VO_2max).[4,5]

■ MEASURES OF CARDIORESPIRATORY EXERCISE INTENSITY *IN REVIEW*

- Intensity for aerobic activities can be determined or estimated by VO_2, HR, RPE, RER, and the talk test.
- Submaximal and maximal graded exercise tests can determine cardiovascular health and provide VO_2max, HR_{max}, and RER values.
- RER values reveal the percentages of the nonprotein fuel sources (carbohydrate and fat) being used.

■ CARDIORESPIRATORY TRAINING CONCEPTS

How are the training concepts for endurance exercise similar to those for strength training, and what are the special considerations for endurance training? The FITT (frequency, intensity, time, type) principle again provides the framework for all physical training. For any adaptations to occur a stimulus must be present. Therefore, the training concepts related to endurance training begin with the overload concept.

Overload in Cardiorespiratory Training

As discussed in previous chapters, in order for muscle or certain other tissues to adapt, some type of stimulus must be presented. The stimulus received must be different from what is normally encountered. It can be something completely new or simply a change in the quantity, duration, or intensity of the activity. For example, if a person typically jogs three times a week for 30 minutes at a moderate intensity, an overload stimulus could simply be increasing the frequency to four times a week, running longer, or running faster. Once this pattern of activity becomes the norm, a new training overload must be introduced in order for additional adaptations and improvements to occur.

Table 13-4

Recommended intensity Levels for the Deconditioned and General Populations

Deconditioned	General Populations
40–49% VO_{2max}	50–80% VO_{2}max
40–49% HRR	50–80% HRR
55–64% HR_{max}	65–90% HRmax
10–12 RPE	13–19 RPE
2–4 METs	5–10 MET

GRADUAL PROGRESSIVE OVERLOAD. The concept of gradual progressive overload (GPO) simply means to overload the body in a systematic and linear fashion. In other words, one gradually increases the frequency (for example, from 2 days/week to 3 days/week), the intensity, or the time during which the activity is performed. This GPO concept provides linear improvements for the specific training variable (FITT) being addressed.

INTENSITY THRESHOLD FOR OVERLOAD. The benefits from a given intensity level of training depend on the individual's age and current fitness level. People who are active and train regularly need higher intensity levels to accomplish overload, whereas older and sedentary individuals need less intensity to accomplish overload. Table 13-4 shows suggested threshold levels of exercise intensity for deconditioned people in the first column. The ranges in the second column apply to the general population.[6] Intensity levels below these thresholds may provide health benefits, but improvements in VO_2 may not be realized. Figure 13-5 illustrates the target heart rate training zones for various ages.

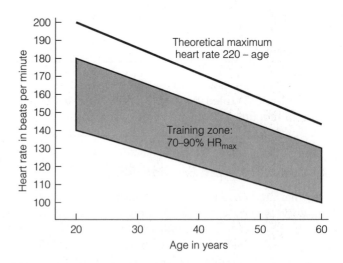

Figure 13-5

Target heart rate training zones. Maximum heart rates and suggested training zones for various age groups.

Table 13-5

FITT Principle for Cardiorespiratory Exercise

Frequency: How often the activity is performed (such as number of runs/week)
Intensity: The level of difficulty the activity is performed (such as %VO$_2$max, %HR$_{max}$)
Time: The duration of the activity (such as 30–60 minutes)
Type: The specific mode of activity (such as running, cycling, swimming)

Key Point

As improvements in fitness level occur, progressively greater intensity levels are required for overload.

FITT Principle

The FITT principle provides a conceptual framework from which all training programs can be designed. FITT is an acronym for the acute training variables: frequency, intensity, time, and type. Manipulating these training variables elicits overload for the physiological adaptations desired. Table 13-5 outlines and defines these components relative to endurance training.

FREQUENCY OF CARDIORESPIRATORY TRAINING. Research studies have shown that improvements in cardiorespiratory function (VO$_2$max) are possible with a training frequency of once a week. However, considerably greater improvements are observed when the training frequency is in the range of 2 to 4 days per week.[6] Further improvements in VO$_2$max can be elicited with increased frequencies of 5 or more days a week, but the relative improvements are diminished, meaning it would take much more effort and time to see similar improvements (the law of diminishing return). For general improvements in cardiorespiratory function with a minimal risk of injury and attainable weight loss goals, a regimen of exercising every other day (exercise on day 1, rest on day 2) has been validated.[6,7] The exercise volume, or distance run per week, has also been linked to reductions in risk factors for coronary heart disease.[8] General fitness endurance programs using more than 4 days/week appear to have higher dropout rates and higher injury rates.[9,10]

Key Point

Optimal frequencies for general fitness endurance training range from 3 to 5 days per week.

INTENSITY OF CARDIORESPIRATORY TRAINING. Competitive endurance events generally require the competitors to perform at an intensity level just below their respective anaerobic thresholds in efforts to improve race times. This intensity level enables the athletes to maximize work output for an extended period while conserving some carbohydrate for the final portion (the "kick") of a race. Thus most of the training intensities should be at the desired intensity level, or "race pace."

However, some of the training bouts should incorporate higher intensity efforts that surpass the anaerobic threshold. The purpose of these training bouts is not only to condition the anaerobic systems but also to elevate or raise the anaerobic threshold. Training bouts at higher intensities force the body to function with elevated lactic acid levels, which should result in an adaptation to process greater amounts of lactic acid and thereby allow greater sustained work rates. Adaptations that allow for greater work rates without excess accumulation of lactic acid elevate the anaerobic threshold. For general fitness purposes the intensity levels do not need to be as high. However, initial threshold levels still need to be reached for improvements to occur.

TIME OF CARDIORESPIRATORY TRAINING. If the goal of the training bout is to accomplish a certain amount of work, expend a certain level of energy, or cover a certain distance, the intensity level determines the duration. For example, if the goal is to expend 300 kcal, less duration (time) would be needed for an intensity level of 90% VO$_2$max versus 70% VO$_2$max because the metabolic demand would be much higher.

Looking back at Figure 13-2, the risk of orthopedic injury begins to rise quite rapidly after 30 minutes of exercise and rises exponentially thereafter. This rise in injury rate also holds true for increasing intensity and frequency levels. Therefore, by maintaining some moderate levels within these components, adequate volumes of work can be accomplished while minimizing the risk of injury. For general fitness purposes the prescribed duration range is 30–60 minutes.

Key Point

The duration of an activity is usually determined by the intensity level at which it is performed.

TYPE AND SPECIFICITY OF CARDIORESPIRATORY TRAINING. As previously mentioned, if VO$_2$max values are calculated for multiple activities, the values are fairly consistent. In other words, the VO$_2$max value for cycling is similar to the VO$_2$max value for running. However, the greatest VO$_2$max values for an athlete are generally found in his or her particular sport. This is due to the specificity (for example, in muscular contraction velocities, tension, ROM, intensity, duration) and adaptations from long hours of practice and performance of that particular activity. Simply stated, a runner is better conditioned for running than for cycling.

Endurance training requires increased oxygen provision to working muscle for aerobic energy metabolism. This increases the demand placed on the pulmonary and cardiovascular systems. Since these systems are enhanced during any form of endurance training, improving general fitness levels can be achieved with any form of cardiorespiratory exercise as long as the initial threshold level (such as 40% VO_2max) is sufficient. However, for improved athletic performance, specific adaptations of the muscle groups used and subsequent blood delivery to tissues must take place. Therefore, the type of activity must be considered in order to receive the desired physiological adaptations for improvements in endurance-related athletic performance.

Key Point

General fitness levels can be achieved with any form of cardiorespiratory exercise.

Periodization of Cardiorespiratory Training

For a runner to successfully compete in a marathon, two factors must be addressed: the distance of the race and the rate at which the distance is covered. These two factors determine the performance level of the athlete. For the best performance, the runner needs to run at the highest velocity that can be maintained while covering the required distance. Using a *periodized* training scheme, the runner concentrates on distance on some days or weeks of training, and speed on others. This accomplishes overload for the specific characteristic (duration, intensity) and associated energy systems.

As discussed in Chapter 3, *periodization* refers to the systematic variation of exercise training, and it provides specific overloads as well as variety to the training. Use of a periodized training program also helps to deter orthopedic injuries. A sample 3-week marathon training schedule is presented in Table 13-6. Notice how intensity and duration are manipulated. This 3-week schedule can be cycled along with 2–3 weeks of lower volumes (less overall mileage) of training to accomplish an appropriate long-term training schedule. Tapering, or decreasing training volume, should occur during the last 3 weeks prior to a race to ensure adequate rest and allow glycogen stores to be maximized.

The training regimens for other endurance sports look similar, with light-intensity days performed with greater duration and higher intensity days performed for

Table 13-6

Sample Marathon Training schedule

	Week 1	Week 2	Week 3
Monday	Active rest, Z1	7–10 mi, Z2	10–12 mi, Z2
Tuesday	4–6 mi, Z1	5–7 mi, Z1	10–12 mi including 10 × ½ mi Z4 intervals
Wednesday	7–9 mi, Z1	10–12 mi, Z2	7–9 mi, Z1
Thursday	10–12 mi, Z2	22–24 mi, Z2	10–12 mi, Z2
Friday	10–12 mi	4–6 mi, Z1	7–9 mi, Z2 including 6 × 1 mi Z4 intervals
Saturday	10–12 mi, Z2	7–9 mi, Z1	6–8 mi, Z2
Sunday	10–12 mi, Z2	10–12 mi, Z2	15 mi, Z3

Note: A training base of 10 miles per week should be achieved prior to beginning this schedule. Z1 = low intensity, Z2 = moderate intensity, Z3 = marathon race pace, Z4 = just over marathon race pace.

shorter amounts of time. For triathletes the training regimen becomes more burdensome because three specific activities must be trained for simultaneously. Despite the number of activities, some type of periodization should be used to enhance performance, minimize the risk of injury, and help alleviate training boredom.

Key Point

Periodized marathon training includes both long, slow runs and shorter, faster paced runs within the overall training regimen.

Cross Training in Cardiorespiratory Exercise

The cross training concepts applied to endurance training are the same as those for strength training. Cross training is the use of exercises and other activities to enhance the performance of one particular activity. The specificity of the activity must be considered for the "transfer of training effect" to be most beneficial. The magnitude of forces, the rate of force development, range of motion, intensity, duration, and so on all need to be similar in order for optimal transfer of training to occur. For example, if improved swimming performance is the objective, then activities such as upper body ergometer (UBE) training and running may help train the appropriate musculature but lack the specificity of the activity. In other words, the activities are not exactly alike and the musculature functions somewhat differently for these activities. Consequently, the adaptations will be specific to the combined training stimuli received.

Cross training can be very helpful in addressing the weaker components of an endurance activity or movement pattern and thus enhance overall performance. This can be especially true for injured athletes involved in a rehabilitation program. For example, if a running athlete injured a knee, swimming may provide the needed stimulus to maintain performance levels while allowing the injury to heal.

In the fitness realm, cross training has been commonly used to describe a program that incorporates a variety of exercises. For example, the use of a treadmill and stationary bike for improving cardiorespiratory function has often been considered cross training. Furthermore, any combination of strength training, endurance training, or other training protocols have been termed cross training among fitness enthusiasts.

Key Point

Cross training is a valuable training concept utilized by many from both the athletic and fitness realms to improve performance and provide variety to training programs.

Interval Cardiorespiratory Training

Interval training is the use of varying intensities throughout an exercise bout. A baseline level of intensity is chosen

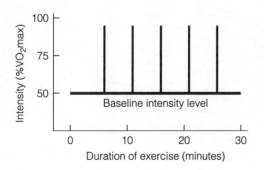

Figure 13-6
Interval training intensity sample for a 30-minute workout. Baseline intensity level is approximately 50% VO_2max with five 1-minute intervals (vertical bars) reaching ~90% VO_2max.

(such as 50% VO_2max) and is followed by periodic intervals of higher intensities (Figure 13-6). The concept is to use the baseline zone as the active recovery zone for the bursts of increased intensity. During these bursts an individual increases exercise intensity and works through the anaerobic threshold (lactate threshold) to the peak zone. When the peak zone is reached, the exerciser need remain at that intensity level for only about 30 seconds. The user then decreases intensity and returns to the baseline zone to recover. Using this method of training accomplishes muscular overload due to increased power outputs while stimulating the anaerobic energy systems. It also forces the aerobic system to recover under duress from the intervals providing an overload stimulus, which enhances adaptation.

EXERCISE EQUIPMENT USED FOR INTERVAL TRAINING. Interval training is common, and most home and commercial cardiovascular exercise machines come equipped with a host of computerized exercise programs that utilize this concept to a certain degree. Program offerings often include programs titled *random*, *rolling hills*, *intervals*, *sprints*, and *steady climb*, all of which vary the intensity levels throughout the exercise period.

The levels of intensity for these interval programs are often broken down into specific zones of intensity. For example, zone 1 is the baseline level and could have an intensity level of 50% VO_2max. Zone 2 could be used as the lactate threshold level and have intensities of 65–75% VO_2max. Zone 3 for this example could be the peak zone, in which the training intensity approaches maximum levels (85%+ VO_2max). Numerous varieties of interval training are used and can boost performance, fitness, and health.

Key Point

Interval training incorporates higher intensity bursts to increase use of anaerobic energy sources, elevate muscular forces and power outputs, and increase caloric expenditures.

Maintaining Cardiorespiratory Performance Levels

Of the acute training variables (FITT), it appears that intensity is the most important for maintaining fitness levels. Researchers have conducted several studies to evaluate the detraining effects from manipulations of training frequencies, intensities, and durations.[11-13] Findings indicated that if intensity levels were maintained, VO_2max remained unchanged despite reductions of up to 66% in frequency and duration. However, when frequency and duration were maintained and intensity levels decreased, significant reductions in VO_2max were observed. Interestingly, similar results occurred for strength training.[14]

Key Point

Intensity is a key factor to maintaining fitness and performance levels.

Genetic Endowment and Cardiorespiratory Capabilities

Approximately a quarter to half of an individual's variation from a VO_2max norm is believed to be predetermined by heredity (genetic predisposition).[15] More specifically, there is a predisposition for muscle fiber type distribution, as described in Chapter 2. The level of training appears to be responsible for the remaining portion. Therefore, if an individual aspires to be an elite endurance athlete, training alone is not enough; he or she needs to have the genetic endowments that enhance endurance performance. It has also been shown that individuals with high VO_2max values due to heredity seem to maintain these values even in a deconditioned state. In other words, their VO_2max values may be better than others with or without training.

■ CARDIORESPIRATORY TRAINING CONCEPTS *IN REVIEW*

- As VO_2max values rise, the training intensity necessary for continued improvements must also rise.
- Endurance training frequencies of 3–5 days per week allow for ample training volumes as well as rest.
- The greatest VO_2max values are found in the particular endurance sport of the athlete.
- Periodized endurance training systematically adjusts frequency, duration, and the intensity of the training bouts.
- Intensity is the key to maintaining fitness and performance levels.
- Interval training is a valuable tool to increase caloric expenditure and incorporate anaerobic systems.
- Heredity plays a significant role in endurance capacity.

■ ACUTE RESPONSES TO CARDIORESPIRATORY EXERCISE

As the body moves from a state of rest to exercise, what changes? Several acute responses occur and they depend on many factors such as current fitness level, sufficiency of

sleep, medications, coffee, alcohol, tobacco, stress, and others. Individuals with higher fitness levels have lower acute responses to exercise than unfit individuals. In other words, exercise is more taxing for the unfit person.

Cardiac Output and Redistribution During Cardiorespiratory Exercise

The demand for energy production rises quickly as the work rate increases. To compensate for the increased energy demand, blood flow to the working muscle tissue increases. This is accomplished primarily through an increase in heart rate and stroke volume, which increase cardiac output. Stroke volume increases in a linear fashion up to 40–50% of VO_2max, where it levels off. Heart rate continues to increase as work rate increases in a positive linear fashion until HR_{max} is achieved (Figure 13-7).

To maximize the volume of blood available to working muscle tissue, some blood is redistributed from the internal organs. For instance, during a resting state approximately 80% of the blood volume occupies the organs of the gut (such as stomach, liver, kidneys). During exercise the blood volume to these organs is reduced to 20% in order to increase the blood flow to the periphery.

Key Point

At the onset of exercise some blood is shunted from the internal organs and redistributed to working muscle tissue.

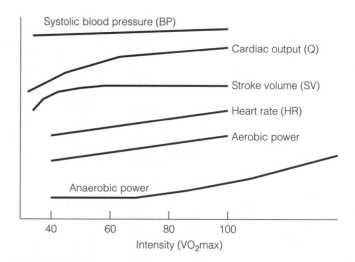

Figure 13-7
Major cardiovascular responses to increasing exercise intensities.

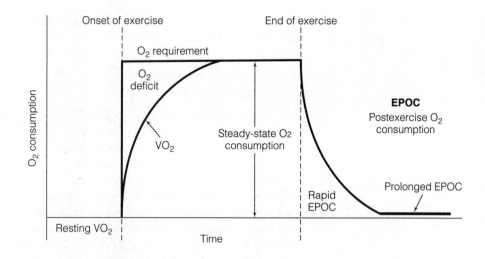

Onset of exercise   End of exercise

O₂ requirement

O₂ deficit

O₂ consumption

VO₂

Steady-state O₂ consumption

EPOC
Postexercise O₂ consumption

Rapid EPOC

Prolonged EPOC

Resting VO₂

Time

Figure 13-8
Oxygen consumption during and after cardiorespiratory exercise. Note that O₂ consumption remains elevated for a time after cessation of the exercise.

BLOOD PRESSURE CHANGES DURING CARDIORES-PIRATORY EXERCISE. Changes in HR, SV, and blood volume all affect blood pressure (BP). Systolic BP, which is the arterial pressure generated during contraction of the left ventricle, increases in proportion to exercise work rate and reaches maximum values between 200 and 249 mm Hg. This provides the driving force for blood delivery to working muscle and other tissue during exercise. Diastolic pressure, which is the arterial pressure during the relaxation of the left ventricle, changes very little, if at all.

Total peripheral resistance to blood flow also influences BP. This is the sum of all factors (HR, SV, blood volume, blood viscosity, blood vessel size, blood vessel resistance, valve function, and so on) affecting the flow of blood. At the onset of exercise, the blood vessels in the active muscles dilate (increase in diameter) to accommodate the increased blood flow and help minimize the pressure needed to deliver the blood.

Key Point

To accommodate increased blood flow demands during cardiorespiratory exercise, changes occur in HR, SV, BP, and blood vessel diameter.

Volume of Oxygen (VO₂) and Ventilation During Cardiorespiratory Exercise

Increased exercise work rates require larger volumes of oxygen (VO₂) in order to aerobically produce ATP in muscle fibers. VO₂ increases linearly with increases in work rate to the maximal level (VO₂max). VO₂max values are influenced by a combination of factors such as heredity, fitness level, sex, physical size, and age.

Minute ventilation (V$_E$) is also increased in response to physical activity. V$_E$ is the number of breaths per minute multiplied by the tidal volume (TV; the volume of air expired or inspired in a breath). In order to properly diffuse oxygen into the bloodstream, plenty of oxygen must be available. Increasing both the tidal volume and frequency allows more oxygen to be delivered and diffused into the blood and more CO₂ to be released from the blood. Both TV and frequency increase in a curvilinear fashion with increasing work rates.

Minute ventilation is positively correlated with VO₂ and increases with work rate in a semilinear fashion. Resting V$_E$ values are generally around 6.0 L/min and can rise as high as 200 L/min in extreme cases. To accompany this increase in V$_E$, the lung diffusion rate (gas exchange) increases by up to three times the resting level, as does the oxygen extraction by muscle tissue.

POSTEXERCISE OXYGEN CONSUMPTION. As demonstrated in Figure 13-8, O₂ consumption remains elevated for a time after completion of an exercise bout. This is referred to as **postexercise oxygen consumption (EPOC)**. As shown in Figure 13-8, EPOC is often broken into two stages or phases. The early phase of EPOC (<1 hour postexercise) is much more dramatic and is often referred to as the *rapid* phase. The later phase of EPOC (>1 hour postexercise) endures and is often referred to as the *prolonged* phase.

Rapid EPOC reflects metabolic efforts to reestablish pre-exercise energy stores, such as creatine phosphate (CP), and oxygen association with myoglobin. The degree of rapid EPOC is related to exercise intensity and duration and thus VO₂ levels and CP recovery efforts.[16] Also, blood lactate levels generated during exercise have been shown to predict EPOC; however, no single measure seems to predict rapid EPOC in all cardiorespiratory exercise situations. Prolonged EPOC remains even more of a puzzle and might reflect enhanced protein turnover and metabolism of energy stores (glycogen and fat) during that time.

Key Point

Minute ventilation (V_E) increases up to five times from a resting to active state.

Body Temperature Regulation in Cardiorespiratory Exercise

After several minutes of exercise, the body core temperature begins to rise. Depending on the intensity and duration of the activity and the environmental conditions, sweating becomes apparent to help dissipate heat. (See Chapter 8 for more detail on sweating.) Blood flow to the skin increases to allow for heat loss by sweating and other means (convection, conduction, and radiation). Efficient body temperature regulation, or **thermoregulation**, is important during exercise, as heat can denature cellular

enzymes and disrupt **homeostasis**, the body's efforts to keep vital functions operating optimally.

Hormonal Response to Acute Cardiorespiratory Exercise

The levels of several hormones change in direct or indirect response to cardiorespiratory exercise bout. As discussed in Chapter 4, the level of insulin released by the pancreas is decreased as exercise intensity surpasses 50% VO_2max. Meanwhile, the levels of norepinephrine and epinephrine increase during exercise. Cortisol levels also increase during prolonged cardiorespiratory exercise at higher submaximal intensity. The net effect of this hormonal balance is to break down energy stores in muscle and adipose tissue as well as the liver. In addition, the liver begins to produce glucose and release it into the blood to help fuel muscle.

Table 13-7 indicates selected hormonal responses to exercise and their related significance, if any. One of the hormonal benefits of endurance training is the increase in β-endorphins during strenuous exercise. This explains the feeling of well-being commonly experienced after exercising, known as "runner's high." As described below, the magnitude of hormonal response to a standard work rate generally declines with endurance training. However, the sensitivity of the *target tissues* increases, so that less hormonal activity is required for a designated purpose.[17]

| Table 13-7 |

Typical Hormonal Responses to Endurance Exercise Training

Hormone	Response
Growth hormone	Lower elevations during exercise
Adrenorticotropin (ACTH)	Higher values during exercise
Thyroxine (T_4)	Increased free thyroxine at rest, increased turnover during exercise
Triiodothyronine (T_3)	Increased turnover during exercise, reduced total levels at rest
Testosterone	Decreases with training due to improved sensitivity
Insulin	Decreases due to improved sensitivity
Glucagon	Adjusts with changes in blood glucose levels
Antidiuretic hormone (ADH)	Increases to conserve water
Endorphins exercise	Increases with long duration
Epinephrine	Increases with heavy volume of exercise
Norepinephrine	Increases with exercise volume
Estrogen	Increases with exercise levels, depending to menstrual phase
Cortisol	Increases with high exercise volumes

■ **ACUTE RESPONSES TO CARDIORESPIRATORY EXERCISE IN REVIEW**

- The onset of exercise elicits several acute responses that include an increase in cardiac output, blood pressure (BP), VO_2, ventilation (V_E), and body temperature.
- Heart rate has a positive linear relationship with work rate.
- Blood is shunted to inner organs and redistributed to working muscle tissue during exercise.
- Minute ventilation (V_E) increases to assist with rising O_2 demands.
- The acute effects of exercise diminish with improved fitness levels.

■ **ENERGY SYSTEMS AND CARDIORESPIRATORY EXERCISE**

What fuel sources are used to maintain ATP levels during cardiorespiratory exercise? How are the relative contributions made by different fuel sources influenced by the intensity and duration of an endurance exercise bout? Also, how do an individual's level of training and energy stores influence energy metabolism? This section summarizes concepts related to the governing factors (intensity, duration, level of training) influencing energy utilization and fuel preference. See Chapters 4–6 for details regarding the metabolism of specific energy nutrients.

Intensity of Cardiorespiratory Exercise and Energy Metabolism

Intensity is a primary determinant of fuel utilization during a cardiorespiratory exercise bout because it determines the hormonal environment during exercise. The hormones of particular interest are insulin, norepinephrine, epinephrine, and cortisol. In addition, intensity level, and thus ATP demand, relative to O_2 availability is a key determinant of fuel utilization, as is the level of training. The expected contributions from various energy resources relative to three different submaximal intensity levels are presented in Figure 13-9 and are the basis for the following discussion.

FUEL SOURCES DURING LOWER INTENSITY CARDIORESPIRATORY EXERCISE.
During lower intensity exercise bouts (25–30% VO_2max), such as walking, fatty acids are the greatest contributor to ATP regeneration. Most of these fatty acids are derived from the blood (see Figure 13-9). At this level of intensity, Type I muscle fibers are recruited and plenty of oxygen is available for aerobic metabolism of fatty acids and carbohydrate. Therefore, very little lactic acid is generated at this level

of intensity, and nonworking muscle fibers and the heart easily metabolize what is produced, so lactate does not accumulate in the blood.

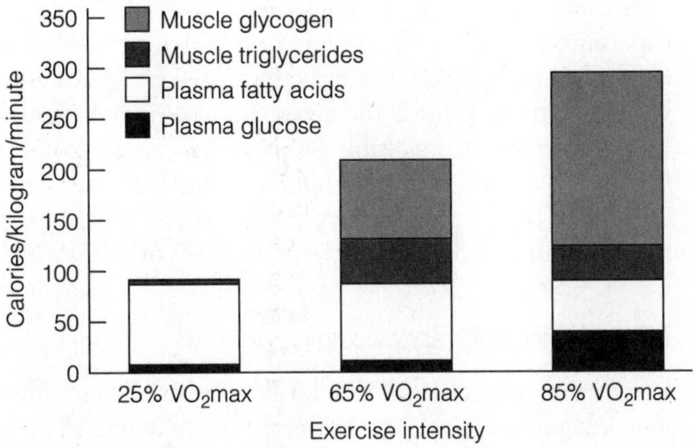

ENERGY SOURCES DURING MODERATE-INTENSITY CARDIORESPIRATORY EXERCISE.
As the level of intensity of cardiorespiratory exercise increases, there is a corresponding increase in reliance on carbohydrate as a fuel source. As shown in Figure 13-9, at an intensity level of 55–65% VO_2max, the use of carbohydrate and fatty acids for fuel is shared fairly evenly. However, at this moderate intensity individuals approach or exceed their lactate threshold, which would indicate an increasing contribution from anaerobic carbohydrate metabolism. Type I muscle fibers, and some Type IIa, are recruited at this intensity level, while insulin release is dampened and the level of epinephrine in the blood is rising. As explained in Chapter 4, this leads to an enhancement of carbohydrate utilization by promoting the breakdown of muscle glycogen stores.

ENERGY SOURCES DURING HIGHER INTENSITY CARDIORESPIRATORY EXERCISE.
As intensity climbs further (≥85% VO_2max), the reliance on carbohydrate is much greater than on fatty acids (see Figure 13-9). More Type IIa and some Type IIb muscle fibers are recruited, dictating greater carbohydrate utilization for ATP regeneration. At this intensity level the ATP demands of the exercise far exceed O_2 availability, so the participant is well above lactate threshold. As a result, the generation of larger amounts of lactic acid becomes a performance-limiting factor.

Key Point

The relative use of the different fuel sources is primarily determined by: (1) intensity and duration of the exercise, (2) level of training, and (3) availability of fuel resources (nutritional state).

Duration of Cardiorespiratory Exercise and Energy Metabolism

The maximal duration of an exercise bout is primarily determined by its intensity. Low- to moderate-intensity exercise bouts can be sustained for several hours, but those of higher intensities cannot. Although a certain intensity level can be maintained throughout one bout of exercise, the contributions made by fuel sources will change. For instance, fat utilization tends to increase throughout a low- to moderate-intensity bout of exercise, as shown in Figure 13-10. This change in energy contribution is largely due to the waning supply of glycogen in muscle. Also, glucose entry into muscle fibers plateaus and might

Figure 13-9
Contribution of different fuel sources to cycling exercise (pre-fasted) at three submaximal intensities. (*Source: Based on data from* Romijn, J.A., Coyle, E.F., Sidossis, L.S., Gastaldelli, A., Horowitz, J.F., Endert, E., Wolfe, R.R. Regulation of endogenous fat and carbohydrate metabolism in relation to exercise intensity and duration. *American Journal of Physiology*, 265: E380–E391 1993.

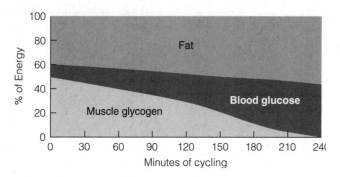

Figure 13-10
Expected fuel sources (as % of energy expended) during a 4-hour cycling bout at 65% VO₂max in pre-exercise fasted young males. (*Source*: Based on data reported in reference 27.)

eventually decrease if blood glucose levels decline during longer endurance bouts. Therefore, for extended bouts of exercise, availability of fuel sources (especially glycogen and blood glucose) becomes an increasingly important performance factor to consider, Carbohydrate availability and performance limitations are discussed in more detail in Chapter 4.

Key Point

Energy contributions change as the duration of the activity is extended.

OXYGEN AVAILABILITY.   Oxygen availability to working muscle tissue increases during cardiorespiratory exercise. Increased cardiac output, redistribution of blood, and minute ventilation serve to increase oxygen delivery. However, these mechanisms are not instantaneous and take time to maximize blood delivery to working muscle. Therefore, O₂ delivery to working muscle tissue increases on a slope, as shown in Figure 13-8. This time lapse can create an oxygen deficit early in an exercise bout (in the first minutes) if the intensity is sufficient (moderate to high). For this reason, cardiorespiratory exercise aimed at fat utilization should last longer than 15 minutes, depending on one's level of training. This period of time allows oxygen delivery to either match oxygen demand or become maximized. During oxygen deficit, aerobic ATP

Key Point

As muscle glycogen decreases during extended activity, the use of blood glucose increases to assist with carbohydrate contribution. However, total carbohydrate contribution wanes as exercise continues.

production pathways are supported by anaerobic energy systems such as CP. In accordance, the rapid EPOC phase largely reflects efforts to restore pre-exercise levels of CP, metabolize lactate, replenish oxygen stores in myoglobin, and pump calcium across cell membranes.

■ ENERGY SYSTEMS AND CARDIORESPIRATORY EXERCISE *IN REVIEW*

- Low-intensity aerobic activity primarily utilizes fat as an energy source.
- Moderate-intensity exercise utilizes a blend of carbohydrate and fat for energy.
- High-intensity endurance activity primarily utilizes carbohydrate for energy.
- Extended bouts of endurance activity require greater contributions from fat, as carbohydrate sources are progressively depleted.

■ ADAPTATIONS FROM ENDURANCE TRAINING

What types of adaptations can be expected from long-term endurance training? Table 13-8 outlines many variables that adapt as a result of endurance training. Several of these variables respond and adapt within 1–3 weeks of training, whereas others may take several months or years. These adaptations are lost as rapidly as they were gained when exercise ceases.

The level or magnitude of adaptation is dependent on the frequency, intensity, time, and type (FITT) of exercise performed and on the person's initial fitness level, genetic capability for improvement, and nutritional status. For the deconditioned person, dramatic increases can be obtained in a short time. But for the active individual, substantially less improvement is possible because there is less potential for improvement. For example, an elite cyclist may need months of training to improve performance levels whereas the beginning cyclist improves in a matter of weeks.

Muscle Fiber Adaptations

Lower intensity exercise primarily recruits Type I fibers. The adaptations observed in these fibers include increases in cross-sectional area (hypertrophy) and vascularity. Vascularity improvements are due to increases in the density of the capillary beds that serve as gas and nutrient exchange sites for working muscle tissue. Within muscle cells, the size and number of mitochondria (Figure 13-11) increase along with oxidative enzymes used in the β-oxidation process or Krebs cycle. These increases help improve aerobic ATP production. To coincide with this, endurance training also elicits an increase in muscle glycogen and fat

Table 13-8

Common Long-Term Adaptations to Endurance Training

Variable		Response
Cardiovascular		
HR_{rest}	Heart rate (rest)	Decreases
HR_{max}	Heart rate (maximum)	Decreases
SV_{rest}	Stroke volume (rest)	Increases
SV_{max}	Stroke volume (max)	Increases
Q_{rest}	Cardiac output (rest)	Increases
Q_{max}	Cardiac output (max)	Increases
Heart volume		Increases
Blood volume		Increases
BP: systolic blood pressure (rest)		Decreases
BP: systolic blood pressure (max)		No change
BP: diastolic blood pressure (rest)		Decreases
BP: diastolic blood pressure (max)		No change
Respiratory		
V_E	Ventilation (rest)	Decreases
V_{Emax}	Ventilation (max)	Increases
TLC	Total lung capacity	No change
TV_{rest}	Tidal volume (rest)	No change
TV_{max}	Tidal volume (max)	Increases
VC	Vital capacity	Increases
RV	Residual volume	Decreases
Metabolic		
A-V O_{2diff}	Arterial-venous O_2 differential (rest)	No change
A-V O_{2diff}	Arterial-venous O_2 differential (max)	Increases
Blood lactate threshold		Increases
RER	Respiratory exchange ratio (submax)	Decreases
RER_{max}	Respiratory exchange ratio (max)	Increases
VO_2rest	Resting oxygen consumption	No change
VO_2max	Maximum oxygen consumption	Increases
Body Composition		
Total body weight		Decreases
Fat weight	Decreases	
Fat-free weight		Increases

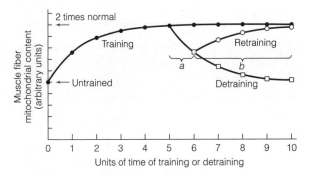

(a) Overall influence of training/detraining

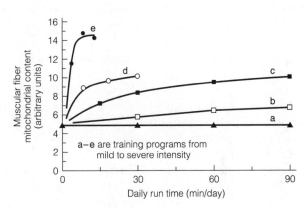

(b) Influence of exercise bout duration

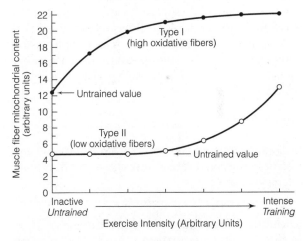

(c) Influence of exercise bout intensity

Figure 13-11

Anticipated influence of training/detraining and exercise intensity duration on mitochondrial content of skeletal muscle. (a) About 50% of the increase of mitochondrial content was lost after 1 week of detraining and all of the adaptation was lost after 5 weeks of detraining. Also, it took four weeks of training to regain the adaptation lost in the first week of detraining. (b) As intensity is increased less time is required for adaptation and the adaptation is greater. (c) At the intensity level of lower cardiorespiratory training, greater adaptation occurs in Type I muscle fibers due to the specific order of motor unit recruitment (see Chapter 3). (*Source*: Based on references 28–30.)

stores, making energy more readily available. Additional glycogen stores and a decreased reliance on carbohydrate can extend performance time prior to fatigue. For example, if the work rate remains constant (such as running at a pace of 7 min/mile), the trained individuals should be able to run for longer periods prior to fatigue than they could prior to training. Another potential adaptation is enhanced fat levels within muscle fibers, particularly Type I, making fatty acids more available as a fuel source for aerobic energy metabolism.

Key Point

The size and number of mitochondria increase with endurance training. These adaptations improve aerobic ATP production.

Endurance Training and Adaptations in the Cardiovascular System

Some of the effects of endurance training that occur within the cardiovascular system include a decreased heart rate at rest and during submaximal exercise. This adaptation is most likely due to the increase in stroke volume (SV), which results from more blood and a greater force exerted by the left ventricle. Endurance training increases blood volume, which can enhance the left ventricle's **preload**, the volume of blood entering the left ventricle to await expulsion into the arterial system by the next contraction of the left ventricle. As SV increases, HR decreases to adjust cardiac output.

Greater preload and transient elevations in systolic pressure during exercise can also stimulate the heart to enlarge. In a manner somewhat similar to skeletal muscle, cardiac muscle hypertrophies when the appropriate stimulus is provided. Also, the capillary density of the heart and skeletal muscle tissue increase, allowing for improved blood delivery and subsequent waste removal. Enlargement of the heart in response to cardiorespiratory training is referred to as **physiological hypertrophy** and is not to be confused with **pathological hypertrophy** associated with chronic high blood pressure (hypertension). Although it is possible for physiological hypertrophy to be pathological, this is extremely rare.

Key Point

Endurance training can improve stroke volume and decrease the heart rate at rest and at submaximal exercise intensities.

Endurance Training and Adaptations in the Respiratory System

Adaptations in the respiratory system include an increase in ventilation due to an expanded TV and increased respiratory rate. Alveolar gas exchange during maximal work rates also improves due to increased pulmonary blood flow and increased pulmonary ventilation. The difference in O_2 content between the arteries and veins (A-V O_2diff) is also affected by training. The primary reason for this is a combination of more efficient extraction of oxygen by the working muscle tissue and increased blood delivery to muscle tissue.

Cardiovascular and ventilation changes coincide with adaptations in muscle energy systems. As a result, endurance training increases the blood lactate threshold, which allows higher work rates to be maintained without the accumulation of lactic acid.

VO$_2$ ADAPTATIONS. VO$_2$ also adapts to endurance training. At rest and during submaximal work, VO$_2$ values remain relatively constant, but during maximal work rates these values increase. The reason for this is a combination of increased oxygen absorption, delivery, extraction, and utilization by muscle tissue. VO$_2$ is considered to be one of the best indicators of cardiorespiratory fitness. Endurance training typically improves VO$_2$ values by 15–20% for an average sedentary person if the training intensity and duration are appropriate.[18]

Key Point

Aerobic training can elevate the lactate threshold, allowing for greater duration of increased intensity efforts.

RESPIRATORY EXCHANGE RATIO (RER) ADAPTATIONS. RER at a given submaximal work rate is lowered as a result of endurance training. This reflects the changes in O_2 availability and the increased capabilities for aerobic energy production in muscle tissue. Quite simply, less carbohydrate and more fat are used during exercise at the same work rate after training in comparison to pretraining, as shown in Figure 13-12. Continued training is necessary to maintain these adaptations.

■ ENERGY SYSTEMS AND CARDIORESPIRATORY EXERCISE *IN REVIEW*

- Regular endurance training elicits changes in the size and number of mitochondria.
- Decreases in resting and submaximal heart rates due to endurance training are often correlated with improvements in stroke volume.
- Respiratory exchange ratios for a given submaximal workload are generally lower after endurance training because of a greater utilization of fat.

■ GENDER- AND AGE-RELATED ASPECTS OF CARDIORESPIRATORY EXERCISE

When comparing the genders in relation to cardiorespiratory endurance, are any differences observed? Does age affect adaptations and responses to this form of training? There is a wide range of values within many of the cardiorespiratory components and this variability may create overlap between the genders. Table 13-9 outlines gender differences in regard to variables involved in endurance training. It should be noted that these differences exist regardless of endurance training. However, training improves many of these variables.

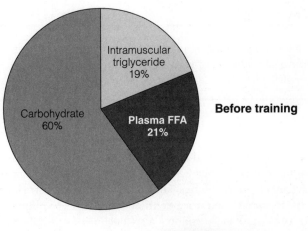

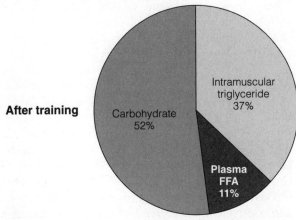

Figure 13-12
Energy substrates used during moderate-intensity exercise. After 12 weeks of cycling training, energy source utilization at the same workload (which was 64% VO₂max pretraining) had changed as shown. The lactate threshold of the young male participants was estimated to be 50–60% VO₂max pretraining and 70% after the training. (*Source*: Based on data from reference 31.)

Cardiovascular Measures and Gender

For a given work rate, women generally have a higher HR. This is most likely due to the HR compensating for the lower SV in an effort to maintain cardiac output. Women generally have a lower SV than men because they tend to have a smaller heart and lower blood volume. Any combination of these decreases the volume of blood within the left ventricle and thus decreases SV.

Another difference in cardiovascular function between the genders is the level of hemoglobin, which carries oxygen in the blood. Women have lower hemoglobin levels and thus lower oxygen saturation than men for a given volume of blood. This reduces the oxygen availability and ultimately affects VO₂max values.

Respiratory Measures and Gender

Respiratory factors also differ between the genders. Because of differences in overall body size and consequently

Table 13-9

Gender-Related Differences of Endurance Exercise Components

Variable	Women Compared to Men
Stroke volume (SV)	lower
Cardiac output (Q)	lower
Blood volume	lower
Hemoglobin levels (Hb)	lower
Tidal volume (TV)	lower
Minute ventilation (VE)	lower
VO₂max	lower

Key Point

The size of the female heart is generally smaller than a male heart, leading to smaller stroke volume and consequently higher heart rates for a given workload.

lung size, women generally have lower maximal minute ventilation (V_E) and tidal volumes (TV). These values represent the maximum volume of air inhaled in a minute's time and per breath, respectively. Greater values allow for better oxygenation of the blood and subsequent delivery to working muscle tissue.

VO₂MAX AND GENDER. When comparing VO₂max values between the genders, women generally have lower values, although there is much overlap. For a given submaximal work rate, the VO₂ values of men and women are equal. This seems reasonable because a certain volume of oxygen is needed for a given amount of work. However, this volume is generally a higher percentage of capacity for a woman than for her male counterpart.

Key Point

Women and men have similar VO₂ values for a given work rate. However, the female is generally at a higher percentage of capacity.

Menstruation and Cardiorespiratory Training

As mentioned in Chapter 12, many types of physical training can cause changes in menstruation. Although menstrual cycles vary considerably among women the effects of physical exercise are significant. The overall training or activity volume appears to be a primary factor in menstrual irregularities, particularly when the exercise volume creates a significant chronic energy deficit.[19,20] If volume is reduced to a point where menstruation becomes "normal," permanent consequences may be avoided. Chronic failure to menstruate, or *amenorrhea*, can result in

Key Point

The overall volume of endurance training may be responsible for changes in menstrual patterns.

decreases in bone density, as discussed in greater detail in Chapter 15.

Cardiorespiratory Performance Across the Life Span

It is apparent that physiological changes occur during aging. Are these changes due to genetics, the environment, particular lifestyles, or a combination of these? As life expectancy continues to increase and because exercise is recommended for all ages, it is important to understand the changes associated with aging. This section provides an introduction to physical and physiological changes associated with different ages and how they affect endurance performance.

VO₂MAX AND AGING Age affects VO_2max values. Young athletes typically have lower values due to their overall size (lungs, blood volume, heart). Also, endurance training in young (prepubescent) children does not provide the same type of responses observed in adolescents or adults. Unlike the marked improvements in performance that result from endurance training in adults, VO_2max values in children do not change in a relative manner. The reason for this has not clearly been established or defined but is most likely due to the development of the heart and subsequent blood delivery to tissues.

As individuals age beyond the young adult years, a decrease of 10% in VO_2max values per decade occurs. This decline begins during the late teens for women and the mid-20s for men.[21] There is no single absolute physiological reason for this, and much of the decline may simply be due to declining activity levels.[22–25]

As people age, the tendency toward physical activity decreases and thus detraining occurs. However, if the intensity of the training can be maintained, VO_2max values can remain relatively consistent. For example, one 80-year-old woman, "Amazing Mavis" Lindgren, had VO_2max values similar to those of a sedentary 20-year-old woman. Mavis had been primarily sedentary until she began training at the age of 62. She began walking and then gradually moved to jogging and subsequently competed in 65 marathons and broke several age bracket records.[26] She finished the New York Marathon at the age of 86. Her VO_2max values began to fall after the age of 80, but she has demonstrated that the human body is adaptable despite some age-related factors.

RESPIRATION AND AGING. Some of the physiological bases for declining VO_2max values (Table 13-10) include certain components of respiration. The **residual volume**

Key Point

If exercise intensity levels are maintained during the aging process, VO_2max values can remain relatively consistent.

(RV) of inspired air increases as people age. This is the volume of air retained in the lungs that cannot be exhaled. An increased RV decreases the volume of air that can be exchanged in the lungs and thus reduces performance. The maximal expiratory ventilation (V_E), the volume of air ventilated in 1 minute, also declines as people age. These changes in RV and V_E have often been attributed to lost elasticity of the respiratory tissues. Much of this needed elasticity seems to be maintainable if exercise intensity levels are sustained throughout the aging process.

BLOOD FLOW AND AGING. Another factor associated with decline in VO_2max is the delivery of oxygen-rich blood. The HR_{max} decreases with age ($HR_{max} = 220 -$ age), thus altering cardiac output in maximal effort exercise. The elasticity of the blood vessels also tends to decrease as people age, which increases peripheral resistance. These two factors combined can have a significant effect on blood delivery, particularly in maximal effort situations, and consequently decrease VO_2max values. Exercise may help deter losses in elasticity but cannot alter HR_{max}. Therefore, although continued training of similar intensity may help maintain VO_2max, the aging process gradually reduces cardiorespiratory function.

The adaptations to endurance training by older participants are similar to those observed in younger adults. Large improvements in aerobic capacity have been reported. The magnitude of improvement is generally based on the beginning fitness level of the individual: those with very low fitness levels generally observe the largest increases in VO_2max. Prescribing endurance exercise for

Table 13-10

Age-Related Changes in Endurance Capacity Variables

Variable	Response with Increasing Age in Adults
Heart rate maximum (HRmax)	↓
Cardiac output (Q)	↓
Residual volume (RV)	↑
Tidal volume (TV)	↓
Minute ventilation (VE)	↓
VO₂max	↓

older individuals should be similar to prescriptions to younger healthy adults with some important considerations. These include lower beginning threshold levels, slower progression of activity levels, and analysis of orthopedic and related medical aspects.

■ GENDER- AND AGE -RELATED ASPECTS OF CARDIORESPIRATORY EXERCISE *IN REVIEW*

- Due to physical size, women generally have lower SV, V_E, blood volume, and subsequently VO_2max values.
- VO_2 values for a given work rate are similar between the genders. However, the female's value is generally at a greater percentage of maximum.
- Extensive endurance training can affect menstruation.
- Intensity is the most significant acute training variable for maintaining fitness levels.
- Loss of elasticity of the lungs, heart, and blood vessels due to the aging process can significantly affect cardiorespiratory endurance variables.

Conclusions

The difference between aerobic and anaerobic exercise is based on the contributions from the various energy systems. These energy contributions are determined by the intensity level of the activity, duration of the activity, and nutritional state of the body. The point where anaerobic systems begin to contribute more energy than aerobic systems is called the lactate or anaerobic threshold.

The intensity level of exercise not only determines the energy sources being utilized but also plays a significant role in maintaining fitness level and determining the duration of exercise. Low- to moderate-intensity levels derive a greater portion of energy needs from the utilization of fat, whereas moderate- to high-intensity levels require greater contributions from carbohydrate sources.

Benefits from endurance training often include decreased resting heart rate, increased stroke volume, and reduced blood pressure. These factors have a significant effect on reducing cardiovascular disease, the number one killer in America. Both men and women respond similarly to endurance training, but due to physical size several differences exist. The aging population also responds positively to endurance training but may be limited by losses in elasticity of the various cardiorespiratory tissues.

 **IN FOCUS**

UNDERSTANDING RESPIRATORY EXCHANGE RATIO (RER) AND RESPIRATORY QUOTIENT (RQ)

The combustion of energy nutrients in cells requires oxygen and produces carbon dioxide. This is referred to as *cellular respiration*. However, it is difficult to measure cellular respiration, and impractical in most situations involving people. Therefore, to assess energy expenditure, researchers measure the changing levels of the gases oxygen and carbon dioxide during breathing. Researchers look at total energy produced as well as the contribution made by different energy molecule classes. This is because the different energy classes utilize different amounts of oxygen relative to the carbon dioxide produced. For instance, when 1 mole of glucose is combusted in cells, 6 moles of O_2 are used and 6 moles of CO_2 are produced. However, when fatty acids are combusted relatively less CO_2 is produced. The balanced chemical equations for the combustion of glucose and palmitic acid, a common fatty acid, are as follows:

$$C_6H_{12}O_6 + 6 O_2 \rightarrow CO_2 + 6 H_2O$$
Glucose
(carbohydrate)

$$C_{16}H_{32}O_2 + 23 O_2 \rightarrow 16 CO_2 + 16 H_2O$$
Palmitic acid
(fatty acid)

The **respiratory exchange ratio (RER)** is generated when the CO_2 in expired gases is divided by the O_2 consumed (CO_2/O_2). For instance, the RER for glucose and palmitic acid is calculated as follows:

$$RER = CO_2/O_2$$

Glucose RER $6 CO_2/6 O_2 = 1.0$
Palmitic acid   RER $16 CO_2/23 O_2 = 0.71$

Once the RER is determined, the fuel mixture can be assessed using Table A. As the scale in Figure A shows, an RER closer to 0.7 means greater fat utilization for fuel, and as the RER approaches 1.0, more carbohydrate is used as fuel. Because protein tends to make only a small contribution to energy expenditure in normal situations, it is not considered. However, during energy restriction and prolonged higher submaximal intensity exercise, amino acids do contribute to energy expenditure. Table A also provides constants for assessing energy expenditure from measurements of gas (O_2 and CO_2) inspired and expired.

Although used interchangeably, the **respiratory quotient (RQ)** and respiratory exchange ratio are not really the same thing. The respiratory quotient

UNDERSTANDING RESPIRATORY EXCHANGE RATIO (RER) AND RESPIRATORY QUOTIENT (RQ) (CONTINUED)

involves the measurement of changing oxygen and carbon dioxide levels associated with a specific cell or isolated tissue of interest such as a muscle. In contrast, the RER is used to assess whole body energy metabolism by measuring changing gas levels in breath. The respiratory exchange ratio is more applicable to the whole body since most of the change in RER from a nonexercise to an exercise state is attributed to contracting muscle fibers and includes other tissue experiencing changes in metabolism during that time (such as the lungs and heart).

Example

A 24-year-old male runner and graduate student was assessed for RER for 1 hour prior to exercise and during a 1-hour training run on a treadmill at a moderate submaximal intensity (67% VO_2max) in an exercise physiology lab. His VO_2 and VCO_2 for the hour prior to running were 20 L of O_2 inspired and 16 L of CO_2 expired. During the run, the values were 165 L of O_2 inspired and 145 L of CO_2 expired. His energy expenditure and RER during those times are shown in Table B, and his RERs are plotted in Figure A.

Table A **RER Constants of O_2 and CO_2 for Nonprotein RER**

RER (Nonprotein)	Energy Value (kcal/L of O_2)	Energy Value (kcal/L of CO_2)	Carbohydrate (%)	Fat (%)
0.707	4.686	6.629	0	100.0
0.71	4.69	6.606	1.1	98.9
0.72	4.702	6.531	4.76	95.2
0.73	4.714	6.458	8.4	91.6
0.74	4.727	6.388	12.0	88.0
0.75	4.739	6.319	15.6	84.4
0.76	4.751	6.253	19.2	80.8
0.77	4.64	6.187	22.8	77.2
0.78	4.776	6.123	26.3	73.7
0.79	4.788	6.062	29.9	70.1
0.80	4.801	6.001	33.4	66.6
0.81	4.813	5.942	36.9	63.1
0.82	4.825	5.884	40.3	59.7
0.83	4.838	5.829	43.8	56.2
0.84	4.85	5.774	47.2	52.8
0.85	4.862	5.721	50.7	49.3
0.86	4.875	5.669	54.1	45.9
0.87	4.887	5.617	57.5	42.5
0.88	4.899	5.568	60.8	39.2
0.89	4.911	5.519	64.2	35.8
0.90	4.924	5.471	67.5	32.5
0.91	4.936	5.424	70.8	29.2
0.92	4.948	5.378	74.1	25.9
0.93	4.961	5.333	77.4	22.6
0.94	4.973	5.29	80.7	19.3
0.95	4.985	5.247	84.0	16.0
0.96	4.998	5.205	87.2	12.8
0.97	5.01	5.165	90.4	9.58
0.98	5.022	5.124	93.6	6.37
0.99	5.035	5.085	96.8	3.18
100	5.047	5.047	100	0

UNDERSTANDING RESPIRATORY EXCHANGE RATIO (RER) AND RESPIRATORY QUOTIENT (RQ) (CONTINUED)

Table B

Example: RER and Other Measures

Measure	Hour Prior to Running (Rest)	Hour of Running (Exercise)
VO2	20 L	165 L
VCO$_2$	16 L	145 L
RER	0.80	0.88
Energy expenditure (EE)	95 kcal	810 kcal
Carbohydrate contribution to EE	33%	61%
Fat contribution to EE	66%	39%
METs	1	8.5

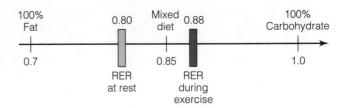

Figure A

Respiratory exchange ratio scale. At a value of 0.7, all nonprotein energy is obtained from fat. At a value of 1.0, all nonprotein energy is obtained from carbohydrate. The rest and exercise values indicated are for the runner in the example.

STUDY QUESTIONS

1. Define the difference between aerobic and anaerobic activity.
2. What are the recommended frequencies, intensities, and durations of endurance exercise for improving general health?
3. List and describe the various methods for determining intensity level.
4. How do intensity level, duration of activity, and nutritional state of the participant affect energy source contributions during exercise?
5. How does the body adjust from a resting to a physically active state?
6. What are the long-term adaptations and benefits of endurance training?
7. What physiological differences relating to endurance exercise exist between the genders?
8. How does age affect the endurance-related variables?

REFERENCES

1. Karvonen M, Kentala K, Mustala O. The effects of training heart rate: a longitudinal study. *Annales Medicinae Experimentalis et Biologae Fenniae* 35:307–315, 1957.
2. Londeree BR, Ames SA. Trend analysis of the %VO2max-HR regression. *Medicine and Science in Sports* 8:122–125, 1976.
3. Borg GA. Psychological basis of physical exertion. *Medicine and Science in Sports & Exercise* 14:377, 1982.
4. Dishman RK. Prescribing exercise intensity for healthy adults using perceived exertion. *Medicine and Science in Sports & Exercise* 26:1087–1094, 1994.
5. Dishman RK, Farquhar R, Cureton KJ. Responses to preferred intensities of exertion in men differing in activity levels. *Medicine and Science in Sports & Exercise* 26:783–790, 1994.
6. American College of Sports Medicine. Position Stand. The recommended quantity and quality of exercise for developing and maintaining cardiorespiratory and muscular fitness, and flexibility in healthy adults. *Medicine and Science in Sports & Exercise* 30 (6), 975–991, 1998.
7. American College of Sports Medicine. Position Stand. The recommended quantity and quality of exercise for developing and maintaining cardiorespiratory and muscular fitness in healthy adults. *Medicine and Science in Sports & Exercise* 22:265–274, 1990.
8. Williams PT. Relationship of distance run per week to coronary heart disease risk factors in 8283 male runners: the national runner's health study. *Archives of Internal Medicine* 157:191–198, 1997.
9. Pollock ML, Gettman LR, Mileses CA, Bah MD, Durstine JL, Johnson RB. Effects of frequency and duration of training on attrition and incidence of injury. *Medicine and Science in Sports* 9:31–36, 1977.

10. Pollock ML. Prescribing exercise for fitness and adherence. In *Exercise Adherence: Its Impact on Public Health* (Dishman RK, ed.), Champaign, IL: Human Kinetics, pp. 259–277, 1988.

11. Hickson RC, Rosenkoetter MA. Reduced training frequencies and maintenance of increased aerobic power. *Medicine and Science in Sports & Exercise* 13:13–16, 1981.

12. Hickson RC, Knakis C, Davis JR, Moore AM, Rich S. Reduced training duration effects on aerobic power, endurance, and cardiac growth. *Journal of Applied Physiology* 53:225–229, 1982.

13. Hickson RC, Foster C, Pollock ML, Galassi TM, Rich S. Reduced training intensities and loss of aerobic power, endurance, and cardiac growth. *Journal of Applied Physiology* 58:492–499, 1985.

14. Graves JE, Pollock ML, Leggett SH, Braith RW, Carpenter DM, Bishop LE. Effects of reduced training frequency on muscular strength. *International Journal of Sports Medicine* 9:316–319, 1988.

15. Bouchard C, Dionne FT, Simoneau JA. Genetics of aerobic and anaerobic performances. *Exercise and Sport Sciences Reviews* 20:27–58, 1992.

16. Tomlin DL, Wenger HA. The relationship between aerobic fitness and recovery from high intensity intermittent exercise. *Sports Medicine* 31(1):1–11, 2001.

17. McArdle WD, Katch FI, Katch VL. *Exercise Physiology, Energy, Nutrition, and Human Performance*, 4th ed., Baltimore: Williams & Wilkins, 1996.

18. Pollock ML. The quantification of endurance training programs. In: *Exercise and Sport Sciences Reviews* (Wilmore JH, ed.), pp.155–188, New York: Academic Press, 1973.

19. Prior JC, Vigna YM, Mckay DW. Reproduction for the athletic female: new understandings of physiology and management. *Sports Medicine* 14:190–99, 1992.

20. Dale E, Gerlock D, Wilhite A. Menstrual dysfunction in distance runners. *Obstetrics and Gynecology* 54:47–53, 1979.

21. Wilmore JH, Costill DL. *Physiology of sport and exercise*. Champaign, IL: Human Kinetics, 1994.

22. Dehn MM, Bruce RA. Longitudinal variations in maximal oxygen intake with age and activity. *Journal of Applied Physiology* 33:805–807, 1972.

23. Kohrt WM, Malley MT, Coggan AR, Spina T, Ogawa AA, et al. Effects of gender, age, and fitness level on response of VO2max to training in 60–71 yr. olds. *Journal of Applied Physiology* 71:2004–2011, 1991.

24. Miyashita M, Haga S, Mizuta T. Training and detraining effects on aerobic power in middle-aged and older men. *Journal of Sports Medicine* 18:131–137, 1978.

25. Pollock ML, Mengelkoh LS, Graves JE, Lowenthal DE, Limacher MC, et al. Twenty year follow-up of aerobic power and body composition of older track athletes. *Journal of Applied Physiology* 82:1508–1516, 1997.

26. Nieman DC. *Fitness and Sports Medicine; A Health Related Approach*, 3rd ed. Mayfield, 1995.

14

SPORT-SPECIFIC TRAINING AND NUTRITION

Personal Snapshot

Steve McAlister/The Image Bank/Getty Images

Coach Johnson is the athletic director at a small college. To maximize the performance of all the athletic programs he has organized a performance committee. The committee includes an exercise physiologist, a sport nutritionist, the strength and conditioning coach, and the head coaches of various sports at the college. Together they made a list of key concepts associated with different sports. These include concepts related to the physical characteristics of the elite athletes in each sport as well as how they should be physically trained to optimize performance. They also listed the most important nutritional considerations for the competitive season and the off-season. As part of the nutrition considerations they addressed eating on the road and nutrition supplementation.

Chapter Objectives

- Provide an overview of the physical characteristics of popular sports.

- Describe the types of training applicable for maximizing performance for different sports.

- Present typical energy expenditures and general recommendations for energy nutrient intake for different sports.

- Discuss nutrition considerations and strategies to enhance athletic performance.

- Identify which nutrition supplements are associated with different sports.

The general nutrition concepts related to exercise and training presented in previous chapters can now serve as the foundation for the sport-specific facets described in this chapter. Each of the sport discussions below focuses on performance characteristics, physical characteristics, training, and nutritional considerations. Sports can seem quite different in the movements involved, the playing fields, and even the uniforms. For instance, at first thought swimming, cycling, wrestling, and football might seem to have little in common. However, the basic nutritional concepts are much the same for all sports, and athletes in sports with comparable training practices and competition and similar environments share special attention to specific nutrients. This chapter summarizes nutritional similarities and identifies sports that have exceptional concerns for specific nutrients.

Eating a diet that is rich in carbohydrate and lower in fat, particularly saturated fat, is recommended for everyone, regardless of athletic status. Whole grain products, fruits, vegetables, legumes, and low-fat dairy products should be the foundation of everyone's carbohydrate intake, barring intolerances, allergies, or diet philosophies. Sport drinks and carbohydrate-dense foods or beverages consumed immediately before, during, and after training and competition can make a significant contribution to carbohydrate consumption and improve performance. Also, eating a variety of foods including fish, fruits, vegetables, and whole grains would help an athlete meet the most basic requirements for vitamins, minerals, and essential fatty acids and provide a variety of health-promoting *nutraceutical* substances. Performance in all sports is optimized by proper hydration efforts. Body weight and composition are also important considerations in many sports, so understanding how to balance energy intake with expenditure is paramount.

As discussed in the opening chapter of this book, the general recommendations for athletic populations are similar to those for the general population in many ways. Differences include athletes' increased need for protein to build and maintain muscle, and for fluid and electrolytes to recover sweat losses. The unique physiological and nutritional aspects of some sports are discussed in the feature In Focus: Nutrition for Exercise at High Altitude and in the Nutrition in Practice features on nutritional considerations for early-morning sports and travel.

■ ICE HOCKEY

Ice hockey has long been popular in countries with cool climates and continues to gain popularity in the United States. In the United States, male and female youth leagues and high school teams are becoming more prevalent along with intercollegiate teams, club sports, and intramurals at colleges and universities. Adult community rinks and ice hockey leagues are also becoming more prevalent and popular.

Photolink/PhotoDisc/Getty Images

Even though ice hockey is an extremely popular sport in many countries, relatively little attention has been focused on the relationship between ice hockey performance and nutrition. Perhaps this is because ice hockey has long been viewed as a game of speed and skill, two attributes that can be viewed more as genetic than inducible through training and diet. Also, ice hockey is one of the least anatomically revealing sports. With the exception of height, very little of a player's physical attributes are noticeable through the padding and uniforms.

There are six positions in ice hockey. Three of these positions are forward and offensive positions (one center and two wings), and three are defensive positions (two defensemen and a goalie). According to one report, defensemen tend to be taller and heavier than forward position players.[1] However, assessment of 27 players in the National Hockey League (NHL) did not reveal a difference in height when participants were separated by position.[2] NHL players such as centers Mario Lemieux (Pittsburgh Penguins) and Eric Lindros (New York Rangers) are certainly testament that taller and heavier players can excel at the forward positions as long as the increased size is coupled with speed and agility. Other data obtained from the assessment of the NHL players revealed that differences in body composition or strength do not exist between the positional players.[2]

Performance Characteristics of Ice Hockey

The sport of ice hockey is played in a cold environment and players wear a full arsenal of gear (helmets, shoulder pads, gloves, and so on). A game consists of three 20-minute periods with two 20-minute intermissions between them. Participation occurs in shifts lasting from 30 seconds to more than a minute with 4–6 minutes of rest between them, depending on shift depth. On-ice performance consists of several sprinting efforts with periods of rest (stoppage of play, face-offs). A player may perform for only 15–20 minutes total in a game.[1] Seasons last approximately 6–8 months, depending on the level, with practices or games on a near daily basis. A professional season may include more than 100 games when counting preseason and playoffs. At the professional level the length of the off-season is shorter than in other sports such as baseball and basketball. Typical ice hockey practices last $1\frac{1}{2}$–2 hours. The sport of ice hockey consists of high-intensity intermittent skating with rapid changes in velocity and frequent body contact.[1]

The sport of ice hockey involves periods of anaerobic performance and aerobic recovery as well as overlapping periods. Peak heart rates during an on-ice shift may reach 90% of HR_{max} with an average of about 85%.[1] Researchers have reported that the average VO_2max of 27 professional ice hockey players in the NHL averaged 53 ml/kg of body weight per minute.[2]

Physical Characteristics of Ice Hockey

Ice hockey is primarily an anaerobic sport. Skating speed, quickness, and agility are essential. Upper body strength and power are necessary for checking and position control. Back and abdominal strength are crucial for the stability and control of the spine. Flexibility is important for all body parts and especially the hamstring and lower back muscles. Training and other protection against injury from contact is needed for the neck and shoulder joint, and injuries to the lower back, hamstring, and adductor muscles from the skating motions must also be addressed.

Key Point

Hockey primarily consists of repetitive high-intensity skating shifts with 4–5 minutes of rest between them.

Physical Training for Ice Hockey

The dynamics of this sport (strength, speed, power) suggest that a periodized program would be the most beneficial. Players benefit from a certain level of mass to attenuate the forces caused by contact (as in checking and positioning). Addressing each of these specific components within a yearly cycle should maximize the training benefits. Strength training in the form of multijoint and Olympic movements mimics on-ice movements and trains the entire body. These exercises improve coordination, stabilization, strength, and power. Specific stabilization exercises for the shoulder joint and shoulder girdle must be included. Several exercises focusing on flexibility, strengthening, and stabilizing of the hamstrings and lower back should also be incorporated. The use of stability, exercise, and medicine balls are also recommended for core stabilization, balance, flexibility, and ballistic movements. Skating and other conditioning drills should simulate game conditions. Performing high-intensity endurance drills of 1–2 minutes with 10- to 20-second maximal burst movements within them should be the primary mode of training.

Key Point

Conditioning should focus on quickness and power drills of 10–15 seconds and high-intensity endurance skates with durations of approximately 2 minutes.

Nutrition for Ice Hockey

There are several nutritional considerations for ice hockey players. These are based on the frequency of training and games, length of season, and travel schedule. One study demonstrated that a well-structured nutrition program improved the performance and in-season weight maintenance of professional players.[3] Another study involving collegiate ice hockey players suggested there are benefits from increasing the athletes' awareness of nutrition and health as it relates to their performance. These athletes identified that making the transition to college and playing ice hockey involved adjusting to new nutritional and athletic environments, increased personal responsibility for food choices, and new meanings for food and eating.[4] They stated that playing ice hockey, health, and taste were major determinants of their food practices. They reported that to them, hockey meant structured schedules, a social network, and performance expectations, and that health meant "feeling good" for hockey and having a lean body composition and a desirable body image.[4]

ENERGY BALANCE AND BODY WEIGHT AND COMPOSITION. Energy expenditure varies among ice hockey players based on the amount of "ice time" and "bench time" as well as resistance training and conditioning off the ice. Professional ice hockey players weighing 180–210 lb (81–94 kg) can expend 12–14 kcal/min of play.[5] Considering all activities, ice hockey players expend 3000–6000 kcal daily.[5] One study involving youth players revealed that they compensated for increased energy expenditure by consuming more energy.[6]

Ice hockey players tend to present a broad range of body compositions, although there is a trend for the players to be bigger and stronger than in the past. Even for youth ice hockey, the players (12- to 13-year-old boys) have been noted to have greater arm mass compared to inactive peers.[6] Additional mass as lean body mass (LBM) can provide greater strength and more opposing force during the struggle in front of the net or against the boards as well as in general "checking." However, additional body mass attributable to enhanced body fat can compromise quickness and stamina. A body fat percentage range of 8–19% might be typical for ice hockey players. For instance, professional ice hockey players assessed in one study had an average body fat percentage of 9.2%.[2] These men also averaged 182.9 cm in height and 85.6 kg in weight. Elite ice hockey players in Sweden averaged 13.9% body fat as assessed by skinfold.[7]

CARBOHYDRATE. Muscle glycogen is a primary fuel source for working muscle in ice hockey players. In one study involving ice hockey players, as much as 60% of the glycogen in the quadriceps muscle was utilized during a single game.[8] Therefore, carbohydrate consumption is important prior to, during, and after a game or practice to optimize glycogen status. Because of the practice and game schedule of ice hockey season, glycogen loading is not practical. However, in an exercise study format, glycogen loading has demonstrated the importance of muscle glycogen to ice hockey performance. Researchers compared muscle glycogen content and performance characteristics in elite Swedish ice hockey players fed either a mixed energy nutrient diet or a glycogen loading diet during two games.[9] As expected, the muscle glycogen concentrations were higher in the glycogen loaded group. In addition, the distance skated, number of shifts skated, amount of time skated within shifts, and skating speed improved with glycogen loading. It was concluded that individual differences in performance were related to muscle glycogen metabolism.

Key Point

Reduction of muscle glycogen stores can reduce ice hockey performance.

The level of carbohydrate in ice hockey player's diet should be at least 60% of energy or 7–10 g/kg of body weight. As many hockey practices and games occur in the late afternoon and early evening, a carbohydrate-containing meal or snack needs to be consumed immediately after exercise, and also later in the evening if more than 2–3 hours elapse between the post-game/practice meal and bedtime. Ice hockey players should strive to consume at least 1.5 g/kg of body weight immediately after a hard practice or game and every 2–3 hours thereafter (see

Chapter 4). Upon waking, carbohydrate should be the foundation of the first meal, providing at least 75–100 g for an adult.

Consuming a sport drink containing 4–8% carbohydrate during games and practices supports muscle carbohydrate utilization. Such consumption during bench time and intermissions might actually allow for partial recovery of glycogen stores. The types of carbohydrate in a pregame/practice meal should also be considered. Ice hockey players should be cautious of any carbohydrate source that may cause intestinal discomfort. Excessive intestinal gas production, bloating, and cramping can negatively impact performance. In youth tournaments, where several games may be played on the same day, carbohydrate-rich snacks or meals should be consumed between games.

Key Point

Ice hockey players should take advantage of bench time to consume fluid and carbohydrate.

PROTEIN. Recommendations for protein intake for hockey players are not firmly established. The primary considerations in determining protein needs include the level and type of training as well as protein requirements for healing injuries. Resistance training can also enhance protein requirements. In addition, the energy level of the diet needs to be adequate to at least maintain nitrogen balance. General recommendations for protein might be 1.2–1.4 g/kg of body weight or 15–20% of energy intake.

Protein should be a component of the post-game/practice meal and can be provided at a level of 0.5 g/kg of body weight or more. This would be at least 42 g for a 185-lb (84-kg) defenseman. There is some support for the notion that protein is supportive of carbohydrate in glycogen recovery. However, perhaps the most outstanding reason for protein consumption is to promote nitrogen balance and possibly positive nitrogen balance as soon after exercise as possible. As for carbohydrate, many sport nutritionists endorse the notion of distributing protein over several smaller meals rather than fewer and larger meals, as discussed in Chapters 5 and 7.

FAT. The level of fat intake reported for elite ice hockey players approximates 40% of their energy consumption.[10,11] If male ice hockey players eat a diet that is higher in fat than the general population, is it due to preference or might it be purposeful? The findings of one study involving collegiate ice hockey players suggest that low-fat foods are viewed generally as healthy, but as not providing taste satisfaction or reward.[4] Despite the encouraging results of glycogen loading trials, an important consideration is the energy density difference between carbohydrate and fat. Replacing an equivalent volume of a higher fat food with a high-carbohydrate food may lead to a reduced energy intake. For instance, when elite Swedish ice hockey players reduced their fat intake from 40% to 30% for 7 months, they also reduced total energy consumption.[11] Therefore, ice hockey players who manipulate their diet composition should also monitor their energy intake and changes in body weight and composition.

VITAMINS AND MINERALS. The information available related to vitamin and mineral intake of ice hockey players is limited. One study involving youth ice hockey players suggests that diet levels of vitamins and minerals may be adequate for this population.[6] Another study involving female ice hockey players suggests that they generally meet recommendations for iron and calcium intakes.[12] Ice hockey players would certainly benefit from a diet that provided adequate levels of vitamins and minerals, which assist in optimal glycogen and energy metabolism and promote general health and disease prevention.

FLUIDS AND HYDRATION. Because ice hockey is associated with cold temperatures, it might seem that sweating would be less significant than in sports such as soccer or football. However, ice hockey has a significant clothing and equipment component, which increases sweating. In fact, sweat volumes of ice hockey players are similar to those of athletes in warm sports, so players need to continuously hydrate before, during, and after games or practices. These athletes can experience a fluid loss of 3–10 lb during a game.[3] They should consume 20 oz (600 ml) for each lb of fluid loss experienced.[13] Players can also gauge their sweat losses by weighing themselves prior to and after play. See the Chapter 8 Nutrition in Practice: Estimating Sweat Loss.

Ice hockey players need to consume water or a sport drink while on the bench and during intermission. The need for carbohydrate is also important, and consuming a sport drink containing 4–8% carbohydrate just before and during a game or practice addresses both issues. Hyperhydration efforts are not recommended because of the availability of fluids during games and practices as well as the problems associated with urinating once in full gear. If fluids are not consumed during a game or practice, players should drink at least 1–2 L of water for each hour of ice time after a practice or game.

NUTRITION SUPPLEMENTATION. Often ice hockey players experiment with various nutrition supplements in an attempt to increase their physical size, power, and speed. Creatine has become fairly popular as players seek an edge to increase these components. Although creatine has not been thoroughly researched with regard to ice hockey, researchers have reported that creatine supplementation increased power and speed aspects of multiple skating bouts.[14] Along with physical attempts to enhance lean body mass comes the temptation to experiment with

nutrition supplements touted to enhance lean body mass development. The IOC and the NCAA as well as some professional sport associations have banned both androstenedione and DHEA. However, other supplements are widely experimented with, including HMB, glutamine, "natural human growth hormone" (arginine, lysine, ornithine), branched chain amino acids (leucine, isoleucine, and valine), and antioxidants. Personal communications with semiprofessional and collegiate ice hockey players have revealed that the interest in supplements to enhance mass is considerable.

■ ICE HOCKEY *IN REVIEW*

- Ice hockey is a game of quickness, speed, power, and agility. A strength training program should address these components. A periodized program using the core lifts, Olympic lifts, and some plyometric exercises is recommended. Conditioning drills should mimic the activities of game play with 1–2 minutes of high intensity skating followed by 3–4 minutes of rest.
- Players should strive to maximize glycogen stores before practices/games and minimize losses during practices/games by eating a higher carbohydrate diet and utilizing a sport drink. Protein requirements of ice hockey players are elevated because of the intensity of practice/games, supportive training, and injury recovery. Sweat rates of ice hockey players can parallel those in warmer sports, so proper hydration before and during a practice/game is paramount.

■ BASEBALL/SOFTBALL

Baseball and softball teams consist of nine players, each playing a position associated with specific characteristics. It is a sport that has a seemingly equal emphasis on individual effort and teamwork. For ease of discussion this assessment focuses on the "position players," which includes everyone except the pitchers and catchers, although it

PhotDisc/Getty Images

addresses pitchers and catchers at certain points. Although baseball and softball have subtle differences, for the most part they are much alike.

Performance Characteristics of Baseball/Softball

The sport is played in a variety of climates but primarily in warm weather, since the active season is in the summer months. Games typically last 2–3 hours at the collegiate and professional level. College baseball's "active" season consists of 50–70 games occurring over 5 months. The off-season for these athletes is also 5 months, leaving approximately 2 months for active rest. Softball has similar characteristics. The major league teams have similar schedules and play 162 games over 7 months, leaving 5 months for an off-season and active rest. Baseball is one of the most competition-dense sports during an active season, when athletes either practice or play just about every day of the week. This heavy schedule demands a continuous emphasis on proper nutrition and rest. In addition, professional baseball players have to adjust to time zone changes during travel.

Physical Characteristics of Baseball/Softball

Baseball is primarily an anaerobic sport. Running speed, quickness, and agility are essential. Lower and upper body strength and power are necessary for throwing motions and batting. Back and abdominal strength are crucial to stabilize and control the spine during throwing and batting motions, and rotator cuff muscles must be strong to accommodate overhand throwing motions. Flexibility is important for all of the musculature, but there is particular concern for the hamstring, lower back, and rotator cuff muscles.

Handedness is perhaps more significant in baseball than any other sport, although arguments could be made for left-side positions in ice and field hockey and lacrosse. For baseball, left-handed pitchers and hitters are hot commodities, along with switch hitters. Although handedness has long been viewed as mere probability, one study observed that the season of birth may be an influential factor; of the baseball players surveyed, those born during the spring or early summer were more likely to be left-handed than those born during the fall and winter.[15]

Physical Training for Baseball/Softball

The dynamics of this sport (speed, power) suggest that a periodized program would be the most beneficial. Power should be developed after a solid foundation of strength has been established. Much of the power for throwing and batting comes from the sequential movements of the large muscle groups of the lower extremity, so adequate volumes of training such as squats should address these muscles. Use of ballistic and plyometric movements (such as medicine balls, box jumps) may prove helpful because they mimic the ballistic nature of throwing, batting, and

sprinting. Specific strengthening and stabilization exercises for the shoulder joint and shoulder girdle must be included. Several exercises focusing on flexibility and strengthening of the hamstrings, lower back, and abdominals should also be included. Most sprinting and other conditioning drills should simulate game conditions: short maximal bursts (3–10 seconds) with a fair amount of rest (30–60 seconds) between repetitions.

In-season training should focus on maintaining strength and power components and address muscular endurance. Off-season training should be devoted to improving strength, power, speed, and in some cases body mass. Some of these gains may be lost during the long active season, but using fairly high intensity levels on occasion helps deter these losses.

Key Point

Power for throwing and batting comes from sequential movements that begin with the hips and legs. Therefore, strength training programs for these sports should focus on power and speed for the lower extremities.

Nutrition for Baseball/Softball

The high-intensity "all-out" movements (running, throwing, batting) in baseball and softball rely heavily on anaerobic energy metabolism. Therefore, during these efforts, creatine phosphate and glycogen are the primary resources for generating ATP. In addition, these sports are most usually played in warm climates, so sweat loss is high. Now more than ever, players possess a great range of height and weight with a general trend toward greater size at perhaps all positions. One nutritional challenge for softball and baseball players is their travel schedule.

ENERGY BALANCE AND BODY WEIGHT AND COMPOSITION. Energy expenditures for baseball and softball players can vary due to different activities related to positions. For instance, a 183-lb pitcher may expend 7.5 kcal/minute, which equates to 450 kcal/hour.[16] A position player weighing the same might expend 5.1 kcal/minute or 306 kcal/hour. Catchers have higher energy expenditures than the position players due to the constant crouching and standing as well as involvement in nearly every pitch and subsequent plays. A catcher squats more than 150 times per game and throws the ball more than any other player. Catchers also back up first base on ground balls and might engage in a few high-impact collisions at the plate. Furthermore, catchers are clad in equipment that increases their mass and also decreases the efficiency of some movements, both of which elevate energy expenditure.

The body composition of baseball players varies depending on performance expectations (speed and power) and personal preferences. This is likely to be tied to field position, and research has suggested that baseball players are better assessed by position.[17] Other estimates for body fat levels of baseball players were 12.6% and for softball players 19.1%.[18,19]

CARBOHYDRATE.   Carbohydrate should provide ≥60% of the total energy consumed by baseball and softball players or 6–10 kcal/kg of body weight. This provides the basis for glycogen recovery following a game or practice. Having greater stored muscle glycogen might be particularly important for pitchers in a rotation as well as relievers, who throw fewer pitches per game but pitch more often. Catchers would also benefit from having greater muscle glycogen levels at the onset of a game. All players might benefit from higher muscle glycogen levels if faced with a doubleheader. These athletes have plenty of time during game and practice situations to consume carbohydrate. Water bottles and coolers containing sport drinks are common on both softball and baseball benches. A sport drink containing 4–8% carbohydrate provides both carbohydrate and water. Recommendations for sport drink consumption are based on fluid needs for recovery of sweat loss, as discussed below.

PROTEIN.   Protein requirements of baseball and softball players are unclear; however, they are higher than the RDA due to the repair of abrasions, contusions, and other musculoskeletal injuries. Also, many players engage in resistance training, and some continue throughout the season, as former MLB slugger Mark McGwire did. The protein needs of these athletes approach those of weightlifters but should be appropriate to the level of training.

Consuming a diet in which 15–20% of the total energy is derived from protein provides adequate protein for baseball and softball players. For instance, a shortstop engaged in minimal amounts of weight training (2–3 days/week) or ballistic training may achieve nitrogen balance closer to the 15% intake versus the 20%. A right fielder or first baseman (traditionally bigger players) who engages in heavy weight training (>3 days/week) to maintain a larger muscle and total body mass might approach the 20% requirement. There is a trend for greater muscularity and general size among many of the positional players. Even pitchers are including more weight training than ever before. Consuming a diet that provides protein at 1.2–1.4 g/kg of body weight should allow baseball players who engage in a moderate level of physical training to achieve nitrogen balance and promote LBM development. If resistance training is heavier during the off-season, protein consumption of 1.6–1.7 g/kg of body weight allows for maximization of gains in strength and mass.

FAT.   Fat intake should be based on achieving energy needs once adequate carbohydrate and protein are provided. Thus, a large left-fielder who derives 60% and 18% of his energy requirements from carbohydrate and protein, respectively, would leave only 22% of the energy to be derived from fat. However, traveling on the road and picking up a fast dinner on the way home from practice or games can lead to poor diet choices. Among the most significant aspects of poor diet planning is consuming foods that provide too much fat. Athletes and their coaches should plan ahead when traveling and bring or have available more nutritious food choices after play. See Nutrition in Practice: Eating on the Road for tips on eating healthy during a heavy traveling schedule.

FLUIDS AND HYDRATION.   Baseball and softball practices and games can last 2–3 hours, and maintaining optimal hydration "might be" the most significant immediate nutrition factor in performance. For instance, baseball players practicing on a warmer day (36.7°C) experienced an average body weight loss of 3% when fluid was not consumed.[20] The sweat rate of these athletes was estimated at 56.5 ml/kg of body weight during practice and the game that followed. Additionally, when these athletes were allowed to consume fluid (sport drink) during the game, most failed to completely compensate for water losses. Proper hydration is also extremely important for youth baseball players whose ballfield does not include a covered dugout.

Lower levels of humidity make sweat less noticeable because evaporation occurs more easily. Baseball players must be aware that they need to drink fluids before they become thirsty and that even subtle reductions in body water can impair performance. This is especially true for pitchers. Significant dehydration can decrease blood volume and circulation through the pitching arm and the legs. Proper hydration is necessary for optimal mental acuity, and baseball is a thinking game; players must be able to quickly comprehend changing situations. Reduced coordination is also related to hydration status, as a rising core temperature may affect the function of the central nervous system. In general, at least 1 cup (~250 ml) of water is recommended every 15 minutes during practices and games. Athletes can gauge their potential water loss during a game situation during the preseason.

Key Point

Maintaining optimal hydration is the most important nutritional concern for baseball and softball players.

VITAMINS AND MINERALS.   Little information exists regarding the vitamin and mineral intake of baseball players. However, since their requirements for these nutrients are not particularly enhanced, a balanced diet would allow for adequate consumption. As demonstrated in the Nutrition in Practice examples in Chapters 9 and 10, an energy-adequate diet containing a variety of natural foods would easily provide at least the RDA for vitamins and minerals. The vitamin and mineral content of the diet of younger baseball players should be of particular concern in light of

When the team bus pulls out for a 5-day road trip that includes three games and one practice, eating a diet that boosts performance becomes a different ballgame than it is at home. An athlete traveling to compete in a triathlon in a city one state away faces the same sort of challenge. How do athletes and coaches deal with the need to eat to prepare for optimal performance as well as recover from training and competition while on the road?

Sadly, fast food is often the choice because of its convenience and low cost. When the travel stop is a convenience store or supermarket, common purchases are candy bars and snack foods (chips, cookies). However, what is saved in time and expense may be lost on the playing fields. The bottom line is that poor food choices on the road can fail to prepare an athlete for competition because they compromise glycogen stores and hydration status. In addition, food choices after competition are an important consideration as an athlete looks forward to the next competition, which may be less than 24 hours away. Following are some suggestions for eating on the road to optimize performance.

PREPARE A PERFORMANCE PACKAGE. Depending on the length of the trip and the competitive level, one can pack or request to have several food and beverage items available; see Table A for some suggestions. Coaches can also provide practical advice to their athletes regarding the travel schedule and what to pack. Simplicity is as important as performance suitability. The length of trip affects the types of foods to pack. For one-day trips, coolers can broaden the range of foods packed. For longer trips, certain foods can be packed in travel bags and sent with the luggage.

BEWARE OF THE UNSEEN. One potential problem athletes face when traveling to an unfamiliar area to compete is microorganisms that can cause diarrhea, malaise, and sickness. The microorganisms can be found in foods, drinking water, and areas of contact. To reduce the risk of contamination athletes can adhere to the following suggestions.

- Drink bottled water instead of tap or fountain water, or boil tap water. Do not drink water while showering and do not brush teeth with tap water. When dining out, drink bottled water, canned or bottled juices, or soft drinks.
- Make sure foods have been thoroughly cooked, especially meats.
- Purchase peelable fruits (oranges, bananas, grapefruit) rather than fruit that does not require peeling (grapes, apples, pears, peaches). If the only choice is unpeeled fruit, wash it thoroughly in bottled or boiled water.
- Do not purchase burritos, sandwiches, or other ready-made food in convenience stores.

SUPERMARKETS. Supermarkets are a better choice than fast-food restaurants. Athletes can purchase a variety of ready-to-eat foods that are higher in carbohydrate and protein and lower in fat. In addition, bottled water and juices can be purchased to maintain optimal hydration. Good food choices include fresh baked products (bagels, breads), yogurt, precooked chicken and turkey, low-fat milk, cereal, and sport bars and drinks.

FAST-FOOD RESTAURANTS. Fast-food restaurants have long been a stop along the road for athletes and teams. Until recently, the food available in these places was for the most part high in fat, but inexpensive and convenient. Over the past few years, many traditional fast-food restaurants have included leaner meats and lower fat preparations and menu items. For instance, depending on an athlete's needs he or she could choose skim milk, low-fat milk, whole milk, or a low-fat or high-fat vanilla milkshake. Salad bars and side orders such as low-fat (air-cooked) fries, garden salads, and rolls can provide carbohydrate-rich, low-fat food choices. Low-fat submarine sandwiches, grilled meat sandwiches, and grilled chicken tacos and burritos can provide a lot of protein without a lot of fat.

Table A   **Travel Foods to Pack or to Buy at Supermarkets**

High-Carbohydrate Foods	High-Protein Foods	Carbohydrate/ Protein Foods	Energy-Dense, High-Fat Foods
Fig bars	Beef jerky	Sport bars and drinks	Peanut butter
Dry cereal (granola)	Precooked chicken	(ready-to-drink, RTD)	Sunflower seeds
Bagels	and turkey (air-sealed)	Blend of nonfat dry milk	Walnuts
Crackers	Soynuts	with chocolate powder	Cashews
Breadsticks	Sport bars	(add water)	Peatnuts
Bananas			
Apples			

growth. Parents should provide nutritious meals and snacks and monitor other foods and beverages consumed by their children.

NUTRITION SUPPLEMENTATION. Baseball is among the sports targeted by some creatine manufacturers. In theory, increased muscle creatine phosphate could enhance ATP-regenerating capabilities such as during all-out sprinting efforts. Applied examples include sprinting around bases (doubles and triples) or trying to snag a fly ball hit in the shallow outfield. In addition, an improved multiple sprint performance might benefit a base stealer who just stole second base and will try for third base on the next pitch. Some of the research involving creatine supplementation and ice hockey players might apply to baseball players (such as base stealers and center fielders).[14] It has been speculated that creatine may benefit pitchers and assist their ability to maintain high pitch velocities despite a climbing pitch count (>100 pitches for a starter). Research is lacking in this area despite the fact that many professional teams have a nutrition consultant. About the only area of abundant research involving baseball players and what they put in their mouth is in regard to chewing tobacco.[21–24]

Other supplements such as DHEA, androstenedione, and related steroid structures are fairly popular among baseball players. However, because researchers have reported that androstenedione and DHEA are not effective in increasing testosterone levels in a controllable manner, these supplements should not be used. Furthermore, the NCAA has banned androstenedione and DHEA. Baseball players use other supplements such as ribose and HMB, more so than softball players. However, these substances have not been thoroughly researched, and recommendations for supplementation cannot be made at this time.

■ BASEBALL/SOFTBALL *IN REVIEW*

- Baseball is a game of quickness, speed, and power, and the strength training program should address these components. A periodized program using the core lifts, Olympic lifts, and some plyometric exercises is recommended. Flexibility is critical for the rotator cuff muscles and others involved in batting and running. Stabilizing the core is also essential, as all throwing and batting motions involve the abdominal and lower back muscles.
- More active players need to maximize glycogen stores before practices/games and minimize losses during practices/games by eating a higher carbohydrate diet and utilizing a sport drink on the bench. Protein requirements are related to supportive training and injury recovery. A warmer and sunnier climate can increase sweating significantly, and players need to properly hydrate between and during practices and games.

Pete Saloutos/Corbis

■ FOOTBALL

The football team has several positions, each with unique physical and performance demands. Linemen need to be big and strong, whereas the receivers and defensive backs must be fast and agile. A high school or college lineman who is not only big and strong but also quick will always get a second look by scouts. The same is true for defensive backs or receivers who are not only quick and agile but also bigger and stronger than most. Running backs and linebackers are hybrids of the players just mentioned. They must be fast, agile, strong, and fairly massive in order to endure the demands of physical contact.

Football players continue to get bigger and stronger.[25] For instance, a survey of High School All-American football teams revealed that the body mass index (BMI) of players increased dramatically during the 1970s and 1980s.[26] Perhaps this is most obvious in the college and professional ranks. In the 1980s, William "The Refrigerator" Perry became one of the first players to surpass the 300-lb mark in professional football, which drew considerable media attention. Today, the average weight of starting offensive or defensive linemen of Division I college and professional football teams is typically 300 lb.

Football is one of the few sports in which most if not all players on the field engage in physical contact during each play. Although the protective equipment is obvious, many question whether it is adequate or utilized correctly.

Many defensive backs and receivers do not use thigh and kneepads, and receivers often wear very small shoulder pads. Head impact can cause the brain to move and slam into the wall of the skull. Depending on the region of impact, trauma to the brain can cause blackouts or decreased orientation and cognitive capabilities. Concussions are a growing concern in football, and many players are not well educated as to the signs and symptoms.[27] Special nutrition considerations for cerebral trauma are completely speculative and require more attention.

Performance Characteristics of Football

Football is played in a variety of environments ranging from severe cold to very warm. Players wear a full complement of pads and gear and a professional game consists of four 15-minute periods or "quarters" with a halftime intermission of 15 minutes after the second quarter. Players are usually associated with either the offense or defense, so a given player is probably active for only half of the total game time or less. In the United States, football seasons typically begin in the hot summer months and continue through the fall and into the beginning of winter. Games are played once a week, with practice sessions occurring during the 5–6 days between games. A practice lasts about 2 hours, and training camp may consist of "two-a-day" practices.

Physical Characteristics of Football

Football is primarily an anaerobic sport. A typical play lasts for 6 seconds with roughly 30–45 seconds of rest between plays. However, the unique activities at different football positions allow a separation of football athletes into three position classes: receivers and defensive backs, running backs and linebackers, and offensive and defensive linemen. Specialty players, namely punters and kickers, are not discussed here.

RECEIVERS AND DEFENSIVE BACKS. Receivers and defensive backs primarily engage in near maximal sprinting with some contact associated with coverage, blocking, and tackling. Because of the physical contact and ballistic nature of the game, players need total body strength and power. Back, neck, and abdominal strength are crucial to stabilize and control the spine. Flexibility is important for all the body parts to enhance speed and function and decrease the incidence of injury.

RUNNING BACKS AND LINEBACKERS. Running backs and linebackers engage in regular contact associated with blocking and tackling. They also perform near maximal sprinting and movements requiring change of direction and quickness. Because of the physical contact and ballistic nature of the game, players need body mass, strength, and power. Back, neck, and abdominal strength are crucial for the stability and control of the spine during contact. Mass development (hypertrophy) is necessary to help create and attenuate the forces of physical contact. Flexibility for all body parts is important to enhance speed and decrease the incidence of injury.

OFFENSIVE AND DEFENSIVE LINEMEN. Linemen are generally the largest and strongest players on the field. Offensive and defensive linemen engage in physical contact on every play in an attempt to control the line of scrimmage. Players line up 2–3 yards from one another, so there is little time or space to accelerate; hence the need for a large mass to produce force. Quickness is important, but it should not be at the expense of losing significant amounts of size or strength.

Key Point

Optimal weight and speed components are different for the various positions in football. Linemen typically need to maintain a large mass at the expense of speed, and receivers need to minimize mass in order to maximize speed.

Physical Training for Football

Physical training for football is based on the position and its associated traits as well as individual desired modifications, if any, to body weight and composition. The degree of muscle hypertrophy, strength, and power depend on the position played and on individual physical traits. Increases in body mass may be helpful for some players and detrimental for others based on changes in speed and agility. These are discussed below relative to certain positions.

RECEIVERS AND DEFENSIVE BACKS. The dynamics of this sport (strength, speed, power) suggest that a periodized program would be the most beneficial. Addressing the specific components within a yearly cycle should maximize the training benefits. Excessive muscular hypertrophy for these players (receivers and defensive backs) may not be beneficial, as the increased mass could slow running speed and agility. Strength training in the form of multijoint and Olympic movements mimics athletic movements and trains the entire body simultaneously. These exercises improve coordination, stabilization, strength, and power. Several exercises focusing on flexibility and strengthening and stabilizing the hamstrings and lower back should also be included. Use of plyometric exercises assists in the development of speed and quickness. Conditioning drills should emphasize footwork and maximize speed. Maximal and high-intensity endurance exercises should make up a majority of the conditioning. For improvement of speed, adequate amounts of rest (about 2 minutes) must be allowed between maximal effort exercises.

RUNNING BACKS AND LINEBACKERS.

Running backs and linebackers need to consider muscle hypertrophy along with strength and power. A specific focus on each of these components suggests the use of a periodized program. Use of multijoint and Olympic movements for strength training mimics athletic movements and trains the entire body simultaneously. These exercises improve coordination, stabilization, strength, and power. Additional isolation exercises may assist in mass development. Use of plyometric exercises assists in the development of speed and quickness. Conditioning drills should emphasize footwork, including change of direction, and maximize quickness. Most drills and conditioning should include 5–20 seconds of near maximal efforts with 20–30 seconds of rest between.

OFFENSIVE AND DEFENSIVE LINEMEN.

Because of the dual need for size and strength, a periodized program addressing both of these components is likely to produce the best results. Also, because of their large size, linemen often have poor flexibility and agility, so using Olympic movements in strength training may assist in developing total body coordination and maintain or improve flexibility. By utilizing the power lifts (squat, deadlift, bench press), one can develop large amounts of both mass and strength, so these should constitute a large portion of the training volume. Conditioning drills for linemen should emphasize footwork and quickness. These drills should be performed in 10–20 seconds covering short distances, with 20–30 seconds of rest between repetitions.

Nutrition for Football

Nutritional concerns and requirements of football players must take into consideration the unique physical attributes and performance requirements at the different position classes. In addition, if changes in body weight and composition are desired to maximize performance, the diet is affected. For instance, a lineman wishing to add weight, primarily LBM, during the off-season needs a calorie intake greater than his energy requirements. Receivers and defensive backs may restrict their energy intake to a degree during the off-season to promote weight loss.

In comparison to many other team sports, football has a shorter season and thus a longer off-season. For American football, summer camp begins in July and the regular season games end in December. Playoffs prolong the season by several weeks for many high schools, some college divisions, and professional leagues. Still, college bowl games end in early January and high school playoffs typically finish prior to winter holiday breaks. This allows for significant time during the off-season to recover from injury and to modify body weight and composition.

ENERGY BALANCE AND BODY WEIGHT AND COMPOSITION.

Body weight and composition can vary in relation to football position.[28] For instance, receivers and defensive backs tend to be leaner and lighter than the other positions; estimations of body fat levels have averaged 9.6%.[18] These positions dictate careful consideration of the relationships between body weight, body composition, agility, and speed. Although play at these positions is sometimes enhanced by increased strength associated with a greater muscle mass (for example, in downfield blocking and tackling) these players are generally not willing to compromise their quickness and speed for extra mass. A possible exception is the strong safety, whose job often involves a good deal of run support. Although defensive backs and receivers may experiment with subtle increases in mass and musculature in the off-season, they are often among the best-conditioned athletes coming into training camp as their off-season training revolves around speed and agility drills.

Running backs and linebackers tend to be more massive than receivers and defensive backs, but smaller than linemen. Here again the relationship between body mass and composition and physical performance must be experimented with and understood completely. Although examples of heavier running backs are easy to recall (such as Ron Dayne and Jerome Bettis of the NFL), most running backs have similar size parameters. Enhanced physical size and weight would clearly offer an advantage during moments of impact and first-down surges. However, the extra mass cannot compromise quickness, speed, or game endurance. The offensive scheme of many teams involves a fullback, who tends to be heavier and not as lean as a halfback. Their body type is ideal for running the ball "up the middle" and blocking. The halfback is the "feature back" on all teams and tends to be geared more toward off tackle and around-the-end quickness and speed. The body fat level of halfbacks can be similar to that of receivers and defensive backs. The body fat level of fullbacks is usually closer to that of linebackers, which seems to average 14%.[18] Most of the qualities that are advantageous for running backs also work well for linebackers. However, linebackers tend to be 10–20 lb (5–10 kg) heavier, especially middle linebackers. Outside linebackers are usually a little lighter and leaner than middle linebackers. Outside linebackers are more involved in pass coverage, blitzing, and tracking down a ball carrier coming around the end. Often teams experiment with rotating their linebackers, so that inside linebackers sporting larger masses might only play first down and short-yardage second and third downs, as these are higher percentage running plays. However, a lighter and quicker linebacker may come in for more passing situations (such as third down and 8).

As mentioned earlier, there is a definite trend over the past years for football players, particularly linemen, to become bigger.[25] In related fashion, energy intakes are staggering and increase relative to player size. For instance, energy consumption of linemen can approximate 5000–6000 kcal daily and researchers have estimated that linemen might consume at least 1500 kcal/day more than defensive backs.[29] Coaches, athletic trainers, and team physicians should monitor players' weights and body compositions during both the off-season and active season and identify those athletes with disordered eating patterns that render them prone to obesity. Recommendations for optimal body compositions for linemen have not been proposed. However, most linemen know the body weight (suggestive of body composition) at which they are "at their best." As mentioned, heavier linemen with a higher percentage of body fat might compromise quickness and agility and may not be able to endure four quarters of high-intensity play.[28] Examples of this are apparent in the professional ranks when a 300+-lb defensive tackle with a higher amount of body fat becomes slow and ineffective in the fourth quarter. In addition, there is evidence to suggest that body fat levels may be related to the incidence of injury in linemen.[30,31] For instance, researchers reported that the high school football players with higher body fat levels they assessed had a $2\frac{1}{2}$ times greater risk of injury than leaner players.[31] The disqualification of extremely overweight football players for reasons of medical risk has also been considered recently.[32] As discussed in earlier chapters, individuals having a higher percentage of body fat (obesity) are at an increased risk of heart disease.

CARBOHYDRATE. Maximal effort sprints separated by limited periods of rest dictate a strong reliance on muscle glycogen. Training drills allow for adaptations leading to greater glycogen stores in both Type I and II muscle fibers. A diet providing at least 60% of its energy as carbohydrates or 7–10 g/kg of body weight is ideal and will allow for appropriate recovery of glycogen stores after practices and games. However, some studies that included nutrition assessment of football players have suggested that they might derive less than half of their energy from carbohydrate.[29,33]

Game times are usually around the noon hour, late afternoon, or early evening and players need to be conscious of their carbohydrate consumption earlier in the day. Football players should eat a meal containing at least 200 g of carbohydrate 3–6 hours prior to a game or strenuous practice. The carbohydrate-containing foods that can evoke undesirable gastrointestinal effects such as bloating and cramping should be avoided prior to practices and on game days. Also, foods that might cause diarrhea should be avoided for 1–2 days, as this situation can hinder optimal hydration. Sport drinks containing 4–8% carbohydrate should be consumed during games and practices in amounts appropriate for maintenance of optimal hydration. For instance, if a running back were to drink 4 cups (~1000 ml) of Gatorade (6% carbohydrate) during practice, he would have consumed approximately 55 g of carbohydrate in addition to the fluid.

PROTEIN. Although protein needs have not been specifically determined for football players, their requirements are higher than those of the general population because of their greater muscularity and training as well as injury recovery. A diet providing 15–20% of its energy from protein would provide ample protein to achieve nitrogen balance and desirable gains in LBM. Expressed another way, protein intake of 1.2–1.7 g/kg of body weight should be adequate and allows for differences in training and conditioning practices among players. For instance, a 195-lb (~$88\frac{1}{2}$-kg) safety eating 4000 kcal daily with 15% protein is getting 150 g of protein or roughly 1.7 g/kg of body weight. Researchers reported that the college defensive backs they surveyed consumed approximately 16% of their energy as protein.[16] An offensive lineman weighing 275 lb (125 kg) and consuming 5000 kcal with 17% derived from protein would receive 212 g of protein, which would equate to 1.7 g/kg of body weight. In the future it will probably be more precise to express protein needs relative to LBM instead of total body mass.

Running backs and linebackers are more centralized in the football continuum as their position requires mass, speed, quickness, strength, and power. Off-season training must involve training for strength and power as well as quickness and speed. These athletes might pump iron with linemen and run drills with the defensive backs and receivers. Their protein needs are based enhanced muscle mass, training protocol, as well as injury recovery (cuts and bruises). Again this level of protein could be met by consuming 15–20% of total energy as protein. For instance, as energy intakes may approximate 4000–4500 kcal daily with 16–17% protein, this would provide a protein intake of at least 1.5–1.6 g/kg of body weight for a 230-lb linebacker.

FLUIDS AND HYDRATION. Hydration status is a primary nutritional consideration for football players. Warmer climates and padding and uniforms can increase sweat losses. Copious water consumption is necessary, especially during preseason camps in the summer months. Also, during high school, college, and professional summer football camps it is common for teams to practice twice a day. Football players must strive to recover water losses between those practices.

Mesh jerseys may be worn during practices and games, and the arms and lower legs are often exposed. This allows for greater heat loss via convection. Plastic and foam protective gear such as helmets, shoulder pads, hip girdles, and arm pads decrease the potential for heat loss via convection. Often, quarterbacks wear a protective "flak jacket" under their jersey. This apparatus greatly decreases heat loss from the torso via convection. Efficient heat removal is imperative for these large athletes, and a few recent deaths of linemen have been attributed to heat intolerance during summer practice. Fans, mist-machines, and other sideline devices can help cool these athletes and decrease sweating reliance.

It would not be surprising for a lineman to have a sweat rate of $1\frac{1}{2}$–2 L/hour on a warm, sunny day, causing a player to lose 3–10 lb by the end of practice. Without question, consuming either water (pure) or sport drinks would benefit a football player's hydration status. These athletes should drink fluids before, during, and after games and practices. They should drink roughly 400–600 ml of fluid 2–3 hours prior to a game. This maximizes hydration status at game time and allows enough time for urination of the excess. While on the sideline and in the locker room, receivers and defensive backs should drink at least 250–300 ml of fluid every 15–20 minutes, more (300–500 ml) if sweating is heavy. A sport drink should be beneficial; however, the carbohydrate should not exceed 8% to allow for rapid emptying from the stomach. These players can estimate fluid losses by weighing themselves prior to and after a practice to estimate body weight change, as discussed in Chapter 8.

College and professional athletes should limit alcohol consumption during the week and completely abstain for 48 hours prior to a game to maximize hydration at game time. Caffeinated beverages such as coffee, teas, and colas must be considered. Although some caffeine prior to a workout, practice, or game may be beneficial for some athletes it can be a hindrance to others. Some athletes may feel lethargic from heavy caffeine consumption a few hours prior to practice or a game. The diuretic effect of caffeine should be considered, and increased fluid consumption is needed to match the increased urination.

VITAMINS AND MINERALS. Nutrition surveys of football players indicate that they tend to consume at least the generally recommended amounts of vitamins and mineral because of the higher volume of food they eat. However, an exhaustive study of vitamin and mineral metabolism in football players remains to be performed. Perhaps the greatest immediate concern is related to mineral losses experienced during sweating. Sport drinks providing sodium at approximately 110–165 mg/8 oz and potassium at 19–46 mg/8 oz should provide ample electrolytes to recover losses when fluid consumption covers sweat loss.

NUTRITION SUPPLEMENTATION. Football players are prime users of nutrition supplements. Creatine is one of the more recognizable supplements used to enhance strength and power and LBM. A body of evidence supports the notion that creatine can promote weight gains, primarily in LBM, which coincides with potential gains in strength. For instance, researchers reported that 15.75 g of creatine monohydrate, as part of a glucose/taurine/electrolyte supplement, promoted greater gains in fat- and bone-free mass, isotonic lifting volume, and sprint performance during intense resistance/agility training in college football players.[33]

Football players also use supplements containing DHEA and androstenedione, although their popularity seems to be declining. Caution should be used with these supplements, as their efficacy has yet to be proven. Substances such as ribose and inosine are popular among football players as isolated supplements or ingredients of multisupplements and sport foods. These substances are touted to enhance ATP levels in skeletal muscle, but there is little supportive information for them so far.[34] Other supplements that are popular among football players include HMB and glutamine, and some receivers and defensive backs might use metabolic enhancement and fat-burning supplements, including ephedrine products, to control body weight and maintain speed.

■ FOOTBALL *IN REVIEW*

- Training for football should include exercises designed to increase strength, speed, power, quickness, agility, and sometimes mass. To address all these components, periodized programs emphasizing the components most important to the particular athlete and position are recommended. Many of the Olympic movements should be incorporated along with the powerlifts. Maintaining or improving flexibility is also a necessity for many of these large athletes. Finally, most of the conditioning drills should mimic game play and last no longer than 20 seconds with 20–40 seconds of rest between them.
- Players need to maximize glycogen stores before practices/games and minimize losses during practices/games by eating a higher carbohydrate diet and utilizing a sport drink on the sideline and at halftime. Protein requirements are elevated relative to the intensity of practice/games, supportive training, and injury recovery. Sweat rates can be excessive due to padding and equipment, especially on warmer and sunnier days, so proper hydration before and during practice/games is extremely important.

■ BASKETBALL

A basketball team consists of several positions, each with specific duties and tasks. However, each position relies heavily on the same basic physical components, so the following discussion addresses all the positions collectively.

Performance Characteristics of Basketball

The sport is primarily played inside gymnasiums and the environment is fairly consistent with temperatures around 70 °F. Players wear a pair of shorts, a tank top, socks, and

Paul A. Souders/Corbis

athletic shoes. College games consist of two 20-minute halves and a 15-minute halftime intermission. The games are basically continuous with stoppage of play for time-outs and fouls. A professional game consists of four 15-minute quarters and a halftime intermission. Basketball seasons begin in November and continue through March for collegiate athletes. College teams play roughly 30 games during this time and most play approximately three games per week. Practice occurs on most of the remaining days within a given week. The NBA schedule consists of 82 games played over 7 months with an average of three games per week. Teams practice on many of the remaining days, with practice averaging 2 hours. Basketball is played by both males and females worldwide and many adult leagues exist for continued recreational competition once high school, collegiate, or professional careers end. John Wooden, one of the most successful college coaches in history, was noted for training his athletes hard while emphasizing good nutrition and resting habits,[35] the cornerstones of any successful athletic program.

Physical Characteristics of Basketball

Basketball is a blend of aerobic and anaerobic activity. Players run, jog, and sprint up and down the court with occasional rest periods for fouls and time-outs. Most players do not play an entire quarter or half but may play a significant portion of it. The game is quick in nature and

has been compared to interval training where a baseline activity level is set with periodic maximal-effort sprints and jumps. The forwards and centers need strength and some mass in order to gain position for rebounding and offensive play.

Basketball players need to be quick, agile, and have good jumping ability. Performing the skills of jumping, shooting, and defensive play often puts these athletes in less than ideal bodily positions. Therefore, athletes need total body strength and power. Back and abdominal strength are crucial for the stability and control of the spine. Flexibility for all body parts is important to enhance speed, function, and decrease the incidence of injury.

Some sports have greater associations between particular anatomical and performance traits than others. Perhaps this is most evident with height in the sport of basketball, particularly at the center and forward positions.

Physical Training for Basketball

Based on the dynamics of the sport a periodized program is recommended. Muscular hypertrophy for basketball players is often important, particularly for the taller players. Gaining mass will improve strength as well as assist in playing inside positions of forward and center. Strength training in the form of multijoint and Olympic movements mimics playing movements and trains the entire body simultaneously. These exercises improve coordination, stabilization, strength, and power. Several isolation exercises may be needed for muscle hypertrophy. Also, several exercises focusing on flexibility, strengthening, and stabilizing of the hamstrings and lower back should be included. Use of plyometric exercises (box jumps, medicine ball throws) assists in the development of speed, quickness, and explosiveness. Conditioning drills should emphasize footwork and maximize speed. Maximal- and high-intensity endurance exercises should make up a majority of the conditioning.

Key Point

Because of the explosive nature of basketball, a strength training program geared around power and strength is necessary. Plyometric exercises and Olympic movements address these components.

Nutrition for Basketball

Special attention must be afforded to diet during the basketball season, as players typically play several games a week and college and professional teams travel. It is not uncommon for players to lose weight during the season. This reflects that the large energy demands of the sport have not been appropriately balanced with food intake.

ENERGY BALANCE AND BODY WEIGHT AND COMPOSITION.

Basketball players tend to be fairly lean; however, at the center and forward positions greater body mass can assist in securing position under the basket for rebounds. Body composition of female and male basketball players has been reported using UWW, DXA, BIA, and skinfold measures. In general the males tend to be leaner than the females with 10–11% and 18–29% body fat, respectively.[36–39] Basketball players can expend considerable energy during practice and games. Energy expenditures during play for a 120-lb (55-kg), a 170-lb (77-kg), and a 210-lb (95-kg) player might be 7.7, 11.2, and 13.8 kcal/min. This expenditure, in addition to physical size and off-court training and conditioning, can lead to energy expenditure ranges of 3000–6000 kcal/day.

CARBOHYDRATE.

A high-carbohydrate diet ($\geq$60% of energy) is recommended to maximize glycogen status during training, practices, and games. Recommendations for daily carbohydrate intake based on body weight are 7–10 g/kg during strenuous practice and games schedules. As many collegiate and professional games are played in the late afternoon to early evening, players should take advantage of pregame meals and beverages to maximize muscle glycogen stores. Many sport nutritionists believe that several smaller carbohydrate-containing meals throughout the day are more beneficial than fewer larger meals. Carbohydrate-containing foods with a history of causing intolerance symptoms (legumes, dairy) should be avoided. Sport drinks containing 4–8% carbohydrate should be consumed during games in amounts appropriate for hydration maintenance. The meal immediately following games or practices should include carbohydrate at a level of 1.0–1.5 g/kg of body weight.

PROTEIN.

Protein intake should come from a variety of sources and provide roughly 1.2–1.4 g/kg of body weight or 15–20% of total energy. However, this recommendation is extrapolated from other sports and not based on research with basketball players. This level would provide a 220-lb (100-kg) basketball player consuming 4500 kcal with 15% of energy from protein, roughly 168 g of protein or 1.68 g/kg of body weight. Higher protein levels are especially important for players involved in weight training to increase muscle mass. Protein should be a component of most if not all meals and snacks and is an important component of the meal immediately after training, practice, or a game.

FLUIDS AND HYDRATION.

One of the largest nutrition concerns for basketball players is the maintenance of proper hydration status during practices and games.

Players substituting in and out need to drink fluids on the bench. A sport drink containing 4–8% carbohydrate can provide performance benefit, especially later in the game. Halftime is also an excellent opportunity to ingest fluids. At least 6–12 oz (150–300 ml) or about $\frac{3}{4}$–$1\frac{1}{2}$ cups of fluid every 15–20 minutes during practice and games is recommended. This allows for approximately 20–40 oz (600–1200 ml) or $2\frac{1}{2}$–5 cups of fluid/hour and can match water losses when sweating is mild or moderate. However, additional fluid is required when the exercise takes place in a very warm environment and sweating is heavier. This might be the case in outside play during the warmer months or when practices and games are played in warm gyms and arenas. Basketball players can gauge their sweat loss during practices and scrimmages to estimate their fluid needs during games (see Chapter 8).

Caffeinated beverages also require special attention, as overconsumption can have a diuretic effect and compromise hydration status at the onset of a game or practice. Often, basketball practices and games occur in the late afternoon and early evening, and caffeinated beverages (such as soda and coffee) might have been consumed throughout the day. Furthermore, as shooting precision is extremely important, players need to understand the effects of either excessive caffeine or withdrawal.

Key Point

Basketball games played outside during the warmer months or in a warm gymnasium can lead to heavy sweating and possibly dehydration.

VITAMINS AND MINERALS.

The intake and status of vitamins and minerals for basketball players has received considerable attention. In a recent study, weighed-food dietary reports revealed that elite women basketball players in Spain failed to achieve that country's minimal intake recommendations for magnesium and zinc, although their copper intake was ample.[40] Other researchers reported that iron intake, as well as heme iron, was higher for women basketball players than for nonathletes.[41] However, their serum ferritin levels were actually lower than the nonathletes'. Elsewhere, researchers reported that the vitamin C intake of some of the basketball players they assessed were below RDA recommendations, although serum ascorbic acid levels were not different from those of sedentary people.[42] Another assessment determined that vitamin supplementation containing vitamin C is prevalent among competitive basketball players.[43] However, when a multivitamin/mineral supplement containing vitamin C was provided to a variety of athletes including basketball players, there was not an improvement in performance.[44]

NUTRITION SUPPLEMENTATION. Basketball players might use a broad range of nutrition supplements. For instance, one study of elite Spanish basketball players revealed that multivitamins and specific vitamins were used by approximately one-half of these athletes.[43] Other supplements provided protein, amino acids, and carbohydrates. In general, basketball players might experiment with supplements such as creatine and other potential mass builders such as HMB, glutamine, and androstenedione. Creatine is touted to enhance strength, power, and speed as well as increase the ability to perform repeated high-intensity exercise bouts, such as sprinting up and down the court. The efficacy of creatine supplementation in basketball players for this purpose has not been well studied even though some basketball players use creatine supplements.

Coaches, trainers, or team physicians should monitor use of supplements by basketball players. Because of the large amount of muscle tissue, basketball players have the potential to enhance their body weight significantly. Thus the relationship between mass, speed, and endurance must be carefully evaluated. For instance, a center or power forward would benefit from an increased muscle mass and power during positioning underneath the basket. Yet, the additional mass could become a detriment if the athlete fatigues during the second half.

■ BASKETBALL *IN REVIEW*

- Basketball is a quick and explosive sport. Use of a periodized program incorporating Olympic and power movements is recommended. Plyometric exercises are recommended to improve jumping ability, quickness, and agility. Flexibility and core stabilization are also important for minimizing injury risks. Conditioning drills should primarily focus on moderate- to high-intensity endurance intervals.
- Basketball players need to maximize glycogen stores before practices/games and minimize losses during practices/games by eating a higher carbohydrate diet and utilizing a sport drink while on the bench and at halftime. Protein recommendations are elevated relative to intensity of practice/games, supportive training, injury recovery, and growth. Sweat rates can be excessive due to intense sprints and the warm environment of gymnasiums or outside courts on warm and sunny days, making proper hydration before and during practices/games extremely important.

■ VOLLEYBALL

High school, collegiate, and Olympic volleyball teams consist of several positions, each with specific duties and tasks. However, most positions rely on the same basic

physical components, so the following discussion addresses all the positions collectively. Competitive amateur and professional sand volleyball uses only two, three, or four players to a team and involves much more activity.

Performance Characteristics of Volleyball

High school, collegiate, and Olympic volleyball is played in an indoor environment with consistent temperatures around 70 °F. Players wear shorts, a shirt, socks and athletic shoes. Sand volleyball is played outside, and the athletes are clad mostly in swimsuits and exposed to the environmental elements. College matches consist of the best three out of five games with 5–10 minutes between games. Each game consists of the first team to reach 30 points in a rally scoring format. If the match requires the play of game five, it is played to 15 points. Matches typically last 1.5–2 hours. Seasons last approximately 4 months with matches 2–3 days per week. Practice occurs on most of the remaining days, and typical practice sessions last 2 hours.

Game play is intermittent with points lasting an average of 6 seconds with 10–20 seconds between points. The game primarily consists of explosive and quick movements with numerous repetitions of jumping and spiking.

Key Point

Due to the excessive jumping in volleyball, strength programs should address the lower extremity muscles that will produce the forces as well as attenuate them.

Physical Characteristics of Volleyball

Volleyball is an anaerobic sport. Players utilize quick-burst movements for offensive and defensive play. The starting players compete for the entire game; however, they may rotate out for a brief rest. Volleyball players need to be quick, agile, and have great jumping ability. This requires strength and power from the total body.

Performing the repetitive skills of jumping and spiking can lead to overuse injuries to the knees (such as patellar tendonitis), lower back, and shoulder muscles (such as rotator cuff injuries). Therefore, athletes need to address the musculature that attenuates the forces created by these movements. Back and abdominal strength are crucial for the stability and control of the spine as well as assisting in the development of power. Flexibility is important for all body parts to enhance speed, function, and decrease the incidence of injury.

Physical Training for Volleyball

Based on the dynamics of the sport a periodized program is recommended. Strength training that uses multijoint and Olympic movements mimics the ballistic nature of volleyball and trains a large portion of the musculature simultaneously. These exercises improve coordination, stabilization, strength, and power. Extended use of plyometric exercises (box jumps, medicine ball throws) assist in the development of speed, quickness, and explosiveness. Specific exercises focusing on the rotator cuff muscles and eccentric loading of the lower extremity should assist in decreasing injuries to the shoulders and knees, respectively.

Conditioning drills should emphasize footwork and explosive jumping. Adequate rest periods should be used when jumping improvement is being addressed. Otherwise, high-intensity endurance activities (5–20 seconds of activity with 10–20 seconds rest) should make up the bulk of training.

Nutrition for Volleyball

Volleyball play primarily engages anaerobic ATP generation, which may provide as much as 90% of ATP.[45] Muscle glycogen is a major fuel provider to working muscle during a volleyball play. Body composition might be a bigger concern for sand versus indoor volleyball players. Sand volleyball players are clad in swimsuits that can be more revealing than indoor uniforms. Furthermore, with fewer players on a court, an outdoor player may be more self-conscious about attention to his or her physical appearance.

ENERGY BALANCE AND BODY WEIGHT AND COMPOSITION.

Energy expenditure during sand volleyball play would be greater than indoor volleyball. First, the reduced player numbers increase the workload for each player. Second, movement in loose sand can provide resistance and inefficient movements, which increases energy expenditure per movement. Some volleyball players might attempt to lose weight prior to and during a competitive season, as weight loss has been noted to potentially improve leaping ability.[46] It is believed that for shorter volleyball athletes, especially high school and college women, extra body weight may be a minor yet competitively significant hindrance during the jumping required for spiking and blocking efforts. This does not mean that weight loss should be encouraged in already lean volleyball players, especially during periods of growth in younger athletes. However, for some volleyball players with excessive body fat, the relationship between the extra body mass and jumping ability should be evaluated. Average body fat percentages for male and female volleyball players have been reported to be 9.8 and 17.9%, respectively.[47]

Estimations of energy expenditure for vigorous volleyball play are based on body weight. For instance, 100-, 160-, and 190-lb volleyball athletes might expend 6.5, 10.4, and 12.4 kcal/minute of play. The average daily energy intake for two groups of competitive adolescent volleyball players was 2013 and 1529 kcal.[48]

CARBOHYDRATE.

Volleyball athletes would benefit from a higher carbohydrate intake (>60% total energy). This would be supportive of greater glycogen storage. Pre-game/practice meals require some consideration. A meal consisting of at least 100 g of carbohydrate should be consumed 3–4 hours prior to a game. Carbohydrate-containing meals and drinks closer to game time should be experimented with during practice. Food sources containing poorly tolerated carbohydrates (fiber and oligosaccharides) should also be avoided. Abdominal bloating and cramping could be exacerbated from jumping efforts. Nutrition assessment of Greek adolescent female volleyball players revealed that carbohydrate accounted for only 46% of the energy in their diet.[48]

PROTEIN.

Very little is known about protein needs for volleyball athletes; however, estimates of protein intake have been between 12 and 20%.[45] Protein needs would largely be based on training requirements with the possible exception of enhanced needs for repair and recovery from injuries. Protein needs for volleyball players might be best expressed as roughly 15–20% of energy requirement or 1.2–1.4 g/kg of body weight. One study involving adolescent female volleyball players estimated protein intake at 16% of energy intake.[48]

FAT.

Recommendations for the level of fat in the diet of volleyball athletes are not different from those for the general population (20–30% of energy).[45] Recommendations for the level of fat are based on achieving carbohydrate and protein requirements first. For instance, a volleyball player's diet providing 60% and 18% of the energy as carbohydrate and protein, respectively, would limit fat energy to 22% of the total. A recent nutrition assessment of adolescent female volleyball players revealed that fat accounted for 37% of the energy in their diet.[48] Some volleyball athletes dramatically restrict their fat intake to achieve a lighter and leaner body. However, fat cannot be completely eliminated from the diet as it provides essential fatty acids.

Key Point

Some volleyball athletes dramatically restrict their fat intake to achieve a lighter and leaner body. However, fat cannot be completely eliminated from the diet as it provides essential fatty acids.

Key Point

The maintenance of optimal hydration is the most significant immediate concern for volleyball players. LBM relative to body weight is an important consideration for jumping performance (spiking, blocking).

VITAMINS AND MINERALS. The general requirements for vitamins and minerals for volleyball players are probably not higher than those for the general population. However, certain vitamins and minerals have been noted as deficient in some volleyball players' diets. For instance, one study involving highly competitive adolescent female volleyball players revealed that their intake of calcium, iron, folic acid, magnesium, zinc, vitamin A, and certain B vitamins was below recommendations.[48] Nutritional assessment of elite female volleyball players in Greece suggests that the intake of vitamins and minerals meet recommendations except for iron.[49] Coaches can use nutrition assessment tools to evaluate the vitamin and mineral composition of their players' diet and assist those athletes in achieving at least minimal recommendations.

Perhaps the most immediate concern for these nutrients is related to mineral losses experienced during excessive sweating in sand volleyball. Sport drinks providing sodium at approximately 110–165 mg/8 oz and potassium at 19–46 mg/8 oz should provide ample electrolytes to recover losses when fluid consumption covers sweat loss. Also, volleyball players following restrictive diets can fail to meet recommendation levels for vitamins and minerals.

FLUIDS AND HYDRATION. Indoor volleyball is typically played in a climate-controlled environment and sweating is not excessive. In addition, players can rotate out of the lineup, which would allow them to consume water. However, dehydration is a primary concern for sand volleyball players. Sand volleyball athletes are exposed to the environment and generally occurs in a warm climate (such as a beach), with significant exposure to sunlight. Sweating tends to be copious and thus special attention should be paid to taking in fluids before and during games. Tournaments often involve several games in a single day. Therefore, conscious efforts should be made to maintain adequate hydration levels. In between games players should seek out shelter or shade and consume sport drinks (4–8% carbohydrate) or even enjoy a carbohydrate-containing snack if tolerated relative to the next game.

Volleyball players should drink at least 2 cups or 16 oz (~500 ml) of water or sport drink 2 hours before practice or games. These athletes should strive to consume at least 1–1½ L of fluid for each hour of play or about ½–1 cup of fluid every 15 minutes. As sweat rate can be greater for the sand volleyball players, they should estimate their sweat volume and plan for optimal fluid consumption. Practice provides an excellent opportunity to gauge sweat volume as described in Chapter 8.

NUTRITION SUPPLEMENTATION. Nutrition supplementation by volleyball players often involves attempts to enhance repeated explosive actions and reduce body fat mass. Supplements geared toward anaerobic energy systems have been the target of some experimentation; however, their use is not as prevalent in volleyball as in other sports. These supplements include ribose, inosine, and creatine. Males are more likely to use these supplements than female volleyball players.

Some volleyball players, especially females, experiment with nutrition supplements touted as metabolic boosters and fat burners. These supplements typically include ephedrine, caffeine, and carnitine. Here the goal may be to increase mental alertness and perhaps jumping actions. Or perhaps the use of these supplements is an attempt to reduce body fat mass to improve jumping efforts and perhaps aesthetic appearance. Also, some volleyball players use diuretics and laxatives to limit body weight.[50]

■ VOLLEYBALL *IN REVIEW*

- Use of multijoint and Olympic movements mimics the ballistic movements of volleyball and improves coordination, stabilization, strength, and power while training a large portion of the musculature simultaneously. Use of plyometric exercises assists in the development of speed, quickness, and explosiveness. Exercises focusing on the rotator cuff muscles and eccentric loading of the lower extremity should assist in decreasing injuries to the shoulders and knees, respectively. Most conditioning drills should be of high intensity and short duration.
- Players should eat a higher carbohydrate diet to support the recovery of energy spent in repeated jumping efforts, and should utilize a sport drink if expenditure will be excessive (as in tournaments and sand volleyball). Protein requirements are elevated relative to the intensity of practice/games, supportive training, and injury recovery. Sweat rates for indoor volleyball may not be as high as for other team sports; however, outdoor games with two to four players on a team lead to higher sweat rates. Proper hydration before and during practice/games helps maximize performance.

PhotoDisc/Getty Images

■ SOCCER

Soccer or "football" outside the United States is the most popular sport worldwide. Soccer teams consist of several types of players, but most need the same physical characteristics to perform at a given level. The discussion below is relevant to all the players with the exception of the goalie. The sport is popular with both genders, and leagues are available for youths as well as adults.

Performance Characteristics of Soccer

The sport of soccer is primarily played in warm environments on a grass field. Players wear shorts, shirts, socks with shin protectors, and cleats. Soccer seasons typically last about 5 months with one to three games a week. Practice and training sessions occur on most of the remaining days of the week. A professional soccer game consists of two continuous 45-minute halves with a 15-minute intermission. Starting players typically participate in most if not all of the game's duration. The running distance covered may approximate 6–6½ miles (~10 km). The travel schedule for most collegiate and professional soccer players is moderate and mainly regional for the college teams.

Physical Characteristics of Soccer

Soccer is an endurance sport with many quick anaerobic bursts. Players are basically on the move the entire game

with sprints and quick moves incorporated in offensive and defensive play. This type of movement is similar to interval training, where a baseline level of intensity is used with periodic bursts of higher intensity lasting for several seconds. Players kick and pass the ball with either foot, requiring strength and coordination from the lower extremity and torso. Because of the constant running, injury concerns are generally associated with muscle pulls and strains.

Key Point

Soccer requires a high degree of endurance with frequent sprints. Normal practices address the endurance component, and conditioning drills should focus on speed and quickness.

Physical Training for Soccer

Soccer players should engage in a periodized program and build power, strength, and speed in the off-season and focus on soccer skills throughout the season. Because daily practices have a high endurance component, other endurance training may not be needed. To improve quickness, interval training is recommended for a large portion of the conditioning. This improves both the aerobic and anaerobic systems for energy production. Other drills should be soccer-related and emphasize quick feet.

The strength training exercises to enhance strength and power may come in the form of both Olympic movements as well as powerlifting movements. Multijoint exercises such as the squat and lunge are valuable in improving lower extremity strength. The eccentric component of these exercises may also prove beneficial in reducing overuse injuries. Incorporating plyometric exercises enhances quickness, agility, and power for the ballistic kicking and sprinting movements.

To reduce the incidence of injury, isolation exercises for the hamstring and lower back should be incorporated. Abdominal and lower back strength can be enhanced through many exercises; however, the use of stability balls and medicine balls are recommended for core stabilization, balance, flexibility, and ballistic movements.

Nutrition for Soccer

Because soccer involves continuous movement of varying intensities for most players, the energy supply comes from a mixture of nutrients and metabolic pathways. For the short-duration and high-intensity sprints, creatine phosphate and glycogen are the primary mechanisms involved in ATP regeneration. During lower intensity jogging and walking, a combination of carbohydrate and fat are oxidized and muscle fibers attempt to recover ATP and creatine phosphate in preparation for the next sprint. Here the primary determinant of the fuel source is O_2 delivery to working muscle. The VO_2max of soccer players tends to approximate 60 ml/kg of body weight per minute.[51]

ENERGY BALANCE AND BODY WEIGHT AND COMPOSITION.
Soccer players vary in energy expenditure depending on their size and position. For instance, a 165-lb (75-kg) soccer player might expend 17 kcal/minute, which can approximate 1500 kcal during a game. Thus, daily energy requirements might approximate 4000 and 3200 kcal for men and women, respectively. Energy requirements expressed by body weight for male and female soccer players might be in the ranges of 47–60 kcal/kg/day and 45–50 kcal/kg/day, respectively.[51] One study estimated the energy intake of professional male soccer players as ranging between 2033 and 3923 kcal/day.[52] Other estimations of energy intake of professional soccer players ranged between 1618 and 4394 kcal/day and averaged 2662 kcal/day.[51] Soccer players tend to be lean. For instance, researchers have reported a body fat level of 9% for male players.[47]

CARBOHYDRATE.
As for most sports, carbohydrate should serve as the foundation of the diet. If multiple games are to be played during the same week, players must seize all eating opportunities to maximize their glycogen stores. A diet that derives at least 60% of its energy from carbohydrate is needed and strenuous training should be minimized between games to enhance glycogen recovery. That is, male and female soccer players should strive to consume 8–10 g/kg body weight throughout a day.[51] The meal following a strenuous practice or game should provide a carbohydrate intake of at least 2 g/kg of body weight. This might provide 125–150 g of carbohydrate for an adult and 50–125 g for a younger player.

To stress the importance of dietary carbohydrate, researchers evaluated the performance of male soccer players during and after a 90-minute 4-on-4 soccer game after consuming a 65% carbohydrate diet or a 30% carbohydrate diet.[53] After consuming the higher carbohydrate diet, the players performed more high-intensity efforts during the game. Other studies have reported that reduced performance during soccer can be related to the capacity of muscle to utilize glycogen.[54,55] Soccer players may benefit from consuming a carbohydrate-containing beverage during practices or games as well. For instance, researchers reported that a 6.9%-carbohydrate drink could reduce glycogen utilization by as much as 22% as compared with a noncarbohydrate placebo in relevant athletes.[56] These athletes were participants in university soccer, hockey, and rugby sports, and the exercise protocol involved 90 minutes of intermittent high-intensity running, which bears a resemblance to soccer.

PROTEIN.
Protein recommendations for soccer players depend on the level of training and performance. Because soccer is extremely popular among children and teens, age is also a consideration with regard to protein recommendations. Younger players may have relatively greater protein requirements (in g/kg of body weight) than adults due to training and competition as well as natural growth processes.

Protein is important during resistance training by soccer players, and their requirements are probably elevated for that training. An intake of roughly 1.2–1.4 g/kg of body weight may be adequate for nitrogen balance, although this has not been substantiated by controlled scientific investigation.[57] Using a percentage of total energy consumed (15–20%) should also allow for adequate dietary protein to maintain nitrogen balance.

FAT.
The level of fat in the diet should be appropriate for body weight maintenance. Recommendations for fat intake are usually based on remaining energy requirements once carbohydrate and protein needs are addressed. For instance, if >60% and 15–20% of energy are derived from carbohydrate and protein, respectively, <25% would come via fat. Despite such recommendations, surveys have suggested that soccer players consume less carbohydrate and more fat than recommended.[58] Consuming a diet higher in fat may be a mechanism for maintaining an adequate energy intake.

FLUIDS AND HYDRATION.
With the exception of indoor leagues, soccer is mostly played outdoors during daylight hours (weekends and early evenings) and is played in warmer climates. In the United States and many other countries, soccer is played in the heat of the summer. For these reasons sweating can be excessive and reduced body water content leading to decreased performance is a concern. Players should strive to drink at least 8–10 oz. (250–300 ml) every 15–20 minutes during practices and games in warm climates when sweating is excessive. More individualized recommendations are based on body weight changes associated with sweating, as described in Chapter 8. Players should drink about $1\frac{1}{2}$–$2\frac{1}{2}$ cups (400–600 ml) of fluid 2–3 hours prior to practice or a game. During game situations, when fluid is infrequently available, players should make sure they begin the practice or game in an optimally hydrated or even hyperhydrated state (see Chapter 8). This requires some individual experimentation because too much fluid prior to a game can result in the need for urination.

Soccer players may drink caffeinated beverages before a game. Although numerous studies have attempted to determine how and when caffeine may best be applied athletically, little work has been conducted with soccer as the athletic model. Therefore, it is difficult to discuss the timing of caffeine intake and quantity of caffeine appropriate for soccer players. Here again, the athlete can experiment in training and practice situations and should be wary of the effects of caffeine excesses and withdrawal. In addition, players must be aware of possible "doping" concerns, as the IOC and NCAA consider caffeine a doping agent at certain levels.

VITAMINS AND MINERALS.   Soccer players can sweat profusely, so the most immediate concern for vitamins and minerals is recovering electrolyte loss in sweat. Sport drinks providing sodium at approximately 110–165 mg/8 oz and potassium at 19–46 mg/8 oz should provide ample electrolytes to recover losses, when fluid consumption covers sweat loss. In addition, a balanced diet at the energy intakes required for soccer players would allow for electrolyte recovery and supply desirable levels of vitamins and minerals.

NUTRITION SUPPLEMENTATION.   Several ergogenic substances have been considered for performance enhancement for soccer players. These are creatine, bicarbonate, caffeine, BCAAs, glycerol, and ginseng. In theory, enhancements of creatine phosphate in skeletal muscle might allow for a prolongation of a sprint and improve performance in repeated bouts of sprinting. Glycerol may enhance hyperhydration efforts when provided within a solution prior to practice or a game. There is evidence that glycerol can enhance water stores in the body by increasing the volume of the blood.[59] Research findings indicate that glycerol supplementation might allow an athlete to maintain a lower performing heart rate and body core temperature during exercise in heat stress situations.[60] However, this research did not involve soccer players, and these players would need to experiment during nongame situations, as there might be side effects (nausea and headaches).

■ SOCCER *IN REVIEW*

- Training the lower extremity is of primary concern and using functional exercises such as the squat and lunge should improve lower extremity strength. General upper body strength without significant mass generation is recommended. Use of plyometric exercises enhances quickness, agility, and power for the ballistic kicking and sprinting movements. Core stabilization can be enhanced through the use of stability balls and medicine balls. Conditioning drills should primarily concentrate on the sprinting components similar to those of game situations.
- Players need to maximize glycogen stores before practice/games and minimize losses during practices/games by eating a higher carbohydrate diet and utilizing a sport drink. Protein requirements are elevated relative to intensity of practice/games, supportive training, and injury recovery. Sweat rates for outdoor soccer can be excessive as games are typically played in warmer climates. Creatine and glycerol might be beneficial supplements.

Photolink/PhotoDisc/Getty Images

■ TENNIS

The sport of tennis involves both singles and doubles play. Both men and women participate in this sport at all ages. The following discussion addresses singles play since this is much more demanding from a physical conditioning perspective.

Performance Characteristics of Tennis

The sport of tennis is played primarily in warmer environments. Venues are a combination of indoor and outdoor, and the surfaces played on include grass, clay, hard court, and the various synthetic indoor surfaces. Players wear shorts or skirts, a top, socks, and athletic shoes. Because tennis season primarily occurs over the summer months, court conditions can be extremely hot and humid. In fact, court temperatures often exceed 100 °F. The professional circuit covers most of the year, but the heart of the season occurs in 8 months. Tournaments run 4 days and are played about every 2 weeks. It is common for players to increase the frequency of play by participating in both singles and doubles play. Practice and training sessions occur on most of the remaining days of the week.

A professional tennis match generally consists of the best of five sets for the male players and best of three for the female players. These matches can last from 45 minutes to more than 3 hours. Players get a short break of 25 seconds between points and 90 seconds after finishing an odd number of games (changeovers). The travel schedule for a professional tennis player is demanding, as events are held all over the world. Players usually spend about a week at the location of a tennis tournament.

Physical Characteristics of Tennis

Tennis requires a player to be quick, strong, explosive, flexible, and have a certain level of endurance to last an entire match. The ballistic acts of serving and overhead smashes occur often and require a full range of motion. Also, play often forces athletes to execute shots in awkward and unbalanced positions that may place them in injurious situations. Players need to cover the entire court efficiently while changing directions an average of four times per point. Typical point durations are often dependent on the surface being used but can be estimated at less than 10 seconds per point. Power for the various strokes is primarily created by the lower extremity. This power is sequentially passed through the torso, shoulders, and arms. Therefore, strength is needed in the lower and upper body and muscles of the core. Also, because tennis players perform high repetitions of certain movements (such as the right-handed forehand shot), muscle imbalances within the core can occur, so proper training of the opposite side of the body is important. Training to help prevent injury to the rotator cuff and lower extremity muscles (such as the adductors) must also be considered.

Key Point

Due to the high demands on one arm and one side of the body, tennis players may develop asymmetry. Strength programs should address this with specific exercises for the nondominant side of the body and for the core.

Physical Training for Tennis

The dynamics of this sport (strength, speed, power, endurance) suggest that a periodized program would be the most beneficial. Also, because of the demanding tournament schedule, certain important tournaments should be chosen for peak performance (such as the U.S. Open, Wimbledon, the French Open). Therefore, the training cycle should be periodized to achieve optimal performance in the tournaments chosen.

Although not critical for most tennis players, increasing muscle mass may prove helpful in the development of strength and power. Many of the Olympic movements and other multijoint free-weight exercises provide the greatest transfer of training effect and are therefore recommended. Several isolation exercises may help enhance muscle hypertrophy for appropriate muscle groups. Also, several exercises need to address the stability, flexibility, and function of the shoulder joint (rotator cuff). Eccentric loading exercises may enhance the ability to attenuate forces and help reduce overuse injuries.

Because tennis is played on a variety of surfaces, the conditioning and tennis drills practiced should be based on the performance characteristics associated with a particular surface. For example, more sliding occurs on clay surfaces, so improving the strength and endurance of the thigh adductor muscles prior to a particular tournament may prove helpful. Conditioning drills should be geared around game conditions with near maximal efforts for 5–30 seconds and rest periods of 25–30 seconds. The use of slide boards, stability balls, and medicine balls are recommended for core stabilization, balance, flexibility, and ballistic movements. The aerobic base should come with consistent practice and play but may need to be augmented with low-impact endurance activities such as cycling.

Nutrition for Tennis

Because of the short duration and high intensity of the points in tennis, movement relies heavily on the use of anaerobic systems to regenerate ATP. In between points, muscle fibers derive more of their energy from aerobic systems. This includes the recovery of creatine phosphate.

ENERGY BALANCE AND BODY WEIGHT AND COMPOSITION.   During tournaments it is not unusual for players to compete in several matches in the same day (both singles and doubles). Also, it is not unlikely that a 175-lb male may expend more than 500 kcal/hour during vigorous play. Competitive male and female tennis players were reported by one study to have average body fat percentages of 11% and 22%, respectively.[47] However, subjective evaluation of elite professional women players yields an even leaner status.

CARBOHYDRATE.   Diets rich in carbohydrate should be used to meet the high glycogen demand of this sport. This is achieved by providing >60% of energy as carbohydrate, or 8–10 g/kg of body weight daily. Although game lengths can vary tremendously, most last a few hours and can potentially result in moderate to nearly total depletion of glycogen stores in the involved muscles. Pregame meals are paramount; eating well-tolerated food(s) containing at least 100–150 g of carbohydrate 3–4 hours prior to match play may help maximize muscle and liver glycogen levels. Consuming a sport drink with 4–8% carbohydrate during play may enhance performance late in the match. Sport drink consumption should be based on fluid needs for the maintenance of optimal hydration. Proper rest is needed between matches if more than one is played in a day. This aids glycogen recovery in skeletal muscle.

After a match, players should strive to consume a carbohydrate-rich meal providing at least 2 g/kg of body weight to begin the recovery of glycogen stores. However, if another game will be played within 1–2 hours, a reduction in food volume may be necessary to avoid stomach discomfort. This might make carbohydrate-dense sport drinks (>8% carbohydrate) and sport bars more desirable. As many tennis practices and matches occur in the late afternoon and early evening, a carbohydrate-containing meal or snack should be considered if more than 2–3 hours elapse between the postexercise meal and bed. Carbohydrate should be a major component of meals leading up to the next practice or game; several smaller meals are recommended.

PROTEIN.   Although tennis players are not typically massive athletes, it does seem that players are becoming bigger and stronger. Many tennis players participate in resistance training programs to enhance strength and power. The recommendations for tennis players might be similar to those discussed above for soccer players (protein intake of 1.2–1.4 g/kg of body weight), and a diet providing 15–20% protein should allow for nitrogen balance. Many sport nutritionists advocate spreading protein intake over several smaller meals to complement carbohydrate intake.

FLUIDS AND HYDRATION.   More often than not, tennis competitions are outdoors and in warmer climates and during the daylight hours. In these cases, sweating is the principal route of heat removal, placing the player at risk of dehydration. As a rule, players need to continuously hydrate before, during, and after games or practices. They should take the opportunity to hydrate and rest during breaks, such as crossovers between sets. Chairs are often provided for brief rests between sets.

Fluid should be consumed in an attempt to replenish water losses during the match. Striving to drink at least 250–300 ml of fluid every 15–20 minutes might allow for the recovery of 1–1.2 L of sweat water loss in the match. However, players should weigh themselves before and after a game-simulating practice to gauge water losses and determine their personal recovery protocol.

In a recent study involving male and female players from two Division I university tennis teams, body weight percentage changes were similar for males and females, approximating a 1% reduction in body weight.[61] In addition, the estimated daily losses of sweat sodium and potassium were recovered by the players' daily dietary intakes of these electrolytes. The researchers concluded that these athletes generally maintained overall fluid-electrolyte balance in response to playing multiple tennis matches on 3 successive days in a hot environment, without the occurrence of heat illness.

Key Point

Tennis players can dehydrate rapidly during outside play on warmer days and need to drink fluids during crossovers and between games.

VITAMINS AND MINERALS.   When tennis players sweat profusely during matches and longer practices, electrolyte imbalances may occur. Sport drinks providing sodium at approximately 110–165 mg/8 oz and potassium at 19–46 mg/8 oz should provide ample electrolytes to recover losses if fluid consumption covers sweat loss.

Collegiate as well as professional tennis players often travel to tournaments and must eat on the road. Therefore eating vitamin- and mineral-dense meals can require planning.

NUTRITION SUPPLEMENTATION.   Tennis is not a sport typically associated with sport supplementation. However, it is an extremely demanding sport and several types of supplements could easily be applied to tennis. For instance, creatine might be experimented with as a possible way to enhance tennis stroke power and repeated sprinting cross-court and to the net. Other supplements touted to enhance ATP levels or anaerobic energy systems may be marketed to tennis players. These include adenine, inosine, and ribose.

Glycerol as a potential hyperhydrating substance could also be applied to tennis. As discussed, during matches players consume fluids infrequently (between sets). Therefore it is typical for water losses to exceed water provision during a match. Tennis players who wish to experiment with glycerol should do so during training situations in the heat.

■ TENNIS *IN REVIEW*

- Because strength, speed, power, and endurance are needed, a periodized program would be the most beneficial. Periodization should revolve around in-season maintenance of fitness levels while trying to peak for major tournaments. The Olympic movements and other multijoint free-weight exercises are recommended to provide the greatest transfer of training effect. Several isolation exercises may help enhance muscle hypertrophy for appropriate muscle groups. Also, several exercises are needed to address the stability, flexibility, and function of the shoulder joint (rotator cuff). Finally, conditioning drills should focus on the quickness and agility components and mimic the surfaces to be played upon.

- Tennis players need to maximize glycogen stores before and minimize losses during practices/matches by eating a higher carbohydrate diet and utilizing a sport drink (6–8% carbohydrate) during crossovers. Protein requirements are primarily related to supportive training and injury recovery. A warmer and sunnier climate can increase sweating significantly and players need to properly hydrate between and during practices/matches.

Photolink/PhotoDisc/Getty Images

■ GOLF

Golf has steadily gained popularity in the United States over the past decade or so. Exciting play by younger athletes such as Sergio Garcia and Tiger Woods and continued brilliant play by seniors such as Lee Trevino and Jack Nicklaus has made golf popular within all age groups. Long viewed as a game of practice and skill, golf is now dominated by athletes who have added an emphasis on physical training and nutrition to optimize their potential.

Performance Characteristics of Golf

The game of golf is played primarily in warm environments. Players wear pants or shorts, a shirt, socks, and spiked shoes. Because golf season is mainly the summer months, course conditions can be extremely hot and humid. The professional circuit covers most of the year, but the heart of the season is in the summer. Tournaments are played about every 2 weeks, although a player could compete nearly weekly, and a round of golf lasts about 3–3½ hours. Practice and training sessions occur before and after rounds of golf, and many hours are spent on the driving range. It is not uncommon for a professional golfer to spend 5–6 hours a day on golf. In collegiate golf, tournament play may involve three rounds of golf in 2 days. Adding practice time, these athletes may be involved in play for 10 or more hours a day.

A round of golf generally consists of playing and walking 18 holes. This translates into 4 miles or more of walking. However, a professional golfer has the luxury of a caddie to carry the golf clubs, making this task much easier. The walking pace is somewhat leisurely and the player has plenty of time to recover from the walk prior to hitting the next shot.

The travel schedule for a professional golfer is demanding. Events are held all over the United States and overseas, and players generally spend an entire week at a given location in order to compete in 4 days of golf. This includes practice rounds and hours at the driving range and chipping and putting greens.

Physical Characteristics of Golf

Golf requires a player to have a generally good level of aerobic conditioning with a fair degree of flexibility and strength. The ballistic nature of the drive and other swings requires a full range of motion (which provides more time to accelerate) for enhanced distance. So strength throughout a full range of motion is important for the golfer. Power is also important for creating club head speed and force to be imparted on the ball.

The power of a golf swing comes from the sequential movements of the lower extremity transferred through the torso and finishing through the shoulders, elbows, and wrists. In order to efficiently transfer the forces created, strong abdominal and back muscles are important. Also, because of the rotational component of the swing, these muscles are important for injury prevention. Along with strong back muscles to prevent injury, golfers should ensure proper flexibility and strength of the rotator cuff muscles.

Physical Training for Golf

Golf requires basic endurance, moderate levels of strength, and a moderately high level of flexibility. Based on these components, a general strength and conditioning program is recommended that incorporates basic multijoint exercises (such as leg press, lunges, lat pull-down) with general aerobic conditioning (30–45 minutes, moderate intensity). The use of free weights and particularly dumbbells may allow for the greatest range of motion and freedom of movement while requiring a large degree of stabilization and control. Although generally not warranted in golfers, muscle hypertrophy may coincide with strength training, and a concerted effort must be made to maintain and/or improve flexibility because changes in muscle mass and strength may affect swing mechanics. Therefore, gain in mass should be minimized unless the individual needs it, and plenty of golf practice should be maintained.

Specific exercises focusing on the abdominal, lower back, and rotator cuff muscles should be incorporated into the training. Use of medicine balls and stability balls and other functional training devices enhance the training by mimicking the ballistic nature, coordination, balance, and dynamics of the golf swing. A variety of both static and dynamic flexibility exercises should also be included in the training regimen.

Key Point

Due to the high degree of precision required of golfers, muscle hypertrophy, strength, and flexibility must be carefully considered with regard to the golf swing.

Nutrition for Golf

Golf is both an anaerobic and aerobic sport. During drives and fairway shots, golf swings can be powerful and predominantly anaerobic. Existing ATP and creatine phosphate are the predominant fuels for these shots. However, pitching, putting and walking between shots are aerobic, with ATP regenerated from both carbohydrate and fat resources.

ENERGY BALANCE AND BODY WEIGHT AND COMPOSITION.   Most of the energy expenditure associated with the sport of golf is associated with walking between shots. Players may walk 4–6 miles to play 18 holes, and a 175-lb male golf athlete may expend about 200–250 additional kilocalories for each hour of play. Golf athletes can carry food in their bags to provide energy during practices and competitive rounds.

Although individuals vary considerably, competitive golf athletes tend to be a little leaner than the general population. Competitive male and female golf athletes typically have a body fat range of 10–16% and 12–20%, respectively.[47] All players should evaluate the influence of body weight and distribution of the weight on their golf mechanics and play performance.

CARBOHYDRATE.   A diet higher in carbohydrates (>60%) is supportive of golf activities and other training, such as resistance and cardiorespiratory exercise used by many golfers to enhance driving distances and improve endurance during long rounds. A carbohydrate-containing food benefits golfers during extended periods of practice and play. Sport bars and drinks as well as fig and granola bars are popular among players.

PROTEIN.   The optimal level of protein in a golfer's diet depends on what supportive training is used. A golfer who does not engage in resistance or aerobic training probably has the same protein needs as the general population. Many golfers today train with resistance equipment and weights to enhance strength and power, and many also run, cycle, or engage in another type of aerobic training program. In these cases, recommendations for protein are based on training type and volume and may be in the range of 1.0–1.4 g/kg of body weight.

FLUIDS AND HYDRATION.   Golf is typically played in warmer environments, so maintaining optimal hydration status is one of the most important nutrition considerations. Players need to continuously hydrate before, during, and after rounds or practices. As a rule golfers should be well hydrated prior to stepping on to the first tee box and should ingest at least a half liter of water for each hour of activity. As hydration status is linked to performance it should be of foremost concern. For instance, even 1–2% dehydration can impair physical and cognitive performance.[62] However, research investigating the relationship between hydration status and golf play is lacking.

Golf allows greater availability of water and opportunity for its ingestion than many other land sports. Golf carts and bags allow players to carry water with them at all times. Water coolers are typically available every couple of tee boxes, and the clubhouse is generally situated at "the turn" from the front nine holes to the back nine holes, allowing players to refill water bottles and pick up sport drinks. Some courses even have a cart driven around to sell beverages. Alcohol should be avoided for 24–48 hours prior to games. However, at the recreational level perhaps no other sport is more associated with drinking alcoholic beverages during play. Here alcohol is consumed more for relaxation and socialization than for ergogenics, although some claim enhanced play.

Key Point

Dehydration is the most significant immediate concern for golf athletes during play.

VITAMINS AND MINERALS.   For these nutrients, perhaps the most immediate concern for golf athletes is electrolyte loss through sweating, as a round of golf (or two) can take several hours. Sport drinks are popular among golf athletes, and those containing sodium (110–165 mg/8 oz) and potassium (9–46 mg/8 oz) should provide ample electrolytes to recover losses if fluid consumption covers sweat loss.

NUTRITION SUPPLEMENTATION.   Some golf athletes experiment with creatine supplements to enhance power in their drives and fairway shots. Other compounds that golf athletes may experiment with are caffeine, taurine, and glucuronolactone, which are components of certain "energy drinks" (such as Red Bull). The combination of these compounds in an "energy drink" system has been reported to increase cognitive capability and mood.[63] The notion is that this could assist golf athletes with maintaining concentration during long rounds of golf and perhaps stress associated with the game. However, some caution should be applied regarding the level of caffeine and restrictions on caffeine as a doping agent.

■ GOLF *IN REVIEW*

- A basic strength training program for golf is generally recommended. The effect of muscle hypertrophy and strength on the golf swing must be considered. Flexibility and the training of the muscles in the body core are paramount for injury prevention and for the generation of force through the torso. Use of multijoint free-weight exercises combined with stability/medicine ball training should provide needed stimuli. A moderate cardiovascular/aerobic base is also warranted, which can be achieved by walking or cycling.
- The primary nutritional consideration for golfers is proper hydration status. Eating a higher carbohydrate diet supports general health and supportive training. Protein requirements may be elevated above nonathlete levels to support peripheral training programs. Golf athletes might experiment with creatine to increase power, and with taurine, caffeine, and glucuronolactone to improve the mental components of the game.

■ WRESTLING

Wrestling has drawn attention for health concerns associated with the weight loss protocols that wrestlers often engage in to wrestle at their lowest possible weight class. Wrestling is also the most challenging individual contact sport and is steeped in history and tradition dating back to the Babylonians, Greeks, and Romans.

Performance Characteristics of Wrestling

The sport of wrestling normally takes place in a gymnasium. Athletes wear body tights, wrestling shoes, and a special head strap to protect the ears. A wrestling match consists of three rounds each lasting 3 minutes, with a minute rest between them. The grappling lasts the entire 3 minutes of each round, with occasional short breaks (15–30 seconds) to return the wrestlers to the middle of the mat if either wrestler ventures outside the ring.

Wrestling season is generally 4–5 months in the winter. Wrestlers consistently need to meet weight requirements prior to a match. Therefore a large degree of dieting is often associated with this sport.

Physical Characteristics of Wrestling

Wrestling requires a unique combination of both anaerobic and aerobic energy sources. Muscular power, quickness, strength, and endurance are necessary for the entire body (upper/lower extremity, the core). Hand and forearm strength is also important for holding and grasping opponents. Because of the variety of positions often encountered, a large degree of flexibility is required of athletes in this sport. This is important for both decreasing the chance of injury and improving functional offensive and defensive moves. Injuries are primarily muscle strains.

Key Point

Wrestling consists of a series of 3-minute rounds involving high-intensity efforts. Conditioning drills and exercises should simulate competition with durations of 1–3 minutes and short rest periods between sets or repetitions.

Physical Training for Wrestling

The dynamics of wrestling (strength, quickness, power, endurance) suggest that a periodized program would be the most beneficial. Athletes need to maximize these components while minimizing body weight. Wrestlers need to manipulate these characteristics and try to find a weight category that is appropriate. Strength training in the form of multijoint and Olympic movements mimics wrestling movements and trains the entire body simultaneously. These exercises improve coordination, stabilization, strength, and power. Conditioning drills should simulate matches with activities lasting up to 3 minutes with periodic bursts of high intensity. However, for improvement in quickness, speed, and power, rest intervals must be long enough to replenish energy stores so that maximal efforts can be made.

Nutrition for Wrestling

Wrestling performance hinges on muscular power, strength, and quickness. The high intensity and short duration of matches make wrestling extremely reliant on muscle glycogen. However, although a match may last only 10 minutes or less, practices usually last $1\frac{1}{2}$–$2\frac{1}{2}$ hours with athletes working on moves, leverage skills, and match simulations as well as conditioning exercises.

ENERGY EXPENDITURE AND BODY WEIGHT AND COMPOSITION. Energy expenditure associated with wrestling depends on the training involved. Weight classes were established to make the sport more competitive. Stronger wrestlers within a given weight class would then

PhotoDisc/Getty Images

hold a competitive advantage. In an attempt to capitalize on this concept, wrestlers often attempt to "cut weight" to drop to the next lower weight class. This was glamorized in the 1980s movie *Vision Quest*. Roughly a third of the wrestlers surveyed in one study wrestled below minimal wrestling weights (MWW).[64] Furthermore, the percentage of those wrestling under the MWW increased as the weight class went down. For instance, in the lightest four weight classes, 62% wrestled below MWW. In the middle four classes 29% wrestled below the MWW, whereas in the heaviest four classes only 6% wrestled below MWW. Many high school and college wrestlers are lean to begin with, so they are unlikely to be able to trim 5–10 lb of fat before the season begins. In fact, not considering the heavyweight class, most elite wrestlers compete at 10–12% body fat or less. Therefore, it would be difficult for weight loss not to include LBM. Some scientists have speculated that the weight maintenance and reduction practices could be detrimental for hormonal balances and appropriate growth in boys.[65] However, some research suggests that normalized levels of anabolic hormones and patterns of growth pick up again in the off-season if energy restriction is ceased.[66,67]

CARBOHYDRATE. Resistance training and intense wrestling practices dictate that muscle tissue relies primarily on glycogen for fuel. Carbohydrates should provide ≥60% of energy in a wrestler's diet, or daily carbohydrate consumption should be 7–10 g/kg of body weight. As a match draws near, many wrestlers attempting a lower body weight restrict their carbohydrate and total energy intake. Following a weigh-in, these wrestlers should eat carbohydrate-dense foods such as a bagel or toast with jelly, a sandwich, or a sport bar and drink. Depending on the degree of depletion and the amount of time between weigh-in and a match, a wrestler may be able to completely recover muscle and liver glycogen stores. For instance, collegiate wrestling matches may take place a day after the weigh-in. However, most high school matches take place in the afternoon after a morning weigh-in. This leaves less time to recover glycogen stores.

Wrestlers attempting to recover glycogen stores between weigh-in and a match should be cautious not to consume too great a volume of food. Foods should be restricted to those that are lower in fat to allow for a more rapid emptying from the stomach. Also, higher glycemic index foods and lower fiber foods might maximize the recovery process and decrease the likelihood of intestinal discomfort.

PROTEIN. Weight training is popular among wrestlers, especially in the off-season as they strive for maximal strength and power. Although protein requirements have not been established for wrestlers, intense weight training and practice increase protein requirements. A diet providing 15–20% of total energy as protein should provide ample protein to allow a wrestler to maximize strength gains. Expressed differently, the protein recommendation for wrestlers is 1.2–1.7 g/kg of body weight, depending on the level of training. During periods of energy restriction, protein intake cannot be compromised and may even need to be higher to help limit the loss of body protein. Typically wrestlers limit carbohydrate and fat during weight-cutting efforts.

FLUIDS AND HYDRATION. Wrestling practices often take place in wrestling rooms or small gymnasiums that may be quite warm and poorly ventilated. This can lead to fairly high sweat rates, and wrestlers should drink fluids during practice sessions relative to the degree of sweating. Water bottles should be brought to practice if water is not available in coolers or fountains. Drinking 250–300 ml every 15–20 minutes can provide about 1.0–1.2 L of fluid in an hour. The difference in weight before and after practice provides a rough estimate of the sweat loss volume and rate and can be used to plan for hydration during practice.

Water consumption during a match may not be necessary, as they are relatively short in duration. Therefore a wrestler who is well hydrated entering a match probably will not experience enough of a sweat loss during the match to compromise strength and performance.

In addition to energy restriction, wrestlers practice hypohydration prior to matches in an attempt to drop weight. However, at times these hypohydration efforts can become critical. For instance, in 1997 a college wrestler died of kidney and heart failure while working out in a wet suit in a 92 °F room.[68] Other deaths have also been associated with dangerous hypohydration practices by wrestlers. Coaches and wrestlers should be well educated on the dangers of severe dehydration and electrolyte imbalances. Recently the American College of Sports Medicine published a position paper on weight loss in wrestlers.[69]

Wrestlers who attempt to reduce body weight by hypohydrating need to focus on consuming fluids immediately after a weigh-in. Here again, if ample time exists between weigh-in and a match, complete rehydration can occur. Water and sport drinks should be consumed liberally. Sport drinks and foods can also help recover electrolytes if they were restricted as well. It is not unlikely for a wrestler to experience a weight gain of 1–3% between weigh-in and the match.[70] Furthermore, wrestlers who gained more weight in the interval were more successful in their matches.

Key Point

Hypohydration and other weight-loss efforts can reduce physical performance.

VITAMINS AND MINERALS. Restrictive diets for weight management and loss can limit the consumption of certain vitamins and minerals. One study that assessed the nutrition of college wrestlers during the season revealed shortcomings in the consumption of vitamins A, C, and B_6 and iron, zinc, and magnesium.[71] Coaches should monitor wrestlers' diets, teach good dietary practices, and possibly encourage vitamin and mineral supplementation.

Excessive sweating during practice and possibly during hypohydration increases the risk of electrolyte imbalance. Sport drinks providing sodium at approximately 110–165 mg/8 oz and potassium at 19–46 mg/8 oz should provide ample electrolytes to recover losses if fluid consumption covers sweat loss.

NUTRITION SUPPLEMENTATION. Several nutrition supplements are popular with wrestlers including creatine, androstenedione, metabolic stimulants, and fat burners. Creatine supplementation might be sought to enhance strength and power needed for takedowns and pins. For weight-cutting wrestlers, creatine supplementation may enhance body weight or make weight reduction more difficult. Like glycogen, muscle creatine phosphate attracts water. Therefore if wrestlers choose to experiment with creatine, they should begin in the off-season so that they can comprehend the potential weight gain.

Fat burners and supplements touted to enhance metabolic rate are popular with some wrestlers, probably more so at the middle and lower weight classes. Research specifically addressing the effects of these supplements on wrestlers is lacking. If an athlete is to use any supplement it should be experimented with in practice in order to predict potential influences on performance during a match.

■ WRESTLING *IN REVIEW*

- Wrestling requires strength, power, speed, and endurance, so a periodized training program is recommended. Core and Olympic movements improve stabilization, power, and strength for most of the body's musculature. Use of plyometric exercises and stability balls also assist in creating explosiveness while training the core for stability. Most of the conditioning drills should simulate competition, using durations of 1–3 minutes with short rest periods between sets or repetitions.

- Wrestlers need to maximize glycogen stores before practices and supportive training by eating a higher carbohydrate diet. Proper hydration maximizes wrestling strength during practices and matches. Protein requirements are elevated relative to the intensity of practices, supportive training, and injury recovery.

- Wrestlers who cut weight by hypohydration and carbohydrate restriction must be aware of potential performance reductions and try to maximally rehydrate and recover carbohydrate before a match.

Photolink/PhotoDisc/Getty Images

■ SWIMMING

Swimming is an extremely popular sport and enjoyed by people of all ages. Many communities have public pools in addition to the pools of local health clubs and high schools. Practice and competition occur in water that may be 15–25 °F below body temperature, which significantly influences heat loss mechanisms compared to land sports. Also, the resistance to physical movement in water is greater than in land sports, influencing the relative workload. However, the buoyancy of the body in the water may counterbalance some of the workload.

Performance Characteristics of Swimming

The sport of swimming involves various strokes and distances covered. The strokes include freestyle (crawl/front), backstroke, breaststroke, and butterfly. Distances range from 50 to 1650 m. The longest races, regardless of stroke, have durations of less than 20 minutes (collegiate). However, most of the races are under 5 minutes and a few (such as the 50- and 100-m freestyle) are under 1 minute.

Collegiate swimming seasons last approximately 6 months. Several big meets occur throughout the season, and individual swimmers may qualify for the national meet during any event. Once qualified, peak performance should be geared around the national meet that occurs at the end of the season. Training for swimming usually requires one or two workouts a day, each lasting an average of 2 hours. Swimmers are generally trained in several or all of the strokes. The distances covered during training for the sprinters and distance swimmers are often similar. One of the underlying premises for the high volumes of training swimmers receive is to improve swimming mechanics and "feel for the water." Water is not the normal environment for humans and thus large volumes of training are considered necessary to become efficient at moving in this environment. The appropriate volume of training for sprinters as well as distance swimmers is still debated. From the perspective of energy systems and muscle fiber types, sprinters should primarily train for short durations using high intensities, as many sprint races last less than a minute.

Physical Characteristics of Swimming

Swimming requires the use of a large portion of the body's musculature. A good portion of the forces needed for the various strokes comes from the large muscles of the upper extremity (including the latissimus dorsi and pectoralis major). The muscles of the legs play a smaller role and function to keep the hips, legs, and lower torso elevated (horizontal). This is more important in reducing drag than providing propulsive force. These large muscle groups need to be strong and high in endurance. Muscles of the abdomen also need to be strong in order to maintain proper body alignment.

Because of the volume of training, swimmers often encounter overuse injuries. Thousands of strokes are accumulated weekly, and injuries such as tendonitis and bursitis are common. This is another debated point when considering the volume of training and the water environment.

Physical Training for Swimming

Because of the long season and high endurance component, a periodized training program is recommended. Cyclic variations in distances covered, intensities used, and particular strokes need to be planned to help maximize training benefits, avoid overtraining, reduce the incidence of injury, and help reduce psychological boredom. Strength and conditioning cycles should be based on major meets and the national event held at the end of the season.

Strength training programs should focus on moderate-intensity muscular endurance and avoid muscle hypertrophy. Increased lean body mass generally affects performance negatively by increasing drag forces. Also, because swimming movements have few or no eccentric components, these should be addressed within the strength program to help reduce the incidence and severity of overuse injuries. Furthermore, because of the high repetition of swimming movements, muscular asymmetry often occurs. Therefore, training opposing muscle groups may help maintain proper joint stability and movement mechanics.

Swimming drills and sessions should include a wide range of intensities. This enhances the various metabolic pathways and trains the musculature to produce a variety of forces. The variety also reduces the boredom often associated with extended training. Although often neglected, specifically training for a chosen stroke within appropriate intensity levels (such as sprints, mid-distance) has the greatest effect on performance and should be a major consideration.

Key Point

Swimming requires many repetitions of movements with few or no eccentric components, increasing the risk of overuse injuries.

Nutrition for Swimmers

During training, swimmers may practice twice a day. The first practice may be at the crack of dawn to get an hour of training in before classes or work. Then another training session may be scheduled for the late afternoon or evening. See Nutrition in Practice: Fueling Up for Exercise at the Crack of Dawn for helpful recommendations on food and beverage consumption before early morning sport training and competition.

ENERGY BALANCE AND BODY WEIGHT AND COMPOSITION. The energy expense of swimming can be tremendous because movement occurs in a medium (water) that offers significant resistance to movement. For instance, researchers estimated that the total energy expenditure (TEE) of female swimmers during high-volume training (17–18 km/day), which reflected two training sessions a day for a total of 5–6 hours, averaged 5600 kcal daily.[72] For these swimmers, their TEE was on average three times as high as their estimated resting energy expenditure (REE). It is expected that on average the energy expenditure for swimmers during training might be 3600–5200 kcal/day; the nature of the practices (times/day, intensity, and number of laps) determines the actual expenditure. Therefore, one concern for swimmers is to consume adequate energy (and carbohydrate) to maintain glycogen levels and body weight during training.

Swimmers often attempt to reduce their body mass to reduce their drag in the water and work output. In addition, swimsuits are small and body-hugging, and certainly among the most physique-revealing uniforms. Therefore, self-consciousness about physical appearance may lead a swimmer to reduce body weight, with or without performance in mind. Coaches and trainers should monitor changes in body weight and composition among swimmers and look for signs of disordered eating. The average body fat percentage for male and female swimmers has been estimated at 8.8% and 15.6–17.2%, respectively.[47]

Key Point

Swimmers often practice early in the morning, making pre-exercise energy consumption a challenge.

CARBOHYDRATE. For shorter "sprint" races, where the duration is less than 2 minutes, glycogen is the predominant fuel resource with early support from creatine phosphate. For these races most (>50%) of the ATP is produced via anaerobic systems, and practices for sprinters can easily reduce the glycogen content below 50–60%

NUTRITION IN PRACTICE   Fueling up for Exercise at the Crack of Dawn

Some athletes have to practice or train at the crack of dawn. For instance, when swimming pools are shared between high school and college programs, swimmers may have to practice at 5 or 6 A.M. Also, many endurance athletes opt to train first thing in the morning during warmer months to minimize their exposure to heat. It is also common for distance runs and cycling events to begin at 7 or 8 A.M. As athletes are often advised to eat 2–3 hours before exercise to allow for stomach emptying, these early morning efforts present a challenge.

Some athletes do not want to eat just before an early practice or competition, as they fear that the stomach contents would be uncomfortable during training. However, skipping food or a meal before exercise, especially after an overnight fast, can lead to compromised glycogen stores. This in turn can impair the ability to train and compete optimally. So what can these athletes do to get in the habit of providing a source of carbohydrate and fluid before they head to the pool, gym, or track or hitting the road?

Early morning athletes need to either wake up well before practice or competition or focus on consuming a tolerable carbohydrate food and/or beverage just before activity; perhaps in the car or walking across campus. Athletes who just will not eat before an early practice should pay special attention to eating in the evening and consume 300–500 ml of a sport drink with 6–8% carbohydrate before starting exercise. Although these early morning fueling practices might not come easily at first, once the athlete gets into the swing of it, it will become habit. Coaches should discuss different strategies with their athletes. Below are some suggestions for fueling before exercise at the crack of dawn.

LATE NIGHT SNACK.   Athletes who do not want to eat in the morning can try the following late-night snacks and consume a sport drink in the morning.

- Eat a small bowl of pasta and drink 8–12 oz of juice.
- Eat apple slices from two medium or large apples spread with peanut butter and drink 8–12 oz of milk.
- Eat one or two slices of homemade pizza topped with tomatoes and peppers.
- Eat a bowl of cereal (>1 cup) with 8–12 oz of juice.

CATNAP SNACK.   With handy snack like the following, athletes can wake up 2–3 hours before practice or competition, reset the alarm, chow down, and go back to sleep.

- Keep a bagel with jelly in a baggie on the nightstand.
- Put a couple of juice boxes or high-carbohydrate ready-to-drink (RTD) shakes in a cooler by the bed.

LIGHT BREAKFAST.   Athletes who are able to get up and eat a small meal before heading out the door can try these.

- 2 pieces of toast and a cup of fruit juice
- 1 bowl of banana and apple slices with a cup of skim milk or juice
- 1 cup of cereal with skim milk and ½ cup of fruit juice and water
- 1 cup of oatmeal with 1 cup of fruit juice

GRAB AND GO.   These early morning snacks are for athletes who want to grab something on the way out the door as they head to the pool, gym, or wherever.

- Baggie of raisins and a juice box
- Bagel and water bottle with 8 oz of sport drink (6–8% carbohydrate)
- Small fig bars (6–8) and 8 oz of juice
- Sport bar that is high carbohydrate/low fat (such as a PowerBar, Gatorade Energy Bar)

And don't forget to drink at least 12–16 oz of water!

---

of maximum. On the other hand, during longer races and endurance swims, swimmers utilize more aerobic mechanisms to power muscle contraction, and muscle glycogen and blood glucose are important fuel sources. So, as for other sports, a higher carbohydrate diet is recommended for swimmers. This can be accomplished by consuming >60% energy as carbohydrate or 8–10 g/kg of body weight daily. Swimmers who practice twice a day need to pay particular attention to their carbohydrate intake between the morning and evening practices and between the

evening and next-morning practices. Swimmers should try to eat a carbohydrate-rich meal after practice with a carbohydrate content of at least 2 g/kg of body weight.

PROTEIN.   The protein requirements for swimmers have not been determined. However, because swimming through water offers a lot of resistance, many swimmers are muscular. In addition, the long practices (several hours a day) may increase amino acid oxidation. However, research is needed in this area to determine if this is indeed the case. In light of the higher energy expenditure associated with swimming, eating 15–20% of total energy as protein should provide enough protein to at least maintain nitrogen balance. For instance, a 170-lb (~80-kg) male swimmer might expend 4300 kcal daily. To match this energy requirement with 15% of the energy being derived from protein would mean consuming 161 g of protein or 2 g/kg of body weight, which would likely be more than enough protein.

FLUIDS AND HYDRATION.   Swimming does not produce as much sweat as land sports due to the cooler water temperature. However, the opportunity for dehydration certainly exists and swimmers should employ the same hydrating practices as other athletes. At least 1 L of water should be consumed for each hour in the pool. Because swimmers get water in their mouth, the sensation of dry mouth is not present as an encouragement to drink water during a practice session. Therefore many swimmers fail to ingest water during practice sessions.

VITAMINS AND MINERALS.   Swimmers consume large amounts of energy and if their diet contains a variety of foods, then they should not have any dietary deficiencies in vitamins and minerals. However, iron deficiency has been noted to be a potential problem for some female swimmers.[49] Electrolyte loss in sweat is not great because of the lower volume of sweat for swimmers than for land athletes with similar work output (see Chapter 8). However, a training-dense schedule and restrictive diet can lead to electrolyte imbalances.

NUTRITION SUPPLEMENTATION.   Creatine supplementation has been purported to potentially benefit sprint-style swimmers. For instance, it was reported that creatine loading (25 g/day for 4 days) improved the average interval swim time for elite swimmers during maximal interval sessions; however, a few of the swimmers did not show improvement.[73] Lower levels of supplementation (5 g/day for 2 months) failed to significantly improve swimming performance. Other researchers determined that creatine supplementation in both males and females failed to improve the swimming velocity during 10×25-yard swims (~30 seconds) for both genders or 6×50-meter interval sets for women.[74] On the other hand, supplementation did improve the performance by males in the 6×50-meter intervals, each lasting about 10–15 seconds.[74] Again, experimentation with any nutrition supplement should be discussed with the coaching staff and properly evaluated for efficacy in the off-season. Some scientists have speculated that creatine supplementation may enhance body mass, thus increasing drag and LBM, which would decrease buoyancy of swimmer. Although this is still speculation, it should be considered and monitored because it could be a detriment, especially to a distance swimmer.

■ SWIMMING *IN REVIEW*

- Swimming is a unique sport with a long season. A periodized program manipulating distances and intensities is warranted. Strength training should focus on joint stability, symmetry, and eccentric loading in efforts to minimize injuries and avoid strength discrepancies. Sprinters should incorporate exercises with higher power outputs to simulate sprinting events. Muscle hypertrophy should be taken into consideration with muscular strength and power since it increases drag forces.
- Swimmers need to maximize glycogen stores before practices and meets by eating a higher carbohydrate diet and utilizing a sport drink, especially when they are in the water more than once a day. Protein requirements are elevated due to intensity of training. Sweat rates are not as great as for land sports, but adequate fluid consumption and proper hydration are still important considerations for swimmers.

■ ENDURANCE CYCLING

The different types of cycling (road, mountain, sprint, velodrome) each have their own unique characteristics. This section is dedicated to endurance road racing, which

Photolink/PhotoDisc/Getty Images

is one of the most popular sports worldwide and includes one of the most recognizable sporting events, the Tour de France.

Performance Characteristics of Cycling

The sport of endurance cycling is performed in a variety of environments from early April through October. Athletes wear helmets, and aerodynamic body Lycra. Races consist of a variety of distances; most are 60 miles. Other common race distances are the century and half-century (100 and 50 miles, respectively). The traditional distance of 60 miles is accomplished in less than 3 hours with average speeds of 27 miles per hour and top speeds that can exceed 50 miles an hour. The average range of cycling cadence in revolutions per minute (RPM) is 80–90. Hence a cyclist performs more than 9600 revolutions during a 2-hour workout or race. Circuit and criteria ("crits") races, 40 and 20 miles respectively, are held just about every weekend for competition as well as preparation for the six to ten major races per season. The terrain covered varies from extreme hills to virtually flat. Training sessions are 2–2½ hours in length and consist of a variety of intensities (such as sprint work and distance rides). Cyclists typically cover 200–300 miles per week in their training schedules.

Physical Characteristics of Cycling

Endurance cycling is an aerobic sport. Cycling speeds and duration are essential to performance levels. Lower extremity strength and aerobic power are critical. Only minimal upper body strength is needed to support the various riding positions. Upper body mass should be kept to a minimum because it increases both air resistance and load. Back and abdominal strength are important to maintain body posture. Injuries tend to be associated with overuse (such as patellar or iliotibial band tendonitis) or with bike collisions and crashes.

Physical Training for Cycling

Because of the distances covered, variety of terrain, and psychological demand of the sport, a periodized program would be the most beneficial. To elicit the most adaptation, athletes focus on distance certain days and weeks, and speed or power on other days and weeks. Varying the distances and intensities provides variety and challenge, and helps deter overtraining and injury rates. Also, addressing the specific components within a yearly cycle based on important races should maximize the training benefits and performance levels.

Strength training exercises should address the lower extremity, lower back, and abdominals. Endurance cycling has a high concentric muscle contraction component with little, if any, eccentric muscle activity and eccentric exercises have proven useful in the prevention and treatment of tendonitis. Therefore, including some eccentric exercises is appropriate. Thorough stretching of the lower extremity also helps reduce the incidence and severity of tendonitis.

Because of the high endurance training of the lower extremity during cycling, strength training the gluteal, quadriceps, and hamstring muscles need only be moderate in intensity. It is difficult to predict if this will improve performance due to the large volume of training already being received from cycling itself. However, strength training appears to assist in the prevention of overuse injuries. Additionally, abdominal and lower back muscles must be strong, flexible, and able to endure the sustained riding positions. Therefore, considerable attention must be given to this body region.

Nutrition for Cycling

Endurance cyclists may cover great distances in a single day. For example, the CoreStates USPRO Championship in Philadelphia held each June covers more than 150 miles in a day. Tour cyclists compete over several days in "stages." For example, in the premier cycling event, the Tour de France, a cyclist covers roughly 4000 km (~2500 miles) in a period of about 3 weeks. The terrain involves mountains and flats. Along with ironman triathlons, the Tour de France may be considered the pinnacle of athletic endurance performance.

Hydration status, eating enough carbohydrate, and maintenance of muscle mass are foremost considerations for endurance cyclists. Cycling has several advantages over other endurance sports such as running and swimming. First, cyclists have a performance apparatus to which they can attach water bottle cages and fanny sacks. Also, cycling jerseys have rear pockets for storing food with minimal effects on drag. The higher velocities of movement may allow for more heat loss via convection in comparison to running, and periodic coasting and drafting in a pack of cyclists provides periods of rest. Also, cycling results in less bouncing of the torso in comparison to running, which may allow cyclists to tolerate solid foods during a ride.

ENERGY BALANCE AND BODY WEIGHT AND COMPOSITION.   Cyclists can expend tremendous amounts of energy during rides. The major factors determining the level of expenditure are the mass of the cyclist and the intensity and duration of the ride. It is within reason to estimate that elite cyclists can expend 600–900 kcal/hour during a competitive ride. Daily expenditure can exceed 6000 kcal during competition. The cyclists assessed in one study ate, on average, 4162 kcal/day during training periods and 4460 kcal on race days. The body fat percentages for elite male and female cyclists tend to approximate 8–12% and 10–15%, respectively.[47]

CARBOHYDRATE. Cyclists competing in the Tour de France have been noted to consume a diet that derives approximately 62%, 15%, and 23% of its energy from carbohydrate, protein, and fat, respectively.[75] As much as 30% of the energy they consume comes in the form of liquid carbohydrate-enriched supplements. As the cyclists may spend 3–5 hours on their bike during a tour, they have to consume a significant quantity of their energy while on the bike.

During touring events, a cyclist must endure efforts over several successive days. This makes glycogen loading impossible, and cyclists must rely on a high-carbohydrate diet (>60% total energy) to maximize glycogen stores prior to the next day's ride. As cyclists might expend in excess of 6000 kcal, they require as much as 8–11 g of carbohydrate/kg of body weight daily.[76] Training and competing at higher intensities is a huge drain on glycogen stores. Although the need for carbohydrate is clear, cyclists may consume lower than recommended levels of carbohydrate in favor of more fat-laden foods to help meet energy requirements. Carbohydrate-dense foods need to be consumed after completing a ride (≥2 g/kg of body weight) and every 2–3 hours thereafter to restore glycogen stores.

During competition, cyclists tend to favor higher carbohydrate foods that are well tolerated (no intestinal discomfort) such as sport drinks, some sport bars, and plain bagels. Sport glucose tablets are also popular to complement sport drinks. Whether the energy of food ingested during competition should be carbohydrate only or a mixture of energy nutrients is unclear. For instance, one recent study suggested that consuming a sport bar containing a mixture of carbohydrate, protein, and fat during a 330-minute endurance ride allowed for greater fat utilization during the effort in comparison to receiving carbohydrate alone.[77] The cyclists performed at 50% of their peak power output (PPO), an average of 203 watts during the ride. However, a third of the riders failed to finish the time trial that followed after receiving the mixed-energy sport bar during an endurance ride. All riders finished the time trial after receiving only carbohydrate during the endurance ride. Therefore individuality is again a key factor and cyclists must experiment during practice rides to determine what will work best during competition.

PROTEIN. Although a reduced body weight would theoretically be a performance advantage by reducing workload, most elite cyclists are not thinking along these lines. Furthermore, weight loss during a tour may be viewed as undesirable as it may include LBM. The higher intensity, longer duration bouts of exercise endured by a cyclist lead to a net breakdown of muscle protein during rides and

perhaps for periods of time thereafter. Protein needs are elevated, and these individuals require a protein intake of at least 1.2–1.4 g/kg of body weight to maintain nitrogen balance during heavy training or multistage competition. Here again, a diet attributing about 15–20% of its energy to protein should meet this requirement.

Today, many cyclists resistance train their legs in an attempt to develop greater strength and power. Some also think the extra muscle might counterbalance some of the catabolic effects of endurance training. Little research is available to speculate on the protein metabolism and dietary requirement for cyclists who engage in both resistance and endurance training within the same general training program.

FLUIDS AND HYDRATION. Water requirements are significant, and cyclists should drink at least 1–2 L of water for each hour on the bike. They will probably reap late performance benefits if the fluid includes a 6–8% carbohydrate sport drink. Most cyclists have water bottle cages attached to the bike frame and professional teams have a team vehicle, which provides riders with fresh supplies of fluids during competition.

Sweat can be pulled off the skin more easily during cycling than running, so cyclists may not think that they are sweating to the extent they really are. If a cyclist waits until thirst is perceived before hydrating, performance may already have been hindered. To win a tour event, a rider must have the lowest combined cycling time throughout the stages. Thus every second counts. Cyclists can weigh themselves before and after a training bout to gauge sweat loss volume and rate. However, cyclists must pay particular attention to accounting for fluid and foods consumed during a cycling bout, as well as urine production as discussed in Chapter 8.

Caffeinated beverages such as Coca-Cola are often the choice of cyclists as a supplemental fluid during rides and as a preride beverage. The ergogenic potential of caffeinated beverages is individualized and requires experimentation. However, one of the detriments associated with caffeine is the potential for immediate increased urine formation. The need to void the bladder during a training ride might not be a problem, but during competition it can be disastrous.

VITAMINS AND MINERALS. Endurance cyclists consume large amounts of energy and if their diet contains a variety of foods, they should not have any diet deficiencies in vitamins and minerals. However, the volume of sweat produced by cyclists makes electrolyte recovery an immediate concern during longer cycling events (>1 hr). Sport drinks providing sodium at approximately 110–165 mg/8 oz and potassium at 19–46 mg/8 oz should provide ample electrolytes to recover losses if fluid consumption covers sweat loss. In addition, cyclists need to at least meet general intake recommendations for vitamins and minerals to support and maintain physiological adaptations such as enhanced aerobic metabolic capabilities and antioxidant systems.

NUTRITION SUPPLEMENTATION. Competitive cyclists need to understand which substances are banned by their sport (see Appendix C for contact information for the United States Cycling Federation/USCF). Several nutrition supplements are targeted to cyclists, including creatine and carnitine. However, the efficacy of creatine and carnitine supplementation to improve distance cycling performance has not been supported.[78] Carnitine is touted to enhance fat utilization, which could spare muscle glycogen and push back fatigue. Creatine supplementation has been speculated to improve endurance cycling performance by increasing strength and endurance. As an anaerobic ATP regenerating resource, it is unlikely to affect submaximal-intensity cycling. However, during breakaway and finishing sprints and in hill and mountain climbs, creatine supplementation may be of help. Can a cyclist ingest creatine in the days just prior to climbing or sprinting stages of a tour and receive a benefit? Researchers are attempting to answer this question.

One argument against creatine supplementation is the potential for weight gain. Carrying extra weight increases a cyclist's workload and energy expenditure, thereby influencing fatigue. Thus, experimentation is required and body weight and performance must be monitored. Certainly sprint cycling (velodrome cycling) is a candidate for testing the efficacy of creatine supplementation.

Glycerol is sometimes marketed to cyclists who train or compete in warmer or more arid environments. Glycerol has the potential to aid in hyperhydrating efforts. Depending on sweat rate a cyclist may be able to maintain adequate hydration with water bottles and a camelback water carrier. Two large water bottles and a larger bladder-containing backpack (camelback) can easily provide 5 L of fluid during a ride. Therefore glycerol supplementation may not be necessary. Cyclists need to experiment with glycerol during practice since intestinal discomfort, bloating, and cramping are possible side effects.

Many cyclists use antioxidant supplements. Free radical production seems to increase during cardiorespiratory exercise, which increases the potential for oxidative damage to muscle tissue. However, antioxidant enzyme systems also appear to be enhanced as an adaptive response. At this time there is not indisputable evidence that antioxidant supplementation allows for greater tissue protection in comparison to naturally adapting systems in well-nourished individuals. In fact, some scientists speculate that supplementation may actually hinder some adaptive processes that may be redux sensitive. For more on this issue see the In Focus: Antioxidant Supplementation by Athletes in Chapter 9.

■ ENDURANCE CYCLING *IN REVIEW*

- Endurance cycling requires a periodized training program. Both distance training and speed training need to be addressed in order to maximize performance. Varying ride distances and intensity add variety to the training and help decrease overuse injuries. Strength training primarily needs to address the lower extremity and core. Eccentric loading and an aggressive flexibility program also help deter overuse injuries. Moderate intensities are recommended, using multijoint lifts to increase basic strength. Use of stability balls help train the abdomen and lower back.
- Cyclists need to strive to increase glycogen stores between training rides and races by eating high-carbohydrate diets. Sport drinks contain carbohydrates and can provide a performance-extending fuel source during longer rides or higher intensity short rides. Protein requirements are elevated relative to intensity and duration of training/races, adaptation demands, and injury recovery. Sweating can be excessive, especially on warmer days, and proper hydration before and during training/races can maximize efforts.

■ DISTANCE RUNNING

Running is one of the most popular sports in the world. It is enjoyed by both genders of all ages. It is perhaps the most simplistic of sports as the movement is so familiar from childhood play. However, running at higher intensities for longer durations is far from play. Running is truly a perpetual dynamic sport, as runners do not enjoy the luxury of coasting as do their wheeled brethren, the cyclists. This reduces the opportunity for brief recovery periods. In addition, consuming fluids and energy during a run is never easy. Distance running is extremely taxing on the musculoskeletal system. Researchers have chronicled the breakdown of damaged muscle tissue after a marathon as discussed in the Chapter 11 feature In Focus: Sport Injury and Nutrition Supplementation.

Karl Weatherly/PhotoDisc/Getty Images

Performance Characteristics

The sport of endurance running is performed in a variety of environments, and peak season is generally from early April through October. Athletes generally wear shorts, tank top, socks, and running shoes. If the weather is particularly cold, runners may wear body polypropylene, spandex, gloves, and a hat. Marathon races are 26.2 miles in length. Shorter races of half marathons, 10-milers, and 10 K (6.2 miles) runs are quite popular. Many marathoners use these shorter races for preparation of longer races. Although it is not clear what distance constitutes an endurance run, many consider the 5 K (3.1 miles) to be at the lower end of the range, whereas others consider it to be the 10 K.

Top marathon runners finish 26.2 miles in less than $2\frac{1}{2}$ hours. Running paces average 5 minutes per mile, or lower, for the entire race! Obviously, to cover 26.2 miles many strides occur. If the average stride length is 5 feet, this translates into 27,456 strides per marathon. With this number of repetitions, overuse injuries must be considered. The terrain covered is generally flat, but many of the shorter races cover hills of varying grades. Training durations are typically 1–2 hours and consist of a variety of intensities (long and slow, and race pace). Endurance runners often cover 50+ miles per week in their training schedules.

Physical Characteristics of Endurance Runners

Endurance running is an aerobic sport. Running speeds and duration are essential to performance levels. Lower extremity strength and aerobic power are critical. Upper body strength requirements are minimal. Upper body mass should be kept to a minimum as this increases the load to be moved. Back and abdominal strength are important to maintain body posture. Injury considerations are those associated with overuse (such as patellar or iliotibial band tendonitis, stress fractures of tarsal bones, shin splints).

Key Point

Endurance running requires a periodized program that fluctuates between distance and speed on a regular basis.

Physical Training for Endurance Runners

Because of the distances covered and psychological demand of the sport, a periodized program would be the most beneficial. To elicit the most adaptation, athletes focus on distance certain days and weeks, and speed on other days and weeks. Varying the distances and intensities provides variety and challenge, and helps deter overtraining and injury rates. Also, addressing the specific components within a yearly cycle based on important races should maximize the training benefits and performance.

Strength training exercises should address the lower extremity, lower back, and abdominals. Endurance running involves a large degree of impact due to the constant contact with the running surface (usually concrete). Attenuating these forces efficiently helps deter and decreases the intensity of overuse injuries. These forces are attenuated with eccentric muscle actions and other mechanical means. Strength training exercises with a moderate eccentric load assists in developing this type of strength. Thorough stretching of the soft tissue associated with the lower extremity also helps reduce the incidence and severity of overuse injuries.

Because of the high endurance training of the lower extremity during running, strength training the gluteal, quadriceps, hamstring, and lower back muscles need only be moderate in intensity. It is difficult to predict if strength training will assist in improving running performance, but many runners feel that it helps prevent overuse injuries. Some runners have felt that strength training improves their "kick" (higher velocity running), but this is difficult to validate.

Nutrition for Endurance Runners

Some of the most significant considerations for runners include their high energy needs, high carbohydrate needs, high fluid needs, and recovery from impact injury of tissue. Many marathoners say that there are actually two races within a marathon, the first 20 miles and the last 6.2. It is in the last 6.2 miles that many marathoners "bonk" or "hit the wall." Once the wall has been hit, performance drops and athletes cope with symptoms ranging from lightheadedness and exhaustion to loss of bowel control and fainting. This bonk is central fatigue, which results from hypoglycemia and/or dehydration, as discussed in Chapter 4. Preventive steps include properly hydrating and consuming carbohydrate during the race.

ENERGY BALANCE AND BODY WEIGHT AND COMPOSITION.   The energy intake of distance runners varies based on size and distance and intensity of training and competition. Some distance runners also engage in resistance training in an attempt to maintain slightly greater muscle mass. Energy expenditures can be 3000–6000 kcal/day. The body fat percentages of competitive male and female runners have been estimated at 4.7% and 14.7%, respectively.[79,80] Restrictive energy consumption to achieve a lighter and leaner body is prevalent in runners, especially females.[81] Restrictive energy consumption and eating disorders limit the intake not only of energy nutrients and energy in general but also of vitamins and minerals.

CARBOHYDRATE.   Runners who push too hard early in a race use muscle glycogen at a faster rate. This increases the reliance on blood glucose earlier in the race. Starting with more muscle glycogen before the onset of a race delays depletion. Proper tapering of training during the last week along with a high-carbohydrate diet (as in glycogen loading) and a prerace carbohydrate source along with carbohydrate ingestion during the marathon decreases the potential for glycogen depletion. For shorter distance endurance runs (5 K and 10 K) glycogen loading is not necessary and may be a detriment if body mass is increased. As the race distance is increased to 15 K and longer, runners can experiment with glycogen loading to see if it is beneficial to their performance.

Because of the high glycogen demand of training and competition, carbohydrate should contribute at least 60% to the total energy of a runner, a carbohydrate intake of about 8–10 g/kg of body weight. For longer distances, runners should strive to ingest 30–60 g of carbohydrate per hour of running; drinking 600–1200 ml of 4–8% carbohydrate sport drink provides this amount. This carbohydrate becomes particularly important as glycogen stores wane later in the run. As training runs are typically scheduled daily (with periodic rest days), the emphasis on glycogen recovery is paramount to prepare for the next run. A posttraining meal with a carbohydrate content of at least 2 g/kg of body weight would begin glycogen recovery.

PROTEIN.   Distance runners are generally quite lean, but their protein needs approximate those of football players and bodybuilders. It is recommended that distance runners maintain a diet with a daily protein content of 1.2–1.4 g/kg of body weight. This is easily obtainable by eating a diet where 15–20% of energy comes in the form of protein. The higher protein need of distance runners is related to increased amino acid oxidation and the significant recovery and repair operations associated with longer runs.

Longer runs at a moderately high intensity level may increase muscle protein breakdown. Therefore the posttraining/race meal may be particularly important with regard to protein metabolism. Although not proven scientifically, it might well be that including protein with carbohydrate in the postrun meal slows protein breakdown and promotes protein synthesis (see Chapter 5). This would allow for a more rapid return to protein balance and subsequent positive protein equilibrium.

FAT.   Fat is an important component of a distance runner's diet. It provides a concentrated energy resource as well as a source of essential fatty acids. Although many runners trying to achieve and maintain a leaner body composition limit the level of fat in their diet, fat cannot be avoided completely. This is especially true in the meal following training or competition. Researchers have reported that the level of fat in the meal following a run corresponds with repletion of fat stores within muscle fibers.[82] This could be an important consideration as a runner prepares for the next training run. One study suggested that increasing the level of fat in the diet might also improve running performance in some runners who have severely restricted their fat intake in the past.[83] These researchers also determined that higher levels of fat in the

diet, up to 40%, did not have a negative influence on the levels of certain factors associated with immune and antioxidant systems.

Key Point

Diets that are too restrictive in fat may decrease longer distance running performance.

FLUIDS AND HYDRATION.   Most training and races occur during daylight hours, exposing runners to heat and other weather elements. On warmer days, significant losses in body water can be experienced due to enhanced sweating. On cooler days, sweating is less severe; however, hydration is still an important factor. It is typical for marathon runners to experience a 3–5% reduction in body weight during races in warmer climates. Runners should strive to drink at least 6–12 oz (150–350 ml) of fluid every 15–20 minutes during longer runs. Weighing in before and after a run can help runners gauge their approximate sweat rate. As part of this process, they account for fluids consumed during the run and urine produced, as explained in the Chapter 8 In Focus feature.

Runners should practice drinking while on the run to decrease the risk of aspiration during a race. Some runners slow their pace or walk for 10–50 yards to allow ease of swallowing. Many runners prefer to drink from a water bottle to increase the control of the fluid. Fluids should be consumed at each water station during a race. This would make fluids available every 12–20 minutes for paces of 6–10 minutes/mile, as water stations are about 2 miles apart.

VITAMINS AND MINERALS.   Endurance runners consume large amounts of energy, and if their diet contains a variety of foods they should not have any dietary deficiencies in vitamins and minerals. However, this is not always the case. Shortcomings in iron consumption have been recognized in many female runners.[49] This is a special concern for women because of menstrual loss of iron. Also, in a study involving female NCAA athletic groups including cross-country runners, the average calcium intake was found to be below recommendations.[84]

At the least, runners need to meet general recommendations for intake of vitamins and minerals to support and maintain physiological adaptations such as enhanced aerobic metabolic capabilities and antioxidant systems as well as to support recovery from injury. However, whether supplementation of antioxidant vitamins (vitamins C and E) and minerals (Cu, Se, Fe, Zn) is necessary is still debated. The volume of sweat produced by runners makes electrolyte recovery an immediate concern during longer runs (>1 hour). Sport drinks providing sodium at approximately 110–165 mg/8 oz and potassium at 19–46 mg/8 oz should provide ample electrolytes to recover losses if fluid consumption covers sweat loss.

NUTRITION SUPPLEMENTATION. Nutrient supplementation is common in distant runners. One survey reported that 48% of the runners reported at least one supplement used within 3 days of running the Los Angeles Marathon.[85] However, there is no convincing evidence that a runner eating an appropriate level of energy and a diet that is rich in a variety of foods and deriving at least 60% of energy from nutrient-dense carbohydrate foods (whole grains, fruits, vegetables) lacks any nutrients.

Many endurance runners experiment with antioxidant supplements, as discussed in the Chapter 9 feature In Focus: Antioxidant Supplementation by Athletes. Other supplements associated with distance running include branched chain amino acids (BCAAs), glutamine, and glycerol. The BCAAs may be used to inhibit the onset of central fatigue later in an endurance bout, although there is not strong support research backing this concept. In addition, BCAAs and glutamine may be used to combat possible reductions in immune system capacity following exhaustive endurance efforts such as marathons and triathlons. Researchers are attempting to determine whether these supplements may help reduce the incidence of infections and illness that are common after these endurance bouts. Glycerol supplementation is an attempt to hyperhydrate before runs in warmer climates, when fluid consumption would not match sweat losses.

■ ENDURANCE RUNNING *IN REVIEW*

- Endurance running requires a periodized training program. Both distance training and speed training must be addressed in order to maximize performance. Varying distances and intensity adds variety to the training and helps decrease overuse injuries. Strength training should focus primarily on the lower body. Eccentric loading and an aggressive flexibility program also help deter overuse injuries often associated with running. Moderate intensities are recommended, using multijoint lifts to increase basic strength. This may also improve the strength needed for the "kick" of a race. Finally, stability balls are recommended for training abdominal and lower back muscles.
- Runners need to recover glycogen stores between training runs and races by eating a higher carbohydrate diet. Carbohydrate sport drinks can provide a performance-extending fuel source during longer runs. Protein requirements are elevated relative to the intensity and duration of training runs and races as well as injury recovery. Sweating can be excessive, especially on warmer days, and proper hydration before and during runs can maximize performance.

Karl Weatherly/PhotoDisc/Getty Images

■ BODYBUILDING

Some of the earliest examples of bodybuilders include Steve Reeves, who dazzled cinema audiences in the 1950s and '60s as Hercules. Bodybuilding grew slowly and then exploded in the 1960s and '70s as the competitors became massive in size. The movie *Pumping Iron* provided an inside look into the world of bodybuilding and may have stimulated interest. Today, the sport still evokes fascination and wonder, but its reputation is scarred by the abuse of anabolic agents and what may be unhealthy practices associated with preparing for competitions (shows).

Performance Characteristics of Bodybuilding

The sport of bodybuilding brings athletes to a staged setting in order to present their physiques. The presentation is generally accomplished through a choreographed series of moves and poses. The athletes are evaluated on several physical components that include muscular size, shape, symmetry, and definition. There are generally only a few major competitions annually, so much of the remaining year is simply spent training.

Physical Characteristics of Bodybuilding

Bodybuilding is primarily an anaerobic sport focused on developing muscular size and symmetry. There is a small endurance component associated with the length of the training sessions, but this is generally addressed by the activity itself. Athletes need to maximize their muscular size, shape, symmetry, and definition in order to be successful. This includes both the large and small muscle groups.

Physical Training of Bodybuilders

Bodybuilding revolves around a few competitions per year, and each of these requires peak physical conditioning. In order to reach peak physical condition for this sport, athletes must have a balance among muscle mass, symmetry, and definition. To accomplish the optimal combination of these components a periodized program is recommended. The phases of training consist of hypertrophy, symmetry, and then definition. During the last phase before competition athletes must decrease body fat in order to present the lean musculature of the body. This phase generally consists of severe dieting, dehydration, and a tapering of training.

Because the musculature is evaluated for several components, a combination of compound and isolation exercises is used. The compound exercises (such as squat, bench, lat-pull) build a solid base for the large volume of training and assist in the development of overall mass and symmetry. However, many isolation exercises are needed to keep the development of all the musculature progressing at similar rates. Bodybuilders are known for trying just about anything in order to create a training stimulus. Therefore a variety of set and repetition schemes as well as exercises are often utilized. Most bodybuilders use a split routine format (4-, 5-, or 6-day split) to address specific body parts several days per week.

Overtraining is common, as these athletes often strength train as much as 4 hours daily. Some signs of overtraining include decreased strength levels and muscular size, poor sleeping patterns, fatigue, cramping, and injury. Several exercises focusing specifically on flexibility are needed for the maintenance of proper joint mobility, which is readily hindered in these athletes because of the large muscle masses involved.

Key Point

Due to the large training demand of muscle hypertrophy, many bodybuilders utilize split routines to enhance the concentration on just a few muscle groups per workout.

Nutrition for Bodybuilders

Bodybuilders strive to achieve a physique that demonstrates a high level of muscle mass and an extremely low level of fat. Achieving an extremely low degree of subcutaneous fat allows the definition of the underlying skeletal muscle to be revealed. Bodybuilders might gain 10–20 lb during the *growth phase* and then lose the same or more weight prior to competition. The purpose of the growth phase is to place the body in an anabolic state by consuming plenty of energy and training hard. Body fat mass is often elevated during this time, but attempts are made to minimize this gain. Also, during this time, perceived exertion is reduced and workouts tend to feel powerful with a better "pump." As competition approaches, bodybuilders strive to lower their body fat mass while trying to maintain as much of the acquired muscle tissue as possible.

Numerous nutrition practices are experimented with and practiced by bodybuilders as a competition or "show" draws near. However, the discussion that follows applies more to bodybuilders who do not compete or who for other reasons maintain greater consistency in nutrition and training habits throughout the year.

ENERGY BALANCE AND BODY WEIGHT AND COMPOSITION. Some bodybuilders compete in only one or two shows per year. In order to achieve a high level of mass and incredible leanness, most professional bodybuilders engage in several months of a growth or *anabolic phase* and then prepare for competition(s) with several weeks of restrictive and carefully planned dieting. The growth phase includes an energy intake well in excess of what is needed to maintain body weight. During the precontest phase these athletes consume an energy-restricted diet to promote body fat loss.

Energy expenditures may approximate 0.085 kcal/kg of body weight per minute during heavy training. Therefore a 220-lb (100-kg) bodybuilder may expend roughly 500 kcal/hour in the gym during heavy training. An enhanced energy expenditure of 100–300 kcal is likely over the next day or so during repair and adaptation processes. The body composition of male and female bodybuilders at competition is around 6–8% and 7–12%, respectively.[86,87]

CARBOHYDRATE. To achieve and then maintain higher quantities of muscle mass, bodybuilders need to consume adequate energy, and their diets should be high in carbohydrate (>60% of energy). A meal containing 50–100 g of carbohydrate should be consumed 2–3 hours before a heavy training session to help maximize glycogen stores in the muscle groups to be trained. Sport drinks containing 4–8% carbohydrate may enhance a training session, especially if muscle groups will be trained long and hard. In addition, a carbohydrate source either during or after a workout might create a more favorable hormonal scenario. Elevated insulin levels after a training session might decrease protein breakdown and support protein synthesis. However, during the precontest phase it is typically for bodybuilders to methodically reduce their carbohydrate intake.

PROTEIN. Protein needs are elevated for bodybuilders to as much as 1.4–1.75 g/kg of body weight. Fat should be restricted to less than 30% of energy. Recent research indicates that bodybuilders and weight trainers may present an earlier positive net protein turnover in trained muscle after a training session if they ingest a source of essential amino acids just prior to training. This concept is based on work by researchers who speculated that increased blood flow to working muscle would allow for a greater uptake of amino acids.[88] The supplementation would result in an augmented amino acid pool, which would then be available for protein synthesis after the conclusion of the training bout.

Many sport nutritionists feel that bodybuilders may benefit from eating numerous smaller meals containing both carbohydrate and protein versus a lesser number of larger meals. Although still speculative, the notion is that a more consistently elevated insulin level throughout the day supports a steady delivery of amino acids to muscle tissue.

Isolated whey protein is popular with bodybuilders, as it is purported to yield greater developments in muscle mass. One reason for its popularity is the relatively high content of BCAAs, which are believed to promote protein synthesis in muscle. Isolated soy protein might be used during energy restriction (precompetition) as a mechanism for counterbalancing the anticipated decrease in thyroid hormone levels. Soy protein has been reported to increase thyroid hormone and thyroid-stimulating hormone (TSH); however, the potential application to athletes remains to be determined.[89]

Key Point

Whey and soy protein are popular protein sources with bodybuilders because of BCAA content and the purported ability to positively influence thyroid hormone levels, respectively.

FLUIDS AND HYDRATION. Gyms and training rooms can be hot and sweating can be copious. Bodybuilders need to maintain optimal hydration during training to ensure maximal strength. Even subtle reductions in hydration status can reduce strength and potentially reduce muscle gains and maintenance. Most gyms have water fountains, which should be utilized frequently during workouts. During a training session, 2 cups ($\frac{1}{2}$L) of water per hour should be the minimum, and up to 4 cups (1 L) per hour or more is needed if sweating is excessive. Sport drinks are a fairly common sight in the gym and can help the bodybuilder maintain hydration status. Many bodybuilders believe that they need to drink extra fluid in association with their higher protein intakes and greater muscle mass. Increased urea and creatinine formation may in turn increase urine production. Although this is not an absolute relationship, a little extra urine production is to be expected.

VITAMINS AND MINERALS. Despite their healthy physical appearance, the diet of bodybuilders is often lacking in some nutrients due to the focus on particular energy sources and the restriction practices before competition. However, bodybuilders consume a variety of supplements and sport foods, which can make up for inadequacies in vitamins and minerals.

NUTRITION SUPPLEMENTATION. Bodybuilding is the largest target market for sport supplement manufacturers. However, many supplements are marketed with little or no scientific backing other than simple speculation or application from other physiological scenarios. For instance, one study evaluated advertisements for nutrition supplements in popular bodybuilding magazines and determined that 42% of product claims could not be substantiated by any scientific findings.[90] In addition, 32% of the product claims were determined to have some related

scientific documentation but were marketed in a misleading manner. Only 21% of advertised products had appropriate scientific documentation to support claims.

Many nutrition supplements have been purported to enhance muscle development and promote leanness, claims that are attractive to bodybuilders. Among these supplements are creatine, HMB, carnitine, arginine, lysine and ornithine, DHEA, and androstenedione. As discussed in Chapter 11, scientific investigation has revealed that creatine probably holds the greatest promise for a positive effect. DHEA and androstenedione, although they are precursors for testosterone, may not elevate blood testosterone levels when taken at manufacturers' recommended doses. Furthermore, some research has suggested that supplementation of DHEA and androstenedione may raise estrogen levels. Only a few studies have tested HMB, so conclusions cannot be made at this time. "Natural growth hormone" (arginine, lysine, ornithine) is still experimental, and glutamine, ribose, and inosine have received even less research.

Metabolic enhancement supplements containing caffeine and ephedrine as well as fat burners such as carnitine and chromium picolinate are still popular. Although chromium picolinate and carnitine have not proved effective independently, research is warranted to determine if there can be a synergistic effect among more complex supplement formulations.

■ BODYBUILDING *IN REVIEW*

- Bodybuilding requires a periodized program focusing on maximizing mass development and symmetry and then body leanness. High training volumes with moderate intensities are used during the mass building stage. During the precompetition phase, high volumes using high-repetition sets are often used to metabolize as much fat as possible. Training includes power, compound, and isolation exercises to maximize size and symmetry. Cardiovascular conditioning is incorporated in the precompetition phase in an effort to metabolize fat and "lean out."
- Eating a higher carbohydrate diet maximizes workouts when muscle groups are to be trained hard (a higher number of sets to fatigue). A sport food before or during training may slow glycogen losses and promote a positive muscle protein balance. Protein requirements are elevated relative to the level of training and adaptive response as well as LBM. Proper hydration supports maximal workouts. Precompetition protocols vary but typically involve lower fat and higher protein intakes. Carbohydrate and fluid are restricted just prior to competition.

Conclusions

The nutritional demands of a sport are based on the physical characteristics of the sport as well as the requirements of training, competition, recovery, and adaptation. In general, all types of athletes would benefit from a diet that supplies more of its energy from natural carbohydrate sources such as whole grains, low-fat dairy, fruits, vegetables, and legumes. Providing protein at a level of 15–20% of the energy required for weight maintenance and desirable weight gain would more than likely be adequate to achieve nitrogen balance. Initiating a training session or competition in an optimally hydrated state should be a consideration for athletes, especially those who hypohydrate to make a weight class (such as wrestlers). Athletes participating in sports with a strong aesthetic component (such as bodybuilding and fitness competitions) probably compete in a slightly or moderately dehydrated state. Athletes experiment with a variety of supplements and each must be appropriately evaluated for scientific basis, efficacy, safety, and permissibility (nonviolation of doping bans) for a sport. All athletes should understand the nature and demands of their sport to help plan their training programs and diet during the off-season, during the season, and prior to, during, and after competition.

 **IN FOCUS**   NUTRITION FOR EXERCISE AT HIGH ALTITUDE

Abrupt exposure to altitudes greater than 10,000 ft (3050 m) is frequently associated with symptoms of altitude illness, a combination of symptoms that include headaches, appetite loss (anorexia), nausea, vomiting, and malaise. The combined effect of these symptoms is usually a profound depression of appetite and reduction of food intake just when the climber needs energy the most. Climbers who anticipate altitude-impaired appetite may at least minimize the consequences: reduced energy intake, depleted muscle glycogen stores, negative nitrogen balances, and loss of critical lean body mass.

Gradual ascent, to acclimatize to progressively higher altitude exposure, is the best preventive medicine for high-altitude sickness. Unfortunately, it is not always practical or possible to delay ascent to altitude. Trekkers and rescue workers frequently must travel abruptly to high altitudes to perform critical tasks. Prior acclimatization is not always possible. Abrupt transportation from sea level to high altitude may be accompanied by debilitating altitude sickness symptoms, including altered mood, appetite, and performance. These uncomfortable symptoms usually increase in intensity for up to 48 hours after altitude exposure and then gradually lessen. Unfortunately, it is usually during the first 48 hours at altitude that critical work must be accomplished. The strenuous activities associated with work or recreation at altitude, plus an initial increase in resting metabolic rate and the lack of adequate food intakes, almost invariably result in an initially negative energy balance. Altitude illness can limit volitional activity, but energy expenditures of experienced and motivated climbers who are acclimatized can be quite high, depending on the activity level achievable under hypoxic conditions.

Effect of Altitude on Energy Balance

Food intakes are typically reduced by 10–50% during acute altitude exposure, depending on the individual and the rapidity of ascent. Scientists observed depressed food intakes and weight loss at altitude even under the controlled hypobaric chamber conditions of Operation Everest II.[91] In this study, work requirements were relatively low and a thermoneutral hypobaric environment with an adequate quantity and variety of palatable food was provided. Reduced food intake under these conditions indicated that hypoxia by itself was a major factor reducing appetite and food intake. Adequate food intake can be achieved at altitude, but it requires a concerted effort for dietary management and forced eating.[92] The combination of anorexia and reduced food intake can potentially exert a negative effect on work performance at even moderate altitude.[93]

Numerous pharmacological attempts to reduce acute mountain sickness have been investigated, with limited success. Caffeine has been reported to enhance relatively short-term, high-intensity work at simulated high altitude, perhaps via an influence on blood glucose availability. High-carbohydrate diets have been recommended by some as a nonpharmacological method to reduce the symptoms associated with acute mountain sickness. As an adjunct for lessening or preventing altitude illness, high-carbohydrate diets should be fed before and during the initial 3- to 4-day critical period of acute altitude exposure. It should be noted that only a limited number of investigators have studied high-carbohydrate diets or carbohydrate supplements for the relief of acute mountain sickness and performance enhancement. Some,[94,95] but not all,[96] have reported some beneficial effects on symptoms, mood, and performance. Most investigators agree that, at the very least, energy balances can be improved by aggressive carbohydrate supplementation at altitude, particularly via the fluid component of the diet. In addition to improving energy balance, carbohydrate supplementation improves nitrogen balance in

NUTRITION FOR EXERCISE AT HIGH ALTITUDE (CONTINUED)

the initial phase of acute altitude exposure. Scientists have confirmed that the negative nitrogen balance encountered at altitude is not due to any hypoxia-related decrease in protein digestibility or absorption but is primarily due to a negative energy balance.[97]

The mechanism by which carbohydrate relieves symptoms of altitude sickness and prolongs endurance at altitude may be related to improving blood oxygenation. Research has shown that blood oxygen tension is increased by a high-carbohydrate diet.[98] In addition, it has been reported that carbohydrate can increase lung pulmonary diffusion capacity at altitude.[99] Recently, scientists have demonstrated that carbohydrate consumption significantly increased oxygen tension and oxyhemoglobin saturation in arterial blood of subjects during simulated altitude (reduced oxygen in inspired air).[100] In addition to improving blood oxygenation, carbohydrate is a more "efficient" energy source at altitude than fat or protein. The energy production per liter of oxygen uptake is greater when carbohydrate is the energy source compared to fat (carbohydrate, 5.05 kcal/L of O_2; fat, 4.69 kcal/L of O_2) regardless of the oxygen tension in the inspired air. Taken together, these different lines of evidence suggest that carbohydrate is a more efficient energy source for work at reduced oxygen tension.

Influence of Altitude on Substrate Utilization and Nutrient Requirements

Research findings suggested that work at altitude in acclimatized individuals may be less reliant on fat metabolism and hence more strongly influenced by carbohydrate availability.[101] However, other researchers contend that the relative contribution of carbohydrate does not increase after altitude acclimatization and, as at sea level, the relative intensity of exercise is the major determinant of metabolic fuel utilization at high altitude.[102]

There is little evidence that chronic or acute altitude exposure increases the requirement for any specific nutrients other than possibly vitamin E and iron.[103] Studies of the effects of cold, energy expenditure, UV light exposure, and the reductive atmosphere at altitude indicates that supplementation of vitamins having an antioxidant function may be desirable at high altitude.[104–106] Supplemental antioxidant vitamins taken during a prolonged stay at high altitude may prevent a deterioration of blood flow and subsequent decrease in physical performance associated with free radical damage to cellular antioxidant defense systems.[104,107] Manipulations that improve oxygen delivery to tissues under the conditions of hypoxia are generally beneficial to work performance. In general, dietary treatments that preserve or enhance the fluidity or deformability of red blood cell (RBC) membranes at altitude are beneficial to oxygen delivery to tissues. Exposure to hypoxia and resultant lipid peroxidation of the unsaturated fatty acids in the red blood cell membrane reduces red cell deformity (the ability of RBCs to bend or flex as they pass through a capillary bed). The improvement of RBC membrane fluidity (increased ability to deform) can be achieved by two dietary mechanisms: supplementing the diet with polyunsaturated fatty acids, and supplementing the diet with antioxidants such as vitamin E to protect existing membrane polyunsaturated fatty acids from free radical peroxidation.

The suggestion that supplemental dietary iron may be beneficial at altitude stems from the observation that there is an increased erythropoietic response and increased hemoglobin synthesis at high altitude. Normal dietary iron intakes are adequate to support increased hemoglobin synthesis for males at high altitude, but females exposed to high altitude may benefit from a dietary iron supplement. All iron-deficient individuals regardless of gender may benefit from iron supplementation prior to going to altitude. Scientists have demonstrated that iron-deficient runners regardless of sex fail to exhibit a normal hematopoietic response upon exposure to altitude.[108] Although oral supplementation of iron (ferrous sulfate, 200–300 mg/day) for 2–3 weeks has been recommended before ascent and for the first 2–4 weeks at altitude, it was cautioned that a simultaneous free radical production might be enhanced by excess free iron.[109]

Fluid Requirements at Altitude

Water requirements at altitude may be greater than those at sea level, due to the low humidity of the atmosphere at altitude and hyperventilation associated with altitude exposure.[93,110] The risk of dehydration is high at altitude due to diuresis, water loss in breath and sweat, and the difficulty of obtaining adequate water. An inappropriate thirst response coupled with an increase in insensible water loss and a transient diuresis during the initial hours of altitude exposure can result in rapid dehydration if adequate fluid is either unavailable or neglected. The rate of respiratory water loss at altitude is about twice the rate of respiratory water loss for an equivalent activity at sea level.[111]

Hypoxia Versus Cold

High altitude and cold environments are often similar with respect to the thermal challenge, tempting one to categorize work in the cold at sea level with work under similar cold conditions at altitude. There are some

NUTRITION FOR EXERCISE AT HIGH ALTITUDE (CONTINUED)

distinct differences, however, which should be considered when planning nutritional support at high altitude. Fat, although tolerated relatively well in the cold at sea level, may not be as well tolerated in diets at high altitude. The symptoms of acute altitude exposure may be exacerbated if fat displaces carbohydrate from the diet. Although high-fat foods are energy dense and reduce the weight/calorie aspect of food carried on climbs, fat requires more oxygen for metabolism than carbohydrate and will place a small, but added, burden on the already overtaxed oxygen economy of the climber. Fat absorption may also be reduced at extremely high elevations; however, elevations commonly reached by recreational skiers, snowshoers, and backpackers are usually not associated with impaired fat or protein or carbohydrate absorption.[97] Another difference between cold exposure at sea level and high altitude is the calorigenic response to cold.[112] Cold exposure during hypoxia results in an increased reliance on shivering for thermogenesis due to a reduction in nonshivering thermogenesis at altitude, perhaps due to a reduction in aerobic catabolism of free fatty acids during hypoxia or an alteration in the neural-hormonal axis thermogenic response.

Recommendations

Inappropriate thirst and appetite responses, together with increased insensible water loss, transient diuresis, and increased energy expenditures, can lead to rapid dehydration and glycogen depletion and weight loss at altitude if adequate food and fluid are neglected. Dehydration may intensify the symptoms of altitude sickness and result in even lower food intakes. One of the most effective and practical performance-sustaining measures that can be adopted upon arrival at high altitude is to consume a minimum of 3–4 L of fluid per day containing 200–300 g of carbohydrate in addition to that contained in the diet. This should prevent dehydration, improve energy balance, improve the oxygen delivery capability of the circulatory system, replenish muscle glycogen, and conserve body protein levels.

The author of this feature is E. Wayne Askew, Ph.D., Professor of Foods and Nutrition at the University of Utah.

STUDY QUESTIONS

1. How are the training concepts for ice hockey and football similar?
2. How do the physical characteristics differ among the different positions in football and baseball?
3. Describe the relationships and training concepts related to body mass, force, and acceleration in different sports?
4. How do training exercises focusing on the body core enhance the performance of throwing, batting, and serving skills?
5. Describe how periodized training programs fit in with the various seasons of the sports.
6. Which sports are particularly sensitive to hydration?
7. What are the ranges of basic protein and carbohydrate requirements for most athletes?
8. The performance of which sports might be improved with creatine supplementation?

REFERENCES

1. Montgomery DL. Physiology of ice hockey. *Sports Medicine* 5(2):99–126, 1998.
2. Agre JC, Casal DC, Leon AS, McNally C, Baxter TL, Serfass RC. Professional ice hockey players: physiologic, anthropometric, and musculoskeletal characteristics. *Archives of Physical Medicine and Rehabilitation* 69(3 Pt 1):188–192, 1998.
3. Burns J, Dugan L. Working with professional athletes in the rink: the evolution of a nutrition program for an NHL team. *International Journal of Sport Nutrition* 4(2):132–134, 1994.
4. Smart LR, Bisogni CA. Personal food systems of male college hockey players. *Appetite* 37(1):57–70, 2001.
5. Langely, S. Hockey. In: *Sports Nutrition: A Guide for the Professional Working with Active People*, 3rd ed. (Rosenbloom C, ed.), Chicago: American Dietetic Association, 2000.
6. Rankinen T, Fogelholm M, Kujala U, Rauramaa R, Uusitupa M. Dietary intake and nutritional status of athletic and nonathletic children in early puberty. *International Journal of Sport Nutrition* 5(2):136–150, 1995.
7. Tegelman R, Aberg T, Eklof R, Pousette A, Carlstrom K, Berglund L. Influence of a diet regimen on glucose homeostasis and serum lipid levels of male elite athletes. *Metabolism* 45(4):435–441, 1996.
8. Houston ME. Nutrition and ice hockey performance. *Canadian Journal of Applied Sport Science* 4(1):98–99, 1979.
9. Akermark C, Jacobs I, Rasmusson M, Karlsson J. Diet and muscle glycogen concentration in relation to physical performance in Swedish elite ice hockey players. *International Journal of Sport Nutrition* 6(3):272–284, 1996.
10. Grandjean AC, Reimers KJ, Ruud JS. Dietary habits of olympic athletes. In: *Nutrition in Exercise and Sport,*

3rd ed. (Wolinsky I, ed.), Boca Raton, FL: CRC Press, 1998.

11. Tegelman R, Aberg T, Pousette A, Carlstrom K. Effects of a diet regimen on pituitary and steroid hormones in male ice hockey players. *International Journal of Sports Medicine* 13(5):424–430, 1992.

12. Nutter J. Seasonal changes in female athletes' diets. *International Journal of Sport Nutrition* 1(4):395–407, 1991.

13. Gonzalez-Alonso J, Heaps CL, Coyle EF. Rehydration after exercise with common beverages and water. *International Journal of Sports Medicine* 13(5):399–406, 1992.

14. Jones AM, Atter T, Georg KP. Oral creatine supplementation improves multiple sprint performance in elite ice-hockey players. *Journal of Sports Medicine and Physical Fitness* 39(3):189–196, 1999.

15. Rogerson PA. On the relationship between handedness and season of birth for men. *Perception and Motor Skills* 79(1 Pt 2):499–506, 1994.

16. Rosenbloom, CA. Baseball. In: *Sports Nutrition: A Guide for the Professional Working with Active People*, 3rd ed. (Rosenbloom C, ed.), Chicago: American Dietetic Association, 2000.

17. Carda RD, Looney MA. Differences in physical characteristics in collegiate baseball players. A descriptive position by position analysis. *Sports Medicine and Physical Fitness* 34(4):370–376, 1994.

18. Wilmore JH. Body composition in sport and exercise: directions for future research. *Medicine and Science in Sports and Exercise* 15(1):21–31, 1983.

19. Withers RT, Whittingham NO, Norton KI, La Forgia J, Ellis MW, Crockett A. Relative body fat and anthropometric prediction of body density of female athletes. *European Journal of Applied Physiology and Occupational Physiology* 56(2):169–180, 1987.

20. Yoshida T, Nakai S, Yorimoto A, Kawabata T, Morimoto T. Effect of aerobic capacity on sweat rate and fluid intake during outdoor exercise in the heat. *European Journal of Applied Occupation Physiology* 71(2–3):235–239, 1995.

21. Ratzin Jackson CG. Overview of human energy transfer and nutrition. In: *Nutrition in Exercise and Sport*, 3rd ed. (Wolinsky I, ed.), Boca Raton, FL: CRC Press, 1998.

22. Greene JC, Walsh MM, Letendre MA. Prevalence of spit tobacco use across studies of professional baseball players. *Journal of the California Dental Association* 26(5):358–364, 1998.

23. Walsh MM, Hilton JF, Masouredis CM, Gee L, Chesney MA, Ernster VL. Smokeless tobacco cessation intervention for college athletes: results after 1 year. *American Journal of Public Health* 89(2):228–234, 1999.

24. Robertson PB, Walsh MM, Greene JC. Oral effects of smokeless tobacco use by professional baseball players. *Advances in Dental Research* 11(3):307–312, 1997.

25. Williford HN, Kirkpatrick J, Scharff-Olson M, Blessing DL, Wang NZ. Physical and performance characteristics of successful high school football players. *American Journal of Sports Medicine* 22(6):859–862, 1994.

26. Parks PS, Read MH. Adolescent male athletes: body image, diet, and exercise. *Adolescence* 32(127):593–602, 1997.

27. Delaney JS, Lacroix VJ, Leclerc S, Johnston KM. Concussions during the 1997 Canadian Football League season. *Clinical Journal of Sports Medicine* 10(1):9–14, 2000.

28. Miller TA, White ED, Kinley KA, Congleton JJ, Clark MJ. The effects of training history, player position, and body composition on exercise performance in collegiate football players. *Journal of Strength Conditioning Research* 16(1):44–49, 2002.

29. Short S, Short WR. Four-year study of university athletes' dietary intake. *Journal of the American Dietetics Association* 82:632–642, 1983.

30. Gomez JE, Ross SK, Calmbach WL, Kimmel Schmidt DR, Dhanda R. Body fatness and increased injury rates in high school football linemen. *Clinical Journal of Sports Medicine* 8(2):115–120, 1998.

31. Kaplan TA, Digel SL, Scavo VA, Arellana SB. Effect of obesity on injury risk in high school football players. *Clinical Journal of Sports Medicine* 5(1):43–47, 1995.

32. Kaplan TA. Obesity in a high school football candidate: a case presentation. *Medicine and Science in Sports and Exercise* 24(4):406–409, 1992.

33. Kreider RB, Ferreira M, Wilson M, Grindstaff P, Plisk S, Reinardy J, et al. Effects of creatine supplementation on body composition, strength, and sprint performance. *Medicine and Science in Sports and Exercise* 30(1):73–82, 1998.

34. Op 'T Eijnde B, Van Leemputte M, Brouns F, Van Der Vusse GJ, Labarque V, et al. No effects of oral ribose supplementation on repeated maximal exercise and de novo ATP resynthesis. *Journal of Applied Physiology* 91(5):2275–2281, 2001.

35. Berardot D. Combined power and endurance sports (Basketball) In: *Nutrition for Serious Athletes*. Champaign, IL: Human Kinetics, 2000.

36. Nowak RK, Knudsen KS, Schulz LO. Body composition and nutrient intakes of college men and women basketball players. *Journal of American Dietetic Association* 88(5):575–578, 1988.

37. Bale P. Anthropometric, body composition and performance variables of young elite female basketball players. *Journal of Sports Medicine and Physical Fitness* 31(2):173–177, 1991.

38. Siders WA, Bolonchuk WW, Lukaski HC. Effects of participation in a collegiate sport season on body composition. *Journal of Sports Medicine and Physical Fitness* 31(4):571–576, 1991.

39. Walsh FK, Heyward VH, Schau CG. Estimation of body composition of female intercollegiate basketball players. *Physics and Sports Medicine* 12:74–79, 1984.

40. Nuviala RJ, Lapieza MG, Bernal E. Magnesium, zinc, and copper status in women involved in different sports. *Internation Journal of Sports Nutrition* 9(3):295–309, 1999.

41. Nuviala RJ, Castillo MC, Lapieza MG, Escanero JF. Iron nutritional status in female karatekas, handball and basketball players, and runners. *Physiology and Behavior* 59(3):449–453, 1996.

42. Rokitzki L, Hinkel S, Klemp C, Cufi D, Keul J. Dietary, serum and urine ascorbic acid status in male athletes. *International Journal of Sports Medicine* 15(7):435–440, 1994.

43. Schroder H, Navarro E, Mora J, Seco J, Torregrosa JM, Tramullas A. The type, amount, frequency and timing of dietary supplement use by elite players in the First Spanish Basketball League. *Journal of Sports Science* 20(4):353–358, 2002.

44. Telford RD, Catchpole EA, Deakin V, Hahn AG, Plank AW. The effect of 7 to 8 months of vitamin/mineral supplementation on athletic performance. *International Journal of Sport Nutrition* 2(2):135–153, 1992.

45. Kundrat S. Volleyball. In: *Sports Nutrition: A Guide for the Professional Working with Active People*, 3rd ed. (Rosenbloom C, ed.), Chicago: American Dietetic Association, 2000.

46. Viitasalo JT, Kyrolainen H, Bosco C, Alen M. Effects of rapid weight reduction on force production and vertical jumping height. *International Journal of Sports Medicine* 8(4):281–285, 1987.

47. Sining WE. Body composition in athletes. In: *Human Body Composition* (Roche AF, Heymsfield SB, Lohman TG, eds.), pp. 257–273, Champaign, IL: Human Kinetics, 1996.

48. Papadopoulou SK, Papadopoulou SD, Gallos GK. Macro- and micro-nutrient intake of adolescent Greek female volleyball players. *International Journal of Sports Nutrition and Exercise Metabolism* 12(1):73–80, 2002.

49. Hassapidou MN, Manstrantoni A. Dietary intakes of elite female athletes in Greece. *Journal of Human Nutrition and Dietetics* 14(5):391–396, 2001.

50. Martin M, Schlabach G, Shibinski K. The use of nonprescription weight loss products among female basketball, softball and volleyball athletes from NCAA Division I institutions: issues and concerns. *Journal of Athletic Training* 33:41–44, 1998.

51. Macedonio M. Soccer. In: *Sports Nutrition: A Guide for the Professional Working with Active People*, 3rd ed. (Rosenbloom C, ed.), Chicago: American Dietetic Association, 2000.

52. Clark K. Nutritional guidance to soccer players for training and competition. *Journal of Sports Science* 12(Spec No): S43–S50, 1994.

53. Balsom PD, Wood K, Olsson P, Ekblom B. Carbohydrate intake and multiple sprints sports: with special reference to football (soccer). *International Journal of Sports Medicine* 20(1):48–52, 1999.

54. Zehnder M, Rico-Sanz J, Kuhne G, Boutellier U. Resynthesis of muscle glycogen after soccer specific performance examined by 13C-magnetic resonance spectroscopy in elite players. *European Journal of Applied Physiology* 84(5): 443–447, 2001.

55. Rico-Sanz J, Zehnder M, Buchli R, Dambach M, Boutellier U. Muscle glycogen degradation during simulation of a fatiguing soccer match in elite soccer players examined noninvasively by 13C-MRS. *Medicine and Science in Sports and Exercise* 31(11):1587–1593, 1999.

56. Kirkendall DT. Effects of nutrition on performance in soccer. *Medicine and Science in Sports and Exercise* 25(12): 1370–1374.

57. Lemon PW. Protein requirements for soccer. *Journal of Sports Science* 12(Spec No):S17–22, 1994.

58. Nicholas CW, Tsintzas K, Boobis L, Williams C. Carbohydrate-electrolyte ingestion during intermittent high-intensity running. *Medicine and Science in Sports and Exercise* 31(9):1280–1286, 1999.

59. Lyons T, Riedesel ML, Meuli LE, Chick TW. Effects of glycerol-induced hyperhydration prior to exercise in the heat on sweating and core temperature. *Medicine and Science in Sports and Exercise* 22:477–483, 1990.

60. Applegate E. Effective nutritional ergogenic aids. *International Journal of Sport Nutrition* (2):229–239, 1999.

61. Bergeron MF, Maresh CM, Armstrong LE, Signorile JF, Castellani JW et al. Fluid-electrolyte balance associated with tennis match play in a hot environment. *International Journal of Sport Nutrition* 5(3):180–193, 1995.

62. Armstrong LE, Epstein Y. Fluid-electrolyte balance during labor and exercise: concepts and misconceptions. *International Journal of Sports Nutrition* 9(1):1–12, 1999.

63. Alford C, Cox H, Wescott R. The effects of Red Bull energy drink on human performance and mood. *Amino Acids* 21(2):139–150, 2001.

64. Wroble RR, Moxley DP. Weight loss patterns and success rates in high school wrestlers. *Medicine and Science in Sports and Exercise* 30(4):625–628, 1998.

65. Horswill CA. Weight loss and weight cycling in amateur wrestlers: implications for performance and resting metabolic rate. *International Journal of Sport Nutrition* 3(3): 245–260, 1993.

66. Roemmich JN, Sinning WE. Weight loss and wrestling training: effects on growth-related hormones. *Journal of Applied Physiology* 82(6):1760–1764, 1997.

67. Roemmich JN, Sinning WE. Weight loss and wrestling training: effects on nutrition, growth, maturation, body composition, and strength. *Journal of Applied Physiology* 82(6):1751–1759, 1997.

68. Berardot D. Power sports (Wrestling). In: *Nutrition for Serious Athletes*. Champaign, IL: Human Kinetics, 2000.

69. Oppliger RA, Case HS, Horswill CA, Landry GL, Shelter AC. American College of Sports Medicine position stand. Weight loss in wrestlers. *Medicine and Science in Sports and Exercise* 28(6):ix–xii, 1996.

70. Wroble RR, Moxley DP. Acute weight gain and its relationship to success in high school wrestlers. *Medicine and Science in Sports and Exercise* 30(6):949–951, 1998.

71. Steen SN, McKinney S. Nutrition assessment of college wrestlers. *Physics and Sports Medicine* 1141:100–116, 1986.

72. Trappe TA, Gastaldelli A, Jozsi AC, Troup JP, Wolfe RR. Energy expenditure of swimmers during high volume training. *Medicine and Science in Sports and Exercise* 29(7):950–954, 1997.

73. Theodorou AS, Cooke CB, King RF, Hood C, Denison T, et al. The effect of longer-term creatine supplementation on elite swimming performance after an acute creatine loading. *Journal of Sports Science* 17(11):853–859, 1999.

74. Leenders NM, Lamb DR, Nelson TE. Creatine supplementation and swimming performance. *International Journal of Sport Nutrition* 9(3):251–262, 1999.

75. Benedot D. Endurance sports. In: *Nutrition for Serious Athletes*. Champaign, IL: Human Kinetics, 2000.

76. Burke LM. Nutritional practices of male and female endurance cyclists. *Sports Medicine* 31(7):521–532, 2001.

77. Rauch HG, Hawley JA, Woodey M, Noakes TD, Dennis SC. Effects of ingesting a sports bar versus glucose polymer on substrate utilization and ultra-endurance performance. *International Journal of Sports Medicine* 20(4):252–257, 1999.

78. Juhn MS, Tarnopolsky M. Oral creatine supplementation and athletic performance: a critical review. *Clinical Journal of Sport Medicine* 8(4):286–297, 1998.

79. Graves JE, Pollock ML, Sparling PB. Body composition of elite female distance runners. *Research Quarterly in Exercise and Sport* 60:239–245, 1987.

80. Pollock ML, Gettman LR, Jackson A, Ayers J, Ward A, Linnerud AC. Body composition of elite class distance runners. *Annals of New York Academy of Science* 301:351–370, 1977.

81. Hulley AJ, Hill AJ. Eating disorders and health in elite women distance runners. *International Journal of Eating Disorders* 30(3):312–317, 2001.

82. Larson-Meyer DE, Newcomer BR, Hunter GR. Influence of endurance running and recovery diet on intramyocellular lipid content in women: a 1H NMR study. *American Journal of Physiology Endocrinology and Metabolism* 282(1):E95–E106, 2002.

83. Venkatraman JT, Feng X, Pendergast D. Effects of dietary fat and endurance exercise on plasma cortisol, prostaglandin E2, interferon-gamma and lipid peroxides in runners. *Journal of the American College of Nutrition* 20(5):529–536, 2001.

84. Leachman Slawson D, McClanahan BS, Clemens LH, Ward KD, Klesges RC, et al. Food sources of calcium in a sample of African-American and Euro-American collegiate athletes. *International Journal of Sport Nutrition and Exercise Metabolism* 11(2):199–208, 2001.

85. Nieman DC, Gates JR, Butler JV, Pollett LM, Dietrich SJ, Lutz RD. Supplementation patterns in marathon runner. *Journal of the American Dietetics Association* 89(11):1615–1619, 1989.

86. Saylor K, Wildman REC, Willard G. Dietary practices, physical assessment and blood chemistry of competitive male bodybuilders. *FASEB Abstract*, 2002.

87. Wildman REC, Saylor K, Willard G. Dietary practices, physical assessment and blood chemistry of competitive female bodybuilders. *FASEB Abstract*, 2001.

88. Wolfe RR. Regulation of muscle protein by amino acids. *Journal of Nutrition* 132(10):3219S–3224S, 2002.

89. Persky VW, Turyk ME, Wang L, Freels S, Chatterton R Jr., et al. Effect of soy protein on endogenous hormones in postmenopausal women. *American Journal of Clinical Nutrition* 75(1):145–153, 2002.

90. Barron RL, Vanscoy GJ. Natural products and the athlete: facts and folklore. *Annals of Pharmacotherapy* 27(5):607–615, 1993.

91. Rose MS, Houston CS, Fulco CS, Coates G, Sutton JR, Cymerman A. Operation Everest. II Nutrition and body composition. *Journal of Applied Physiology* 65:2545–2553, 1988.

92. Butterfield GE. Maintenance of body weight at altitude: in search of 500 kcal/day. In: *Nutritional Needs in Cold and in High-Altitude Environments* (Marriott BM, Carlson SJ, eds.), Washington, D.C.: National Academy Press, 1996.

93. Askew EW. Cold weather and high altitude nutrition: overview of the issues, In: *Nutritional Needs in Cold and in High-Altitude Environments* (Marriott BM, Carlson SJ, eds.), Washington, D.C.: National Academy Press, 1996.

94. Consolazio CF, Matoush LO, Johnson HL, Krzywicki HJ, Daws TA, Isaac GJ. Effects of high-carbohydrate diets on performance and clinical symptomatology after rapid ascent to high altitude. *Federal Proceedings* 28:937–945, 1969.

95. Askew EW. Nutrition and performance in hot, cold and high altitude environments, In: *Nutrition in Exercise and Sport*, 3rd ed. (Wolinsky I, ed.), pp. 597–619, Boca Raton, FL: CRC Press, 1997.

96. Swenson ER, MacDonald A, Vatheuer M, Maks C, Treadwell A, et al. Acute mountain sickness is not altered by a high carbohydrate diet nor associated with elevated circulating cytokines. *Aviation and Space Environmental Medicine* 68:499–503, 1997.

97. Butterfield GE, Gates J, Fleming S, Brooks GA, Sutton IR, Reeves JT. Increased energy intake minimizes weight loss in men at high altitude. *Journal of Applied Physiology* 72:1741–1748, 1992.

98. Hansen JE, Hartley LH, Hogan RP. Arterial oxygen increase by high-carbohydrate diet at altitude. *Journal of Applied Physiology* 33:441–445, 1972.

99. Dramise JG, Inouye CM, Christensen BM, Fults RD, Canham JE, Consolazio CF. Effects of a glucose meal on human pulmonary function at 1600 m and 4300 m altitudes. *Aviation and Space Environmental Medicine* 46:365–340, 1975.

100. Lawless NP, Dillard TA, Torrington KG, Davis HQ, Kamimori G. Improvement in hypoxemia at 4600 meters simulated altitude with carbohydrate ingestion. *Aviation and Space Environmental Medicine* 70:874–878, 1999.

101. Roberts AC, Butterfield GE, Cymerman A, Reeves JT, Wolfel EE, Brooks GA. Acclimatization to 4,300-m altitude decreases reliance on fat as a substrate. *Journal of Applied Physiology* 81:1762–1771, 1996.

102. McClelland GB, Hochachka PW, Weber JM. Carbohydrate utilization during exercise after high altitude acclimation: A new perspective. *Proceedings of the National Academy of Sciences* 95:10288–10293, 1998.

103. Marriott BM, Carlson SJ (eds.). *Nutritional Needs in Cold and in High-Altitude Environments*, Washington, D.C.: National Academy Press, 1996.

104. Simon-Schnass I. Oxidative stress at high altitudes and effects of vitamin E. In: *Nutritional Needs in Cold and in High-Altitude Environments* (Marriott BM, Carlson SJ, eds.), p. 393, Washington, D.C.: National Academy Press, 1996.

105. Pfeiffer JM, Askew EW, Roberts DE, Wood SM, Benson JE, et al. Effect of antioxidant supplementation on urine and blood markers of oxidative stress during extended moderate altitude training, *Wilderness and Environmental Medicine* 10:66–74, 1999.

106. Chao W, Askew EW, Roberts DE, Wood SM, Perkins JB. Oxidative stress in humans during work at moderate altitude. *Journal of Nutrition* 129:2009–2012, 1999.

107. Askew EW. Environmental and physical stress and nutrient requirements, *American Journal of Clinical Nutrition* 61:631S–637S, 1995.

108. Stray-Gundersen J, Alexander C, Hochstein A, deLomos D, Levine BD. Failure of red cell volume to increase to altitude exposure in iron deficient runners. *Medicine and Science in Sports and Exercise* 24:S90, 1992.

109. Berglund B. High-altitude training aspects of haematological adaptation. *Sports Medicine* 14:289–294, 1992.

110. Hoyt RW, Honig A. Body fluid and energy metabolism at high altitude. In: *Handbook of Physiology, Section 4: Environmental Physiology* (Blatteis CM, Frealy MJ, eds.), p. 1277, New York: Oxford University Press, 1996.

111. Milledge J. Respiratory water loss at altitude. *Newsletter of the International Society of Mountain Medicine* 2(3):5, 1992.

112. Giesbrecht GG, Fewell JE, Megirian D, Brant R, Remmers JE. Hypoxia similarly impairs metabolic responses to cutaneous and core cold stimuli in conscious rats. *Journal of Applied Physiology* 77:726–732, 1994.

CHAPTER 15

NUTRITION, EXERCISE, AND SPECIAL POPULATIONS

Chapter Objectives

- Discuss nutrition and exercise considerations across the life span, including pregnancy and lactation.

- Describe the nature of types 1 and 2 diabetes mellitus and discuss the roles of diet and exercise in prevention and treatment.

- Discuss nutrition and exercise considerations for vegetarians.

- Describe eating disorders among athletes and other active people.

Personal Snapshot

© Fotografia/Corbis

Amy's friends call her the die-hard gym rat because she spends about 10 hours a week at her local fitness club. She loves to pump iron and hit the elliptical trainer and cannot imagine a day without exercise. Amy and her husband want to start a family and she often talks with her mom about what to expect during pregnancy. According to her mother, she will need to give up exercise when she is pregnant and while she is breastfeeding. But Amy has seen pregnant women working out in the gym, and wonders if what is known about exercise and pregnancy has changed in recent years. What should she do about working out during pregnancy and breastfeeding?

The bulk of this book addresses exercise and nutrition concepts for healthy young adults. From time to time unique considerations based on age and gender are included, but they have not received as much attention. In addition, exercise and nutrition recommendations during pregnancy have not been discussed yet, nor have certain endocrinological and psychological scenarios such as diabetes mellitus and eating disorders. Also, vegetarian diets for athletes warrant further consideration. This chapter provides an overview of topics in exercise and nutrition for special populations not covered in the earlier chapters as well as a synopsis of key concepts related to young and older active people.

■ EXERCISE AND NUTRITION DURING PREGNANCY AND LACTATION

A couple of decades ago, women were discouraged from exercising during pregnancy. Today, active women are encouraged to continue exercise during pregnancy. In fact, many believe that women who maintain their activity throughout pregnancy may experience an easier delivery. Special nutritional and exercise considerations for pregnancy are reviewed in this section.

Key Point

Contrary to advice in years gone by, exercise is now encouraged during pregnancy.

Exercise During Pregnancy and Lactation

Exercise involves special considerations for women who are pregnant. Although a low to moderate level of exertion can have a favorable influence on the mother's physical and mental state during pregnancy, higher intensity exercise is not recommended. Exercise overloads and trains the body of the mother, but the effects of physical training on the fetus are not completely understood. Because of the shift in blood supply from the inner organs to the working muscle tissue during heavy exercise, a decrease in the O_2-rich and nutrient-loaded blood supply serving the fetus is possible. Research has not shown a clear relationship between exercise and fetal disorders, but several studies have indicated that vigorous exercise during pregnancy reduces infant birth weights.

Using a moderate intensity for exercise during pregnancy seems to benefit the mother, and the redistributed blood flow is kept to a minimum and thus does not reduce uterine flow. The American College of Obstetricians and Gynecologists (ACOG) released its guidelines for exercise during and after pregnancy. It suggested moderate exercise with heart rates of approximately 140 beats/minute or less. It was also recommended that pregnant women not engage in exercise in the supine position (flat on back) after the first trimester of pregnancy because of the reduced blood supply to the uterus during exercises in this position. Prolonged periods of motionless standing were also discouraged. Activities that are high impact or ballistic in nature were discouraged because of joint laxity that occurs during pregnancy (which allows the pelvic bones to spread as needed) and the possibility of fetal trauma. Figure 15-1 highlights the recommendations for exercise during and after pregnancy.

Along with the specific recommendations, the ACOG urged individuals to use common sense. So when addressing exercise and pregnancy, a prescription should consider the woman's health, previous activity level, pregnancy complications, and goals of the exercise program. Thus, if a woman was physically active prior to pregnancy, maintaining activity levels with some modifications should not complicate the pregnancy.

The ACOG further stated that the physiological and morphological changes associated with pregnancy persist for 4 to 6 weeks postpartum. Therefore, activity levels should be increased back to prepregnancy levels gradually. Finally, the ACOG stated several contraindications to exercise during pregnancy. In other words, a woman should consult with a physician or not exercise if any of these conditions are present:

- Pregnancy-induced hypertension
- Preterm rupture of membrane
- Preterm labor during the prior or current pregnancy
- Incompetent cervix
- Persistent bleeding in the second to third trimester
- Intrauterine growth retardation

Key Point

The American College of Obstetricians and Gynecologists recommend that a woman's heart rate not exceed 140 beats/minute during pregnancy.

Nutrition During Pregnancy and Lactation

The extra energy requirements of exercise and of pregnancy must be accounted for when the mother assesses her energy needs. During the first trimester the daily needs for the embryo and fetus are smaller than in the second and third trimesters, in which a woman should be eating roughly 300 more kcal daily. Excessive exercise

DO	DON'T

DO

Do exercise regularly (at least three times a week).

Do warm up with 5 to 10 minutes of light activity.

Do exercise for 20 to 30 minutes at your target heart rate.

Do cool down with 5 to 10 minutes of slow activity and gentle stretching.

Do drink water before, after, and during exercise.

Do eat enough to support the additional needs of pregnancy plus exercise.

© Michael Salas/The Image Bank

Pregnant women can enjoy the benefits of exercise.

DON'T

Don't exercise vigorously after long periods of inactivity.

Don't exercise in hot, humid weather.

Don't exercise when sick with fever.

Don't exercise while lying on your back after the first trimester of pregnancy or stand motionless for prolonged periods.

Don't exercise if you experience any pain or discomfort.

Don't participate in activities that may harm the abdomen or involve jerky, bouncy movements.

Figure 15-1
Exercise guidelines during pregnancy.

and inadequate energy intake may yield inadequate maternal weight gain and impede fetal growth. Total weight gain during pregnancy of 25–35 lb is recommended; the variations depend on body weight at the initiation of pregnancy. Little research information is available regarding the specific metabolism of vitamins and minerals during pregnancy for a woman who maintains aspects of training. However, as displayed in the U.S. Dietary Reference Intake levels in Appendix A, the recommended (RDA and AI) levels for several vitamins and minerals are increased during pregnancy and lactation.

Lactating women who exercise expend more energy and tend to be leaner than nonexercising lactating women. In addition, if their energy intake appropriately accounts for exercise as well as lactation, they may also produce slightly more milk. The energy recommendation is increased by 500 kcal daily for lactation alone, so an active woman needs to pay special attention to her energy intake. It is particularly important that lactating women maintain proper hydration by taking into account the increased fluid needs for both exercise and lactation. For more information on exercise and nutrition during pregnancy and lactation, see Appendix C for contact information for several professional and private organizations and educational information.

■ EXERCISE AND NUTRITION DURING PREGNANCY AND LACTATION *IN REVIEW*

- Most women can safely exercise during pregnancy.
- Pregnant women should not engage in exercise in the supine position (flat on back) after the first trimester of pregnancy.
- High-impact or ballistic activities are discouraged due to joint laxity during pregnancy and the possibility of fetal trauma.
- Pregnant and lactating women have increased needs for energy, protein, and vitamins and minerals.
- Pregnant and lactating women who exercise must account for water loss as sweat in conjunction with increased water requirements.

■ EXERCISE AND NUTRITION FOR CHILDREN AND TEENS

Childhood and the teen years present special considerations for exercise and nutrition. Among the most significant matters is growth. As shown in Figure 15-2, growth spurts

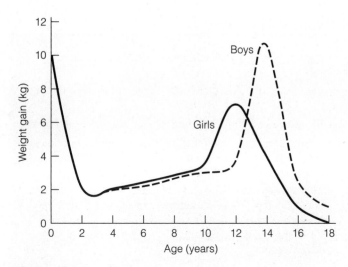

Figure 15-2
Average annual weight gain of boys and girls.

Table 15-1
Suggestions and Recommendations for Physical Activity and Children • Regular walking, bicycling, and outdoor play with other children. • Limit TV and computer use to 1–2 hours a day. • Weekly participation in organized sports, lessons, and clubs. • Regular physical education that includes at least 20 minutes of coordinated large-muscle exercise. • Physical activities should be fun, improve confidence in participating in physical activity, and involve friends and peers. • Regular family outings that involve walking, cycling, swimming or other recreational activities. • Positive role modeling for a physically active lifestyle by parents, teachers, and coaches.

are experienced at different time points. Children and adolescents are particularly sensitive to peer and self-evaluation regarding appearance. Individuals who develop an altered self-image are prone to experience disordered eating patterns. The incidence and characteristics of disordered eating with special attention to more active people and athletes are discussed in the feature In Focus: Eating Disorders and the Female Athlete Triad at the end of this chapter.

Exercise for Children and Teens

Strength and endurance training concepts and their importance to children were detailed in Chapters 12 and 13. Healthy activity patterns often begin in the early childhood years, so it is important that children learn to enjoy and participate in a variety of physical activities. A healthy level of physical activity requires regular participation in activities that increase energy expenditure above resting levels. Active children should participate in regular physical education classes, play sports, perform regular and routine household chores, play recreationally both indoors and preferably outdoors, and regularly travel by foot, bicycle, skateboard, scooter, rollerblades, or other human-powered means. Health professionals prescribing exercise for children should consider the child's maturational age, gender, and inclination toward activity. The season of the year, popular likes and dislikes, and parental attitudes should also be considered when making these recommendations. Tables 15-1 and 15-2 provide suggestions and guidelines for children and parents to increase physical activity and help establish lifelong behavior patterns.

Physical activity is important for all children, including those who are less coordinated or physically impaired. For children with a higher than typical body weight, physical activity is particularly helpful for both their physical and psychological health. Research consistently demonstrates that active children are more likely to remain active throughout adulthood. Therefore, encouraging activity in children can be an important component of weight control during adult years. The type of physical activity should be geared around play rather than exercise and the activity choices should be consistent with the child's interests and skill level, and within the family's budget.

Key Point

Because active adults are more likely to have been active when they were children and teens, physical activity needs to be encouraged early in life.

Nutrition for Children and Teens

Children experience periods of rapid growth that cause their requirements for energy and protein to be higher than those for adults. For instance, the energy requirements of growth may be about 5 kcal/g of tissue gained. This translates to approximately 1825 kcal for each pound of tissue gained. Increased energy needs relative to body weight during childhood are displayed in Figure 15-3. Protein needs are also enhanced during childhood and adolescence in comparison to adulthood. For instance, the RDA for protein is 25% higher (relative to body weight) for a boy or girl 7–18 years old than for an adult. However, after 2 years of age, children have the same recommendations as adults for limiting fat (<30% of energy) and saturated fat (<10% of energy).

Table 15-2   Parent Tips for Rearing Active and Healthy Children

- Help your children develop good physical activity habits at an early age by setting a good example yourself. Practice heart-healthy habits.
- Limit the amount of television, movies, videos, and computer games to less than 2 hours a day. Fill the rest of leisure time with physical activity.
- Plan family outings and vacations so that they involve vigorous activities such as hiking, bicycling, skiing, swimming, and so on.
- Give your children some household chores that require physical exertion, keeping in mind their levels of strength, coordination, and maturity. Mowing lawns, raking leaves, scrubbing floors, and taking out the garbage not only teach responsibility but can be good exercise.
- Observe what sports and activities appeal to your children, then find out about lessons and clubs. Some children thrive on team sports; other children prefer individual activities. Some activities, like tennis and swimming, can be enjoyed for a lifetime and are much easier to learn during childhood.
- If it is safe to walk or bike rather than drive, do so. Use stairs instead of elevators and escalators.
- Increase the distances you and your children walk.
- At daycare, make sure the kids exercise at least 30 minutes a day.
- Stay involved in your child's physical education classes at school. Ask about frequency of classes and activity, class size, curriculum (instruction in lifetime fitness activities as well as team sports should be emphasized), physical fitness assessments, qualifications of the teacher (should hold appropriate certification in physical education and be an appropriate role model for students). Physical fitness should be measured at the beginning and end of each year, and goals should be established for each child. Encourage your school board to put emphasis on skills students can use for the rest of their lives.
- Discourage doing homework immediately after school to allow children to find some diversion from the structure of the school day. Children should be active after school and before dinner.
- Choose fitness-oriented gifts—a jump rope, mini-trampoline, tennis racket, baseball bat, a youth membership at the local YMCA or YWCA. Select the gift with your child's skills and interests in mind.
- Take advantage of your city's recreation opportunities, from soccer leagues to fun runs. Check out the various camps or organizations like the Sierra Club that sponsor outdoor activities such as camping, hiking trips, and bird watching.
- Spring your infant from mechanical restraints as much as possible. Strollers and playpens are high on convenience but low on activity potential. Try to unleash your diapered dynamo whenever and wherever he or she can safely move around.
- When your children are bored, suggest something that gets them moving, like playing catch or building a snowman in the yard.

Other nutrients of primary concern for children and teens include calcium, vitamin D, vitamin C, copper, zinc and iron for proper bone development. Physical activity and proper nutrition allow for optimal bone development during childhood and adolescence. Many researchers feel that the years just before puberty may be extremely important for bone density development and that weight-bearing physical activity along with optimal nutrition intake are the keys in achieving maximal bone mineral density, which tends to occur around age 30. Iron is also of

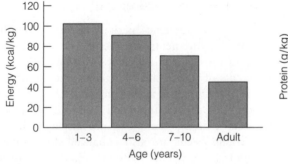
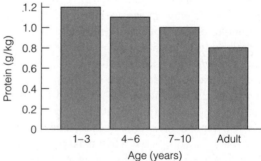

Figure 15-3

Approximate requirements for energy (in kcal/kg of body weight) and protein (in g/kg of body weight) for children and adults. Energy and protein requirements are greater for children than for adults when expressed relative to body weight.

concern for females once they begin to menstruate, because iron is lost along with blood. Therefore, nutrition requirements for children and teen athletes vary from those for adults in light of normal growth and maturation processes. Much more research information is available for adult athletic populations than for younger ones.

■ EXERCISE AND NUTRITION FOR CHILDREN AND TEENS *IN REVIEW*

- Physical activity needs to be encouraged during childhood and adolescence.
- Regular physical activity is important for youth to provide healthy energy expenditure, assist with bone mass development, and hopefully provide the basis for a physically active adult lifestyle.
- Protein, calcium, phosphate, vitamins C and D, copper, zinc and iron are crucial to the proper formation of the skeleton.
- General dietary recommendations for children tend to be higher than for adults when they are based on body mass.
- Females have an additional need for iron once they begin to menstruate.

■ EXERCISE AND NUTRITION FOR ADVANCING AGE

At what age do recommendations for nutrition and exercise end? It is extremely important for the senior population to maintain or increase their activity levels to support physical health and enhance activities of daily life. However, individuals of advancing age have unique considerations for exercise. Advancing age also places emphasis on certain nutrients due to decreased absorption and metabolism. This section provides an overview of some of the most significant considerations related to exercise and nutrition for people of advancing age.

Exercise for People of Advancing Age

People generally become less physically active as they get older. Nearly 40% of people over the age of 55 report no leisure time physical activity. This is a serious problem because as one becomes older the need for regular exercise may increase. Many studies have shown that increased levels of physical activity are associated with a reduced incidence of heart disease, hypertension, type 2 diabetes mellitus, depression, and anxiety. Regular physical activity also improves muscular strength and endurance, which are important to balance and coordination and thus reduce

Table 15-3

Exercise Tips for Older Americans

- If you have a family history of heart disease, check with your doctor first. It's a good idea to have a physical examination and take a graded exercise test before you start an exercise program.
- Pick rhythmic, repetitive activities that challenge the circulatory system and exercise at an intensity appropriate for you.
- Pick activities that are fun, that suit your needs, and that you can do year-round.
- Wear comfortable clothing and footwear appropriate for the temperature, humidity, and activity.
- If you decide that walking is a great activity for you, choose a place that has a smooth, soft surface; that does not intersect with traffic; and that is well-lighted and safe. Many older Americans walk at area shopping malls.
- Find a companion to exercise with you if it will help you stay on a regular schedule and add to your enjoyment.
- Because muscular adaptation and elasticity are generally slowed with age, take more time to warm up and cool down while exercising. Make sure you stretch slowly.
- Start exercising at a low intensity, especially if you have been mostly sedentary, and progress gradually.
- If you plan to be active more than 30 minutes, try to drink some water every 15 minutes, especially when exercising in hot, humid conditions. As you age, your sense of thirst tends to decrease and you can't completely rely on your internal sense of thirst.

the likelihood of falling and possible fracture. Also, basic daily life skills of carrying groceries, performing yard work, and household chores can be enhanced, giving rise to an improved sense of independence. Therefore, active people generally improve their lives both qualitatively and quantitatively. Tips for achieving a more active lifestyle are presented in Table 15-3.

Inactive people lose muscle mass at the rate of 1–2% per year starting between the ages of 20 and 35. This can be a very significant loss (25% or more) in muscle mass by the age of 60 and leads to a large and increasing number of elderly people living at or below functional levels. In this condition, a minor illness or mishap could make them completely dependent on others for their daily care. For improvements in health, fitness, and function to be observed in a person who has been sedentary, physical activities need only be low to moderate on the intensity scale. Over time, continual improvements in these areas will require greater volumes of physical activity. However, maintenance of fitness levels may be more important than improving them. Therefore, when a certain functional level of fitness has been reached, efforts should be focused on keeping it. There are a couple of precautions for the senior exerciser to consider:

- Use an extensive warm-up to ensure the body is properly prepared for work.
- Progress slowly in regard to intensity, duration, and volume.
- Use rest periods of 2–3 minutes between strength training exercises to ensure recovery from the previous workload.
- Begin with exercises that are simple in nature and gradually progress to more complex movements.
- Avoid ballistic and high-impact exercises.

Elderly individuals often have physical disabilities that hinder participation in normal physical activities. However, they can still benefit tremendously from exercise. Alterations in standard exercise practices may have to be made, but finding creative ways for these people to be active is extremely important. People with disabilities often enjoy the use of swimming pools for much of their physical activity. Many dumbbell and strength exercises can be incorporated for those confined to wheelchairs and walkers for physical activity. Finding exercises that are challenging, useful, and enjoyable will provide the greatest benefits.

Key Point

During advancing age, diet and exercise become increasingly important in reducing the risk of diseases and improving the quality of life.

Nutrition for Advancing Age

Sometimes advancing age is viewed as an opposite scenario to growing. For example, energy expenditure is higher relative to body weight during infancy and childhood than in adulthood, and it progressively slows during the waning decades of life. However, this is an inaccurate view of nutritional needs later in life. In fact, the nutritional requirements for some vitamins and minerals actually increase. Among these may be calcium and vitamin D. Considerably more research is needed to understand the changes in nutritional requirements.

Nutritional considerations during advancing age may be related to reductions in optimal performance of digestive and absorptive activities as well as decreased metabolic and storage capabilities. For instance, it seems that the production of vitamin D via exposure to sunlight is not as efficient for a 60-year-old man as it was when he was 20. In addition, the presence of disease processes can necessitate particular nutrition adjustments. For instance, individuals who have experienced a heart attack may need to adhere to a restricted diet.

■ EXERCISE AND NUTRITION FOR ADVANCING AGE *IN REVIEW*

- Physical activity for the aging population is extremely important in order to maintain muscular strength, muscular endurance, flexibility, and aerobic capacity.
- Increased levels of physical activity are associated with a reduced incidence of heart disease, hypertension, type 2 diabetes mellitus, depression, and anxiety.
- Nutritional requirements for some vitamins and minerals may actually increase in the last two to three decades of life.

■ EXERCISE AND NUTRITION FOR DIABETES MELLITUS

Diabetes mellitus is a broad metabolic disorder whose common characteristic is hyperglycemia. Approximately 17 million people in the United States, or 6.2% of the population, have diabetes. Of these, an estimated 11.1 million have been diagnosed and another 5.9 million people (or one-third) are unaware that they have the disease. There are different classifications of diabetes mellitus, and nutrition and exercise are important considerations for both types. This section provides an overview of the two principal types of diabetes mellitus along with their relationships to exercise and nutrition.

Key Point

Hyperglycemia is the underlying diagnostic indicator for diabetes mellitus, although the metabolism of all energy nutrients is affected.

Types of Diabetes Mellitus

Diabetes mellitus has been recognized for thousands of years and today it is a leading cause of blindness, need for amputation, and renal failure and a significant contributor to the development of heart disease and stroke. Even though diagnosis is founded on hyperglycemia, diabetes mellitus is not simply a condition of abnormal carbohydrate metabolism. It should be viewed as a metabolic scenario in which lipids and protein are also considered. In type 1 diabetes mellitus, there is a critical reduction in insulin production by the β cells of the islet of Langerhans within the pancreas. These cells are actually destroyed in autoimmune events; however, the trigger(s) remain elusive. Insulin therapy is absolutely necessary in type 1 diabetes mellitus, making its treatment different from type 2.

Approximately 95% of people diagnosed with diabetes mellitus are classified as having type 2. This form differs from type 1 in that insulin is produced, often at levels that are higher than normal. However, the potency of the insulin is reduced at the cellular level. This results in hyperglycemia. The risk of developing type 2 diabetes mellitus is increased when there is a family history of the disease or for individuals of Native American, Hispanic, African, or Pacific Islander descent. Compared with Caucasian and Asian Americans, African Americans are twice as likely to be diagnosed with type 2 diabetes mellitus, Hispanic Americans are two to three times more likely, and Native Americans are five times more likely. Other risk factors include age and a sedentary lifestyle.

Above all, obesity is the single most important risk factor for the development of type 2 diabetes mellitus, as roughly 9 out of 10 diagnosed individuals are at least 20% above their ideal body weights. The prevalence of type 2 diabetes mellitus is approximately three times greater in obese individuals compared to average weight individuals. By reducing obesity levels, individuals can often significantly improve the regulation of blood glucose and insulin levels. Swollen adipocytes release factors into the blood that may decrease insulin sensitivity and thus glucose tolerance. Insulin sensitivity returns relative to body fat reduction.

Key Point

Obesity is the most significant risk factor for type 2 diabetes mellitus.

Exercise and Diabetes Mellitus

The benefits of regular exercise for the diabetic or any other individual are well known. Exercise can serve as both a preventive factor and a treatment therapy. As for the general population, heart disease is the foremost killer of individuals with diabetes mellitus. Adults with diabetes mellitus have a heart disease death rate that is two to four times higher than adults that do not have diabetes mellitus. Regular cardiovascular exercise strengthens heart muscle tissue and improves vascularization. This type of exercise potentially decreases total cholesterol levels, primarily by reducing LDL-cholesterol, and increases HDL-cholesterol levels. Exercise can also improve other risk factors such as hypertension, obesity, and stress as well as improve one's psychological outlook. Before a regular exercise regimen begins, a complete physical examination by a diabetic specialist is recommended along with nutritional consultation provided by a registered dietitian (R.D.).

Exercise adds a new dimension in blood glucose control. Before an individual with diabetes mellitus begins an exercise program, the diabetic athlete should be aware of the need to pay closer attention to his or her diet and monitor blood glucose levels more frequently. During muscle contraction, glucose can enter muscle cells in an insulin-independent fashion. As discussed in Chapter 4, events related to muscle contraction can increase the number of glucose transport proteins (GluT4) on the plasma membrane (sarcolemma) of working cells. Also, it seems that exercise allows for an enhanced insulin-stimulated translocation of GluT4 to the sarcolemma for several hours after exercise. This can decrease insulin requirements, so too much insulin during or after exercise can result in hypoglycemia. Not only would hypoglycemia reduce exercise performance, it can progress to life-threatening stages.

Exercise also affects the rate at which injected insulin is absorbed from subcutaneous tissue, enhancing the actions of insulin during exercise. This is primarily attributable to increased blood flow to the skin, allowing for greater absorption of injected insulin from subcutaneous adipose tissue. Also, exercising in a warmer environment has an additive effect. Therefore individuals with type 1 diabetes mellitus who are involved in sports played in warmer climates, such as soccer, football, and baseball, must consider not only the exercise effect but the ambient temperature as well. General tips for the diabetic athlete are presented in Table 15-4.

Table 15-4   Tips for the Diabetic Athlete

- Exercise regularly, preferably at the same time each day to help determine and stabilize insulin and energy requirements.
- Exercise with a partner who understands the hypoglycemic risks and knows what to do in case of emergency.
- If participating in a spontaneous bout of activity (such as an unexpected baseball game), eat before and reduce insulin dose as appropriate.
- Always exercise approximately 1 hour after eating, when blood glucose levels are on the rise.
- Never begin exercise with a low blood glucose level.
- Always carry glucose tablets or some form of sugar. Always carry change for a phone call or a vending machine.
- On a long day trip, such as hiking or cycling, eat six small meals containing both carbohydrates and protein. Be overprepared with extra emergency food, in case of an unexpected delay.
- Drink plenty of fluids before and during exercise, to prevent dehydration.
- An athlete who exhibits diabetic complications such as neuropathy (nerve damage) and peripheral circulation damages may not be able to train safely and should consult his or her physician.

Key Point

Exercise can decrease insulin requirements by increasing glucose uptake into muscle, both during and after exercise.

Nutrition and Diabetes Mellitus

The primary goal for "diabetes control" is to limit meal-induced hyperglycemia and to maintain fasting (postabsorptive) euglycemia (70–110 mg/100 ml blood). The diet recommended for individuals with diabetes mellitus is basically the same as healthy eating recommendations for the general population. These individuals must also balance their meals with insulin dosages, oral medication, and physical activity to avoid hypoglycemic or hyperglycemic episodes. The help of an R.D. or a diabetic educator plays an important role in making wise food choice plans for the diabetic patient.

As recommended by the American Diabetes Association (ADA), a typical diet for the person with type 1 diabetes mellitus includes 50–60% of the energy to be derived from carbohydrates and about 12–20% from protein sources. In accordance with general recommendations by the American Heart Association, individuals with diabetes mellitus should limit their fat intake to less than 30% of total energy and saturated fat to less than 10%. Individuals with diabetes have the same requirements for vitamins and minerals as the general population. More information regarding diabetes can be found at the website of the American Diabetes Association (www. diabetes.org).

■ EXERCISE AND NUTRITION FOR DIABETES MELLITUS *IN REVIEW*

- Diabetes mellitus is a broad metabolic disorder
- Exercise is important in promoting control of hyperglycemia.
- Exercise can help prevent and treat obesity, the predominant risk factor for type 2 diabetes mellitus.
- Dietary recommendations for individuals managing diabetes mellitus are the same as for the general population.
- Because obesity is strongly associated with the development of type 2 diabetes mellitus, reduction of body fat through diet and exercise is strongly recommended.
- Moderate amounts of physical activity can assist with weight management for persons with diabetes.
- Special consideration of the intensity and duration of physical activity are needed when prescribing exercise for the diabetic.

Two major concerns are associated with type 2 diabetes mellitus. First, chronic hyperglycemia is believed to cause cellular damage and increase tissue aging. Second, the elevated insulin levels associated with type 2 diabetes mellitus promote increased blood pressure as well as total and LDL-cholesterol levels. So as expected, heart disease is the most common cause of death in individuals who fail to control type 2 diabetes mellitus. In addition, insulin is an anabolic hormone and is believed to possibly accelerate the progression of some types of cancer.

Key Point

Type 2 diabetes mellitus is strongly related to obesity, and when adipose tissue is reduced the severity of this disorder often decreases in a related manner.

■ VEGETARIANISM FOR ACTIVE PEOPLE AND ATHLETES

Many people choose to follow a vegetarian diet, in which the foundation of the foods eaten are plants or plant-derived. However, vegetarians have significant variation in food choices. For instance, a *vegan* eats only foods of plant origin such as fruits, vegetables, legumes (dried beans and peas), grains, seeds, and nuts. A *lacto-vegetarian* also includes dairy products, and an *ovo-lacto-vegetarian* (or lacto-ovo-vegetarian) includes eggs and dairy products. The so-called semi-vegetarians do not eat red meat but consume chicken and fish to complement plant foods, dairy products, and eggs.

Because they are lower in animal (especially mammalian) products, vegetarian diets are generally lower than nonvegetarian diets in the content of total fat, saturated fat, and cholesterol. Diets containing higher amounts of these items are associated with increased risk of obesity, high blood pressure, heart disease, type 2 diabetes mellitus, and some types of cancer. Does this mean a diet that is vegetarian or resembles that style is the way to go? There really is not an absolute answer for this question. For instance, although red meat is often viewed as a "bad guy," very lean red meat is an excellent source of protein, iron, zinc, and vitamin B_{12}. Also, the more restrictive a vegetarian diet is, the more care one must take to make it a nutritionally balanced diet.

Protein Quality and Quantity

One concern long associated with vegetarian diets is related to the protein quantity and quality of a diet plan. However, plant proteins alone can provide enough of the essential and nonessential amino acids as long as sources of dietary protein are fairly varied and energy intake is

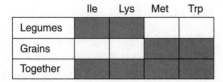

Figure 15-4
Complementary amino acids in the diet. In general, legumes provide plenty of isoleucine (Ile) and lysine (Lys), but fall short in methionine (Met) and tryptophan (Trp). Grains have the opposite strengths and weaknesses, making them a perfect match for legumes.

adequate to at least meet expenditure. As discussed in Chapter 5, whole grains, legumes, vegetables, seeds, and nuts contain both essential and nonessential amino acids. However, the quantity of certain essential amino acids may be below human protein levels. For instance, legumes are somewhat limited in methionine and tryptophan but provide plenty of isoleucine and lysine (Figure 15-4). Whole grains are rich in methionine and tryptophan but limited in isoleucine and lysine. This means that if legumes and grains are part of the same meal they will complement each other. The essential amino acid limitations in one plant protein source can be complemented by the abundance in another plant protein source. It probably is not essential that vegetarians complement foods during the same meal, but complementary plant proteins should be well represented in the diet at some point daily.

Vegetarian athletes need to estimate their protein quantity and quality. For those who merely increase food volume, a proportionate amount of their energy consumption will be attributable to protein. However, if low-protein foods (sport drinks, fruits, and juices) are consumed to increase energy intake, then a protein imbalance can be created. Vegetarian athletes can work with an R.D. to assess their diet and plan their meals appropriately.

Vitamins and Minerals

Vegetarians tend to eat more vitamin C and B-complex vitamins than nonvegetarians. However, vitamins A and B_{12} are generally lacking in plant foods, and vitamin D is found most abundantly in fortified milk and dairy products. Vegans can find reliable sources of vitamin B_{12} in fortified breakfast cereals, fortified soy beverages, some brands of nutritional (brewer's) yeast, and vitamin supplements. Vegans can also obtain vitamin D in margarine, fortified breakfast cereals, supplements, and via sunlight as discussed in Chapter 9. The conversion of carotenoids (such as β-carotene) to vitamin A provides adequate levels of this vitamin for vegetarians who do not drink fortified milk or eat dairy foods.

Deriving enough iron from the diet is a potential problem often associated with a vegetarian diet. Iron-fortified and whole grain cereals, dried beans, dark green leafy vegetables (such as spinach and kale), brewer's yeast, and dried fruits are all good plant sources of iron. There is not convincing evidence that vegetarians are more prone to develop iron deficiency than nonvegetarians. Vegetarians may have greater nonheme iron absorption because of a higher vitamin C intake. Zinc deficiency has also been noted as a concern, as it is more abundant in animal foods. Also, zinc (like iron) from plant sources is not well absorbed due to hindering substances such as phytate in plant tissue. As with iron, most vegetarians are not zinc deficient, and individual zinc supplementation is not recommended unless medically warranted. Calcium may also be a consideration for vegans, but for lacto-vegetarians the dietary calcium content is similar to that for the general population.

■ VEGETARIANISM FOR ACTIVE INDIVIDUALS AND ATHLETES *IN REVIEW*

- Vegetarians are often classified by the degree of restriction of animal-based foods.
- The amount and type of protein can be a consideration for vegan athletes with enhanced protein requirements.
- The major concerns for vitamin inadequacies can be addressed by eating fortified foods, taking supplements, eating carotenoids (which are converted to vitamin A), and using exposure to sunlight to promote vitamin D production in the body.
- The major concerns for minerals include calcium, iron, and milk.
- Nutritional assessment of the diet of more restrictive vegetarians is recommended, and a registered dietitian (R.D.) can provide assistance in planning a balanced diet.

Conclusions

Exercise and nutrition are important considerations for health regardless of age. For children, the increased nutrition demands for exercise are in addition to increased need based on growth. Although all essential nutrients are important, among the most significant for children are energy, protein, calcium, and vitamins A, C, and D. Physical activity during childhood and adolescence promotes the development of healthy habits that are likely to endure into adulthood. On the other end of the aging progression, older people benefit tremendously from regular physical activity to minimize the loss of bone material, improve cardiovascular health and performance, and promote a desirable body composition.

Certain eating philosophies, such as vegetarianism, can lead to unique nutrient concerns such as for protein, vitamins A, D, and B_{12}, and iron and zinc. However, there is not strong evidence of poor status of these nutrients among more active people and athletes who are vegetarians.

Certain metabolic circumstances can also bring about unique nutritional concerns. For individuals who are diagnosed with type 1 diabetes mellitus, exercise, nutrition, and insulin therapy must be precisely coordinated to maintain blood glucose control. For most individuals who are diagnosed with type 2 diabetes mellitus, the most important considerations might be energy imbalance and obesity. Exercise is an important factor in prevention as well as treatment of this metabolic disorder.

 **IN FOCUS** EATING DISORDERS AND THE FEMALE ATHLETE TRIAD

The female athlete triad is a very serious health problem among female athletes and encompasses three interrelated health issues that include disordered eating, amenorrhea, and osteoporosis (Figure A). Endurance athletes, as well as athletes in appearance based sports such as gymnastics, ballet, figure skating, diving, etc. may be particularly prone to disordered eating and the associated health risks. This phenomena as well as the potential deficiencies and long-term health problems that may result have a serious impact on the athlete's performance and overall health.

Many who suffer from this condition do not meet the strict DSM IV (Diagnostic and Statistical Manual for Mental Disorders) criteria for the diagnosis of an eating disorder. However, they do engage in disordered eating behavior which may include severe dietary restriction. When severely restricting calories there is the potential for deficiency in any nutrient, depending on the severity of restriction and the types of foods restricted or omitted from the diet. The extreme demands of training and competing may increase the nutrient needs of many athletes and thus put them at an even greater risk of deficiency and long-term health problems than a restricting non-athlete.

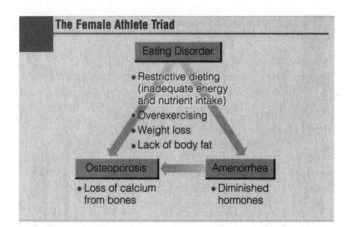

Figure A

Perhaps the most symptomatic nutrient deficiency associated with dietary restriction is **iron deficient anemia**. In an attempt to cut calories many athletes may decrease their intake of meats and other iron rich foods. Heme-iron is the most bioavailable form of iron and it is found in animal sources, non-heme iron is found predominantly in plant sources of iron. Therefore, if an athlete is restricting meats, the majority of her iron intake is from a source with low bioavailability, non-heme iron.

Iron losses may also be greater in athletes. In environments where sweat loss is great, iron loss in the sweat may be increased. Although, when sweat rate increases iron lost in sweat usually decreases. There is also an increase in red blood cell breakdown during exercise and iron lost through the GI tract may be increased in endurance athletes. If the athlete is still menstruating iron will also be lost in the menstrual blood.

Symptoms of anemia include fatigue, lethargy and decreased oxygen delivery to the working muscle. With a decrease of oxygen delivery there will be a decrease in energy production. A decrease in energy production may also occur due to iron's role in the various stages in the energy pathways. Iron is also required for cytochromes in the electron transport chain and for succinate dehydrogenase and aconitase in the citric acid cycle. In addition to fatigue, and decreased energy production, iron deficiency may also decrease VO_2 max and decrease onset of blood lactate (OBLA). The decrease in VO_2 max and OBLA may decrease performance during prolonged intense activity, or during anaerobic activity when it is beneficial to delay onset of blood lactate.

Athletes restricting calories and meats and other iron rich foods may also have inadequate intakes of **Zinc**. Zinc levels in athletes may be low due to decreased intake, increase loss in sweat, and an increase excretion in urine seen after exercise. Zinc is a cofactor for many functions in cell metabolism and may therefore decrease athletic performance, immune function, etc.

Although a decrease in performance as a result of inadequate intake of the **B Vitamins** is unlikely unless

EATING DISORDERS AND THE FEMALE ATHLETE TRIAD (CONTINUED)

these are entirely omitted from the diet, omission of the B vitamins will result in a decrease in VO_2 max and a decrease in OBLA. Which will result in a decrease in time to exhaustion and a decrease in max aerobic capability. A rebound can be seen after two weeks of adequate intake after deficiency.

Athletes restricting calories may also restrict **carbohydrate** intake, especially with the current focus on low carbohydrate high protein/high fat diets. A decrease in carbohydrate intake will inevitably lead to a decrease in glycogen stores especially when glycogen is not replaced after prolonged and/or consecutive days of training. This would be of particular concern for athletes competing in events lasting longer than one hour when reliance on blood glucose decreases and there is an increased usage of glycogen stores. When glycogen stores are depleted the body will rely more on fats and protein as an energy source and amino acids will be used for gluconeogenesis. Overtime this could lead to an increased overall body protein catabolism especially if restriction is so severe that protein intake is inadequate. Additionally, when the body relies more on fat for energy efficiency during exercise is decreased. A low carbohydrate intake may not be as concerning for the athlete participating in events requiring the ATP/PC system, such as 200 meter sprints and power lifting however the endurance athlete's will most likely be impaired.

Many women in the US do not meet the RDA for **calcium**. In an attempt to cut fat and calories, many athletes restrict their intake of the most nutrient dense source of calcium, dairy products An inadequate intake of calcium is one of the major risk factors for osteoporosis. The athlete with disordered eating is at an increased risk compared to a non-athlete as they often have the confounding risk factor of amenorrhea. Amenorrhea is accompanied by low blood estrogen levels and estrogen is an important hormone involved in the development and formation of bone. When estrogen levels are decreased, as during amenorrhea, bone development is impaired. The longer the amenorrhea lasts or the more frequently it occurs, the greater risk. Low levels of estrogen can negate the benefits that physical activity has on the formation of bone. In many athletes low levels of estrogen result from a delayed onset of menses. This delay is sometimes intentional in sports where a female physique may not be considered optimal for performance, such as gymnastics. Athletes participating in these sports may be at an increased risk for disordered eating and other associated problems.

The inadequate intake and amenorrhea may prevent the young athlete form ever reaching peak bone mass, which puts her at greater risk for developing osteoporosis in general, and puts her at risk for developing the disease at an earlier age. Most women don't lose a significant amount of bone mass until about age 50-60 when estrogen levels decrease due to menopause, if a woman never reached her peak bone density she will be starting at a disadvantage and may see a more rapid decline and therefore onset of the disease at an earlier age than expected.

The acute affects of an inadequate intake of calcium on performance are negligible since blood calcium levels are tightly maintained by vitamin D, parathyroid hormone, and calcitonin. This also makes symptoms of calcium deficiency difficult to detect. Symptoms are often not detected until a disease state has occurred and may be irreversible. DEXA Scans can detect bone density abnormalities and identify those at risk for osteoporosis. Most changes occur first in the lumbar spine, then the wrist, then lastly in the hip. Women with inadequate calcium intakes are at an increased for Type 1 or post menopausal osteoporosis, and for hip fractures.

An inadequate intake of the **antioxidants** may increase oxidative damage by free radicals produced during exercise. This could lead to an increased risk of cancer. Vitamin E intake may be inadequate on a very low fat diet. If selenium levels are low this may impair the function of Glutathione Peroxidase. If vitamin C levels are diminished it may impair the function of vitamin E and further decrease reduction reactions.

The Female Athlete Triad does not just lead to a decrease in performance but may impair the long-term health of the athlete. The best way to address this problem is through prevention and education (see Table A).

Table A

Tips for Combating Eating Disorders

General Guidelines

- Never restrict food servings to below the numbers suggested for adequacy by the Food Guide Pyramid.
- Eat frequently. Include healthy snacks between meals. The person who eats frequently never gets so hungry as to allow hunger to dictate food choices.
- If not at a healthy weight, establish a reasonable weight goal based on a healthy body composition.
- Allow a reasonable time to achieve the goal. A reasonable loss of excess fat can be achieved at the rate of about 10 percent of body weight in six months.
- Establish a weight-maintenance support group with people who share interests.

Specific Guidelines for Athletes and Dancers

- Remember that eating disorders impair physical performance. Seek confidential help in obtaining treatment if needed.
- Focus on proper nutrition as an important facet of your training, as important as proper technique.

EATING DISORDERS AND THE FEMALE ATHLETE TRIAD (CONTINUED)

Athletes, coaches, and parents need to be aware of the serious health problems associated with restricting calories, amenorrhea, and delaying menses. Healthy nutrition and weight loss techniques should be taught to parents and athletes at an early age to establish positive well-informed healthy attitudes as well as the practices that go along with them. Coaches need to encourage healthy eating, and discourage severe restricting and weight loss. It should be noted that any disordered eating should be addressed since there is the potential for the development of a serious eating disorder in the future. The ultimate risk, of course, being death.

The author of this feature is Susan Hewlings, Ph.D., R.D., Department of Integrated Health Sciences, Stetson University, Orlando, Florida.

STUDY QUESTIONS

1. What are some of the most important nutrients during childhood and adolescence?
2. Why is exercise important during childhood and adolescence, and are there any special exercise considerations for children?
3. What are the key nutrients for bone development?
4. What are the exercise limitations for pregnant women?
5. What are the major nutritional considerations for pregnant and lactating women?
6. Does the senior population have any special nutritional considerations?
7. Are there any limitations to consider when prescribing exercise to older people?
8. Why is exercise a special consideration for individuals with type 1 diabetes mellitus?
9. Although their diet might be limited in vitamin A, why might this not be a concern for vegetarians?

SUGGESTED READINGS

Brown W. The benefits of physical activity during pregnancy. *Journal of Science and Medicine in Sport* 5(1):37–45, 2002.

Kramer MS. Aerobic exercise for women during pregnancy. *Cochrane Database Systematic Reviews.* 2000(2):CD000180, 2000.

APPENDIX A

Dietary Reference Intakes

Dietary Reference Intakes: RDA, AI, and UL

Recommended Dietary Allowances (RDA) and Adequate Intakes (AI)

Age (yr)	Thiamin RDA (mg/day)	Riboflavin RDA (mg/day)	Niacin RDA (mg/day)[a]	Biotin AI (µg/day)	Pantothenic acid AI (mg/day)	Vitamin B6 RDA (mg/day)	Folate RDA (µg/day)[b]	Vitamin B12 RDA (µg/day)	Choline AI (mg/day)	Vitamin C RDA (mg/day)[c]	Vitamin A RDA (µg/day)[d]	Vitamin D AI (µg/day)[e]
Infants												
0–0.5	0.2	0.3	2	5	1.7	0.1	65	0.4	125	40	400	5
0.5–1	0.3	0.4	4	6	1.8	0.3	80	0.5	150	50	500	5
Children												
1–3	0.5	0.5	6	8	2	0.5	150	0.9	200	15	300	5
4–8	0.6	0.6	8	12	3	0.6	200	1.2	250	25	400	5
Males												
9–13	0.9	0.9	12	20	4	1.0	300	1.8	375	45	600	5
14–18	1.2	1.3	16	25	5	1.3	400	2.4	550	75	900	5
19–30	1.2	1.3	16	30	5	1.3	400	2.4	550	90	900	5
31–50	1.2	1.3	16	30	5	1.3	400	2.4	550	90	900	5
51–70	1.2	1.3	16	30	5	1.7	400	2.4	550	90	900	10
>70	1.2	1.3	16	30	5	1.7	400	2.4 *	550	90	900	15
Females												
9–13	0.9	0.9	12	20	4	1.0	300	1.8	375	45	600	5
14–18	1.0	1.0	14	25	5	1.2	400	2.4	400	65	700	5
19–30	1.1	1.1	14	30	5	1.3	400	2.4	425	75	700	5
31–50	1.1	1.1	14	30	5	1.3	400	2.4	425	75	700	5
51–70	1.1	1.1	14	30	5	1.5	400	2.4	425	75	700	10
>70	1.1	1.1	14	30	5	1.5	400	2.4	425	75	700	15
Pregnancy												
≤18	1.4	1.4	18	30	6	1.9	600	2.6	450	80	750	5
19–30	1.4	1.4	18	30	6	1.9	600	2.6	450	85	770	5
31–50	1.4	1.4	18	30	6	1.9	600	2.6	450	85	770	5
Lactation												
≤18	1.4	1.6	17	35	7	2.0	500	2.8	550	115	1200	5
19–30	1.4	1.6	17	35	7	2.0	500	2.8	550	120	1300	5
31–50	1.4	1.6	17	35	7	2.0	500	2.8	550	120	1300	5

NOTE: For all nutrients, values for infants are AI.

[a] Niacin recommendations are expressed as niacin equivalents (NE), except for recommendations for infants younger than 6 months, which are expressed as preformed niacin.

[b] Folate recommendations are expressed as dietary folate equivalents (DFE).

[c] Values are for nonsmokers. Smokers require an additional 35 milligrams per day.

[d] Vitamin A recommendations are expressed as retinol activity equivalents (RAE).

[e] Vitamin D recommendations are expressed as cholecalciferol and assume an absence of adequate exposure to sunlight.

Tolerable Upper Intake Levels (UL)

Age (yr)	Niacin (mg/day)[a]	Vitamin B6 (mg/day)	Folate (µg/day)[a]	Choline (mg/day)	Vitamin C (mg/day)	Vitamin A (µg/day)[b]	Vitamin D (µg/day)	Vitamin E (mg/day)[c]	Calcium (mg/day)	Phosphorus (mg/day)	Magnesium (mg/day)[d]	Iron (mg/day)
				Vitamins						**Minerals**		
Infants												
0–0.5	—	—	—	—	—	600	25	—	—	—	—	40
0.5–1	—	—	—	—	—	600	25	—	—	—	—	40
Children												
1–3	10	30	300	1000	400	600	50	200	2500	3000	65	40
4–8	15	40	400	1000	650	900	50	300	2500	3000	110	40
9–13	20	60	600	2000	1200	1700	50	600	2500	4000	350	40
Adolescents												
14–18	30	80	800	3000	1800	2800	50	800	2500	4000	350	45
Adults												
19–70	35	100	1000	3500	2000	3000	50	1000	2500	4000	350	45
>70	35	100	1000	3500	2000	3000	50	1000	2500	3000	350	45
Pregnancy												
≤18	30	80	800	3000	1800	2800	50	800	2500	3500	350	45
19–50	35	100	1000	3500	2000	3000	50	1000	2500	3500	350	45
Lactation												
≤18	30	80	800	3000	1800	2800	50	800	2500	4000	350	45
19–50	35	100	1000	3500	2000	3000	50	1000	2500	4000	350	45

[a] The UL for niacin and folate apply to synthetic forms obtained from supplements, fortified foods, or a combination of the two.

[b] The UL for vitamin A applies to the preformed vitamin only.

[c] The UL for vitamin E applies to any form of supplemental α-tocopherol, fortified foods, or a combination of the two.

[d] The UL for magnesium applies to synthetic forms obtained from supplements or drugs only.

Vitamins							Minerals						
Vitamin E RDA (mg/day)[e]	Vitamin K AI (µg/day)	Calcium AI (mg/day)	Phosphorus RDA (mg/day)	Magnesium RDA (mg/day)	Iron RDA (mg/day)	Zinc RDA (mg/day)	Iodine RDA (µg/day)	Selenium RDA (µg/day)	Copper RDA (µg/day)	Manganese AI (mg/day)	Fluoride AI (mg/day)	Chromium AI (µg/day)	Molybdenum RDA (µg/day)
4	2.0	210	100	30	0.27	2	110	15	200	0.003	0.01	0.2	2
5	2.5	270	275	75	11	3	130	20	220	0.6	0.5	5.5	3
6	30	500	460	80	7	3	90	20	340	1.2	0.7	11	17
7	55	800	500	130	10	5	90	30	440	1.5	1.0	15	22
11	60	1300	1250	240	8	8	120	40	700	1.9	2	25	34
15	75	1300	1250	410	11	11	150	55	890	2.2	3	35	43
15	120	1000	700	400	8	11	150	55	900	2.3	4	35	45
15	120	1000	700	420	8	11	150	55	900	2.3	4	35	45
15	120	1200	700	420	8	11	150	55	900	2.3	4	30	45
15	120	1200	700	420	8	11	150	55	900	2.3	4	30	45
11	60	1300	1250	240	8	8	120	40	700	1.6	2	21	34
15	75	1300	1250	360	15	9	150	55	890	1.6	3	24	43
15	90	1000	700	310	18	8	150	55	900	1.8	3	25	45
15	90	1000	700	320	18	8	150	55	900	1.8	3	25	45
15	90	1200	700	320	8	8	150	55	900	1.8	3	20	45
15	90	1200	700	320	8	8	150	55	900	1.8	3	20	45
15	75	1300	1250	400	27	13	220	60	1000	2.0	3	29	50
15	90	1000	700	350	27	11	220	60	1000	2.0	3	30	50
15	90	1000	700	360	27	11	220	60	1000	2.0	3	30	50
19	75	1300	1250	360	10	14	290	70	1300	2.6	3	44	50
19	90	1000	700	310	9	12	290	70	1300	2.6	3	45	50
19	90	1000	700	320	9	12	290	70	1300	2.6	3	45	50

[e] Vitamin E recommendations are expressed as α-tocopherol.

SOURCE: Adapted with permission from the *Dietary Reference Intakes* series, National Academy Press. Copyright 1997, 1998, 2000, 2001, by the National Academy of Sciences. Courtesy of the National Academy Press, Washington, D.C.

Minerals									
Zinc (mg/day)	Iodine (µg/day)	Selenium (µg/day)	Copper (µg/day)	Manganese (mg/day)	Fluoride (mg/day)	Molybdenum (µg/day)	Boron (mg/day)	Nickel (mg/day)	Vanadium (mg/day)
4	—	45	—	—	0.7	—	—	—	—
5	—	60	—	—	0.9	—	—	—	—
7	200	90	1000	2	1.3	300	3	0.2	—
12	300	150	3000	3	2.2	600	6	0.3	—
23	600	280	5000	6	10	1100	11	0.6	—
34	900	400	8000	9	10	1700	17	1.0	—
40	1100	400	10,000	11	10	2000	20	1.0	1.8
40	1100	400	10,000	11	10	2000	20	1.0	1.8
34	900	400	8000	9	10	1700	17	1.0	—
40	1100	400	10,000	11	10	2000	20	1.0	—
34	900	400	8000	9	10	1700	17	1.0	—
40	1100	400	10,000	11	10	2000	20	1.0	—

NOTE: An Upper Limit was not established for vitamins and minerals not listed and for those age groups listed with a dash (—) because of a lack of data, not because these nutrients are safe to consume at any level of intake. All nutrients can have adverse effects when intakes are excessive.

SOURCE: Adapted with permission from the *Dietary Reference Intakes* series, National Academy Press. Copyright 1997, 1998, 2000, 2001, by the National Academy of Sciences. Courtesy of the National Academy Press, Washington, D.C.

Dietary Reference Intakes:
Energy, Carbohydrate, Fiber, Essential Fatty Acids, and Protein

Age (yr)	Energy EER[a] (cal/day[b])	Carbohydrate RDA (g/day)	Total Fiber AI (g/day)	Linoleic Acid AI (g/day)	α-Linolenic Acid AI (g/day)	Protein RDA (g/kg/day)
Males						
0-0.5	570	60	—	4.4	0.5	1.52
0.5-1	743	95	—	4.6	0.5	1.5
1-3[c]	1,046	130	19	7	0.7	1.1
4-8[c]	1,742	130	25	10	0.9	0.95
9-13	2,279	130	31	12	1.2	0.95
14-18	3,152	130	38	16	1.6	0.85
19-30	3,067[d]	130	38	17	1.6	0.8
31-50	3,067[d]	130	38	17	1.6	0.8
>51	3,067[d]	130	30	14	1.6	0.8
Females						
0-0.5	520	60	—	4.4	0.5	1.52
0.5-1	676	95	—	4.6	0.5	1.5
1-3[c]	992	130	19	7	0.7	1.1
4-8[c]	1,642	130	25	10	0.9	0.95
9-13	2,071	130	26	10	1.0	0.95
14-18	2,368	130	36	11	1.1	0.85
19-30	2,403[d]	130	25	12	1.1	0.8
31-50	2,403[d]	130	21	12	1.1	0.8
>51	2,403[d]	130	21	11	1.1	0.8
Pregnancy						
14–18, *1st trimester*	2,368	175	28	13	1.4	1.1
2nd trimester	2,708	175	28	13	1.4	1.1
3rd trimester	2,820	175	28	13	1.4	1.1
19-50, *1st trimester*	2,403[e]	175	28	13	1.4	1.1
2nd trimester	2,743[e]	175	28	13	1.4	1.1
3rd trimester	2,855[e]	175	28	13	1.4	1.1
Lactation						
14-18, *1st 6 months*	2,698	210	29	13	1.3	1.1
2nd 6 months	2,768	210	29	13	1.3	1.1
19-50, *1st 6 months*	2,733[e]	210	29	13	1.3	1.1
2nd 6 months	2,803[e]	210	29	13	1.3	1.1

NOTE: For all nutrients, values for infants are AI; AI is not equivalent to RDA (see Chapter 2). Dashes indicate that values have not been determined.

[a]Estimated Energy Requirement (EER) is the average dietary energy intake predicted to maintain energy balance and is consistent with good health in healthy adults. EER values are determined at four physical activity levels; the values above are for the "active" person.

[b]Kilocalories per day.

[c]For energy, the age groups for young children are 1–2 years and 3–8 years.

[d]Subtract 10 calories per day for males and 7 calories per day for females for each year of age above 19.

[e]Subtract 7 calories per day for each year of age above 19.

GLOSSARY

acetylcholine Chemical neurotransmitter released by some nerve endings.

actin Intracellular thin contractile protein of muscle tissue.

action potential Rapid and specific change in the electrical properties of a plasma membrane in response to a stimulus.

active transport Energy-requiring movement of a substance across a membrane, via a transport protein, against a concentration or electrical gradient.

adipocytes Cells that mainly store fat.

adipose tissue Tissue consisting of adipocytes (fat cells) held together by connective tissue.

aerobic Refers to the requirement of oxygen for energy metabolism.

agonist Muscle primarily responsible for a specific joint action.

albumin The most abundant plasma protein, which transports many substances including free fatty acids (FFAs), bile acids, and several minerals.

aldosterone Hormone produced by the adrenal glands; decreases sodium loss in the urine by increasing sodium reabsorption.

amenorrhea An absence or marked suppression of menstruation.

amino acid pool The free amino acids found dissolved in the intracellular and extracellular fluids, which reflect protein turnover in cells and amino acid exchange between tissues.

amino acids Nitrogen-containing molecules, many of which can serve as building blocks for protein.

amylase Enzyme found in saliva and pancreatic juice that digests amylose and straight chains in amylopectin.

amylopectin Starch consisting of straight chains with branching.

amylose Straight chains of glucose in starch.

anabolism The building or synthesis operations in metabolism.

anaerobic Refers to energy systems that do not have an oxygen requirement.

analgesic A pain reliever.

androgenic Promoting the development and maintenance of primary and secondary sexual characteristics; testosterone is the main androgen.

anemia Hemoglobin levels that are too low to supply the body's demand for oxygen; often defined as hemoglobin levels of <7 g/100 ml of blood. Has many possible causes.

anion An ion containing negative charge (such as chloride, Cl^-; bicarbonate, HCO_3^-).

anorexia nervosa An eating disorder involving severe dietary restriction, refusal to maintain a minimally normal body weight, and a distorted perception of body image.

antagonist Muscle opposing the agonist in a specific joint action.

antagonistic Describes competitive or opposing actions.

antidiuretic hormone (ADH) Hormone produced by the hypothalamus and released by the posterior pituitary gland; stimulates the kidneys to reabsorb more water, reducing blood volume and conserving plasma volume.

apolipoprotein Any protein found in a lipoprotein shell.

aquaporins Water channels in cell membranes.

arrhythmia Irregular heart rate.

atherosclerosis The development of lipid, fibrous protein, and mineral deposits within the wall of an artery.

basal metabolic rate (BMR) The measurement of basal metabolism during a specific period of time (such as an hour or a day).

basal metabolism The energy expended during nonactive rest at a comfortable environmental temperature and at least 12 hours after the consumption of a meal.

basal Refers to lowest or foundational level.

bile Solution produced in the liver and delivered to the small intestine to coat lipids, allowing lipases to break them down more readily.

bioelectrical impedance analysis (BIA) A method of estimating body fat using low-intensity electric current passed through the body from one electrode to another.

biological value (BV) Describes how well the amino acid composition of a food protein meets human needs. High biological value indicates the protein has all essential amino acids in proportion to human cell needs.

biopsies Extraction of a sample of tissue for laboratory assessment.

body composition The relative contributions to a person's mass made by different substances or tissues. Body composition can be broken down in various ways, such as fat mass and fat-free mass, or as water, bone mineral mass, other fat-free mass, and fat mass.

body mass index (BMI) An index of a person's weight in relation to height; determined by dividing the weight (in kilograms) by the square of the height (in meters).

body weight The total mass of a person, expressed in pounds (lb) or kilograms (kg).

bone mineral content (BMC) The total mass of all minerals in bone including calcium, phosphorus, magnesium, sodium, and fluoride.

bonking Common term used to describe extreme fatigue during endurance exercise.

bronchodilation Dilation of the bronchioles and smaller airways in the lungs.

brown adipose tissue Specialized adipose tissue that can generate additional heat to maintain body temperature.

bulimia nervosa An eating disorder characterized by repeated episodes of binge eating followed by purging (self-induced vomiting, misuse of laxatives, fasting, or excessive exercise).

calcium oxalate Crystals that are one type of kidney stone.

carbon skeleton An amino acid after the amine group has been removed. The carbon skeleton is usually metabolized for energy.

carbonic anhydrase Enzyme that catalyzes the formation of carbonic acid (H_2CO_3) from CO_2 and H_2O, and vice versa.

cardiac output Product of stroke volume and heart rate for a given period (for example, 5 L/minute).

cardiogenic shock General organ system failure resulting from decreased cardiac perfusion.

carnitine palmitoyl transferase I (CPT-I) An enzyme complex that transports longer chain fatty acids across the inner membrane of mitochondria.

carotenoids Lipid molecules produced by plants; many have antioxidant properties and/or can be converted to vitamin A.

casein Predominant protein component in milk

catabolism The breakdown operations in metabolism.

cation An ion containing a positive charge (such as sodium, Na^+; potassium, K^+).

cell differentiation The process by which cells acquire certain characteristics and properties to form the various body tissues.

central fatigue Problems with the central nervous system resulting from the effects of hypoglycemia on the brain and nerves (see *neuroglucopenia*).

ceruloplasmin Copper transport protein that is also responsible for oxidizing iron, thus allowing it to be transported aboard transferrin.

chelated Electrically bound, such as the binding of a positive Ca^{2+} ion to a negatively charged phosphate (PO_4^{3-}) ion.

chloride shift Movement of chloride into a red blood cell in exchange for an exiting bicarbonate ion. This serves to electrically balance the movement of bicarbonate out of the cell.

cholesterol esters Cholesterol molecules linked to fatty acid by an ester bond.

chondroitin sulfate Glycosaminoglycan found in joint connective tissue such as cartilage.

chylomicron A form of lipoprotein that carries triglycerides and cholesterol from the small intestine to the general circulation via the lymphatic system.

clinical trials Objective process of testing the outcome of a certain treatment or situation.

clotting factors A group of more than 30 substances required for blood to properly coagulate and form a clot.

collagenase Enzyme that digests collagen.

complex training Strength training that combines high-force and high-velocity movements successively in a training session.

computerized axial tomography (CAT or CT scan) Sophisticated form of a low intensity X-ray received by a detector across the tissue that produces a "slice" image of the body. Slices can be pieced together using a computer to provide a three-dimensional image.

concentric Contraction while shortening a muscle.

conduction The transfer of heat between objects in contact.

convection The transfer of heat between an object and a fluid environment (air or liquid).

core body temperature Temperature in and around vital organs.

cross bridge The union of the contractile proteins actin and myosin in muscle.

cross-sectional area Diameter of muscle tissue.

deamination Enzymatic removal of an amine group of an amino acid, thereby generating a carbon skeleton and ammonia.

dehydration A reduced body water content.

denature To break peptide bonds that hold a protein in its three-dimensional conformation, thus straightening out the protein; stomach acid denatures protein early in digestion.

Dietary Supplement Health and Education Act (DSHEA) A 1994 U.S. law requiring the manufacturers of nutrition supplements to ensure that a product is safe; the FDA acts only after an unsafe supplement reaches the market.

direct calorimetry A method to assess whole body energy expenditure; heat energy released by the body warms a layer of fluid surrounding a specialized room (metabolic chamber), and the change in fluid temperature is the energy expended.

disaccharides Carbohydrate molecules composed of two monosaccharides linked by a chemical bond.

diverticulosis Presence of herniations of the wall of the large intestine (diverticula) caused by increased pressure.

docosahexaenoic acid (DHA) An essential fatty acid that can be used to make eicosanoids.

doubly labeled water (DLW) Tool used to assess energy expenditure based on estimations of CO_2 production; uses water molecules with isotopes of hydrogen ^2H and ^{18}O.

dual-energy X-ray absorptiometry (DXA or DEXA) Assesses body composition using low levels of X-ray; includes assessment of bone tissue.

dynamic constant external resistance (DCER) Strength training exercises in which the external resistance remains constant and is not affected by mechanical advantage (for example, using barbells and dumbbells).

eccentric Contraction while lengthening a muscle.

eicosanoids Hormonelike molecules derived from fatty acids and involved in the regulation of local events such as inflammatory processes.

eicosapentaenoic acid (EPA) An essential fatty acid that can be used to make eicosanoids.

electron transport chain Series of enzymatic complexes associated with the inner membrane of the mitochondria that serve as the site of oxidative phosphorylation.

emulsifier A substance that has a detergent effect, coating lipids to promote their breakdown.

energy balance The balance between energy consumed (absorbed) and energy expended from the body.

enrichment The process of adding certain nutrients to a grain product that were removed in processing efforts.

enterocytes Cells lining the wall of the small intestine involved in the absorption of nutrients.

epidemiological research Medical studies of disease causes, distribution, and controls in populations.

erythrocytes Red blood cells.

erythropoietin Hormone produced in the kidneys in response to reduced O_2 content of the blood; stimulates red blood cell formation in the bone marrow.

essential amino acid Amino acid not made in tissue either at all or in sufficient quantity to meet human requirements for maintenance and growth.

essential fat Body fat associated with bone marrow, the central nervous system, and internal organs, and in women the mammary glands and pelvic region; considered indispensable.

essential nutrients Nutrients that must be provided by the diet to prevent the development of deficiency disorders.

euglycemia Fasting blood glucose level achieved by hormonal regulation.

excitable tissue Tissue that possesses the ability to change the electrical properties of its plasma membrane in response to a stimuli; excitable cells are muscle cells and neurons.

exercise Formalized training to address one or more physical characteristics.

extracellular fluid (ECF) Water-based fluid outside of cells; includes the plasma of the blood and the interstitial fluid between cells.

facilitated diffusion Movement of a substance across a membrane, via a transport protein, down a gradient (e.g. concentration, electrical).

fat-free mass (FFM) The mass of the body that is not fat, including muscle, bone, skin, and organs.

fatigue Reduction in athletic performance that stems from physiological events in muscle (see *peripheral fatigue*) and/or the central nervous system (see *central fatigue*).

fatty acids Organic molecules that can be attached to glycerol to form glycerides (e.g., triglycerides) or to other molecules.

Federal Trade Commission (FTC) U.S. agency that regulates nutrition supplement advertising.

female athlete triad A syndrome affecting some female athletes; it has three principal aspects: disordered eating, menstrual irregularities, and osteoporosis.

fenestration A small opening in a capillary cell membrane that allows fluid and small solutes to pass.

ferritin Principal iron storage protein found in tissue; serum ferritin is used as an indicator of iron status.

ferroxidation Oxidation of iron converting it from ferrous to ferric; necessary for iron to interact with its transport protein (transferrin) and circulate.

fiber Indigestible fibrous structural molecules found in plant tissue; includes indigestible straight and branched polysaccharides as well as lignin.

fitness Level of physical condition: cardiorespiratory endurance, muscular strength, muscular endurance, flexibility, and body composition.

fluid and electrolyte replacement (FER) Describes a sport drink in which the carbohydrate content is relatively low, which is better during exercise than a higher carbohydrate drink used before and after exercise.

Food and Drug Administration (FDA) U.S. agency that regulates food labeling, drugs, and nutrition supplements.

force Ability to do work or cause physical change; energy, strength.

fortification The process of adding a nutrient to a food to increase its level above what is already present.

free radical A particle with an unpaired electron, which makes it highly reactive and destructive to cells.

frequency How often a specific activity is performed, such as number of running sessions per week.

fructose Monosaccharide found in fruits and honey and half of the disaccharide sucrose.

functional foods Foods that contain one or more nutraceutical substances.

functional training Strength training exercises that simulate real-world activities; heavy emphasis on postural control.

galactose Monosaccharide found to a limited degree in foods, and as half of the disaccharide lactose.

gastroesophageal reflux (GER) Expulsion of stomach content into the esophagus.

globular protein Any protein whose final three-dimensional shape is globular, such as hemoglobin and the enzyme hexokinase.

glucogenic Describes energy molecules, particularly amino acids, that can be used to make glucose in the liver.

glucosamine A basic building block of glycosaminoglycans; consists of glucose with an attached amine group.

glucose The most abundant monosaccharide in food carbohydrate and the primary carbohydrate that circulates in the blood.

glucose tolerance The ability to appropriately reestablish euglycemia following meal-induced hyperglycemia.

glucose transporter (GluT) Family of proteins that transport glucose across cell membranes.

glucose-6-phosphate Created in the first reaction of glycolysis and can be used to make glycogen or enter the pentose phosphate pathway or continue glycolysis.

glutathione A tripeptide of glutamic acid, cysteine and glycine involve in the antioxidant activities.

glycemic (glycemia) Term that refers to the level of glucose in the blood.

glycemic index The glycemic response of a food relative to a standard carbohydrate-containing food.

glycemic response Describes the degree and duration of blood glucose elevation produced by eating a specific food.

glycogen Carbohydrate storage molecule consisting of branching chains of glucose.

glycosaminoglycans (GAGs) Structurally unique polysaccharides that include amino sugar (glucosamine or galactosamine); found in joints and connective tissue.

gold standard or **criterion method** Refers to a tool or technique that has the highest level of accuracy and precision; used to determine the accuracy of other tools.

graded exercise test (GXT) Exercise test assessing physiological response, such as VO_2, BP, HR, RPE, to increasing workload.

health General term referring to an overall condition of well-being. Health can be viewed as either good or bad or on a relative scale.

healthy weight For gender, height, and build, the body weight associated with lower risk of disease.

heart disease Refers to pathological alterations in the structure and function of the heart. The most common cause of heart disease is atherosclerosis.

heart rate reserve (HRR) Difference between maximum and resting heart rate. It is the number of heartbeats available for work.

hematocrit The percentage of total blood volume occupied by erythrocytes.

hemoconcentration Occurs when the water content of the blood is reduced, thus concentrating the remaining fluid.

hemodilution Occurs when the water content of the blood is increased, which reduces the concentration of dissolved and suspended substances in the extracellular fluid.

hemoglobin Oxygen-transporting protein in erythrocytes.

hemosiderin Secondary iron storage protein found in cells; derived from transferrin. The level of hemosiderin increases with iron status.

hepatocytes Highly specialized liver cells that are involved in the metabolism of energy nutrients including the storage of glycogen, gluconeogenesis and urea production.

heredity Genetic transmission of traits to offspring.

high-density lipoproteins (HDLs) Lipoproteins that accumulate cholesterol in the circulation and return it to the liver.

high-energy phosphate molecules Molecules that have a high-energy phosphate bond that can be split to provide energy (ATP) and molecules that can quickly release phosphate for ATP production (GTP, CP).

high-fructose corn syrup Partially digested corn starch in which approximately half of the carbohydrate becomes glucose and fructose. HFCS is used as a commercial sweetener and is popular in sport foods.

homeostasis The maintenance of constant internal conditions by the body's control systems.

hormone-sensitive lipase (HSL) The most important enzyme involved in the release of fatty acids from adipose tissue.

humidity The level of moisture in the air, which influences the ability of sweat water to vaporize and cool the body.

hyaluronic acid Independent glycosaminoglycan found in joints.

hydration Water status of the body.

hydrogenation Food-processing technique that adds hydrogen to double bonds of unsaturated fatty acids, which allows an oil to solidify.

hydrostatic pressure The force a fluid applies to the surface of its container.

hyperglycemia Elevated blood glucose.

hyperinsulinemia Elevation of the level of insulin in circulation in response to an increase in blood glucose.

hyperkalemia Elevated potassium level in the blood; can disrupt the function of excitable tissue and fluid balance.

hypernatremia Elevated sodium level in the blood; can disrupt the function of excitable tissue and fluid balance.

hyperplasia The growth of tissue through cell reproduction.

hyperthermia A rise in core body temperature.

hypertrophy Increase in muscle cell size.

hypochromic microcytic anemia Anemia involving small red blood cells with a lowered content of hemoglobin; results from poor iron status.

hypoglycemia Blood glucose concentration that is below fasting level.

hypohydration Deliberately ingesting less water than expended; also, the state of low body water (body water deficit) resulting from an imbalance of water intake and loss. Hypohydration efforts purposely lower body weight or alter body appearance.

hyponatremia Low sodium level in the blood; can disrupt the function of excitable tissue and fluid balance.

hypotonic Describes a fluid that has an osmolality lower than blood serum (300 mOsm/kg of water).

hypovolemia A reduction in total blood volume; can disrupt blood pressure regulation and effective circulation.

hypoxia A reduced level of oxygen in the blood.

indirect calorimetry A method to estimate whole body energy expenditure based on measuring the amount of oxygen utilized and carbon dioxide produced during energy metabolism.

insomnia Inability to sleep well.

intensity The level of difficulty; how hard one performs an activity. For strength training, measured in repetitions of maximum; for aerobic activity, measured as VO_2max, heart rate, or other measures.

intercellular cleft A small opening between cells in a capillary wall that allows fluid and small solutes to pass.

interstitial fluid Found between cells in tissue. This fluid is exchanged with the plasma, although their composition is not the same.

intracellular fluid (ICF) Water-based fluid inside cells bathing organelles.

ion An element or molecule containing a net positive or negative charge.

isokinetic Describes strength training equipment in which the movement speed or velocity is controlled.

isomer One form of a molecule that can take two or more forms.

isometric Contraction with no movement of a muscle.

isotope One of two or more atomic forms of the same element. Isotopes vary only in the number of neutrons they contain.

isozyme One form of an enzyme that has more than one form.

ketogenic Describes energy molecules, particularly amino acids, that can be used to make ketone bodies in the liver.

ketosis A physiological state in which the ketone body production exceeds metabolism, which results in an accumulation of ketone bodies in the blood and tissue.

kilocalories (kcal) Measurement of energy; the amount of energy required to raise 1 kg of water by 1°C; equal to 4.184 kilojoules (kJ).

lactase Digestive enzyme produced by the small intestine that splits the disaccharide lactose.

lactate threshold (anaerobic threshold) Level of intensity associated with a significant increase in the blood level of lactate, reflecting greater reliance on anaerobic glycolysis.

lacteal A small lymphatic vessel.

lactose intolerance Symptoms such as bloating, cramping and diarrhea experience by people who produce insufficient amount of lactase.

lactose Disaccharide found in milk and dairy; it is digested by the enzyme *lactase*.

lean body mass (LBM) The sum of the body's fat-free mass and essential fat.

leukocytes White blood cells, which function in the immune system.

limiting amino acid The essential amino acid found in lowest proportion to human tissue needs in a food protein of low biological value.

lipase Enzymes that break down lipids.

lipid Class of chemicals, including fat and cholesterol, that do not dissolve well into water and are largely composed of hydrogen and carbon (hydrocarbon).

lipolysis Process of triglyceride store breakdown in adipose tissue involving hormone sensitive lipase.

lipoprotein Lipid-carrying shuttles made in the small intestine and liver; a shell that includes protein, phospholipids, and cholesterol encases lipids.

long chain fatty acid (LCFA) A fatty acid chain with 12–22 carbons.

low-density lipoproteins (LDLs) Lipoproteins that carry cholesterol to various tissues.

macrophage A nonlymphocyte white blood cell specialized for phagocytosis.

magnetic resonance imaging (MRI) Method for determining tissue density by using a large electromagnet that creates a magnetic field. After radio waves are introduced the magnetic field realigns hydrogen atoms, which give off their excess energy at different rates depending on the type of tissue. Emitted energy is then analyzed by a computer that constructs an image.

maltase Digestive enzyme produced by the small intestine that splits the disaccharide maltose.

maltodextrin Consist of small links of glucose with a branch point. Can be produced by partially digesting starch.

maltose Disaccharide found in limited amounts in the diet; produced when starch is broken down. Maltose is digested in the small intestine by the enzyme *maltase*.

margarine Hydrogenated oil that has been fortified with vitamins A and D and used as a substitute for butter.

medium chain fatty acid (MCFA) A fatty acid chain with 6–12 carbons.

menstruation The periodic discharge of bloody fluid from the uterus occurring at regular intervals between puberty and menopause.

metabolic cart Instrument that measures the volume of gas exchanged between an individual and the environment; measures O_2 utilized and CO_2 produced.

metabolic chamber Insulated room used in direct calorimetry.

metabolic rate The energy expended in a certain period of time (such as kcal/hour)

metabolism Refers to energy-releasing processes of cells, tissues, or the human body; may be measured in kilocalories expended.

metallothionein Mineral-binding protein produced in several tissues including the small intestinal wall and the liver.

minerals Inorganic substances, many of which are essential nutrients.

minute ventilation (V_E) The volume of air in a breath (tidal volume) multiplied by the frequency of breaths in a minute.

monocyte A circulating white blood cell that ingests foreign bodies.

monosaccharide The simplest carbohydrate molecules.

monounsaturated fatty acid (MUFA) A fatty acid that has just one point of unsaturation or double bond.

morbidity Rate of incidence of a disease.

mortality Death or death rate.

motor cortex Region of the brain that initiates voluntary skeletal muscle contraction.

motor unit A motor neuron and all of the muscle fibers it innervates.

muscle fiber A skeletal muscle cell, called a fiber because it is long relative to its diameter.

muscle spindle The spindle-shaped sensory receptor located within skeletal muscle that senses muscle "stretch."

mutation A change in the sequence of nucleotides within a gene.

myelin Lipid-rich insulating layer that is wrapped around neurons to increase signal transmission rate.

myofibril Bundle of myofilaments forming an internal subdivision of skeletal and cardiac muscle.

myosin Intracellular thick contractile protein of muscle tissue.

neuroglucopenia Hypoglycemic effects on the central nervous system including lightheadedness, lethargy, and nausea; the hallmarks of central fatigue.

neuromuscular junction The anatomical structure (synapse) intersecting a muscle fiber and nerve fiber.

nitrogen balance A balance between nitrogen lost from the body and ingested in the diet for a given period, to measure whether protein is being lost, gained, or kept in balance for the body as a whole.

nonessential fat Body fat located in subcutaneous and visceral stores; considered dispensable.

nonessential nutrients Nourishing substances that do not need to be part of the diet.

Northern blotting Laboratory technique used to determine the quantity of specific mRNA.

nutraceutical A nutrient that can promote health by hindering disease processes or is supportive in the treatment of existing disease.

nutrients Substances that nourish the human body.

nutrition The science pertaining to the nourishment of the human body.

obesity A body composition with fat mass accounting for more than 25% and 33% of body weight for men and women, respectively.

oligosaccharides Short indigestible chains of monosaccharides found mostly in legumes.

osmolality Measure of the concentration of particles in a solution, expressed as number of milliosmoles (mOsm) per kilogram of water.

osmosis The movement of water down its concentration gradient through a semipermeable barrier such as a cell membrane or capillary wall.

osmotic pressure The water-pulling force generated by substances dissolved in fluid.

osteomalacia Bone disorder in adulthood resulting from a reduction in bone mineral content; usually caused by poor vitamin D status and/or calcium intake.

osteoporosis Refers to a clinical situation in which bone integrity has been reduced, increasing the risk of fracture. Osteoporosis is diagnosed when bone density is less than two standard deviations below the average for a young adult.

overhydration Elevated body water content, typically created experimentally.

overload Physical stimulus above normal level.

oxalate A principal metabolite of vitamin C; also found as a component of some kidney stones.

oxidation Removal of electron(s) from a reactant.

oxidative phosphorylation Process of ATP synthesis that uses the energy of electrons removed from reactants to attach a phosphate group.

pathological hypertrophy Enlargement of the heart induced by chronic hypertension or other entity that is disruptive to effective cardiac activity.

peak bone mass Point in the development of bone in which it reaches its greatest mass and density; typically occurs in the late 20s to early 30s. Also known as maximal bone density.

peptide bonds The covalent bonds that link amino acids together.

periodization Systematic variation applied to a strength training program.

peripheral fatigue Reduced muscle performance resulting from muscle cells' inability to meet ATP demands of activity.

phosphate Form of phosphorus typically found in nature; an inorganic complex of phosphorus and oxygen (HPO_4^{2-} and $H_2PO_4^-$).

physiological hypertrophy Enlargement of the heart induced by cardiorespiratory training allowing the heart to function more efficiently per beat.

phytate Inositol hexaphosphate (Figure 10-3) made by plants; it contains numerous negative charges and can electrically interact in the digestive tract with positively charged minerals such as calcium, magnesium, zinc, copper, and iron and reduce their absorption.

phytochemical Plant-derived substance (nutraceutical) that may help prevent or treat diseases such as cancer, atherosclerosis, and osteoporosis.

platelet Formed element of blood, aids in clotting.

plate-loaded equipment Strength training equipment in which weight plates are manually loaded on the machine.

plethysmography A method of estimating body composition based on densitometry. Body volume is based on air displacement; often referred to as Bod Pod.

plyometric Exercises that utilize the stretch-shortening cycle in which muscle spindles are forcefully stretched, the reflexive response can provide additional motor recruitment. Examples include depth jumps and medicine ball catching and throwing.

polar Unbalanced electrical charge that is separated.

polysaccharides Complex carbohydrates containing hundreds of monosaccharides; polysaccharides include starch, fiber, and glycosaminoglycans.

polyunsaturated fatty acid (PUFA) A fatty acid that has more than one point of unsaturation or double bond.

porphyrin Nonprotein building block of heme, which is the iron-containing part of hemoglobin.

postexercise oxygen consumption (EPOC) Enhanced O_2 consumption after completion of an exercise bout; reflects recovery and adaptive efforts.

posttranslational modification The specific modification of an amino acid after it has been incorporated into a protein. Examples include hydroxylysine in collagen and 3-methyl histidine in actin.

power The amount of work in a given period of time: $P = w/t$.

preload The volume of blood that enters and collects in the left ventricle prior to contraction.

primary amenorrhea Delayed menarche; the absence of menstruation by age 16.

primary sweat The initial sweat produced by the exocrine cells at the foundation of sweat glands.

protease Any enzyme that splits peptide bonds between amino acids, thereby dismantling or digesting a protein. Proteases are important for digestion of food protein as well as protein breakdown in tissue.

protein fiber (fibrous protein) In reference to proteins, any protein whose final three-dimensional shape is long and straight, such as collagen and actin.

protein turnover The processes of protein breakdown and synthesis considered collectively.

provitamin A Carotenoids produced by plants that can be converted to vitamin A.

pseudoanemia Reduction in blood hemoglobin levels associated with hemodilution.

radiation The emission of energy (heat) from the surface of a warm object in the form of electromagnetic waves.

radioactive Refers to an atom that contains degrading neutrons, which result in the emission of energy.

range of motion (ROM) The displacement of body segments for a given joint.

rating of perceived exertion (RPE) A subjective assessment of intensity level by performer. Often used with a chart (Borg scale) to correlate with heart rate.

reabsorption Absorption of water or a solute back into the blood after it has been filtered out, for example, in the renal tubule system and sweat glands.

ready-to-drink (RTD) Term for sport shakes that need no added fluid.

renin-angiotensin system Enzyme/hormone system involved in regulating blood pressure. Renin activates angiotensin, which triggers the release of aldosterone and promotes vasoconstriction.

repetitions of maximum (RM) Maximum amount of weight that can be lifted for a specified number of repetitions. A 1 RM is the maximum amount of weight that can be lifted.

residual volume (RV) The volume of air remaining in the lungs after maximal exhalation efforts.

resorption Process of extracting minerals such as calcium and phosphorus (phosphate) from bone tissue in an effort to balance blood levels.

respiratory exchange ratio (RER) The ratio of carbon dioxide expired to oxygen consumed at the level of the lungs, which correlates to the ratio of carbohydrate to fat used for energy. Also known as respiratory quotient (RQ).

respiratory quotient (RQ) The measurement of changing oxygen and carbon dioxide levels associated with a specific cell or isolated tissue, which indicates the proportion of energy from carbohydrate versus fat.

resting metabolic rate (RMR) Level of energy expenditure during nonactive rest at a com-fortable environmental temperature and 4 hours after the consumption of a meal; tends to be about 10% greater than basal metabolic rate.

retinoids Active forms of vitamin A found in animal foods.

ribose A five-carbon monosaccharide that serves as a component of RNA, DNA, and ATP.

risk factors Elements that contribute to a condition or process.

S-adenosyl methionine (SAM) A principal donor of methyl groups during the construction of certain molecules.

salt A compound composed of a positive ion other than H^+ and a negative ion other than OH^-.

sarcolemma Plasma membrane of a muscle fiber.

sarcomere Functional contractile unit of a muscle cell.

sarcoplasmic reticulum Specialized form of the endoplasmic reticulum in skeletal muscle fibers that stores calcium.

saturated fatty acid (SFA) See *saturation*.

saturation Describes a fatty acid in which all carbon molecules between the ends are bonded to two hydrogen molecules and there are no double bonds.

secondary amenorrhea The absence of three or more menstrual cycles after menarche.

selectorized machines Strength training equipment that utilizes a weight stack in which the amount of resistance is selected by the use of a weight pin.

semistarvation Energy level approximating half of energy needs for weight maintenance.

serotonin Neurotransmitter derived from tryptophan.

short chain fatty acid (SCFA) A fatty acid with four carbons or fewer.

simple sugars Common terminology for naturally occurring monosaccharides and disaccharides.

sinoatrial node (SA node) Highly specialized region of the right atrium responsible for generating action potentials that pace the heart rate.

sliding filament theory Theory describing the mechanics of muscle contraction.

sonography Ultrasound; imaging method utilizing sound waves to reconstruct tissue.

specific heat The amount of heat (in kilocalories of energy) required to raise the temperature of 1 kg of the specific substance (such as water or fat) by 1°C.

split routine Strength training program that divides the body into specific muscle groups or regions. These specific units are then organized into a training schedule.

stable isotope Nonradioactive isotope; its neutrons are not degrading.

starch Digestible polysaccharides consisting of straight chains of glucose that can have branching.

stereoisomer Nonsuperimposable, mirror-image form of a molecule.

steroids Common terminology for molecules derived from cholesterol; include testosterone and estrogens.

sterols Class of molecules made by plants and animals that have a particular four-ring design and include cholesterol.

strength The amount of force that can be generated.

stroke volume Volume of blood ejected from a ventricle of the heart in one heartbeat.

subcutaneous Describes location beneath the skin.

subunits Protein components of a larger functional protein complex. For instance, hemoglobin contains four subunits that are held together to form the final functional hemoglobin structure.

sucrase Digestive enzyme produced by the small intestine that splits the disaccharide sucrose

sucrose Disaccharide derived from the sugar cane plant and used as a sweetener in recipes; it is digested in the small intestine by the enzyme *sucrase*.

summation Phenomenon of combined effects of multiple twitches of a muscle fiber.

sweat gland Exocrine organs dispersed in the skin and responsible for producing and regulating the composition of sweat.

sympathomimetic Mimicking the effects of the sympathetic nervous system, including central nervous system (CNS) stimulation, vasoconstriction, elevated blood pressure, and bronchodilation.

synergist Muscle functioning to assist the antagonist in a specific joint action.

synergistic Two or more factors whose combined effects are greater than the sum of their separate effects.

tachycardia Too-fast heart rate.

tendons Connective tissue that connects muscle to bone.

thermoregulation Physiological efforts (such as sweating) aimed at maintaining body temperature within a narrow range that allows for optimal function.

thirst A symptom of dehydration that prompts water ingestion.

tidal volume (TV) Volume of air inspired and expired during normal breathing.

time In describing training, refers to the duration of an activity. For strength training, usually refers to weeks or months; for endurance training, the time spent in aerobic activity for a session.

total energy expenditure Sum of all energy-releasing processes of the human body.

total iron binding capacity (TIBC) The availability of space on transferrin for iron; elevated TIBC is associated with decreased iron status in tissue.

total peripheral resistance The sum of all factors (HR, SV, blood volume, blood viscosity, blood vessel size, blood vessel resistance, valve function, and so on) affecting the flow of blood through the cardiovascular system.

training volume Amount of training stress, or work done. Usually calculated for strength training exercises by multiplying sets × repetitions × weight lifted. For endurance activities the volume may be determined by distance covered or by duration of activity.

transaminase Enzyme that transfers amine groups from amino acids to α-keto acids, such as pyruvate and α-ketoglutarate.

transamination Transfer of an amine group from an amino acid to an α-keto acid. Creates a nonessential amino acid and generates a carbon skeleton, which can be further broken down for energy.

transferrin Principal iron transport protein in the blood.

translation The process of linking amino acids together in a specific sequence as coded for in DNA. Translation occurs on ribosomes, to which tRNA delivers amino acids.

transverse tubule (T-tubule) Tubular extension of plasma membrane of a muscle cell that conducts action potentials.

triacylglycerol Same as *triglyceride*.

triglyceride Fat; consists of a glycerol backbone with three fatty acids attached. The fatty acids vary in structure in different triglycerides.

tropomyosin Regulatory contractile protein in muscle.

troponin Protein assisting tropomyosin in regulating muscle fiber contraction.

twitch One cycle of contraction and relaxation of a muscle fiber.

Type I muscle fiber Slow-twitch muscle fiber deriving energy from aerobic sources.

Type II muscle fiber Fast-twitch muscle fiber deriving most energy from anaerobic sources.

type In describing training, refers to the specific modality or exercise performed; each type of exercise elicits specific adaptations.

underwater weighing (UWW) Widely used procedure for measuring body density. The difference of weight on land and weight underwater (corrected for water density) equals body volume. Further corrections for residual air must also be considered.

unsaturated fatty acid A fatty acid in which adjacent carbons at one or more points are linked to only one hydrogen atom each. See *saturation*.

variable resistance Strength training equipment in which a noncircular pulley or cam provides varying resistance designed to try to match the weight load to the strength curve for a particular movement.

vasoconstriction The constriction of blood vessels, which can raise blood pressure, reduce blood flow to specific tissues, and restrict heat loss.

vertigo Dizziness.

very long chain fatty acid (VLCFA) A fatty acid chain with 24 carbons or more.

very low density lipoproteins (VLDLs) Lipoproteins that carry fat to tissues, primarily fat cells.

vitamins Organic essential nutrients that do not provide energy.

VO$_2$ Volume of oxygen consumed.

VO$_2$max Maximum volume of oxygen consumed in a given time period. Also termed peak VO$_2$.

Western blotting Laboratory technique used to determine the quantity of specific proteins.

whey Serum protein fraction in milk.

work The product of the force exerted on an object and the distance the object moves $(w = f \times d)$.

PHOTO CREDITS

INDEX